Monographs in Electrical and Electronic Engineering

Series editors: P. Hammond, T. J. E. Miller, and S. Yamamura

Monographs in Electrical and Electronic Engineering

---

10. *The theory of linear induction machinery* (1980) Michel Poloujadoff
12. *Energy methods in electromagnetism* (1981) P. Hammond
15. *Superconducting rotating electrical machines* (1983) J. R. Bumby
16. *Stepping motors and their microprocessor controls* (1984) T. Kenjo
17. *Machinery noise measurement* (1985) S. J. Yang and A. J. Ellison
18. *Permanent-magnet and brushless d.c. motors* (1985) T. Kenjo and S.Nagamori
19. *Metal–semiconductor contacts: Second edition* (1988) E. H. Rhoderick and R. H. Williams
20. *Introduction to power electronics* (1988) Eiichi Ohno
21. *Brushless permanent-magnet and reluctance motor drives* (1989) T. J. E. Miller
22. *Vector control of a.c. machines* (1990) Peter Vas
23. *Brushless servomotors: fundamentals and applications* (1990) Y. Dote and S. Kinoshita
24. *Semiconductor devices, circuits, and systems* (1991) Albrecht Möschwitzer
25. *Electrical machines and drives: a space-vector theory approach* (1992) Peter Vas
26. *Spiral vector theory of a.c. circuits and machines* (1992) Sakae Yamamura
27. *Parameter estimation, condition monitoring, and diagnosis of electrical machines* (1993) Peter Vas
28. *An introduction to ultrasonic motors* (1993) S. Sashida and T. Kenjo
29. *Ultrasonic motors: theory and applications* (1993) S. Ueha and Y. Tomikawa
30. *Linear induction drives* (1993) J. F. Gieras
31. *Switched reluctance motors and their control* (1993) T. J. E. Miller
32. *Numerical modelling of eddy currents* (1993) Andrzej Krawczyk and John A. Tegopoulus
33. *Rectifiers, cycloconverters, and a.c. controllers* (1994) Thomas H. Barton

Rectifiers, Cycloconverters, and AC Controllers

Thomas H. Barton

Emeritus Professor of Electrical Engineering
University of Calgary

CLARENDON PRESS • OXFORD
1994

Oxford University Press, Walton Street, Oxford OX2 6DP

Oxford New York Toronto
Delhi Bombay Calcutta Madras Karachi
Kuala Lumpur Singapore Hong Kong Tokyo
Nairobi Dar es Salaam Cape Town
Melbourne Auckland Madrid

and associated companies in
Berlin Ibadan

Oxford is trade mark of Oxford University Press

Published in the United States
by Oxford University Press Inc., New York

A catalogue record for this book is available from the British Library

Library of Congress Cataloging in Publication Data
(Data available)
ISBN 0 19 856163 6

Typeset by the author using LaTeX
Printed in Great Britain by
Bookcraft (Bath) Ltd, Midsomer Norton, Avon

To my dear wife
for her unfailing
patience, understanding,
and encouragement

PREFACE

Introduction

The concept of electronic control of industrial processes is commonplace. We are all aware that the combustion process in automobiles is increasingly controlled by a microprocessor, that whole petrochemical plants are controlled by computers, that the rolling of steel and the manufacture of cars is controlled by a variety of electronic devices. Electronics is ubiquitous in the modern world. Unfortunately, one of the greatest merits of electronic devices, very low power consumption, severely limits their direct application to industry whose processes are inevitably associated with high power levels. The manufacture of the staples of commerce, paper, plastics, soap, steel, sugar, etc. is measured in thousands of tons and their production requires commensurate electrical inputs measured in thousands of kilowatts. How then are we to control say the rolling of steel, involving numerous electric motors of 1000 and more horsepower with a computer operating at a few watts? There are a number of answers to this question involving either compressed air or hydraulic fluid under pressure which are excellent in specific circumstances. However, the most generally applicable and most flexible solution lies in the use of power electronics, electronic devices which bridge the gap between the milliwatt power levels of the control devices and the megawatt levels of the process.

Power electronics is the general term applied to the study of high power electronic devices and the circuits which employ them. However, since there are some high power applications which are not generally regarded as within the purview of power electronics and there are some modest power applications which are, this definition requires some expansion. While it is obvious that the devices which control the motors driving a paper making machine are an application of power electronics and that the computer which controls the process is a conventional electronic device, in other circumstances it is less easy to draw the boundary between conventional and power electronics. Power level is obviously important but is not the only consideration. The power output of an audio amplifier may well be in the kilowatt range and yet it is considered to be a part, admittedly a highly specialized part, of conventional electronics. The output of the amplifiers which launch TV signals onto the transmitting antenna are measured in tens or hundreds of kilowatts and yet they are part, again specialized, of conventional electronics. On the other hand the devices which control the speed and torque of a quarter horsepower motor will usually fit without question into the scope of power electronics. It is the combination of a number of factors, power level, voltage level, and operating frequency which point to the use of power electronic devices and which place the circuits within the field of power electronics.

Power electronic devices

The basic power electronic device is the silicon diode, a rectifying element which, in its larger sizes, can conduct thousands of amperes and block thousands of volts. The diode is a passive, two terminal, device whose operating condition depends solely on conditions in the circuit employing it. If the voltage across it is positive, i.e. its anode terminal is positive with respect of its cathode terminal, it conducts — current entering the anode terminal and leaving the cathode terminal. If the voltage across it is negative, it blocks. The next commonest device is the thyristor, which introduces the concept of control via a third terminal, the gate. When its gate terminal is at the same potential as its cathode, a thyristor blocks both positive and negative voltage. If the gate is made momentarily positive so that current flows from the gate to the cathode, the behavior switches to that of a diode, conducting in the forward biased state, from anode to cathode, but still blocking in the reverse biased state. A development of the thyristor, the triac, blocks in either direction before signal is applied to its gate and conducts in either direction after signal is applied.

While the utility of thyristors is enormous they have the disadvantage that, once switched from the blocking to the conducting state, they remain in the conducting state as long as current is flowing through them. They do not recover the blocking state until the current flow stops. In a.c. circuits where current flow naturally stops and reverses twice per cycle, this is not a problem but in d.c. circuits thyristors can only be employed if an external agency forces the current to zero for a brief period, *forced commutation* as against the *natural commutation* of the a.c. circuits. This demerit is repaired by the GTO, the gate turn-off thyristor, which is returned to the blocking mode by extracting current from its gate.

Thyristors, triacs, and GTOs share a common line of descent from diodes and indeed the thyristor is sometimes referred to as an SCR, a silicon controlled rectifier, in recognition of this. Two other major classes of power electronic device have arisen directly from the mainstream of low power electronic devices, the power bipolar junction transistor and the power field effect transistor. These devices are fully controllable, they can be turned off as well as on, and they respond much more quickly. While their power handling capability is an order of magnitude less than that of thyristors, the power bipolar transistors can control substantial power measured in hundreds of kilowatts. However, they have low current gain and consequently require base drive current of the same order as the collector current. The power field effect devices have very high power gains and are consequently easily controlled but their power handling capability is an order of magnitude less than that of bipolar devices. Much current research and development effort is directed towards devices which combine the merits of both classes of device. One such device in commercial production is the insulated gate bipolar junction transistor, the IGBT.

Structure and applicability of the book

The book is devoted to naturally commutated circuits using only diodes, thyristors, and triacs, and can be broadly divided into four sections. The first of these discusses rectifiers and occupies Chapters 2 through 10. The second, Chapter 11, is a review of cycloconverter theory mainly in the context of a suppressed circulating current device but with a brief consideration of a device with circulating current. The third part, Chapter 12, is a review of the transformers used in rectifier and cycloconverter circuits and the special analytical problems which they pose. The fourth and final part, Chapters 13 and 14, is devoted to a.c. voltage regulators. In addition, topics and analytic expressions whose consideration would disrupt the main text are dealt with in five appendices, A through E.

The aim of the book is to be comprehensive and it is not intended to be read from cover to cover. Rather those sections which are relevant to the task in hand can be picked out. Without carrying it to the excess of endless repetition of earlier material, I have tried to make each chapter an independent entity. For example the chapter on cycloconverters can be understood by anyone with an introductory knowledge of rectifiers. Chapters 13 and 14 on a.c. voltage regulators stand on their own and Chapter 11 can be followed by anyone with an introductory knowledge of transformers.

Use as a course text

I have used the material for many years at many teaching levels ranging from a final year undergraduate elective course, through graduate level courses to postgraduate and industrial courses. In an undergraduate course of 13 weeks duration with 39 hours of formal instruction and 39 hours of combined computational–experimental laboratory I find that I can cover the whole of Chapters 2 through 4, about half of Chapter 5, a brief qualitative review of Chapter 6, about half of Chapter 8 and Chapter 13. This leaves about five lecture hours for a review of the circuits with forced commutation found in inverters and choppers.

For Chapter 5, the Effects of Finite Impedance, I feel it is sufficient at the undergraduate level to limit discussion to a description of commutation overlap and the derivation of the rectifier equivalent circuit without entering into the details of waveform analysis.

At the undergraduate level it is sufficient to cover only the latter portion of Chapter 8 which deals with the use of Fourier analysis as a prime analytic tool. This should be restricted to the derivation of the spectra of output voltage and current of an ideal rectifier and should omit the complexities of multiplying Fourier series to obtain input quantities.

I have always felt that, despite its great practical importance, the case of capacitor smoothing, Chapter 7, is too analytically complex for inclusion in an introductory course. However, if the emphasis of the course were on single phase rectifiers this chapter could appropriately replace the material of Chapters 5 and 6.

Cycloconverters, Chapter 11, and transformers, Chapter 12, are specialized topics and would consequently be rarely included in an undergraduate course.

At the undergraduate level a guided tour of the power electronics laboratory is highly desirable to provide students with an idea of the physical nature of the devices and circuits.

I have found that problem assignments requiring computer programming take up too much of an undergraduate student's time and latterly have restricted problems to those which can be solved on a programmable calculator.

In a postgraduate course of 39 lecture hours devoted to naturally commutated circuits I have found that I can cover most of the material in the book. As at the undergraduate level, problem solving is essential and, as graduate students are generally competent programmers, my problem assignments usually involve the use of a computer. However, great care in the design of the problems is essential and much support must be given to prevent the student from being bogged down in a morass of debugging.

Mathematical requirements

While a qualitative treatment based on words and diagrams can serve the needs of those requiring only a passing acquaintance with the subject, analysis is essential to a full understanding. The treatment here is highly analytical with the flow of algebra relieved and illuminated at frequent intervals by numerical examples. As far as possible these are interrelated so that by following the examples the reader should gain a profound quantitative understanding of the circuits. However, do not be put off by the mass of mathematics.

It is important in approaching the text to make a clear distinction between the intellectual difficulty of the mathematics and the complexity of the manipulations. The necessary mathematics is well within the scope of that required of undergraduate students of electrical engineering. The required mathematical toolbox need only contain a knowledge of trigonometry as used by electrical engineers, elementary differential equations, and Fourier analysis. I have occasionally made use of eigenvalue analysis but only in situations such as drive dynamics in Chapter 10 where the reader could be expected to have familiarity with this topic.

Mathematical complexity is another matter. There is no doubt that care and patience are required to follow much of the analysis. The minimum equipment for numerical computation is a scientific calculator, preferably with the ability to store several intermediate results. However, it must be stressed that this is a minimum. Once past the first four chapters, the reader will quickly perceive a need for programing ability. Much can be done with a programmable calculator but a full understanding requires access to a computer. I have made all the calculations and have drawn the diagrams using an IBM compatible PC, initially a 286 machine, then a 386 machine, and now a 486 machine. All my computational programs are written in Fortran which I find very suited to calculations. However, the solutions could equally well be obtained using any of the other computational programming languages such as Pascal and C. The figures are created in Postscript with the aid of auxiliary programs written in C and Fortran. The text is created in the LaTeX version of TeX.

Modeling assumptions

Here, as in so many situations, the complexities of the real world make true mathematical modeling impossible. However, a few judicious and in general unimportant assumptions make this topic analytically accessible. The assumptions are clearly stated in the text. Probably the most important is the use of a piecewise linear model for the power electronic devices, with zero conduction when blocking and a constant small voltage drop when conducting. Three phase sources are assumed to be balanced with sinusoidal voltage waveforms and sometimes a small, inductive line impedance. DC sources are assumed to have pure d.c. e.m.fs free of voltage ripple. These assumptions introduce very little error into the analytical results but, where the devices are concerned, are unsuited to the design process and more complex considerations come to the fore when creating essential auxiliary components such as gating circuits and snubbers, providing adequate fusing, and creating adequate means for dissipating the heat generated within the devices. The text is devoted to analysis and these design considerations are not considered.

Computing and engineering accuracy

I am very conscious of the limits of analysis, that if the results of calculations are within a 1 or 2% of the real life results it is cause for satisfaction. Nevertheless, I have given numerical results to a much higher order of accuracy, often to six significant figures. This has been done for two reasons. First, many of the computations involve the small differences of large quantities and so require highly accurate intermediate computations to attain reasonable accuracy in the final result. Second, it is helpful to readers to have precise answers against which to check their own computations.

Many of the algebraic expressions are of great complexity and, while a great deal of cross checking has been done to ensure accuracy, it is inevitable that errors will have been made. I would appreciate hearing of any errors that readers may find. I have high confidence in the accuracy of the numerical results since they are taken directly from the output of the computer programs which have numerous built-in cross checks. Hence, if an algebraic expression fails to give the numerical result, the expression probably is in error rather than the calculation.

Nomenclature

The number of variables and subscripts is so great that I feel that a table of symbols would be of little value. I have instead defined variables as and when required and do not expect the reader to recollect, at least not without prompting, what was defined several chapters earlier. That being said there are some basic principles of nomenclature which I have followed and which are in general use.

I have of course used such standard nomenclature as v for voltage, i for current, R for resistance, etc. All time-varying quantities are represented by lower case symbols and all constant quantities are represented by upper case symbols unless the symbol is a Greek letter. Thus, a voltage having a sinusoidal waveform, a radian frequency of ω, and an r.m.s. value of V is represented by $v = \sqrt{2}V\cos(\omega t)$. The instantaneous voltage is represented by the lower case v. Its r.m.s. value is represented by the upper case V. The circular frequency, although a constant like the r.m.s. value, is represented by the lower case Greek ω. The time variable is represented by the lower case t. Occasionally, in order to emphasize the time invariant nature of the quantity, I have used an upper case Greek letter. Thus, in Chapter 10 when considering small perturbations in motor speed about a quiescent point I have used Ω for the speed at the quiescent point.

I have used subscripts freely to distinguish between similar quantities, my practice being guided by the need to name variables and parameters in computer programs. Within the great variety of subscripts there are some guiding principles. I have restricted the single subscript s to supply quantities and the single subscript o to output quantities. Thus v_s would be a supply voltage and i_o would be an output current. The subscript rms is obvious and the subscript pk indicates a peak value. Thus, for the sinusoidal voltage referred to in the preceding paragraph $V_{pk} = \sqrt{2}V_{rms}$. Where three phase quantities require greater clarification I have distinguished the three lines as red, yellow, and blue with subscripts r, y, and b. I assume that the three phase supply provided by a utility is grounded somewhere and often reference potentials to this point even though the user may have a three wire system with no access to the neutral. In the more general three phase context I have used the subscripts ll for line to line and ln for line to neutral.

Simple numerical quantities, whether time varying or constant, are represented by italic letters, e.g. v, i, R, L, C. Complex quantities such as phasors and vectors are represented by upper case, bold face, sanserif symbols such as $\mathsf{\mathbf{V}}$, $\mathsf{\mathbf{I}}$, and $\mathsf{\mathbf{Z}}$. A complex quantity can be represented by its real and imaginary components, e.g. as $\mathsf{\mathbf{Z}} = R + jX$ or by its modulus and argument. Rather than use the exponential form for the polar expression I have used the form $Z\underline{/\zeta}$ if the phase angle ζ is in radians or $Z\underline{/\zeta^\circ}$ if it is in degrees.

In power electronic problems it is often important to keep track of which a.c. line is currently connected to the d.c. system. To aid in this, when a circuit diagram contains a representation of a transformer, windings on the same limb of the transformer core are always drawn with their axes parallel and, where necessary, I have used the dot notation to indicate polarity.

Further reading

Since the invention of the grid-controlled mercury arc rectifier in about 1930, dozens of books and thousands of articles have been written on power electronics. The majority of the analyses presented in this book were developed during the 1930s and the book by H. Rissik, *The Fundamental Theory Of Arc Converters*, Chapman and Hall Ltd., 1939, is an excellent guide to that literature. The invention of the thyristor in 1957 led to numerous new books on the topic and the most accessible sources in English of current literature are the *Transactions of the Industry Applications Society of the Institute of Electrical and Electronic Engineers*, New York, and the *Proceedings of the Institution of Electrical Engineers*, London.

Acknowledgements

Finally I wish to recognize my indebtedness to many people and institutions. The first of these is my wife who has with great forbearance and good humor put up with being a computer widow for more years than either of us care to remember. Fortunately this problem has been somewhat relieved by the development of personal computers and by the tolerance of my university so that I can do most of my work at home. I am indebted to generations of graduate students with whom I have explored this topic and in particular I wish to recognize my own mentors, Dr T. F. Wall and Dr O. I. Butler who, many years ago set me on this path with a Ph.D. study of commutation in grid controlled mercury arc rectifiers at the University of Sheffield. I most gratefully acknowledge the financial support of the Natural Sciences and Engineering Research Council of Canada through three decades of my research in this and allied areas. I am deeply grateful to McGill University, where I was a professor from 1957 to 1975, and since then to the University of Calgary who have provided the policies and infrastructure which made this work possible. In particular I am most grateful for the award by the University of Calgary of a Killam Resident Fellowship which freed me to concentrate on writing during the second half of 1988. Finally, I am most grateful to Oxford University Press for their great patience through the many years this book has been in the writing and especially to their Engineering Editor, Richard Lawrence, who throughout all this time has unfailingly provided most valuable advice and encouragement.

Calgary, Canada
July 1993 T. H. B.

CONTENTS

1

PROLOGUE

1.1 Introduction

Electronic equipment has become an indispensable aspect of the lives of all of us. Everyone is familiar with the television receiver, the video cassette recorder, the personal computer, the portable telephone, etc. One important feature of such equipment is its low power consumption, in the range of a few watts to a few hundreds of watts. Far less familiar is electronic equipment which operates at the power levels of interest to industry, in the range of a few kilowatts to many megawatts. Such equipment, its study, design, manufacture, and utilization, is known under the general heading of *power electronics.*

Large amounts of electric power are used in the manufacture of the basic products of industry: aluminum, cement, paper, plastics, steel, etc. An aluminum smelter will typically use hundreds of megawatts in the electrolytic production of the metal. A plant devoted to the production of paper or steel will employ many electric motors whose power ranges from a few hundred to a few thousand kilowatts as well as a much larger number of lower power.

Manufacturing industry which converts the basic materials into the multitude of products, such as automobiles, refrigerators, television sets, and telephones, which we use in our daily lives employs large numbers of electric motors, from a few tens to a few hundreds of kilowatts, and a wide variety of other equipment such as heaters, coolers, and ventilators that are also prodigious consumers of electric power.

Of the numerous reasons for the importance of power electronics, one is pre-eminent: controllability. Power electronic equipment is highly controllable. Its energy throughput and the voltage level at which it is delivered can be controlled with the expenditure of very little power and with speeds which, by comparison with the speeds of industrial processes, are very high. The power gain of a power electronic device, by which is meant the rate of change of output power with control power, $\mathrm{d}p_o/\mathrm{d}p_c$, is typically within the range 10^6–10^8 so that the output power can be increased from zero to a megawatt by changing the control power by a fraction of a watt. This can typically be accomplished in a time comparable with the cycle time of the a.c. supply, i.e. 10–20 ms.

1.2 The link between the computer and the process

Its very high power gain makes power electronic equipment the ideal intermediary between the computer and the process, between the computer which is overseeing and controlling the industrial process and the process itself. The low control power requirements consequent on the high power gain are well suited to the power levels at which control computers and their associated equipment operate. The fast response facilitates and improves the control process.

1.3 Power electronic devices

Like most other electronic devices, power electronic devices are based on the properties of extremely pure, monocrystalline silicon, the active part of the device being a wafer of this material suitably doped with acceptor and donor elements for the task it is to perform. The major difference between the low voltage, low power devices normally encountered and power electronic devices lies in the wafer dimensions. The wafer must be thick to support hundreds and possibly thousands of volts, and it must have a large area to carry the current, tens to hundreds of amperes. To give some idea of scale, the wafer thickness will be of the order of 0.5 mm and its area will be of the order of 1 $mm^2\,A^{-1}$. Thus the wafer for a 100 A device will be a disk of about 12 mm diameter and 0.5 mm thick and the wafer for a 1000 A device will be a disk of about 35 mm diameter of similar thickness.

There are several types of power electronic device in common use: diodes, thyristors, triacs, gate turn-off thyristors, power bipolar junction transistors, power field effect transistors, and insulated gate power bipolar junction transistors. In order to function efficiently they are all employed in the on–off mode, either permitting the almost unimpeded flow of current or almost perfectly blocking it. All the devices are extremely efficient, indeed in comparison to their nonsilicon predecessors (the mercury arc rectifier, the copper oxide rectifier, and the selenium rectifier) they are almost perfect. In the on mode they conduct current and absorb very little voltage, of the order of 2 V. In the off mode the leakage current, i.e. the current they allow through, is extremely small, of the order of one-millionth of their rated current.

1.3.1 *Packaging*

The size of the silicon wafer is a small portion of the total size of the package containing it. This package has provision for current to be led into and away from the wafer and for control signals to be applied to it. It protects the wafer from mechanical damage and atmospheric corrosion. It dissipates the heat generated within the wafer so that the wafer temperature is maintained within the operating limit.

Although these devices have very high efficiency, substantial electric power is lost within them as heat. As an example consider a device which blocks 500 V in the off state and conducts 1000 A in the on state. It is controlling $500 \times 1000 \times 10^{-3} = 500$ kW and, since it will absorb about 2 V, it will dissipate about 2 kW in the on state, 0.4% of the power it is controlling. This heat loss, while a very small part of the total power, is twice the power of a typical domestic electric kettle but instead of being put into 1 l or more of water it is being released into about 0.5 cm^3 of silicon which will quickly melt unless the packaging provides some very effective means for dissipating the heat.

While the plastic packaging, used for many electronic devices found in such familiar items as television sets and computers, is employed for devices rated below a few kilowatts the more usual package is a combination of metal for current and heat conduction and porcelain for electrical insulation, the whole forming a hermetically sealed, mechanically strong enclosure. This, since the active piece of material inside it is a thin wafer of silicon, has much the same appearance no matter what type of device is enclosed.

1.3.2 *Diodes*

Silicon diodes have no control function but are universally employed in power electronic equipment where they perform a variety of roles, principally as rectifiers and to steer current flow into desired paths. A diode conducts current in one direction while absorbing very little voltage and blocks voltage in the other direction while allowing very little current flow. The diode symbol and the conventional voltage polarity and current directions for the diode voltage, v_d and the diode current, i_d are shown in Fig. 1.1(a). Positive current enters at the anode, labelled A, and leaves at the cathode, labelled C.

The static voltage–current characteristic showing the relationship between v_d and i_d is given in Fig. 1.1(b). This shows the typical shape of the relationship but is not drawn to scale. When the diode voltage is negative the leakage current is very small and is negligible for almost all practical purposes. When the voltage is positive, current flows and the diode voltage is positive and small, of the order of 2 V.

For the majority of purposes, including all those covered in this book, this is all the knowledge about diodes that is needed. However, it is appropriate to mention some aspects of extreme behavior since these are crucial to the viability of the device. As the magnitude of the negative voltage

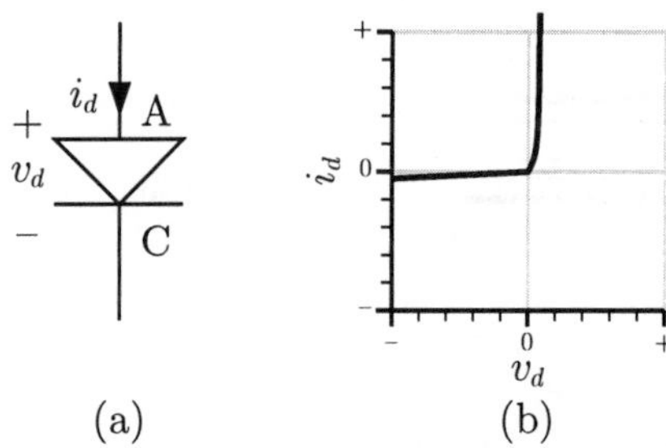

FIG. 1.1. (a) The diode symbol and (b) its static volt-ampere characteristic.

applied to a diode is increased, a point is reached beyond which the leakage current through it increases rapidly. This happens because the voltage stress in the silicon wafer has become so high that the silicon can no longer support it and experiences an avalanche breakdown. The high voltage gradient within the material causes large numbers of electrons to be freed from the outer layers of the silicon atoms and the reverse leakage current is enormously increased. The combination of high voltage and high leakage current results in a very high heat dissipation in the wafer and causes failure within milliseconds unless the condition is relieved. The voltage level at which this blocking failure occurs is well above the rated operating voltage, typically two or more times greater, and the equipment is designed so that the limit is never exceeded. This breakdown region is well outside the normal operating voltage range shown in Fig. 1.1(b).

In the on state, the voltage absorbed by the device is not constant as has hitherto been implied, but varies somewhat with current. It is true that the voltage absorbed by the forward biased P–N junction within the silicon wafer is more or less constant over a very wide range of current but as the current becomes small and approaches zero the voltage absorbed also tends to zero.

The package enclosing the wafer presents a very small resistance to current flow which results in a modest increase in voltage drop with increase in current. However, for currents up to the rated value this effect is small and it is adequate for most purposes to model the device in the on state by a small voltage drop of about 2 V. After all, a device voltage drop of 2 V is only 0.4% of the output voltage of a 500 V rectifier and a 10% error in its estimation would have only a 0.04% effect on computations. Indeed the device voltage drop is so small that it is sometimes ignored altogether, a course followed here in Chapters 13 and 14.

Throughout the text it is assumed that a diode has zero reverse leakage current and has a forward voltage drop whose value is about 2 V and is independent of the forward current flowing though the device.

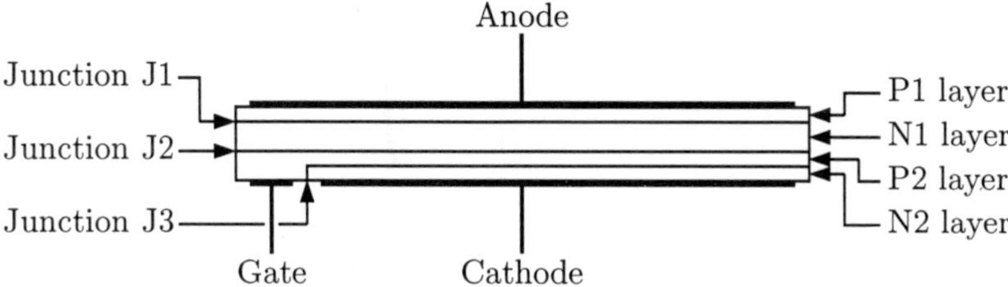

FIG. 1.2. Diagrammatic representation of the silicon wafer of a thyristor showing four P–N–P–N layers and the three junctions J1, J2, and J3.

1.3.3 *Thyristors*

Thyristors, alternatively called silicon controlled rectifiers or SCRs, are built on the same principles as diodes. A diagrammatic cross-section of the silicon wafer of a thyristor is shown in Fig. 1.2. There are four layers in the wafer so that the structure is P–N–P–N providing three junctions, two P–N junctions, J1 and J3, surrounding one N–P junction, J2, as shown in Fig. 1.2. The anode lead is connected to the outer positive layer, P1, the cathode lead is connected to the outer negative layer, N2, and the control electrode or gate lead is connected to the inner positive layer, P2. The series combination of the first P–N junction, J1, and the central N–P junction, J2, means that the device blocks current flow in both directions. This not very useful characteristic is turned into something of enormous value by the presence of the second P–N junction, J3. The P2 and N2 layers are formed so that a minute amount of control power will cause junction J3 to break down and flood junction J2 with so many carriers that it too collapses and permits almost unimpeded current flow.

The thyristor acts as a switch which is normally open but which can be closed by the momentary application of a small positive voltage to its gate terminal. The very low power required to initiate turn on and the rapidity of the process are due to a regenerative process which occurs in the wafer. As soon as the turn-on is initiated, the main power circuit ensures that it rapidly progresses towards completion. However, this desirable behavior has a down side: control is lost once the device is turned on and current is flowing through the wafer. Control can only be regained if the current becomes zero for a brief period.

The symbol for a thyristor is shown in Fig. 1.3(a) with the conventional polarity for the thyristor voltage, v_{th} and the conventional direction of the thyristor current, i_{th}. Positive current enters the anode, A, and leaves at the cathode, C. The control or gate terminal is marked G.

The static volt-ampere characteristic showing the relationship between v_{th} and i_{th} is displayed in Fig. 1.3(b). This shows the typical shape of the

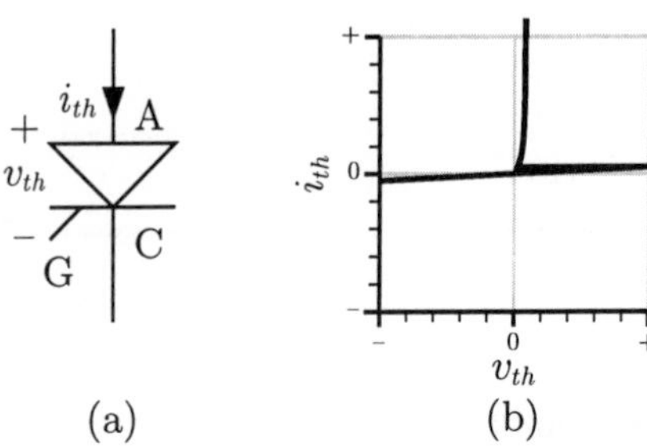

FIG. 1.3. (a) The thyristor symbol and (b) its static volt-ampere characteristic.

relationship but is not drawn to scale. The static characteristic shows the same type of blocking behavior as a diode but now in both directions. There is negligible leakage current until the applied voltage approaches the breakdown value which is well beyond the operating voltage range shown in Fig. 1.3(a). For negative voltages, breakdown is exactly like that described for a diode. For positive voltages, breakdown initiates the regenerative process mentioned above and the thyristor switches to the on mode.

Exceeding the forward voltage breakdown limit is not a useful way of controlling the thyristor and voltage is normally held well below the breakdown limit, with control being exercised by the application of positive voltage and current to the gate terminal.

For most purposes, including those of this book, the thyristor can be treated as a switch with precisely controllable turn-on capability and with a constant voltage drop in the on state of about 2 V. However, a very brief discussion of its dynamic characteristics is appropriate here.

While the power required to initiate turn-on is very small, it is finite and too small a gate signal will not produce turn-on or will cause the turn-on process to be slow. It is therefore important to have a turn-on signal power well above the device manufacturer's minimum and to have a rapid build up to the full signal, rapid in this context meaning within a microsecond or less. Given a gate signal of adequate power, turn-on will be accomplished in a few microseconds, a period which is brief enough to be regarded as negligible in the context of the usual power system period. However, it is possible, in circuits of low inductance combined with large driving voltage, for the anode to cathode current to rise faster than the rate of growth of the turned-on region can accommodate. This excessive di/dt causes overheating of the wafer which may lead to failure. The thyristor data sheets supplied by the device manufacturer will specify a maximum rate of rise of current for the device.

Two ways in which turn-on can be initiated have already been mentioned: by applying a gate signal and by applying a high anode to cathode voltage. Additionally, a high rate of increase of anode to cathode voltage can initiate turn-on. This is the dv/dt phenomenon and, as for its counterpart the di/dt effect, the thyristor manufacturer will specify a maximum value. Both this dv/dt effect and the di/dt effect are of little importance for the majority of applications at the normal power frequencies of 50 and 60 Hz.

Throughout the text the assumption is made that a thyristor in the off state conducts no current and that in the on state it can be represented by a voltage drop of the order of 2 V which is independent of the current passing through it.

1.3.4 *Triacs*

Triacs have a more complex internal wafer structure, with five layers and four P–N junctions which provide them with thyristor-like behavior in both directions, as illustrated in Fig. 1.4. This shows the device symbol (a) and its static volt-ampere characteristic (b). Since a triac conducts current in both directions it cannot strictly be said to have an anode and a cathode. Nevertheless, the standard practice has been followed in Fig. 1.4(a) by labelling the side remote from the gate the anode and the side closest to the gate the cathode.

A triac blocks current flow in both directions until a signal, a small voltage and current, is applied to its gate. Then it breaks down and freely conducts current in either direction. The comments made above regarding the limitations of diodes and thyristors apply. Too high a voltage in the blocking mode will cause the triac to switch to the conducting mode as will too high a dv/dt. Too high a rate of increase of current through it will destroy it.

Triacs are extremely useful for the control of a.c. currents, as described in Chapters 13 and 14. Many hundreds of millions are employed the world over in the familiar domestic lamp dimmer. However, their complex structure means that they cannot be manufactured to the high voltage and current levels common for diodes and thyristors so that their use is limited to relatively low power levels of a few tens of kilowatts. At higher power levels their function is fulfilled by a pair of back to back connected thyristors as illustrated in Fig. 13.1. This is extremely effective but the convenience of a single gate is lost and two electrically isolated gate driver circuits must be employed

In the text it is assumed that a triac in the off state conducts no current and that in the on state it can be represented by a voltage drop of about +2 V when the current is positive and about −2 V when the current is negative.

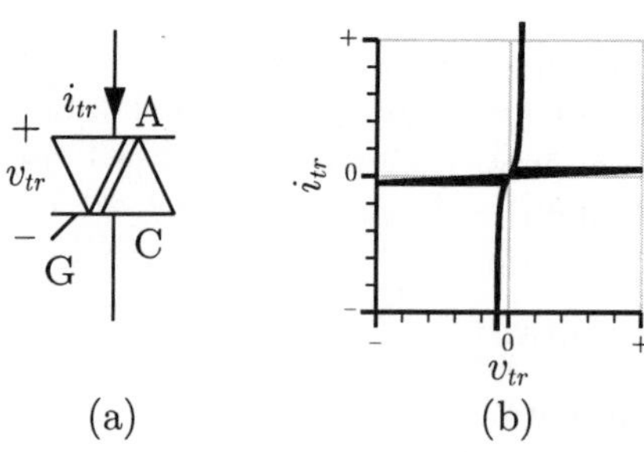

FIG. 1.4. (a) The triac symbol and (b) its static volt-ampere characteristic.

1.3.5 *GTOs, BJTs, FETs, and IGBTs*

The remaining devices, gate turn-off thyristors or GTOs, power bipolar junction transistors or BJTs, power field effect transistors or FETs, and insulated gate bipolar junction transistors or IGBTs, have turn-off as well as turn-on capability. Like the thyristor from which it is derived, the GTO is an on–off device, it either conducts current while absorbing very little voltage or blocks voltage while conducting very little current. It is like a switch that can be opened as well as closed at will. The transistors, like their low power counterparts, can exercise continuous control of current but are always used in the on–off mode as continuous control under power electronic conditions would result in destructively high internal device losses.

These devices are not used in the type of circuits described in this book and will not be considered further.

1.4 Natural commutation

Turn-off capability comes at high cost, both monetarily and in device characteristics, and turn-off devices are used only where essential, in inverters and choppers. When the power electronic equipment is operated from an a.c. supply, the supply itself contains the turn-off mechanism in that its voltage reverses direction twice per cycle. Under these circumstances thyristors can be employed and the circuits can exploit their positive features—very high voltage handling capability in the off state, and very high current carrying capacity and low voltage drop in the on state.

Circuits which exploit the a.c. supply to turn off thyristors are said to be *naturally commutated.* They are the subject of this book and the devices with turn-off capability are not further mentioned except very briefly at the end of Chapter 9. There it is noted that a rectifier with controllable turn-off capability in addition to controllable turn-on capability has a number of very attractive features. Unfortunately such rectifiers are not yet feasible for the general run of applications for a number of reasons: cost, limited power range, and limited switching frequency capability. No doubt as existing devices are improved and new devices are developed this situation will change and the newer and improved devices may well challenge the present dominant role of diodes and thyristors in the naturally commutated circuits.

1.5 Naturally commutated circuits

There are three types of naturally commutated circuit: rectifiers cycloconverters and a.c. voltage regulators. They have one thing in common, one side of them is connected to an a.c. supply, usually the power system. One side of a rectifier is connected to an a.c. supply and the other to a d.c. system. Usually power flows from the a.c. side to the d.c. side, but if the d.c. side contains a power source, e.g. a d.c. generator, the power flow can reverse, from the d.c. side to the a.c. side. In this event the rectifier is said to operate in the inverter mode, a usage of the word inverter which must not be confused with the more common application to an apparatus in which turn-off devices are employed to convert d.c. into a.c. without the aid of an existing a.c. supply.

A cycloconverter draws input power from the a.c. fixed frequency, fixed voltage system and converts it to a.c. output power at another frequency and voltage, both these output quantities usually being variable.

An a.c. voltage regulator draws input power from the a.c. supply at fixed frequency and voltage and converts it to a.c. output power of the same frequency but variable voltage.

Rectifiers are by far the most important of these three types of apparatus, a fact reflected by the devotion to them of the majority of this book. Rectifier topics occupy Chapters 2 through 10, an overview of cycloconverters is given in Chapter 11, and a.c. voltage regulators are dealt with in Chapters 13 and 14. Rectifier transformers, which are as important for cycloconverters as for rectifiers, are considered in Chapter 12.

1.6 Three phase and single phase systems

In this book the word *power* in the phrase *power electronics* is emphasized. The concern is with equipment of interest to industry which will typically operate at power levels of a kilowatt or more from a three phase system. This is not to say that industry does not use single phase power, it does, especially below one kilowatt, but to recognize that the overwhelming majority of industrial power is supplied and used as three phase. Consequently there is very little mention of single phase in the book. Chapter 13 is entirely devoted to single phase a.c. regulators and, although written in a three phase context, substantial segments of the remainder can be applied to single phase situations. For example a study of Chapters 2 through 8 would leave the reader well equipped for the analysis of any single phase rectifier problem. Nevertheless it must be recognized that single phase rectifiers do have problems and techniques peculiar to themselves.

Most single phase rectifiers operate at low power levels of the order of 100–200 W, e.g. the primary power supply for a TV receiver or a personal computer. In such applications capacitor smoothing is almost universally employed and the coverage of this topic in Chapter 8 is readily applied to the single phase case. At lower power levels of a few watts, e.g. the battery chargers for portable telephones and pocket calculators, special circuits are employed which sacrifice efficiency to obtain other advantages such as low manufacturing cost and reduced volume and weight. Articles and texts specializing in these applications are available.

It has been implied up to now that single phase applications operate at low power levels. This is true for most such equipment but is not true for one very important application, electric traction. While some electric railways still operate from a d.c. excited third rail or catenary cable, the preferred method today is to supply the trains with standard 50 or 60 Hz power from a relatively high voltage catenary, e.g. 25 kV. On the locomotive, transformers reduce this voltage to an appropriate level and single phase rectifiers rated at hundreds of kilowatts convert it to d.c. for further processing, usually to variable frequency, variable voltage, three phase a.c., to feed the propulsion motors. These rectifiers and their associated circuits have special configurations and operating modes whose principal aim is to minimize as far as possible the problems associated with the use of single phase at such high power levels. These are the subject of special articles and texts.

2

SIMPLE SINGLE WAY RECTIFIERS

2.1 Introduction

A single way rectifier is one in which the output current to the load passes through a single thyristor or diode, or a single group of series and/or parallel connected thyristors or diodes, on its way from the a.c. to the d.c. system. A simple rectifier is one which is composed solely of thyristors and/or diodes with the possibility of a transformer interposed between the rectifier and the a.c. system. Because of their basic simplicity, simple single way rectifiers, while relatively rare in practice, provide a good starting point for the study of the more popular but more complex bridge rectifiers whose study begins in Chapter 3.

2.2 The simple three phase rectifier

The circuit diagram of a simple, single way, three phase rectifier is shown in Fig. 2.1. It is fed from a three phase voltage source, terminals R, Y, and B at the left, of sinusoidal waveform, frequency f_s, r.m.s. line to line voltage V_s, via a delta/wye transformer. The wye connected secondary provides the three phase output at terminals 1, 2, and 3 and also provides a neutral terminal, N, which is necessary for the functioning of the rectifier. Terminals 1, 2, and 3 are connected to the anodes of three thyristors, T1, T2, and T3 whose cathodes are joined together to form the positive pole, K, of the d.c. output.

The transformer diagram, as well as indicating the windings also indicates their location on the core and polarity by the following convention. Windings whose axes are parallel are on the same limb of the transformer and coils wound in the same direction have the same polarity. Thus, the primary winding which is connected between the R and Y terminals and the secondary winding connected between terminal 1 and terminal N have parallel axes and are therefore on the same limb of the transformer core. Further, they are wound in the same direction so they have the same polarity, a current entering the secondary at the terminal 1 end having a similar magnetic effect as a current entering the primary at the terminal R end. This convention is followed throughout the book.

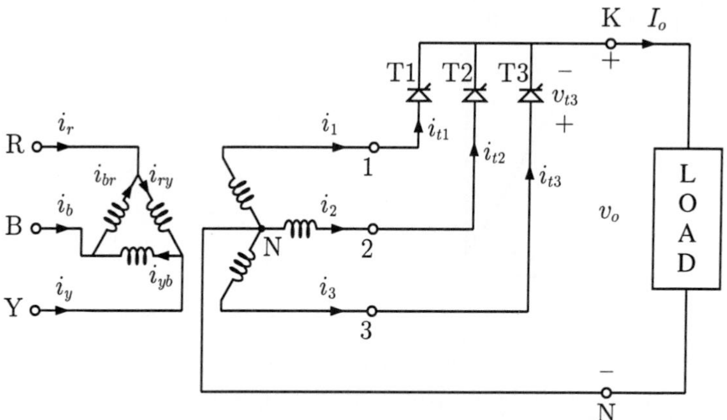

FIG. 2.1. The circuit diagram of a single way, three phase rectifier.

The simple, single way, three phase rectifier is rarely encountered in practice but when it is used a more complex transformer, a zigzag transformer, is employed to avoid certain difficulties concerned with core magnetization. This type of transformer is discussed in Section 12.4. For the time being, a magnetically linear core will be assumed so that this problem does not arise.

The load is assumed to be active, e.g. a battery or d.c. motor armature, with a pure d.c. e.m.f. E_o, i.e. a time invariant, ripple free e.m.f. The assumption is also made that the load has sufficient inductance to effectively smooth the output current which is therefore the time invariant, ripple free quantity I_o. The transformer is assumed to be ideal with a turns ratio, secondary winding turns to primary winding turns per phase, of N_{tsp}. All these assumptions are removed in succeeding chapters.

2.3 The secondary voltages

The secondary phase voltage has an r.m.s. value $N_{tsp}\, V_s$ and the time origin will be chosen so that the voltages of terminals 1, 2, and 3 relative to the transformer neutral are

$$v_1 = \sqrt{2} N_{tsp} V_s \cos(\theta) \tag{2.1}$$

$$v_2 = \sqrt{2} N_{tsp} V_s \cos(\theta - \frac{2\pi}{3}) \tag{2.2}$$

$$v_3 = \sqrt{2} N_{tsp} V_s \cos(\theta + \frac{2\pi}{3}) \tag{2.3}$$

where $\theta = \omega_s t$ is the angle corresponding to time t and ω_s is the radian frequency of the supply, $2\pi f_s$.

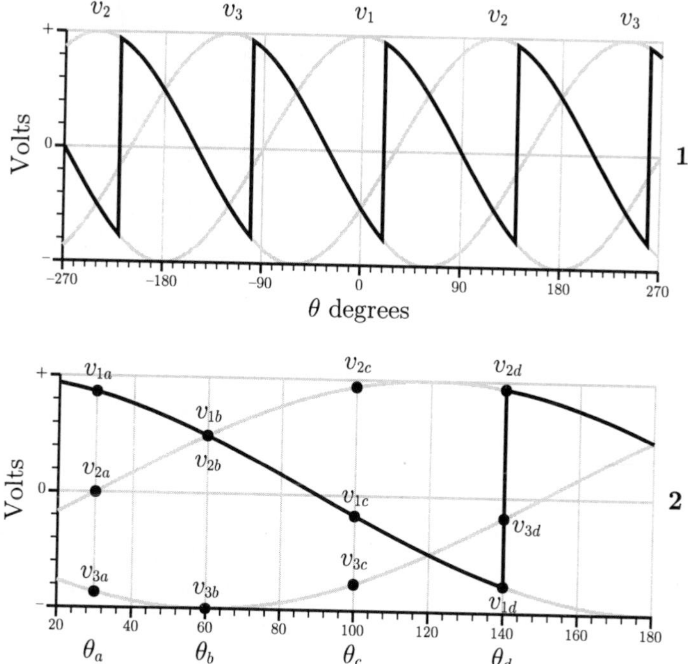

FIG. 2.2. Voltage waveforms in a simple, three phase rectifier.

The three line to neutral voltages v_1, v_2, and v_3 are shown by gray lines in Fig. 2.2 and the rectifier output voltage is shown by a black line. The graphs are identified by a number to their right. Graph 1 covers the range $-270° < \theta < 270°$ and an expanded portion of the θ axis covering the range $20° < \theta < 180°$ is shown in graph 2. The waves are identified in graph 1 by the notations v_1, v_2, and v_3 above their positive peaks. Specific values of $\theta, \theta_a, \theta_b, \theta_c$, and θ_d corresponding to specific times t_a, t_b, t_c, and t_d are indicated below the θ axis of graph 2. Specific voltages v_{1a}, v_{2a}, v_{3a}, etc. corresponding to these value of θ are indicated by annotated black dots on graph 2.

Referring to graph 2, at θ_a the voltage of terminal 1 relative to the transformer neutral is v_{1a}, the voltage of terminal 2 relative to the transformer neutral is v_{2a} and the voltage of terminal 3 relative to the transformer neutral is v_{3a}. Thus terminal 1 is the most positive of the three and thyristor T1 conducts. Thyristor T2 is reversed biased by the voltage $(v_{1a} - v_{2a})$ and therefore does not conduct. Thyristor T3 is reverse biased by the voltage $(v_{1a} - v_{3a})$ and does not conduct. The positive output terminal, K, is connected to terminal 1 by T1 so that, except for a small voltage drop in the thyristor, v_k is equal to v_1. Since v_1 is the potential of terminal 1 relative to the neutral and, since the neutral forms the negative pole of the output, the output voltage, v_o, is equal to v_1 less the small voltage drop in the conducting thyristor.

As time passes, study of graph 2 shows that the potential of terminal 1 falls, the potential of terminal 2 rises and the reverse bias across thyristor T2 decreases until at θ_b the two potentials are equal and the reverse bias is zero. The potential of terminal 1 continues to fall and the potential of terminal 2 continues to rise until at θ_c there is a substantial forward bias, $v_{2c} - v_{1c}$, across T2. The third thyristor T3 is still reverse biased by the voltage $v_{1c} - v_{3c}$.

T2 can withstand the forward voltage until it receives a gate signal at $\theta = \theta_d$ which converts it to the conducting mode. Terminal 2 is immediately connected to K and a reverse bias, $v_{2d} - v_{1d}$, is applied across T1 which therefore ceases to conduct, recovers its blocking capability and prevents further conduction from terminal 1. It will be assumed that each thyristor is fired at a similar instant in its cycle. Thus T3 is fired at $\theta_d + 120°$, T1 at $\theta_d + 240°$, T2 at $\theta_d + 360°$, etc. The output voltage is then as indicated by the black line of graph 1, being comprised of 120° segments of the three phase voltages.

2.4 Delay angle

If diodes were used instead of thyristors, they would be unable to support any positive voltage and commutation from one to the next would occur as soon as the incoming diode was forward biased, i.e. D2 would commence to conduct at $\theta_b = 60°$, D3 would commence to conduct at $\theta_b + 120° = 180°$, D1 would commence to conduct at $\theta_b - 120° = -60°$. These points, $\theta_b - 120°$, θ_b, and $\theta_b + 120°$, are the *natural commutation instants* for T1, T2, and T3 and commutation is said to be delayed by the *delay angle*, α, where $\alpha = \theta_d - \theta_b$.

2.5 The output voltage

The waveform of the output voltage is shown by the black lines in Fig. 2.3 for delay angles of 20°, 50°, 80°, 110°, and 140°. The waveforms of the three phase voltages v_1, v_2, and v_3 are shown for reference by the gray lines. Each graph is identified by a number to its right and each graph is annotated with its delay angle α.

The output voltage has a mean or d.c. value with a variation or *ripple* superimposed upon it, the waveform of the ripple repeating every 120° so that the ripple frequency is three times that of the supply. Shortly, quantitative measures of the mean value and ripple will be determined but for the moment it is evident from an inspection of Fig. 2.3 that the

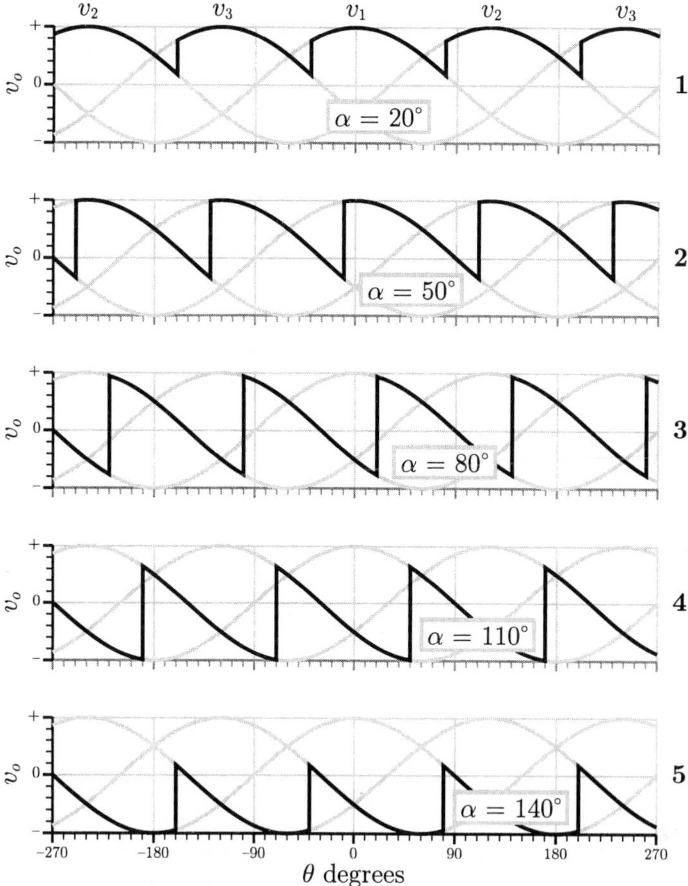

FIG. 2.3. Output voltage waveforms for a simple, single way, three phase rectifier for delay angles of 20, 50, 80, 110, and 140°.

mean value decreases as the delay angle increases and indeed is negative for graphs 4 and 5 drawn for delay angles of 110° and 140°. To maintain constant current flow in the load in these circumstances requires that the load be active, i.e. that it has an e.m.f. capable of maintaining positive current flow against the negative mean output voltage of the rectifier.

As already noted, the output voltage waveform is comprised of 120° segments of the three phase waveforms so that for $-60° + \alpha < \theta < 60° + \alpha$

$$v_o = v_1 = \sqrt{2} N_{tsp} V_s \cos(\theta) \tag{2.4}$$

and for $60° + \alpha < \theta < 180° + \alpha$

$$v_o = v_2 = \sqrt{2} N_{tsp} V_s \cos(\theta - \frac{2\pi}{3}) \tag{2.5}$$

and for $180° + \alpha < \theta < 300° + \alpha$

$$v_o = v_3 = \sqrt{2} N_{tsp} V_s \cos(\theta + \frac{2\pi}{3}) \tag{2.6}$$

etc.

2.6 The mean output voltage

The shape of each 120° segment is the same as that of all others so that the mean value, V_o, of the output voltage is given by

$$V_o = \frac{3}{2\pi} \int_{-\pi/3+\alpha}^{\pi/3+\alpha} \sqrt{2} N_{tsp} V_s \cos(\theta)\, d\theta \tag{2.7}$$

i.e.

$$V_o = \frac{3}{\pi} \sqrt{\frac{3}{2}} N_{tsp} V_s \cos(\alpha) \tag{2.8}$$

or

$$V_o = V_{om} \cos(\alpha) \tag{2.9}$$

where V_{om} is the maximum value of the output voltage, obtained when the delay angle is zero. It is

$$V_{om} = \frac{3}{\pi} \sqrt{\frac{3}{2}} N_{tsp} V_s = 1.1695 N_{tsp} V_s \tag{2.10}$$

It is seen that the output voltage is a cosinusoidal function of the delay angle, varying from $+V_{om}$ at $\alpha = 0$, through zero at $\alpha = 90°$, to $-V_{om}$ at $\alpha = 180°$, as shown in Fig. 2.4.

The output voltage is not pure d.c. but has a substantial ripple component. In Chapter 8 the Fourier components of the ripple are derived. Here only its frequency and r.m.s. value are noted.

2.7 The output voltage ripple

Since the output voltage waveform repeats every 120°, the ripple frequency, f_{rip}, is three times the supply frequency

$$f_{rip} = 3 f_s \tag{2.11}$$

The ripple voltage, v_{rip}, is the difference between the actual voltage and the mean value. Thus for the interval $-60° + \alpha < \theta < 60° + \alpha$

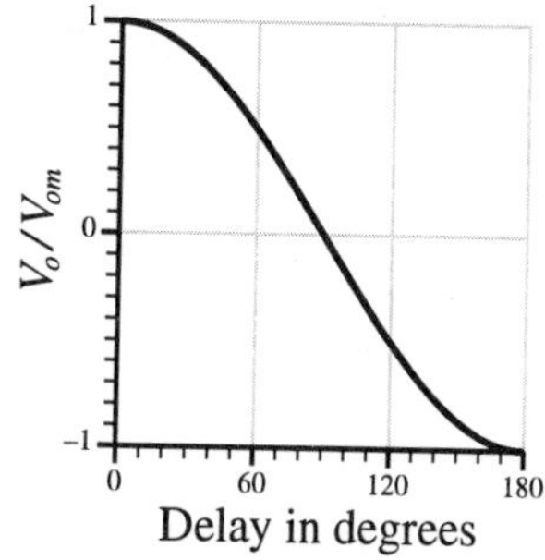

FIG. 2.4. The mean output voltage as a function of delay angle.

$$v_{rip} = \sqrt{2}N_{tsp}V_s\cos(\theta) - V_o$$

It is convenient to convert this to a function of V_{om}

$$v_{rip} = V_{om}\left[\frac{2\pi}{3\sqrt{3}}\cos(\theta) - \cos(\alpha)\right]$$

The r.m.s. value, V_{rip}, of v_{rip} is obtained in the usual way by squaring this expression, averaging, and taking the square root

$$V_{rip} = V_{om}\left[\left(\frac{2\pi^2}{27} - \frac{1}{2}\right) - \left(\frac{1}{2} - \frac{\pi\sqrt{3}}{18}\right)\cos(2\alpha)\right]^{1/2} \tag{2.12}$$

or expressed numerically

$$V_{rip} = V_{om}\sqrt{0.2311 - 0.1977\ \cos(2\alpha)} \tag{2.13}$$

Thus the ripple voltage is a function of 2α, increasing from a minimum of $0.1827V_{om}$ when the delay angle is zero to a maximum of $0.6548V_{om}$ when the delay angle is 90° and then decreasing back to the same minimum, $0.1827V_{om}$, when the delay angle is 180°.

The r.m.s. value of the ripple voltage is shown as a function of delay angle in Fig. 2.5.

Example 2.1

A single way, three phase rectifier is to be fed from a 440 V 60 Hz three phase supply. It must be capable of producing up to 300 V d.c. across a 200 kW load. Determine

- the rated output current
- the ratio of the transformer

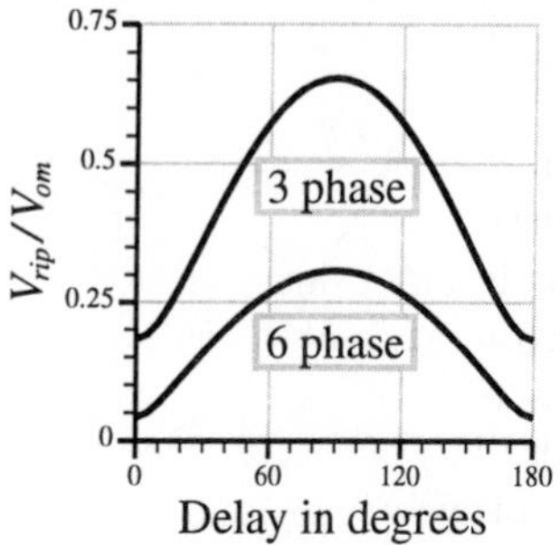

FIG. 2.5. The r.m.s. ripple voltage for single way, three and six phase rectifiers as a function of the angle of delay.

- the frequency of the output voltage ripple
- the mean value of the output voltage at delay angles of 0°, 40°, and 80°
- the r.m.s. value of the output voltage ripple at these delay angles

The rated output current of the rectifier is

$$I_{orat} = \frac{200 \times 1000}{300} = 666.7 \text{ A}$$

In eqn 2.10 it is known that V_s is 440 V and that V_{om} is 300 V . From these the transformer ratio is

$$N_{tsp} = 0.5830$$

The ripple frequency is three times the supply frequency

$$f_{rip} = 180 \text{ Hz}$$

Since V_{om} is 300 V

$$\begin{aligned} \alpha &= 0^\circ, & V_o &= 300.0 \text{ V} \\ \alpha &= 40^\circ, & V_o &= 300.0 \cos(40^\circ) = 229.8 \text{ V} \\ \alpha &= 80^\circ, & V_o &= 300.0 \cos(80^\circ) = 52.1 \text{ V} \end{aligned}$$

The r.m.s. value of the voltage ripple is given by eqn 2.13.

$$\begin{aligned} \alpha &= 0^\circ, & V_{rip} &= 54.8 \text{ V} \\ \alpha &= 40^\circ, & V_{rip} &= 133.1 \text{ V} \\ \alpha &= 80^\circ, & V_{rip} &= 193.7 \text{ V} \end{aligned}$$

****

2.8 The output current

The waveform of the output current is shown in graph 1 of Fig. 2.8. Since it is assumed that the inductive component of the load impedance is very large, the ripple component is negligible and the wave appears as a horizontal straight line. The output current wave has been subdivided by gray vertical lines into segments of 120° width corresponding to the periods when the thyristors conduct. The figure is drawn for a delay angle of 75°.

The mean value of the output current, I_o, is given by

$$I_o = \frac{V_o - V_{thy} - E_o}{R_o} \tag{2.14}$$

where V_o is the mean value of the output voltage, V_{thy} is the voltage drop across a conducting thyristor, E_o is the load e.m.f. and R_o is the load resistance.

Example 2.2

If, for the single way rectifier circuit described in Example 2.1, the load circuit resistance is 0.042 Ω, the thyristor forward voltage drop is 1.8 V and the delay angle is 40°, determine the load current when the load e.m.f. is 210 V.

$$I_o = \frac{300.0\cos(40°) - 1.8 - 210.0}{0.042} = 428.9 \text{ A}$$

****

2.9 The thyristor voltage

Considering thyristor T1, the voltage across it, v_{t1}, has a small positive value during the time it conducts, for $-60° + \alpha < \theta < 60° + \alpha$. While this voltage is somewhat dependent on the thyristor current, it is more or less constant at about 2 V and it is being approximated by a small constant quantity whose symbol is V_{thy}.

When the next thyristor in the sequence, T2, conducts, the anode of thyristor T1 is at the potential v_1 and its cathode is at the potential v_2 (apart from the small voltage drop in T2, of the about 2 V). Thus the voltage across T1 is $v_1 - v_2$.

When T3 conducts, the anode potential of T1 is still v_1 but its cathode potential is now v_3 and the thyristor voltage is $v_1 - v_3$. By drawing

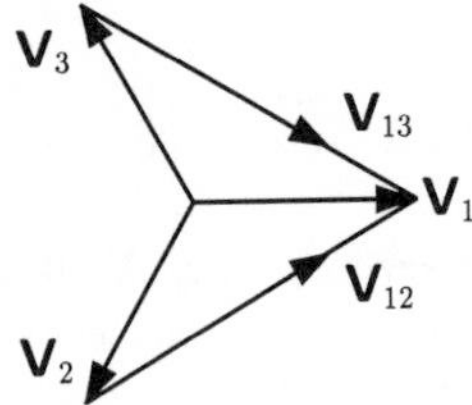

FIG. 2.6. Voltage phasor diagram for the single way, three phase rectifier.

the voltage phasor diagram, Fig. 2.6, the thyristor voltage can readily be identified. Thus for $-60° + \alpha < \theta < 60° + \alpha$

$$v_{t1} = V_{thy} \tag{2.15}$$

for $60° + \alpha < \theta < 180° + \alpha$

$$v_{t1} = v_1 - v_2 = \sqrt{6} N_{tsp} V_s \cos(\theta + \pi/6) \tag{2.16}$$

and for $180° + \alpha < \theta < 300° + \alpha$

$$v_{t1} = v_1 - v_3 = \sqrt{6} N_{tsp} V_s \cos(\theta - \pi/6) \tag{2.17}$$

The thyristor voltage is comprised of three segments each of 120° duration. The first segment extends from $-60° + \alpha$ to $60° + \alpha$, the thyristor is conducting and the voltage across it is a small positive and essentially constant voltage of about 2 V. The second segment extends from $60° + \alpha$ to $180° + \alpha$ and is a section of the wave v_{12}. The third segment extends from $180° + \alpha$ to $300° + \alpha$ and is a section of the wave v_{13}. These waves are shown by gray lines in Fig. 2.7 together with thyristor voltage waves, shown by the black lines for delay angles of zero, 45°, 90°, and 135°. Clearly the thyristor must be capable of repeatedly withstanding the peak value of the secondary line to line voltage, $\sqrt{6} N_{tsp} V_s$, in both positive and negative directions, i.e. the maximum thyristor voltage, V_{tmax} is

$$V_{tmax} = \sqrt{6} N_{tsp} V_s \tag{2.18}$$

or, in terms of V_{om},

$$V_{tmax} = \frac{2\pi}{3} V_{om} = 2.0944 V_{om} \tag{2.19}$$

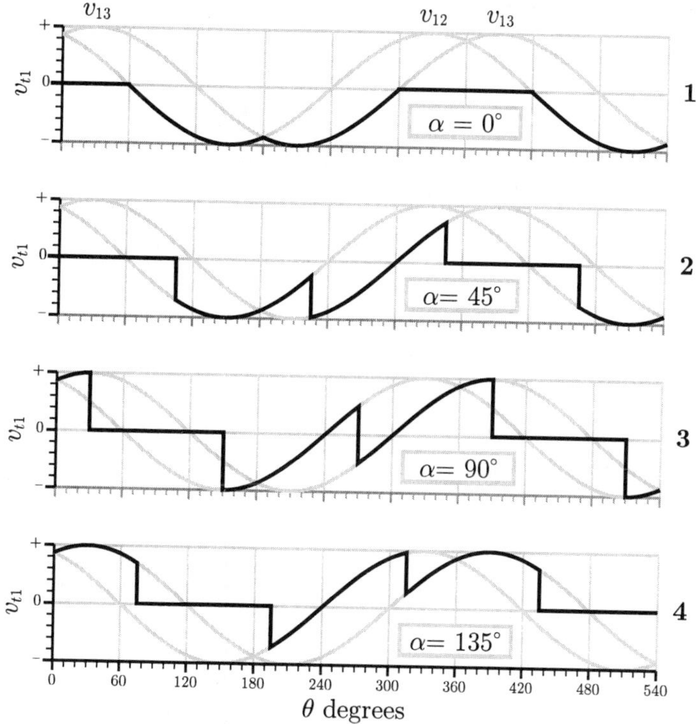

FIG. 2.7. Thyristor voltage waveforms for various delay angles.

2.10 The thyristor current

The waveforms of the thyristor currents are shown in graphs 2, 3, and 4 of Fig. 2.8. The waveforms are rectangular with each thyristor carrying the output current, I_o, for one third of a cycle. The mean thyristor current, I_t, is given by

$$I_t = I_o/3 \tag{2.20}$$

The r.m.s. thyristor current, I_{trms}, is given by

$$I_{trms} = I_o/\sqrt{3} \tag{2.21}$$

2.11 Thyristor power loss

Representing a thyristor in the on state by a constant voltage drop is an excellent approximation for the type of computations which are made in this book. There is however one situation where a better model is required,

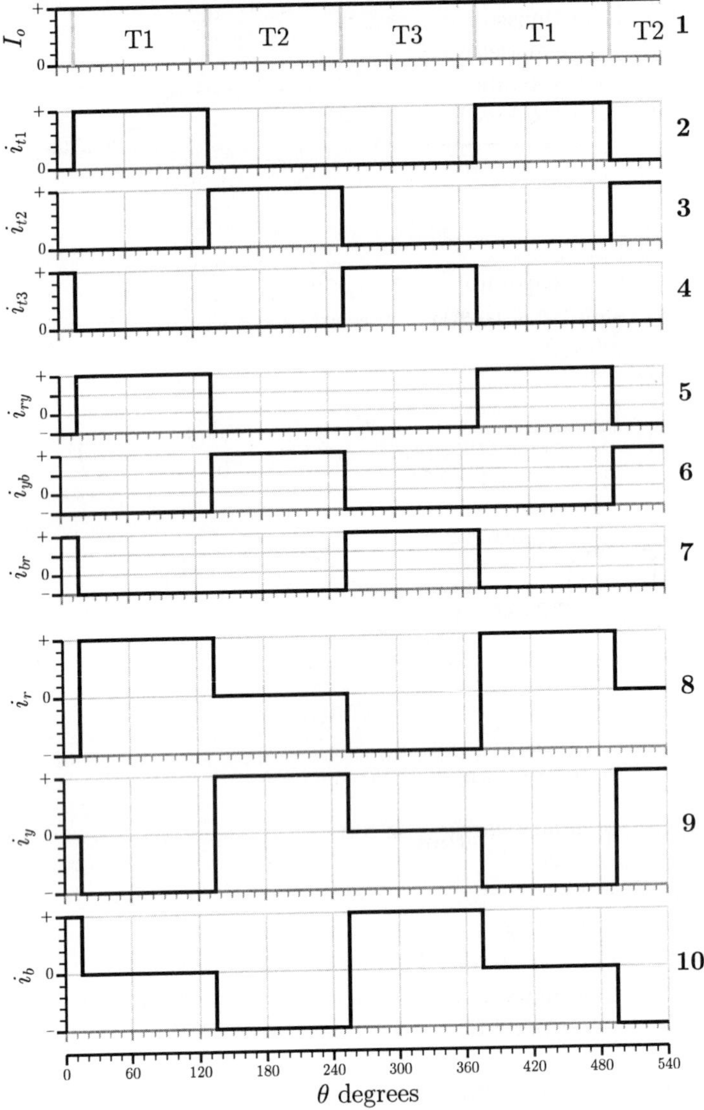

FIG. 2.8. Current waveforms for various delay angles.

the determination of the mean power loss in the thyristor. This is important when designing the heat sink which will maintain the thyristor wafer temperature below the permissible maximum.

If the thyristor in the on state could be accurately modeled by a constant voltage drop, then the mean power loss would be the product of that voltage drop and the mean thyristor current. If on the other hand the thyristor in the on state could be modeled by a resistor, then the mean power loss would the product of that resistance and the square of the thyristor r.m.s. current. In reality the thyristor loss lies somewhere between these two values and manufacturer's data sheets normally contain information for the accurate determination of loss.

Example 2.3

Determine for the example the maximum voltage stress on a thyristor and the mean and r.m.s. thyristor currents when the load current has the rated value of 666.7 A.

From eqn 2.19

$$\text{Max thyristor reverse voltage} = \text{Max forward voltage} = 628.3 \text{ V}$$

From eqn 2.20

$$\text{Thyristor mean current} = 222.2 \text{ A}$$

From eqn 2.21

$$\text{Thyristor r.m.s. current} = 384.9 \text{ A}$$

****

2.12 The transformer and supply currents

A transformer secondary carries the same current as its thyristor so that its r.m.s. value, I_{sec}, is

$$I_{sec} = \frac{1}{\sqrt{3}} I_o = 0.5774 I_o \tag{2.22}$$

A transformer primary would be expected to carry a current $I_{prim} = N_{tsp} I_{sec}$ which would provide ampere–turn balance. However, such a current would have a mean value of $N_{tsp} I_{sec}/3$ and, since there are no sources of d.c. on the primary side, this is not possible. The primary current waveform is therefore as shown in Fig. 2.8, graphs 5, 6, and 7. The waveform is rectangular with an amplitude of $+2N_{tsp}I_o/3$ for one third of a cycle and $-N_{tsp}I_o/3$ for two thirds of a cycle giving a mean value of zero. The r.m.s. value, I_{prim}, is

$$I_{prim} = \frac{\sqrt{2}}{3} N_{tsp} I_o = 0.4714 N_{tsp} I_o \tag{2.23}$$

The discrepancy between the primary and secondary m.m.fs results in d.c. magnetization of the transformer core which is highly undesirable. In practice a zigzag transformer, described in Section 12.4, would be employed to avoid this situation.

An a.c. line current is the difference between two transformer primary currents. For example the current in the red line is given by

$$i_r = i_{ry} - i_{br} \tag{2.24}$$

The waveforms of the a.c. line currents are shown in Fig. 2.8, graphs 8, 9, and 10. They comprise positive and negative pulses of amplitude $N_{tsp}I_o$ and duration 120° separated by periods of zero current of duration 60°. The mean value of the line current is zero and its r.m.s. value, I_s, is

$$I_s = \sqrt{\frac{2}{3}}N_{tsp}I_o = 0.8165N_{tsp}I_o \tag{2.25}$$

2.13 The transformer size

The size of a transformer is proportional to the product of its voltage and r.m.s. current and is commonly expressed as its kVA rating. Rectifier transformer differ from normal a.c. system transformers in that the waveforms are non-sinusoidal. This requires that individual kVA ratings for the various windings must be determined. In the present case a primary winding sustains an r.m.s. voltage V_s and, under rated conditions, carries an r.m.s. current $\sqrt{2}N_{tsp}I_{orat}/3$, where I_{orat} is the rated d.c. output current. Hence the primary rating, kVA_{prim}, for all three phases is, using eqn 2.23

$$kVA_{prim} = 3V_s\sqrt{\frac{2}{3}}N_{tsp}I_{orat}10^{-3}$$

Expressing V_s in terms of the d.c. output voltage according to eqn 2.8 this becomes

$$kVA_{prim} = \frac{2\pi}{3\sqrt{3}}V_{om}I_{orat}10^{-3} = 1.2092V_{om}I_{orat}10^{-3} \tag{2.26}$$

A secondary winding supports a voltage $N_{tsp}V_s$ and carries a current $I_o/\sqrt{3}$ so that the secondary rating, kVA_{sec}, is

$$kVA_{sec} = 3N_{tsp}V_s\frac{1}{\sqrt{3}}I_{orat}10^{-3}$$

which, again expressing V_s in terms of V_{om}, gives

$$kVA_{sec} = \frac{\sqrt{2}\pi}{3}V_{om}I_{orat}10^{-3} = 1.4810V_{om}I_{orat}10^{-3} \tag{2.27}$$

The secondary rating is about 22% greater than the primary rating, a situation which will be manifested by a greater weight of copper in the winding.

2.14 The power factor

The size and cost of an a.c. supply system are determined by the voltage and the r.m.s. current, i.e. by the load volt-amperes. The power which is delivered to the customer is smaller than this by the power factor. In systems having sinusoidal waveforms, the power factor is the cosine of the phase difference between the current and the voltage. In non-sinusoidal systems such as the ones under discussion, defining the power factor is more difficult since there are harmonics to be considered in addition to the fundamental components.

The power factor when the waveforms are not sinusoidal is not a well-defined quantity. However, there are two definitions in fairly general use both of which will be employed as the circumstances indicate. They are the *fundamental power factor* and the *conventional power factor*. The fundamental power factor is the cosine of the phase difference between the fundamental components of voltage and current. This will be used when discussing the Fourier analysis of rectifier waveforms in Chapter 8. Here the conventional power factor will be used. This is the ratio of the input power, P_s, to the input volt-amperes, VA_s, as conventionally measured, i.e. $\sqrt{3}$ multiplied by the product of the r.m.s. line to line voltage and the r.m.s. line current. In this case, since zero losses are assumed, the a.c. input power is equal to the d.c. output power so that

$$P_s = V_o I_o = V_{om} I_o \cos(\alpha)$$

and the input kVA is

$$kVA_s = \sqrt{3} V_1 \frac{\sqrt{2}}{3} N_{tsp} I_o 10^{-3}$$

Expressing V_s in terms of V_{om}, this gives

$$kVA_s = \frac{2\pi}{3\sqrt{3}} V_{om} I_o = 1.2092 V_{om} I_o 10^{-3}$$

the same as the primary kVA, eqn 2.26.

Thus the power factor as conventionally measured is

$$Pf = \frac{3\sqrt{3}}{2\pi} \cos(\alpha) = 0.8270 \cos(\alpha) \tag{2.28}$$

The fundamental power factor is in fact equal to $\cos(\alpha)$ and the conventional power factor is about 17% less because of the non-sinusoidal input current waveform.

Example 2.4

At rated power determine the r.m.s. values of the secondary, primary, and line currents, the transformer primary and secondary kVA and the input power factor as conventionally measured when the delay angle is 40°.

For an output current of 666.7 A

$$\begin{aligned}
\text{Secondary current} &= 384.9 \text{ A r.m.s.} \\
\text{Primary current} &= 183.2 \text{ A r.m.s.} \\
\text{Supply line current} &= 317.4 \text{ A r.m.s.} \\
\text{Secondary kVA} &= 296.2 \\
\text{Primary kVA} &= 241.8 \\
\text{Supply kVA} &= 241.8 \\
\text{Input power factor} &= 0.6335 \text{ as conventionally measured}
\end{aligned}$$

The kVA ratings should be compared with the rated output power of 200 kW.

****

2.15 The common anode connection

Before leaving the topic of three phase rectifiers an alternative to the common cathode connection which has been discussed should be noted. This is the common anode circuit shown in Fig. 2.9. The three thyristor anodes are connected together to form the negative terminal of the d.c. supply and the transformer neutral provides the positive terminal. In general the differences between this circuit and the common cathode circuit are minor. However, a most significant difference, of which particular note is taken in the next chapter, is that the output voltage relative to the transformer neutral is negative and is displaced in phase by 60° relative to that for the common cathode connection. The voltage of output terminal A relative to terminal N is shown for delay angles of 20, 50, 80, 110, and 140° by black lines in Fig. 2.10, the potentials of terminals 1, 2, and 3 relative to terminal N being shown by gray lines. It should be noted that the voltage shown by the black line, v_{an}, is -1 times the output voltage, v_o as defined in Fig. 2.9.

The common cathode circuit provides a common terminal to which all three thyristor gate driver circuits can be referenced. The common anode circuit does not have this convenience and requires three electrically independent gate drivers. It is therefore rarely encountered on its own. However, it forms a part of the extremely popular bridge circuits which are discussed in Chapter 3.

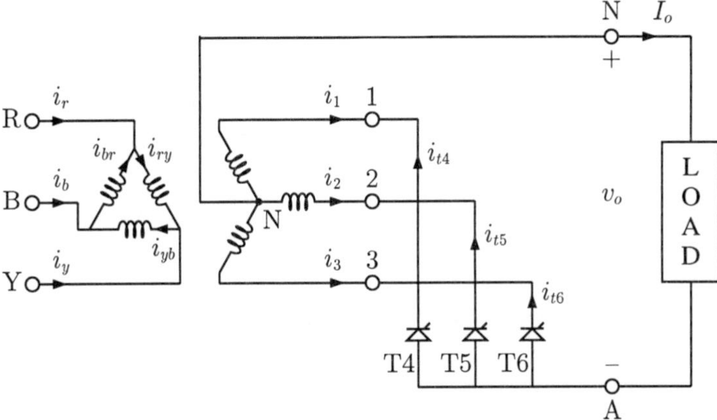

FIG. 2.9. Circuit diagram for a single way, three phase rectifier using the common anode connection.

2.16 Review

The study of the simple three phase rectifier begins to illustrate the dilemmas facing the rectifier designer who wishes to produce d.c., i.e. ripple free voltage and current, from a standard a.c. source of sinusoidal waveform. It has been seen that this is not possible, the output voltage inevitably has substantial ripple superimposed on the d.c. and if measures are taken to minimize the associated current ripple, the currents on the a.c. side are far from sinusoidal. The rectifier also requires a substantially larger transformer than might be expected from the rated load power and a substantial reduction in power factor from the value to be expected from the delay angle.

Some of these problems are relieved by increasing the number of phases and in the balance of this chapter a simple six phase rectifier is reviewed, six being a phase number readily derived from a standard three phase supply. However, the study will also show that the gains are accompanied by losses which make this alternative equally undesirable.

2.17 The simple six phase rectifier

A six phase supply can be obtained from the standard three phase supply by center tapping the windings of the secondary of a three phase transformer and connecting the center taps together to form a neutral as illustrated in Fig. 2.11. The primary windings must be delta connected as shown in the figure so that *triplen* currents, i.e. Fourier components whose order is a multiple of three times the supply frequency, can flow round the mesh.

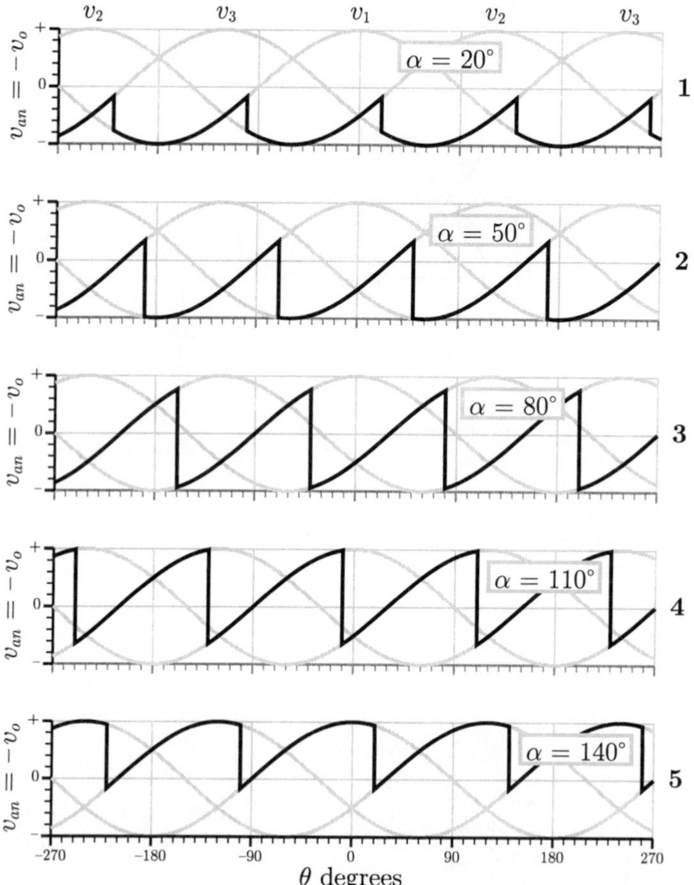

FIG. 2.10. Output voltage waveforms in a single way, three phase rectifier using the common anode connection, for delay angles of 20, 50, 80, and 110°.

These currents must flow in order to balance the m.m.fs of corresponding components in the secondary windings and yet they cannot flow in a three wire, three phase system such as that supplying the transformer. Delta connection of the primary provides a local path round which these currents can circulate.

The transformer secondary neutral forms the negative pole of the d.c. supply. The thyristors are connected to the ends of the six secondary phases and their common cathode connection is the positive pole of the d.c. supply.

The potentials of the six secondary phase ends relative to the neutral are shown by gray curves in Fig. 2.12. Again the time origin has been

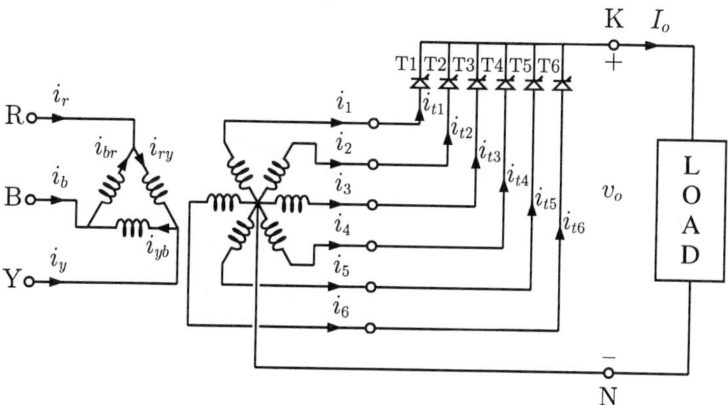

FIG. 2.11. Circuit diagram of a single way, six phase rectifier.

chosen so that

$$v_1 = \sqrt{2}N_{tsp}V_s\cos(\theta) \tag{2.29}$$

Following the reasoning already employed for the simple three phase rectifier, the natural conduction period of T1 is from $-30°$ to $30°$ after which conduction transfers to T2 etc. Each thyristor conducts for one sixth of a cycle.

If conduction is delayed by α, the conduction period for T1 becomes $-30° + \alpha$ to $30° + \alpha$. The output voltage waveforms for delay angles of $20°$, $50°$, $80°$, $110°$, and $140°$ are shown by black lines in Fig. 2.12 and the secondary voltages are shown for reference in gray. Each graph is identified by a number to its right, each graph is annotated with its delay angle and the secondary voltages are identified by v_1 etc. over the appropriate voltage peaks of graph 1.

2.18 The mean output voltage

Since each conducting period is similar to the others, attention will be concentrated on T1 and its conduction period $-30° + \alpha < \theta < 30° + \alpha$. The potential of the thyristor anode relative to the neutral is

$$v_1 = \sqrt{2}N_{tsp}V_s\cos(\theta)$$

where N_{tsp} is the ratio of the number of turns of one of the six secondary windings to the turns of a primary winding. The voltage v_1, less a small amount for thyristor voltage drop, is the output voltage for the conducting period. Thus the mean output voltage is

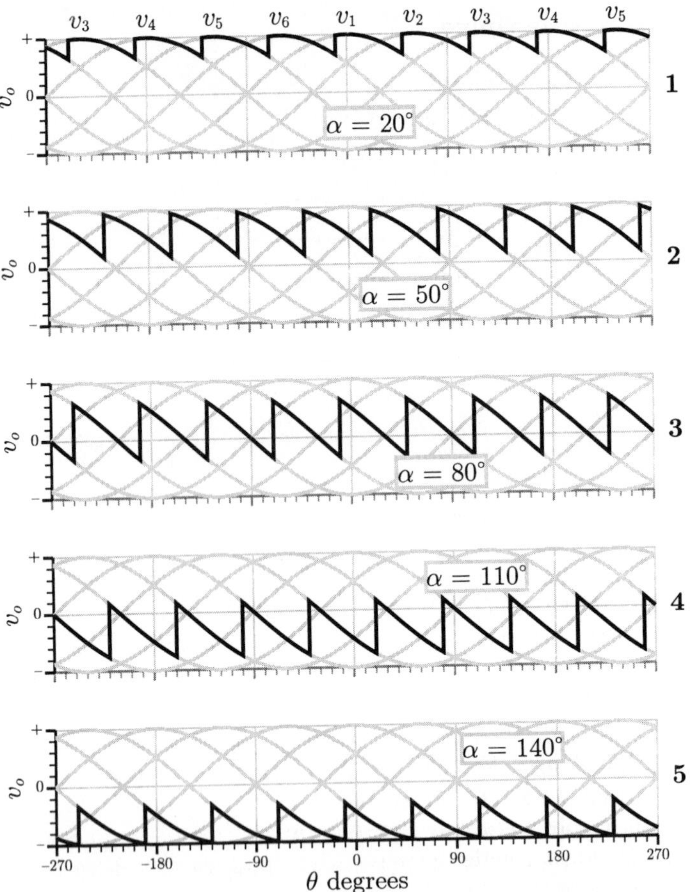

FIG. 2.12. Output voltage waveforms for a single way, six phase rectifier at delay angles of 20, 50, 80, 110, and 140°.

$$V_o = \frac{3}{\pi} \int_{\pi/6+\alpha}^{-\pi/6+\alpha} \sqrt{2} N_{tsp} V_s \cos(\theta)\, d\theta$$

i.e.

$$V_o = \frac{3\sqrt{2}}{\pi} N_{tsp} V_s \cos(\alpha) \tag{2.30}$$

or

$$V_o = V_{om} \cos(\alpha) \tag{2.31}$$

where

$$V_{om} = \frac{3\sqrt{2}}{\pi} N_{tsp} V_s = 1.3505 N_{tsp} V_s \tag{2.32}$$

2.19 The voltage ripple

The output voltage waveform, shown for various delay angles in Fig. 2.12, repeats six times per input cycle so that the ripple frequency is six times the supply frequency

$$f_{rip} = 6f_s \tag{2.33}$$

During the interval $-30° + \alpha < \theta < 30° + \alpha$, T1 conducts and the expression for the ripple voltage is

$$v_{rip} = \sqrt{2}N_{tsp}V_s\cos(\theta) - V_o$$

Referring all quantities to V_{om}

$$v_{rip} = V_{om}\left[\frac{\pi}{3}\cos(\theta) - \cos(\alpha)\right]$$

Following the procedure used in obtaining eqn 2.13, the r.m.s. value of the ripple voltage is

$$V_{rip} = V_{om}\left[(\frac{\pi^2}{18} - \frac{1}{2}) - (\frac{1}{2} - \frac{\pi}{4\sqrt{3}})\cos(2\alpha)\right]^{1/2} \tag{2.34}$$

or in numerical form

$$V_{rip} = V_{om}\sqrt{0.04831 - 0.04655\cos(2\alpha)} \tag{2.35}$$

The r.m.s. ripple voltage is shown as a function of the delay angle in Fig. 2.5. As the delay angle increases, the ripple voltage increases from a minimum of 4.2% of V_{om} at zero delay to a maximum of 30.8% of V_{om} at 90° delay and then decreases back to 4.2% at 180° delay, values substantially less than the values for the three phase rectifier which are shown on the same graph.

Example 2.5

A six phase rectifier has the same input and output specification as the three phase rectifier of Example 2.1, namely 440 V r.m.s. line to line, 60 Hz input, and 300 V maximum, 200 kW rated output. Determine its ripple frequency and the r.m.s. value of its ripple voltage for the delay angles of Fig. 2.12, namely 20, 50, 80, 110, and 140°. Compare the values with those obtained for the three phase rectifier.

The required values, found by applying eqns 2.32, 2.33, and 2.35, are given below. The corresponding values for the three phase rectifier are given in parentheses after the six phase number.

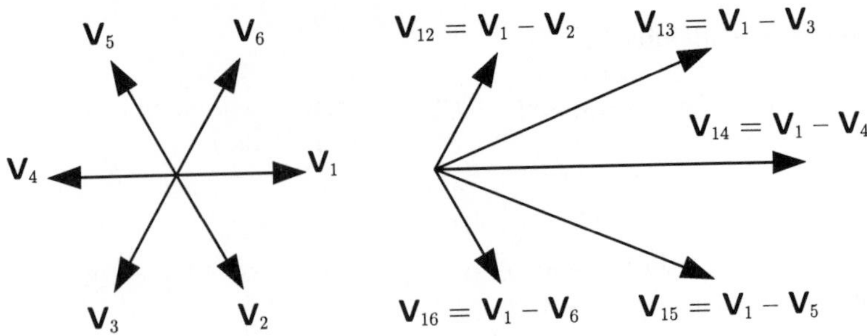

FIG. 2.13. Phasor diagrams for voltages in a single way, six phase rectifier.

The turns ratio is smaller,

$$\text{Turns ratio, } N_{tsp} = 0.5049 \quad (0.5830)$$

The ripple frequency is doubled.

$$\text{Ripple frequency, } f_{rip} = 360 \text{ Hz} \quad (180)$$

The ripple voltage is substantially reduced at all delay angles.

$$\begin{array}{lll} \alpha = 20°, & V_{rip} = 33.7 \text{ V r.m.s.} & (84.7) \\ \alpha = 50°, & V_{rip} = 71.2 \text{ V r.m.s.} & (154.6) \\ \alpha = 80°, & V_{rip} = 91.1 \text{ V r.m.s.} & (193.7) \\ \alpha = 110°, & V_{rip} = 86.9 \text{ V r.m.s.} & (185.5) \\ \alpha = 140°, & V_{rip} = 60.2 \text{ V r.m.s.} & (133.1) \end{array}$$

****

2.20 Output current

Equation 2.14 for the output current is unaffected by the phase number and for the rated load condition and the same thyristor voltage drop, the current in the example will still be 666.7 A.

2.21 The thyristor voltage

The thyristor voltage waveform is more complex than for the three phase case. Referring to the phasor diagram of Fig. 2.13, when T1 is conducting, its anode is slightly positive, by about 2 V, relative to its cathode. When T2 conducts, the anode of T1 is at v_1 and its cathode is at v_2 (less about 2 V for the drop in T2) so that the voltage across T1 is v_{12}. Similarly when T3 conducts, the voltage across T1 is v_{13} etc. Thus for $-30° + \alpha < \theta < 30° + \alpha$

$$v_{t1} = V_{thy} \tag{2.36}$$

for $30° + \alpha < \theta < 90° + \alpha$

$$v_{t1} = \sqrt{2} N_{tsp} V_s \cos(\theta + \frac{\pi}{3}) \tag{2.37}$$

for $90° + \alpha < \theta < 150° + \alpha$

$$v_{t1} = \sqrt{6} N_{tsp} V_s \cos(\theta + \frac{\pi}{6}) \tag{2.38}$$

for $150° + \alpha < \theta < 210° + \alpha$

$$v_{t1} = 2\sqrt{2} N_{tsp} V_s \cos(\theta) \tag{2.39}$$

for $210° + \alpha < \theta < 270° + \alpha$

$$v_{t1} = \sqrt{6} N_{tsp} V_s \cos(\theta - \frac{\pi}{6}) \tag{2.40}$$

for $270° + \alpha < \theta < 330° + \alpha$

$$v_{t1} = \sqrt{2} N_{tsp} V_s \cos(\alpha - \frac{\pi}{3}) \tag{2.41}$$

The five voltage waves, v_{12} through v_{16} are shown by gray lines in Fig. 2.14 and the thyristor voltage is shown by the black lines for delay angles of zero, 45°, 90°, and 135°.

Clearly the maximum voltage stress on a thyristor is $2\sqrt{2} N_{tsp} V_s$ in both the positive and negative directions.

$$V_{tmax} = 2\sqrt{2} N_{tsp} V_s = 2.8284 N_{tsp} V_s \tag{2.42}$$

or, in terms of V_{om}

$$V_{tmax} = \frac{2\pi}{3} V_{om} = 2.0944 V_{om} \tag{2.43}$$

the same expression as for the three phase rectifier, see eqn 2.19.

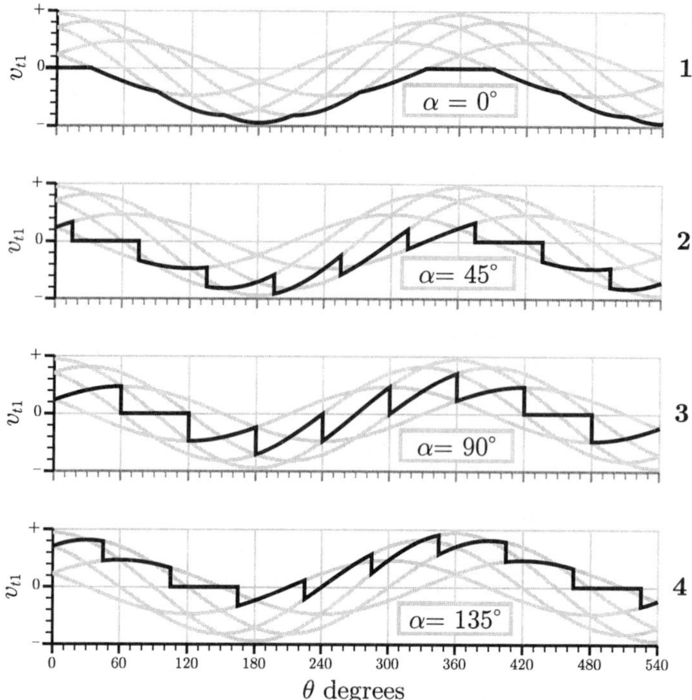

FIG. 2.14. Thyristor voltage waveforms in the single way, six phase rectifier for various delay angles.

2.22 The thyristor current

The current waveforms are shown in Fig. 2.15 where the graphs are identified by a number to their right. Graph 1 is the output current, subdivided by vertical gray lines into 60° segments for each conducting thyristor as indicated. The next six graphs, 2 through 7, show the thyristor currents. Each thyristor carries the load current for one sixth of a cycle. Thus the mean and r.m.s. thyristor currents are

$$I_t = I_o/6 = 0.1667I_o \tag{2.44}$$

$$I_{trms} = I_o/\sqrt{6} = 0.4082I_o \tag{2.45}$$

Example 2.6

Determine for the rectifier of Example 2.5 the values of the maximum thyristor voltage and the mean and r.m.s. values of the thyristor current. Compare them with the corresponding values for the three phase rectifier.

The solutions are as follows with the three phase rectifier values given in parentheses. From eqn 2.43 the maximum thyristor voltage is

$$V_{tmax} = 628.3 \text{ V} \quad (628.3)$$

From eqn 2.44 the thyristor mean current is

$$I_t = 111.1 \text{ A} \quad (222.2)$$

From eqn 2.45 the thyristor r.m.s. current

$$I_{trms} = 272.2 \text{ A} \quad (384.9)$$

****

2.23 The transformer and supply currents

A transformer secondary carries a thyristor current, i.e. one unidirectional pulse per supply cycle of amplitude I_o and duration one sixth of a cycle, as shown in Fig. 2.15, graphs 2 through 7. The r.m.s. value, I_{sec}, is the same as that for a thyristor.

$$I_{sec} = I_o/\sqrt{6} = 0.4082 I_o \tag{2.46}$$

The waveforms of the transformer primary currents are shown in graphs 8, 9, and 10 of Fig. 2.15. A primary current balances the m.m.fs of two secondary currents, for example i_{ry} balances the m.m.fs of i_1 and i_4. The waveform comprises alternate positive and negative pulses of amplitude $N_{tsp}I_o$ and duration one sixth of a supply cycle. The mean value of a primary current is zero and the d.c. magnetization problem which occurs in the three phase transformer is not encountered. The primary r.m.s. current, I_{prim}, is

$$I_{prim} = N_{tsp}I_o/\sqrt{3} = 0.5774 N_{tsp}I_o \tag{2.47}$$

The primary will also carry the transformer magnetizing current which, since an ideal transformer is assumed, is zero.

The waveforms of the three a.c. supply line currents are shown in graphs 11, 12, and 13 of Fig. 2.15. The supply line current is the difference between two primary currents, for example $i_r = i_{ry} - i_{br}$. A line current's waveform comprises positive and negative pulses of amplitude $N_{tsp}I_o$ and duration one third of a cycle. The mean value is zero and the r.m.s. value, I_s, is

$$I_s = \sqrt{\frac{2}{3}} N_{tsp}I_o = 0.8165 N_{tsp}I_o \tag{2.48}$$

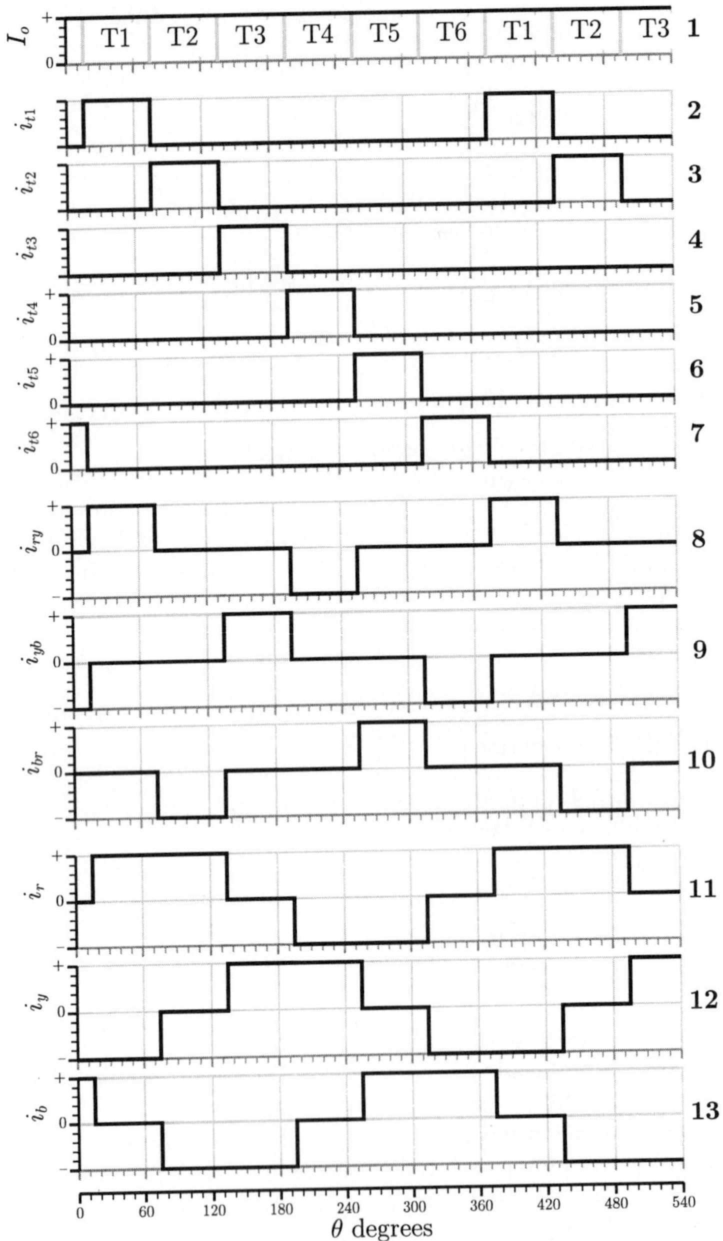

FIG. 2.15. Current waveforms in a single way, six phase rectifier.

2.24 The transformer size

The total kVA rating of the six secondary windings is

$$kVA_{sec} = 6N_{tsp}V_s\frac{I_o}{\sqrt{6}}$$

Using eqn 2.32 this becomes

$$kVA_{sec} = \frac{\pi}{\sqrt{3}}V_{om}I_o10^{-3} = 1.8138V_{om}I_o10^{-3} \tag{2.49}$$

The kVA rating of the primary is

$$kVA_{prim} = 3V_s\sqrt{1/3}\,N_{tsp}I_o10^{-3}$$

which, by use of eqn 2.32 becomes

$$kVA_{prim} = \frac{\pi}{\sqrt{6}}V_{om}I_{om}10^{-3} = 1.2825V_{om}I_o10^{-3} \tag{2.50}$$

Thus, there is substantially more material, 22%, in the six phase secondary than in the three phase secondary and somewhat more, 6%, in the primary. Add to this the extra complication of the six phase secondary and it is seen that the transformer is substantially larger and more expensive.

2.25 The input kVA and the power factor

The input kVA as conventionally measured is

$$kVA_s = \sqrt{3}V_s\sqrt{\frac{2}{3}}N_{tsp}I_o10^{-3}$$

which by use of eqn 2.32 becomes

$$kVA_s = \frac{\pi}{3}V_{om}I_o10^{-3} = 1.0472V_{om}I_o10^{-3} \tag{2.51}$$

This value is significantly less than the primary kVA, eqn 2.50, because the primary windings carry *triplen* harmonic currents, i.e. current harmonics whose order is a multiple of 3, which circulate round the delta but do not get into the a.c. lines.

Since losses are neglected, the input power is $V_oI_o = V_{om}I_o\cos(\alpha)$ so that the input power factor as conventionally measured is

$$Pf_s = \frac{3}{\pi}\cos(\alpha) = 0.9549\cos(\alpha) \tag{2.52}$$

This is significantly greater than for the three phase case, indicating a lower harmonic content in the input current.

Example 2.7

Determine for the six phase rectifier example supplying the rated load current of 666.7 A, the r.m.s. values of the secondary, primary, and supply line currents, the kVA of the secondary and primary windings, the input kVA, and the input power factor as conventionally measured. Compare the values with those obtained for the three phase rectifier.

Using eqns 2.46 through 2.52, the various values are as follows, the corresponding values for the three phase rectifier being shown in parentheses.

$$\begin{aligned} I_{sec} &= 272.2 \text{ A} \quad (384.9) \\ I_{prim} &= 194.3 \text{ A} \quad (183.2) \\ I_s &= 274.8 \text{ A} \quad (317.3) \\ kVA_{sec} &= 362.8 \quad (296.2) \\ kVA_{prim} &= 256.5 \quad (241.8) \\ Pf_s &= 0.7315 \quad (0.6335) \end{aligned}$$

****

2.26 Conclusions

The increase in phase number from three to six has improved the quality of both the d.c. output and the a.c. input in that the ripple content of the output voltage is decreased and the harmonic content of the input current has also decreased. The price of this improvement has been an increase in rectifier complexity, six thyristors and their associated electronics as against three, and an increase in size and complexity of the transformer. The transformer has six secondary windings as against three and, for a given magnetic core, there is 22% more copper in the winding. The primary of the six phase transformer is somewhat larger, for the same core it has 6% more copper.

These conclusions are generally true. As the phase number is increased, the waveforms on both the d.c. and a.c. sides improve but the circuit complexity and cost increase. Substantial improvements on this situation can be made by the use of modified circuits which place rectifiers in series or in parallel. The simple rectifiers discussed here are therefore rarely used in practice. Two series arrangements, the fully controlled bridge and the semi controlled bridge, dominate the current market and are discussed in the next chapter.

3

THREE PHASE BRIDGE RECTIFIERS

3.1 Introduction

It was seen in Chapter 2 that increasing the number of phases of a simple rectifier improves the waveforms on both the d.c. and a.c. sides but also increases the transformer size and complexity. It was stated that the advantages can be retained and the disadvantages reduced by series and/or parallel connection of rectifiers. This chapter discusses two particular series arrangements, the fully controlled and semi-controlled bridge rectifiers, which constitute the overwhelming majority of power rectifiers in commercial use. Again, perfect components and smooth, ripple-free output current are assumed.

3.2 Series connection of single way rectifiers

In the previous chapter, Figs 2.1 and 2.9 show single way, three phase rectifiers in the common cathode and common anode configurations. These rectifiers can be connected in series to form a three phase bridge rectifier whose advantages are so great that it dominates the current market for rectifiers of substantial size. The processes by which the bridge circuit is derived from the two single way circuits is illustrated in Fig. 3.1. Figure 3.1(a) shows the two rectifiers sharing a common transformer secondary so that the negative d.c. terminal of the common cathode rectifier is at the same potential as the positive d.c. terminal of the common anode rectifier. Clearly if the two loads are drawing the same current, the transformer neutral current, I_n, is zero and the neutral connection can be removed. This has been done in Fig. 3.1(b). Since the rectifier is now supplied by only the three a.c. lines, it no longer needs the transformer and this has been omitted from Fig. 3.1(c).

The circuit without transformer is the one which will be studied in the remainder of this chapter. However, it must be recognized that a transformer may be required in practice in order to attain one or more of the following three objectives.

- To obtain a specific d.c. output voltage unrelated to the a.c. supply voltage.

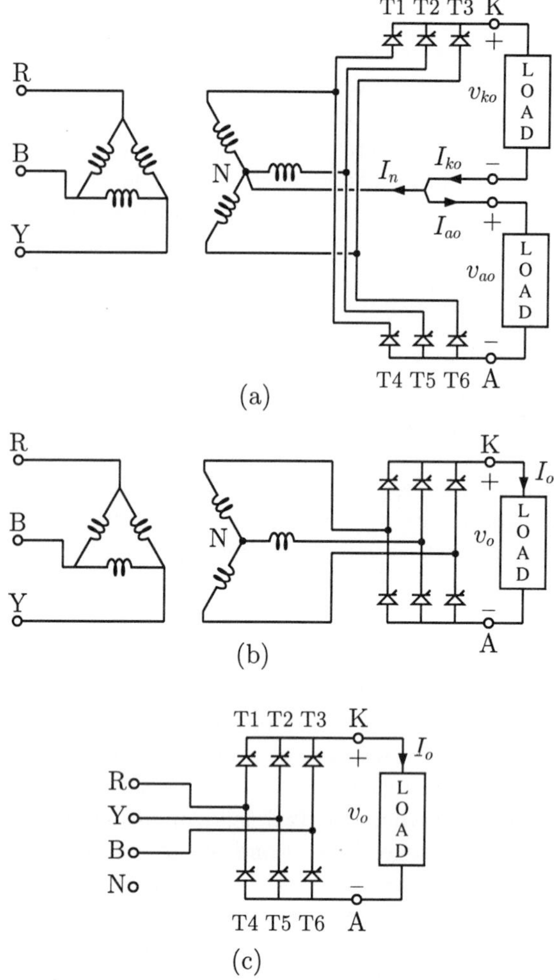

FIG. 3.1. The evolution of the three phase bridge rectifier from the series connection of two single way, three phase rectifiers, one in the common cathode and the other in the common anode connection.

- To reduce the effects of the rectifier on the a.c. system voltage waveform.
- To provide electrical isolation of the d.c. output from the a.c. system.

Figure 3.1(c) retains the neutral terminal for reference purposes only. The practice of the preceding chapter of referring many voltages to this point will be continued. All single subscript voltages are referenced to this point so that v_k is the potential of terminal K relative to N and v_a is the potential of terminal A relative to N, but v_{ka} is the potential of terminal K relative to that of terminal A, i.e. it is $v_k - v_a$.

For the study, the neutral has no significance other than as a reference point for potentials, but before leaving the topic it must be recognized that the neutral has considerable practical significance. The overwhelming majority of three wire, three phase a.c. systems do in fact have a grounded neutral somewhere back in the system. This ensures that the system is always referenced to ground potential. For the bridge rectifier this means that the positive d.c. terminal will be positive relative to ground and that the negative d.c. terminal will be negative relative to ground.

3.3 Nomenclature

Omission of the transformer requires two changes in the nomenclature. The r.m.s. a.c. input line to line voltage will continue to be denoted by the symbol V_s so that the r.m.s. value of the secondary r.m.s. line to neutral voltage which was the compound symbol $N_{tsp}V_s$ in the previous chapter is now $V_s/\sqrt{3}$. This substitution will be made in all the formulae obtained from Chapter 2.

The output voltage, v_o, instead of being relative to the neutral, N, is the voltage of the positive output terminal, K, relative to the negative output terminal, A. The delay angle of the common cathode half will be denoted by α_k and that of the common anode half by α_a. These delay angles are totally independent and can have any values but only three particular cases are of practical importance. The first, and by far the most important, is the case where the two delay angles are equal. This provides the fully controlled bridge. The second is the case where one part of the bridge, usually the common anode half, is made from diodes rather than thyristors and therefore has a delay angle of zero. This provides the semi-controlled bridge. The third is the case where the two delay angles are varied sequentially. This provides sequential control which is of significance in large rectifiers and is discussed in Chapter 9.

3.4 The fully controlled bridge rectifier

The fully controlled bridge rectifier, in which the two delay angles are equal, is by far the most popular of the three phase bridges. Its circuit is that of Fig. 3.1(c). Since $\alpha_k = \alpha_a$, the suffix will be omitted and the delay will be denoted by α. The output voltage waveforms of a fully controlled bridge are shown in Fig. 3.2 which has been drawn for a delay angle of 20°. Graph 1 shows the potentials of terminals K and A relative to the system neutral, the solid black line being the instantaneous value, the dashed line the mean value and the potentials of terminals 1, 2, and 3 being shown by

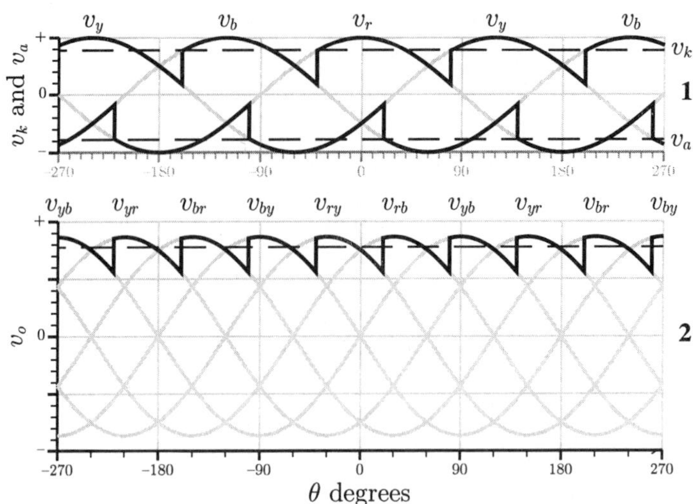

FIG. 3.2. Output voltage waveforms of the fully controlled, three phase bridge rectifier.

gray lines. The potential of terminal K, v_k, is the upper black line and the potential of terminal A, v_a, is the lower black line, both being identified at the right-hand edge of the graphs. Graph 2 shows the output voltage, v_{ka}.

Making the substitution of $V_s/\sqrt{3}$ for $N_{tsp}V_s$ in eqn 2.8, the mean potentials, relative to the neutral, of the positive and negative d.c. output terminals are $\pm 3V_s \cos(\alpha)/(\pi\sqrt{2})$ so that the mean output voltage, i.e. the voltage of terminal K relative to terminal A, is

$$V_o = \frac{3\sqrt{2}}{\pi} V_s \cos(\alpha) \tag{3.1}$$

or

$$V_o = V_{om} \cos(\alpha) \tag{3.2}$$

where

$$V_{om} = \frac{3\sqrt{2}}{\pi} V_s = 1.3505 V_s \tag{3.3}$$

Thus there is the same cosinusoidal relationship between the mean output voltage and the delay angle as was found for the single way circuits. However, for a given line to line voltage at the rectifier terminals, the maximum value of the mean voltage is doubled to $3\sqrt{2}V_s/\pi$ as expected from the series connection of two rectifiers. The values of V_{om} for various common a.c. system voltages are given in Table 3.1. If these values are unsuited to the application or if a.c. system voltage waveform quality is a problem

Table 3.1 *The maximum values of mean output voltage and thyristor voltage when a bridge rectifier is directly connected to a three phase system*

Line to line r.m.s. voltage	V_{om} V	V_{tmax} V
220	297.1	311
400	540.2	566
440	594.2	622
600	810.3	849
3300	4456.6	4667

or if safety requirements demand electrical isolation, a transformer must be interposed between the supply and the rectifier.

3.5 The output voltage ripple

Unfortunately the ripple voltage cannot be determined from the results of Chapter 2. It must be obtained directly. Defining the time origin so that the phasor voltage of the red terminal relative to the system neutral is the datum, as indicated in the phasor diagram, Fig. 3.3, the potentials of the red, R, yellow, Y, and blue, B, terminals relative to the neutral are, following the phasor diagram of Fig. 3.3,

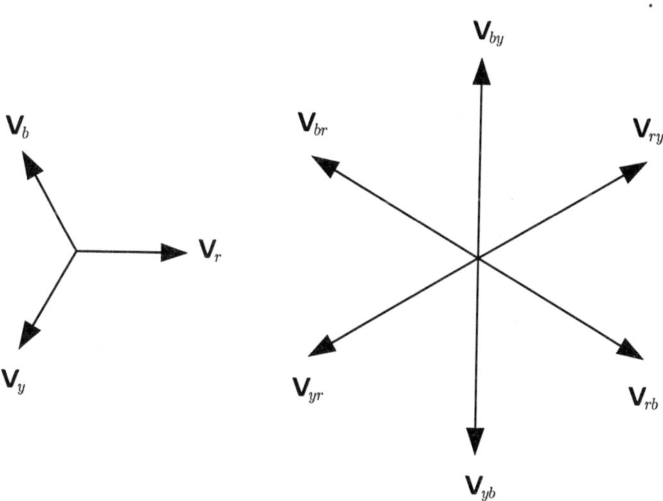

FIG. 3.3. Phasor diagrams for the line to system neutral and the line to line voltages of the a.c. system supplying the bridge rectifier.

$$v_r = \frac{\sqrt{2}V_s}{\sqrt{3}}\cos(\theta) \tag{3.4}$$

$$v_y = \frac{\sqrt{2}V_s}{\sqrt{3}}\cos(\theta - \frac{2\pi}{3}) \tag{3.5}$$

$$v_b = \frac{\sqrt{2}V_s}{\sqrt{3}}\cos(\theta + \frac{2\pi}{3}) \tag{3.6}$$

The line to line voltages are

$$v_{ry} = \sqrt{2}V_s\cos(\theta + \frac{\pi}{6}) = -v_{yr} \tag{3.7}$$

$$v_{yb} = \sqrt{2}V_s\cos(\theta - \frac{\pi}{2}) = -v_{by} \tag{3.8}$$

$$v_{br} = \sqrt{2}V_s\cos(\theta + \frac{5\pi}{3}) = -v_{rb} \tag{3.9}$$

The thyristor conduction periods and the output voltage during those periods are listed in Table 3.2.

The output voltage ripple waveform repeats six times per supply cycle so that the ripple frequency is

$$f_{rip} = 6f_s \tag{3.10}$$

During the 60° interval α to $\alpha + 60°$ the output voltage is v_{rb} so that the ripple voltage during this interval is

$$v_{rip} = \sqrt{2}V_s\cos(\theta - \frac{\pi}{6}) - V_{om}\cos(\alpha) \tag{3.11}$$

which in terms of V_{om}, eqn 3.3, becomes

$$v_{rip} = V_{om}\left[\frac{\pi}{3}\cos(\theta - \frac{\pi}{6}) - \cos(\alpha)\right] \tag{3.12}$$

Squaring, integrating, and averaging this expression gives the r.m.s. value of the ripple voltage as

$$V_{rip} = V_{om}\sqrt{(\frac{\pi^2}{18} - \frac{1}{2}) - (\frac{1}{2} - \frac{\pi}{4\sqrt{3}})\cos(2\alpha)} \tag{3.13}$$

which in numerical terms reduces to

$$V_{rip} = V_{om}\sqrt{0.04831 - 0.04655\cos(2\alpha)} \tag{3.14}$$

Thus, the ripple frequency and ripple voltage are the same as the values for a single way, six phase rectifier having the same rectifier mean output voltage.

Table 3.2 *Conduction periods and output voltage for the thyristors in the fully controlled bridge rectifier*

Thyristor		Conduction period		Output
Top	Bottom	Start	End	voltage
T1	T5	$-60° + \alpha$	$0° + \alpha$	v_{ry}
T1	T6	$0° + \alpha$	$60° + \alpha$	v_{rb}
T2	T6	$60° + \alpha$	$120° + \alpha$	v_{yb}
T2	T4	$120° + \alpha$	$180° + \alpha$	v_{yr}
T3	T4	$180° + \alpha$	$240° + \alpha$	v_{br}
T3	T5	$240° + \alpha$	$300° + \alpha$	v_{by}
T1	T5	$300° + \alpha$	$360° + \alpha$	v_{ry}

3.6 The thyristor voltage and current

The thyristor voltages and currents are identical with those found in the preceding chapter for the single way, three phase rectifier provided the substitution of $V_s/\sqrt{3}$ is made for $N_{tsp}V_s$. Thus the thyristor voltage waveforms are as shown in Fig. 2.7 with the maximum voltage stress on a thyristor being

$$V_{tmax} = \sqrt{2}V_s \tag{3.15}$$

or, in terms of the output voltage from eqn 3.3,

$$V_{tmax} = \frac{\pi}{3}V_{om} = 1.047V_{om} \tag{3.16}$$

For a given maximum d.c. output voltage, the voltage stresses on the thyristors are half those on the thyristors of the single way, six phase rectifier. Table 3.1 lists, in the third column, the maximum value of the thyristor voltage. Allowing a factor of safety of 2, direct connection of a bridge rectifier to systems up to 600 V is well within the capabilities of commercially available thyristors which have blocking capabilities up to 2000 V. Direct connection to a 3300 V system would require, for the same safety factor, that four or more thyristors be connected in series in each leg of the bridge.

The thyristor currents are identical with those in the single way, three phase rectifier, each thyristor carries the load current for one-third of a cycle and the mean and r.m.s. thyristor currents are

$$I_t = I_o/3 \tag{3.17}$$

and

$$I_{trms} = I_o/\sqrt{3} \tag{3.18}$$

3.7 The a.c. line current, volt-amperes, and power factor

The a.c. line current cannot be directly derived from the results for the single way rectifier since there are now two thyristors connected to each line, one drawing current from the line and the other returning current to it. Thus

$$i_r = i_{t1} - i_{t4} \tag{3.19}$$

$$i_y = i_{t2} - i_{t5} \tag{3.20}$$

$$i_b = i_{t3} - i_{t6} \tag{3.21}$$

The waveforms of the a.c. line currents are shown in graphs 3, 4, and 5 of Fig. 3.4 which also shows, in graphs 1 and 2, the division of the output current among the six thyristors. Each line current is a rectangular wave with positive and negative pulses, each of amplitude I_o and duration 120°. The mean value of the line current is zero and the r.m.s. value is

$$I_s = \sqrt{\frac{2}{3}} I_o \tag{3.22}$$

The input volt-amperes as conventionally measured is therefore

$$VA_s = \sqrt{3} V_s I_s$$

which, after substituting $\pi V_{om}/(3\sqrt{2})$ for V_s and $\sqrt{2/3}\, I_o$ for I_s, yields

$$kVA_s = \frac{\pi}{3} V_{om} I_o 10^{-3} = 1.0472 V_{om} I_o 10^{-3} \tag{3.23}$$

Since an ideal rectifier with zero losses is assumed, the input power equals the output power, i.e.

$$P_s = V_o I_o = V_{om} I_o \cos(\alpha) \tag{3.24}$$

The power factor as conventionally measured is then

$$Pf = \frac{P_s}{VA_s} = \frac{3}{\pi} \cos(\alpha) = 0.9549 \cos(\alpha) \tag{3.25}$$

a value identical with that for the simple six phase rectifier, eqn 2.52.

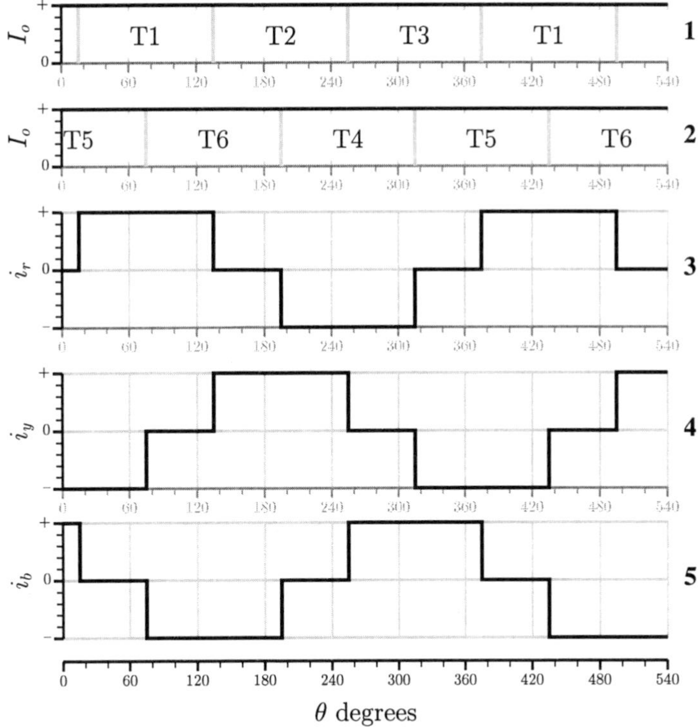

FIG. 3.4. Current waveforms in a fully controlled, three phase bridge rectifier. The figure has been drawn for a delay angle of 75°.

3.8 Summary

The fully controlled bridge rectifier provides a performance on the a.c. and d.c. sides identical with that of the simple six phase rectifier while behaving internally as a single way, three phase rectifier thus combining the merits of both. Further, it becomes possible to eliminate the transformer provided the d.c. load can accommodate the resulting voltage and can be floated relative to ground. These are not onerous conditions and even in those situations where a transformer is necessary it is simpler and significantly smaller than the transformer for the six phase system.

For a given d.c. voltage, the maximum voltage stress on the thyristors is half that for the simple six phase unit while the mean thyristor current is doubled and the r.m.s. current is increased by a factor of $\sqrt{2}$.

The losses in the thyristors are approximately doubled but since the loss is a very small proportion of the input power, less than 1%, this is not a serious problem.

Example 3.1

Consider the load of Chapter 2, i.e. 200 kW d.c., redesigned to be suitable for direct connection to the 440 V 60 Hz system via a three phase bridge rectifier. Determine all salient values at zero and a 40° delay.

Direct connection via a bridge rectifier to the 440 V system produces a maximum d.c. output of 594.2 V. For the rated load power of 200 kW, the rated current is then 336.6 A.

The output voltage ripple frequency is 360 Hz, the minimum ripple voltage is 24.9 V r.m.s. at 0° delay and the maximum ripple voltage is 183.0 V r.m.s. at 90° delay.

The maximum thyristor voltage stress is 622.3 V and, at rated load, the mean thyristor current is 112.2 A and the r.m.s. thyristor current is 194.3 A.

At rated load and zero delay the input line current is 274.8 A, the input kVA as conventionally measured is 209.4 and the power factor is 0.9549. At 40° delay the current and kVA are unchanged and the power factor is reduced to 0.7315.

Thus the same result as for the simple six phase rectifier has been obtained by direct connection to a normal three wire, three phase system without the use of a transformer. The price paid is the redesign of the load to suit the new d.c. voltage and some additional thyristor loss. An idea of the insignificance of this loss is obtained by assuming the forward voltage drop of a thyristor to be 2 V. The extra loss is then $2 \times 3 \times 112.2 = 673.2$ W, i.e. 0.34% of the rated output.

****

3.9 The semi-controlled bridge rectifier

Despite its simplicity and many excellent characteristics, the fully controlled rectifier shares a disadvantage with all fully controlled rectifiers, a large kVA input and poor power factor at reduced voltage. It has been seen that

$$V_o = V_{om}\cos(\alpha)$$

and

$$Pf = \frac{3}{\pi}\cos(\alpha)$$

so that the power factor is reduced in the same proportion as the output voltage. At full voltage the power factor is 0.9549, at half voltage it is 0.4775, and at zero voltage it is zero.

Further, for a given output current, I_o, the input volt-amperes is constant at $1.0472V_{om}I_o$ even though the output power is reduced by increasing the angle of delay.

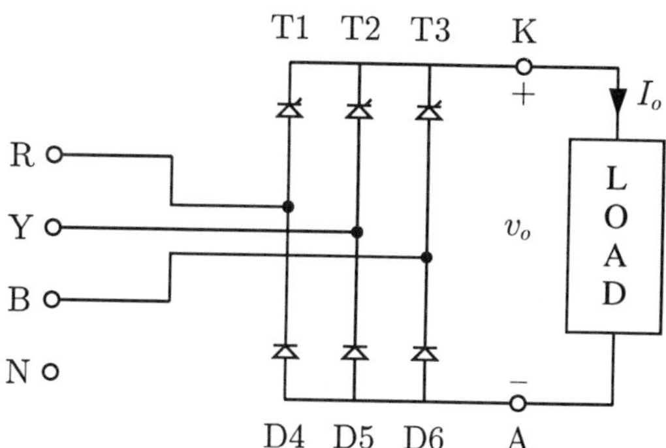

FIG. 3.5. The three phase, semi-controlled, bridge rectifier.

These problems can be relieved by sequential control, a topic discussed in some detail in Chapter 9. Full application of this technique is normally reserved for rectifiers of higher power and phase number than the three phase units. However, a simplified version is employed to some extent at this level. This is the semi-controlled bridge in which the three common anode thyristors in the lower half of the rectifier are replaced by diodes. The delay angle, α_a, of the lower half is then fixed at zero and results in an improvement in power factor and a reduction in kVA demand at the larger delay angles. Unfortunately a substantial price is paid for these improvements in that the ripple of both the input current and output voltage are increased and the rectifier will not permit reversal of power flow, from the d.c. to the a.c. side.

The semi-controlled rectifier is shown in Fig. 3.5. The waveforms of key variables are shown in Figs 3.6 and 3.7. Each of these latter two figures shows, by black lines, in graph 1 the waveforms of the potentials of terminals K and A relative to the system neutral, in graph 2 the output voltage v_{ak}, in graph 3 the current in thyristor T1, in graph 4 the current in diode D4 and in graph 5 the current in the red line. Figure 3.6 has been drawn for a delay of 40° and Fig. 3.7 for a delay of 80°. The figures also show, by dashed lines, the mean values of v_k, v_a, and v_{ak} and by gray lines, the waveforms of the line to system neutral voltages in graph 1 and the line to line voltages in graph 2. These voltages are identified above their positive peaks.

Figures 3.6 and 3.7 illustrate that the delay angle of 60° is a critical point for the semi-controlled rectifier. When the delay is less than 60°, T1 and D4 conduct in separate and distinct periods. When the delay exceeds 60°, the conduction periods of T1 and D4 overlap and for a brief period the load is short circuited through the simultaneously conducting thyristor and diode. This condition, in which the load current circulates through a

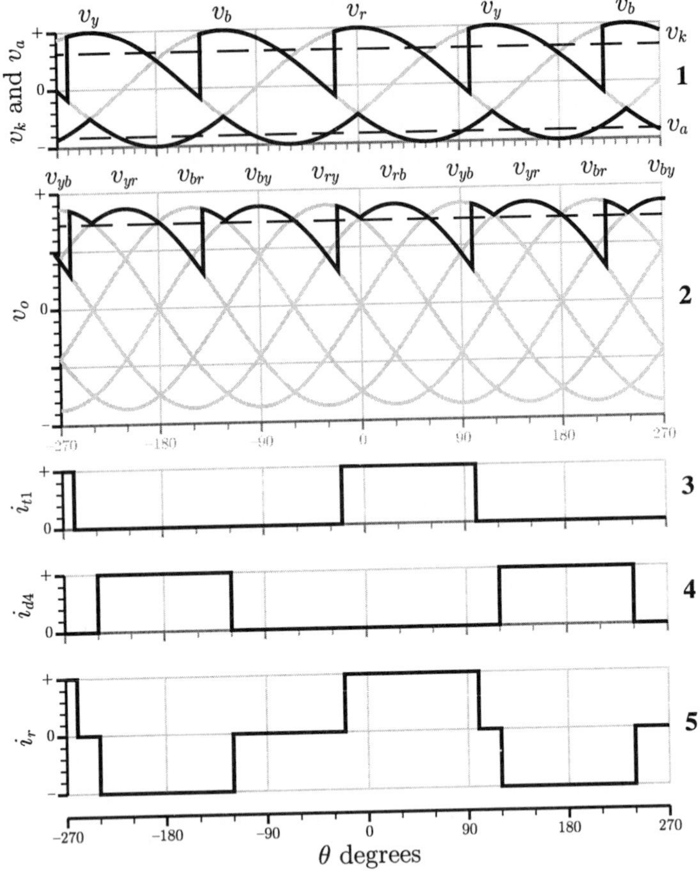

FIG. 3.6. Waveforms for the three phase, semi-controlled bridge rectifier for a delay angle of 40°.

local path within the rectifier rather than being drawn from the a.c. supply system, is known as *freewheeling*. It occurs during the shaded areas of graph 5 of Fig. 3.7.

3.10 The output voltage

The mean potential of terminal K is

$$V_k = \frac{3}{\pi\sqrt{2}} V_s \cos(\alpha)$$

and the mean potential of terminal A is

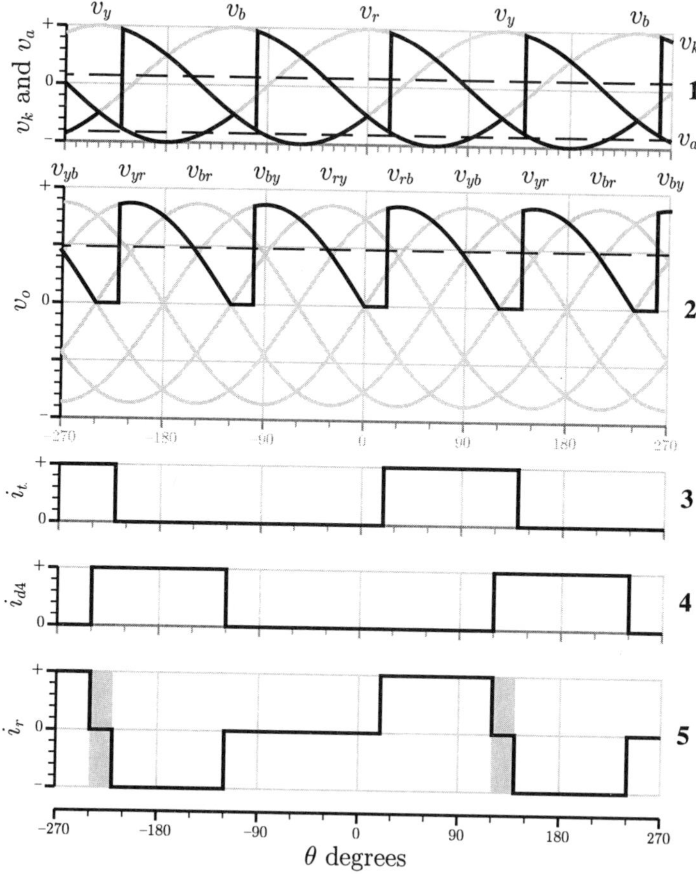

FIG. 3.7. Waveforms for the three phase, semi-controlled bridge rectifier for a delay angle of 80°.

$$V_a = \frac{3}{\pi\sqrt{2}} V_s$$

so that the mean output voltage is

$$V_o = \frac{3}{\pi\sqrt{2}} V_s[1 + \cos(\alpha)] \tag{3.26}$$

or

$$V_o = V_{om}\frac{1 + \cos(\alpha)}{2} \tag{3.27}$$

where

$$V_{om} = \frac{3\sqrt{2}}{\pi} V_s = 1.3505 V_s \tag{3.28}$$

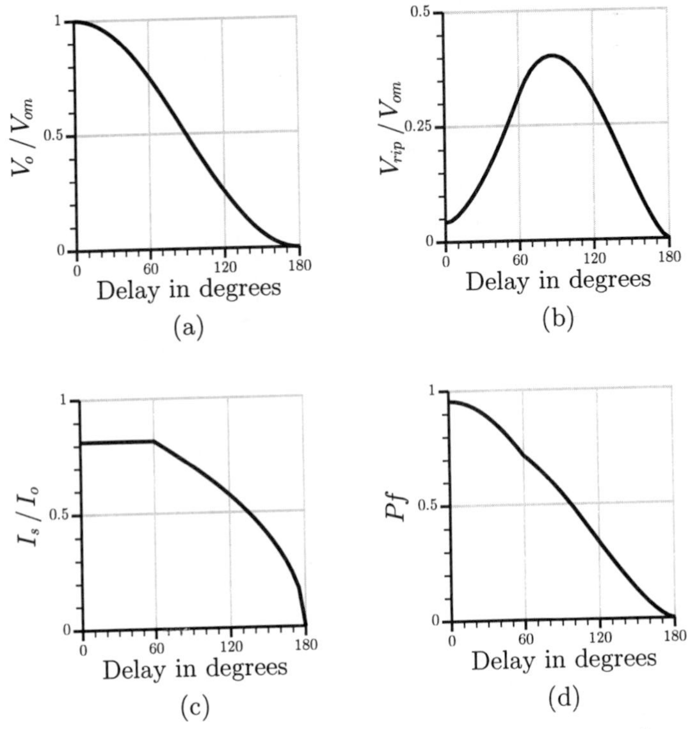

FIG. 3.8. Output voltage, ripple voltage, supply current, and power factor of the semi-controlled rectifier, as functions of delay angle.

Figure 3.8(a) shows the mean output voltage as a function of delay. It highlights the fact that the mean output voltage is unidirectional, starting from the same value as the fully controlled bridge at zero delay, reducing to half this value at 90° delay, and becoming zero at 180° delay.

Inspection of Figs 3.6 and 3.7 shows that the ripple frequency is three times the supply frequency

$$f_{rip} = 3f_s \tag{3.29}$$

The ripple voltage is the difference between the instantaneous and mean output voltages

$$v_{rip} = v_o - V_o \tag{3.30}$$

Since all 120° intervals are identical, the interval when T1 conducts, $-60° + \alpha < \theta < 60° + \alpha$ will be considered. This must be done for the two cases of $\alpha \leq 60°$ and for $\alpha > 60°$.

3.10.1 *Delay less than or equal to 60°*

For $\alpha \leq 60°$, T1 and D5 conduct for the period $-60° + \alpha$ to $0°$ and T1 and D6 conduct for the period $0°$ to $60° + \alpha$. Hence, during the period $-60° + \alpha < \theta < 0°$

$$v_o = v_{ry} = \sqrt{2}V_s \cos(\theta + \frac{\pi}{6}) \tag{3.31}$$

and, using eqn 3.28

$$v_{rip} = V_{om}\left[\frac{\pi}{3}\cos(\theta + \frac{\pi}{6}) - \frac{1 + \cos(\alpha)}{2}\right] \tag{3.32}$$

For the period $0° < \theta < 60° + \alpha$

$$v_o = v_{rb} = \sqrt{2}V_s \cos(\theta - \frac{\pi}{6}) \tag{3.33}$$

and

$$v_{rip} = V_{om}\left[\frac{\pi}{3}\cos(\theta - \frac{\pi}{6}) - \frac{1 + \cos(\alpha)}{2}\right] \tag{3.34}$$

Squaring eqns 3.32 and 3.34, integrating with respect to θ over the appropriate interval, averaging and taking the square root, yields the r.m.s. value of the ripple voltage

$$V_{rip} = V_{om}\left[\frac{\pi^2}{18} + \frac{\pi}{8\sqrt{3}} - \frac{3}{8} - \frac{1}{2}\cos(\alpha) + (\frac{\pi}{8\sqrt{3}} - \frac{1}{8})\cos(2\alpha)\right]^{1/2} \tag{3.35}$$

i.e. in numerical form

$$V_{rip} = V_{om}\sqrt{0.4 - 0.5\cos(\alpha) + 0.1017\cos(2\alpha)} \tag{3.36}$$

Evaluating eqn 3.36 at the extremes of its applicable range shows that there is an increase in the r.m.s. ripple voltage from $0.0420V_{om}$ at zero delay to $0.3149V_{om}$ at 60° delay.

3.10.2 *Delay greater than 60°*

When $\alpha > 60°$, T1 and D6 conduct for the period $-60° + \alpha$ to $120°$ and T1 and D4 conduct for the period $120°$ to $60° + \alpha$. During the first of these periods, $-60° + \alpha < \theta < 120°$

$$v_o = v_{rb} = \sqrt{2}V_s \cos(\theta - \frac{\pi}{6}) \tag{3.37}$$

and

$$v_{rip} = V_{om}\left[\frac{\pi}{3}\cos(\theta - \frac{\pi}{6}) - \frac{1+\cos(\alpha)}{2}\right] \tag{3.38}$$

During the second period $120° < \theta < 60° + \alpha$

$$v_o = 0 \tag{3.39}$$

and

$$v_{rip} = -V_{om}\frac{1+\cos(\alpha)}{2} \tag{3.40}$$

Squaring eqns 3.38 and 3.40, integrating with respect to θ over the appropriate interval, averaging, and taking the square root, yields the r.m.s. value of the ripple voltage

$$V_{rip} = V_{om}\left[\frac{\pi^2}{12} - \frac{3}{8} - \frac{\pi}{12}\alpha - \frac{1}{2}\cos(\alpha) + \frac{\pi}{24}\sin(2\alpha) - \frac{1}{8}\cos(2\alpha)\right]^{1/2} \tag{3.41}$$

i.e. in numerical form

$$V_{rip} = V_{om}\big[0.4475 - 0.2618\alpha - 0.5\cos(\alpha) + 0.1309\sin(2\alpha) - 0.1250\cos(2\alpha)\big]^{1/2} \tag{3.42}$$

The r.m.s. ripple voltage is shown as a function of the delay angle in Fig. 3.8(b). It is greater than the value for the fully controlled bridge over much of the delay angle range.

The maximum ripple voltage of $0.4022V_{om}$ occurs at a delay of 87.36°. As the delay is increased beyond this point the ripple decreases and becomes zero at 180° delay.

3.11 The thyristor and diode currents

The thyristor and diode current waveforms are shown in graphs 3 and 4 of Figs 3.6 and 3.7. Apart from the fact that the diode current waves are fixed in phase relative to the supply voltage at a position corresponding to zero delay, the waveforms are identical to those for the fully controlled bridge.

The mean and r.m.s. values of the thyristor and diode currents are affected by the freewheeling diode, an invariable addition to the bridge which is discussed in Section 3.15. However, the freewheeling diode does not affect the bridge currents when the delay is less than 60° so that for $\alpha \le 60°$

$$I_t = I_d = I_o/3 \tag{3.43}$$

and

$$I_{trms} = I_{drms} = I_o/\sqrt{3} \tag{3.44}$$

3.12 The a.c. line current

The currents in the three a.c. lines have similar waveforms, differing only by a mutual phase displacement of 120°. Considering the red line, the current i_r is the difference between the currents in thyristor T1 and diode D4, i.e.

$$i_r = i_{t4} - i_{d4} \tag{3.45}$$

The waveform of i_r is shown in graph 5 of Figs 3.6 and 3.7. The current wave is comprised of rectangular positive and negative current pulses of amplitude I_o and equal duration. Thus the mean value of the line current is zero as expected. The r.m.s. value depends on the duration, δ, of the pulses and is

$$I_s = I_o\sqrt{\frac{\delta}{\pi}} \tag{3.46}$$

For delay angles less than 60°, δ is 120° so that

$$I_s = \sqrt{\frac{2}{3}}I_o \tag{3.47}$$

When the delay angle exceeds 60° the thyristor and diode currents overlap and there is a period of freewheeling when no current flows in the a.c. line. This period is indicated in Fig. 3.7, graph 5, by the shaded area. The thyristor finishes conducting at $60° + \alpha$ and the diode starts conducting at 120° so that the overlap angle is $60° + \alpha - 120°$, i.e. $\alpha - 60°$. The width of the line current pulses is 120° less this overlap angle, i.e. $180° - \alpha$. Thus for $\alpha > 60°$, $\delta = \pi - \alpha$ and

$$I_s = \sqrt{\frac{\pi - \alpha}{\pi}}I_o \tag{3.48}$$

The ratio I_s/I_o is shown as a function of delay in Fig. 3.8(c).

3.13 The input kVA and power factor

The input volt-amperes as conventionally measured is

$$VA_s = \sqrt{3}V_s I_s$$

This, when expressed in terms of the load current and voltage, becomes, when $\alpha \leq 60°$

$$VA_s = \frac{\pi}{3}V_{om}I_o \tag{3.49}$$

and when $\alpha > 60°$

$$VA_s = \sqrt{\frac{\pi(\pi - \alpha)}{6}}\, V_{om} I_o \tag{3.50}$$

The variation of the input volt-amperes with delay angle is the same as the variation of r.m.s. input current so that Fig. 3.8(c) also shows the variation with delay of the ratio of volt-amperes to $V_{om}I_o$.

The input power factor as conventionally measured is the ratio of the input power to the input volt-amperes and, since the ideal lossless case is being considered, this is equal to the ratio of output power to input volt-amperes. Thus for $\alpha \leq 60°$

$$Pf = \frac{3}{\pi}\frac{1 + \cos(\alpha)}{2} = 0.4775[1 + \cos(\alpha)] \tag{3.51}$$

and for $\alpha > 60°$

$$Pf = \sqrt{\frac{6}{\pi(\pi - \alpha)}}\frac{1 + \cos(\alpha)}{2} = \sqrt{\frac{0.4775}{\pi - \alpha}}\,[1 + \cos(\alpha)] \tag{3.52}$$

The power factor is shown as a function of delay in Fig. 3.8(d). The power factor of the semi-controlled rectifier is slightly better than that of its fully controlled counterpart.

Example 3.2

Repeat Example 3.1 but using a semi-controlled bridge directly connected to the same 440 V, 60 Hz system.

The rated load current is unchanged at 336.6 A but the output voltage ripple frequency is halved to 180 Hz and its magnitude increases from 24.5 V at zero delay to 187.1 V at 60° delay and then decreases to zero at 180° delay.

At rated output current, the a.c. line current is 274.8 A for delay angles less than 60°. As the delay is increased beyond 60° the a.c. line current steadily decreases, being 238.0 A at 90° delay, 168.3 A at 135° delay and zero at 180° delay.

The input power factor is given by eqn 3.50 when the delay is less than 60° and eqn 3.51 for larger delays.

****

3.14 Operation at large angles of delay

The essential function of a controlled rectifier is the production of variable voltage d.c. In many applications this requirement includes the production, from time to time, of a very low voltage. As an example, if the load is the armature of a d.c. motor, the machine may be required to run at very low speed, i.e. very low armature voltage, for part of its operating cycle, e.g. when it is decking an elevator, setting up a machine tool or threading paper through a printing press. This duty presents no problem to a fully controlled bridge but the semi-controlled bridge as described so far has a serious problem, the possibility of commutation failure.

As the delay approaches 180° the output voltage waveform becomes a series of brief triangular pulses as indicated in graph 1 of Fig. 3.9 which has been drawn for a delay of 165°. Graphs 2 and 3 show the distribution of the output current between the thyristors and the diodes. Attention will be concentrated on the commutation from T3 to T1 which occurs at $\theta = -60° + \alpha = 105°$. Hitherto, an ideal situation with zero impedance in the a.c. circuit and the supply system has been assumed. Now it must be recognized that in reality the lines carrying current to the thyristors will have inductance. Voltage will be absorbed in removing the stored magnetic energy associated with the flow of i_{t3} in the line inductance and in building up the magnetic energy associated with i_{t1}. The commutation voltage, i.e. the voltage which forces i_{t3} down to zero and i_{t1} up to I_o, is v_{rb}. Its waveform is shown by a gray line in graph 1 of Fig. 3.9 where it is identified by v_{rb} above its positive peaks. All the line to line voltages are similarly shown and identified in graphs 1 and 4.

At large delay angles the commutation voltage becomes very small and is zero at 180° delay. Under these circumstances it may well be that the current transfer is not completed at the critical point, $\theta = 120°$, when the commutation voltage reverses direction and opposes rather than aids the transfer. If this occurs, T3 does not recover its forward voltage blocking capability and a disastrous chain of events commences.

At $\theta = 120°$ the commutation voltage reverses, T1 stops conducting and T3 continues to conduct. The output voltage is now $v_{br} = \sqrt{2}V_s \cos(\theta - \pi/6)$. The applicable portion of it is shown by a black line in graph 4. T2 is fired at $\theta = 225°$ but the commutation voltage, v_{yb}, is negative and conduction through T3 continues. Transfer of current from D4 to D5 occurs at $\theta = 240°$ and the output voltage then becomes $v_{by} = \sqrt{2}V_s \cos(\theta + \pi/2)$. The applicable portion of this voltage is also shown by a black line. This voltage peaks at $\theta = 270°$ and then decreases to zero at $\theta = 360°$ when current transfers from D5 to D6, the rectifier goes into the freewheeling state and the output voltage is zero. This continues until $\theta = 465°$ when T1 is again fired.

The very large voltage pulse which the load receives as a result of the commutation failure produces a very considerable increase in load current and, since the circuit was unable to commutate the much smaller normal current, it will not be able to commutate this large current so that the fault

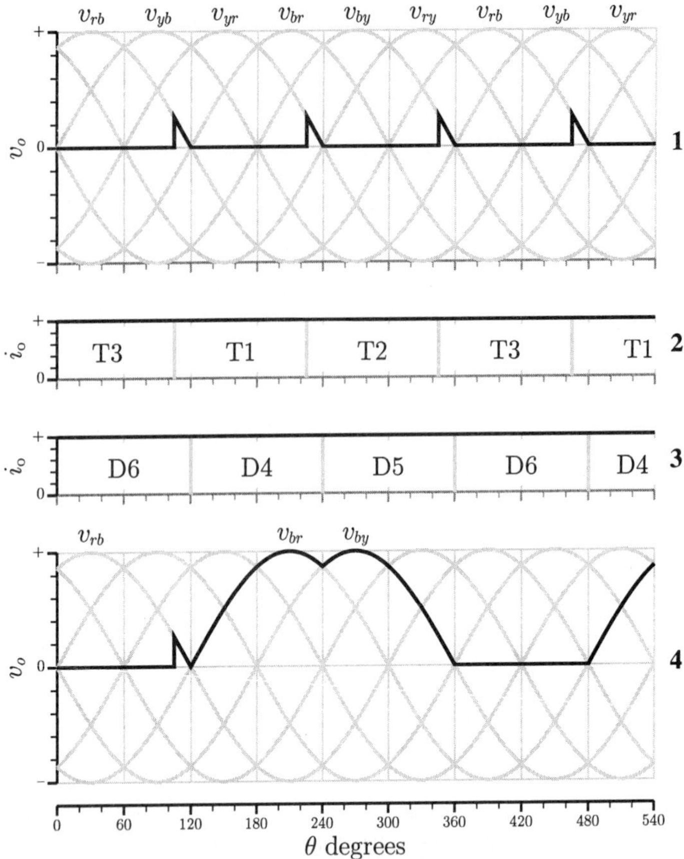

FIG. 3.9. Operation at 165° delay. Graph 1 shows normal operation. Graph 2 shows a commutation failure during the transfer from T3 to T1.

condition will continue until the protective devices are able to interrupt the connection between the d.c. load and the rectifier.

So far, a fault condition which might occur in any circuit has been described. However, considering the nature of the situation which led to it, it is seen that the consequences are likely to be far more serious than the blowing of a fuse or opening of a circuit breaker. The motor supplied by the rectifier will be accelerated fiercely and lives may well be jeopardized.

The remedy which is employed with the fully controlled rectifier, of limiting the delay angle to a value at which successful commutation is ensured, would excessively limit the field of application of the semi-controlled rectifier and the situation is relieved by the introduction of a fourth diode, the *freewheeling diode.*

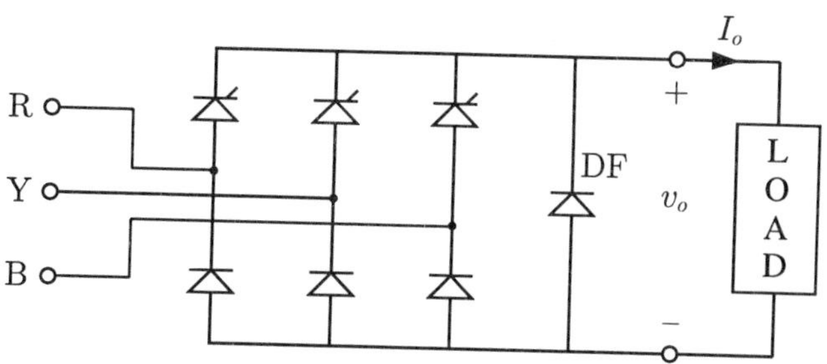

FIG. 3.10. The semi-controlled bridge rectifier with freewheeling diode.

3.15 The freewheeling diode

The freewheeling diode DF is connected between the output terminals of the bridge as shown in Fig. 3.10. Provided the output voltage is positive, the diode is reverse biased and takes no part in the operation. However, when freewheeling occurs much of the load current is diverted through the freewheeling diode, the current in the thyristor is greatly reduced, and the chances of a successful commutation are correspondingly increased.

The operation of the freewheeling diode is illustrated by Fig. 3.11 which shows an approximation to the nonlinear conduction characteristics of the thyristors and diodes, the vertical axis being device current and the horizontal axis the device forward voltage, v_{thy}. The conduction characteristics of the three devices, the thyristor, its diode, and the freewheeling diode, are assumed to be the same and are given by curve A. The conduction characteristic of the series combination of the thyristor and diode is shown by curve B which, for the same current, has a voltage drop twice that of curve A. The conduction characteristic of the parallel combination of the feedback diode and the bridge thyristor and diode is shown in curve C. On this curve, for a given voltage drop, the output current, i_o, is the sum of the currents through the freewheeling diode, i_{fwd}, and the series combination of thyristor and diode, i_{td}.

When freewheeling, the output current, I_o, is made up of I_{fwd} through the freewheeling diode and I_{td} through the thyristor and diode, these three currents being indicated in Fig. 3.11 by annotated horizontal dashed lines. It is evident that the device conduction nonlinearity causes most of the current to be diverted away from the thyristor through the feedback diode. The actual degree of nonlinearity will be substantially more than is shown in Fig. 3.11 and in practice a much smaller proportion of the current than is shown flows through the thyristor.

In addition to its primary function of assisting commutation, the freewheeling diode provides a modest improvement in efficiency by reducing the voltage drop during freewheeling by a factor of about two.

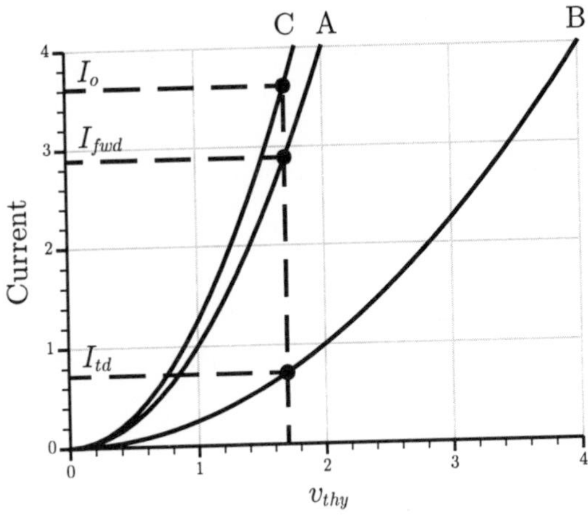

FIG. 3.11. Device conduction characteristics showing how the nonlinearity increases the diversion of current through the freewheeling diode.

During the freewheeling period most of the load current is diverted from the bridge to the freewheeling diode and, making the extreme assumption that all of the current is so diverted, the following expressions can be derived for the mean and r.m.s. values of the currents in a bridge thyristor, a bridge diode, and the freewheeling diode. The thyristor and its associated diode conduct once per supply cycle and their currents are averaged over 2π while the freewheeling diode operates three times per supply cycle and its current is averaged over $2\pi/3$. When $\alpha \geq 60°$

$$I_t = I_d = I_o \frac{\pi - \alpha}{2\pi} \tag{3.53}$$

$$I_{fwd} = I_o \frac{\alpha - \pi/3}{2\pi/3} \tag{3.54}$$

$$I_{trms} = I_{drms} = I_o \sqrt{\frac{\pi - \alpha}{2\pi}} \tag{3.55}$$

$$I_{fwdrms} = I_o \sqrt{\frac{\alpha - \pi/3}{2\pi/3}} \tag{3.56}$$

3.16 Comparison of the two bridges

The salient difference between the two bridges is the inability of the semi-controlled version to reverse power flow. This limits it to applications where regeneration, for example to provide rapid speed reduction of a d.c. motor,

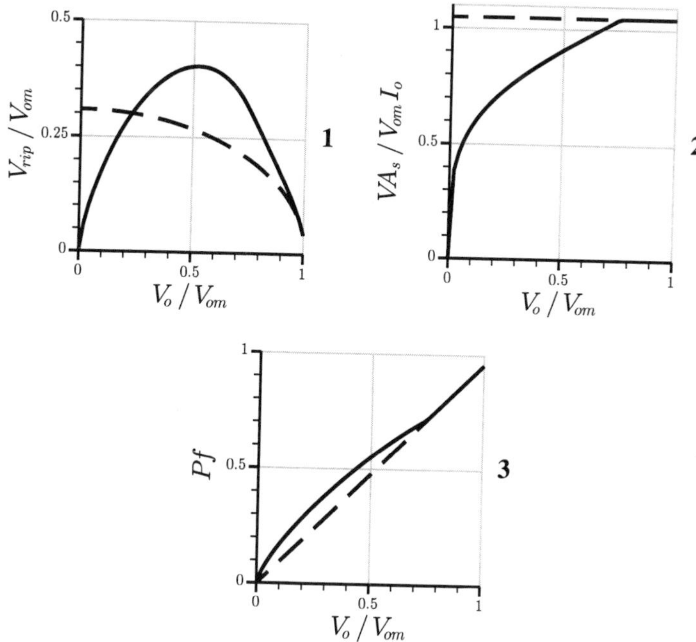

FIG. 3.12. Comparison of the characteristics of the semi-controlled rectifier, shown by the solid line, with those of the fully controlled rectifier, shown by the dashed lines.

is not required. If this condition is met, Fig. 3.12 provides a valid comparison of the two bridges.

Figure 3.12 shows the r.m.s. ripple voltage, the input volt-amperes, and the power factor as functions of output voltage for the two bridges. The characteristics of the semi-controlled bridge are shown by solid lines and of the fully controlled bridge by dashed lines. All values are normalized to the maximum value of the output voltage, V_{om}, or the product $V_{om}I_o$ as appropriate. At low output voltages, say below 50% of V_{om}, The kVA demand of the semi-controlled bridge is substantially lower than that of the fully controlled bridge. This could be a significant factor if operation at low voltage forms an appreciable proportion of the system duty cycle.

Below 87% of V_{om} the power factor of the semi-controlled bridge is slightly better than that of the fully controlled bridge but the improvement is not of the same order as the reduction in kVA just noted.

Except for considerable voltage reductions to below 25% of V_{om}, the output voltage ripple of the semi-controlled bridge is greater than that of the fully controlled circuit. When it is recollected that the ripple frequency is also half that of the fully controlled bridge, it is seen that the semi-controlled device is inferior in regard to ripple.

In many applications the need for inversion of power flow dictates the fully controlled bridge. When this is not a factor there is no unanimity as to the relative merits of the two circuits. However, as the number of rectifiers connected to supply systems increases and the problems of harmonic pollution increase, use of the semi-controlled version appears to be diminishing.

4

GATING CONSIDERATIONS AND OPERATION AT LARGE ANGLES OF DELAY

4.1 Introduction

The basic theory of simple single way and bridge rectifiers operating under idealized conditions was reviewed in Chapters 2 and 3 without too much concern for details. In this and succeeding chapters many of those details will be considered. In this chapter a number of matters concerned with thyristor gating and operation at large delay angles are investigated.

4.2 Delay angle limitations

In the preceding two chapters it has been tacitly assumed that the delay angle is limited to the range zero to 180°. Why this is so and why in reality the range is somewhat more limited, from zero to about 165°, are the topic of this section.

Consider the two thyristors TA and TB shown in Fig. 4.1(a). The inductances in series with the thyristors are the input line inductances. They are small but have a vital effect on commutation, the transfer of current from one thyristor to the next, since the energy in the magnetic field of one must be removed and the energy of the other must be built up. This process absorbs voltage and takes time. Assuming that TA is conducting so that the voltage of its cathode is equal to, except for the small thyristor voltage drop, the voltage, v_a, of its anode. The anode voltage of TB is v_b so that the voltage, v_{tb}, across TB is $v_b - v_a$. This voltage must be positive if commutation from TA to TB is to occur when TB is fired.

Figure 4.1(b) shows the two anode voltages. It has been drawn for a six phase rectifier but is valid for any number of phases P. The voltage v_b lags the voltage v_a by $360°/P = 60°$ in Fig. 4.1(b). The time origin is placed at the natural commutation instant so that

$$v_a = V_{pk}\cos(\theta + \frac{\pi}{P}) \tag{4.1}$$

and

$$v_b = V_{pk}\cos(\theta - \frac{\pi}{P}) \tag{4.2}$$

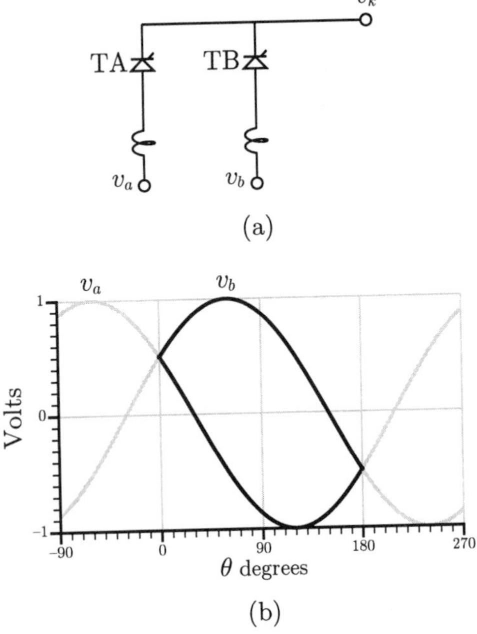

FIG. 4.1. Commutation from thyristor TA to thyristor TB.

It follows that

$$v_{tb} = v_b - v_a = 2V_{pk}\sin(\frac{\pi}{P})\sin(\theta) \tag{4.3}$$

The voltage v_{tb} across thyristor TB is positive over the range $0 < \theta < 180°$ which is the region on Fig. 4.2(b) where the waveforms of v_a and v_b are shown in black. This is the range for which natural commutation is possible. The adjective *natural* is used here to signify that commutation resulting from the normal operation of the circuit is being referred to as distinct from *forced* commutation when external means are employed to make the current transfer. All circuits described in this book are naturally commutated.

If the gate pulse is delivered at a negative delay angle, what happens depends on the width of the pulse. If the width is greater than the magnitude of the delay angle, gate drive is still applied when v_{tb} becomes positive and current transfer occurs at an effective gate angle of zero. If on the other hand the gate pulse width is less than the magnitude of the delay angle, commutation will not occur. Thus, $\alpha = 0$ is the phase forward limit.

At the other extreme, if the firing pulse is delivered at a delay of 180°, the commutation voltage is zero and immediately becomes negative. Thus, current transfer does not occur and conduction continues through TA. To ensure a successful commutation, the firing pulse must be delivered to TB at a delay angle somewhat less than 180° when there is sufficient commutation voltage to ensure a successful transfer. The precise maximum effective value of delay depends on the values of the load current and of the supply impedance. However, a reasonable estimate of the phase back limit is about 90% of the theoretical limit of 180°, i.e. about 162°.

Thus, the practical range of delay angle for a naturally commutated rectifier is

$$0 < \alpha < K \times 180^\circ \tag{4.4}$$

where K depends on the input impedance and can be assumed to be of the order of 0.9.

4.3 Firing pulse width

A number of considerations influence the width of the firing pulses applied to a rectifier thyristor. First, a thyristor can be switched into the conducting state by a very brief current pulse to its gate, of the order of 10 to 20 μs duration. Second, it is poor practice to apply gate current to a reverse biased thyristor since this substantially increases the reverse leakage current and increases the device losses. The remaining factors, which are discussed below, are functions of the type of rectifier and the external circuits.

To ensure stability when the delay is small, the load is active and the output current is small, the width of the firing pulse must be at least $180^\circ/P$ for a P phase rectifier as discussed in Section 6.24. Thus, for a fully controlled three phase bridge, which acts as a six phase rectifier as far as the input and output circuits are concerned, the width of the firing pulses must be at least 30°.

In any two way rectifier, such as the three phase bridge, in which two thyristors always operate in series, it is possible that the thyristor which is nominally conducting will in reality have ceased to conduct and have recovered its blocking capability when the incoming thyristor is fired, a matter discussed in detail in Chapter 6. Under these circumstances conduction does not occur unless gate signal is applied simultaneously to both thyristors. One of two arrangements can be employed, double pulsing or continuous gating.

In the double pulsing arrangement, firing pulses are delivered simultaneously to the gates of both the incoming and the nominally conducting thyristor. Thus each thyristor receives two gate pulses, the first when it is due to start conduction and the second when the next thyristor starts to conduct. For the three phase bridge, the two pulses are 60° apart.

With continuous gating, gate signal is applied to the thyristor for at least the period from the commencement of conduction to the gating of

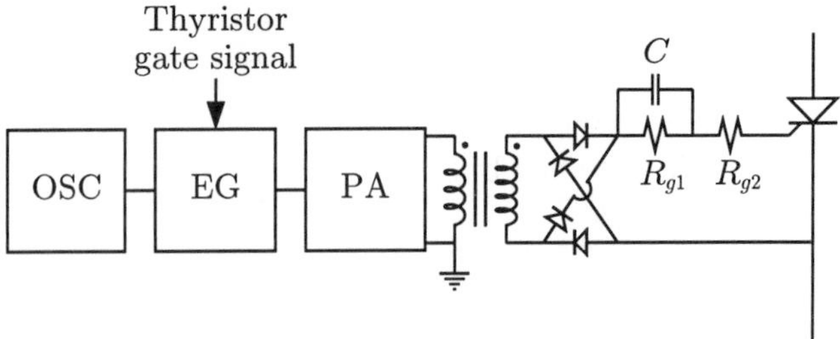

FIG. 4.2. A thyristor gating circuit.

the next thyristor in the other half of the bridge. Thus, for a three phase bridge rectifier, the width of the gate pulses would be somewhat more than 60°, say 70–75°.

In view of the requirement noted above for firing pulses at least 30° wide, continuous gating is more common as fulfilling both needs.

4.4 Firing circuits

A wide variety of firing circuits is employed but the circuit of Fig. 4.2 will serve to illustrate the principles involved in their design. The oscillator, OSC, produces a frequency of several hundred times the supply frequency, say 50 kHz. It is connected to the firing circuit of each thyristor by an electronic gate, EG, one for each thyristor, which is opened by a low level logic signal derived from the system controller.

The low level oscillator signal passed by the open gate serves as the input to a power amplifier, PA, which produces the required power level for firing the thyristor, of the order of 10 W. The power amplifier output is delivered to the thyristor gate via a high frequency transformer, rectifier, and limiting resistors, R_{g1} and R_{g2}.

The rectifier control circuits operate at very low power levels and at voltage levels close to ground potential and the transformer serves to isolate them from the thyristor cathode potential which makes large and rapid excursions relative to ground, typically hundreds of volts in microseconds. The isolation can be improved by providing optical isolation between the gate and the power amplifier and/or having an electrostatic shield between the primary and secondary windings of the transformer to minimize disturbances transmitted from the thyristor via the inter-winding capacitance. To the same end, the power supply for the power amplifier is separate and well isolated from the power supply for the system controller.

The rectifier ensures that negative voltages are not applied to the thyristor gate. A full wave rectifier such as is shown ensures that the transformer windings carry only a.c. and thus avoids problems due to d.c. saturation of its core.

The gate resistance ensures that the nonlinear and very variable gate to cathode resistance is swamped by a linear resistance thus ensuring reliable and repeatable operation.

The capacitor which shunts part of the gate resistance gives the initial portion of the gate pulse a higher current than the later portion, to give a condition called *hard gating.* A high initial gate current, approaching the maximum permissible value, ensures that the thyristor turns on quickly, reliably and with its maximum di/dt capability. Once the thyristor is turned on, a much lower level of gate drive will serve to keep it in the on state, and the gate current can be reduced. The capacitor initially short circuits R_{g1}. As it charges, the full value of R_{g1} becomes effective and reduces the gate current. The time constant will be such that high gate current is applied for only a few microseconds.

An idealized example in which the ripple in the rectified gate drive supply, the internal impedance of that supply, and the gate to cathode impedance are all neglected will suffice to illustrate the principle.

Example 4.1

For the gate driver circuit of Fig. 4.2, determine the form of the gate current if the d.c. gate drive source is 10 V, R_{g1} is 30 Ω, R_{g2} is 10 Ω and C is 1 μF. Assume that the gate to cathode impedance is negligible.

If t is in microseconds, the expression for the gate current in amperes is

$$i_g = 0.25 + 0.75 \exp(-\frac{t}{7.5})$$

Thus, the gate current is initially 1 A and decays to a steady value of 0.25 A in about 25 μs as illustrated in Fig. 4.3.

Assuming the gate drive to be maintained for a 75° interval, i.e. for about 4000 μs in the typical 50 or 60 Hz rectifier, the power delivered to each gate circuit would, except for the brief initial high current period, be about 2.5 W. For a three phase bridge rectifier the average power per gate would be about 0.52 W and the mean power delivered to all six gates would be about 3.1 W. Such a bridge rectifier could supply hundreds of kilowatts so that the power gain would be of the order of 10^5.

****

4.5 The controller

The controller which regulates the thyristor gate currents by opening and closing the electronic gate typically incorporates the following features,

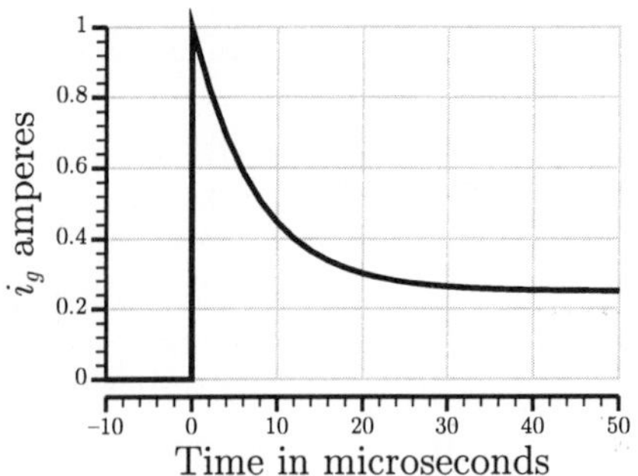

FIG. 4.3. Thyristor gate current.

- it linearizes the rectifier steady state transfer function,
- it provides protection against excessive phase advance and phase retard,
- it provides an inhibit feature which on command blocks all firing pulses.

The transfer characteristic of a fully controlled rectifier follows the cosine law

$$V_o = V_{om}\cos(\alpha) \tag{4.5}$$

thus, if the controller is to provide a linear relationship between a control voltage v_c and the mean output voltage V_o, it must produce a delay angle which obeys the relationship

$$\alpha = \arccos(\frac{v_c}{V_{cm}}) \tag{4.6}$$

where V_{cm} is the maximum effective value of v_c. By taking the cosine of each side, this translates to

$$\cos(\alpha) = \frac{v_c}{V_{cm}} \tag{4.7}$$

Substituting the relationship of eqn 4.7 into eqn 4.5 yields

$$V_o = \frac{V_{om}}{V_{cm}}\, v_c \tag{4.8}$$

i.e. the mean output voltage is proportional to the control voltage, the steady state gain being V_{om}/V_{cm}.

The desired linear relationship between control voltage and delay angle is obtained using the biased cosine principle which can be accomplished by analog or by digital means. Although digital controllers are now more popular than the analog type, the analog technique will first be described since it more clearly illustrates the principles involved.

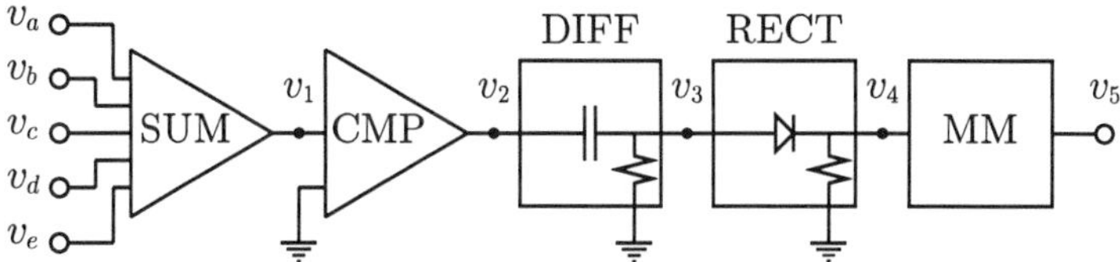

FIG. 4.4. A biased cosine controller channel.

4.6 The biased cosine controller

The essentials of a controller channel for one thyristor are shown in block diagram form in Fig. 4.4. For the fully controlled bridge rectifier, the controller has six such channels, one for each thyristor. The control channel starts with a five input summer, SUM, whose output, v_1, is the inverse sum of the five input voltages, v_a through v_e. The output of the summer is taken to the comparator, CMP, whose output, v_2, feeds a differentiator, DIFF. The negative pulses in the differentiator output, v_3, are removed by the rectifier, RECT. The resulting positive pulses, v_4, flip a monostable circuit, MM, which flops back to its original state after a preset time. It is the output of this monostable, v_5, which opens the electronic gate of Fig. 4.2 to send a firing signal to the thyristor. The waveforms of the voltages are shown in Fig. 4.5 which has been drawn assuming that the channel is controlling thyristor T1 at a delay angle of 60°.

The first input to the summer, v_a, is a sinusoidal wave whose negative peak coincides with the natural commutation point of the thyristor being controlled

$$v_a = V_{cm} \cos(\theta - \psi) \tag{4.9}$$

where the phase angle, ψ, is in this case $2\pi/3$. This waveform is shown in graph 1 of Fig. 4.5.

The second input, v_b, is a d.c. bias voltage which will for the time being be assumed to be zero.

The third input, v_c, is the control voltage. It is shown in graph 4 of Fig. 4.4 by the horizontal dashed line.

The fourth input, v_d, is a negative rectangular pulse train, one pulse per supply cycle. Its waveform is shown in graph 2 of Fig. 4.5. The trailing edge of each pulse coincides with a negative peak of v_a and defines the zero delay point. The pulse amplitude is approximately equal to V_{cm}.

The fifth input, v_e, is a positive rectangular pulse train, one pulse per supply cycle. Its waveform is shown in graph 3 of Fig. 4.5. The leading edge of each pulse defines the maximum delay point and occurs somewhat before a positive peak of v_a. The pulse amplitude is approximately equal to V_{cm}. In Fig. 4.5, the leading edge has been set at $\theta = 105°$.

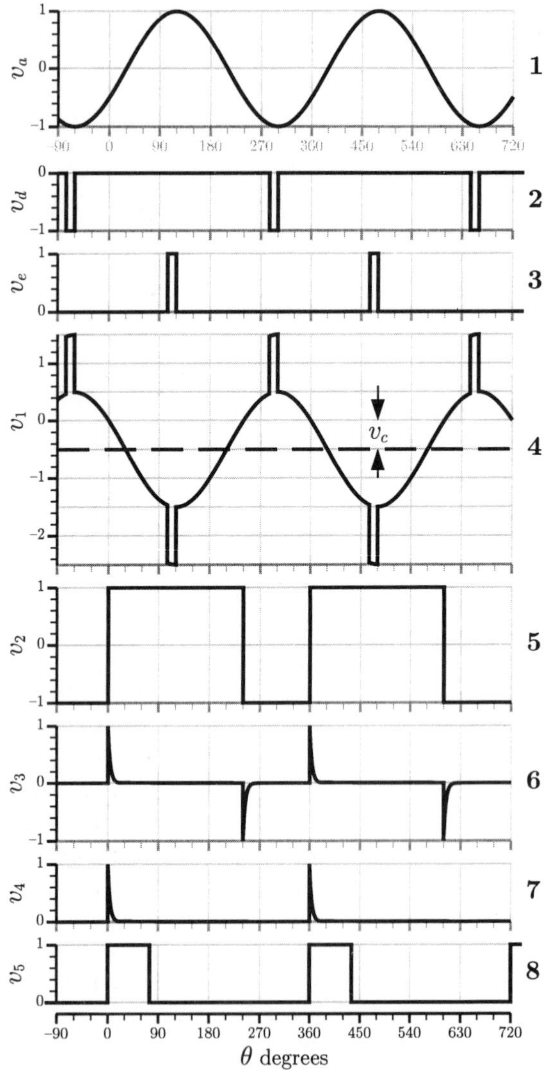

FIG. 4.5. Waveforms in a biased cosine controller.

By making the pulse widths 15–25° as appropriate, the fourth and fifth inputs can be combined into a single symmetrical pulse train if desired.

The waveform of the inverted sum of the input voltages, v_1, is given in graph 4 of Fig. 4.5. For the moment the rectangular pulses can be ignored and attention can be concentrated on the sinusoidal portions of the wave biased by the control input, v_c. The voltage v_1 is then

$$v_1 = V_{cm} \cos(\theta - \psi) - v_c \tag{4.10}$$

It has a negative going zero crossing at θ_1 where

$$\cos(\theta_1 - \psi) = \frac{v_c}{V_{cm}} \tag{4.11}$$

and $0 < \theta_1 - \psi < \pi$. It has a positive going zero crossing at θ_2 where

$$\theta_2 = 2\psi - \theta_1 \tag{4.12}$$

If the phase shift ψ has been correctly chosen, the angle $\theta_1 - \psi$ is the delay angle. In the case shown in Fig. 4.5 with $v_c = V_{cm}/2$ and $\psi = 2\pi/3$, $\theta_1 = 0$, $\theta_2 = 4\pi/3 = 240°$ and the delay, for thyristor T1, is 60°.

Equation 4.11 provides the desired cosinusoidal relationship, see eqn 4.7, between θ_1 and the control voltage, v_c, so that if the negative going zero crossing of the summer output, v_1, is used to trigger the thyristor the desired linearization will have been achieved. It is the function of the remaining components of Fig. 4.4 to do this.

As the output, v_1, of the summer passes through zero, the output, v_2, of the comparator switches rapidly from one extreme to the other as shown in graph 5 of Fig. 4.5. When v_1 is positive, v_2 has its maximum negative value, when v_1 is negative, v_2 has its maximum positive value.

The differentiator, which can be a simple RC combination, generates positive and negative steep fronted pulses at θ_1 and θ_2.

The negative pulses at θ_2 are blocked by the rectifier and the positive pulses at θ_1 cause the monostable flip-flop to flip from the quiescent to the active state. The monostable remains in this state with its output high for a preset period and then flops back to the quiescent state. For a fully controlled bridge rectifier the monostable timer would maintain the output high for about 75°.

Provided the timing of θ_1 is correctly arranged relative to the thyristor voltage, the output of the monostable can be used to open the electronic gate of the firing circuit, Fig. 4.2, to deliver to the thyristor a correctly timed gate pulse of correct width. The question of correct timing will shortly be discussed, but for the moment we must remain with the control channel to review the functions of the phase forward and phase back limits.

4.6.1 *Phase forward and phase back limits*

As the control voltage approaches $+V_{cm}$, the intercept with the cosine wave becomes shallower and shallower and so less well defined. When $v_c = V_{cm}$, v_c is tangential to v_a, the delay angle is zero and the rectifier output voltage is at its maximum value. If the control voltage is further increased, it no longer intersects with v_a and, if there were no phase forward barrier, v_1 would not reach the zero level, the comparator output would not switch, no firing signal would be sent to the thyristor, and the rectifier output would collapse, usually with disastrous consequences. From a practical viewpoint it is impossible to limit the control voltage so precisely that this does not occur. The phase forward limit pulse, v_d, provides a realistic solution to this problem.

If the control voltage becomes too high, the negative going intercept occurs at the trailing edge of the phase forward limit pulse and firing pulses continue to be delivered to the thyristor at a delay angle of zero. By making the height of the limit pulse approximately equal to the amplitude, V_{cm}, of the cosine wave, it becomes a simple and entirely practicable matter to limit the control voltage so as to maintain correct firing.

The phase back limit, v_e, works in a similar way. The leading edge of the phase back limit pulse prevents the delay angle from exceeding the value which the system designer decides is prudent, 165° in the case of Fig. 4.5.

4.6.2 *Setting the phase relationships*

Correct firing requires that the negative maximum of the cosine wave, v_a, coincide with the zero delay point of the thyristor to be fired. Considering for example the three phase bridge rectifier of Fig. 3.1(c), the zero delay point of thyristor T1 is at $\theta = -\pi/3$ so that the required controller voltage is given by eqn 4.13.

$$v_a = V_{cm} \cos(\theta - \frac{2\pi}{3}) \tag{4.13}$$

Inspection of the phasor diagrams of Fig. 4.6(a) and 4.6(b) shows that a voltage of the correct phase is obtained by summing the two line to line voltages, v_{yr} and v_{yb}. This can be done by series connection of two transformer secondaries as shown in Fig. 4.6(c). However, the replacement of the series connection by a summing amplifier, S, as shown in Fig. 4.6(d) provides much greater flexibility. The output voltage in this event is

$$\mathbf{V}_o = N_{tsp} R_3 (\frac{\mathbf{V}_{yr}}{R_1} + \frac{\mathbf{V}_{yb}}{R_2}) \tag{4.14}$$

where $\mathbf{V}_o$, $\mathbf{V}_{yr}$, and $\mathbf{V}_{yb}$ are phasor representations of the voltages and N_{tsp} is the ratio of secondary to primary turns of the transformer. Separate adjustment of the resistances R_1 and R_2 provides control of the phase of the

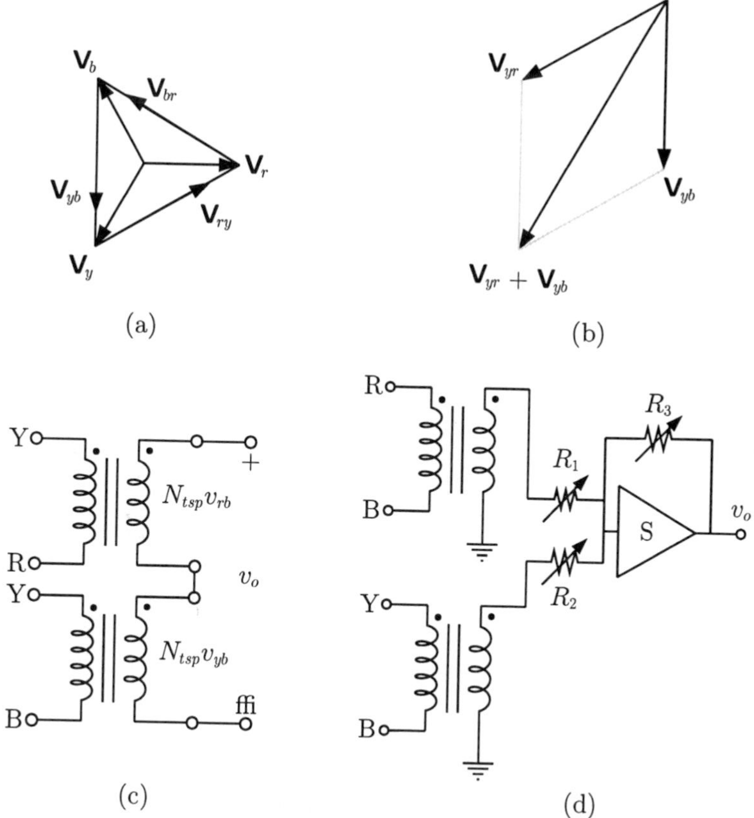

FIG. 4.6. Creation of a voltage having a specific phase.

output voltage and adjustment of R_3 provides control of its amplitude. Indeed, by the appropriate use of inverting amplifiers preceding the summing amplifier, the output voltage can be given any phase angle between zero and 360°.

Unfortunately, this simple solution is not practical because of noise in the three phase supply. The major component of this noise is notches associated with commutation in the rectifier, a topic considered in Chapter 5. Such waveform disturbances transmitted to the controller cause misfiring and failure of the rectifier. To eliminate this, the outputs of the transformers of Fig. 4.6 must be passed through low pass filters to remove the noise components.

Fortunately high voltages are available from the control transformers and very substantial filtering using simple components can be employed. As an example a simple, single stage, RC low pass filter with its break

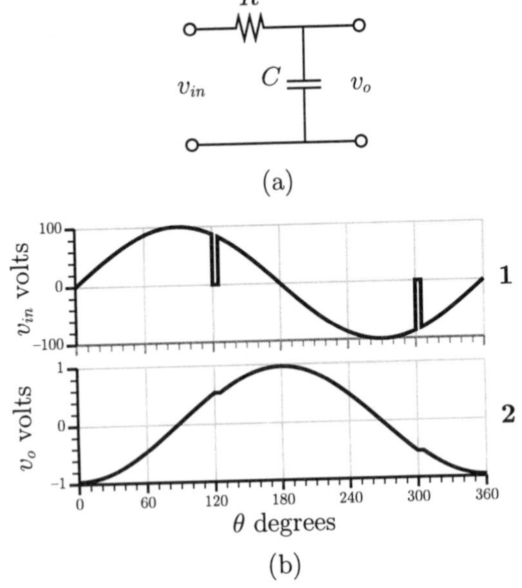

FIG. 4.7. Filtering the reference wave.

frequency set two decades below the supply frequency will attenuate the supply frequency component by a factor of 100 and retard its phase by 89.43°. A noise component at ten times the supply frequency is attenuated by a factor of 1000 and is retarded in phase by 89.94°. Thus, the phase shift of such a filter is insensitive to the filter parameters, being close to $-90°$, and it attenuates the noise relative to the signal by substantial amounts.

The efficacy of such a filtering system is illustrated by Fig. 4.7 which shows in part (a) a simple RC low pass filter and in graph 2 of part (b) the effect of that filter on an input voltage with deep notches, graph 1. The notches are reduced to minor level changes in the output voltage. The figure has been drawn for the case of Example 4.2 and exaggerates the area of the notches so as to make the effect of the filter clearly visible.

Example 4.2

Consider a three phase bridge rectifier connected to a 440 V, 60 Hz, three wire, three phase supply. Taking the red line to neutral voltage as datum, i.e.

$$v_r = 359.3\cos(\theta)$$

derive sinusoidal voltages, amplitude 10 V, for use in a biased cosine controller for firing thyristor T1. Use simple RC filters to remove noise on the system voltages.

Since the filters will introduce a phase lag of 90°, a filter input voltage which lags v_r by 30° is required. Inspection of Fig. 4.6(a) shows that this can be obtained directly from the line to line voltage v_{rb}. An isolating transformer, ratio $N_{tsp} = 0.1607$, will therefore be connected between the red and blue lines. The output of this transformer will be a voltage, amplitude 100 V, lagging v_r by 30°

Selecting a filter capacitance of 4 microfarads and a filter resistance of 66.3 kΩ gives a cutoff frequency of 0.6 Hz, an attenuation of 100 at 60 Hz and a phase lag of 89.43°. Thus, after filtering, the output has almost the desired phase. The loss in the filter will be 75 mW, a negligible amount.

Should the phase angle error of 0.57° be too large it can be eliminated by adding to the voltage derived from v_{rb} filtered and suitably scaled signals derived from the other two line to line voltages using circuits similar to that shown in Fig. 4.6(d).

The effect of this filter is illustrated by Fig. 4.7 which shows an input voltage with deep notches reaching to the zero voltage line and 5° wide. Such notches are severe and are unlikely to be encountered in practice. The effect of these notches is to vertically shift portions of the output voltage by 0.038 V.

****

4.6.3 *The common anode half*

Nothing has so far been said about the biased cosine waves required for firing the bottom half thyristors, T4, T5, and T6. Since these are fired in antiphase with T1, T2, and T3, inverting the outputs of the summing amplifiers in the control channels for the common cathode thyristors provides the necessary additional cosine waves for the control channels for the common anode thyristors.

4.6.4 *Rectifier gain*

With the biasing signal, v_b, set to zero, the transfer characteristic is linear with the mean output voltage proportional to the control voltage up to V_{om} which occurs at zero delay, $v_c = V_{cm}$, and down to $V_{om}\cos(\alpha_m)$ where α_m is the maximum value of delay angle defined by the phase back limit. Figure 4.8 shows such a characteristic labelled B and assuming $\alpha_m = 160°$. The steady state gain of the converter–controller combination is the slope of the operating part of the characteristic. It depends on the particular design but some idea of the values involved can be obtained by considering a rectifier directly connected to a 440 V supply and using standard operational amplifiers in its controller. For this, $V_{om} = 595$ V and V_{cm} will typically be about 10 V. Thus a gain of about 60 can be expected.

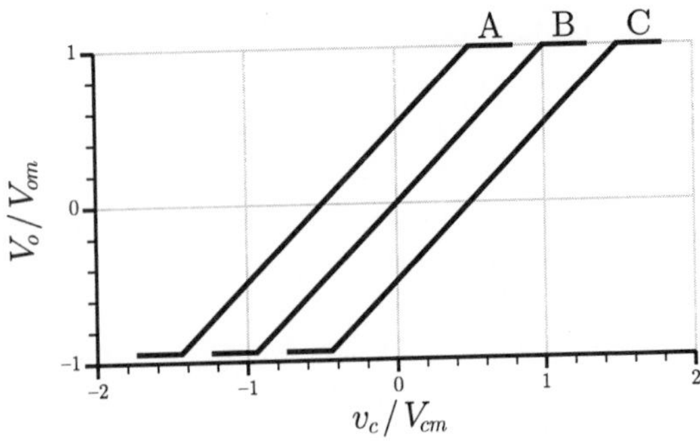

FIG. 4.8. Steady state transfer characteristics of the biased cosine controlled rectifier.

4.6.5 *Biasing*

The symmetrical transfer characteristic referred to above may not suit the needs of the system designer or the characteristics of the control signal. A d.c. signal, v_b of Fig. 4.4, can be used to bias the control characteristic into the desired region. Figure 4.8 shows the effects of making the bias voltage equal to $V_{cm}/2$ in which case the characteristic is moved to the left by $v_c = V_{cm}/2$, characteristic A, or $-V_{cm}/2$ in which case it is moved to the right by the same amount, characteristic C.

4.6.6 *Stability*

No mention has been made of stabilization against supply voltage variations. This is deliberate. If the d.c. sources which provide the control and bias voltages are obtained from the same a.c. source from which the cosine waves are derived, any variation in a.c. supply voltage will affect all equally and the control relationships will be maintained.

4.6.7 *Protection*

A rectifier controller should include a gate pulse inhibitor which on demand will prevent firing pulses being delivered to the thyristors. This can be accomplished by removing an enable signal from the electronic gate, EG, of Fig. 4.2. This feature is useful in protecting the rectifier from at least some of the detrimental effects of faults.

If the rectifier currents become excessive, protective devices such as overload relays will cause it to be disconnected from the supplies. The gate pulse inhibit feature provides a backup which operates extremely quickly, within a fraction of a cycle of fault detection.

Rectifiers typically feed active loads such as d.c. motors and secondary batteries which produce an e.m.f. and must be protected from fault currents supplied from the d.c. side as well as from the a.c. supply. The interruption of a d.c. current is a far more onerous task than the interruption of an a.c. current. An alternating current automatically becomes zero twice per cycle so that its interruption only requires inhibition of the restarting of current flow. A direct current in contrast must be forcibly extinguished. These differing characteristics of a.c. and d.c. which have such a marked effect on circuit breaker design also impinge upon the effectiveness of the gate pulse inhibit circuit. It interrupts currents derived from the a.c. system but cannot stop the flow of currents created by a d.c. side e.m.f. The thyristor which is conducting continues to conduct current derived from the load and the interruption of this current becomes a task for the thyristor fuse and/or the d.c. side circuit breaker.

4.7 Digital control

Rectifier controllers are no exception to the trend towards digital control. The variety of digital controllers encountered in practice is as great as for analog controls. The basic functions which such a controller must provide are shown as blocks in Fig. 4.9. Assuming that the control voltage is analog, it is first converted to digital form by an analog to digital converter A/D. The numerical output of this is converted and limited to an appropriate range corresponding to the phase forward and phase back limits, and the resulting number, which is proportional to $\cos(\alpha)$, is converted to the delay angle by a decoder. This is most probably a read-only memory whose addresses are defined by the output of the limiter and whose stored numbers are the values of delay angle. The delay angle is stored in a counter on receipt of a load signal. The counter is then placed in the ready to count state by an enable signal. The counter starts to count clock pulses as soon as the start signal is given and as soon as the required count is reached it emits a pulse which opens the electronic gate of the thyristor gating circuit of Fig. 4.2.

The counter starting points are obtained by creating correctly phased sinusoidal waves by transforming, filtering, and summing as in Figs 4.6 and 4.7. These waves are then applied to comparators, differentiators, and rectifiers, as in Fig. 4.4, to produce properly timed activating pulses for the counters.

With a digital system, control is discrete, in contrast to the continuous control of the analog system. However, quite simple systems can give good accuracy as Example 4.3 indicates.

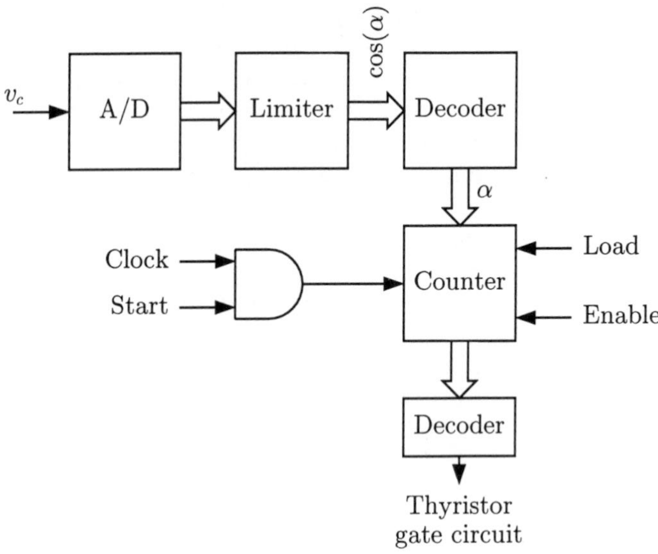

FIG. 4.9. A digital controller channel.

Example 4.3

Determine the accuracy with which the delay angle can be specified if the A/D converter of Fig. 4.9 has an output word comprised of eight bits.

An eight bit A to D converter provides 256 discrete measures of the control voltage and so can address 256 memory locations. If each memory location contains an eight bit word defining the delay angle, the output range from V_{om} to V_{omin} is divided into 256 equal steps. The accuracy is plus or minus half a step and is therefore about 0.4%.

****

4.8 Operation into active loads

In this and the preceding two chapters it has been assumed that the delay angle can have any value between zero and the maximum value which is approximately 165°. We have been able to do this by the implicit assumption that the output is a smooth d.c. current source. It has been seen that, as the delay is increased from zero to 90°, the mean output voltage and hence the power delivered to the load diminish correspondingly from their maximum values down to zero. As the delay is increased beyond 90° the mean output voltage becomes negative and the power delivered to the load is negative. The power flow has reversed, d.c. power produced by the load is being delivered to the a.c. system through the rectifier.

This reversal of power flow implies an active load and in reality many rectifier loads are of this type, for example, the armature of a d.c. machine or a secondary battery. Under steady state conditions such loads can be modeled as a series combination of a d.c. e.m.f. and a resistance. The e.m.f. is nearly equal to the mean rectifier output voltage and is associated with the energy conversion process: electrical to mechanical energy in the case of the motor, electrical to chemical energy in the case of the battery. The resistance represents the sources of energy loss and is small, typically it will absorb less than 5% of the rated voltage at rated current. The load circuit inductance plays an important role in an operation of this type. The energy stored in the magnetic field of the inductor helps bridge the gap between the rectifier output voltage varying at a multiple of the supply frequency and the steady load e.m.f. as illustrated by Example 4.4.

Example 4.4

A fully controlled three phase bridge rectifier is directly connected to a 440 V supply. Determine the operating conditions when its delay angle is 130° and it is connected to a load whose e.m.f. is −400 V. The total resistance in the output circuit is 0.08 Ω and voltage drops in the conducting thyristors can be ignored.

Figure 4.10 shows the waveform of the output voltage by a black line and the load e.m.f. by the lower horizontal dashed line. The mean output voltage, −382.0 V, is shown by the upper horizontal dashed line.

The mean load current, I_o, is

$$I_o = \frac{-382.0 - (-400.0)}{0.08} = 225 \text{ A}$$

Using the nomenclature of Fig. 3.1, thyristors T1 and T5 conduct for $70° < \theta < 130°$. At the start of this 60° segment, the output voltage of the rectifier is

$$v_o = \sqrt{2} \times 440 \times \cos(\theta + \pi/6) = -108.1 \text{ V}$$

so that the voltage driving the current through the load impedance is

$$V_z = -108.1 + 400.0 = 291.9 \text{ V}$$

This is far greater than the 18.0 V necessary to maintain the mean current of 225 A in the load resistance so that the current will tend to increase and in the process to store additional energy in the load inductance. This continues until the output voltage equals −382.0 V, i.e. until $\theta = 97.9°$,

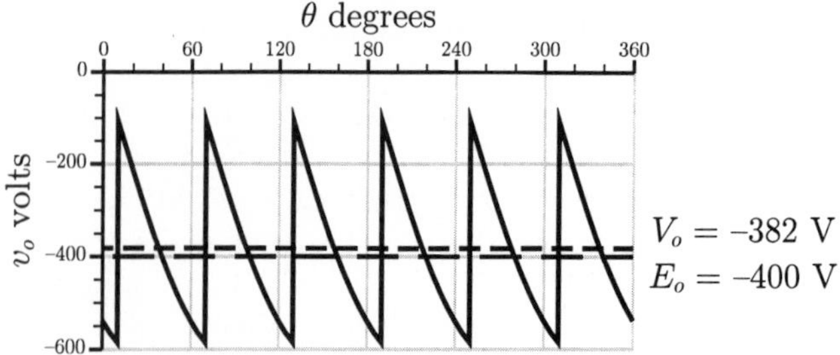

FIG. 4.10. Operation with an active load and with power flowing from the d.c. to the a.c. side.

at which point the 18 V difference between it and the load e.m.f. is just sufficient to force 225 A through the load resistance. Thereafter the difference voltage is inadequate, the load current tends to diminish and energy is drawn from the store in the load inductance.

****

The load inductance acts as an energy bank in which energy is stored when the instantaneous rectifier output voltage is algebraically greater than the load e.m.f. and which delivers energy when the rectifier output voltage is less than the load e.m.f. The inductance bridges the difference between the rectifier output voltage which has considerable ripple and the load e.m.f. which is effectively pure d.c. Up to now it has been assumed that the inductance is so large that variations in the output current are negligible. In reality the current may have a significant ripple, a matter which will be considered in Chapter 5.

4.9 Two and four quadrant operation

The rectifier requires that load current flow in only one direction so that reversal of power flow can only occur when the output voltage reverses. It is usual to describe this situation in the diagram of Fig. 4.11 which represents the load voltage on the horizontal axis and the load current on the vertical axis. The rectifier operates with positive current and positive or negative voltage depending on the delay angle. An operating point such as P1 in the first quadrant corresponds to power flowing from the a.c. side to the d.c. side. An operating point such as P2 in the second quadrant corresponds to inverted power flowing from the d.c. to the a.c. side. The rectifier is said to operate in the either the first quadrant or the second quadrant.

A typical rectifier load, such as a d.c. machine armature or battery, reverses power flow by reversing the direction of current flow and thus operates in the first and fourth quadrants. A d.c. machine can in fact operate

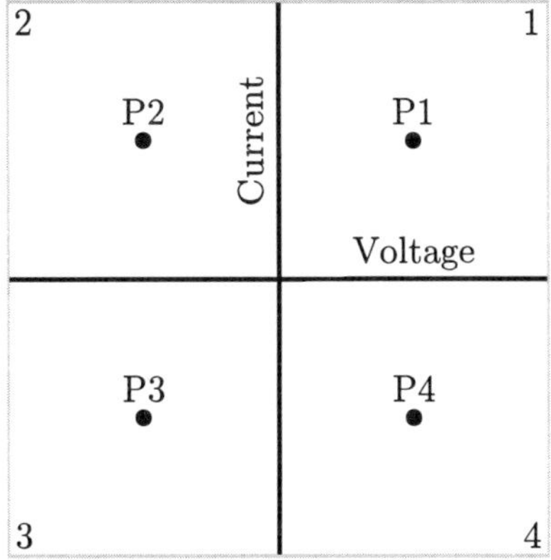

FIG. 4.11. The four operating quadrants.

in all four quadrants if its speed reverses. In this event its e.m.f. reverses so that it then motors in the third quadrant and generates in the second.

The operating characteristics of a rectifier are therefore ill-matched to those of its typical loads and a simple rectifier–load combination can operate in only one quadrant, the first. Such a system can be fed by a semi-controlled rectifier with improvements in input power factor.

Special measures must be taken for multiquadrant operation. Either the load e.m.f. must be reversed, or the rectifier's output voltage must be reversed, or dual rectifiers must be employed. These alternatives are discussed in detail in Chapter 10.

4.10 The controlled rectifier as an a.c. source

Up to now the rectifier has been considered as a load on an a.c. system. It draws current from the system and, if the input power is positive, energy flows from the a.c. to the d.c. side. If the input power is negative, energy flows from the d.c. to the a.c. side. This viewpoint is satisfactory for those cases, the majority, where the rectifier represents a relatively small load on a large system which can accept regenerated power in any form. However, when the rectifier capacity is significant compared to the system capacity, as is the case with the rectifiers at the load end of a d.c. transmission line, problems can be encountered with the negative power flow from the d.c. to the a.c. side, the normal case for this situation. To see why this is so, consider Fig. 4.12(a) which shows the rectifier, a load, and something which will be called a *compensator*, on the a.c. side. The sign conventions already adopted will be used so that the rectifier is shown as delivering power to the d.c. side. However, the rectifier will be operating with a delay angle greater

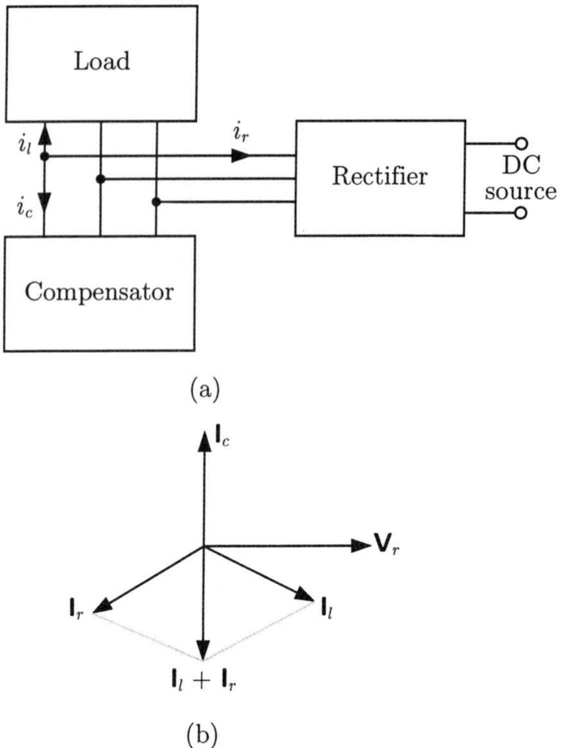

FIG. 4.12. A rectifier supplying an a.c. load with energy derived from the d.c. side.

than 90° and the d.c. output voltage will be negative so that the power flow is negative, being taken from the d.c. side and delivered to the a.c. side. To simplify the situation it is assumed that there is no generating capacity on the a.c. side, that all the power consumed by the load is supplied by the rectifier and that the compensator is a purely reactive element capable of storing but not consuming energy.

Applying Kirchhoff's current law to any one of the three a.c. nodes yields

$$i_r + i_l + i_c = 0 \tag{4.15}$$

Considering only the fundamental components of the three currents, this translates into the phasor equation

$$\mathbf{I}_r + \mathbf{I}_l + \mathbf{I}_c = 0 \tag{4.16}$$

The phasor diagram corresponding to eqn 4.16 is shown in Fig. 4.12(b). Everything on this diagram is referenced to the phase potential, $\mathbf{V}_r$. The load is shown as having a lagging power factor of 0.9, typical of a large load such as a city. The rectifier is shown as operating at a large delay, the value used for the diagram being 150°, and the fundamental component of its input current lags the reference voltage by the delay angle. Since the power consumed by the load is supplied by the rectifier, the real component of the rectifier current is equal and opposite to the real component of the load current. In order to meet the requirements of eqn 4.16, the current consumed by the compensator must be equal and opposite to the sum of the imaginary components of the load and rectifier current. Since the latter are both negative, the compensator must draw a substantial leading current from the a.c. system.

In brief, the rectifier, when operating at large delay angles in the inverter mode, can only supply current to an a.c. load of leading power factor. Although the majority of loads have a lagging power factor, this normally is not a problem since the system generators can play the role of compensator and generate the volt-amperes reactive to balance the system. Where this is not possible a compensator must be installed in the form of a capacitor bank or a synchronous phase modifier, an over excited synchronous motor with zero mechanical load. The necessary installation can be substantial as Example 4.5 illustrates.

Example 4.5

Estimate the leading reactive volt-amperes which must be supplied at the a.c. terminals of an inverting rectifier if it is operating at a delay angle of 150°, and is supplying 500 MW of power to an a.c. load whose power factor is 0.9 lagging.

Since its power factor is 0.9 lagging, the current consumed by the load lags by $\arccos(0.9) = 25.84°$ on the voltage. The load therefore absorbs $500\tan(25.84°) = 242.2$ MVARS lagging. The rectifier, operating at a delay angle of 150° requires a load which must absorb $500\tan(150) = -288.7$ MVARS lagging. The difference between these amounts, $242.2-(-288.7) = 530.8$ MVARS leading must be provided at the terminals of the rectifier by either capacitors or over excited synchronous machines.

****

4.11 Rectifiers with forced commutation

Many of the problems which rectifiers pose for the a.c. system are connected with the fact that their commutation requires that they operate at a lagging power factor. If commutation is forced by external means rather than by using the supply system voltages, the rectifier need no longer operate at a lagging power factor, it can operate at unity or at a leading power factor. This method of operation is technically possible today but is not yet economically viable. However, as power electronics grows to represent a significant portion of the a.c. system load and as devices improve in capability and fall in cost it seems likely that this mode of operation will become attractive. A brief account of such systems is give at the end of Chapter 9.

5

THE EFFECTS OF FINITE IMPEDANCE

5.1 Introduction

In the initial studies of rectifiers contained in the previous three chapters it has been assumed that the a.c. supply impedance is negligible, the load inductance is infinite, and the transformer, if used, is ideal. With these assumptions, the voltages at the rectifier input terminals are pure sinusoids, commutation, the transfer of current from one thyristor to the next, is instantaneous, and the load current is pure, ripple free, d.c. In reality the supply will have impedance and the load inductance will be finite. The transformer, when used, will have impedance and will require magnetizing current and the rectifier may contain a smoothing choke. The waveforms of the rectifier input voltage and output current will be affected and commutation will take a finite time. These effects are investigated in this chapter.

The chapter can be subdivided into four parts. The first part, Sections 5.1 through 5.9, defines the problem and formulates it in mathematical terms. The analysis is simple, involving no more than the development of the differential equations describing the circuit behavior. The second part, Section 5.10, introduces some simplifying approximations which lead to an equivalent circuit model for the rectifier. The third part, Sections 5.11 through 5.17, solves without approximations the equations derived in the first part and develops expressions for all the voltage, current, and power waveforms. The fourth and last part, Section 5.18 compares solutions obtained using the approximations of the second part with those obtained in the third part.

While the level of mathematical difficulty is not high, being limited to the solution of first order, linear, constant coefficient differential equations, the level of complexity in the third part is high and it would be appropriate to terminate a first reading of this chapter at the end of the second part, i.e. the end of Section 5.10.

Considerable use is made of Thévenin's theorem, the a.c. supply being modeled as an infinite bus behind an impedance. This is extended to the thyristor input terminals by modeling the combination of rectifier transformer and supply as an infinite bus behind an impedance.

Accurate analytical treatment of the transformer magnetizing current is extremely difficult. The problem is twofold: first, the relationship between magnetizing current and voltage is nonlinear because of the ferromagnetic core and second, the non-sinusoidal current waveforms cause voltages away from the infinite bus to have non-sinusoidal waveforms. These problems arise whenever transformers are encountered and it is usual to approximate the magnetizing impedance by a linear equivalent which, under normal operating conditions, will consume the same r.m.s. current as the actual impedance. This leads to a Tee type equivalent circuit which is often converted to an analytically simpler L type circuit by moving the magnetizing branch to the input terminals. In this chapter this approximation is carried a stage further by shifting the magnetizing impedance to the infinite bus. Moving the impedance in this way greatly simplifies the analysis and normally involves very little error because the a.c. line and transformer leakage impedances are very small compared to other circuit impedances. These matters are expanded upon in Chapter 12.

The analysis is illuminated by a numerical example which runs throughout the chapter and whose solution is used for all waveform plots.

In general, waveforms will be non-sinusoidal and impedances will therefore, unless otherwise stated, be in the operational form, $Z = R + L\,\mathrm{d}/\mathrm{d}t$. The analysis has been kept as simple as possible by the substitution, where appropriate, of single symbols for complex expressions. In general an upper case V represents the value of a voltage be it mean, r.m.s., or peak, etc., an upper case I represents similar quantities for a current and, for sinusoidal quantities, ϕ_v is the phase of a voltage, ϕ_c is the phase of a current and ζ is the phase angle of a complex impedance. The various parameters are identified by appropriate subscripts.

5.2 The generalized circuit

Figure 5.1 shows a generalized circuit of the type which is studied in this chapter. The potentials of the infinite three phase bus, terminals R, Y, and B, are balanced three phase voltages of positive phase sequence and sinusoidal waveform. The supply system connecting the bus to the rectifier transformer has an inductive impedance, $Z_s = R_s + L_s\mathrm{d}/\mathrm{d}t$. The input terminals of the transformer are labelled R′, Y′, and B′. The windings of the three phase transformer will be connected in a configuration appropriate to the rectifier. The rectifier is a three phase bridge requiring only a three-wire input, terminals 1, 2, and 3. The smoothing choke, impedance $Z_{ch} = R_{ch} + L_{ch}\mathrm{d}/\mathrm{d}t$, is part of the rectifier. The output terminals of the rectifier are labelled K and A, K being the terminal from which current flows to the load and A the terminal to which it returns. The load is active and inductive, producing a d.c. e.m.f. of E_o V and having an impedance $Z_o = R_o + L_o\,\mathrm{d}/\mathrm{d}t$.

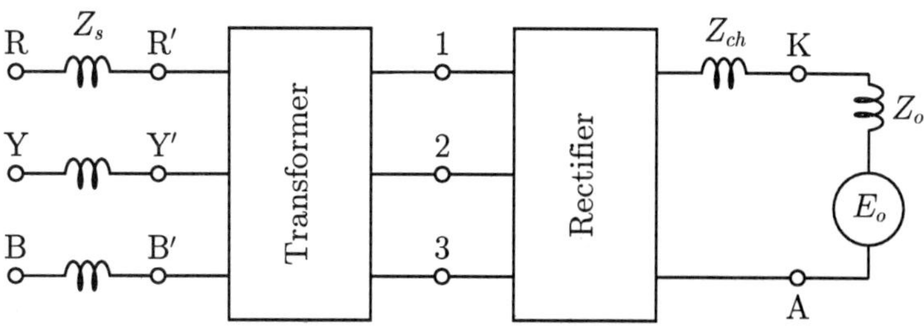

FIG. 5.1. A rectifier fed from a three phase system via a transformer.

5.3 The Thévenin equivalent circuit

The derivation of the Thévenin parameters will be performed in the context of the standard transformer connection, a delta primary and a wye secondary. A similar method of analysis with only minor variations can be employed for any other connection.

The equivalent circuit of the supply system as seen from the rectifier terminals is shown in Fig. 5.2. In Fig. 5.2(a) the transformer is modeled by the Tee equivalent circuit, $Z_{p\sigma}$ being its primary leakage impedance, $Z_{s\sigma}$ its secondary leakage impedance, and Z_m its magnetizing impedance. The number of turns on a primary phase is N_{tp}, the number on a secondary phase is N_{ts}, the ratio of primary to secondary turns per phase is N_{tps}, and the ratio of secondary to primary turns per phase is N_{tsp}. The supply impedance is Z_s. The potentials of the infinite bus relative to the system neutral are v_r, v_y, and v_b. The potentials at the transformer input terminals relative to the system neutral are v_r', v_y', and v_b'. The potentials of the transformer output terminals, which are also the rectifier input terminals, relative to the secondary neutral are v_1, v_2, and v_3. The supply line currents are i_r, i_y, and i_b. The transformer primary currents are i_{ry}, i_{yb}, and i_{br}. The transformer secondary currents are i_1, i_2, and i_3. Positive current directions are indicated in Fig. 5.2 by arrows.

The analysis of any circuit containing a transformer is greatly complicated by the presence of the magnetizing impedance between the primary and secondary leakage impedances. For the majority of problems the effort of dealing with this complication results in only a minor improvement in accuracy since the magnetizing current is small compared to the rated current. It is therefore common practice to transfer the magnetizing impedance to the primary input terminals so as to convert the Tee to a more easily analyzed L. For most problems the errors thus introduced are entirely negligible, a matter dealt with at length in Chapter 12.

Unfortunately, in the case represented by Fig. 5.2(a), this approximation would still leave a Tee type circuit, the supply impedance, Z_s, being

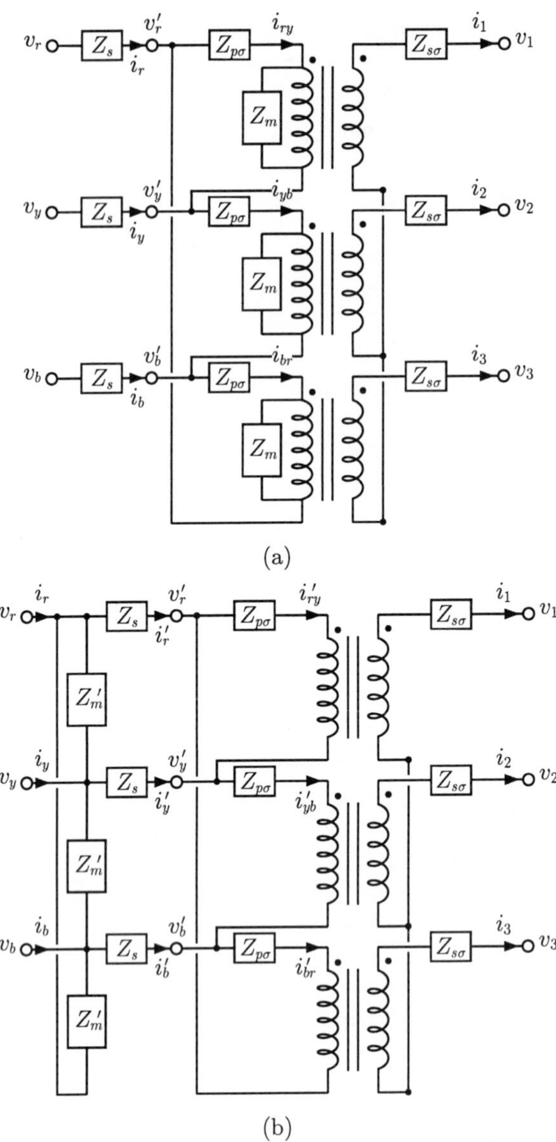

FIG. 5.2. Equivalent circuits of the transformer and supply.

interposed between the transformer input terminals and the infinite bus. Since the supply impedance is usually small compared to the transformer leakage impedances, this difficulty will be circumvented by taking the further step of transferring the magnetizing impedance to the infinite bus to give the circuit of Fig. 5.2(b). The consequent analytical simplification is enormous and the errors thus introduced are small and can be reduced by augmenting the magnetizing impedance by the delta equivalent of the line impedance, $3Z_s$, and the primary leakage impedance so that the magnetizing impedance shown in Fig. 5.2(b) as Z'_m becomes $Z_m + 3Z_s + Z_{p\sigma}$.

The transformer primary voltages may be expressed in terms of the output voltages and currents as follows

$$v'_{ry} = v'_r - v'_y = Z_{p\sigma} i'_{ry} + N_{tps} Z_{s\sigma} i_1 + N_{tps} v_1 \tag{5.1}$$
$$v'_{yb} = v'_y - v'_b = Z_{p\sigma} i'_{yb} + N_{tps} Z_{s\sigma} i_2 + N_{tps} v_2 \tag{5.2}$$
$$v'_{br} = v'_b - v'_r = Z_{p\sigma} i'_{br} + N_{tps} Z_{s\sigma} i_3 + N_{tps} v_3 \tag{5.3}$$

The requirement for m.m.f. balance in the transformer core provides the current relationships

$$N_{tp} i'_{ry} = N_{ts} i_1 \tag{5.4}$$
$$N_{tp} i'_{yb} = N_{ts} i_2 \tag{5.5}$$
$$N_{tp} i'_{br} = N_{ts} i_3 \tag{5.6}$$

The transformer input voltages may also be expressed in terms of the bus voltages and the supply line currents thus

$$v'_{ry} = v_r - Z_s i'_r - (v_y - Z_s i'_y) \tag{5.7}$$
$$v'_{yb} = v_y - Z_s i'_y - (v_b - Z_s i'_b) \tag{5.8}$$
$$v'_{br} = v_b - Z_s i'_b - (v_r - Z_s i'_r) \tag{5.9}$$

Applying Kirchhoff's current law to the transformer input terminals yields the current relationships

$$i'_r = i'_{ry} - i'_{br} \tag{5.10}$$
$$i'_y = i'_{yb} - i'_{ry} \tag{5.11}$$
$$i'_b = i'_{br} - i'_{yb} \tag{5.12}$$

Furthermore, since the input and output systems are three-wire

$$i'_r + i'_y + i'_b = 0 \tag{5.13}$$
$$i_1 + i_2 + i_3 = 0 \tag{5.14}$$

and, since it has been seen in Chapter 3 that there are no zero sequence currents circulating round the transformer primary,

$$i'_{ry} + i'_{yb} + i'_{br} = 0 \tag{5.15}$$

Equating the right-hand sides of eqns 5.1 through 5.3 to the right-hand sides of eqns 5.7 through 5.9 and using the current relationships of eqns 5.4 through 5.6 and 5.10 through 5.15 yields

$$v_1 = N_{tsp}(v_r - v_y) - (3N_{tsp}^2 Z_s + N_{tsp}^2 Z_{p\sigma} + Z_{s\sigma}) i_1 \tag{5.16}$$

$$v_2 = N_{tsp}(v_y - v_b) - (3N_{tsp}^2 Z_s + N_{tsp}^2 Z_{p\sigma} + Z_{s\sigma})i_2 \tag{5.17}$$
$$v_3 = N_{tsp}(v_b - v_r) - (3N_{tsp}^2 Z_s + N_{tsp}^2 Z_{p\sigma} + Z_{s\sigma})i_3 \tag{5.18}$$

These three equations show that the combination of supply system and transformer as seen from the rectifier input terminals can be modeled by the three voltages $v_{st1} = N_{tsp}(v_r - v_y)$, $v_{st2} = N_{tsp}(v_y - v_b)$, and $v_{st3} = N_{tsp}(v_b - v_r)$, each behind an impedance $Z_{st} = 3N_{tsp}^2 Z_s + N_{tsp}^2 Z_{p\sigma} + Z_{s\sigma}$. The subscript *st* signifies *System Thévenin*, the Thévenin equivalent of the supply system and transformer. This circuit model is shown in Fig. 5.3.

The equivalent circuit of Fig. 5.3 permits the determination of the output current, the thyristor currents, and the transformer secondary currents. Multiplying the secondary currents by the turns ratio N_{tsp} gives the primary current components i'_{ry}, i'_{yb}, and i'_{br} of Fig. 5.2(b). The line current components, i'_r, i'_y, and i'_b of Fig. 5.2(b) can then be obtained using eqns 5.10 through 5.12. Adding the currents drawn by the magnetizing impedances, Z'_m, of Fig. 5.2(b) yields the currents drawn from the a.c. supply, i_r, i_y, and i_b of Fig. 5.2(b). These currents will be a close approximation to the actual currents.

To maintain continuity with the analysis of Chapter 3 the time origin is chosen so that

$$v_{st1} = \sqrt{2}V_{st}\cos(\theta) \tag{5.19}$$
$$v_{st2} = \sqrt{2}V_{st}\cos(\theta - \frac{2\pi}{3}) \tag{5.20}$$
$$v_{st3} = \sqrt{2}V_{st}\cos(\theta - \frac{4\pi}{3}) \tag{5.21}$$

where V_{st} is the r.m.s. value of the line to line voltage at the infinite bus, V_{bll}, multiplied by the transformation ratio N_{tsp}.

The phase relationships of eqns 5.19 through 5.21 imply that the bus potentials are

$$v_r = \sqrt{\frac{2}{3}}V_{bll}\cos(\theta - \frac{\pi}{6}) \tag{5.22}$$
$$v_y = \sqrt{\frac{2}{3}}V_{bll}\cos(\theta - \frac{5\pi}{6}) \tag{5.23}$$
$$v_b = \sqrt{\frac{2}{3}}V_{bll}\cos(\theta + \frac{\pi}{2}) \tag{5.24}$$

The application of these expressions is illustrated by Example 5.1 in which the basic data for the example for this chapter are give and the Thévenin parameters are derived.

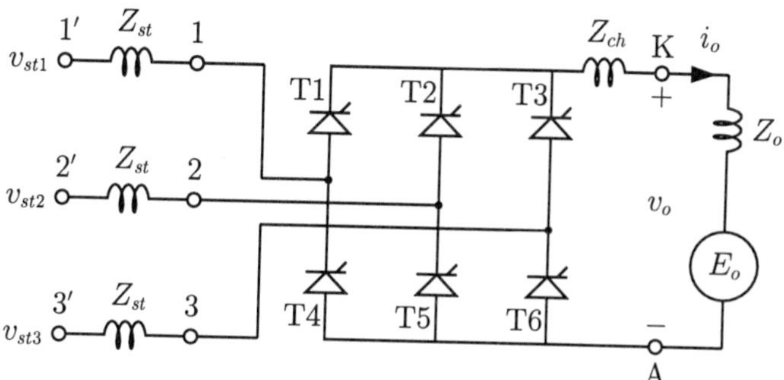

FIG. 5.3. The rectifier and the Thévenin equivalent of the supply and transformer.

Example 5.1

A fully controlled bridge rectifier is fed from a three phase, 60 Hz, 3.3 kV supply via a transformer.

The rectifier load is active with an e.m.f. E_o and an impedance which can be represented by a series combination of a resistance of 0.062 Ω and an inductance of 0.32 mH.

The rectifier includes a smoothing choke whose impedance can be represented by a resistance of 0.029 Ω in series with an inductance of 0.66 mH.

The voltage drop in a conducting thyristor can be taken as 1.6 V.

The transformer primary, which is connected to the 3.3 kV system, is delta connected. Its resistance and leakage inductance are 0.45 Ω and 6.3 mH per phase.

The transformer secondary is wye connected. Its resistance and leakage inductance are 0.0025 Ω and 0.041 mH per phase.

The magnetizing impedance can be modeled as a resistance in series with an inductance, their values referred to a primary phase being 97 Ω and 2.3 henries.

There are 13 times as many turns on a primary phase as on a secondary phase.

The internal impedance of the three phase supply may be represented by a resistance of 0.034 Ω in series with an inductance of 0.6 mH per phase.

Determine the Thévenin parameters of the system as seen from the input terminals of the rectifier.

Using eqns 5.16 through 5.24 the following values for the Thévenin parameters are obtained.

$$V_{st} = 253.8 \text{ V}$$

$$R_{st} = 0.005766\ \Omega$$
$$L_{st} = 0.08893\ \text{mH}$$

****

5.4 Normal conduction and commutation

During normal conduction, two thyristors always conduct the current, one in the upper half of the bridge and the other in the lower half. The finite inductance in the load circuit means that there is some ripple in the output current in response to the ripple in the output voltage. This will be called the *normal mode.*

The a.c. side currents corresponding to the load current store energy in the magnetic fields associated with the a.c. side inductors. This energy must be rearranged during a commutation to correspond to the new a.c. side current configuration, a process which takes time. This period is called the *overlap period* and during it three thyristors conduct, the outgoing thyristor, the incoming thyristor, and the thyristor in the other half of the bridge. This will be called the *commutation mode.*

Since, from the viewpoint of the output current, all 60° periods are identical, only one such period need be analyzed to gain a complete picture of rectifier operation. Attention will be concentrated on the period $60°+\alpha < \theta < 120° + \alpha$. At the start of this period there is a commutation from thyristor T1 to T2 in the upper half of the bridge while T6 conducts in the lower half. The period when T1 and T2 are simultaneously conducting is called the *overlap period* and its duration is designated by the symbol u. Hence, during the commutation interval, $60° + \alpha < \theta < 60° + \alpha + u$, the current in T1 is diminishing towards zero and the current in T2 is increasing towards the load current value.

At the end of commutation, $\theta = 60° + \alpha + u$, the current in T1 is zero, T1 ceases to conduct, and all the load current is carried by T2. The two thyristors T2 and T6 conduct the load current until $\theta = 120° + \alpha$ when another commutation, from T6 to T4, is initiated by firing T4.

The analysis of this behavior requires circuit equations for the two cases, three thyristors conducting and two thyristors conducting. These are obtained by applying Kirchhoff's laws to the equivalent circuit of Fig. 5.4 for the three thyristor commutation mode and Fig. 5.5 for the two thyristor normal mode.

5.5 Commutation mode circuit equations

The commutation process is analyzed by applying Kirchhoff's voltage law to the loops from an a.c. input terminal of Fig. 5.3, through a conducting

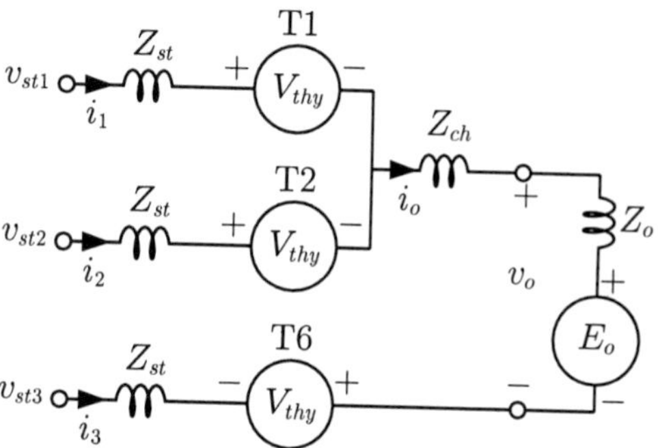

FIG. 5.4. Equivalent circuit during the commutation mode.

thyristor, through the load, and back to another input terminal via another conducting thyristor. For the interval $60° + \alpha < \theta < 60° + \alpha + u$ the obvious paths are from terminal 1′ to terminal 3′ via T1 and T6, and from terminal 2′ to terminal 3′ via T2 and T6. The equivalent circuit for this process is shown in Fig. 5.4. In writing the equations, the fact that Kirchhoff's current law requires that $i_3 = -i_o$ is kept in mind.

For the loop from terminal 1′ through T1, the load, and T6 to terminal 3′,

$$v_{st1} - Z_{st}i_1 - V_{thy} - Z_{ch}i_o - Z_oi_o - E_o - V_{thy} - Z_{st}i_o - v_{st3} = 0 \quad (5.25)$$

For the loop from terminal 2′ through T2, the load, and T6 to terminal 3′,

$$v_{st2} - Z_{st}i_2 - V_{thy} - Z_{ch}i_o - Z_oi_o - E_o - V_{thy} - Z_{st}i_o - v_{st3} = 0 \quad (5.26)$$

Adding these two equations and remembering that $i_1 + i_2 = i_o$ and $v_{st1} + v_{st2} + v_{st3} = 0$, yields the following first order differential equation for the load current,

$$R_1i_o + L_1\frac{di_o}{dt} = -\frac{3}{2}v_{st3} - 2V_{thy} - E_o \quad (5.27)$$

where

$$R_1 = \frac{3}{2}R_{st} + R_{ch} + R_o \quad (5.28)$$

and

$$L_1 = \frac{3}{2}L_{st} + L_{ch} + L_o \quad (5.29)$$

The solution to eqn 5.27 is the sum of three parts, the steady state response to the sinusoidal driving function, $-3v_{st3}/2$, the steady state response to the d.c. driving function, $-(2V_{thy} + E_o)$, and the transient response.

The sinusoidal driving function is, from eqn 5.20,

$$-\frac{3}{2}v_{st3} = -\frac{3}{2}\sqrt{2}V_{st}\cos(\theta - \frac{4\pi}{3}) = \frac{3}{2}\sqrt{2}V_{st}\cos(\theta - \frac{\pi}{3})$$

and the steady state response to it is, by the application of a.c. circuit theory,

$$i_o = I_1 \cos(\theta + \phi_{c1}) \tag{5.30}$$

where

$$I_1 = \frac{3}{2}\frac{\sqrt{2}V_{st}}{Z_1} \tag{5.31}$$

$$Z_1 = \sqrt{R_1^2 + X_1^2} \tag{5.32}$$

$$X_1 = \omega_s L_1 \tag{5.33}$$

$$\phi_{c1} = -\frac{\pi}{3} - \zeta_1 \tag{5.34}$$

$$\zeta_1 = \arctan(X_1/R_1) \tag{5.35}$$

The steady state response to the d.c. driving function is, by the application of d.c. circuit theory,

$$i_o = -I_2 \tag{5.36}$$

where

$$I_2 = \frac{2V_{thy} + E_o}{R_1} \tag{5.37}$$

The transient response is

$$i_o = I_3 \exp(-\frac{\theta}{\tan(\zeta_1)}) \tag{5.38}$$

where the more usual time function, $t/(L_1/R_1)$ has been replaced by the more convenient angle function $\theta/(X_1/R_1) = \theta/\tan(\zeta_1)$.

Thus, the complete solution is

$$i_o = I_1 \cos(\theta + \phi_{c1}) - I_2 + I_3 \exp(-\frac{\theta}{\tan(\zeta_1)}) \tag{5.39}$$

The value of I_3 depends on the initial conditions. If i_o has the value I_{o1} when $\theta = \pi/3 + \alpha$, then

$$I_3 = \left[I_{o1} - I_1 \cos(\frac{\pi}{3} + \alpha + \phi_{c1}) + I_2\right] \exp(\frac{\pi/3 + \alpha}{\tan(\zeta_1)}) \tag{5.40}$$

The value of i_o at the end of the overlap period will be I_{o2} where

$$I_{o2} = I_1 \cos(\frac{\pi}{3} + \alpha + u + \phi_{c1}) - I_2 + I_3 \exp(-\frac{\pi/3 + \alpha + u}{\tan(\zeta_1)}) \tag{5.41}$$

The application of these expressions is illustrated by Example 5.2.

Example 5.2

For the example, determine an expression for the output current when the delay angle is 45°, the e.m.f. E_o is 370 V, and I_{o1} is 400 A.

Using eqns 5.27 through 5.40 it is found that

$$i_o = 1248.22\cos(\theta - 2.3849) - 3745.13 + 4762.69\exp(-\frac{\theta}{4.2122})$$

****

5.6 Differential equations for thyristor currents

The output current expression of eqn 5.39 can be substituted into eqns 5.25 and 5.26 to produce the following first order differential equations in the currents i_1 and i_2

$$R_{st}i_1 + L_{st}\frac{di_1}{dt} = V_{14}\cos(\theta + \phi_{v14}) - V_5 + V_6\exp(-\frac{\theta}{\tan(\zeta_1)}) \quad (5.42)$$

$$R_{st}i_2 + L_{st}\frac{di_2}{dt} = V_{24}\cos(\theta + \phi_{v24}) - V_5 + V_6\exp(-\frac{\theta}{\tan(\zeta_1)}) \quad (5.43)$$

where

$$V_{14} = \sqrt{A_1^2 + B_1^2} \quad (5.44)$$

$$\phi_{v14} = \arctan(-B_1/A_1) \quad (5.45)$$

$$V_{24} = \sqrt{A_2^2 + B_2^2} \quad (5.46)$$

$$\phi_{v24} = \arctan(-B_2/A_2) \quad (5.47)$$

$$A_1 = \frac{3\sqrt{2}}{2}V_{st} - R_2I_1\cos(\phi_{c1}) + X_2I_1\sin(\phi_{c1}) \quad (5.48)$$

$$B_1 = \sqrt{\frac{3}{2}}V_{st} + R_2I_1\sin(\phi_{c1}) + X_2I_1\cos(\phi_{c1}) \quad (5.49)$$

$$A_2 = -R_2I_1\cos(\phi_{c1}) + X_2I_1\sin(\phi_{c1}) \quad (5.50)$$

$$B_2 = \sqrt{6}V_{st} + R_2I_1\sin(\phi_{c1}) + X_2I_1\cos(\phi_{c1}) \quad (5.51)$$

$$R_2 = R_{st} + R_{ch} + R_o \quad (5.52)$$

$$X_2 = X_{st} + X_{ch} + X_o \quad (5.53)$$

$$V_5 = 2V_{thy} + E_o - R_2I_2 \quad (5.54)$$

$$V_6 = \frac{R_1X_2 - R_2X_1}{X_1}I_3 \quad (5.55)$$

Example 5.3

For the example, determine the values of the parameters in eqns 5.42 and 5.43.

Using eqns 5.44 through 5.55

$$R_2 = 0.09677\ \Omega$$
$$X_2 = 0.4030\ \Omega$$
$$V_{14} = 312.95\ \text{V}$$
$$\phi_{v14} = 0.4558\ \text{radians}$$
$$V_{24} = 310.29\ \text{V}$$
$$\phi_{v24} = -2.5497\ \text{radians}$$
$$V_5 = 10.80\ \text{V}$$
$$V_6 = -5.22\ \text{V}$$
$$\tan(\zeta_1) = 4.2122$$

****

5.7 The thyristor currents

The right-hand sides of eqns 5.42 and 5.43 show three driving functions. The solutions for i_1 and i_2 are therefore the sums of four terms, the steady state responses to each driving function and the transient response. The first driving function is an a.c. voltage at the supply frequency and the steady state response to it is obtained by the application of a.c. circuit theory. The second driving function is a d.c. voltage and the steady state response to it is obtained by the application of d.c. circuit theory. The third function is an exponentially decaying voltage and the steady state response to it is obtained by the application of a.c. circuit theory but with the exponential coefficient, $-1/\tan(\zeta_1)$ replacing $j\omega_s$. The transient response is an exponentially decaying term which bridges the gap between the steady state solution and the actual response. Thus,

$$i_1 = I_{14}\cos(\theta + \phi_{c14}) - I_5 + I_6 \exp(-\frac{\theta}{\tan(\zeta_1)}) + I_{17}\exp(-\frac{\theta}{\tan(\zeta_{st})}) \quad (5.56)$$

and

$$i_2 = I_{24}\cos(\theta + \phi_{c24}) - I_5 + I_6 \exp(-\frac{\theta}{\tan(\zeta_1)}) + I_{27}\exp(-\frac{\theta}{\tan(\zeta_{st})}) \quad (5.57)$$

where

$$I_{14} = \frac{V_{14}}{Z_{st}} \quad (5.58)$$

$$\phi_{c14} = \phi_{v14} - \zeta_{st} \tag{5.59}$$

$$I_{24} = \frac{V_{24}}{Z_{st}} \tag{5.60}$$

$$\phi_{c24} = \phi_{v24} - \zeta_{st} \tag{5.61}$$

$$Z_{st} = \sqrt{R_{st}^2 + X_{st}^2} \tag{5.62}$$

$$\zeta_{st} = \arctan(X_{st}/R_{st}) \tag{5.63}$$

$$I_5 = \frac{V_5}{R_{st}} = \frac{1}{2} I_2 \tag{5.64}$$

$$I_6 = V_6 \frac{X_1}{R_{st}X_1 - X_{st}R_1} = \frac{1}{2} I_3 \tag{5.65}$$

driving

The amplitudes, I_{17} and I_{27}, of the transient components depend on the initial conditions but, since $i_1 + i_2 = i_o$ it follows that

$$I_{17} = -I_{27} \tag{5.66}$$

Since $i_2 = 0$ when $\theta = \alpha + 60°$

$$I_{27} = -\left[I_{24}\cos(\alpha + \frac{\pi}{3} + \phi_{c24}) - I_5 + I_6 \exp(-\frac{\alpha + \pi/3}{\tan(\zeta_1)})\right] \times \exp(\frac{\alpha + \pi/3}{\tan(\zeta_{st})}) \tag{5.67}$$

Expressions having a form similar to that of i_o, i_1, and i_2, eqns 5.39, 5.56, and 5.57, occur frequently in power electronics analysis, so frequently that it is appropriate to standardize it. This has been done in Appendix A where the expression is called the *basic power electronics expression.* Various useful functions of the basic expression are derived in that appendix.

Example 5.4

For the example, determine expressions for the currents in thyristors T1 and T2 during the commutation period and the values of the currents at the start and finish of commutation given that the overlap angle is 3.3709°.

The current in thyristor T1 is given by eqn 5.56 and the current in thyristor T2 is given by eqn 5.57. Determining numerical values for the coefficients of these expressions with the aid of eqns 5.58 through 5.65 it is found that

$$i_1 = 9199.52\cos(\theta - 0.9446) - 1872.57 + 2381.34\exp(-\frac{\theta}{4.2122}) - 6953.39\exp(-\frac{\theta}{5.8141})$$

and

$$i_2 = 9121.41\cos(\theta - 3.9501) - 1872.57 + 2381.34\exp(-\frac{\theta}{4.2122}) + 6953.39\exp(-\frac{\theta}{5.8141})$$

A check on the accuracy of the calculations is obtained by determining the output current at the start and end of commutation and confirming that these values are, within the limits of computational accuracy, equal respectively to the initial value of i_1 and the final value of i_2. At the start of commutation, $\theta = 105°$, the values of the currents are

$$i_o = 400.0 \qquad i_1 = 400.0 \qquad i_2 = 0.0$$

At the end of commutation, $\theta = 108.3709°$, the values are

$$i_o = 393.91 \qquad i_1 = 0.0 \qquad i_2 = 393.90$$

These values indicate a typical difficulty with current equations in power electronic problems. The currents are the relatively small differences between large quantities and high accuracy must be maintained in the calculations in order to have reasonable accuracy in the final results.

****

5.8 The normal mode, two thyristors conducting

Commutation ends at $\theta = \pi/3 + \alpha + u$ when the current in T2 equals the output current and T1 has ceased to conduct. The period from this point to $\theta = 2\pi/3 + \alpha$ is the normal mode when only two thyristors conduct, in this case T2 and T6. The analysis of this stage is started by writing down Kirchhoff's voltage law for the loop from terminal $2'$ of Fig. 5.5 through the load to terminal $3'$, recognizing that $i_2 = i_o$ and $i_3 = -i_o$. This yields the first order differential equation in i_o

$$R_3 i_o + L_3 \frac{di_o}{dt} = v_{st2} - v_{st3} - 2V_{thy} - E_o \tag{5.68}$$

where

$$R_3 = 2R_{st} + R_{ch} + R_o \tag{5.69}$$

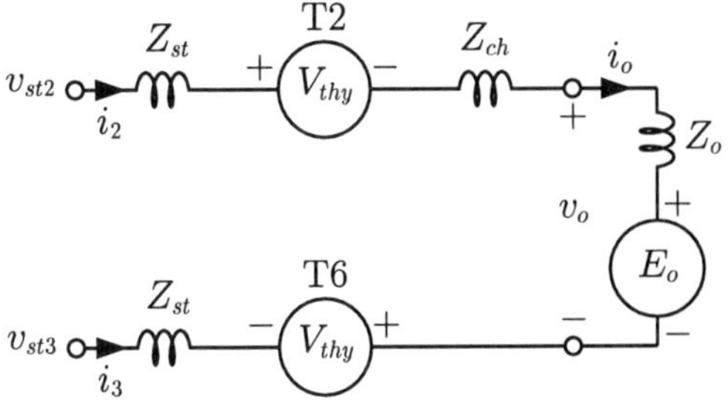

FIG. 5.5. Equivalent circuit during the normal mode.

$$L_3 = 2L_{st} + L_{ch} + L_o \tag{5.70}$$

and, using the expressions given in eqns 5.19 through 5.21

$$v_{st2} - v_{st3} = \sqrt{6}V_{st}\cos(\theta - \frac{\pi}{2}) \tag{5.71}$$

This equation is very similar to eqn 5.27 and has a similar solution, again in the form of the general power electronics expression of Appendix A

$$i_o = I_8\cos(\theta + \phi_{c8}) - I_9 + I_{10}\exp(-\frac{\theta}{\tan(\zeta_3)}) \tag{5.72}$$

where

$$I_8 = \frac{\sqrt{6}V_{st}}{Z_3} \tag{5.73}$$

$$Z_3 = \sqrt{R_3^2 + X_3^2} \tag{5.74}$$

$$X_3 = \omega_s L_3 \tag{5.75}$$

$$\phi_{c8} = -\frac{\pi}{2} - \zeta_3 \tag{5.76}$$

$$\zeta_3 = \arctan(X_3/R_3) \tag{5.77}$$

$$I_9 = \frac{2V_{thy} + E_o}{R_3} \tag{5.78}$$

and I_{10} is determined by the initial conditions.

Since there must be continuity of current in an inductive circuit, the current at the start of the normal mode is the current, I_{o2}, at the end of the commutation period, $\theta = \pi/3 + \alpha + u$. Substituting this value in eqn 5.72 yields I_{10} as

$$I_{10} = \left[I_{o2} - I_8\cos(\frac{\pi}{3} + \alpha + u + \phi_{c8}) + I_9\right]\exp(\frac{\pi/3 + \alpha + u}{\tan(\zeta_3)}) \tag{5.79}$$

The value of the output current at the end of the normal mode is the value of the current at the start of the next commutation, from T6 to T4, at $\theta = 2\pi/3 + \alpha$. Denoting this value by I_{o3},

$$I_{o3} = I_8 \cos(2\pi/3 + \alpha + \phi_{c8}) - I_9 + I_{10} \exp(-\frac{2\pi/3 + \alpha}{\tan(\zeta_3)}) \tag{5.80}$$

Example 5.5

For the example, determine the expression for the output current during the normal mode, i.e. for $108.37° < \theta < 165.00°$.

It has just been found that the output current at the end of commutation is 393.90 A so that

$$i_o = 1386.75 \cos(\theta - 2.9109) - 3939.82 + 5157.35 \exp(-\frac{\theta}{4.2572})$$

This equation gives the value of I_{o2} as 393.91 A and the value of I_{o3} as 368.35 A. The value of I_{o2} is very close to the value obtained in Example 5.4, a strong indication that the computations are correct.

****

5.9 The steady state

In the steady state, the 60° segments of the output current are identical so that, for the steady state, I_{o3} must equal I_{o1}. It has just been found that this condition is not satisfied by the example, that the output current at the end of the 60° segment, 368.35 A, is less than the value at the start, 400.0 A. The system is not in the steady state but is making a transition to a lower value of output current.

The transient behavior could be followed through succeeding commutations until the system reaches the steady state. However, this is a tedious process and the computations can be greatly speeded up.

This completes the end of the first part and before continuing the analysis in the third part, which starts with Section 5.11, a useful and relatively simple approximate solution to the problem will be developed.

5.10 Approximate solution

There are two reasons for developing an approximate treatment of the effects of finite impedance on rectifier behavior. First, the exact analysis is complicated and, while this may present no difficulty once it has been incorporated in a computer program, it is useful to have a simpler method available for hand calculation. Second, the speed with which a computer program determines the steady state solution is improved by having a good initial estimate of the solution. This is provided by the approximate solution.

The approximate solution requires that three assumptions regarding the rectifier be made. None of these is true but all are in practice nearly true. The errors introduced by these assumptions are in large measure compensatory and the final result is a good approximation to the exact analytic solution.

5.10.1 *The first assumption*

The first assumption is that the commutation and output current ripples are independent phenomena. The example has shown that the change in output current during commutation is small, from 400.0 to 393.90 A, a change of 1.5%, and that the commutation process occupies only a small part, 3.37° or 5.6% of the 60° conduction segment.

This assumption permits the development of separate, independent solutions for commutation and output current ripple.

5.10.2 *The second assumption*

During commutation it is assumed that the rate of change of the output current is so small relative to the rates of change of the thyristor currents that it can be ignored. In the example it has been found that, during the overlap period of 3.37° the output current decreases from 400.0 to 393.90 A, an average rate of change of −1.81 A per degree. During the same period i_1 decreases from 400.0 to 0.0 A for an average rate of change of −118.7 A per degree and i_2 increases from 0.0 to 393.90 for an average rate of change of 116.9 A per degree.

From this it is concluded that little error will be introduced if it is assumed that the output current during commutation is constant at the mean value.

5.10.3 *The third assumption*

During commutation, the rates of change of the thyristor currents are so great that the resistance voltage drops are very small in comparison to the inductance voltage drops. In the example the Thévenin circuit resistance is $R_{st} = 0.005766\ \Omega$ and the inductance is $L_{st} = 0.08893$ mH. The average voltage drop in the resistance is about $R_{st}I_{o1}/2 = 1.15$ V. The average rate of change of current is about 2576 A ms^{-1} so that the average voltage drop in the inductance is about 229 V.

From this it is concluded that little error will be introduced if R_{st} is omitted from the commutation calculations.

5.10.4 *Approximate solution to commutation*

It is assumed that the output current is constant at the mean value, I_o, and that the effects of circuit resistance are negligible.

Subtracting eqn 5.26 from 5.25, replacing Z_{st} by $L_{st}\mathrm{d}/dt$ or its equivalent, $X_{st}\mathrm{d}/\mathrm{d}\theta$, and remembering that $i_1 + i_2 = I_o$, where I_o is the mean value of the output current, yields the following equation for i_1

$$X_{st}\frac{\mathrm{d}i_1}{\mathrm{d}\theta} = \frac{v_{st1} - v_{st2}}{2} \tag{5.81}$$

Integrating this expression using the values of v_{st1} and v_{st2} given in eqns 5.19 through 5.21 yields

$$i_1 = I_{11}\cos(\theta - \frac{\pi}{3}) + I_{12} \tag{5.82}$$

where I_{11} is

$$I_{11} = \sqrt{\frac{3}{2}}\frac{V_{st}}{X_{st}} \tag{5.83}$$

and I_{12} is the constant of integration.

Since $i_1 = I_o$ at the start of commutation when $\theta = 60° + \alpha$

$$I_{12} = I_o - I_{11}\cos(\alpha) \tag{5.84}$$

so that the current in the outgoing thyristor is

$$i_1 = I_o - I_{11}\cos(\alpha) + I_{11}\cos(\theta - \frac{\pi}{3}) \tag{5.85}$$

and, since $i_1 + i_2 = I_o$, the current in the incoming thyristor is

$$i_2 = I_{11}\cos(\alpha) - I_{11}\cos(\theta - \frac{\pi}{3}) \tag{5.86}$$

Commutation ends at $\theta = \pi/3 + \alpha + u$ when i_1 is zero. Substitution of this angle in eqn 5.85 yields the following equation

$$\cos(\alpha + u) - \cos(\alpha) = -\frac{I_o}{I_{11}} \tag{5.87}$$

from which the overlap angle is obtained as

$$u = \arccos\left[\cos(\alpha) - \frac{I_o}{I_{11}}\right] - \alpha \tag{5.88}$$

Example 5.6

With the delay angle of 45° and load e.m.f. of 370 V, determine by the approximate method expressions for the currents in the outgoing and incoming thyristors and the overlap angle.

Using eqns 5.83 through 5.88 it is found that the commutation period starts at $\theta = 105.0°$ and ends at $\theta = 107.9552°$. During this period the thyristor currents are given by

$$i_1 = -6210.55 + 9273.46\cos(\theta - 1.04720)$$

$$i_2 = 6557.33 - 9273.46\cos(\theta - 1.04720)$$

and the overlap angle is

$$u = 2.9552°$$

****

5.10.5 *The voltage lost during commutation*

In Fig. 5.3, considering the path from terminal 2′ through the load to terminal 3′, the voltage, δv, absorbed by the inductance L_{st} in series with thyristor T2 is given by

$$\delta v = L_{st}\frac{\mathrm{d}i_2}{\mathrm{d}t} = X_{st}\frac{\mathrm{d}i_2}{\mathrm{d}\theta} \tag{5.89}$$

The mean value, ΔV, of the voltage lost in this way, over an output cycle of 60° duration, is

$$\Delta V = \frac{3}{\pi}\int_{\pi/3+\alpha}^{\pi/3+\alpha+u} X_{st}\frac{\mathrm{d}i_2}{\mathrm{d}\theta}\mathrm{d}\theta = \frac{3}{\pi}\int_0^{I_o} X_{st}\mathrm{d}i_2 = \frac{3}{\pi}X_{st}I_o \tag{5.90}$$

Thus, as far as the d.c. output is concerned, the system inductance, L_{st}, appears as an additional internal resistance of $3X_{st}/\pi\ \Omega$, making the total *rectifier Thévenin resistance*, R_{rt}, in the path of the mean output current

$$R_{rt} = 2R_{st} + \frac{3}{\pi}X_{st} + R_{ch} \tag{5.91}$$

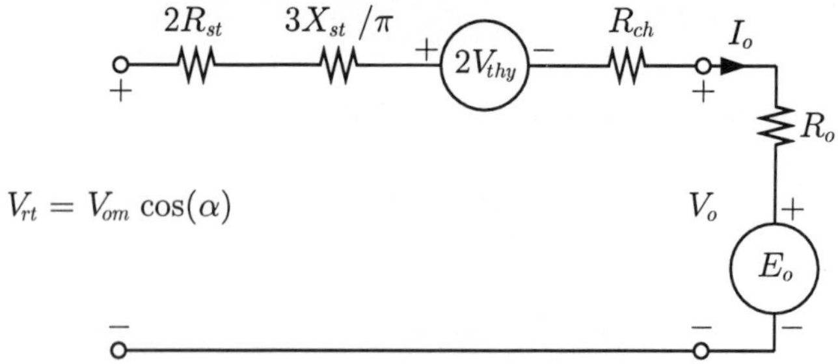

FIG. 5.6. The d.c. equivalent circuit of the rectifier.

5.10.6 *Output voltage and the equivalent circuit*

It is necessary to distinguish the actual output voltage of the rectifier, which shows the effect of the internal voltage losses, from the voltages v_o and V_o which have been found in Chapter 3 assuming an ideal circuit. The latter are renamed the *rectifier instantaneous Thévenin voltage* and the *mean rectifier Thévenin voltage* and they are given the symbols v_{rt} and V_{rt}. The rectifier mean Thévenin voltage and the mean output current are related by

$$V_{rt} = V_{om}\cos(\alpha) = E_o + 2V_{thy} + (R_{rt} + R_o)I_o \tag{5.92}$$

The equivalent circuit of the rectifier, as seen from its d.c. side, can now be drawn. It is shown in Fig. 5.6 and provides a quick and, in view of the approximations, remarkably accurate method for determining the mean output voltage under any conditions of load current and delay angle.

Example 5.7

Determine, using the approximate method, the mean output voltage of the example when the delay angle is 45° and the load e.m.f. is 370 V.

Since $V_{st} = 253.84$ V, the value of V_{om} from eqn 3.3 is

$$V_{om} = 1.3505 \times \sqrt{3} \times 253.84 = 593.77 \text{ V}$$

At 45° delay the mean output voltage is

$$V_{rt} = 593.77\cos(45°) = 419.86 \text{ V}$$

From eqn 5.90, the internal resistance of the rectifier as seen from its d.c. terminals is

$$R_{rt} = 2 \times 0.005766 + \frac{3}{\pi}0.03353 + 0.029 = 0.07255\ \Omega$$

It is noteworthy that a substantial part, $0.03202\ \Omega = 44\%$, is contributed to this resistance by the commutation component, $3X_{st}/\pi$.

The mean output current is then given by

$$I_o = \frac{419.86 - 2 \times 1.6 - 370.0}{0.07255 + 0.062} = 346.78 \text{ A}$$

The mean output voltage is

$$V_o = 370.0 + 0.062 \times 346.78 = 391.50 \text{ V}$$

This explains why, when the exact analysis of the example was made with a mean output current of 400 A it was found that the system was in a transitional state to a lower mean output current whose value is now seen to be 346.8 A.

****

5.10.7 *The load characteristics*

From the d.c. equivalent circuit it is seen that the load characteristics of the rectifier, i.e. its mean output voltage as a function of mean output current for a given angle of delay, are parallel straight lines whose slopes are $-R_{rt}$ V A^{-1} and which intersect with the voltage axis at $V_{om}\cos(\alpha) - 2V_{thy}$. The characteristics for the example are shown in Fig. 5.7. It will be seen later, in Section 5.18, that these characteristics are remarkably accurate despite the approximations.

5.10.8 *The waveform of the output current*

The waveform of the output current can now be determined within the limitations of the approximations, the relevant approximation being that the overlap angle can be neglected in this calculation. The analysis starts from eqn 5.26 which is modified in several ways to suit the approximations. First, because instantaneous commutation is now assumed, i_2 is set equal to i_o. Second, to allow for the effects of commutation on the mean value of the output current, an e.m.f. whose value is $3X_{st}I_o/\pi$ is placed in series with the load e.m.f. This yields the first order differential equation

$$R_3 i_o + L_3 \frac{di_o}{dt} = v_{st2} - v_{st3} - E_o - \frac{3}{\pi} X_{st} I_o - 2V_{thy} \tag{5.93}$$

where R_3 and L_3 have been defined in eqns 5.69 and 5.70.

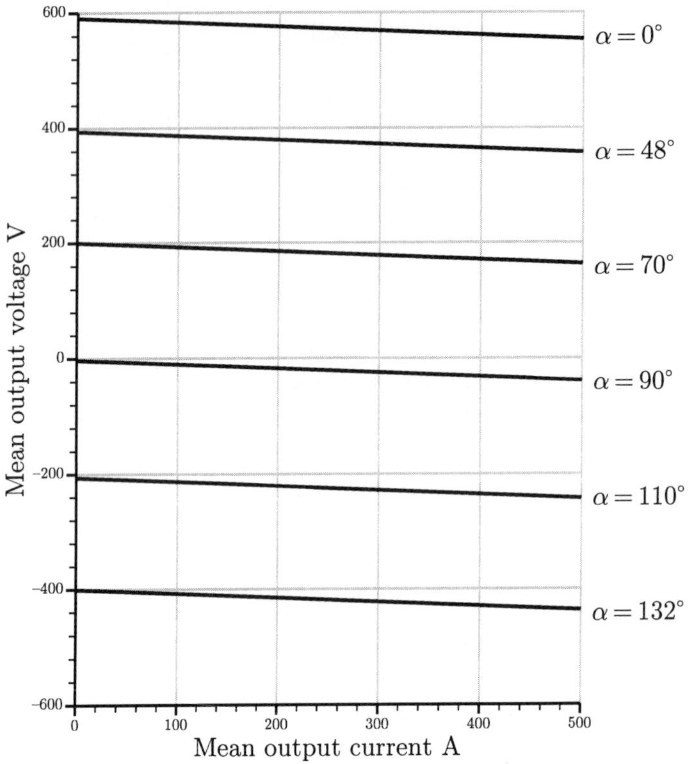

FIG. 5.7. Rectifier load characteristics at various angles of delay.

From eqns 5.19 through 5.21, $v_{st2} - v_{st3} = \sqrt{6}V_{st}\cos(\theta - \pi/2)$ and, following the pattern established for the solution of eqn 5.68, the solution of eqn 5.93 is

$$i_o = I_8\cos(\theta + \phi_{c8}) - I_{13} + I_{14}\exp(-\frac{\theta}{\tan(\zeta_3)}) \tag{5.94}$$

where I_8, ϕ_{c8}, and ζ_3 have been defined by eqns 5.73 through 5.77, I_{13} is given by

$$I_{13} = \frac{E_o + \frac{3}{\pi}X_{st}I_o + 2V_{thy}}{R_3} \tag{5.95}$$

and I_{14} depends on the boundary conditions.

In determining the value of I_{14}, the initial value, I_{o1}, of the output current at $\theta = 60° + \alpha$ is not known but it is known that, in the steady state, the final value, I_{o3}, at $\theta = 120° + \alpha$ will be the same. Substituting these values of θ in eqn 5.94 yields the following equality

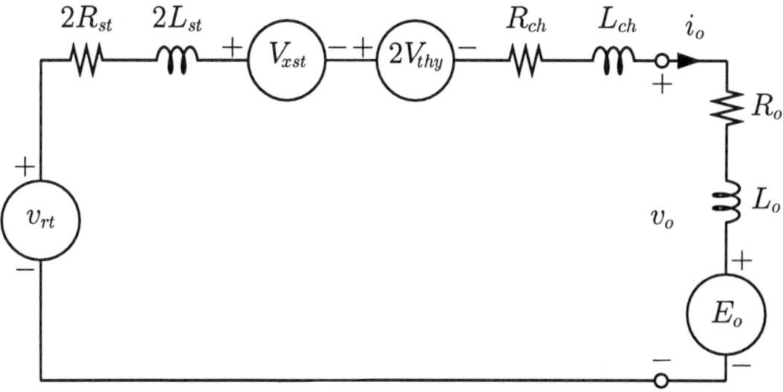

FIG. 5.8. The a.c. equivalent circuit of the rectifier. The inclusion of the voltage source $V_{xst} = (3/\pi)X_{st}I_o$ ensures that the mean value of the output current derived from this circuit is in agreement with that derived from the d.c. equivalent circuit, Fig. 5.6.

$$\begin{aligned} I_8 \cos(\frac{\pi}{3} + \alpha + \phi_{c8}) - I_{13} + I_{14} \exp(-\frac{\pi/3 + \alpha}{\tan(\zeta_3)}) &= \\ I_8 \cos(\frac{2\pi}{3} + \alpha + \phi_{c8}) - I_{13} + I_{14} \exp(-\frac{2\pi/3 + \alpha}{\tan(\zeta_3)}) & \end{aligned} \tag{5.96}$$

from which it is found that I_{14} is given by

$$I_{14} = \frac{-I_8 \cos(\alpha + \phi_{c8})}{\exp(-\frac{\pi/3+\alpha}{\tan(\zeta_3)}) - \exp(-\frac{2\pi/3+\alpha}{\tan(\zeta_3)})} \tag{5.97}$$

5.10.9 *The a.c. equivalent circuit*

The a.c. equivalent circuit of the rectifier, as seen from its d.c. terminals, can now be drawn. It is shown in Fig. 5.8. As noted in Section 5.10.6, the input voltage to this circuit is called the *rectifier instantaneous Thévenin voltage*, symbol v_{rt}. The effects of commutation cannot be included as an equivalent resistance $3X_{st}/\pi$ as was done in the d.c. equivalent circuit since the circuit resistance would then be too high and the damping of transient components of current too great. However, if it is not included, the mean value of the output current as derived from the a.c. equivalent circuit will be significantly greater than that derived from the d.c. circuit. This dilemma is resolved by modeling the mean voltage drop due to commutation by a voltage source V_{xst} whose strength is $I_o \times 3X_{st}/\pi$ as indicated in Fig. 5.8.

The waveform of the rectifier Thévenin voltage repeats at six times the supply frequency, every 60° segment being like every other. The value of this voltage over the segment $60° + \alpha < \theta < 120° + \alpha$ is

$$v_{rt} = \sqrt{6}V_{st}\cos(\theta - \pi/2) = \sqrt{6}V_{st}\sin(\theta) \tag{5.98}$$

If this segment is designated as Segment 1 and the segment after it as 2, etc., the general expression for v_{rt} is

$$v_{rt} = \sqrt{6}V_{st}\sin(\psi) \tag{5.99}$$

where

$$\psi = \theta - (K-1)\frac{\pi}{3} \tag{5.100}$$

and K is the number of the segment.

Example 5.8

For the example, using the approximate method, determine an expression for the output current when the delay angle is 45° and the load e.m.f. is 370 V.

In determining the expression for the waveform it is found from eqn 5.69 that $R_3 = 0.1025\ \Omega$, from eqn 5.70 that $X_3 = 0.4365\ \Omega$, that $Z_3 = 0.4484\ \Omega$, and that $\zeta_3 = 76.7811°$.

From eqn 5.94 and using eqns 5.73, 5.76, 5.95, and 5.97 it is found that the required expression is

$$i_o = 1386.75\cos(\theta - 2.9109) - 3748.10 + 5151.18\exp(-\frac{\theta}{4.2572})$$

At the start of the segment, $\theta = 105°$, this equation gives $i_o = 256.93$ A. The same value is obtained at the end of the segment, $\theta = 165°$, indicating that the correct solution has been obtained.

The output current waveform derived from this expression is shown by a gray line in Fig. 5.15. It will be seen that it is a reasonable approximation to the actual waveform shown by the black line.

****

5.11 The exact solution for the steady state

It has been seen how the exact solution requires knowledge of two of the system variables, the load e.m.f. and the value of the output current, I_{o1}, at the start of commutation. The commutation progresses from this point in two stages covering the commutation mode and the normal mode. Expressions for the currents during commutation are derived, the overlap angle is found and the output current at the end of commutation, I_{o2}, is calculated. Knowledge of this current permits the determination of an expression for the output current and its value, I_{o3}, at the end of the normal mode which is the value at the start of the next commutation. In the steady state this value is equal to the starting value, I_{o1}, and the computation must be repeated until convergence is attained. Iteration can follow the natural process, the calculations being repeated with the starting current of the new cycle being equal to the value at the end of the old one. However, this is a slow process and, unless the transient process is of interest, it unnecessarily uses up computer time.

There are two requirements for the minimization of computation time. First, the computation must start from values as close as possible to the steady state. A good approximation to the overlap angle is provided by the approximate method. A good approximation to the initial current, I_{o1}, is provided by the calculation described below. Second, methods which force convergence to the steady state solution should be used.

5.11.1 *The initial values*

The approximate method is used to determine the overlap angle and, with the delay angle and load e.m.f. specified, the set of eqns 5.40, 5.41, 5.79, and 5.80 for I_{o1} for the steady state, i.e. assuming $I_{o3} = I_{o1}$, are solved. The resulting expression for I_{o1} is

$$I_{o1} = \frac{C_1 I_1 + C_2 I_2 + C_8 I_8 + C_9 I_9}{C_{o1}} \tag{5.101}$$

where

$$C_1 = \left[\cos(\frac{\pi}{3} + \alpha + u + \phi_{c1}) - \cos(\frac{\pi}{3} + \alpha + \phi_{c1}) \exp(-\frac{u}{\tan(\zeta_1)})\right] \times \exp(-\frac{\pi/3 - u}{\tan(\zeta_3)}) \tag{5.102}$$

$$C_2 = -\left[1 - \exp(\frac{-u}{\tan(\zeta_1)})\right] \exp(-\frac{\pi/3 - u}{\tan(\zeta_3)}) \tag{5.103}$$

$$C_8 = \cos(\frac{2\pi}{3} + \alpha + u + \phi_{c8}) - \cos(\frac{\pi}{3} + \alpha + \phi_{c1}) \exp(-\frac{\pi/3 - u}{\tan(\zeta_3)}) \tag{5.104}$$

$$C_9 = -1 + \exp(-\frac{\pi/3 - u}{\tan(\zeta_3)}) \tag{5.105}$$

$$C_{o1} = 1 - \exp(\frac{-u}{\tan(\zeta_1)}) \exp(-\frac{\pi/3 - u}{\tan(\zeta_3)}) \tag{5.106}$$

These values of u and I_{o1} are the starting values for the exact solution.

The expression for the current in the outgoing thyristor is derived, eqn 5.56, and the time at which commutation is completed is found as the angle at which this current becomes zero. Unfortunately, this is a transcendental equation and its zero must therefore be found by successive approximation. Since the approximate solution provides a good starting point, the Newton–Raphson technique is excellent for this purpose.

5.11.2 *Overlap by the Newton–Raphson method*

For the Newton–Raphson method both eqn 5.56 and its derivative with respect to θ are required.

$$\frac{\mathrm{d}i_1}{\mathrm{d}\theta} = -I_{c14}\sin(\theta+\phi_{c14}) - \frac{I_6}{\tan(\zeta_1)}\exp(-\frac{\theta}{\tan(\zeta_1)}) - \frac{I_{c17}}{\tan(\zeta_{st})}\exp(-\frac{\theta}{\tan(\zeta_{st})}) \tag{5.107}$$

If θ_1 is an estimate of the correct value, then an improved estimate, θ_2, is

$$\theta_2 = \theta_1 - \frac{i_1(\theta_1)}{\frac{\mathrm{d}i_1}{\mathrm{d}\theta}(\theta_1)} \tag{5.108}$$

The starting value of θ_1 is the value given by the approximate solution, $\pi/3+\alpha+u$, and the process is repeated, with the new value of θ serving as the starting value, until an acceptable level of accuracy is attained.

5.11.3 *Iteration for I_{o1}*

The value of θ just found allows the value of the overlap angle to be updated and with this the calculations of I_{o1} and u can be reiterated. These calculations are repeated until the accuracy of I_{o1} is acceptable. The number of major loop iterations for 0.1% accuracy is of the order of five and, within this loop, the number of Newton–Raphson iterations is of the order of two per major loop for 0.0001 radian accuracy.

The value of the output current at the end of commutation is now found using eqn 5.39. This is the current I_{o2} needed for determination of the output current during the normal mode, eqns 5.72 through 5.79.

The value of the output current at the end of commutation, I_{o3} eqn 5.80, can now be found and a check of convergence is made by testing that $I_{o3} = I_{o1}$ within acceptable limits of accuracy.

5.11.4 *Some programming requirements*

It is possible for the above program to go into an infinite major or Newton–Raphson loop so that loop counters with provision for exiting with an error message are essential. Setting the maximum number of iterations to 9 is satisfactory.

The initial calculation of the overlap angle by the approximate method produces errors when the delay approaches 180°. The argument of the arc cosine function of eqn 5.88 becomes less than -1 and an error message is produced. Many programming languages automatically deal with this problem by returning the angle π. However, this is not effective for the calculation as it results in the computational equivalent of a rectifier commutation failure and the Newton–Raphson calculation of overlap does not converge. It is therefore essential to test the value of $\cos(\alpha) - I_o/I_{11}$ and, if it is less than a number X close to -1, to replace it by X. The value, $X = -0.99$, corresponding to $\alpha + u = 171.9°$, seems to be adequate for the resolution of this difficulty.

Sometimes values of load e.m.f. and rectifier delay will be selected which result in commutation failure. With the loop counters described above this will result in the program stopping and producing error messages. However, it is better to monitor this situation and produce a more precise error message. This is done by noting if the value of $\alpha + u$ exceeds π. If this occurs, commutation failure results and the program should be stopped and an appropriate message should be printed.

Example 5.9

Determine steady state expressions for the output current and the currents in the incoming and outgoing thyristors for the example when the delay angle is 45° and the load e.m.f. is 370 V.

Starting from the values of overlap angle and initial current found in the previous example, the following steady state expressions are obtained after eight major loop iterations, the final correction to I_{o1} being 0.0008 A.

$$\text{Overlap angle} = 2.4372°$$

During commutation, $105° < \theta < 107.4372°$

$$\begin{aligned}
i_o &= 1248.21\cos(\theta - 2.3849) - 3745.13 + 4586.97\exp(-\frac{\theta}{4.2122}) \\
i_1 &= 9199.50\cos(\theta - 0.9446) - 1872.57 + 2293.48\exp(-\frac{\theta}{4.2122}) - \\
&\qquad 7031.32\exp(-\frac{\theta}{5.8141}) \\
i_2 &= 9121.41\cos(\theta - 3.9501) - 1872.57 + 2293.48\exp(-\frac{\theta}{4.2122}) - \\
&\qquad 7031.32\exp(-\frac{\theta}{5.8141})
\end{aligned}$$

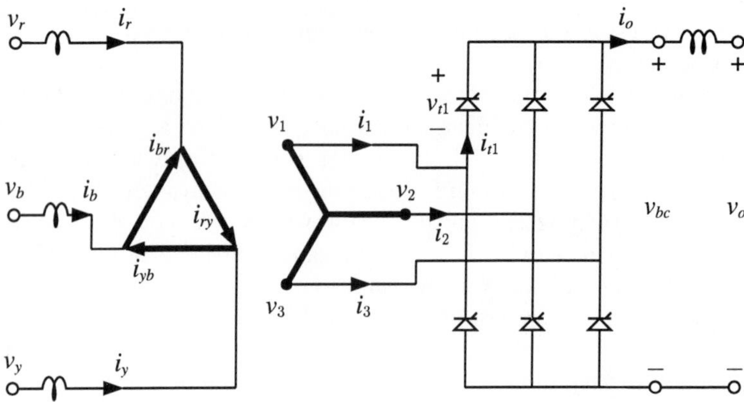

FIG. 5.9. The rectifier circuit showing all the currents and voltages.

During the normal mode, $107.4372° < \theta < 165°$

$$i_o = 1386.75\cos(\theta - 2.9109) - 3639.82 + 4995.91\exp(-\frac{\theta}{4.2572})$$

Key values derived from these equations are

$$\begin{aligned}\text{Output current at the start of commutation} &= 286.27 \text{ A}\\ \text{Output current at the end of commutation} &= 283.32 \text{ A}\\ \text{Output current at the end of the normal mode} &= 286.27 \text{ A}\end{aligned}$$

The value of current at the end of the normal mode is sufficiently close to the value at the start to confirm that the steady state has been reached.

****

5.12 The current waveforms

Expressions for the currents i_{t1}, i_{t2}, and i_o over the 60° range $60° + \alpha < \theta < 120° + \alpha$ have now been obtained. From these the waveforms of the sixteen currents identified in Fig. 5.9 can be determined. To avoid overloading this figure with information, the transformer windings have been represented by thick lines and for the thyristors only the current through and the voltage across thyristor T1 have been shown.

To add flexibility to the waveform expressions, the time dependant variable θ will be changed to ψ and some linear combinations of the expressions will be derived.

5.12.1 *The basic current expressions*

The basic current expressions are the expressions already derived with the variable θ changed to ψ, ψ being the linear function of θ

$$\psi = \theta - \delta \tag{5.109}$$

where δ is a constant.

This change will, by using the appropriate value of δ, allow the current expressions to be moved to any position on the θ axis to correspond with commutation from any thyristor to the next.

With this change of variable, for the commutation mode, $60° + \alpha < \psi < 60° + \alpha + u$, the basic current expressions are, from eqns 5.56, 5.57, and 5.39

$$\begin{aligned} i_a = {} & I_{14}\cos(\psi + \phi_{c14}) - I_5 + I_6 \exp(-\frac{\psi}{\tan(\zeta_1)}) + \\ & I_{17}\exp(-\frac{\psi}{\tan(\zeta_{st})}) \end{aligned} \tag{5.110}$$

$$\begin{aligned} i_b = {} & I_{24}\cos(\psi + \phi_{c24}) - I_5 + I_6 \exp(-\frac{\psi}{\tan(\zeta_1)}) + \\ & I_{27}\exp(-\frac{\psi}{\tan(\zeta_{st})}) \end{aligned} \tag{5.111}$$

$$i_c = I_1\cos(\psi + \phi_{c1}) - I_2 + I_3\exp(-\frac{\psi}{\tan(\zeta_1)}) \tag{5.112}$$

Since frequent reference to these expressions will be made, they will be called expressions **A**, **B**, and **C**. It is noted that $\mathbf{C} = \mathbf{A} + \mathbf{B}$.

It will be convenient to have the following linear combinations of expressions **A**, **B**, and **C** which will be designated as **D**, **E**, and **F**.

$$i_d = \mathbf{D} = \mathbf{A} - \mathbf{B} \tag{5.113}$$

$$i_e = \mathbf{E} = \mathbf{A} + \mathbf{C} \tag{5.114}$$

$$i_f = \mathbf{F} = \mathbf{B} + \mathbf{C} \tag{5.115}$$

For the normal mode, $60° + \alpha + u < \psi < 120° + \alpha$,

$$i_g = I_4\cos(\psi + \phi_{c4}) - I_5 + I_6\exp(-\frac{\psi}{\tan(\phi_4)}) \tag{5.116}$$

This expression will be called **G**.

The basic current expression names, **A** through **G**, carry the implication of the region over which the expression applies. The waveforms of these

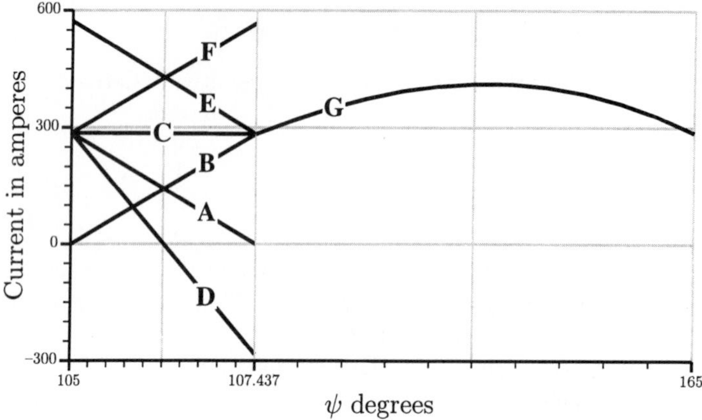

FIG. 5.10. The waveforms of the basic current expressions. Note that the scale of ψ during the commutation mode is ten times that during the normal mode.

functions are shown in Fig. 5.10 which has been drawn using the numerical expression obtained in Example 5.10. For clarity the angle scale for the commutation mode has been made ten times that for the normal mode.

It is important to note that the basic expressions all have the form of the basic power electronics expression discussed in Appendix A. This is very helpful in developing subroutines which will manipulate the basic expressions in various ways.

Example 5.10

Determine, for the example with a delay angle of 45° and an e.m.f. of 370 V, the seven basic current expressions.

Expressions **A**, **B**, **C**, and **G** have already been derived. From these expressions **D**, **E**, and **F** are determined by the use of eqns 5.113, 5.114,, and 5.115.

For the commutation mode, $105° < \psi < 107.44°$

$$\mathbf{A} = 9199.52\cos(\psi - 0.9446) - 1872.57 + 2293.48\exp(-\frac{\psi}{4.2122}) - 7031.32\exp(-\frac{\psi}{5.8141})$$

$$\mathbf{B} = 9121.41\cos(\psi - 3.9501) - 1872.57 + 2293.48\exp(-\frac{\psi}{4.2122}) + 7031.32\exp(-\frac{\psi}{5.8141})$$

$$\mathbf{C} = 1248.22\cos(\psi - 2.3849) - 3745.13 + 4586.97\exp(-\frac{\psi}{4.2122})$$

$$\mathbf{D} = 18278.52\cos(\psi - 0.8769) - 14062.65\exp(-\frac{\psi}{5.8141})$$

$$\mathbf{E} = 9443.42\cos(\psi - 1.0761) - 5617.70 + 6880.45\exp(-\frac{\psi}{4.2122}) - 7031.32\exp(-\frac{\psi}{5.8141})$$

$$\mathbf{F} = 9213.31\cos(\psi - 3.8142) - 5617.70 + 6880.45\exp(-\frac{\psi}{4.2122}) + 7031.32\exp(-\frac{\psi}{5.8141})$$

During the normal mode, $107.44° < \psi < 165.0°$

$$\mathbf{G} = 1386.75\cos(\psi - 2.9109) - 3639.82 + 4995.91\exp(-\frac{\psi}{4.2572})$$

****

5.12.2 *Time shifting the basic expressions*

The change of variable from θ to ψ enables the basic expressions to be shifted to any location on the θ axis. Thus, if $\delta = 120°$, $\theta = \psi + 120°$, and the commutation mode starts at $\theta = 180° + \alpha$ and ends at $180° + \alpha + u$. The normal mode starts at $\theta = 180° + \alpha + u$ and ends at $240° + \alpha$. The 60° segment that follows the firing of thyristor T_3 has therefore been created.

5.12.3 *The basic segments*

The application of the basic expressions is made easier if a supply cycle is divided into twelve basic segments, six commutation mode segments, and six normal mode segments. Any 360° period can be selected and the one starting with firing thyristor T1 at $-60° + \alpha$ and ending at $300° + \alpha$ when this thyristor is again fired will be chosen.

The twelve segments are identified in graph 1 of Fig. 5.11 with reference to the current i_1 in the transformer secondary number 1. This figure also shows the waveform of the current in the corresponding primary winding which will be referred to later.

5.12.4 *The current waveforms*

The tools for assembling the current waveforms are now available in the concept of the basic current expressions and basic segments of Fig. 5.11. These are summarized in Tables 5.1 through 5.3.

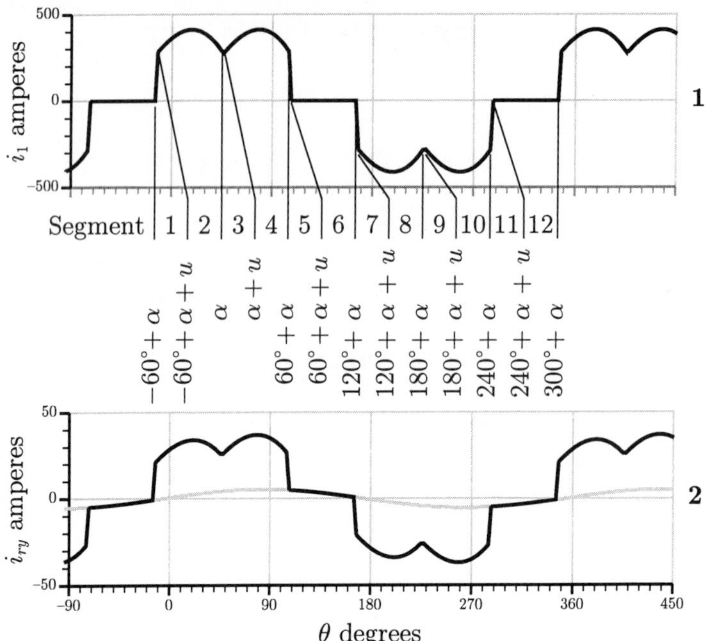

FIG. 5.11. Waveforms of the transformer winding currents. Graph 1 shows the current in secondary winding number 1 and the 12 basic segments. Graph 2 shows the current in the corresponding primary winding, RY, with the magnetizing component shown by the gray line. The graphs have been drawn for a delay angle of 45°.

Each of Tables 5.1, 5.2, and 5.3 is a rectangular array with 19 rows and 5 columns. Each defines the composition of the 16 rectifier currents during one third of a supply cycle, Table 5.1 covering the first third of the cycle, Table 5.2 the middle third of the cycle, and Table 5.3 the last third of the cycle. The period covered by the three tables begins with the firing of T1 at $\theta = -60° + \alpha$ and ends 360° later, at $300° + \alpha$ the next time T1 is fired. The 16 currents are identified in the first column of each table, the one headed 'Current'.

Eleven of the 16 currents are identified in Fig. 5.9, the remaining five being the currents in thyristors T2 through T6. Columns 2 through 5 of each table, the ones headed 'Segment number', define the current during one of the 12 segments identified in Fig. 5.11, the segment number being given in row 2. Row 3 gives the value of the segment shift angle δ in degrees. Rows 4 through 19 list the function defining the current during that segment. In the case of currents on the secondary side of the transformer, rows 4 through 13, the functions are **A** through **G**.

To obtain expressions for the currents on the primary side, rows 14 through 19 of the tables, the functions must be multiplied by the transformation ratio N_{tsp} and the magnetizing component must be added to the result. Scaling by the transformation ratio is indicated by appending primes to the function symbols which become $\mathbf{A}'$ through $\mathbf{G}'$. Rows 14 through 16 define the transformer primary currents and the magnetizing current components are the currents flowing in the delta connected magnetizing impedances of Fig. 5.2(b).

Rows 17 through 19 of the tables define the a.c. line currents. The magnetizing current components in the lines are the wye equivalents of those used for the primary winding currents whose magnitudes must therefore be multiplied by $\sqrt{3}$ and whose phase angles must be advanced by 30°. Thus, $\mathbf{I}_{mr} = \sqrt{3}\,\mathbf{I}_{mry}\underline{/30^\circ}$.

5.12.5 *The output current*

From the point of view of the output, conditions repeat every 60° so that the output current waveform comprises the basic expression **C** in the odd numbered segments and basic expression **G** in the even numbered segments. These are shown in the fourth row of Tables 5.1, 5.2, and 5.3 and the waveform for the example is shown in graph 2 of Fig. 5.12.

5.12.6 *The thyristor currents*

Consider the current i_{t1} in thyristor T1 which is defined by the fifth row of Tables 5.1 through 5.3. During segment 1 this thyristor is commutating on so that its current is described by basic current expression **B** with a shift of −120°. During segment 2 it is in the normal mode and with its current defined by basic expression **G** with a shift of −120°. During segment 3 it is carrying the output current while a commutation from T5 to T6 occurs in the lower half of the bridge. Its current is therefore defined by basic expression **C** with a shift of −60°. During segment 4 the thyristor is again in the normal mode and its current is defined by basic expression **G** with a shift of −60°. During segment 5 the thyristor is commutating off and its current is defined by basic expression **A** with zero shift. During the remainder of the cycle, segments 6 through 12, the thyristor current is zero.

The currents in the other five thyristors are obtained by phase shifting i_{t1} by the appropriate amount. Thus, i_{t2} lags i_{t1} by 120°, i_{t3} lags by 240°, i_{t4} lags by 180°, i_{t5} leads by 60°, and i_{t6} lags by 60°. This has been done to create rows 6 through 10 of the tables.

The waveform of i_{t1} is shown in graph 4 of Fig. 5.12.

Table 5.1 *Composition of the rectifier currents*

Current	Segment number			
	1	2	3	4
$\delta°$	−120	−120	−60	−60
i_o	**C**	**G**	**C**	**G**
i_{t1}	**B**	**G**	**C**	**G**
i_{t2}	0	0	0	0
i_{t3}	**A**	0	0	0
i_{t4}	0	0	0	0
i_{t5}	**C**	**G**	**A**	0
i_{t6}	0	0	**B**	**G**
i_1	**B**	**G**	**C**	**G**
i_2	−**C**	−**G**	−**A**	0
i_3	**A**	0	−**B**	−**G**
i_{ry}	$\mathbf{B}' + i_{mry}$	$\mathbf{G}' + i_{mry}$	$\mathbf{C}' + i_{mry}$	$\mathbf{G}' + i_{mry}$
i_{yb}	$-\mathbf{C}' + i_{myb}$	$-\mathbf{G}' + i_{myb}$	$-\mathbf{A}' + i_{myb}$	i_{myb}
i_{br}	$\mathbf{A}' + i_{mbr}$	i_{mbr}	$-\mathbf{B}' + i_{mbr}$	$-\mathbf{G}' + i_{mbr}$
i_r	$-\mathbf{D}' + i_{mr}$	$\mathbf{G}' + i_{mr}$	$\mathbf{F}' + i_{mr}$	$2\mathbf{G}' + i_{mr}$
i_y	$-\mathbf{F}' + i_{my}$	$-2\mathbf{G}' + i_{my}$	$-\mathbf{E}' + i_{my}$	$-\mathbf{G}' + i_{my}$
i_b	$\mathbf{E}' + i_{mb}$	$\mathbf{G}' + i_{mb}$	$\mathbf{D}' + i_{mb}$	$-\mathbf{G}' + i_{mb}$

5.12.7 *The transformer secondary currents*

The transformer secondary winding currents are linear combinations of the thyristor currents, $i_1 = i_{t1} - i_{t4}$, $i_2 = i_{t2} - i_{t5}$, and $i_3 = i_{t3} - i_{t6}$. This has been done to create rows 11, 12, and 13 of the tables.

The waveform of the current i_1 is shown in graph 6 of Fig. 5.12

5.12.8 *The transformer primary currents*

Each transformer primary current is the sum of the reflection into the primary of the corresponding secondary current and the appropriate magnetizing current. A reflected secondary current component is the appropriate secondary current just found in the preceding section multiplied by the transformation ratio N_{tsp}. A magnetizing current component is the current flowing in the corresponding phase of the delta connected magnetizing impedance of Fig. 5.2(b).

With the bus potentials given in eqns 5.22 through 5.24, the bus line to line voltages are

Table 5.2 *Composition of the rectifier currents*

Current	Segment number			
	5	6	7	8
$\delta°$	0	0	60	60
i_o	**C**	**G**	**C**	**G**
i_{t1}	**A**	0	0	0
i_{t2}	**B**	**G**	**C**	**G**
i_{t3}	0	0	0	0
i_{t4}	0	0	**B**	**G**
i_{t5}	0	0	0	0
i_{t6}	**C**	**G**	**A**	0
i_1	**A**	0	$-\mathbf{B}$	$-\mathbf{G}$
i_2	**B**	**G**	**C**	**G**
i_3	$-\mathbf{C}$	$-\mathbf{G}$	$-\mathbf{A}$	0
i_{ry}	$\mathbf{A}' + i_{mry}$	i_{mry}	$-\mathbf{B}' + i_{mry}$	$-\mathbf{G}' + i_{mry}$
i_{yb}	$\mathbf{B}' + i_{myb}$	$\mathbf{G}' + i_{myb}$	$\mathbf{C}' + i_{myb}$	$\mathbf{G}' + i_{myb}$
i_{br}	$-\mathbf{C}' + i_{mbr}$	$-\mathbf{G}' + i_{mbr}$	$-\mathbf{A}' + i_{mbr}$	i_{mbr}
i_r	$\mathbf{E}' + i_{mr}$	$\mathbf{G}' + i_{mr}$	$\mathbf{D}' + i_{mr}$	$-\mathbf{G}' + i_{mr}$
i_y	$-\mathbf{D}' + i_{my}$	$\mathbf{G}' + i_{my}$	$\mathbf{F}' + i_{my}$	$2\mathbf{G}' + i_{my}$
i_b	$-\mathbf{F}' + i_{mb}$	$-2\mathbf{G}' + i_{mb}$	$-\mathbf{E}' + i_{mb}$	$-\mathbf{G}' + i_{mb}$

$$v_{ry} = \sqrt{2}V_{bll}\cos(\theta) \tag{5.117}$$
$$v_{yb} = \sqrt{2}V_{bll}\cos(\theta - \frac{2\pi}{3}) \tag{5.118}$$
$$v_{br} = \sqrt{2}V_{bll}\cos(\theta - \frac{4\pi}{3}) \tag{5.119}$$

where V_{bll} is the r.m.s. value of the line to line voltage at the bus.

The magnetizing currents are therefore approximated by

$$i_{mry} = \sqrt{2}I_m\cos(\theta - \zeta'_m) \tag{5.120}$$
$$i_{myb} = \sqrt{2}I_m\cos(\theta - \zeta'_m - 2\pi/3) \tag{5.121}$$
$$i_{mbr} = \sqrt{2}I_m\cos(\theta - \zeta'_m - 4\pi/3) \tag{5.122}$$

where

$$I_m = V_s/Z'_m \tag{5.123}$$

the equivalent magnetizing impedance at the bus is

Table 5.3 *Composition of the rectifier currents*

Current	Segment number			
	9	10	11	12
δ°	120	120	180	180
i_o	**C**	**G**	**C**	**G**
i_{t1}	0	0	0	0
i_{t2}	**A**	0	0	0
i_{t3}	**B**	**G**	**C**	**G**
i_{t4}	**C**	**G**	**A**	0
i_{t5}	0	0	**B**	**G**
i_{t6}	0	0	0	0
i_1	−**C**	−**G**	−**A**	0
i_2	**A**	0	−**B**	−**G**
i_3	**B**	**G**	**C**	**G**
i_{ry}	$-\mathbf{C}' + i_{mry}$	$-\mathbf{G}' + i_{mry}$	$-\mathbf{A}' + i_{mry}$	i_{mry}
i_{yb}	$\mathbf{A}' + i_{myb}$	i_{myb}	$-\mathbf{B}' + i_{myb}$	$-\mathbf{G}' + i_{myb}$
i_{br}	$\mathbf{B}' + i_{mbr}$	$\mathbf{G}' + i_{mbr}$	$\mathbf{C}' + i_{mbr}$	$\mathbf{G}' + i_{mbr}$
i_r	$-\mathbf{F}' + i_{mr}$	$-2\mathbf{G}' + i_{mr}$	$-\mathbf{E}' + i_{mr}$	$-\mathbf{G}' + i_{mr}$
i_y	$\mathbf{E}' + i_{my}$	$\mathbf{G}' + i_{my}$	$\mathbf{D}' + i_{my}$	$-\mathbf{G}' + i_{my}$
i_b	$-\mathbf{D}' + i_{mb}$	$\mathbf{G}' + i_{mb}$	$\mathbf{F}' + i_{mb}$	$2\mathbf{G}' + i_{mb}$

$$Z'_m = \sqrt{(R_m + 3R_s + R_{p\sigma})^2 + (X_m + 3X_s + X_{p\sigma})^2} \tag{5.124}$$

and its phase angle ζ'_m is

$$\zeta'_m = \arctan(\frac{X_m + 3X_s + X_{p\sigma}}{R_m + 3R_s + R_{p\sigma}}) \tag{5.125}$$

The composition of the primary currents is indicated in rows 14, 15, and 16 of the Tables 5.1 through 5.3.

The waveform of the primary current i_{ry} is shown in graph 2 of Fig. 5.11 with the magnetizing current component shown by the gray line.

5.12.9 *The input line current*

The input line currents are linear combinations of the transformer primary currents, $i_r = i_{ry} - i_{br}$, $i_y = i_{yb} - i_{ry}$, and $i_b = i_{br} - i_{yb}$.

The composition of these currents is indicated in rows 17, 18, and 19 of the tables. Because of lack of space in the tables, the combination of two

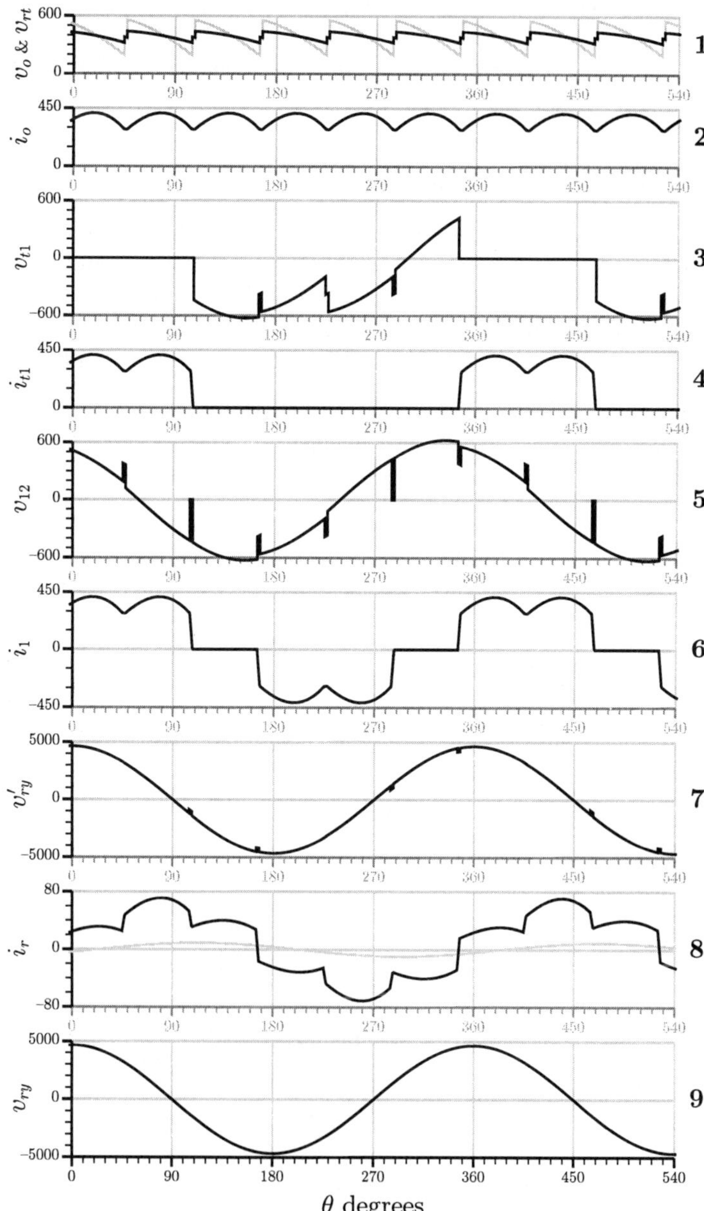

FIG. 5.12. The waveforms of various currents and voltages in the rectifier. The nine graphs are identified by numbers at the right-hand end of the θ axis.

primary phase magnetizing currents is shown as the equivalent line current. Thus, $i_{mry} - i_{mbr}$ is represented by i_{mr} etc.

The waveform of the current i_r is shown in graph 8 of Fig. 5.12 with the magnetizing component shown by a gray line.

5.13 The voltage waveforms

Derivation of the voltage waveforms is similar to the derivation of the current waveforms in that it is based on the basic expressions and the basic segments. However, it is more complex because it includes not only the resistance voltage drops, a simple scaling of the basic expressions by the appropriate resistance, but also the inductance voltage drops which involve new functions, the derivatives of the basic expressions. Since $\theta = \omega_s t$ is being used instead of time, it is convenient to replace inductance by reactance at the supply frequency and $\mathrm{d}i/\mathrm{d}t$ by $\mathrm{d}i/\mathrm{d}\theta$, i.e.

$$L\frac{\mathrm{d}i}{\mathrm{d}t} = X\frac{\mathrm{d}i}{\mathrm{d}\theta} = X\frac{\mathrm{d}i}{\mathrm{d}\psi}$$

where $X = \omega_s L$

The derivatives of the basic expressions are designated by the corresponding bold face italic letter. Thus, $\boldsymbol{F}$ represents $\mathrm{d}\mathbf{F}/\mathrm{d}\theta$ and $\boldsymbol{F}'$ represents $\boldsymbol{F}$ referred to the primary side of the transformer, i.e. $\boldsymbol{F}' = \boldsymbol{F} N_{tsp}$.

Example 5.11

Determine for the example, the derivatives with respect to ψ of the basic expressions $\boldsymbol{A}$ through $\boldsymbol{G}$.

The numerical values for the basic expressions appropriate to the example are

$$\boldsymbol{A} = -9199.52\sin(\psi - 0.9446) - 544.49\exp(-\frac{\psi}{4.2122}) + 1209.37\exp(-\frac{\psi}{5.8141})$$

$$\boldsymbol{B} = -9121.41\sin(\psi - 3.9501) - 544.49\exp(-\frac{\psi}{4.2122}) - 1209.37\exp(-\frac{\psi}{5.8141})$$

$$\boldsymbol{C} = -1248.22\sin(\psi - 2.3849) - 1088.98\exp(-\frac{\psi}{4.2122})$$

$$\boldsymbol{D} = -18278.52\sin(\psi - 0.8769) + 2418.73\exp(-\frac{\psi}{5.8141})$$

$$\boldsymbol{E} = -9443.42\sin(\psi - 1.0761) - 1633.47\exp(-\frac{\psi}{4.2122}) + 1209.37\exp(-\frac{\psi}{5.8141})$$

$$\boldsymbol{F} = -9213.31\sin(\psi - 3.8142) - 1633.47\exp(-\frac{\psi}{4.2122}) - 1209.37\exp(-\frac{\psi}{5.8141})$$

$$\boldsymbol{G} = -1386.75\sin(\psi - 2.9109) - 1173.52\exp(-\frac{\psi}{4.2572})$$

****

The derivation of the voltages must start with known voltages and, as these are available in the busbar voltages and the load e.m.f., the analysis can start from either end of the system. The start will be at the infinite bus and will work towards the load, the final derivation being that of the load e.m.f. This provides a good test of a complex process in that the e.m.f. so derived should match with the actual e.m.f.

5.13.1 *The busbar voltages*

With the Thévenin voltages of eqns 5.19 through 5.21, the potentials of the busbars relative to the system neutral are

$$v_r = \sqrt{\frac{2}{3}}V_{bll}\cos(\theta - \pi/6) \tag{5.126}$$

$$v_y = \sqrt{\frac{2}{3}}V_{bll}\cos(\theta - \frac{5\pi}{6}) \tag{5.127}$$

$$v_b = \sqrt{\frac{2}{3}}V_{bll}\cos(\theta + \frac{\pi}{2}) \tag{5.128}$$

and the voltages between busbars are

$$v_{ry} = \sqrt{2}V_{bll}\cos(\theta) \tag{5.129}$$

$$v_{yb} = \sqrt{2}V_{bll}\cos(\theta - \frac{2\pi}{3}) \tag{5.130}$$

$$v_{br} = \sqrt{2}V_{bll}\cos(\theta + \frac{2\pi}{3}) \tag{5.131}$$

The waveform of v_{ry} is shown in graph 9 of Fig. 5.12. The other voltages are similar but with the appropriate phase shift as indicated by eqns 5.126 through 5.131 and in the case of the bus potentials of eqns 5.126 through 5.128 scaled by the factor $1/\sqrt{3}$.

5.13.2 *The transformer input voltages*

The transformer input voltages are obtained by subtracting the line voltage drops from the infinite bus voltages. Thus, the potentials of the transformer primary terminals relative to the system neutral are given by

$$v'_r = v_r - R_s i_r - X_s \frac{\mathrm{d}i_r}{\mathrm{d}\theta} \tag{5.132}$$

$$v'_y = v_y - R_s i_y - X_s \frac{\mathrm{d}i_y}{\mathrm{d}\theta} \tag{5.133}$$

$$v'_b = v_b - R_s i_b - X_s \frac{\mathrm{d}i_b}{\mathrm{d}\theta} \tag{5.134}$$

The line to line voltages at the transformer primary terminals are then

$$v'_{ry} = v'_r - v'_y \tag{5.135}$$

$$v'_{yb} = v'_y - v'_b \tag{5.136}$$

$$v'_{br} = v'_r - v'_r \tag{5.137}$$

The waveform of the voltage v'_{ry} is shown in graph 7 of Fig. 5.12. The waveforms of v'_{yb} and v'_{br} are similar but with phase lags of 120° and 240°.

5.13.3 *The transformer output voltages*

The potential of a transformer output terminal relative to the secondary winding neutral is the voltage across the corresponding primary winding referred to the secondary, less the voltage drop in the transformer leakage impedances. Thus,

$$v_1 = \left[v'_{ry} - R_{p\sigma}i_{ry} - X_{p\sigma}\frac{\mathrm{d}i_{ry}}{\mathrm{d}\theta}\right]N_{tsp} - R_{s\sigma}i_1 - X_{s\sigma}\frac{\mathrm{d}i_1}{\mathrm{d}\theta} \tag{5.138}$$

$$v_2 = \left[v'_{yb} - R_{p\sigma}i_{yb} - X_{p\sigma}\frac{\mathrm{d}i_{yb}}{\mathrm{d}\theta}\right]N_{tsp} - R_{s\sigma}i_2 - X_{s\sigma}\frac{\mathrm{d}i_2}{\mathrm{d}\theta} \tag{5.139}$$

$$v_3 = \left[v'_{br} - R_{p\sigma}i_{br} - X_{p\sigma}\frac{\mathrm{d}i_{by}}{\mathrm{d}\theta}\right]N_{tsp} - R_{s\sigma}i_3 - X_{s\sigma}\frac{\mathrm{d}i_3}{\mathrm{d}\theta} \tag{5.140}$$

The secondary line to line voltages are then obtained as

$$v_{12} = v_1 - v_2 \tag{5.141}$$

$$v_{23} = v_2 - v_3 \tag{5.142}$$

$$v_{31} = v_3 - v_1 \tag{5.143}$$

The waveform of v_{12} is shown in graph 5 of Fig. 5.12. The waveforms of v_{23} and v_{31} are similar but with phase lags of 120° and 240°.

5.13.4 *The output voltage behind the choke*

For brevity the output voltage on the rectifier side of the smoothing inductor will be called the *output voltage behind the choke* with the symbol v_{bc}. Its location is indicated on Fig. 5.9. Its waveform in the steady state is an endless repetition of the 60° section of the waveform of v_{12} when thyristors T1 and T5 are conducting, less the voltage drops in the conducting thyristors. This section of v_{12} covers basic segments 1 and 2, i.e. from $-60° + \alpha$ to $-60° + \alpha + u$ and from $-60° + \alpha + u$ to α. The voltage drop in the thyristors is 2 V_{thy} whether the rectifier is in the commutating or the normal mode.

The waveform of v_{bc} is shown by the gray line in graph 1 of Fig. 5.12.

5.13.5 *The output voltage*

The output voltage is the output voltage behind the choke less the voltage drop in the choke. Thus,

$$v_o = v_{bc} - R_{ch} i_o - X_{ch} \frac{\mathrm{d}i_o}{\mathrm{d}\theta} \tag{5.144}$$

The waveform of v_o is shown by the black line of graph 1 of Fig. 5.12.

5.13.6 *The thyristor voltage.*

The only remaining voltage of interest is the voltage across a thyristor. For the upper half of the bridge this is equal to the potential of the line connected to the thyristor anode less the potential of the cathode of the conducting thyristor. A similar definition holds for a thyristor in the bottom half of the bridge with the words anode and cathode interchanged.

Considering thyristor T1, it conducts from $-60° + \alpha$ to $60° + \alpha + u$ and during this period the voltage across it is V_{thy}. For the period $60° + \alpha + u$ to $180° + \alpha$, T2 conducts and the voltage across T1 is $v_{12} - V_{thy}$. For the period $180° + \alpha$ to $300° + \alpha$, thyristor T3 conducts and the voltage across T1 is $v_{13} - V_{thy}$.

The voltage across T1 is shown in graph 5 of Fig. 5.12. The voltages across the other thyristors are similar but with appropriate phase shifts.

5.13.7 *The load e.m.f.*

Although the load e.m.f. is known, its computation provides a valuable check on the accuracy of the waveform calculations which are very complicated. The e.m.f. is the output voltage less the drop in the load impedance, i.e.

$$E_o = v_o - R_o i_o - X_o \frac{\mathrm{d}i_o}{\mathrm{d}\theta} \tag{5.145}$$

5.14 Commentary on the waveforms

Figure 5.12 shows waveforms of the principal voltages and currents associated with the rectifier. The given sinusoidal busbar voltages are represented by the voltage v_{ry} shown in graph 9. After rectification this provides the output voltage, v_o, shown by the black line of graph 1 and the output current of graph 2. The notch in the output voltage due to commutation is evident in graph 1.

The smoothing action of the rectifier choke is made evident by comparing the voltage behind the choke, v_{bc}, shown by the gray line of graph 1, with the output voltage, the black line.

The thyristor current is represented by graph 4 which shows the current i_{t1} in thyristor T1. When the thyristor is fired there is a steep rise in current during the commutation period which in the case shown extends for 2.44°. This is followed by a normal mode period, a commutation period in the other half of the bridge, and a second normal mode period. During these three periods the thyristor current waveform is the same as that of the output current. Finally, 120° after the thyristor was fired, there is a commutation to the next thyristor and the current falls rapidly. Thereafter the thyristor current is zero until it is fired once again, 360° after the first firing.

The waveform of an input line current to the rectifier is typified by graph 6 which shows the current i_1 in line 1. This is the difference between the currents in thyristors T1 and T4 and exhibits the characteristics discussed in the preceding paragraph.

The flow of the current i_1, and its primary reflection, i_{ry}, in the transformer cause voltage drops in the transformer leakage impedances which result in the transformer output voltage typified by graph 5 which shows v_{12}. The most notable feature of this waveform is the deep notches caused by the interaction of the high rate of change of current during commutation with the transformer leakage and supply line inductances.

A commutation notch occurs at each commutation so that there are six per cycle, two deep ones and four less deep ones. The deep notches take the wave down to 0 V. The deep notches in v_{12} occur when there is a commutation from T1, connected to line 1, to T2, connected to line 2 and when there is a commutation from T4, connected to line 1, to T5, connected to line 2. Similarly deep notches occur in the waveform of v_{23} when T2 commutates to T3 and when T5 commutates to T6, and in that of v_{31} when T3 commutates to T1 and when T6 commutates to T4.

The other four notches have about half this depth because only one of the two lines is involved. Thus, for v_{12}, the notch which occurs at $\theta = 165°$ is associated with commutation from T6 to T4. Only the rising current in T4, which flows in line 1, affects the voltage v_{12}, the falling current in T6 flowing in line 3. Since the current in T4 is rising, the transformer leakage inductance absorbs voltage and the notch reduces the terminal voltage.

During the next commutation notch, at $\theta = 225°$, the inverse behavior is seen. This notch is associated with commutation from T2 to T3. The current, which flows in line 2, is decreasing rapidly and the transformer inductance voltage tries to maintain the flow by boosting v_{12}. The current in the incoming thyristor, T3, flows in line 3 and therefore does not affect v_{12} so that this also is a half height notch.

The thyristor voltage waveform, typified by graph 3 for thyristor T1, shows the effects of commutation by two notches associated with commutations in the other half of the bridge and a commutation step associated with the commutation in the same half of the bridge which does not involve the thyristor in question. Where the notches show a positive rate of change of voltage, as at the leading edge of the notch at 165° and the trailing edge of the notch at 285°, are potential problem points for the thyristors. Thyristors have a maximum $\mathrm{d}v/\mathrm{d}t$ limit which cannot be exceeded without risk of them losing their hold-off capability. The thyristors are protected against this potentially disastrous eventuality by snubber circuits. Thyristors also have a $\mathrm{d}i/\mathrm{d}t$ limit but the rates of change of current during commutation, while large compared to those normally experienced on the a.c. system, are typically well within the thyristor capability and are, in any case taken care of by the snubbers.

The waveform of the supply line currents is typified by graph 8 which shows the current, i_r, in the red line. This shows the rapid changes associated with commutation, the slower changes which occur during normal mode operation and the small, sinusoidal effect of the transformer magnetizing current which is shown by the gray line.

The rapid changes in line current associated with commutation introduce notches into the transformer input voltages as illustrated by graph 7 which shows v'_{ry}. These notches reflect, but are much smaller than, those in the transformer output voltages. However, only four notches are apparent, all about the same size, notches at 225° and 405° not being apparent. The explanation for this lies in the fact that a delta–wye transformer is being considered so that the voltage v'_{ry} is the primary reflection of v_1 which does not experience the commutations from T2 to T3 and from T5 to T6.

A most significant feature of graph 7 is the small size of the notches. Roughly speaking it can be said that the supply line impedance and the leakage impedance of the transformer form an inductive voltage divider during the commutation periods so that the total depth of the notch as seen at the secondary terminals is reduced by a factor of $(3\, X_s \times {N_{tsp}}^2)/X_{st}$ at the primary terminals. For the example, this factor is 0.12. Considering the notch at $\theta = 165°$, this has a depth of about 237 V on the secondary side and 237 V on the primary side. Referring the primary value to the secondary side by multiplying by N_{tsp} yields an equivalent secondary notch of 26.3 V. The ratio of the two notches is therefore $26.3/237 = 0.111$, close to the inductance ratio.

The frequency spectrum of these narrow commutation notches extends far into the audio frequency spectrum and can be a source of interference with other equipment connected to the same a.c. system, not least of which is the rectifier controller itself. Direct connection of the rectifier to the a.c. system exacerbates these problems in two ways. First, the full depth of the notches is experienced by the a.c. system with no attenuation in the transformer. Second, since the inductance seen by the rectifier is considerably reduced, commutation is much faster, the notches are much narrower and their potential for interference is correspondingly greater. In sensitive environments it may be necessary to connect the rectifier to the system distribution bus via a transformer simply to reduce the commutation notch noise.

5.15 The power flow

Ideally, for a three phase system the power flows from the a.c. system into the rectifier and from the rectifier into the d.c. load should be time invariant. Due to the ripple in the rectified voltage, this is not the case. Graph 1 of Fig. 5.13 shows, by a black line, the waveform of the power, $p_o = v_o \times i_o$, flowing into the load and, by the gray line, that of the power behind the choke, $p_{bc} = v_{bc} \times i_o$. The waveform of the input power from the a.c. system, $p_{in} = v_{ry}i_{ry} + v_{yb}i_{yb} + v_{br}i_{br}$, is shown in graph 2 of Fig. 5.13. The ripple in all three powers is substantial but, in the output power, is significantly reduced by the smoothing effect of the rectifier choke. The effect of the commutation notches is evident in p_o and p_{bc} but is not nearly so apparent in p_{in} due to the smoothing effect of the transformer leakage inductance.

The product of a voltage expression and a current expression which yields the power is complicated, involving the sum of up to 13 separate terms with 21 coefficients. The form of this expression is given in eqn A.5.

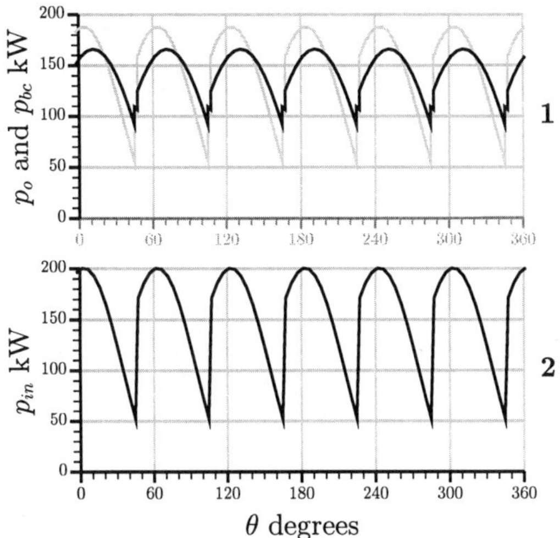

FIG. 5.13. Waveforms of power flows. Graph 1 shows the output power by the black line and the power behind the choke by the gray line. Graph 2 shows the total a.c. input power to the rectifier transformer.

Obtaining the mean power requires the integral of the power expression with respect to θ and this is given in eqn A.19.

Example 5.12

Derive expressions for the output power as a function of θ during the commutation and normal modes and determine the average values of the output power, the power behind the choke, and the input power for the example when the delay angle is 45° and the load e.m.f. is 370 V.

Multiplying the appropriate basic voltage and current expressions yields the following expressions for the output power in kilowatts.

During the commutation mode, $105° < \theta < 107.44°$, i.e. segment 5 of Fig. 5.11,

$$p_o = -467.79 + 105.66\cos(2\theta - 3.6738) + 701.90\exp(-\frac{\theta}{2.1061}) + 576.12\cos(\theta - 4.1617) + 59.0121\exp(-\frac{\theta}{4.2122}) + 880.45\cos(\theta - 1.4830)\exp(-\frac{\theta}{4.2122})$$

During the normal mode, $107.44° < \theta < 165°$, i.e. segment 6 of Fig. 5.11,

$$p_o = -465.72 + 130.42\cos(2\theta - 4.7257) + \\ 840.20\exp(-\frac{\theta}{2.1286}) + 619.28\cos(\theta - 4.6648) + \\ 108.93\exp(-\frac{\theta}{4.2572}) + \\ 1066.67\cos(\theta - 2.0105)\exp(-\frac{\theta}{4.2472})$$

The mean output power is 143.84 kW, the mean power behind the choke is 147.77 kW, and the mean input power is 154.58 kW. Thus, the rectifier efficiency, including the losses in its transformer and smoothing choke, is 93.05%.

****

5.16 Mean and r.m.s. values of the voltages and currents

The mean values of the output voltages and currents and of the thyristor current are of interest. They are obtained by summing the integrals of the basic expression over the appropriate intervals and dividing by the total integration interval. For the output quantities integration is required only over a 60° interval comprising a commutation mode segment and a normal mode segment. The waveform of a thyristor current only repeats once per supply system cycle and its integral must be carried out over a complete cycle. The integral of the basic power electronics expression, of which the expressions for the voltages and currents are subsets, is given in eqn A.3.

The r.m.s. values of all the voltages and currents are of interest. The basic expressions must be squared and integrated, a process facilitated by eqn A.19.

The ripple content of the output quantities is easily obtained as the square root of the difference between the squares of the r.m.s. value and the mean value. A similar process, substituting the fundamental component for the mean value, gives the ripple content of the a.c. side quantities and is covered in Chapter 8.

Example 5.13

For the example, operating with a delay angle of 45° and a load e.m.f. of 370 V, determine the mean and r.m.s. values of the currents and voltages.

Using the methods just described and with the aid of the functions of the basic power electronics expression given in Appendix A, the various quantities are

Mean output voltage = 392.69 V

$$\begin{aligned}
\text{R.m.s. output voltage} &= 394.23 \text{ V}\\
\text{R.m.s. ripple in the output voltage} &= 34.78 \text{ V}\\
\text{Mean output current} &= 366.01 \text{ A}\\
\text{R.m.s. output current} &= 368.33 \text{ A}\\
\text{R.m.s. ripple in the output current} &= 41.23 \text{ A}\\
\text{Mean voltage behind the choke} &= 403.31 \text{ V}\\
\text{R.m.s. voltage behind the choke} &= 417.07 \text{ V}\\
\text{R.m.s. ripple in the voltage behind the choke} &= 106.29 \text{ V}\\
\text{Mean thyristor current} &= 122.00 \text{ A}\\
\text{R.m.s. thyristor current} &= 212.22 \text{ A}\\
\text{Transformer secondary line to line voltage} &= 430.41 \text{ V r.m.s.}\\
\text{Transformer secondary current} &= 300.12 \text{ A r.m.s.}\\
\text{Transformer primary voltage} &= 3288.11 \text{ V r.m.s.}\\
\text{Transformer primary current} &= 26.08 \text{ A r.m.s.}\\
\text{AC line current} &= 45.17 \text{ A r.m.s.}
\end{aligned}$$

Note that the ripple content of the output voltage is reduced by the choke from 106.29 to 34.78 V.

****

5.17 The input kVA and power factor

As conventionally measured, the input kVA is $\sqrt{3}$ times the product of the r.m.s. input line to line voltage and the r.m.s. line current divided by 1000 and the power factor is the ratio of the mean input power to this kVA.

Thus, for the example, the input kVA is $\sqrt{3} \times 3288.11 \times 45.17 \div 1000 = 257.27$.

The power factor is $154.58 \div 257.27 = 0.6009$.

The value of the power factor is substantially smaller than the value for an ideal system which is $0.9549 \cos(45°) = 0.6752$, see eqn 3.25. The difference has a variety of causes. The ripple increases the r.m.s. quantities, commutation overlap increases the lag of the a.c. currents and the transformer magnetizing current, because of its low lagging power factor, increases the lag of the a.c. side currents.

5.18 Comparison of the exact and approximate solutions

Figures 5.14 and 5.15 provide a comparison between the approximate and exact methods of analysis. Figure 5.14 shows load characteristics, mean

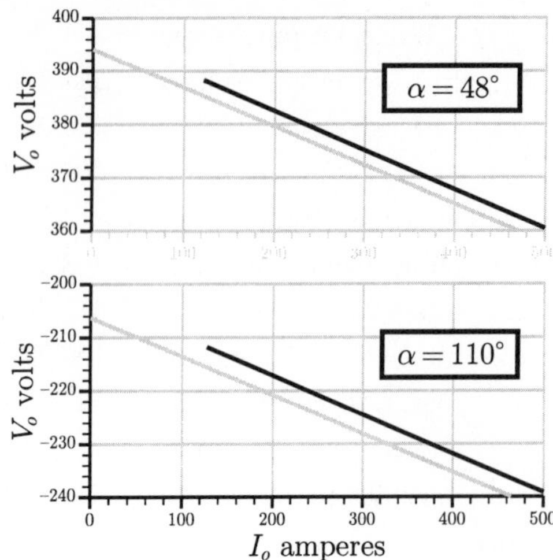

FIG. 5.14. Load characteristics by the exact method, black line, and by the approximate method, gray line.

output voltage as a function of mean output current, for delay angles of 48° and 110°, values used in drawing Fig. 5.7. The black lines were obtained by the exact method and the gray lines by the approximate method. The voltage scale has been expanded to emphasize the differences. The characteristics for the exact method have been stopped short of zero current because a new phenomenon, discontinuous conduction, occurs at low load current and is the subject of Chapter 6.

The voltage by the approximate method is 2–3 V lower than that by the exact method. For a given delay angle, the reduction is constant, the two characteristics being parallel, but it depends on the delay angle, being small, and even negative, when the delay angle is small and largest when the delay angle is in the vicinity of 90°. The difference is due to the assumption in the approximate method that the mean output current is being commutated. In fact, due to the ripple, the actual current being commutated is generally smaller although at small delay angles it may be larger. In the case of the example at 45° delay and 370 V e.m.f., the current being commutated is 286.27 A, 78.2% of the mean output current. This means that the overlap period is shorter and the voltage absorbed during commutation is less than the value obtained by the approximate method. Since the difference is so closely linked to the output current ripple any reduction in ripple by having more inductance in the load circuit diminishes the error.

Figure 5.15 compares the output current waveforms, the black line being obtained by the exact method and the gray line by the approximate method. The major difference is due to the difference in mean values noted

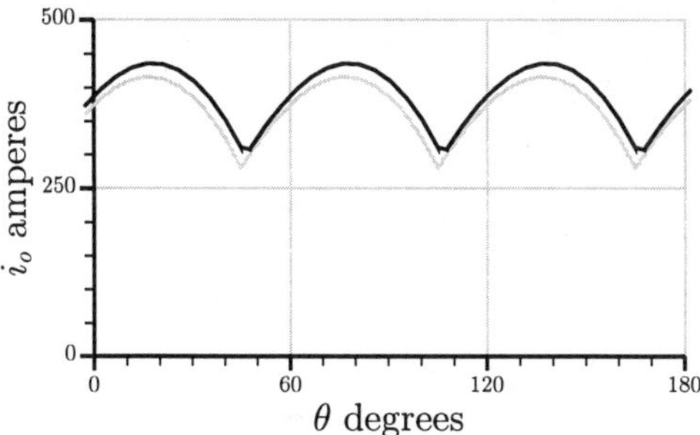

FIG. 5.15. Comparison of the output current waveforms computed by exact method, black line, and by the approximate method, gray line.

above with relatively minor differences in waveform due to the neglect of commutation by the approximate method.

The errors of the approximate method are small, and are normally well within acceptable accuracy limits especially when it is considered that substantial uncertainly attaches to most of the quantities which have been used with such great accuracy. The approximate method, and especially the equivalent circuits of Figs 5.7 and 5.8, is a useful tool for the analysis of rectifier problems. However, once the problem has been programmed for computer solution, the difference between the two methods amounts to no more than a second or so of computer time. Thus, the exact technique, while it may not be either necessary or warranted by the accuracy with which the problem can be posed, has, once it has been incorporated in a computer program, the aesthetic merit of being accurate in so far as its inputs are accurate.

6

DISCONTINUOUS CONDUCTION

6.1 Introduction

In Chapter 5, in which the effects of finite impedances in the rectifier circuit were studied, it has been tacitly assumed that conduction is continuous; that the output current never falls to zero. There are two situations in which this is not the case, when the load on the rectifier is capacitive, typically when capacitor smoothing is employed, and, with an inductive load, at low levels of output current. Under these circumstances conduction is discontinuous, there being a period in each 60° output segment when no output current flows. The onset of this condition profoundly alters the behavior of the rectifier. This mode of behavior as it applies to an inductive load at low output currents is reviewed in this chapter and the case of the capacitive load is considered in Chapter 7.

Rectifier behavior when conduction is discontinuous is very complex with numerous alternative modes of operation depending on the conditions. In this chapter three distinct modes are identified, mode 1 when the delay angle is small, mode 2 for medium values of delay, and mode 3 when the delay is large. Further, mode 1 behavior divides into two submodes, modes 1-1 and 1-2. Unlike in Chapter 5 where the approximate solution could be easily separated in the text from the exact solution, the modes must be considered concurrently, and this makes for complicated reading. Some relief can be gained at a first reading by using the fact that modes 1 and 3 are relatively unimportant so that those parts which deal with these modes can be omitted and yet a good understanding of rectifier operation when conduction is discontinuous can still be obtained.

Generally, a rectifier in discontinuous conduction will operate in mode 2 and the analysis is straightforward. When the delay is small, below about 10°, and the load current is very small, below about 5% of the rated value, the rectifier may operate in either of submodes 1-1 or 1-2. When the delay is large, greater than about 170°, and the load current is very small, it may operate in mode 3. Such large delay angles are beyond the normal phase back limit and run grave risk of commutation failure at larger load currents. Mode 3 is therefore extremely unlikely in practice and it is included in this analysis mainly for the sake of completeness.

Figure 6.1 shows the output current waveform for the rectifier defined

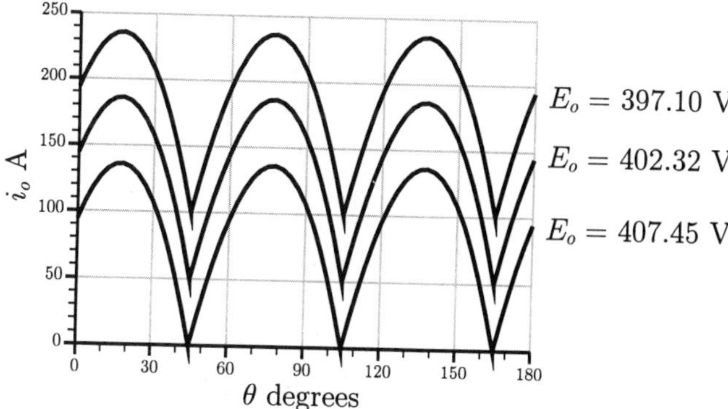

FIG. 6.1. Output current waveforms for the rectifier of Chapter 5 operating at a delay of 45° and with the indicated values of load e.m.f. The critical e.m.f. is 407.45 V.

in Example 5.1 when the delay angle is 45° and at various values of load e.m.f. It is evident that, as the load e.m.f. is raised keeping the delay angle constant, the mean load current decreases and with it the separation between the current waveform and the zero current axis. At some critical value of e.m.f., in this particular case 407.45 V, the lowest points in the output current waveform touch the θ axis, the current at these points being zero. This is the boundary between continuous and discontinuous conduction and for load e.m.fs greater than this critical value conduction will be discontinuous.

This problem is discussed in the context of the fully controlled three phase bridge rectifier and the combination of rectifier, transformer, and load defined in Example 5.1 will be used as an illustrative example. The nomenclature of Chapter 5 is used and attention is focused on the segment $\alpha + \pi/3 < \theta < \alpha + 2\pi/3$ when T2 and T6 conduct. The situation for other rectifier circuits, such as the semi-controlled bridge, can be analyzed by similar techniques.

6.2 The continuous–discontinuous conduction boundary

At the boundary between continuous and discontinuous conduction, known as the *critical boundary*, the output current is normally zero at the commutation instant so that the overlap angle is zero. In rare situations, for delay angles approaching zero or 180°, the current is not zero at this point and the commutation process occurs. However, even under these conditions the current to be commutated is extremely small, the commutation period is very short, and there is very little error incurred in neglecting it. This is done and the system then has the simple equivalent circuit of

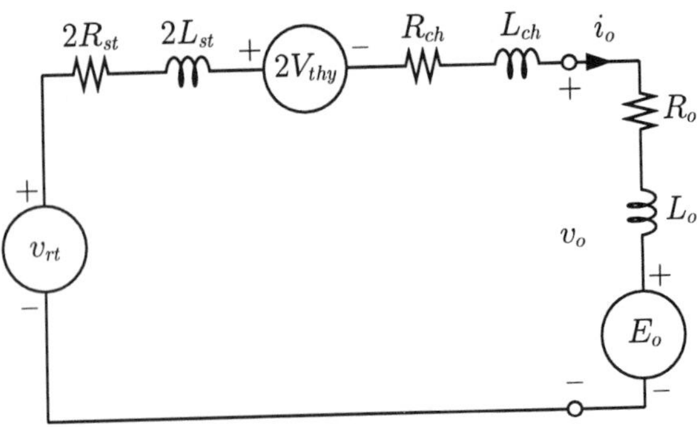

FIG. 6.2. The equivalent circuit of the rectifier when conduction is discontinuous.

Fig. 6.2. This is similar to the a.c. equivalent circuit of Fig. 5.8 but omits the voltage source $V_{xt} = I_o 3X_{st}/\pi$ which represents the voltage absorbed by the system inductance during commutation. The input voltage v_{rt} is the Thévenin voltage of the rectifier and comprises successive 60° segments of the bus bar line to line voltages referred to the secondary and appropriate to the delay angle, and is defined for the general case by eqns 5.99 and 5.100. It is defined by eqn 5.98 for the specific segment under consideration as $\sqrt{6}V_{st}\cos(\theta - \pi/2) = \sqrt{6}V_{st}\sin(\theta)$.

The transition point between continuous and discontinuous conduction is designated the *critical point*, the locus of such points for all delay angles being the *critical boundary*. All quantities at the critical point, e.g. currents and voltages, are described as critical values and their symbols are identified by the subscript *crit*. The critical value of mean output current, I_{ocrit}, is small, typically less than 10% of the rated current.

6.3 The critical point

The critical point marks the boundary between continuous and discontinuous operation. For a given angle of delay, it is defined by the value of load e.m.f. for which the output current just becomes zero at some point in a 60° output segment. Any increase in e.m.f. and therefore reduction in current will cause the output current to try to go negative, which it cannot do because of the unidirectional conductivity of the rectifying elements, and conduction therefore ceases.

The key to defining the critical point is the expression for the output current during the period $\alpha + \pi/3 < \theta < \alpha + 2\pi/3$ which is defined by eqn 5.94 as

$$i_o = I_8 \cos(\theta + \phi_{c8}) - I_{13} + I_{14} \exp(-\frac{\theta}{\tan(\zeta_3)}) \tag{6.1}$$

From eqns 5.73 through 5.77 and eqn 5.97, it is seen that I_8, ϕ_{c8}, and I_{14} are

$$I_8 = \frac{\sqrt{6}V_{st}}{Z_3} \tag{6.2}$$

$$\phi_{c8} = -\frac{\pi}{2} - \zeta_3 \tag{6.3}$$

$$I_{14} = \frac{-I_8 \cos(\alpha + \phi_{c8})}{\exp(-\frac{\pi/3+\alpha}{\tan(\zeta_3)}) - \exp(-\frac{2\pi/3+\alpha}{\tan(\zeta_3)})} \tag{6.4}$$

where

$$R_3 = 2R_{st} + R_{ch} + R_o \tag{6.5}$$

$$X_3 = 2X_{st} + X_{ch} + X_o \tag{6.6}$$

$$Z_3 = \sqrt{R_3^2 + X_3^2} \tag{6.7}$$

$$\zeta_3 = \arctan(X_3/R_3) \tag{6.8}$$

All three quantities defined by eqns 6.2 through 6.4 depend solely on the input voltage, the circuit parameters, and the delay angle and are independent of the load e.m.f. so that the search for the critical point must concentrate on the value of I_{13}, which will be called I_{13crit}, and the corresponding value of load e.m.f. E_{ocrit}.

In determining the critical point, a problem arises since the circuit will behave in one of three modes, each of which necessitates a different treatment. These modes must therefore be identified before continuing further with the analysis.

6.4 Modes of operation

The three possible modes of operation at the critical point are illustrated by Fig. 6.3 which shows, in a pair of graphs for each of four different angles of delay, the waveforms of the rectifier Thévenin voltage, v_{rt}, the sum of the load e.m.f. and thyristor forward voltage drops, $E_o + 2V_{thy}$, and the load current, i_o, for the example at the critical point. The mode number and delay angle are listed in a block on each graph pair and the graphs are drawn for the segment when thyristors T2 and T6 conduct so that the θ scale starts at $\alpha + 60°$ and extends to $\alpha + 120°$. Note that the voltage and current scales change considerably from graph to graph in order that they are appropriate to the displayed data.

At any time the *net circuit driving voltage* is the difference between v_{rt}, shown by the black curve, and $E_o + 2V_{thy}$, shown by the horizontal dashed

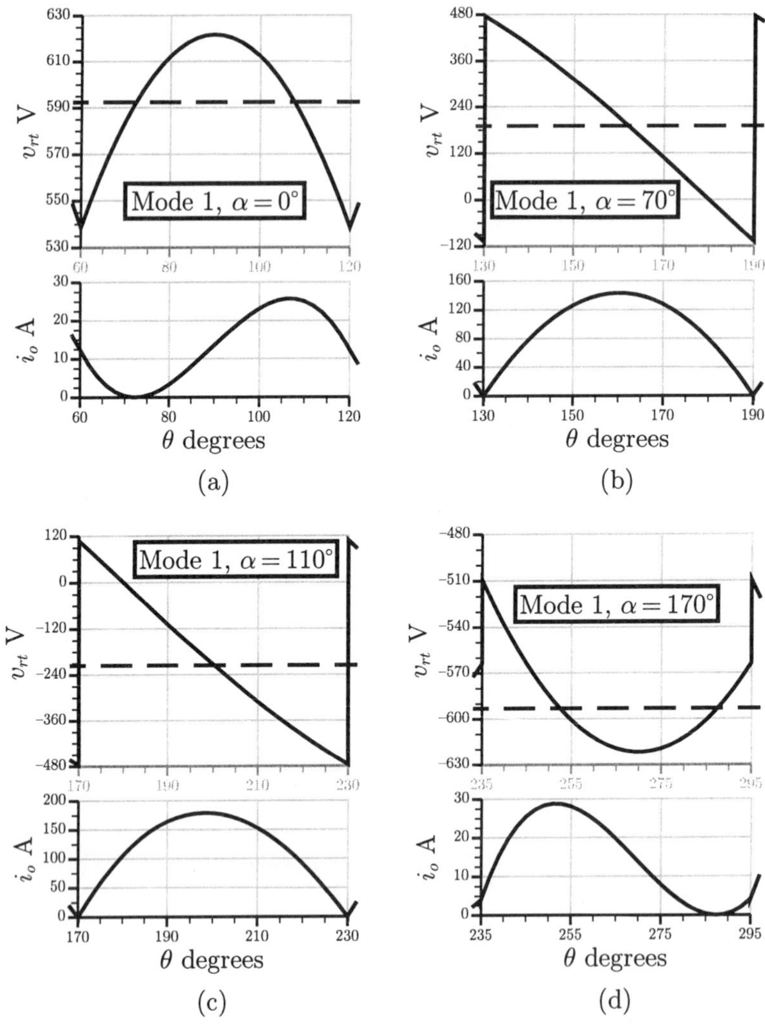

FIG. 6.3. Waveforms of the rectifier Thévenin voltage and output current at the critical operating point for various values of delay angle.

line in the upper graph of each pair. The rate of change of output current is the net circuit driving voltage less the voltage absorbed by the circuit resistance all divided by the circuit inductance, i.e.

$$\frac{\mathrm{d}i_o}{\mathrm{d}t} = \frac{v_{rt} - E_o - 2V_{thy} - R_{st}i_o}{L_{st}} \tag{6.9}$$

6.4.1 *Mode 1*

When the delay angle is small, the net circuit driving voltage at the start of a rectifier segment is negative. The rate of change of current is consequently negative, the current decreases and becomes zero later in the segment. This condition characterizes mode 1 operation. It is typified by Fig. 6.3(a) which has been drawn for the critical point at zero delay.

The e.m.f. for Fig. 6.3(a) has the critical value of 589.24 V and the dashed line is at this value plus 3.2 V, the drop in the two conducting thyristors, i.e. at 592.44 V. The rectifier Thévenin voltage, v_{rt}, is the 60° segment of 621.79 $\cos(\theta - \pi/2)$ which runs from 60° to 120° and is shown by the solid line of the upper graph. At the point when T2 is fired, $\theta = 60°$, $i_o = 12.42$ A, and $v_{rt} = 538.49$ V. Under the negative net circuit driving voltage of −53.95 V, the current decreases, becoming zero after 12.32° at $\theta = 72.32°$. After just touching the zero axis, the current rises to a maximum of 25.7 A at $\theta = 106.86°$ and then decreases to 12.42 A at the end of the segment, $\theta = 120°$, the value at the start of the segment. The mean value of the output voltage is 590.04 V and the mean value of the output current is 12.97 A.

6.4.2 *Mode 2*

If the net circuit driving voltage is positive at the firing instant, the output current has a positive rate of change at that point, a situation characteristic of modes 2 and 3. Mode 2 behavior results when the net circuit driving voltage decreases with time, becomes zero and then goes negative and remains negative during the balance of the segment. For the example, the transition from mode 1 to mode 2 occurs at a delay angle of 9.68° and mode 2 behavior is illustrated by Fig. 6.3(b) drawn for a delay of 70° and Fig. 6.3(c) drawn for a delay of 110°. During mode 2, at the critical point the current starts from zero at the firing instant, rises to a maximum, and then falls back to zero at the end of the segment.

For Fig. 6.3(b), 70° delay, the critical value of load e.m.f. is 187.67 V and the dashed line is drawn 3.2 V higher than this, at 190.87 V. At the point at which T2 is fired, $\theta = 130°$, $v_{rt} = 476.32$ V so that the net circuit driving voltage is 285.45 V and the output current initially rises from zero at the rate of 246.5 $\mathrm{A\,ms^{-1}}$, 11.41 A per degree. The current reaches its maximum value of 179.5 A at $\theta = 160.33°$ just before the net circuit driving voltage becomes zero at $\theta = 162.12°$, the difference between these two angles being due to the voltage absorbed by the circuit resistance. After the maximum, the falling current just reaches zero at the end of the segment, $\theta = 190°$. The mean output voltage is 195.05 V and the mean output current is 119.07 A.

For Fig. 6.3(c), $\alpha = 110°$, the critical value of load e.m.f. is -218.45 V and the dashed line is drawn 3.2 V higher than this, at -215.25 V. At the point at which T2 is fired, $\theta = 170°$, $v_{rt} = 107.97$ V so that the net circuit driving voltage is 323.22 V and the output current initially rises from zero at the rate of 279.2 $\mathrm{A\,ms^{-1}}$, 12.92 A per degree. The current reaches its maximum value of 179.2 A at $\theta = 198.46°$ just before the net circuit driving voltage becomes zero at $\theta = 200.25°$, the difference between these two angles being due to the 18.38 V absorbed by the circuit resistance at the current maximum. After the maximum, the falling current just reaches zero at the end of the segment, $\theta = 230°$. The mean value of output voltage is -211.09 V and the mean value of output current is 118.69 A.

6.4.3 *Mode 3*

As the delay angle approaches 180°, the possibility arises of the net circuit driving voltage becoming positive towards the end of the conducting segment. This leads to mode 3 behavior and can occur for the example when θ exceeds 169.50°.

Such a delay angle is imprudently large, running grave risk of a commutation failure. However, so as to present a complete picture, Fig. 6.3(d) shows mode 3 operation for $\alpha = 175°$ and the critical load e.m.f. of -596.21 V so that the horizontal dashed line is at -593.01 V — the e.m.f. plus the drop in two thyristors. At the firing point, $\theta = 235°$, $v_{rt} = -509.43$ V and the output current is 4.12 A. Under the positive net circuit driving voltage the current rises to a maximum of 28.88 A at $\theta = 251.61°$ and then falls to just become zero at $\theta = 287.50°$. At this point the net circuit driving voltage again becomes positive and the output current rises back to its starting value of 4.12 A at the end of the segment, $\theta = 295°$. The mean value of the output voltage is -595.30 V and the mean value of the output current is 14.6 A.

6.4.4 *The mode 1 to mode 2 transition*

Inspection of Fig. 6.3 shows that the transition between mode 1 and mode 2 at the critical point occurs when the slope of the current waveform is zero at the start of the 60° segment. From eqn 6.1

$$\frac{\mathrm{d}i_o}{\mathrm{d}\theta} = -I_8 \sin(\theta + \phi_{c8}) - \frac{I_{14}}{\tan(\zeta_3)} \exp(-\frac{\theta}{\tan(\zeta_3)}) \qquad (6.10)$$

At the commutation point, $\theta = \alpha + \pi/3$, and substituting the value of I_{14} from eqn 6.4, this equation defines the value of delay angle at the transition, α_{t12}, by

$$-I_8 \sin(\alpha_{t12} + \pi/3 + \phi_{c8}) + \frac{I_8 \cos(\alpha_{t12} + \phi_{c8})}{\tan(\zeta_3)[1 - \exp(-\frac{\pi/3}{\tan(\zeta_3)})]} = 0$$

i.e.

$$\alpha_{t12} = \arctan\left[\frac{-\cos(\phi_{c8}) + A\sin(\pi/3 + \phi_{c8})}{-\sin(\phi_{c8}) - A\cos(\pi/3 + \phi_{c8})}\right] \quad (6.11)$$

where

$$A = \tan(\zeta_3)[1 - \exp(-\frac{\pi/3}{\tan(\zeta_3)})] \quad (6.12)$$

The Fortran or C function *atan2* will return the correct value of delay when the signs given in eqn 6.11 are used. For the example $\alpha_{t12} = 9.68°$

An approximate value for the transition delay can be obtained by neglecting the circuit resistance. Then $\zeta_3 = \pi/2$ and $\phi_{c8} = -\pi$ so that eqn 6.11 becomes

$$\alpha_{t12} = \arctan(\frac{6/\pi - \sqrt{3}}{1}) = 10.08° \quad (6.13)$$

For the example the approximate solution is very close to the exact solution. This is generally the case so that the transition at the critical point from mode 1 to mode 2 can be expected to occur in any rectifier at a delay angle of about 10°.

6.4.5 *Transition between modes 2 and 3*

The transition between mode 2 and mode 3 occurs when the slope of the output current waveform is zero at the end of the 60° segment. By substituting $\theta = \alpha + 2\pi/3$ into eqn 6.10, replacing I_{14} with the expression of eqn 6.4 and equating the result to zero it is found that

$$\alpha_{t23} = \arctan\left[\frac{-\cos(\phi_{c8}) - A\sin(2\pi/3 + \phi_{c8})}{-\sin(\phi_{c8}) + A\cos(2\pi/3 + \phi_{c8})}\right] \quad (6.14)$$

where A is defined by eqn 6.12

Again the Fortran or C functions *atan2* will return the correct delay angle when the signs of eqn 6.14 are used.

For the example $\alpha_{t23} = 169.50°$.

If resistance is neglected, $\zeta_3 = \pi/2$ and $\phi_{c8} = \pi$ and eqn 6.14 reduces to

$$\alpha_{t23} = \arctan(\frac{6\ \pi - \sqrt{3}}{-1}) = 169.92° \quad (6.15)$$

Again this is extremely close to the exact solution and indicates that the transition from mode 2 to mode 3 occurs in any rectifier at a delay of about 170°. This is beyond a prudent limit so that mode 3 operation is unlikely in practice.

6.4.6 *The critical point for mode 2*

The critical operating point, i.e. the transition between continuous and discontinuous conduction, is determined by the value of $I_{13} = I_{13crit}$, which just makes the output current zero at its minimum point. For mode 2 this value is easily found since the minimum occurs at the commutation point. For modes 1 and 3 the minimum occurs at a later point, $\theta = \delta$, which must first be determined

Starting with the simplest case, mode 2, if i_o is zero when $\theta = \alpha + \pi/3$

$$I_{13crit} = I_8 \cos(\alpha + \frac{\pi}{3} + \phi_{c8}) + I_{14} \exp(-\frac{\alpha + \pi/3}{\tan(\zeta_3)}) \quad (6.16)$$

and

$$E_{ocrit} = R_3 I_{13crit} - 2V_{thy} \quad (6.17)$$

where I_8, ζ_3, ϕ_{c8}, and I_{14} are defined by eqns 6.2 through 6.8.

Integrating eqn 6.1 over the range $\alpha + \pi/3 < \theta < \alpha + 2\pi/3$ and averaging yields, after some simplification, the mean output current as

$$I_{ocrit} = \frac{3}{\pi} \frac{I_8 \cos(\alpha)}{\cos(\zeta_3)} - I_{13crit} \quad (6.18)$$

The mean value of the output voltage, V_{ocrit}, is then

$$V_{ocrit} = E_{ocrit} + R_o I_{ocrit} \quad (6.19)$$

Example 6.1

Determine the critical values for the example operating at a delay angle of 45°.

First, eqns 6.3 through 6.8 are used to find that $\zeta_3 = 1.3401$ rad, $I_8 = 1386.75$ A, $\phi_{c8} = -2.9109$ rad, and $I_{14} = 5151.18$ A.

Since the delay angle is 45° the system is in mode 2 so that eqns 6.16 through 6.19 may be used to find the critical values.

$$\begin{aligned} I_{13crit} &= 4005.03 \text{ A} \\ E_{ocrit} &= 407.45 \text{ V} \\ I_{ocrit} &= 89.84 \text{ A} \\ V_{ocrit} &= 413.02 \text{ V} \end{aligned}$$

****

6.4.7 *The critical point for modes 1 and 3*

For modes 1 and 3 the minimum point in the current wave must first be found. The value of θ at this location will be called δ and, since $\mathrm{d}i_o/\mathrm{d}\theta$ is zero at the minimum point, from eqn 6.10,

$$-I_8\sin(\delta+\phi_{c8})-\frac{I_{14}}{\tan(\zeta_3)}\exp(-\frac{\delta}{\tan(\zeta_3)})=0 \tag{6.20}$$

Using the Newton–Raphson method of successive approximation to find δ in this transcendental equation,

$$f(\delta)=-I_8\sin(\delta+\phi_{c8})-\frac{I_{14}}{\tan(\zeta_3)}\exp(-\frac{\delta}{\tan(\zeta_3)}) \tag{6.21}$$

and

$$f'(\delta)=-I_8\cos(\delta+\phi_{c8})+\frac{I_{14}}{\tan^2(\zeta_3)}\exp(-\frac{\delta}{\tan(\zeta_3)}) \tag{6.22}$$

A better approximation δ_2 is obtained from an approximation δ_1 by

$$\delta_2=\delta_1-f(\delta_1)/f'(\delta_1) \tag{6.23}$$

An appropriate initial value for δ for mode 1 is the start of the segment, i.e.

$$\delta_{st}=\alpha+\frac{\pi}{3} \tag{6.24}$$

For mode 3, the initial value of δ should be the end of the output segment, i.e.

$$\delta_{st}=\alpha+\frac{2\pi}{3} \tag{6.25}$$

The value of I_{13crit} is now obtained from eqn 6.16 but with δ substituted for $\alpha+\pi/3$, i.e.

$$I_{13crit}=I_8\cos(\delta+\phi_{c8})+I_{14}\exp(-\frac{\delta}{\tan(\zeta_3)}) \tag{6.26}$$

The values of E_{ocrit}, I_{ocrit}, and V_{ocrit} are given by eqn 6.17 through eqn 6.19.

Example 6.2

Determine the critical values for the example when operating at zero delay and 175° delay.

At $\alpha = 0$, the value of δ is obtained after a few Newton–Raphson iterations as 72.32°. The critical values are then

$$\begin{aligned} I_{13crit} &= 5778.07 \text{ A} \\ E_{ocrit} &= 589.24 \text{ V} \\ I_{ocrit} &= 12.97 \text{ A} \\ V_{ocrit} &= 590.04 \text{ V} \end{aligned}$$

At $\alpha = 175°$, the value of δ is 287.50°. The values at the critical point are then

$$\begin{aligned} I_{13crit} &= -5783.60 \text{ A} \\ E_{ocrit} &= -596.21 \text{ V} \\ I_{ocrit} &= 14.60 \text{ A} \\ V_{ocrit} &= -595.30 \text{ V} \end{aligned}$$

****

6.5 The cutoff voltage

When the load e.m.f. exceeds the critical value, conduction is discontinuous, i.e. there is a period during each output cycle when the bridge is reverse biased and the d.c. and a.c. sides are isolated from each other. During this period the output voltage is equal to the load e.m.f. This is responsible for profound modifications to the rectifier characteristics.

As the e.m.f. is increased, the period of non conduction extends until eventually it equals one sixth of a cycle and the rectifier is totally cut off. The minimum value of e.m.f. at which this occurs is the cutoff voltage which will be given the symbol V_{co}.

The cutoff voltage is clearly the largest value of the rectifier Thévenin voltage so that for $0 \leq \alpha \leq 30°$

$$V_{co} = \sqrt{6} V_{st} \tag{6.27}$$

and for $30° \leq \alpha \leq \alpha_{max}$

$$V_{co} = \sqrt{6} V_{st} \cos(\alpha - \frac{\pi}{6}) \tag{6.28}$$

Example 6.3

Determine for the example the values of load e.m.f. which will just cut off the rectifier for delay angles of 0, 45°, and 175°.

At 0° eqn 6.27 is used to find that a load e.m.f. of 621.79 V will just cut off the rectifier.

At 45° delay eqn 6.28 yields the cutoff voltage as 600.61 V.

At 175° delay eqn 6.28 yields the cutoff voltage as −509.34 V.

****

6.6 E.m.fs between the critical and cutoff values

In discussing the cases where the load e.m.f. is between the critical and cutoff values it will be convenient to define some additional quantities. Within the operating segment under review when thyristors T2 and T6 conduct and which covers the period $\alpha + 60° < \theta < \alpha + 120°$, the value of θ at which current ceases to flow will be called *the extinction angle*, symbol γ. In modes 1 and 3 current flow may recommence at *the initiation angle*, symbol δ. The total voltage driving current round the loop of Fig. 6.2 will be described as the *net circuit driving voltage*, symbol v_{net}

$$v_{net} = v_{rt} - 2V_{thy} - E_o \tag{6.29}$$

At this point the assumption that the voltage drop across a conducting thyristor is constant comes into question since the currents are small. This is a small problem whose resolution does not merit heroic measures such as accurately modeling a conducting thyristor. The solution that has been adopted is to consider the voltage drop to be constant at any discontinuous conduction operating point, with a value linearly dependant on the load e.m.f. relative to the critical and cutoff values according to eqn 6.30, where V_{thyc} is the value of the thyristor forward voltage drop during continuous conduction

$$V_{thy} = V_{thyc} \frac{V_{co} - E_o}{V_{co} - E_{ocrit}} \tag{6.30}$$

This approach is simple and eliminates discontinuities in the computations, and errors associated with it are completely negligible in the context of rectifier behavior as a whole.

6.7 The operating point

For the given delay angle, having ascertained that the load e.m.f. is between the critical and cutoff values, the next task is the determination of the operating mode. There are four of these, modes 1, 2, and 3 as already described but with mode 1 subdivided into two modes, 1-1 and 1-2. These four modes are illustrated by Fig. 6.4(a) through (d), each of which has two graphs, the upper showing voltage waveforms and the lower the output current waveform. The rectifier output voltage and current are shown as black lines, the waveform of the rectifier Thévenin voltage by a gray line, and the value of the load e.m.f. plus the voltage drop in two conducting thyristors, by a dashed line, during the segment $\alpha + 60° < \theta < \alpha + 120°$ for the four modes.

6.7.1 *Distinguishing between the modes*

Figure 6.4(a) is drawn for a delay angle of zero and at an output current level at which the rectifier operates in mode 1-1. The load e.m.f. is 595 V and, according to eqn 6.30 with $V_{thyc} = 1.6$ V, the thyristor forward drop is $V_{thy} = 1.32$ V. At the start of the segment, $\theta = 60°$, the net circuit driving voltage, $v_{net} = -59.14$ V and the output current $i_{ost} = 3.57$ A. The flow of current is maintained until the extinction angle, $\gamma = 61.62°$, by the energy stored in the circuit inductance. The initiation point, $\theta = \delta$, occurs at 74.00°. The net circuit driving voltage then becomes positive and, if the two thyristors involved, in this case T2 and T6, are still in the conducting state, conduction recommences with the output current having the value i_{ofn} at the end of the segment, $\theta = 120°$. In the steady state $i_{ost} = i_{ofn} = 3.57$ A.

Figure 6.4(b) illustrates mode 1-2 behavior. The delay angle is still zero but the load e.m.f. is higher at 610 V and the load current is smaller. At this e.m.f. eqn 6.30 gives the thyristor forward drop as 0.58 V. At the start of the segment, $\theta = 60°$, the net circuit driving voltage is -72.67 V and the output current is zero. The net circuit driving voltage becomes positive at the initiation point, $\theta = \delta = 79.39°$, and conduction begins if the thyristors are still in the conducting mode. Conduction ends before the end of the segment at the extinction angle, $\gamma = 110.92°$.

Figure 6.4(c) illustrates mode 2 behavior. The delay angle is 70°, the load e.m.f. is 280 V and the thyristor forward drop is 1.09 V. Conduction begins as soon as the incoming thyristor is fired, in this case T2 at $\theta = 130°$, and continues until the extinction point, $\theta = \gamma = 173.31°$.

Figure 6.4(d) illustrates mode 3 behavior and again the proviso must be made that operation at the very large delay angle of 175° would be most imprudent, inviting a commutation failure. The delay angle is 175°, the e.m.f. is -580 V, and the thyristor forward drop is 1.30 V. When thyristor T2 is fired at $\theta = 235°$, the load current is 0.86 A. It rises to a maximum of 16.1 A and then falls to zero at the extinction point, $\gamma = 264.06°$. The net circuit driving voltage becomes positive again at the initiation point,

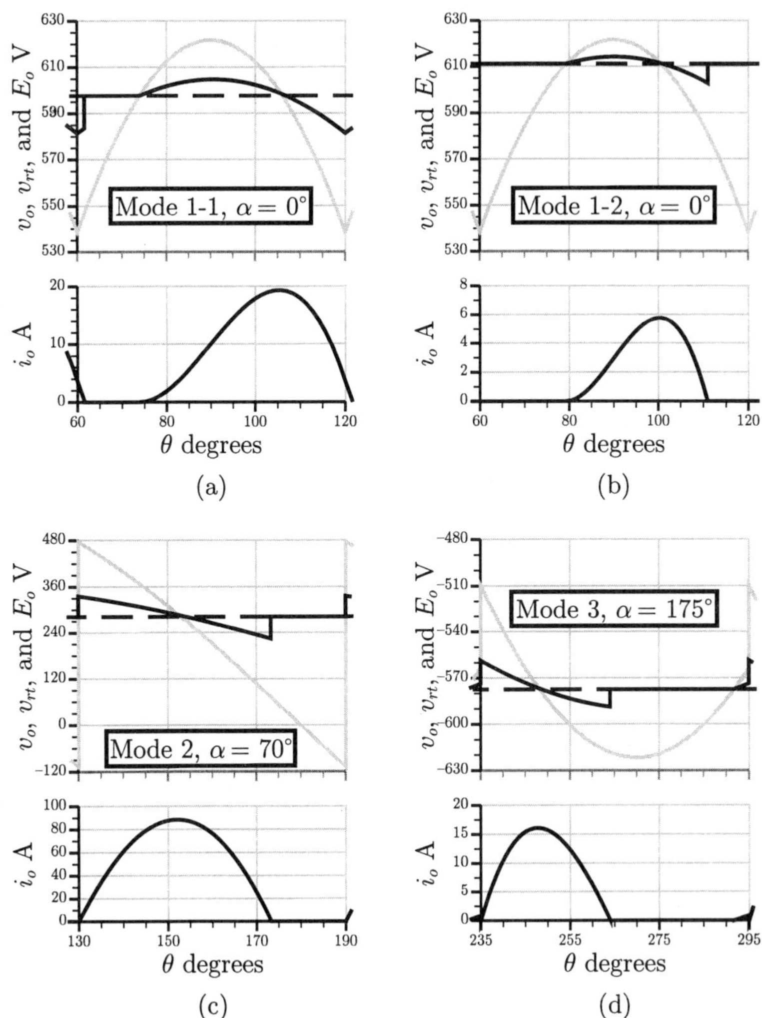

FIG. 6.4. Voltage and current waveforms during discontinuous conduction.

$\delta = 291.78°$ and, if the two thyristors concerned are in the conducting state, conduction recommences. The current rises to i_{ofn} at the end of the segment. In the steady state $i_{ost} = i_{ofn} = 0.87$ A.

6.7.2 *Computational procedure*

The computational procedure is now clear. Having ascertained that the system is in the discontinuous conduction state, the value of the net circuit driving voltage at the start of the segment is determined.

$$v_{net} = \sqrt{6}V_{st}\cos(\alpha - \frac{\pi}{6}) - E_o - 2V_{thy} \tag{6.31}$$

If v_{net} is negative the system is in mode 1, if it is zero or positive the system is in either mode 2 or 3.

If the system is in mode 1, the distinction must be made between modes 1-1 and 1-2. This is done by determining the initiation angle and the expression for the output current from this point onwards. The extinction angle is found from this expression and if it is greater than the value of θ at the end of the segment the system is in mode 1-1 otherwise the system is in mode 1-2.

If the system is not in mode 1, the distinction must be made between modes 2 and 3. This is done by determining the value of the net circuit driving voltage at the end of the conducting segment. If this voltage is positive, the system is in mode 3, if it is zero or negative the system is in mode 2.

Once the mode is known, the computational procedure is straightforward.

6.7.3 *Modes 1-1 and 1-2*

The first task in distinguishing between modes 1-1 and 1-2 is the determination of the initiation angle, δ, as the point at which the net circuit driving voltage becomes positive, i.e.

$$\delta = \arcsin(\frac{E_o + 2V_{thy}}{\sqrt{6}V_{st}}) \tag{6.32}$$

Knowing δ, I_{13} is found as the response of the equivalent circuit of Fig. 6.2 to the d.c. driving voltage, i.e.

$$I_{13} = \frac{E_o + 2V_{thy}}{R_3} \tag{6.33}$$

The current I_{14} is now dimensioned so that the output current is zero at the initiation point, i.e.

$$I_{14} = [I_{13} - I_8\cos(\delta + \phi_{c8})]\exp(\frac{\delta}{\tan(\zeta_3)}) \tag{6.34}$$

The extinction angle, γ, is now found as the point at which the current is again zero, i.e. when

$$I_8\cos(\gamma + \phi_{c8}) - I_{13} + I_{14}\exp(-\frac{\gamma}{\tan(\zeta_3)}) = 0 \tag{6.35}$$

This transcendental equation must be solved by successive approximation and again the Newton–Raphson method is excellent with

$$f(\gamma) = I_8 \cos(\gamma + \phi_{c8}) - I_{13} + I_{14} \exp(-\frac{\gamma}{\tan(\zeta_3)}) \tag{6.36}$$

and

$$f'(\gamma) = -I_8 \sin(\gamma + \phi_{c8}) - \frac{I_{14}}{\tan(\zeta_3)} \exp(-\frac{\gamma}{\tan(\zeta_3)}) \tag{6.37}$$

A suitable starting value for the iteration is $\gamma = \alpha + 2\pi/3$.

If γ is found to be greater than $\alpha + 2\pi/3$, the system is in mode 1-1, otherwise it is in mode 1-2.

If the system is in mode 1-2, no additional computations are necessary. The system is cut off from $\alpha + \pi/3$ to δ and from γ to $\alpha + 2\pi/3$ and current flows from δ to γ with the conduction angle $\beta = \gamma - \delta$.

If the system is in mode 1-1, the value of γ just found is discarded and the value of the output current at $\alpha + 2\pi/3$, i_{ofn}, is found. For the steady state this is also the value, i_{ost}, at the start of the output segment. The parameters of eqn 6.1 which fit these new conditions from $\alpha + \pi/3$ to the extinction point must now be found. The only parameter which changes is I_{14} and its value is given by

$$I_{14} = [i_{ost} - I_8 \cos(\alpha + \pi/3 + \phi_{c8}) + I_{13}] \exp(\frac{\alpha + \pi/3}{\tan(\zeta_3)}) \tag{6.38}$$

The extinction angle is now found using the technique already described in eqns 6.35 through 6.37, but with $\gamma = \alpha + \pi/3$ as the starting point. Current flows from $\alpha+\pi/3$ to γ and from δ to $\alpha+2\pi/3$, and the system is cut off from γ to δ with the conduction angle $\beta = (\gamma - \alpha - \pi/3) + (\alpha + 2\pi/3 - \gamma) = \gamma - \delta + \pi/3$.

Example 6.4

Show for the example that the operating point defined by $\alpha = 0°$ and $E_o = 595$ V is of the mode 1-1 type. Determine the extinction, initiation and conduction angles and the parameters of eqn 6.1 which apply to the two conduction periods within the output segment from $\alpha + 60°$ to $\alpha + 120°$. This is the case illustrated by Fig. 6.4(a).

At zero delay, the critical e.m.f. is 589.24 V and the cutoff voltage is 621.79 V. Taking the thyristor forward voltage drop during continuous conduction as 1.6 V, eqn 6.30 yields the thyristor voltage drop to be used here as 1.32 V.

Following the above procedure, it is found that thyristors T2 and T6 conduct from 60.0° to 61.62° and from 73.98° to 120.0° for a total conduction angle of 47.64°.

The equations for the output current are for the range $60.0^\circ < \theta < 61.62^\circ$

$$i_o = 1386.75\cos(\theta - 2.9109) - 5828.72 + 7970.81\exp(-\frac{\theta}{4.2572})$$

for the range $61.62^\circ < \theta < 73.98^\circ$

$$i_o = 0$$

and for the balance of the segment, $73.98^\circ < \theta < 120.0^\circ$

$$i_o = 1386.75\cos(\theta - 2.9109) - 5828.72 + 7985.71\exp(-\frac{\theta}{4.2572})$$

****

6.7.4 *Mode 2*

The analysis of mode 2 behavior is straightforward since conduction starts with zero current at the commutation point, $\theta = \alpha + \pi/3$. The current I_{14} of eqn 6.1 is then

$$I_{14} = [I_{13} - I_8\cos(\alpha + \pi/3 + \phi_{c8})]\exp(\frac{\alpha + \pi/3}{\tan(\zeta_3)}) \tag{6.39}$$

where I_{13} is given by eqn 6.33.

The extinction angle is found using the Newton–Raphson technique, eqns 6.36 and 6.37, with the starting value $\gamma = \alpha + 2\pi/3$. Current flows from $\alpha + \pi/3$ to γ and the system is cut off from γ to $\alpha + 2\pi/3$. The conduction angle is $\beta = \gamma - \alpha - \pi/3$.

Example 6.5

Determine for the example at the mode 2 operating point, $\alpha = 70^\circ$, $E_o = 280$ V, the extinction and conduction angles and the parameters of eqn 6.1 for the conduction period during the output segment from $\alpha + \pi/3$ to $\alpha + 2\pi/3$. This is the case illustrated by Fig. 6.4(c).

The critical value of e.m.f. for this case is 187.67 V and the cutoff voltage is 476.32 V. With these values, eqn 6.30 gives the thyristor forward voltage drop as 1.09 V.

Using the above procedure for mode 2, it is found that the extinction angle is 173.31°, the conduction angle is 43.31° and the expressions for the output current are, during the period $130° < \theta < 173.31°$,

$$i_o = 1386.75\cos(\theta - 2.9109) - 2752.07 + 2796.87\exp(-\frac{\theta}{4.2572})$$

and, during the period $173.31° < \theta < 190.0°$,

$$i_o = 0$$

****

6.7.5 *Mode 3*

Mode 3 is a continuation of the mode 3 that has already been recognized at the boundary of continuous and discontinuous conduction. It only occurs at delay angles greater than α_{t23} defined by eqn 6.15, i.e. for delay angles greater than about 170°. Mode 3 is characterized by the net circuit driving voltage becoming positive again towards the end of a 60° output segment. If firing pulses of sufficient width are employed, approaching 120°, the active thyristors will again conduct. It will be assumed that this occurs and that conduction commences at δ as indicated by Fig. 6.4(d). The output current rises to i_{ofn} at the end of the 60° segment and for the steady state this is also the current at the start of the segment. The current during the first part of the segment is then defined and its extinction angle γ can be found.

The initiation angle, δ, for mode 3 is found as the point at which the net circuit driving voltage becomes positive. Nominally this is done using eqn 6.32 but the typical arcsine routine would place the angle in the fourth quadrant with a value such as $-80°$. The value required for the segment under consideration is obtained by adding 360° to this, i.e.

$$\delta = \arcsin(\frac{E_o + 2V_{thy}}{\sqrt{6}V_{st}}) + 2\pi \tag{6.40}$$

The value of I_{14} which gives zero output current at δ is now found using eqn 6.34. The output current i_{ofn} at $\alpha + 2\pi/3$ is determined. For the steady state this is also the value, i_{ost}, at $\alpha + \pi/3$ and a new value of I_{14} is found using eqn 6.38.

The extinction angle, γ, is determined using the Newton–Raphson technique, eqns 6.36 and 6.37. So far this is a repetition of the method employed for mode 1-1. However, a different starting point for the Newton–Raphson iteration must be used if convergence to the correct point is to be obtained. A suitable initial value is $\gamma = \alpha + \pi/2$. Conduction occurs from $\alpha + \pi/3$ to γ and from δ to $\alpha + 2\pi/3$.

Example 6.6

Determine for the example at the mode 3 operating point, $\alpha = 175°$, $E_o = -580$ V, the extinction, initiation, and conduction angles and the parameters of eqn 6.1 for the two conduction periods within the output segment from $\alpha + 60°$ to $\alpha + 120°$. This is the case illustrated by Fig. 6.4(d).

At this operating point the critical value of load e.m.f. is -596.21 V and the cutoff voltage is -509.34 V. Equation 6.30 then gives the thyristor forward drop as 1.30 V.

Using the method described above, it is found that conduction occurs from the start of the output segment at 235° to 264.06° and restarts at 291.78° and continues to the end of the segment at 295° for a total conduction angle of 32.74°. The expressions for the output current are for the period $235° < \theta < 264.06°$

$$i_o = 1386.75\cos(\theta - 2.9109) + 5631.35 - 16106.45\exp(-\frac{\theta}{4.2572})$$

for the period $264.06° < \theta < 291.78°$

$$i_o = 0$$

and for the period $291.78° < \theta < 295°$

$$i_o = 1386.75\cos(\theta - 2.9109) + 5631.35 - 15995.29\exp(-\frac{\theta}{4.2572})$$

****

6.8 The output voltage waveform

Expressions for the output current waveform in the various operating modes have been obtained and the next step in the analysis is the derivation of corresponding expressions for the output voltage waveform.

The instantaneous output voltage is equal to E_o during the cutoff period and v_{rt} less the voltage drops in the line, transformer, thyristors, and smoothing choke impedances during conduction, i.e.

$$v_o = \sqrt{6}V_{st}\cos(\theta - \frac{\pi}{2}) - 2V_{thy} - R_6 i_o - X_6\frac{di_o}{d\theta} \qquad (6.41)$$

where

$$R_6 = 2R_{st} + R_{ch} \tag{6.42}$$

and

$$X_6 = 2X_{st} + X_{ch} \tag{6.43}$$

Thus

$$v_o = V_8 \cos(\theta + \phi_{v8}) + V_{13} - V_{14} \exp(-\frac{\theta}{\tan(\zeta_3)}) \tag{6.44}$$

where

$$V_8 = \sqrt{A^2 + B^2} \tag{6.45}$$
$$A = -Z_6 I_8 \cos(\phi_{c8} + \zeta_6) \tag{6.46}$$
$$B = -\sqrt{2}V_s - Z_6 I_8 \sin(\phi_{c8} + \zeta_6) \tag{6.47}$$
$$\phi_{v8} = \arctan(\frac{B}{A}) \tag{6.48}$$
$$V_{13} = R_6 I_{13} - 2V_{thy} \tag{6.49}$$
$$V_{14} = (R_6 - \frac{X_6}{\tan(\zeta_3)}) I_{14} \tag{6.50}$$
$$Z_6 = \sqrt{R_6^2 + X_6^2} \tag{6.51}$$
$$\zeta_6 = \arctan(\frac{X_6}{R_6}) \tag{6.52}$$

These equations suffice for all modes provided the appropriate values of γ, δ, and I_{14} are employed. They have been used to derive the voltage waveforms of Fig. 6.5.

Example 6.7

Derive expressions for the output voltage for the mode 1-1, 2, and 3 operating points specified in the Examples 6.4 through 6.6.

For the mode 1-1 operating point, $\alpha = 0$, $E_o = 595$ V, $V_{thy} = 1.32$ V, the expression for the output voltage during the period $60.00° < \theta < 61.62°$ is

$$v_o = 188.09 \cos(\theta - 1.8148) + 236.25 + 268.32 \exp(-\frac{\theta}{4.2572})$$

during the period $61.62° < \theta < 73.98°$

$$v_o = 595.0$$

and during the period $73.98° < \theta < 120°$

$$v_o = 188.09 \cos(\theta - 1.8148) + 236.25 - 1.57 \exp(-\frac{\theta}{4.2572})$$

For the mode 2 operating point, $\alpha = 70°$, $E_o = 280$ V, the value of thyristor forward voltage drop is $V_{thy} = 1.09$ V, and for the period $130° < \theta < 173.31°$

$$v_o = 188.09\cos(\theta - 1.8148) + 111.55 - 94.15\exp(-\frac{\theta}{4.2572})$$

For the period $173.31° < \theta < 190.0°$

$$v_o = 280.0$$

For the mode 3 operating point, $\alpha = 175°$, $E_o = -580.0$ V, the appropriate value of thyristor forward voltage drop is $V_{thy} = 1.30$ V, and during the period $235° < \theta < 264.06°$

$$v_o = 188.09\cos(\theta - 1.8148) - 228.25 - 542.19\exp(-\frac{\theta}{4.2572})$$

During the period $264.06° < \theta < 291.78°$

$$v_o = -580.0$$

and during the period $291.78° < \theta < 295.0°$

$$v_o = 188.09\cos(\theta - 1.8148) - 228.25 - 538.45\exp(-\frac{\theta}{4.2572})$$

****

6.9 The thyristor, transformer, and a.c. line currents

The derivation of the remaining currents in the circuit follows the procedure established in Chapter 5. There the idea of a set of basic current expressions which could readily be shifted to any appropriate point in the rectifier operating cycle was developed. The task here is simpler in that there are fewer basic current expressions. The method will be illustrated by reference to mode 2 operation since this is the simplest case and covers the majority of operating situations.

In mode 2 there is only one basic current expression, that defined by eqn 6.1 with I_{13} defined by eqn 6.33 and I_{14} defined by eqn 6.38. Thus the basic current expression is

$$\mathbf{G}(\psi) = I_8\cos(\psi + \phi_{c8}) - I_{13} + I_{14}\exp(-\frac{\psi}{\tan(\phi_4)}) \tag{6.53}$$

where

$$\psi = \theta - \delta_{sh} \tag{6.54}$$

This expression can be moved to any point by using the appropriate value of the shift angle δ_{sh}.

Following the procedure of Chapter 5, the supply system cycle which starts with the application of gate signal to thyristor T1 at $\theta = -60° + \alpha$, is divided into twelve subsegments which correspond to the conduction and cutoff of each pair of thyristors. Tables 6.1 through 6.3 define all the circuit currents in terms of the function **G** over a complete supply system cycle for the range $\alpha - 60°$ when T1 is fired through to $\alpha + 300°$ when T1 is fired again. Each of the three tables covers one-third of the cycle.

Each of Tables 6.1 through 6.3 has 21 rows and 5 columns. The first column in each table defines the current and the following four columns give the expression for the current, a zero indicating that no current flows during that subsegment. Row 3 gives the start of the subsegment and row 4 gives its finish. Row 5 gives the shift angle δ_{sh}. Row 6 defines the output current, rows 7 through 12 the thyristor currents, rows 13 through 15 the transformer secondary winding currents, rows 16 through 18 the transformer primary winding currents, and rows 19 through 21 the a.c. line currents. Conduction occurs in odd numbered segments and the bridge is cut off in even numbered segments. Segments 5 and 6 are the ones analyzed in this chapter and the results are applied to the other segments.

In the tables $\mathbf{G}'$ is the function **G** referred to the primary winding, i.e. $\mathbf{G}'(\psi) = \mathbf{G}(\psi)N_{tsp}$. A transformer magnetizing current component is indicated by the subscript m. Thus i_{mry} is the magnetizing component of the current in the red–yellow primary phase and i_{mr} is the magnetizing component of the current flowing in the red line.

Example 6.8

Derive the basic current expression for the example operating at a delay of 75° and with a load e.m.f. of 290 V. Use this expression to derive expressions for the current in thyristor T1.

Under the specified conditions the thyristor forward drop is 0.79 V and the basic current expression is

$$\mathbf{G}(\psi) = 1386.75\cos(\psi - 2.9109) - 2843.85 + 2895.91\exp(-\frac{\psi}{4.2572})$$

with conduction starting at $\psi = 135.0°$ and finishing at $\psi = 167.52°$ for a conduction angle of 32.52°

From Tables 6.1 through 6.3 it is seen that, during the supply system cycle which starts at $-60° + \alpha = 15°$ and ends at $300° + \alpha = 375°$, thyristor T1 conducts during subsegments 1 and 3. During subsegment 1 the shift angle, δ_{sh}, is from row 5 of Table 6.1, $-120°$ and for subsegment 3 it is $-60°$. Elsewhere the current is zero. Applying these shift angles to the basic current expression it is found that, for subsegment 1, $15° < \theta < 47.52°$

Table 6.1 *Composition of the rectifier currents*

Current	Subsegment number			
	1	2	3	4
Start	$\alpha - 60°$	$\gamma - 120°$	α	$\gamma - 60°$
Finish	$\gamma - 120°$	α	$\gamma - 60°$	$\alpha + 60°$
δ_{sh}	-120	-120	-60	-60
i_o	$\mathbf{G}$	0	$\mathbf{G}$	0
i_{t1}	$\mathbf{G}$	0	$\mathbf{G}$	0
i_{t2}	0	0	0	0
i_{t3}	0	0	0	0
i_{t4}	0	0	0	0
i_{t5}	$\mathbf{G}$	0	0	0
i_{t6}	0	0	$\mathbf{G}$	0
i_1	$\mathbf{G}$	0	$\mathbf{G}$	0
i_2	$-\mathbf{G}$	0	0	0
i_3	0	0	$-\mathbf{G}$	0
i_{ry}	$\mathbf{G}' + i_{mry}$	i_{mry}	$\mathbf{G}' + i_{mry}$	i_{mry}
i_{yb}	$-\mathbf{G}' + i_{myb}$	i_{myb}	i_{myb}	i_{myb}
i_{br}	i_{mbr}	i_{mbr}	$-\mathbf{G}' + i_{mbr}$	i_{mbr}
i_r	$\mathbf{G}' + i_{mr}$	i_{mr}	$2\mathbf{G}' + i_{mr}$	i_{mr}
i_y	$-2\mathbf{G}' + i_{my}$	i_{my}	$-\mathbf{G}' + i_{my}$	i_{my}
i_b	$\mathbf{G}' + i_{mb}$	i_{mb}	$-\mathbf{G}' + i_{mb}$	i_{mb}

$$i_{t1} = 1386.75\cos(\theta - 0.8165) - 2843.85 + 1770.63\exp(-\frac{\theta}{4.2572})$$

and for subsegment 3, $75.0° < \theta < 107.52°$

$$i_{t1} = 1386.75\cos(\theta - 1.8637) - 2843.85 + 2264.42\exp(-\frac{\theta}{4.2572})$$

During subsegment 2, $47.45° < \theta < 75°$, and subsegments 4 through 12, $107.52° < \theta < 375°$

$$i_{t1} = 0.0$$

The basic current expression has the form of the basic power electronics expression of Appendix A. Applying eqns A.3, and A.19, which give the integrals of the basic power electronics expression and its square, it is found that the mean and r.m.s. thyristor currents are

$$I_{thy} = 6.01 \text{ A}$$

Table 6.2 *Composition of the rectifier currents*

Current	Subsegment number			
	5	6	7	8
Start	$\alpha + 60°$	$\gamma°$	$\alpha + 120°$	$\gamma + 60°$
Finish	γ	$\alpha + 120°$	$\gamma + 60°$	$\alpha + 180°$
δ_{sh}	0	0	60°	60°
i_o	$\mathbf{G}$	0	$\mathbf{G}$	0
i_{t1}	0	0	0	0
i_{t2}	$\mathbf{G}$	0	$\mathbf{G}$	0
i_{t3}	0	0	0	0
i_{t4}	0	0	$\mathbf{G}$	0
i_{t5}	0	0	0	0
i_{t6}	0	0	0	0
i_1	0	0	$-\mathbf{G}$	0
i_2	$\mathbf{G}$	0	$\mathbf{G}$	0
i_3	$-\mathbf{G}$	0	0	0
i_{ry}	i_{mry}	i_{mry}	$-\mathbf{G}' + i_{mry}$	i_{mry}
i_{yb}	$\mathbf{G}' + i_{myb}$	i_{myb}	i_{myb}	i_{myb}
i_{br}	$-\mathbf{G}' + i_{mbr}$	i_{mbr}	i_{mbr}	i_{mbr}
i_r	$\mathbf{G}' + i_{mr}$	i_{mr}	$-\mathbf{G}' + i_{mr}$	i_{mr}
i_y	$\mathbf{G}' + i_{my}$	i_{my}	$2\mathbf{G}' + i_{my}$	i_{my}
i_b	$-2\mathbf{G}' + i_{mb}$	i_{mb}	$-\mathbf{G}' + i_{mb}$	i_{mb}

$$I_{thyrms} = 15.50 \text{ A}$$

****

6.10 The voltages

The system voltages are obtained by the methods already employed in Chapter 5. Starting from one end of the system, either the a.c. bus bars or the load e.m.f., where the voltage is known, progress is made towards the other end, subtracting or adding, as appropriate, the voltages absorbed by the system impedances. A check on this very complex calculation is provided by the fact that the voltage is known at both ends. Thus, by working from one end to the other, the final voltage can be compared with its known value, any difference beyond the accumulated computer roundoff error indicating an error in the calculations.

Table 6.3 *Composition of the rectifier currents*

Current	Subsegment number			
	9	10	11	12
Start	$\alpha + 180°$	$\gamma° + 120°$	$\alpha + 240°$	$\gamma + 180°$
Finish	$\gamma + 120°$	$\alpha + 240°$	$\gamma + 180°$	$\alpha + 300°$
δ_{sh}	120°	120°	180°	180°
i_o	**G**	0	**G**	0
i_{t1}	0	0	0	0
i_{t2}	0	0	0	0
i_{t3}	**G**	0	**G**	0
i_{t4}	**G**	0	0	0
i_{t5}	0	0	**G**	0
i_{t6}	0	0	0	0
i_1	$-\mathbf{G}$	0	0	0
i_2	0	0	$-\mathbf{G}$	0
i_3	**G**	0	**G**	0
i_{ry}	$-\mathbf{G}' + i_{mry}$	i_{mry}	i_{mry}	i_{mry}
i_{yb}	i_{myb}	i_{myb}	$-\mathbf{G}' + i_{myb}$	i_{myb}
i_{br}	$\mathbf{G}' + i_{mbr}$	i_{mbr}	i_{mbr}	i_{mbr}
i_r	$-2\mathbf{G}' + i_{mr}$	i_{mr}	$-\mathbf{G}' + i_{mr}$	i_{mr}
i_y	$\mathbf{G}' + i_{my}$	i_{my}	$-\mathbf{G}' + i_{my}$	i_{my}
i_b	$\mathbf{G}' + i_{mb}$	i_{mb}	$2\mathbf{G}' + i_{mb}$	i_{mb}

Example 6.9

For the example operating at a delay angle of 75° and a load e.m.f. of 290 V, determine an expression for the voltage between the red and yellow line at the transformer input during subsegment 7. As in the previous example, the thyristor forward drop is 0.79 V.

The voltage of the red bus bar is 2694.44 $\cos(\theta - 0.5236)$ and that of the yellow bus bar is 2694.44 $\cos(\theta - 2.6180)$

Segment 7 runs from 195.0° to 227.52° and has a shift angle of 60°, 1.0471 rad. From Table 6.1, the current in the red line is $i_r = -\mathbf{G}'(\theta - 1.0472) + i_{mr}$ and in the yellow line is $i_y = 2\mathbf{G}'(\theta - 1.0472) + i_{my}$. Now

$$\mathbf{G}'(\theta - 1.0472) = 106.67\cos(\theta - 3.9580) + 218.76 - 284.89\exp(-\frac{\theta}{4.257})$$
$$i_{mr} = 9.265\cos(\theta - 1.9830)$$

$$i_{my} = 9.265\cos(\theta + 2.2058)$$

so that

$$i_r = 110.65\cos(\theta - 0.8935) + 218.76 - 284.89\exp(-\frac{\theta}{4.2572})$$
$$i_y = 222.55\cos(\theta + 2.3199) - 437.52 + 569.77\exp(-\frac{\theta}{4.2572})$$

The flow of these currents in the line impedances, $0.034 + j0.2262\ \Omega$ per line, absorb some voltage and the potentials of the transformer input terminals are

$$v'_r = v_r - R_s i_r - X_s \frac{\mathrm{d}i_r}{\mathrm{d}\theta}$$

and

$$v'_y = v_y - R_s i_y - X_s \frac{\mathrm{d}i_y}{\mathrm{d}\theta}$$

so that

$$v'_{ry} = v_r - v_y - R_s(i_r - i_y) - X_s \frac{\mathrm{d}(i_r - i_y)}{\mathrm{d}\theta}$$

With the aid of the derivative of the basic power electronics expression given in eqn A.2 of Appendix A, and using the numerical values given above

$$v'_{ry} = 4603.22\cos(\theta - 0.0090) - 22.31 - 16.35\exp(-\frac{\theta}{4.2572})$$

****

6.11 The waveforms

The waveform calculations are most tedious to perform manually but it is a relatively minor matter to produce a computer program which will perform them for every voltage of interest and for every subsegment. This has been done to produce the waveforms of Fig. 6.5. The operating conditions for this figure are a delay angle of 75°, a load e.m.f. of 290 V, and a thyristor forward drop of 0.79 V. The conduction angle is 32.52°, slightly more than half a rectifier output segment. Each graph is identified by a number at its right hand end.

Graph 1 shows the waveform of the output voltage as a black line and that of the voltage behind the choke as a gray line. During conduction, the voltage behind the choke follows the sinusoidal input voltage less a small amount for the voltage drops in the line and transformer impedances and in the thyristors. The smoothing effect of the choke is still very apparent.

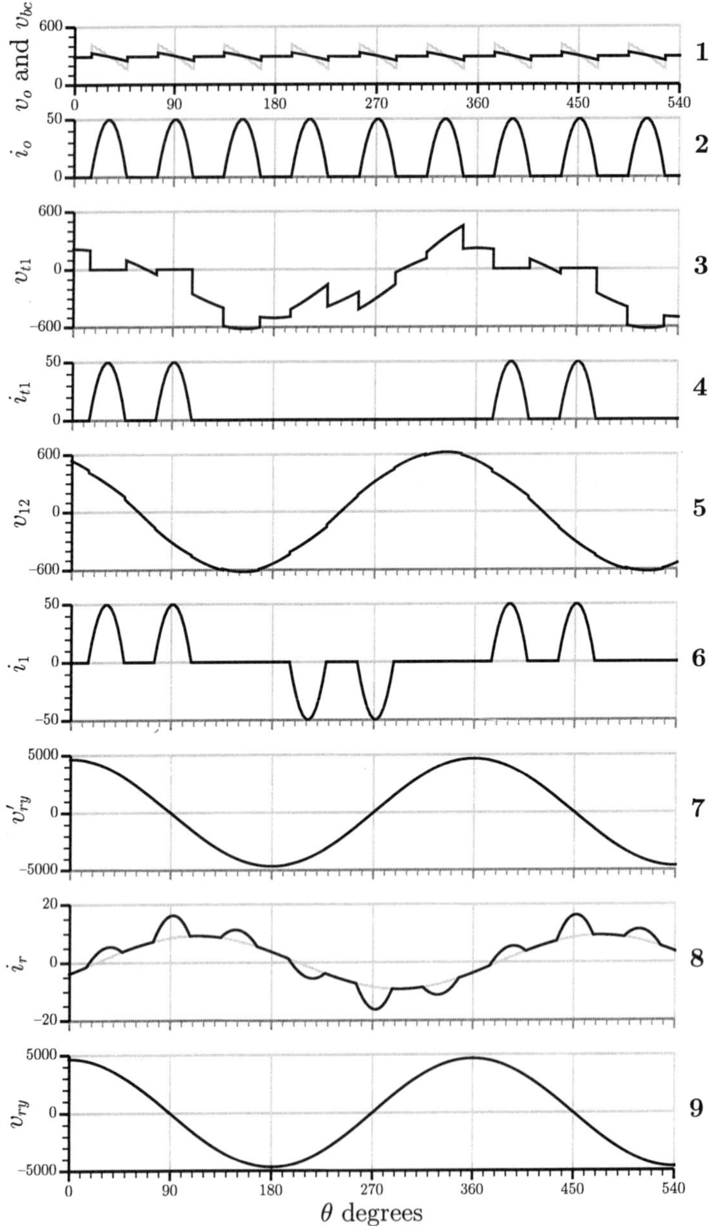

FIG. 6.5. Waveforms in the rectifier of the example operating in mode 2 with a delay angle of 75°, a load e.m.f. of 290 V, and a thyristor forward drop of 0.79 V. Currents are in amperes and voltages in volts.

During the cutoff period, which occupies 27.48°, the output voltage and the voltage behind the choke are both equal to the load e.m.f. of 290 V.

Graph 2 shows the output current as a series of pulses of amplitude 49.7 A, duration 32.52°, one pulse every 60°.

Graph 3 shows the voltage across thyristor T1 and should be compared with Graph 3 of Fig. 5.12. When the thyristor conducts, the voltage across it is the small forward voltage drop which, for these calculations, has been taken as 0.79 V. When T2 conducts, the voltage across T1 is $v_1 - v_2$ and when T3 conducts it is $v_1 - v_3$. During cutoff, the voltage across a thyristor is determined by the blocking impedances of all the thyristors in the bridge and a rigorous analysis results in extremely complex expressions for the thyristor voltages. However, for the special case of equal blocking impedances, the expressions are enormously simplified, the one for thyristor T1 being

$$v_{t1} = v_1 - \frac{E_o}{2} \tag{6.55}$$

Expression 6.55 has been used in determining the waveform of graph 3. At first glance it is surprising to find that the thyristor is forward biased during part of the cutoff subsegment number 2. At the start of this segment $\theta = 47.52°$ and $v_{t1} = 97.82$ V. The thyristor voltage has decreased to zero at $\theta = 66.18°$ and is -52.09 V at the end of this subsegment, $\theta = 75°$. The difficulty is resolved by looking at the voltage across the complementary thyristor, T6, in the other half of the bridge. The voltage across this, assuming equal blocking impedances, is

$$v_{t6} = -v_3 - \frac{E_o}{2} \tag{6.56}$$

At $\theta = 47.52°$, $v_{t6} = -495.51$ V and at $\theta = 75°$, $v_{t6} = -491.76$ V. Thus, even though one of a pair of thyristors may be forward biased, the other is heavily reverse biased and the pair cannot conduct.

Graph 4 shows the current in thyristor T1. There are two identical pulses per supply cycle, the first when the thyristor is fired, in this case at $\theta = 15°$, and the next 60° later when the next thyristor in the other half of the bridge is fired, in this case T6 at 75°. The pulses are identical in shape to the pulses which make up the output current. (This assumes that the thyristor is still in the conducting state when the next incoming thyristor is fired. This does not naturally occur and must be ensured by the gating circuit, a matter reviewed later in Section 6.16)

Graph 5 shows the waveform of the voltage between the rectifier input lines 1 and 2. Although the effects of current flow in the system and transformer impedances can be distinguished, these effects are small and the waveform is essentially sinusoidal. The most notable difference from the continuous conduction case, Graph 5 of Fig. 5.12, is the absence of commutation notches.

Graph 6 shows the waveform of the current in transformer secondary number 1. Since this supplies current to one thyristor, T1, and receives current from another, T4, the current is composed of four pulses per supply cycle, a positive pair when T1 conducts and a negative pair when T4 conducts. Each pulse is identical in shape to an output current pulse.

Graph 7 shows the waveform of the voltage across the red–yellow transformer primary winding. The voltage absorbed by the line impedances is too small to be distinguished and the voltage is essentially the sinusoidal voltage between the red and yellow bus bars.

Graph 8 shows, by the black line, the current in the red supply line and, by the gray line, its transformer magnetizing current component. The line current has six pulses per supply cycle superimposed on the transformer magnetizing current. The pulses are equally spaced at 60° intervals and the center one of a group of three is of double amplitude as indicated in Tables 6.1 through 6.3.

Graph 9 shows the sinusoidal voltage between the red and yellow bus bars and provides a reference for all the other graphs.

6.12 The operating point

The operating point, I_o, V_o, can now be found. The mean output current, I_o, is found by integrating and averaging the output current over the output subsegment. For modes 1-2 and 2, when there is only one conducting period in an output subsegment,

$$I_o = \frac{3}{\pi}\int_{\delta}^{\gamma} i_o \mathrm{d}\theta$$

i.e.

$$I_o = \frac{3}{\pi}\Big\{I_8[\sin(\gamma+\phi_{c8}) - \sin(\delta+\phi_{c8})] - I_{13}(\gamma-\delta) - \tan(\zeta_3)\, I_{14}[\exp(-\frac{\gamma}{\tan(\zeta_3)}) - \exp(-\frac{\delta}{\tan(\zeta_3)})]\Big\} \tag{6.57}$$

For mode 2, the initiation angle δ is $\alpha+\pi/3$.

For modes 1-1 and 3, there are two periods of conduction in each output segment from the trigger point, $\alpha+\pi/3$, to the extinction point, γ, and from the initiation point, δ, to the end of the output subsegment, $\alpha+2\pi/3$. It follows from this that

Table 6.4 *Operating point data for Fig. 6.4*

Mode	Delay degrees	E_o V	V_{thy} V	I_o A	V_o V
1-1	0.0	595.00	1.32	8.63	595.54
1-2	0.0	610.00	0.58	1.71	610.11
2	70.0	280.00	1.09	42.52	282.64
3	175.0	−580.00	1.30	5.13	−579.68

$$I_o = \frac{3}{\pi}\left\{\int_{\alpha+\pi/3}^{\gamma} i_o \mathrm{d}\theta + \int_{\delta}^{\alpha+2\pi/3} i_o \mathrm{d}\theta\right\} \tag{6.58}$$

Each of these two integrals produces an expression similar to that of eqn 6.57 but with the appropriate value of I_{14} being used in each case.

The mean output voltage is found from the mean output current by using the following relationship between the d.c. components

$$V_o = E_o + R_o I_o \tag{6.59}$$

The operating point data for the four conditions which are shown in Fig. 6.4 are summarized in Table 6.4.

Example 6.10

Determine the operating point data for the example when the delay angle is 75° and the load e.m.f. is 290 V.

It has already been found that the output current during subsegment 5 is

$$i_o = 1386.75\cos(\theta - 2.9109) - 2843.85 + 2895.91\exp(-\frac{\theta}{4.2572})$$

Conduction begins at $\theta = 135°$ and ends at $\theta = 167.52°$. Thus, for the mode 2 type operation eqn 6.57, with $\delta = 135° = 2.3562$ rad and $\gamma = 167.52° = 2.9238$ rad, is used to obtain the mean output current as $I_o = 18.03$ A. This current, the load e.m.f. of 290 V and the load resistance of 0.062 Ω inserted into eqn 6.59 give the mean output voltage as $V_o =$ 291.12 V.

****

6.13 The output characteristics

Output characteristics, which show the mean output voltage as a function of the mean output current for fixed delay angles, are generally used to display the discontinuous conduction data. Such characteristics have already been used in Figs 5.7 and 5.14. The latter figure demonstrated the accuracy of the Thévenin equivalent circuit for the continuous conduction mode. It was seen that, with very little error, the continuous conduction output characteristics can be represented by parallel straight lines, having the equation

$$V_o = V_{om}\cos(\alpha) - 2V_{thy} - R_{rt}I_o \tag{6.60}$$

It has been seen that this expression is only true down to the critical output current, I_{ocrit}, defined by eqn 6.18. For output currents below the critical value, conduction is discontinuous and the shape of the output characteristic which spans the gap between the critical point and the cutoff point is defined by the pair of parametric eqns 6.57 and 6.59, with eqn 6.58 used where necessary.

The output characteristics for the example, modified to include the effects of discontinuous conduction, are shown in Fig. 6.6. Eighteen characteristics covering the delay angle range of zero to 170° in 10° steps are shown. Alternate characteristics are identified by the delay angle at their right. The locus of the critical point is shown by a gray line and the point where cutoff occurs by a dot. In comparing Fig. 6.6 with Fig. 5.6 it should be noted that, in order to show the discontinuous conduction phenomena to maximum effect, the horizontal axis has been expanded to cover only the range zero to 200 A. It will be seen that the critical current reaches a maximum, I_{ocm}, equal to 126.51 A for the example, at a delay angle of 90° and that the voltage rise from the critical point to the cutoff point is also a maximum at this delay. Substituting $\alpha = \pi/2$ in eqn 6.57

$$I_{ocm} = -I_{13crit} \tag{6.61}$$

and

$$V_{co} = \frac{V_{st}}{\sqrt{2}} = \frac{\pi}{6}V_{om} = 0.5236V_{om} \tag{6.62}$$

6.14 The voltage rise at light load

The rise in voltage at light loads can present serious problems. As an example, if the load is the armature of a d.c. motor, the motor speed will be more or less proportional to the mean output voltage and there will therefore be a substantial rise in speed at light loads. Taking an extreme

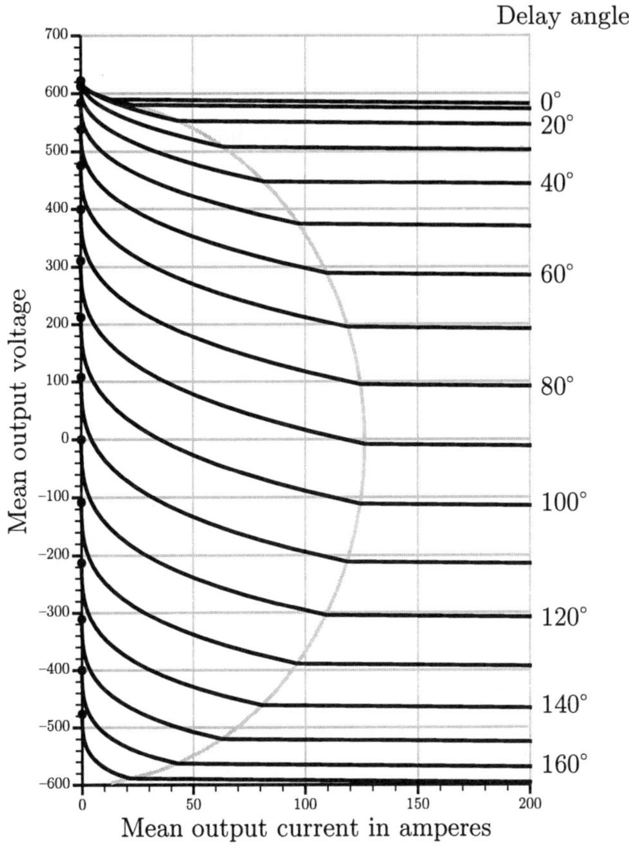

FIG. 6.6. Load characteristics of a rectifier at light load. The load characteristics are shown by the black line. The boundary between continuous and discontinuous conduction is shown by the gray line. The cut off points are indicated by black dots on the voltage axis.

example, if the motor is operating against a substantial load torque and the delay is 90°, the motor speed will be almost zero. If now load torque is entirely removed from the motor, the speed will rise and approach half the maximum speed, disconcerting at the least and potentially dangerous.

The critical current can be reduced by increasing the circuit inductance. However, this is a palliative, not a cure, which introduces new problems in that the system speed of response falls. While systems vary widely, an approximate idea of the scale of this phenomenon in practice is obtained by considering the critical current level to be about one-tenth of the rated current. A satisfactory cure for the problem is provided by the use of a feedback system to control the speed. This topic is considered in somewhat more detail in Chapter 10.

In order to illustrate all the various phenomena, Fig. 6.6 has been drawn showing delay angles up to 170°. It has been seen that mode 3 can only occur for delay angles greater than about 170° so that this mode does not occur in the figure. Again the warning is necessary that it is most imprudent to approach so near to the absolute delay limit because of the danger of loss of control so that mode 3 is in reality extremely rare.

6.15 The thyristor voltage during cutoff

When a thyristor is conducting, the voltage across it is the small, positive conducting voltage drop.

When the thyristor is blocking and another thyristor is conducting, the voltage across it is the appropriate line to line voltage.

When all the thyristors are blocking, the situation is not well defined because the circuit impedances under these circumstances, such as the impedances presented by thyristors and the stray capacitances, are not known. However, assuming that the blocking impedances are equal, the following simple expressions for the thyristor voltages are obtained.

$$v_{t1} = v_1 - E_o/2 \tag{6.63}$$

$$v_{t2} = v_2 - E_o/2 \tag{6.64}$$

$$v_{t3} = v_3 - E_o/2 \tag{6.65}$$

$$v_{t4} = -v_1 - E_o/2 \tag{6.66}$$

$$v_{t5} = -v_2 - E_o/2 \tag{6.67}$$

$$v_{t6} = -v_3 - E_o/2 \tag{6.68}$$

The voltage across thyristor T1 during the cutoff periods shown in Graph 4 of Fig. 6.5, has been computed using eqn 6.63.

6.16 Narrow gate pulses

The foregoing analysis has been based on the assumption of wide firing pulses. It has already been seen that successful operation of a bridge rectifier requires either double pulsing or pulses at least 60° wide. Operation in mode 1 requires pulses at least 90° wide and operation in mode 3 requires pulses at least 120° wide. Wide firing pulses are therefore standard. However, it is interesting to note the effects of narrow firing pulses.

If double, narrow, firing pulses are employed, operation in mode 2 will be as described here. However, operation in modes 1 and 3 is not possible since the thyristors will typically have recovered by the time the initiation point, at $\theta = \delta$, is reached and will therefore block further conduction. Subsequent action depends on the nature of the load. Assuming it to be a separately excited d.c. motor, it will go into a limit cycle. Instead of operating in mode 1, power flow to the motor is cut off. Its speed falls and with it, its e.m.f. At some point the e.m.f. becomes low enough for mode 2 type conduction to begin and the machine is accelerated back to a speed at which it is again cut off. The motor speed therefore oscillates over a small range, 1 or 2% of rated speed, at a low frequency depending upon the load inertia.

7

CAPACITOR SMOOTHING

7.1 Introduction

Hitherto, smoothing of the rectifier output current has been considered to be achieved by a combination of the load inductance and a smoothing inductor. This is an effective technique up to the highest load levels with the relatively minor drawback of discontinuous conduction at light loads and the accompanying rise in output voltage. An alternative at lower load levels, say below 100 kW, is provided by capacitor smoothing for which the inductor in series with the load is replaced by a capacitor in parallel with it. Like inductor smoothing, this is a most effective technique and, like inductor smoothing, it has disadvantages. The rectifier is normally in the discontinuous conduction mode, the mean output voltage varies with changes in load, and power is delivered from the a.c. system in brief current pulses whose amplitude is much greater than the d.c. current being delivered to the load.

The capacitance needed for effective smoothing is so large that it can only be economically obtained by the use of electrolytic capacitors whose polarity requirement prevents reversal of the rectifier output voltage and thereby prohibits reversal of power flow. The rectifier output voltage can be controlled down to zero by varying the delay angle, but typically capacitor smoothing is employed with uncontrolled rectifiers, some other technique such as a chopper being employed for voltage control.

In its physical realization, the capacitor smoothed rectifier is a very simple device but from a theoretical viewpoint it is highly complex, being described by differential equations of higher order than inductor smoothing and with many more possible modes of operation. Indeed a whole book could be devoted to it. Here the treatment is generalized as much as possible, consistent with keeping within the bounds of a single chapter.

The rectifier model already developed in Chapter 5 and used in Chapter 6, Fig. 6.2 will be used. The equivalent circuit is shown in Fig. 7.1(b). Like Fig. 6.2, it differs from Fig. 5.8 by the omission of the voltage source V_{xst}, which represents the voltage absorbed by the a.c. side inductance during commutation, an omission which tacitly recognizes that conduction is normally discontinuous so that the commutation process associated with continuous conduction does not occur. The smoothing capacitor, C, is connected across the d.c. output terminals in parallel with the load. For generality the series inductor, R_{ch} and L_{ch}, is retained but its function is changed from smoothing the output current to controlling the capacitor charging current and it is as a consequence much smaller. Indeed, if the a.c. side inductance is sufficient, the series inductor is not required.

Again for generality a controlled rectifier is considered although capacitor smoothing is usually employed with uncontrolled rectifiers. The exploration of performance is also carried into the negative output region, an operating area normally prohibited because of the use of an electrolytic smoothing capacitor which cannot sustain reversed voltage.

The analytical results are illustrated by reference to a new example with a lower rated power than the one used for Chapters 5 and 6. As the occasion requires, three variants of the example will be examined, a low supply impedance case, a high supply impedance case, and a high smoothing capacitance case. In order that the low supply impedance case be plausible, a circuit without a transformer will be considered. This is illustrated by Fig. 7.1(a).

The rectifier Thévenin voltage is defined by eqns 5.99 and 5.100. It comprises 60° segments of the a.c. input voltages. In the steady state, during part of a voltage segment the rectifier is forward biased and current flows into the smoothing capacitor and the load, the capacitor component increasing the capacitor voltage. This will be called the *charging period.*

During other portions of a rectifier voltage segment the capacitor voltage is sufficient to reverse bias the rectifier, isolating the load from the a.c. supply. The load current is then supplied solely by the capacitor. This will be called the *discharge period.*

As in Chapter 6 the value of θ at which conduction ceases will be called the *extinction angle.*

As in Chapters 5 and 6, attention is concentrated on the segment from $\alpha + 60°$ to $\alpha + 120°$ when thyristors T2 and T6 would be expected to conduct. During this segment the rectifier Thévenin voltage is

$$v_{rt} = \sqrt{2}V_s \cos(\theta - \frac{\pi}{2}) = \sqrt{2}V_s \sin(\theta) \tag{7.1}$$

where the voltage V_s is the r.m.s. value of the a.c. system line to line voltage.

Figure 7.1 identifies the various currents and voltages which will be needed for the analysis. In Fig. 7.1(a) are shown the infinite bus voltages, v_r, v_y, and v_b, the voltages at the rectifier input terminals, v'_r, v'_y, and v'_b, the a.c. line currents, i_r, i_y, and i_b, and the voltage across thyristor T2, v_{t2} and the current through it, i_{t2}. In Fig. 7.1(b) are shown the rectifier output voltage, v_o, which is the same as the capacitor voltage, v_c, the voltage behind the choke, v_{bc}, the rectifier current, i_{rec}, the output current to the load, i_o, and the capacitor charging current, i_c.

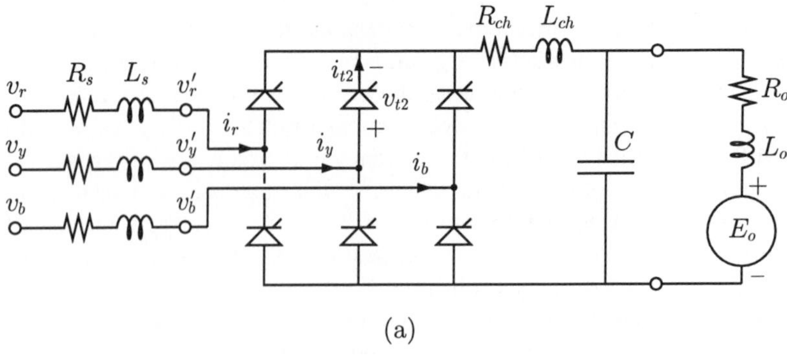

(a)

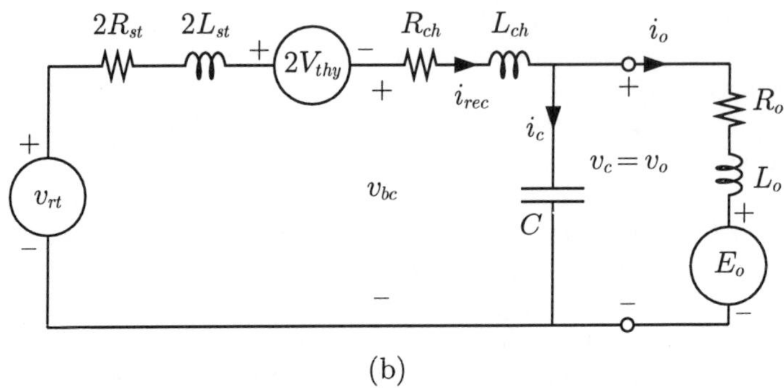

(b)

FIG. 7.1. A three phase bridge rectifier (a) with capacitor smoothing and (b) its equivalent circuit.

The combination of more complex mathematics and numerous alternative modes of behavior make this one of the most difficult chapters in the book. This is unavoidable; while very popular in practice, the workings of the capacitor smoothed rectifier are analytically complex. A first reading could concentrate on the principles involved and avoid the analytical detail. In particular, familiarity with the numerous waveforms given in the figures is most helpful to understanding the workings of this rectifier.

7.2 The example

For the example it will be assumed that the rectifier is directly connected to a balanced three phase, 60 Hz system of sinusoidal waveform and with an equivalent infinite busbar line to line voltage of 440 V r.m.s. Two alternative supply systems are considered, one of low impedance and the other of high impedance. The low impedance supply has a line resistance, R_s, of 0.002 Ω and a line inductance, L_s, of 16 μH. This might be the case when the rectifier load is a small part of the total load on the supply system. The high impedance supply has ten times the impedance, $R_s = 0.02\ \Omega$ and $L_s = 160\ \mu$H. This might be the case when the rectifier is connected to the supply via its own transformer.

Each rectifier element, be it a thyristor or diode, will be assumed to absorb 1.6 V when conducting.

An output inductor is included in the analysis for completeness. However, for all examples its impedance is assumed to be zero, i.e. it is assumed that there is no inductor.

Two smoothing capacitors are considered. One is somewhat on the low side for the load and the other is large. The low capacitance is 4000 μF and the high capacitance is ten times this value, 40 000 μF. The capacitor losses are assumed to be negligible.

The load is active and inductive, e.g. the armature of a d.c. machine. It is rated at 180 A, 100 kW giving a rated load voltage of 555.5 V, about 10% below the peak value of the a.c. line to line voltage. The load resistance, R_o, is 0.17 Ω and the load inductance, L_o, is 1.4 mH. Thus, at rated current the load resistance absorbs 30.6 V, 5.5% of the rated voltage.

The duration of a rectifier segment is 1/360 s and the small capacitor can maintain rated load current for this period with a voltage drop of 125 V. Under the same circumstances the large capacitor would discharge by one tenth of this amount, 12.5 V.

7.3 Operating modes

Unfortunately the expressions describing the behavior of the rectifier are too complicated to permit direct identification of the steady state. The analysis must start from a plausible operating point and progress through a series of transient segments until the steady state is reached. Under these circumstances the rectifier can operate in a variety of modes, a fact which complicates computer programming. For the analysis, the seven types of rectifier segment illustrated in Fig. 7.2 must be recognized. This figure shows by vertical lines the extent of a rectifier segment starting at θ_{st} and finishing 60° later at θ_{fn}, these angles being indicated at the top and bottom of the lines. A rectifier segment is indicated by a horizontal line of which there are seven identified at the left as types 1 through 7. The eighth and lowest line is identified as the basic segment, the one from which all the others can be derived.

During a segment the rectifier can either be conducting, in which case it is supplying energy to the load from the a.c. system, or it can be cut off, in which case connection between the a.c. and d.c. systems is severed and the energy consumed by the load is supplied by the capacitor whose voltage as a consequence diminishes. Those portions of a rectifier segment

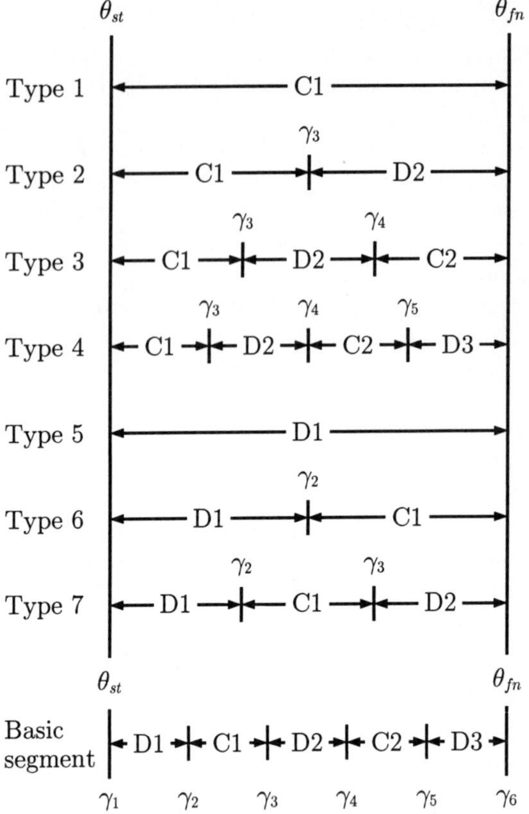

FIG. 7.2. The seven operating modes and the basic segment.

when it is conducting are identified by a C for *charging*, and those portions when it is cut off by a D for *discharging*.

For computing purposes it is convenient to consider the seven types of segment shown in Fig. 7.2 as particular examples of the basic segment shown at the bottom of the figure. This has five subsegments, three discharging and two charging. The first subsegment is of the discharging type and will be called D1. This is followed by alternate charging and discharging type subsegments, C1, D2, C2, and D3. The starting and finishing points of the subsegment are identified by an array of six angles, γ_1 through γ_6, so that D1 extends from γ_1 to γ_2, C1 from γ_2 to γ_3, etc.

A type 1 segment has only one subsegment, C1, extending from the start of the segment at γ_2 to the end at γ_3. This mode is rare in the steady state. It occurs only for the high supply impedance example at small delay angles and is illustrated by Fig. 7.3 which shows key voltage and current

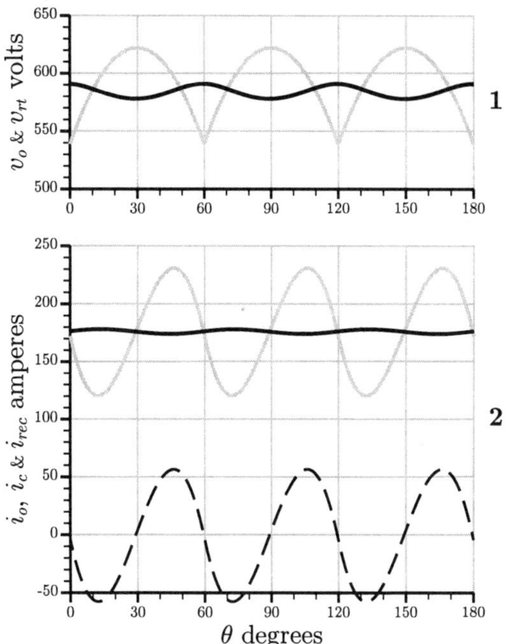

FIG. 7.3. Steady state waveforms of the output voltage, output current, capacitor current, and rectifier current for the large supply inductance case at zero delay. Graph 1 shows v_o by a black line and v_{rt} by a gray line. Graph 2 shows i_o by a black line, i_c by a dashed line, and i_{rec} by a gray line.

waveforms at zero delay. Graph 1 shows by a black line the rectifier output voltage v_o and by a gray line the rectifier Thévenin voltage. Graph 2 shows by a black line the current delivered to the load, i_o, by a gray line the current delivered to the capacitor, i_c, and by a dashed line the sum of these two currents, i_{rec}, the current delivered by the rectifier. The figure is drawn for a delay angle of 0° and a load e.m.f. of 554 V. Under these circumstances the mean output current is 176.2 A and the mean output voltage is 584.0 V.

A type 2 segment has two subsegments, C1 extending from γ_2 to γ_3 and D2 extending from γ_3 to γ_4. This is the steady state type for a rectifier operating at a delay angle of more than about 10° and is illustrated by Fig. 7.4 which is drawn for the low supply impedance, low capacitance case with a delay angle of 45° and a load e.m.f. of 524 V. During the segment from 105° to 165° the extinction angle is 125.74°, the mean output current is 178.0 A, and the mean output voltage is 554.3 V.

A type 3 segment has three subsegments, C1 from γ_2 to γ_3, D2 from γ_3

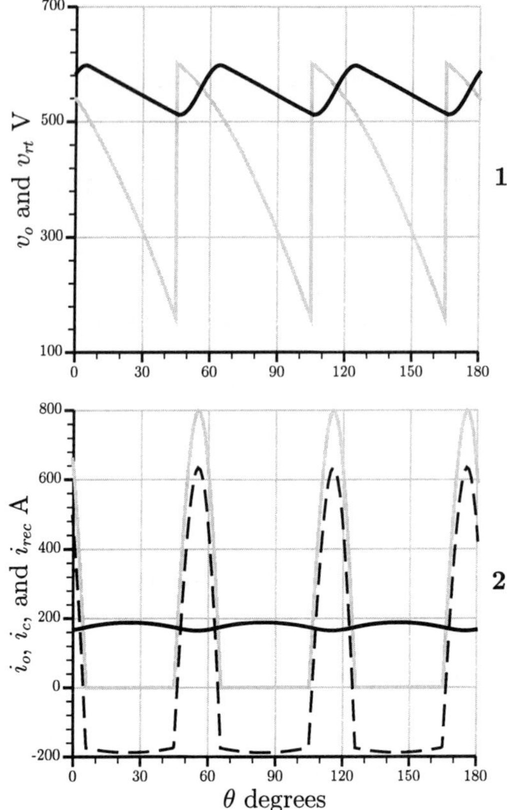

FIG. 7.4. Steady state waveforms of the output voltage, output current, capacitor current, and rectifier current for the low supply impedance case at 45° delay. Graph 1 shows v_o by a black line and v_{rt} by a gray line. Graph 2 shows i_o by a black line, i_c by a dashed line, and i_{rec} by a gray line.

to γ_4, and C2 from γ_4 to γ_5. In the steady state this type of operation is rare.

A type 4 segment has four subsegments, C1 from γ_2 to γ_3, D2 from γ_3 to γ_4, C2 from γ_4 to γ_5, and D3 from γ_5 to γ_6. This is a transient type of operation and cannot occur during steady state operation.

A type 5 segment has only one subsegment, D1 from γ_1 to γ_2. This type only occurs under transient conditions.

A type 6 segment has two subsegments, D1 from γ_1 to γ_2 and C1 from γ_2 to γ_3. This is usually a transient type of operation but it can occur in the steady state when the delay angle is less than 30°.

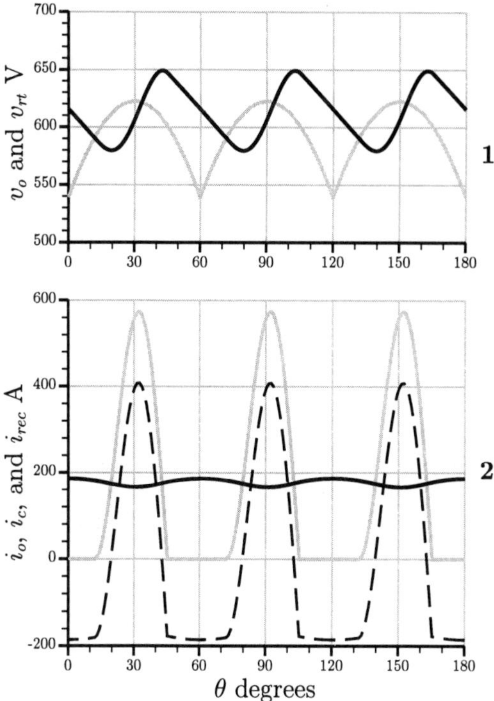

FIG. 7.5. Steady state waveforms of the output voltage, output current, capacitor current, and rectifier current for the low supply impedance case at zero delay. Graph 1 shows v_o by a black line and v_{rt} by a gray line. Graph 2 shows i_o by a black line, i_c by a dashed line, and i_{rec} by a gray line.

A type 7 segment has three subsegments, D1 from γ_1 to γ_2, C1 from γ_2 to γ_3, and D2 from γ_3 to γ_4. This is the normal type of steady state operation when the delay angle is small, below about 10°. It is illustrated for the low supply impedance, low capacitance case by Fig. 7.5 which is drawn for a delay angle of 0° and a load e.m.f. of 583 V. Under these conditions, for the segment from 60° to 120° the re-entry angle is 72.24°, the extinction angle is 105.33°, the mean output current is 176.8 A, and the mean output voltage is 613.1 V.

7.4 The charging mode

Referring to Fig. 7.1 and examining the rectifier segment from $\alpha + 60°$ to $\alpha + 120°$, during the charging period, thyristors T2 and T6 are forward biased and conduct. Using the output current, i_o, the rectifier current,

i_{rec}, and the capacitor voltage, v_c, as state variables, the state variable relationships of eqn 7.2 are obtained by applying Kirchhoff's Voltage Law to the load and the supply branches of the equivalent circuit and Kirchhoff's Current Law to a capacitor terminal of the equivalent circuit

$$\begin{bmatrix} X_o & & \\ & X_r & \\ & & \frac{1}{X_c} \end{bmatrix} \frac{d}{dt} \begin{bmatrix} i_o \\ i_{rec} \\ v_c \end{bmatrix} + \begin{bmatrix} R_o & & -1 \\ & R_r & 1 \\ 1 & -1 & \end{bmatrix} \begin{bmatrix} i_o \\ i_{rec} \\ v_c \end{bmatrix} = \begin{bmatrix} -E_o \\ v_{rt} - 2V_{thy} \\ 0 \end{bmatrix} \tag{7.2}$$

where, for convenience, the equation is expressed in terms of the angle θ rather than time using the relationship, $\theta = \omega_s t$, so that inductances are replaced by reactances at the supply frequency. The resistance, R_r, and the reactance, X_r, are given by

$$R_r = 2R_s + R_{ch} \tag{7.3}$$

and

$$X_r = 2X_s + X_{ch} \tag{7.4}$$

The analysis can progress from this point by several alternative paths. By elimination, a single, third order equation in one of the state variables could be obtained whose solution, by back substitution, would provide expressions for the other two state variables. This is a tedious process involving a great deal of algebra and requiring, for the natural frequencies, the solution of a cubic equation. Alternatively the parameter matrices of the state equation could be handed to a computer to obtain directly the eigenvalues and eigenvectors of the system. A middle of the road approach will be adopted which owes much to the eigenvector method but keeps more closely in touch with the problem.

The solution to the state matrix equation 7.2, is the sum of the steady state responses to the driving functions of its right-hand side and the natural or transient behavior. The driving functions are simple, a sinusoidal a.c. voltage, v_{rt}, and two d.c. driving functions, E_o and $2V_{thy}$. The response to v_{rt} is obtained by the application of a.c. circuit theory and the response to E_o and $2V_{thy}$ by the application of d.c. circuit theory. The analysis starts with the simplest part of the solution, the steady state d.c. response, and progress through the a.c. response to the transient response. The solution for the output current will be obtained and expressions for i_{rec} and v_c will be derived from it.

Examination of the equivalent circuit of Fig. 7.1(b) shows that the d.c. component of i_o is

$$I_{co1} = -\frac{E_o + 2V_{thy}}{R_o + R_r} \tag{7.5}$$

The three characters of the subscript to the current, co1, indicate in reverse order that this is the first coefficient for the output current expression during the charging mode.

The a.c. response is similarly given by the phasor equation

$$\mathbf{I}_{co2} = \frac{\mathbf{V}_{rt}\mathbf{Z}_c}{\mathbf{Z}_c\mathbf{Z}_o + \mathbf{Z}_o\mathbf{Z}_r + \mathbf{Z}_r\mathbf{Z}_c} \tag{7.6}$$

where

- $\mathbf{V}_{rt} = \sqrt{2}V_s\underline{/-\pi/2}$ is the phasor form of the rectifier Thévenin voltage, v_{rt}, during this segment
- $\mathbf{Z}_c$ is the capacitor impedance, $-jX_c$
- $\mathbf{Z}_o$ is the load impedance, $R_o + jX_o$
- $\mathbf{Z}_r$ is the equivalent rectifier impedance, $R_s + R_{ch} + j(2X_s + X_{ch})$
- $\mathbf{I}_{co2}$ is the phasor, magnitude I_{co2}, phase ϕ_{co2}, representing the output current produced by $\mathbf{V}_{rt}$.

The time domain expression for this component of the output current is

$$i_{o2} = I_{co2}\cos(\theta + \phi_{co2}) \tag{7.7}$$

where both the amplitude, I_{co2}, and phase, ϕ_{co2} are readily obtained from eqn 7.6 for any specific case.

Since there are three state equations, the system is of third order and has three natural frequencies which are determined numerically as either the eigenvalues of the state equation or the roots of the characteristic equation. In general, one of these techniques would have to be employed but in this case the concepts of modal analysis provide a direct analytical solution. Figure 7.6 shows the rectifier equivalent circuit with the driving functions removed and therefore subject only to its intrinsic natural behavior. It is seen that there are only two possible modes of behavior, the *series mode* in which the current i_3 circulates round the outer loop formed by the series combination of Z_o and Z_r, and the *parallel mode* in which the current i_4 circulates between the capacitor and the parallel combination of the impedance Z_o and the impedance Z_r.

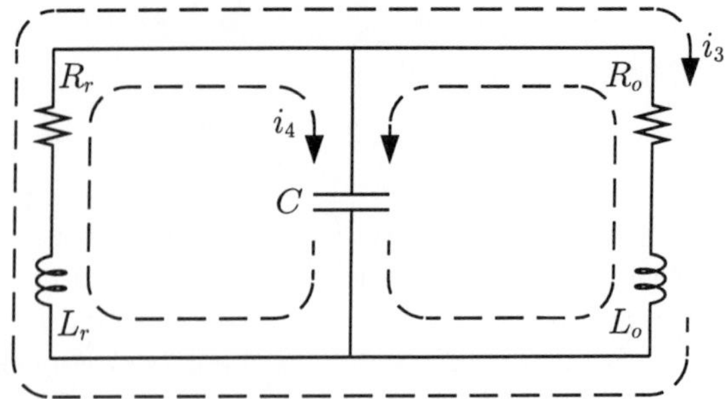

FIG. 7.6. The series and parallel modal circuits.

The series mode is of first order with natural frequency

$$s_{c3} = -\frac{R_o + R_r}{X_o + X_r} \tag{7.8}$$

If $Z_p = R_p + sX_p$ represents the parallel combination of Z_o and Z_r, the parallel mode represents the circulation of current between the capacitor and the inductive impedance, Z_p, and is therefore of second order with two natural frequencies, s_{c4} and s_{c5}, given by the solution of the quadratic equation

$$X_p s^2 + R_p s + X_c = 0 \tag{7.9}$$

i.e.

$$s_{c4} = -\frac{R_p}{2X_p} + \sqrt{\left(\frac{R_p}{2X_p}\right)^2 - \frac{X_c}{X_p}} \tag{7.10}$$

and

$$s_{c5} = -\frac{R_p}{2X_p} - \sqrt{\left(\frac{R_p}{2X_p}\right)^2 - \frac{X_c}{X_p}} \tag{7.11}$$

Depending on the system damping, eqn 7.9 can have two separate real roots, two equal real roots or a pair of conjugate complex roots. Because these are high power circuits which, for reasons of efficiency must be lightly damped, and because the capacitance must be large to have a significant smoothing effect, it is highly probable that the system will be underdamped. It will be assumed that this is the case so that the root s_{c5} is the conjugate of s_{c4}, and root s_{c4} has a real part s_{c4r} and an imaginary part s_{c4i} given by

$$s_{c4r} = -\frac{R_p}{2X_p} \tag{7.12}$$

and

$$s_{c4i} = \sqrt{\frac{X_c}{X_p} - \left(\frac{R_p}{2X_p}\right)^2} \tag{7.13}$$

so that

$$s_{c4} = s_{c4r} + js_{c4i} \tag{7.14}$$

The expression for the output current during the charging mode then has the form

$$\begin{aligned} i_o = I_{co1} + I_{co2}\cos(\theta + \phi_{co2}) + I_{co3}\exp(s_{c3}\theta) + \\ I_{co4}\cos(s_{c4i}\theta + \phi_{co4})\exp(s_{c4r}\theta) \end{aligned} \tag{7.15}$$

where the amplitude I_{co1} is given by eqn 7.5, the amplitude I_{co2} and phase ϕ_{co2} are given by eqn 7.6, the exponent s_{c3} is given by eqn 7.8, the frequency s_{c4i} is given by eqn 7.13, and the exponent s_{c4r} is given by eqn 7.12. The amplitudes I_{co3} and I_{co4} and the phase ϕ_{co4} depend on the initial conditions.

All the expressions encountered in this chapter are variants on the expression on the right-hand side of eqn 7.15. The expression is more complex than the generalized power electronics expression which was developed in Chapter 5 and it will be treated in a similar way, calling it the *basic capacitor smoothing expression.* A number of useful functions of this expression are given in Appendix B.

Example 7.1

Determine the generalized capacitor smoothing expression for the output current during the charging mode for the low supply impedance, low smoothing capacitance example when the load e.m.f. is 583 V—the e.m.f. used for Fig. 7.5.

For a load e.m.f. of 583 V, the amplitude of the current I_{co1} is given by eqn 7.5 as

$$I_{co1} = -586.2/0.174 = -3368.97 \text{ A}$$

The magnitude and phase, I_{co2} and ϕ_{co2}, are obtained from eqn 7.6 as

$$I_{co2} = 1116.91 \text{ A} \qquad \text{and} \qquad \phi_{co2} = -2.83579 \text{ rad.}$$

The exponent s_{c3} is obtained from eqn 7.8 as

$$s_{c3} = -0.322311$$

The frequency s_{c4i} and exponent s_{c4r} are obtained from eqns 7.12 and 7.13 as

$$s_{c4i} = 7.49611 \quad \text{and} \quad s_{c4r} = 0.165680$$

It must be remembered that the frequency s_{c4i} is relative to the a.c. supply frequency and is not an absolute value. Thus, in this case the frequency of the transient component is 7.50 times the a.c. supply frequency, 449.8 Hz or 2826 rad s^{-1}.

With these values, during the charging mode for the segment from $\alpha +$ $60°$ to $\alpha + 120°$, the expression for the output current is

$$i_o = -3368.97 + 1116.91\cos(\theta - 2.83579) + I_{co3}\exp(-0.322311\theta) + I_{co4}\cos(7.49661\theta + \phi_{co4})\exp(-0.165680\theta)$$

****

7.4.1 *The other circuit variables*

Having determined the output current, attention must be turned to the other state variables, the capacitor voltage, v_c, and the rectifier current, i_{rec}. Also, it will be convenient to have two additional variables derived from the state variables, the capacitor current, i_c, and the rate of change of the rectifier current. The capacitor voltage, which is also the rectifier output voltage, is the sum of the load e.m.f. and the voltage absorbed by the load impedance

$$v_c = v_o = E_o + R_o i_o + X_o \frac{di_o}{d\theta} \tag{7.16}$$

The capacitor current is equal to the rate of change of capacitor voltage with respect to θ divided by the reactance of the capacitor at the supply frequency

$$i_c = \frac{1}{X_c}\frac{dv_c}{d\theta} \tag{7.17}$$

The rectifier current is the sum of the output and capacitor currents

$$i_{rec} = i_c + i_o \tag{7.18}$$

Complicated as it may seem, expressions are required for these variables in terms of the basic capacitor smoothing expression for the output current. To facilitate their derivation, the Cartesian form of the basic expression must be created by eliminating the phase angles, ϕ_{co2} and ϕ_{co4}. This is done by expanding the cosine terms of eqn 7.15 to obtain

$$i_o = I_{co1} + I_{co2c}\cos(\theta) + I_{co2s}\sin(\theta) + I_{co3}\exp(s_{c3}\theta) + \\ I_{co4c}\cos(s_{c4i}\theta)\exp(s_{c4r}\theta) + I_{co4s}\sin(s_{c4i}\theta)\exp(s_{c4r}\theta) \tag{7.19}$$

where

$$I_{co2c} = I_{co2}\cos(\phi_{co2}) \tag{7.20}$$
$$I_{co2s} = -I_{co2}\sin(\phi_{co2}) \tag{7.21}$$
$$I_{co4c} = I_{co4}\cos(\phi_{co4}) \tag{7.22}$$
$$I_{co4s} = -I_{co4}\sin(\phi_{co4}) \tag{7.23}$$

and the final subscript, c or s, in the symbol for an amplitude indicates a cosine or sine component, i.e. I_{co2s} is the amplitude of the sine component of the second term of the basic capacitor smoothing expression for the output current during the charging mode.

The capacitor voltage is obtained by substituting the expression just derived for output current into eqn 7.16

$$v_c = V_{cc1} + V_{cc2c}\cos(\theta) + V_{cc2s}\sin(\theta) + V_{cc3}\exp(s_{c3}\theta) + \\ V_{cc4c}\cos(s_{c4i}\theta)\exp(s_{c4r}\theta) + V_{cc4s}\sin(s_{c4i}\theta)\exp(s_{c4r}\theta) \tag{7.24}$$

where

$$V_{cc1} = E_o + R_o I_{co1} \tag{7.25}$$
$$V_{cc2c} = R_o I_{co2c} + X_o I_{co2s} \tag{7.26}$$
$$V_{cc2s} = R_o I_{co2s} - X_o I_{co2c} \tag{7.27}$$
$$V_{cc3} = (R_o + s_{c3}X_o)I_{co3} \tag{7.28}$$
$$V_{cc4c} = (R_o + s_{c4r}X_o)I_{co4c} + s_{c4i}X_o I_{co4s} \tag{7.29}$$
$$V_{cc4s} = (R_o + s_{c4r}X_o)I_{co4s} - s_{c4i}X_o I_{co4c} \tag{7.30}$$

The capacitor current is obtained from this expression for the capacitor voltage by the use of eqn 7.17

$$i_c = I_{cc2c}\cos(\theta) + I_{cc2s}\sin(\theta) + I_{cc3}\exp(s_{c3}\theta) + \\ I_{cc4c}\cos(s_{c4i}\theta)\exp(s_{c4r}\theta) + I_{cc4s}\sin(s_{c4i}\theta)\exp(s_{c4r}\theta) \tag{7.31}$$

where

$$I_{cc1} = 0 \tag{7.32}$$

$$I_{cc2c} = -\frac{X_o}{X_c} I_{co2c} + \frac{R_o}{X_c} I_{co2s} \tag{7.33}$$

$$I_{cc2s} = -\frac{X_o}{X_c} I_{co2s} - \frac{R_o}{X_c} I_{co2c} \tag{7.34}$$

$$I_{cc3} = s_{c3} \frac{R_o + s_{c3} X_o}{X_c} I_{co3} \tag{7.35}$$

$$\begin{aligned} I_{cc4c} = & \frac{s_{c4r} R_o - (s_{c4i}^2 - s_{c4r}^2) X_o}{X_c} I_{co4c} + \\ & s_{c4i} \frac{R_o + 2 s_{c4r} X_o}{X_c} I_{co4s} \end{aligned} \tag{7.36}$$

$$\begin{aligned} I_{cc4s} = & \frac{s_{c4r} R_o - (s_{c4i}^2 - s_{c4r}^2) X_o}{X_c} I_{co4s} - \\ & s_{c4i} \frac{R_o + 2 s_{c4r} X_o}{X_c} I_{co4c} \end{aligned} \tag{7.37}$$

The rectifier current is obtained by summing the output and capacitor currents as already noted in eqn 7.18. The result is

$$\begin{aligned} i_{rec} = & I_{cr1} + I_{cr2c} \cos(\theta) + I_{cr2s} \sin(\theta) + I_{cr3} \exp(s_{c3}\theta) + \\ & I_{cr4c} \cos(s_{c4i}\theta) \exp(s_{c4r}\theta) + I_{cr4s} \sin(s_{c4i}\theta) \exp(s_{c4r}\theta) \end{aligned} \tag{7.38}$$

where

$$I_{cr1} = I_{co1} \tag{7.39}$$

$$I_{cr2c} = \frac{X_c - X_o}{X_c} I_{co2c} + \frac{R_o}{X_c} I_{co2s} \tag{7.40}$$

$$I_{cr2s} = \frac{X_c - X_o}{X_c} I_{co2s} - \frac{R_o}{X_c} I_{co2c} \tag{7.41}$$

$$I_{cr3} = \frac{s_{c3} R_o + X_c + s_{c3}^2 X_o}{X_c} I_{co3} \tag{7.42}$$

$$\begin{aligned} I_{cr4c} = & \frac{s_{c4r} R_o + X_c - (s_{c4i}^2 - s_{c4r}^2) X_o}{X_c} I_{co4c} + \\ & \frac{s_{c4i}(R_o + 2 s_{c4r} X_o)}{X_c} I_{co4s} \end{aligned} \tag{7.43}$$

$$\begin{aligned} I_{cr4s} = & \frac{s_{c4r} R_o + X_c - (s_{c4i}^2 - s_{c4r}^2) X_o}{X_c} I_{co4s} - \\ & s_{c4i} \frac{R_o + 2 s_{c4r} X_o}{X_c} I_{co4c} \end{aligned} \tag{7.44}$$

The rate of change of rectifier current will be required when the various circuit voltages are derived and it is convenient to obtain it now by differentiating eqn 7.38

$$\frac{\mathrm{d}i_{rec}}{\mathrm{d}\theta} = I_{crp1} + I_{crp2c}\cos(\theta) + I_{crp2s}\sin(\theta) + I_{crp3}\exp(s_{c3}\theta) + I_{crp4c}\cos(s_{c4i}\theta)\exp(s_{c4r}\theta) + I_{crp4s}\sin(s_{c4i}\theta)\exp(s_{c4r}\theta) \tag{7.45}$$

where

$$I_{crp1} = 0 \tag{7.46}$$

$$I_{crp2c} = -\frac{R_o}{X_c}I_{co2c} - \frac{X_c - X_o}{X_c}I_{co2s} \tag{7.47}$$

$$I_{crp2s} = -\frac{R_o}{X_c}I_{co2s} + \frac{X_c - X_o}{X_c}I_{co2c} \tag{7.48}$$

$$I_{crp3} = \frac{s_{c3}(s_{c3}R_o + X_c + s_{c3}^2X_o)}{X_c}I_{co3} \tag{7.49}$$

$$I_{crp4c} = \frac{-({s_{c4i}}^2 - s_{c4r}^2)R_o + s_{c4r}X_c - s_{c4r}(3s_{c4i}^2 - s_{c4r}^2)X_o}{X_c}I_{co4c} + s_{c4i}\frac{2s_{c4r}R_o + X_c - (s_{c4i}^2 - 3s_{c4r}^2)X_o}{X_c}I_{co4s} \tag{7.50}$$

$$I_{crp4s} = \frac{-(s_{c4i}^2 - s_{c4r}^2)R_o + s_{c4r}X_c - s_{c4r}(3s_{c4i}^2 - s_{c4r}^2)X_o}{X_c}I_{co4s} - s_{c4i}\frac{2s_{c4r}R_o + X_c - (s_{c4i}^2 - 3s_{c4r}^2)X_o}{X_c}I_{co4c} \tag{7.51}$$

7.4.2 *Initial conditions and the charging mode*

Equations 7.19 through 7.51, provide expressions for all the rectifier variables during the charging mode in terms of the six parameters, I_{co1}, I_{co2c}, I_{co2s}, I_{co3}, I_{co4c}, and I_{co4s}. Of these, the first three are known, being defined by eqns 7.5, 7.6, 7.20, and 7.21. The remaining three, the amplitudes of the transient components, I_{co3}, I_{co4c}, and I_{co4s}, are determined by the initial conditions, the values of the state variables at the start of the charging mode.

It will be assumed that the charging mode commences at θ_{cst} with the output current equal to I_{cost}, the capacitor voltage equal to V_{ccst}, and the rectifier current equal to I_{crst}. Note that although the rectifier current is normally zero at the start of the charging mode, the possibility must be considered that it is not zero in order to be able to deal with the continuous conduction case, Fig. 7.2, type 1, and transient, type 3, behavior.

Strictly speaking, when the initial value of the rectifier current is not zero the effects of commutation should be included in the analysis. However, this adds very considerably to the difficulty of an already complex problem and is not justified by the small improvement in accuracy in those rare cases where it is indicated. Commutation is therefore ignored by assuming it to be instantaneous.

Substituting θ_{cst} for θ in the state variable eqns 7.19 through 7.44, and rearranging the terms results in the following three simultaneous equations in the three unknown amplitudes

$$\begin{bmatrix} A_{11} & A_{12} & A_{13} \\ A_{21} & A_{22} & A_{23} \\ A_{31} & A_{32} & A_{33} \end{bmatrix} \begin{bmatrix} I_{co3} \\ I_{co4c} \\ I_{co4s} \end{bmatrix} = \begin{bmatrix} B_1 \\ B_2 \\ B_3 \end{bmatrix} \tag{7.52}$$

where

$$A_{11} = \exp(s_{c3}\theta_{cst}) \tag{7.53}$$

$$A_{12} = \cos(s_{c4i}\theta_{cst})\exp(s_{c4r}\theta_{cst}) \tag{7.54}$$

$$A_{13} = \sin(s_{c4i}\theta_{cst})\exp(s_{c4r}\theta_{cst}) \tag{7.55}$$

$$A_{21} = (R_o + s_{c3}X_o)A_{11} \tag{7.56}$$

$$A_{22} = (R_o + s_{c4r}X_o)A_{12} - s_{c4i}X_oA_{13} \tag{7.57}$$

$$A_{23} = (R_o + s_{c4r}X_o)A_{13} + s_{c4i}X_oA_{12} \tag{7.58}$$

$$A_{31} = \frac{s_{c3}R_o + X_c + s_{c3}^2X_o}{X_c}A_{11} \tag{7.59}$$

$$A_{32} = \frac{s_{c4r}R_o + X_c - (s_{c4i}^2 - s_{c4r}^2)X_o}{X_c}A_{12} - \frac{s_{c4i}(R_o + 2s_{c4r}X_o)}{X_c}A_{13} \tag{7.60}$$

$$A_{33} = \frac{s_{c4r}R_o + X_c - (s_{c4i}^2 - s_{c4r}^2)X_o}{X_c}A_{13} + \frac{s_{c4i}(R_o + 2s_{c4r}X_o)}{X_c}A_{12} \tag{7.61}$$

$$B_1 = I_{cost} - I_{co1} - I_{co2c}\cos(\theta_{st}) - I_{co2s}\sin(\theta_{st}) \tag{7.62}$$

$$B_2 = V_{ccst} - V_{cc1} - V_{cc2c}\cos(\theta_{st}) - V_{cc2s}\sin(\theta_{st}) \tag{7.63}$$

$$B_3 = I_{crst} - I_{cr1} - I_{cr2c}\cos(\theta_{st}) - I_{cr2s}\sin(\theta_{st}) \tag{7.64}$$

Once in numerical form, eqn 7.52 is readily solved.

Table 7.1 *Coefficients of eqns 7.19, 7.24, 7.31, 7.38, and 7.45*

Variable	Coefficient					
	1	2c	2s	3	4c	4s
i_o	-3368.97	-1065.09	336.250	5321.20	-13.0648	2.78052
v_c	10.2758	-3.59704	619.305	-0.594444	9.92289	51.9220
i_c	0.0	933.889	5.42421	0.288920	584.479	-125.147
i_{rec}	-3368.97	-131.202	341.674	5321.49	571.414	-122.366
$\mathrm{d}i_{rec}/\mathrm{d}\theta$	0.0	341.674	131.202	-1715.17	-1012.00	-4263.40

Example 7.2

The initial conditions for the charging mode portion of the previous example, i.e. the low supply impedance, low smoothing capacitance case with a delay angle of zero, and a load e.m.f. of 583 V used to construct Fig. 7.5, are

$$\begin{aligned}\theta_{cst} &= 72.2365^\circ\\ I_{cost} &= 181.143 \text{ A}\\ V_{ccst} &= 589.387 \text{ V}\\ I_{crst} &= 0.0 \text{ A}\end{aligned}$$

Determine the numerical form of eqn 7.52 and expressions for the output current, capacitor voltage, capacitor current, rectifier current, and rate of change of rectifier current.

The numerical form of eqn 7.52 is

$$\begin{bmatrix} 6.66071\times10^{-1} & -8.11201\times10^{-1} & -2.16514\times10^{-2}\\ -7.44084\times10^{-5} & 1.86968\times10^{-2} & -3.21140\\ 6.66108\times10^{-1} & 3.54878\times10^{1} & 0.99204 \end{bmatrix}\begin{bmatrix} I_{co3}\\ I_{co4c}\\ I_{co4s}\end{bmatrix}$$

$$= \begin{bmatrix} 3.55484\times10^{3}\\ -9.56858\\ 3.08361\times10^{3}\end{bmatrix}$$

The solution of this equation gives the values of I_{co3}, I_{co4c}, and I_{co4s} and from these, using eqns 7.24 through 7.51, the coefficients of the expressions for the other circuit variables are obtained. All these coefficients are listed in Table 7.1.

Transformation from the Cartesian form of Table 7.1 to the polar form of eqn 7.15 then yields the expressions for the output, capacitor, and rectifier currents, the rate of change of rectifier current, and the capacitor voltage listed below

$$\begin{aligned} i_o &= 3368.97 + 1116.91\cos(\theta - 2.83579) + \\ &\quad 5321.20\exp(-0.322311\theta) + \\ &\quad\quad 13.3574\cos(7.49661\theta - 2.93190)\exp(-0.165680\theta) \\ v_c &= 10.2758 + 619.315\cos(\theta - 1.57660) - \\ &\quad 0.594444\exp(-0.322311\theta) + \\ &\quad\quad 52.8617\cos(7.49661\theta - 1.38196)\exp(-0.165680\theta) \\ i_c &= 933.905\cos(\theta - 0.00580812) + 0.288920\exp(-0.322311\theta) + \\ &\quad 597.727\cos(7.49661\theta + 0.210932)\exp(-0.165680\theta) \\ i_{rec} &= -3368.97 + 365.999\cos(\theta - 1.93743) + \\ &\quad 5321.49\exp(-0.322311\theta) + \\ &\quad\quad 584.370\cos(7.49661\theta + 0.210960)\exp(-0.165680\theta) \\ \frac{\mathrm{d}i_{rec}}{\mathrm{d}\theta} &= 365.999\cos(\theta - 0.366636) - 1715.17\exp(-0.322311\theta) + \\ &\quad 4381.86\cos(7.49661\theta + 1.80385)\exp(-0.165680\theta) \end{aligned}$$

****

7.4.3 *Termination of the charging mode*

The charging mode ends either at the extinction angle or at the end of the segment, whichever occurs sooner, and the first task is therefore the determination of the extinction angle. This is the first root after θ_{cst} of the expression formed by equating the right-hand side of eqn 7.38 to zero. This expression is transcendental and the root must be found by a successive approximation technique. Unfortunately the parameter range is so great that the direct use of the powerful Newton–Raphson method is unreliable. A suitable method comprises two stages, a step by step search to localize the required root followed by a Newton–Raphson refinement.

In the two stage method the value of i_{rec} for $\theta = \theta_{cst} + n\Delta$ is found where Δ is a small angle of the order of 3°, and n successively takes on the values 1, 2, 3, etc. Initially the value of i_{rec} increases, reaches a maximum, and then decreases to become negative. If this occurs when $n = N$, the required root lies between $\theta_{cst} + (N-1)\Delta$ and $\theta_{cst} + N\Delta$. A highly accurate approximation to the true value of γ is then quickly found using the Newton–Raphson method whereby an improved approximation, γ_2, is obtained from an approximation, γ_1, by

$$\gamma_2 = \gamma_1 - \frac{i_{rec}(\gamma_1)}{\frac{di_{rec}}{d\theta}(\gamma_1)} \tag{7.65}$$

Having found the extinction angle to the required accuracy, say 0.0001 rad, a check is made as to whether it occurs before or after the end of the rectifier segment, θ_{fn}. If it occurs before the end of the segment the computation proceeds. If this is not the case and γ equals or is greater than the θ_{fn}, the behavior is of type 1, 3, or 6 and γ is set to θ_{fn}.

Example 7.3

Check the correctness of the solution obtained in Example 7.2 by showing that the value of the rectifier current at θ_{cst} is zero. Then determine the extinction angle and find the values of the output current, capacitor voltage, capacitor current, and rectifier current at this point.

Substitution of 1.26076 rad (72.2365°) for θ in the expression for the rectifier current of Example 7.2 yields $i_{rec} = -0.0001$, acceptably close to zero.

The step by step determination of the zero of i_{rec} shows that it occurs between 105.07° and 108.06°. Three Newton–Raphson refinements yield the extinction angle as 105.328° with an accuracy of better than 0.006°. Substitution of this value for θ in the expressions for the currents and voltage gives

$$\begin{aligned} i_o &= 177.727 \text{ A} \\ v_c &= 646.613 \text{ V} \\ i_c &= -177.727 \text{ A} \\ i_{rec} &= -0.000 \text{ A} \end{aligned}$$

The facts that $i_c = -i_o$ and i_{rec} is so small are further indications of the accuracy of the results.

****

7.5 The discharging mode

During the discharging mode, the rectifier is reverse biased and does not conduct. The load current is maintained by the capacitor whose voltage falls as a result of the charge drawn from it. Referring to Fig. 7.1(b), it is seen that the equivalent circuit reduces to the simple series combination of the smoothing capacitor and the load impedance with the load e.m.f. as driving function. Thus the capacitor voltage and the output current are related by

$$v_c = E_o + R_o i_o + X_o \frac{di_o}{d\theta} \tag{7.66}$$

The capacitor current is related to the capacitor voltage by

$$i_c = \frac{1}{X_c}\frac{\mathrm{d}v_c}{\mathrm{d}\theta} \tag{7.67}$$

and, by applying Kirchhoff's Current Law to a rectifier output terminal,

$$i_c = -i_o \tag{7.68}$$

7.5.1 *The output current*

Differentiating eqn 7.66 with respect to θ and using eqns 7.67 and 7.68 to replace $\mathrm{d}v_c/\mathrm{d}\theta$ by $-X_c i_o$ yields the differential equation defining the output current during the discharging mode.

$$\frac{\mathrm{d}^2 i_o}{\mathrm{d}\theta^2} + \frac{R_o}{X_o}\frac{\mathrm{d}i_o}{d\theta} + \frac{X_c}{X_o} i_o = 0 \tag{7.69}$$

The characteristic equation is

$$s^2 + \frac{R_o}{X_o}s + \frac{X_c}{X_o} = 0 \tag{7.70}$$

whose roots s_{d4} and s_{d5} are

$$s_{d4} = -\frac{R_o}{2X_o} + \sqrt{\left(\frac{R_o}{2X_o}\right)^2 - \frac{X_c}{X_o}} \tag{7.71}$$

and

$$s_{d5} = -\frac{R_o}{2X_o} - \sqrt{\left(\frac{R_o}{2X_o}\right)^2 - \frac{X_c}{X_o}} \tag{7.72}$$

As in the case of the charging mode, it is highly probable that the discharging mode will be underdamped and there seems little point in complicating further an already complex task by considering the over and critically damped cases. The analysis will therefore continue on the assumption of under damping so that s_{d5} is the conjugate of s_{d4} and the natural frequency, s_{d4}, of the discharging mode has real and imaginary components, s_{d4r} and s_{d4i}, given by

$$s_{d4r} = -R_o/(2X_o) \tag{7.73}$$

$$s_{d4i} = \sqrt{\frac{X_c}{X_o} - \left(\frac{R_o}{2X_o}\right)^2} \tag{7.74}$$

and

$$s_{d4} = s_{d4r} + js_{d4i} \tag{7.75}$$

The expression for the output current during the discharging mode has the form

$$i_o = I_{do4}\cos(s_{d4i}\theta + \phi_{do4})\exp(s_{d4r}\theta) \tag{7.76}$$

where the frequency s_{d4i} is given by eqn 7.74 and the exponent s_{d4r} is given by eqn 7.73. The amplitude I_{do4} and the phase ϕ_{do4} depend on the initial conditions.

Equation 7.76 is a particular case of the basic capacitor smoothing expression with the amplitudes of the first three terms set to zero.

Example 7.4

Determine the basic capacitor smoothing expression for the output current during the discharging mode for the low supply impedance, low smoothing capacitance example.

The amplitudes of the I_{do1}, I_{do2}, and I_{do3} are zero.

The frequency s_{d4i} and exponent s_{d4r} are obtained from eqns 7.74 and 7.73 as

$$\begin{aligned} s_{d4i} &= 1.10929 \\ s_{d4r} &= -0.16105 \end{aligned}$$

Remember that the frequency s_{d4i} is relative to the a.c. supply frequency and is not an absolute value. Thus, in this case the frequency of the transient component is 1.11 times the a.c. supply frequency, 66.56 Hz or 418.2 rad s^{-1}.

Thus, during the discharging mode, for the segment from $\alpha + 60°$ to $\alpha + 120°$, the expression for the output current is

$$i_o = I_{do4}\cos(1.109290\theta + \phi_{do4})\exp(-0.161050\,\theta)$$

****

7.5.2 *The other circuit variables*

The rectifier current is zero during the discharging mode and the capacitor current is -1 times the load current. The only other variable which is required at this point is the capacitor voltage which is the same as the output voltage. The capacitor voltage is obtained by the application of eqn 7.16, the result having a similar form to eqn 7.24 but with V_{cc2c}, V_{cc2s}, and V_{cc3} equal to zero.

$$v_c = V_{dc1} + V_{dc4c}\cos(s_{d4i}\theta)\exp(s_{d4r}\theta) + V_{dc4s}\sin(s_{d4i}\theta)\exp(s_{d4r}\theta) \quad (7.77)$$

where

$$V_{dc1} = E_o \quad (7.78)$$
$$V_{dc4c} = (R_o + s_{d4r}X_o)I_{do4c} + s_{d4i}X_oI_{do4s} \quad (7.79)$$
$$V_{dc4s} = (R_o + s_{d4r}X_o)I_{do4s} - s_{d4i}X_oI_{do4c} \quad (7.80)$$

For given initial conditions, $i_o = I_{dost}$ and $v_c = V_{dcst}$ at $\theta = \theta_{dst}$, the values of I_{do4c} and I_{do4s} are obtained in a similar manner to that used for the corresponding amplitudes in the charging mode

$$\begin{bmatrix} C_{11} & C_{12} \\ C_{21} & C_{22} \end{bmatrix} \begin{bmatrix} I_{do4c} \\ I_{do4s} \end{bmatrix} = \begin{bmatrix} D_1 \\ D_2 \end{bmatrix} \quad (7.81)$$

where

$$C_{11} = \cos(s_{d4i}\theta_{dst})\exp(s_{d4r}\theta_{dst}) \quad (7.82)$$
$$C_{12} = \sin(s_{d4i}\theta_{dst})\exp(s_{d4r}\theta_{dst}) \quad (7.83)$$
$$C_{21} = (R_o + s_{d4r}X_o)C_{11} - s_{d4i}X_oC_{12} \quad (7.84)$$
$$C_{22} = (R_o + s_{d4r}X_o)C_{12} + s_{d4i}X_oC_{11} \quad (7.85)$$
$$D_1 = I_{dost} \quad (7.86)$$
$$D_2 = V_{dcst} - E_o \quad (7.87)$$

Considering type 7 operation, i.e. the type of the example, each rectifier segment contains two sections where the rectifier operates in the discharging mode. The first precedes the charging mode section and extends from θ_{st} to the initiation angle γ_2. The second follows the charging mode section and extends from the extinction angle γ_3 to θ_{fn}. The initial conditions for the first discharging mode section must be specified. The solution for this section provides the initial conditions for the charging mode section. The solution for this in turn provides the initial conditions for the second discharge mode section etc.

Example 7.5

For the second discharging mode section of Fig. 7.5, determine the numerical form of eqn 7.81 and, by solving it, determine expressions for the output current and capacitor voltage. From these expressions determine the values of the output current and capacitor voltage at the start and end of this section.

In Example 7.3 it was found that the extinction angle is 105.328° = 1.83831 rad. Substituting this value for θ_{dst} in eqn 7.81 yields

$$\begin{bmatrix} -0.335788 & 0.663627 \\ -0.417075 & -0.140185 \end{bmatrix} \begin{bmatrix} I_{do4c} \\ I_{do4s} \end{bmatrix} = \begin{bmatrix} 177.727 \\ 63.6126 \end{bmatrix}$$

The solution of this equation yields

$$\begin{aligned} I_{do4c} &= -207.284 \text{ A} \\ I_{do4s} &= 162.929 \text{ A} \end{aligned}$$

from which, using eqns 7.78 through 7.80,

$$\begin{aligned} V_{dc1} &= 583.0 \text{ V} \\ V_{dc4c} &= 77.771 \text{ V} \\ V_{dc4s} &= 135.207 \text{ V} \end{aligned}$$

Converting to polar form

$$\begin{aligned} i_o &= 263.652\cos(1.10929\theta - 2.47543)\exp(-0.161050\theta) \\ v_c &= 583.0 + 155.979\cos(1.10929\theta - 1.04881)\exp(-0.161050\theta) \end{aligned}$$

Substituting the extinction angle, 1.83831 rad, for θ in these equations gives 177.727 A for the output current and 646.613 V for the capacitor voltage. These values are in agreement with the values obtained in Example 7.3 at the end of the charging mode section, an indication that the solution is correct.

This section ends at the end of the segment, when $\theta = \theta_{fn} = 120° = 2.09440$ rad. At this point the output current is 185.993 A and the capacitor voltage is 615.506 V.

****

7.6 The first discharging mode

Still concentrating on type 7 behavior, the first discharging mode section starts at the start of the rectifier segment and ends at the initiation angle. When the initial values of the rectifier current and capacitor voltage are specified, the determination of the expressions for the currents and voltages during this section follows the above procedure, i.e. eqn 7.81 is solved and the resulting amplitudes, I_{do4c}, I_{do4s}, V_{dc4c}, and V_{dc4s}, are converted to polar form. The initiation angle must then be determined. It occurs when the rectifier driving voltage is sufficient to forward bias the thyristors, i.e. when $v_{rt} > v_c + 2V_{thy}$. Thus the first zero after θ_{st} of the expression $f(\theta) = v_{rt} - 2V_{thy} - v_c$ must be found. Combining eqns 7.1 and 7.77, the appropriate zero of

$$f(\theta) = \sqrt{2}V_s \sin(\theta) - 2V_{thy} - E_o - V_{dc4c} \cos(s_{d4i}\theta) \exp(s_{d4r}\theta) - V_{dc4s} \sin(s_{d4i}\theta) \exp(s_{d4r}\theta) \tag{7.88}$$

is required. Allowance is made for the voltage drop in two conducting thyristors to maintain consistency with the thyristor model. Strictly speaking, the $2V_{thy}$ term should be omitted but this would create discontinuities with the on state results. Whichever method is deemed the most aesthetically satisfactory, putting the term in or leaving it out, can be used, the difference between the two calculations being small, well within the limits of engineering accuracy.

Again the parameter variability in eqn 7.88 is so great that the Newton–Raphson technique cannot on its own locate the correct zero. Again the solution is a step by step determination of the approximate location of the zero followed by a Newton–Raphson refinement. As before, a step size of about 3° for θ should give satisfactory results.

The Newton–Raphson method requires the derivative of $f(\theta)$ with respect to θ

$$\frac{df}{d\theta}(\theta) = \sqrt{2}V_s \cos(\theta) - V_{dc4c}[s_{d4r} \cos(s_{d4i}\theta) - s_{d4i} \sin(s_{d4i}\theta)] \exp(s_{d4r}\theta) - V_{dc4s}[s_{d4r} \sin(s_{d4i}\theta) + s_{d4i} \cos(s_{d4i}\theta)] \exp(s_{d4r}\theta) \tag{7.89}$$

Example 7.6

For the first discharge mode section, determine the numerical form of eqn 7.81 given that at the start of the section the values of load current and capacitor voltage are those found in Example 7.5 at the end of the second discharge mode section, namely

$$\theta = 60^\circ$$
$$i_o = 185.983 \text{ A}$$
$$v_c = 615.506 \text{ V}$$

The numerical form of eqn 7.81 is

$$\begin{bmatrix} 0.336088 & 0.775073 \\ -0.425215 & 0.262651 \end{bmatrix} \begin{bmatrix} I_{do4c} \\ I_{do4s} \end{bmatrix} = \begin{bmatrix} 185.993 \\ 32.506 \end{bmatrix}$$

The solution is

$$\begin{aligned} I_{do4c} &= 56.610 \text{ A} \\ I_{do4s} &= 215.408 \text{ A} \end{aligned}$$

and by inserting these values into eqns 7.79 and 7.80

$$\begin{aligned} V_{dc4c} &= 130.927 \text{ V} \\ V_{dc4s} &= -14.834 \text{ V} \end{aligned}$$

Converted to the polar form of eqns 7.76 and 7.77, these values give the following expressions for the output current and capacitor voltage

$$\begin{aligned} i_o &= 222.723\cos(1.109290\theta - 1.31380)\exp(-0.161050\theta) \\ v_c &= 583.0 + 131.765\cos(1.109290\theta - 0.112818)\exp(-0.161050\theta) \end{aligned}$$

The function $f(\theta)$ and its rate of change, eqns 7.88 and 7.89, are

$$\begin{aligned} f(\theta) &= 622.254\sin(\theta) - 586.2 - \\ &\qquad 131.765\cos(1.109290\theta - 0.11282)\exp(-0.161050\theta) \\ \frac{\mathrm{d}f}{\mathrm{d}\theta}(\theta) &= 622.254\cos(\theta) + \\ &\qquad 147.698\cos(1.109290\theta - 1.3138)\exp(-0.161050\theta) \end{aligned}$$

The step by step evaluation of $f(\theta)$ locates the initiation angle between 72° and 75°. Four Newton–Raphson iterations then locate it at 72.2365° with an accuracy of 0.006°.

The values of output current, capacitor voltage, and rectifier voltage at the initiation point are

$$\begin{aligned} i_o &= 181.143 \text{ A} \\ v_c &= 589.387 \text{ V} \\ v_{rt} &= 592.587 \text{ V} \end{aligned}$$

The difference between the rectifier Thévenin voltage and the capacitor voltage is 3.204 V, close enough to twice the thyristor forward drop, 3.2 V, to indicate that the result is very close to the correct value of the initiation angle.

The values of the output current and capacitor voltage are extremely close to the initial values, 181.143 A and 589.387 V, used in Example 7.2 where the charging mode section was determined.

Thus, within the limits of computer roundoff error, the three sections of the rectifier segment found in Examples 7.1 through 7.6 form a continuous sequence representing a steady state condition. They have in fact been used to plot Fig. 7.5.

****

7.7 Determination of the steady state

The rectifier segment which has been examined in Examples 7.1 through 7.6 starts with an output current of 185.993 A and a capacitor voltage of 615.506 V and ends with the values 185.993 A and 615.506 V. Thus the starting and finishing values are about the same, indicating that the rectifier is very close to the steady state. The means of getting to this point must now be addressed and for this there is no short cut as there was in Chapter 5; transient operation must be considered.

Steady state operation is indicated by the values of the state variables at the end of a rectifier segment being equal to the values at the start of the segment. The transcendental nature of the expressions which describe the state variables mean that a closed form solution cannot be obtained. Initial values for the state variables are proposed, the problem is solved and the final values are found. If the differences between the final and initial values are sufficiently small, the solution is an acceptable approximation to the steady state. If the differences are excessive, the problem must be repeated with new initial values, this process continuing until the results are acceptable.

In this iterative process there are numerous ways in which the new initial values can be derived from the old in order to maximize the rate of convergence. An obvious method is to imitate real life by using the final values as the initial values for the next iteration. In well behaved problems, speedup factors can be employed with considerable success. The final values of the state variables are multiplied by factors which experience shows to be appropriate to provide the new initial conditions. In badly behaved systems one of the numerous hill climbing algorithms can be employed. The present problem fits between these extremes, the problem is well behaved for any particular operating point but the type of behavior depends greatly on the circuit parameters and on the operating point, delay angle, and load e.m.f. My own experience has been that the natural solution, in which the final values from one iteration are the initial values for the next and which models the transient behavior of the system, is the best approach and it is this method which will be described.

7.8 Convergence criteria

Before commencing, the criteria by which convergence to the steady state is reached must be decided. An r.m.s. error criterion is employed which is based on only two of the three state variables, the output current and the capacitor voltage, since the rectifier current only influences the problem during the charging mode and reaches the steady state when the first two attain that condition. The output current error is defined as the difference between the initial and final values divided by the rated load current, and the capacitor voltage error as the difference between the initial and final values divided by the rated load voltage. The *residual error* is the square root of the sum of the squares of these two errors and iteration until the residual is less than 0.0005 on two successive iterations has been employed for all numerical calculations. The two successive iterations requirement is necessary because the transient can under certain circumstances be oscillatory and a small residual from one iteration may be followed by a large residual from the next. Insisting that two successive residuals shall be less than the criterion ensures convergence to the steady state.

7.9 The transient solution

The seven rectifier segment types shown in Fig. 7.2 seem to be adequate for a wide range of circuit parameters and operating points and the determination of transient behavior will be described on this basis with the aid of the flow diagram of Fig. 7.7.

The first task is to define the operating conditions, the values of delay angle and load e.m.f. The second task is the definition of the initial values of the state variables. The third and final requirement for a fully defined problem is the initial value of θ. This has been arbitrarily taken to be the start of a rectifier segment and it is assumed that the thyristor gate pulses are sufficiently wide to maintain the conducting mode until the thyristor becomes forward biased.

The calculation commences with $\theta = \theta_{st}$, $i_o = I_{ost}$, $v_c = V_{cst}$, and $i_{rec} = I_{rst}$ and the first step is to determine if the rectifier is in the charging or discharging mode. There are three possibilities.

- If the initial rectifier current is greater than zero, the first rectifier subsegment is of the charging type.
- If the initial rectifier current is zero and the rectifier is forward biased, the first rectifier subsegment is of the charging type.
- If the initial rectifier current is zero and the rectifier is reverse biased, the first rectifier subsegment is of the discharging type.

In testing for forward or reverse bias, the reverse voltage acting on the circuit is taken as the sum of the load e.m.f. and two thyristor forward drops, i.e. for the conditions of Fig. 7.5 where the e.m.f. is 583 V the reverse voltage is 586.2 V. The first *if* block of Fig. 7.7 which contains the symbol D? makes this decision. If the answer is D for discharge, the computation proceeds through block D1. If not, a jump is made to block C1.

Block D1 solves for the discharging mode exactly as was done above. Equation 7.81 is determined and solved to provide expressions for the output current and capacitor voltage, eqns 7.76 and 7.77. The function $f(\theta)$ and its derivative with respect to θ, eqns 7.88 and 7.89, are then found and the termination of the subsegment is determined by the successive approximation technique described above, i.e. the termination point is first localized by a step by step exploration of $f(\theta)$ and is then refined by the Newton–Raphson technique. If the termination value of θ lies beyond the end of the rectifier segment, the operation is type 5, the termination angle is set equal to θ_{fn} and the values of i_o and v_c are found at this point and are used as the initial values for the next rectifier segment. If the termination value of θ is within the rectifier segment, the values of i_o and v_c are found at this point and are used as the initial values for the next, charging type, subsegment. This process is indicated by the second *if* block with one possible output pointing to *Type 5*.

The next block marked C1 determines conditions during the charging mode in the manner described in Section 7.4. Equation 7.52 is determined and solved to give the expressions for the state variables, i_o, v_c, and i_{rec}, eqns 7.19, 7.24, and 7.38. The end of the subsegment is found as the point at which i_{rec} becomes zero by the successive approximation method followed by Newton–Raphson refinement. If the end occurs after the end of the rectifier segment, the segment is of type 1 or 6. In this event the end of the subsegment is set equal to θ_{fn} and the values of the state variables are found at this point and are used as the starting point for the next rectifier segment. If the subsegment ends before the end of the rectifier segment, the values of the state variables are found at this point and are used as initial values for the next, discharge type, subsegment. This process is indicated in Fig. 7.7 by the third *if* block with the exit arrow marked *Type 1 or 6*.

This procedure continues through successive discharging and charging type subsegments until the end of the rectifier segment is reached, the appropriate type number being attached to the solution. The residual for the segment is calculated and is tested for the steady state. If the steady state has not been reached, the whole process is repeated for the next rectifier segment.

In the computation, each rectifier segment is defined by a type number, arrays of coefficients for the state variable expressions, and an array of values of θ marking the starts and ends of the subsegments.

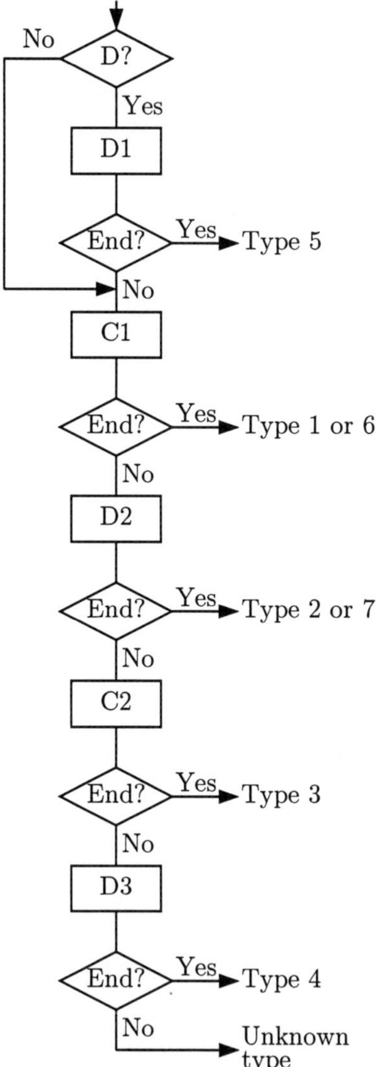

FIG. 7.7. Flow diagram showing the sequence of computations for a rectifier segment.

The final block of Fig. 7.7 is marked *Unknown type* and is a very necessary precaution to deal with unusual operating conditions. If this point is reached, a suitable error message should be printed and the program should gracefully terminate.

The steady state condition illustrated in Fig. 7.5 was attained after 43

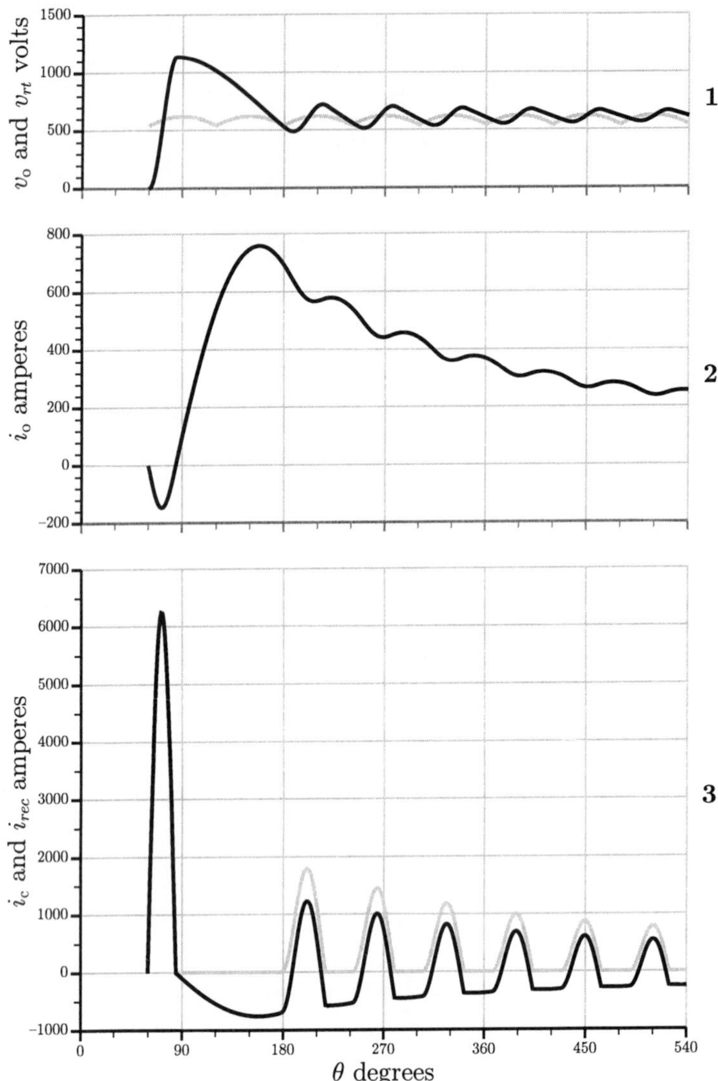

FIG. 7.8. Start up transient for the low supply inductance case.

iterations from the initial condition of all the state variables equal to zero. This is a rather extreme set of initial values which would not normally occur in practice. However, it has the merit of providing a very demanding test of the program's ability to converge. The first nine rectifier segments of this transient are shown in Fig. 7.8. Graph 1 shows by a black line the output voltage and by a gray line the rectifier Thévenin voltage. Graph 2 shows the load current. Graph 3 shows by a black line the capacitor current and by a gray line the rectifier output current.

While, as already noted, the initial conditions are extreme they are none the less possible and as well as testing the validity of the program, the result illustrates a potential problem of capacitor smoothed rectifiers, which must be minimized by the rectifier controller.

When the rectifier is connected to the a.c. system with the smoothing capacitor discharged, as it would normally be in the off state, there is an enormous inrush of current which charges the capacitor to about twice the normal voltage, in the case of Fig. 7.8 to about 1200 V. The extra charge is then dissipated in the load over a few transient rectifier output cycles. In the case of Fig. 7.8, this process takes about nine rectifier cycles, i.e. about 50 ms. If the load were not connected and so could not absorb the excess energy, the capacitor would sit at the high initial charging voltage for a prolonged period, slowly discharging to its normal operating voltage through its own leakage resistance and through any other stray leakage current path that was present.

The initial inrush current can be enormous. In the case illustrated in Fig. 7.8, the initial charging current is a pulse of about 1.5 ms duration whose amplitude is over 6000 A, about 34 times the rated load current! Such a huge pulse would be intolerable on most a.c. systems. It can be reduced by introducing inductance in the charging path either by installing the choke represented in Fig. 7.1 by R_{ch} and L_{ch} or interposing a transformer with the appropriate leakage impedance between the rectifier and the a.c. supply. Roughly speaking, the height of the current pulse will be inversely proportional to the inductance so that for the high supply inductance case cited in Section 7.2 the amplitude of the pulse would be reduced to about 600 A and would probably be acceptable to the supply authorities.

However, control of the current pulse does not solve the other problem, capacitor overvoltage. This can of course be resolved by using a capacitor of adequate voltage rating but this effectively means having a capacitor which is twice as big and twice as expensive as required by the normal operating conditions. When a thyristor rectifier is employed this difficulty is overcome by having the thyristors fully phased back on starting and then phasing them forward at a controlled rate so that the capacitor charge is built up without excessive current flow. This elegant solution is not available with the more usual diode rectifier. A simple solution is then to temporarily connect, during the initial charging process, sufficient resistance in series with the capacitor to critically damp the charging circuit.

Example 7.7

Determine for the low supply impedance, low smoothing capacitance case, the transient condition illustrated in Fig. 7.8 assuming that the capacitor is initially discharged and that the load and rectifier currents are initially zero. Throughout the transient the load e.m.f. is 583 V and the delay angle is 0°. The first firing pulses are delivered to thyristors T2 and T6.

Firing thyristors T2 and T6 connects the yellow and blue a.c. lines to the load so that the rectifier Thévenin voltage during the first segment is

$$v_{rt} = 622.254\cos(\theta - \frac{\pi}{2})$$

At the instant of firing, $\theta = 60°$, v_{rt} has the value 538.888 V, which is greater than the sum of the terminal voltage at this time, $v_c = 0$, and two thyristor forward drops, 3.2 V. The first *if* block of Fig. 7.7 therefore leads the program directly to block C1. The initial rate of increase of the rectifier current in A per radian is the difference voltage, 535.688 V, divided by the reactance of two supply lines, 0.0120637 Ω, i.e. 44404.9 A rad^{-1}. The load e.m.f. also feeds current into the uncharged capacitor at an initial rate equal to the load e.m.f. divided by the load reactance, i.e. $583 \div 0.527788 = 1104.61$ A rad^{-1}. Since initially no current flows into the capacitor the initial rate of rise of its voltage is zero.

The first section of the first segment is therefore of the charging type, starting at $\theta = 60°$ with

$$\begin{aligned} i_o &= -3368.97 + 1116.91\cos(\theta - 2.83579) + 5044.73\exp(-0.322311\theta) + \\ &\quad 164.424\cos(7.49661\theta - 0.0740375)\exp(-0.165680\theta) \end{aligned}$$

$$\begin{aligned} v_c &= 10.2758 + 619.315\cos(\theta - 1.57660) - 0.563559\exp(-0.322311\theta) + \\ &\quad 650.703\cos(7.49661\theta + 1.47590)\exp(-0.165680\theta) \end{aligned}$$

$$\begin{aligned} i_{rec} &= -3368.97 + 365.999\cos(\theta - 1.93743) + 5045.01\exp(-0.322311\theta) + \\ &\quad 7193.33\cos(7.49661\theta + 3.06882)\exp(-0.165680\theta) \end{aligned}$$

This section lasts until $\theta = 84.7186°$ when the rectifier current is zero, the output current is 2.06 A, and the capacitor voltage is 1135.58 V.

The termination of the charging mode before the end of the rectifier segment moves the program to block D2 of Fig. 7.7. During the discharging type section the rectifier current is zero, the capacitor current is -1 times the output current and

$$\begin{aligned} i_o &= 1197.22\cos(1.10929\theta + 3.07435)\exp(-0.161050\theta) \\ v_c &= 583.0 + 708.285\cos(1.10929 - 1.78221)\exp(-0.161050\theta) \end{aligned}$$

This section continues to the end of the rectifier segment at $\theta = 120°$ so that the first segment is of type 2 with one charging and one discharging type section. At the end of the segment the output current is 540.76 A and the capacitor voltage is 1016.29 V.

The second rectifier segment begins at $\theta = 120°$ with thyristors T2 and T4 nominally conducting and with the rectifier Thévenin voltage

$$v_{rt} = 622.254 \cos(\theta - \frac{5\pi}{6})$$

The rectifier is initially reverse biased by the large capacitor voltage and is therefore in the discharging mode—block D1 of Fig. 7.7. Since nothing has changed as far as the load is concerned, the expressions for the state variables are those just given for the discharging mode. Discharging continues until the rectifier becomes forward biased at $\theta = 176.998°$ when the output current is 710.73 A and the capacitor voltage is 551.24 V, 3.2 V less than the rectifier Thévenin voltage.

Assuming that the nominally conducting thyristors, T2 and T4, are still in the conducting mode, the charging mode section, C1, has

$$\begin{aligned} i_o &= -3368.97 + 1116.91 \cos(\theta + 2.40019) + 8904.77 \exp(-0.322311\theta) + \\ &\quad 12.1224 \cos(7.49661\theta + 2.42993) \exp(-0.165680\theta) \\ v_c &= 10.2578 + 619.315 \cos(\theta - 2.62380) - 0.994773 \exp(-0.322311\theta) + \\ &\quad 47.9740 \cos(7.49661\theta - 2.30332) \exp(-0.165680\theta) \\ i_{rec} &= -3368.97 + 365.999 \cos(\theta - 2.98463) + 8905.25 \exp(-0.322311\theta) + \\ &\quad 530.337 \cos(7.49661\theta - 0.710398) \exp(-0.165680\theta) \end{aligned}$$

At the end of the rectifier segment the rectifier current is still positive at 19.63 A so that this segment is of type 6 and the computation goes forward to the third rectifier segment with $i_o = 694.51$ A and $v_c = 527.1$ V.

During the third segment, which lasts from 180° to 240°, the rectifier Thévenin voltage is

$$v_{rt} = 622.254 \cos(\theta - \frac{7\pi}{6})$$

Since the rectifier current is not zero the computation moves directly to block C1 of Fig. 7.7 and determines that

$$i_o = -3368.97 + 1116.91\cos(\theta + 1.35300) + 11779.2\exp(-0.322311\theta) + 43.9178\cos(7.49661\theta + 1.77038)\exp(-0.165680\theta)$$

$$v_c = 10.2758 + 619.315\cos(\theta + 2.61219) - 1.31588\exp(-0.322311\theta) + 173.804\cos(7.49661\theta - 2.96287)\exp(-0.165680\theta)$$

$$i_{rec} = 3368.97 + 365.999\cos(\theta + 2.25136) + 11779.8\exp(-0.322311\theta) + 1921.35\cos(7.49661\theta - 1.36995)\exp(-0.165680\theta)$$

This charging mode section ends at $\theta = 217.991°$ with $i_o = 577.98$ A and $v_c = 707.93$ V.

The following discharging mode section has

$$i_o = 1093.08\cos(1.10929\theta + 1.84235)\exp(-0.161050\theta)$$

$$v_c = 583.0 + 646.676\cos(1.10929\theta - 3.01421)\exp(-0.161050\theta)$$

This mode continues to the end of the segment so that the segment is of type 2. The values of the state variables at the end are $i_o = 545.03$ A, $v_c = 562.73$ V, and $i_{rec} = 0$.

The rectifier segments continue in this manner, all succeeding segments being of type 7.

****

7.10 Steady state, d.c. side, currents, and voltages

The d.c. side currents and voltages, which have been found for the steady state during the rectifier segment which extends from $\alpha+60°$ to $\alpha+120°$ and during which thyristors T2 and T6 are nominally conducting, are repeated during successive segments. The derivation of the expression for another segment only requires the introduction of the appropriate phase shift into the expression already derived.

The expressions are all variants of the basic expression given in eqn 7.15 and listed here, in terms of a dummy angle ψ, in eqn 7.90 for the range $\alpha + 60° < \psi < \alpha + 120°$

$$X_1 + X_2\cos(\psi + X_3) + X_4\exp(X_5\psi) + X_6\cos(X_7\psi + X_8)\exp(X_9\psi) \quad (7.90)$$

This expression is converted to apply to a segment covering the range $\alpha + 60° + \delta < \theta < \alpha + 120° + \delta$, by replacing ψ by $\theta - \delta$ to obtain

$$Y_1 + Y_2\cos(\theta + Y_3) + Y_4\exp(Y_5\theta) + Y_6\cos(Y_7\theta + Y_8)\exp(Y_9\theta) \quad (7.91)$$

where

$$Y_1 = X_1 \quad (7.92)$$

$$Y_2 = X_2 \tag{7.93}$$
$$Y_3 = X_3 - \delta \tag{7.94}$$
$$Y_4 = X_4 \exp(-X_5\delta) \tag{7.95}$$
$$Y_5 = X_5 \tag{7.96}$$
$$Y_6 = X_6 \exp(-X_9\delta) \tag{7.97}$$
$$Y_7 = X_7 \tag{7.98}$$
$$Y_8 = X_8 - X_7\delta \tag{7.99}$$
$$Y_9 = X_9 \tag{7.100}$$

Example 7.8

From the steady state expressions found in Examples 7.1 through 7.6, determine expressions for the steady state output current during the rectifier segment when thyristors T3 and T5 conduct.

When the delay angle is zero, thyristors T3 and T5 conduct during the segment which extends from 240° to 300°. Generalizing the steady state expressions for i_o already found by substituting the dummy angle ψ for θ produces for the range $60.0° < \psi < 72.24°$

$$i_o = 222.721 \cos(1.10929\psi - 1.31381) \exp(-0.161050\psi)$$

for the range $72.24° < \psi < 105.33°$

$$i_o = -3368.97 + 1116.91 \cos(\psi - 2.83580) + 5321.20 \exp(-0.322311\psi) + 13.3573 \cos(7.49661\psi - 2.9319) \exp(-0.165680\psi)$$

and for the range $105.33° < \psi < 120.0°$

$$i_o = 263.65 \cos(1.10929\psi - 2.47544) \exp(-0.161050\psi)$$

The new segment lags by 180° so that $\delta = 180°$ and $\theta - \pi$ is substituted for ψ in these expressions to produce for the range $240.0° < \theta < 252.24°$

$$i_o = 369.397 \cos(1.109290\theta - 4.79875) \exp(-0.1610500\theta)$$

for the range $252.24° < \theta < 285.33°$

$$i_o = -3368.97 + 1116.91 \cos(\theta - 5.97739) + 14647.5 \exp(-0.3223110\theta) + 22.4785 \cos(7.496610\theta - 26.4832) \exp(-0.1656800\theta)$$

and for the range $285.33° < \theta < 300.0°$

$$i_o = 437.281 \cos(1.109290\theta - 5.96038) \exp(-0.1610500\theta)$$

The phase angles can be left unchanged or can be normalized to the range $-\pi$ to π by subtracting or adding integral multiples of 2π. Thus the

Table 7.2 *Composition of the thyristor and a.c. line currents*

Segment	1	2	3	4	5	6
Start	α	$\alpha+60°$	$\alpha+120°$	$\alpha+180°$	$\alpha+240°$	$\alpha+300°$
Finish	$\alpha+60°$	$\alpha+120°$	$\alpha+180°$	$\alpha+240°$	$\alpha+300°$	$\alpha+360°$
Conducting thyristors	T1-T6	T2-T6	T2-T4	T3-T4	T3-T5	T1-T5
Conduction starts	$\gamma_2-60°$	γ_2	$\gamma_2+60°$	$\gamma_2+120°$	$\gamma_2+180°$	$\gamma_2+240°$
ends	$\gamma_3-60°$	γ_3	$\gamma_3+60°$	$\gamma_3+120°$	$\gamma_3+180°$	$\gamma_3+240°$
δ	$-60°$	$0°$	$60°$	$120°$	$180°$	$240°$
i_{rec}	C1	C1	C1	C1	C1	C1
i_{t2}	0	C1	C1	0	0	0
i_r	C1	0	–C1	–C1	0	C1
i_y	0	C1	C1	0	–C1	–C1
i_b	–C1	–C1	0	C1	C1	0

phase angle of -26.4832 rad in the second of the three new expressions can be changed to -1.35046 by adding 8π without changing the numerical values produced by the expression.

****

7.11 Thyristor and a.c. side currents in the steady state

Having reached the steady state and obtained the load variables, attention focuses on the other circuit variables, the voltages and currents of the thyristors, and the a.c. side currents and voltages. These are obtained in a manner similar to that employed in Chapters 5 and 6. A supply cycle is divided into six rectifier segments, each of 60° duration. Segments in the steady state can be of either types 1, 2, or 7, examples being shown in Figs 7.3, 7.4, and 7.5. It is convenient to divide each steady state segment into three sections, a discharging section of type D1 extending from γ_1 to γ_2, a charging section of type C1 extending from γ_2 to γ_3, and a discharging section of type D2 extending from γ_3 to γ_4. In the case of a type 1 segment, the extent of the discharging type sections is zero, i.e. $\gamma_2 = \gamma_1$ and $\gamma_4 = \gamma_3$. In the case of a type 2 segment the extent of the D1 segment is zero, i.e. $\gamma_2 = \gamma_1$.

Table 7.2 shows the composition, for type 7 segments, of the current in thyristor T2, i_{t2}, and the currents in the red, yellow, and blue lines, i_r, i_y, and i_b, in terms of the rectifier current, i_{rec}. Row 1 of the table gives the segment number, 1 through 6, the six segments occupying a supply cycle. Row 2 gives the angle at which the segment starts and row 3 gives the angle at which it ends. Row 4 lists the nominally conducting thyristors. Rows 5 and 6 give the start and finish of conduction. Row 7 shows the angle, δ, by which the steady state expression for the rectifier current which has been found during segment 2, must be shifted to give the correct expression for the segment. Row 8 lists, by the symbol C1 for charging mode number 1, the expression for the rectifier current during the conducting section. This expression is shifted by the angle δ so as to start at $\gamma_2 + \delta$ and finish at $\gamma_3 + \delta$. Rows 9 through 12 list the composition of the currents in thyristor T2 and in the three a.c. lines in terms of the rectifier current function, C1, appropriately shifted by the angle δ.

Example 7.9

Determine for the steady state operating condition investigated in Examples 7.1 through 7.6, expressions for the current in thyristor T2 and in the red input a.c. line over a supply cycle extending from zero to 360°.

It has been found that the rectifier segments are of type 7 with a discharging type section occupying the first 12.2366°, Example 7.6, a charging type section occupying the next 33.091°, Example 7.3, and a discharging type section occupying the final 14.6724°, Example 7.5. During the discharging type sections the rectifier is cut off and no current flows in the thyristors or a.c. lines.

During the second segment, from 60° to 120° it has been found, in Example 7.2, that the expression for the rectifier current, labelled C1 in Table 7.2, is for the range $72.2366° < \theta < 105.3276°$

$$\begin{aligned} i_{rec} = {} & -3368.97 + 365.999\cos(\theta - 1.93743) + 5321.49\exp(-0.322311\theta) + \\ & 584.366\cos(7.49661\theta + 0.210952)\exp(-0.165680\theta) \end{aligned}$$

From Table 7.2 it is seen that this expression, with appropriate phase shifts, describes the current in thyristor T2 during Segments 2 and 3 and that the current is zero throughout the remaining segments, 1, 4, 5, and 6. During Segment 2, no phase shift is required so that for the range $72.2366° < \theta < 105.3276°$

$$\begin{aligned} i_{t2} = {} & -3368.97 + 365.999\cos(\theta - 1.93743) + 5321.49\exp(-0.322311\theta) + \\ & 584.366\cos(7.49661\theta + 0.210952)\exp(-0.165680\theta) \end{aligned}$$

To derive the corresponding expression for Segment 3 requires a phase lag of 60° to obtain the expression for the range $132.2366° < \theta < 165.3276°$

$$\begin{aligned} i_{t2} = {} & -3368.97 + 365.999\cos(\theta - 2.98463) + 7457.91\exp(-0.322311\theta) + \\ & 695.08\cos(7.49661\theta - 7.63948)\exp(-0.165680\theta) \end{aligned}$$

Elsewhere throughout the cycle this thyristor current is zero.

From Table 7.2, during the charging section of Segment 1 the current in the red line is C1 with a phase lead of 60°, i.e. for the range $12.2366° < \theta < 45.3276°$

$$i_r = -3368.97 + 365.999\cos(\theta - 0.89023) + 3797.07\exp(-0.322311\theta) + 491.286\cos(7.496610 + 8.06138)\exp(-0.165680\theta)$$

The current in the red line is zero throughout Segment 2.

During Segment 3, the current is -1 times C1 with a phase lag of 60°, i.e. for the range $132.2366° < \theta < 165.3276°$

$$i_r = 3368.97 - 365.999\cos(\theta - 2.98463) - 7457.91\exp(-0.322311\theta) - 695.08\cos(7.496610 - 7.63948)\exp(-0.165680\theta)$$

During Segment 4, the current is -1 times C1 with a phase lag of 120°, i.e. during the range $192.2366° < \theta < 225.3276°$

$$i_r = 3368.97 - 365.999\cos(\theta - 4.03183) - 10452.1\exp(-0.322311\theta) - 826.77\cos(7.496610 - 15.4899)\exp(-0.165680\theta)$$

The current in the red line is zero throughout Segment 5.

During Segment 6, the current is C1 with a phase lag of 240°, i.e. for the range $312.2366° < \theta < 345.3276°$

$$i_r = -3368.97 + 365.999\cos(\theta - 6.12622) + 20529.1\exp(-0.322311\theta) + 1169.73\cos(7.496610 - 31.1908)\exp(-0.165680\theta)$$

****

7.12 Thyristor and a.c. side voltages in the steady state

As already discussed in Chapters 5 and 6, the system voltages can be computed by starting from either the known voltage at the a.c. busbars or the known load e.m.f. As in those chapters, a start will be made at the a.c. bus and will work towards the load e.m.f. so as to provide a check on this lengthy and complex computation since the final expressions should coincide with the constant value of the e.m.f.

With the red phase voltage as datum, the line to line voltages at the system busbars are

$$v_{ry} = \sqrt{2}V_s\cos(\theta + \frac{\pi}{6}) \tag{7.101}$$

$$v_{yb} = \sqrt{2}V_s \cos(\theta - \frac{\pi}{2}) \tag{7.102}$$

$$v_{br} = \sqrt{2}V_s \cos(\theta + \frac{5\pi}{6}) \tag{7.103}$$

The a.c. line to line voltages at the rectifier input terminals are obtained by subtracting the line voltage drops from the bus bar voltages.

$$v'_{ry} = v_{ry} - R_s(i_r - i_y) - X_s\frac{\mathrm{d}(i_r - i_y)}{\mathrm{d}\theta} \tag{7.104}$$

$$v'_{yb} = v_{yb} - R_s(i_y - i_b) - X_s\frac{\mathrm{d}(i_y - i_b)}{\mathrm{d}\theta} \tag{7.105}$$

$$v'_{br} = v_{br} - R_s(i_b - i_r) - X_s\frac{\mathrm{d}(i_b - i_r)}{\mathrm{d}\theta} \tag{7.106}$$

Alternatively v'_{yb} can be obtained by introducing a phase lag of $2\pi/3$ into v'_{ry} and v'_{br} can be obtained by introducing a phase lead of $2\pi/3$ into v'_{ry} following the procedure already described by eqns 7.90 and 7.91, and used in Examples 7.8 and 7.9.

In determining the voltage across a thyristor, a problem is encountered similar to that faced in Chapter 6—the difficulty of assigning a value to this voltage during cutoff. The same solution is used; assuming that the impedances of the thyristors are equal during cutoff. The voltage across a thyristor is then the appropriate a.c. line to neutral voltage. Thus, the voltage across thyristor T2 during cutoff is taken to be $v_y = \sqrt{2/3}V_s \cos(\theta - 2\pi/3)$. This has the unfortunate result of appearing to give, from time to time, a forward biased device during cutoff. The resolution of this difficulty is provided by remembering that two thyristors in the bridge must conduct in series and that the companion thyristor will always be reverse biased by a greater voltage.

During the conducting mode there is no such problem, the thyristor voltages being always well defined. When the thyristor is conducting, the voltage across it is its forward drop, V_{thy}. When another thyristor in the same half of the bridge is conducting, the voltage across the thyristor in question is the appropriate line to line voltage at the rectifier terminals plus the voltage drop in the conducting thyristor. As an example, thyristor T2 conducts during Segments 2 and 3, from γ_2 to γ_3 and from $\gamma_2 + 60°$ to $\gamma_3 + 60°$, see Table 7.2. During these two periods $v_{t2} = V_{thy}$.

Thyristor T1 conducts during Segments 1 and 6, from $\gamma_2 - 60°$ to $\gamma_3 - 60°$ and from $\gamma_2 + 240°$ to $\gamma_3 + 240°$ and during these periods $v_{t2} = v'_{yr} + V_{thy}$.

Thyristor T3 conducts during Segments 4 and 5, from $\gamma_2 + 120°$ to $\gamma_3 + 120°$ and from $\gamma_2 + 180°$ to $\gamma_3 + 180°$ and during these periods $v_{t2} = v'_{yb} + V_{thy}$.

During cutoff, the voltage behind the choke, v_{bc}, is the capacitor voltage. During the charging mode it is the appropriate line to line voltage at the rectifier input terminals less the forward conducting drops in two thyristors. Considering Segment 1 of Table 7.2, thyristors T1 and T6 conduct from $\gamma_2 - 60°$ to $\gamma_3 - 60°$ connecting the red and blue a.c. lines to the rectifier output. During this period $v_{bc} = v'_{rb} - 2V_{thy}$. During Segment 5, T3 and T5 conduct from $\gamma_2 + 180°$ to $\gamma_3 + 180°$, connecting the blue and yellow a.c. lines to the rectifier output. During this period $v_{bc} = v'_{by} - 2V_{thy}$.

The output voltage, which is also the capacitor voltage, is obtained by subtracting the voltage drop in the choke from the voltage behind the choke, i.e.

$$v_o = v_c = v_{bc} - R_{ch}i_{rec} - X_{ch}\frac{di_{rec}}{d\theta} \tag{7.107}$$

Finally the load e.m.f. is obtained by subtracting the voltage drop in the load impedance from the output voltage, i.e.

$$e_o = v_o - R_o i_o - X_o\frac{di_o}{d\theta} \tag{7.108}$$

Example 7.10

Derive steady state expressions for the voltage across and the current through thyristor T2, the current in the red a.c. line and the voltage between the red and yellow rectifier input terminals during Segment 4 of Table 7.2. Use the low supply impedance, low smoothing capacitance case considered in all the preceding examples with $\alpha = 0$ and $E_o = 583$ V.

Under these conditions, the rectifier exhibits type 7 behavior with a charging section sandwiched between two discharge sections. During Segment 4 of Table 7.2, which covers the range $180° < \theta < 240°$, thyristors T3 and T4 conduct during the charging section from 192.24° to 225.33° and are reverse biased during the two discharge sections from 180° to 192.24° and from 225.33° to 240°.

The current through thyristor T2 is zero throughout this segment.

During cutoff the currents in all three a.c. lines are zero. During the charging mode the rectifier current flows in the positive direction in the blue line and in the negative direction in the red line and there is no current in the yellow line. Thus for the range $180.0° < \theta < 192.24°$

$$i_r = 0.0$$

for the range $192.24° < \theta < 225.33°$

$$i_r = 3368.97 - 365.999\cos(\theta - 4.03183) - 10452.1\exp(-0.322311\theta) - 826.77\cos(7.49661\theta - 15.4899)\exp(-0.165680\theta)$$

and for the range $225.33° < \theta < 240.0°$

$$i_r = 0.0$$

During the whole segment the voltage between the red and yellow busbars is

$$v_{ry} = 622.254\cos(\theta + \frac{\pi}{6})$$

During cutoff this is the voltage between the red and yellow input terminals of the rectifier.

During the charging type section of this segment the rectifier current flows positively in the blue line and negatively in the red line so that the voltage absorbed by the red line impedance, $Z_l i_o$, must be added to the red to yellow busbar voltage. Thus, during the range $180.0° < \theta < 192.24°$

$$v'_{ry} = 622.244\cos(\theta + \frac{\pi}{6})$$

during the range $192.24° < \theta < 225.33°$

$$\begin{aligned} v'_{ry} = &-6.74 + 619.96\cos(\theta + 0.524208) - 0.58\exp(-0.322311\theta) + \\ &37.39\cos(7.496610\theta - 1.37487)\exp(-0.165680\theta) \end{aligned}$$

and during the range $225.33° < \theta < 240.0°$

$$v'_{ry} = 622.244\cos(\theta + \frac{\pi}{6})$$

For the first and third sections it is assumed that the voltage across T2 is the voltage v_y. During the second section T3 conducts, connecting the blue a.c. line to the rectifier positive terminal. The voltage across T2 is then $v'_{yb} + V_{thy}$. Hence, for the range $180.0° < \theta < 192.24°$

$$v_{t2} = 359.26\cos(\theta - 2.09440)$$

for the range $192.24° < \theta < 225.33°$

$$\begin{aligned} v_{t2} = &-5.14 + 619.96\cos(\theta - 1.57019) - 1.15\exp(-0.322311\theta) + \\ &52.91\cos(7.496610\theta - 17.0757)\exp(-0.165680\theta) \end{aligned}$$

and for the range $225.33° < \theta < 240.0°$

$$v_{t2} = 359.26\cos(\theta - 2.09440)$$

The voltage before the choke is the capacitor voltage during cutoff and is $v'_{br} - 2V_{thy}$ during the charging mode. The capacitor voltage during cutoff must be obtained from the expressions already obtained in Examples 7.5 and 7.6 but shifted by 120°. Thus, during the range $180.0° < \theta < 192.24°$

$$v_{bc} = 583.0 + 184.62\cos(1.10929\theta - 2.21048)\exp(-0.161050\theta)$$

during the range $192.24° < \theta < 225.33°$

$$v_{bc} = 10.28 + 619.31\cos(\theta - 2.61219) - 1.17\exp(-0.322311\theta) + 74.79\cos(7.49661\theta + 14.3331)\exp(-0.165680\theta)$$

and during the range $225.33° < \theta < 240.0°$

$$v_{bc} = 583.0 + 281.55\cos(1.10929\theta - 3.37211)\exp(-0.161050\theta)$$

Since the example does not have an output choke, the output voltage is the same as the voltage before the choke. When the drop in the load impedance is subtracted from the expression for $v_{bc} = v_o$ just derived, the result is the load e.m.f. When this computation is carried out it yields, within the limits of computer roundoff error, the constant value of 583 V for the e.m.f. during all three sections of the segment. Since this is the known value of the e.m.f. the result confirms the accuracy of the computations.

****

The steady state voltage and current waveforms for the example are shown in Fig. 7.9 where each graph is identified by a number at its right-hand side. Graph 7 shows the sinusoidal voltage of r.m.s. value 440 V between the red and yellow busbars and all other waveforms use the same θ scale.

Graph 1 shows the rectifier output voltage by a black line and the load e.m.f. by a dashed line.

Graph 2 shows the output current to the load. The ripple at six times the supply frequency is apparent.

Graph 3 shows the waveform of the capacitor current by a solid line and of the rectifier current by a dashed line. The rectifier current is comprised of positive pulses at six times the supply frequency which reach a maximum value of 573 A. The capacitor current, also at six times the supply frequency, is both positive and negative with a mean value of zero. During the discharging periods it supplies the output current and its value is -1 times the output current. During the charging periods, except for brief periods at the start and finish, the capacitor current is positive to restore the capacitor charge lost during the discharging period.

Graph 4 shows the current in thyristor T2. It has two pulses per supply cycle, each pulse being identical with a rectifier output current pulse. The

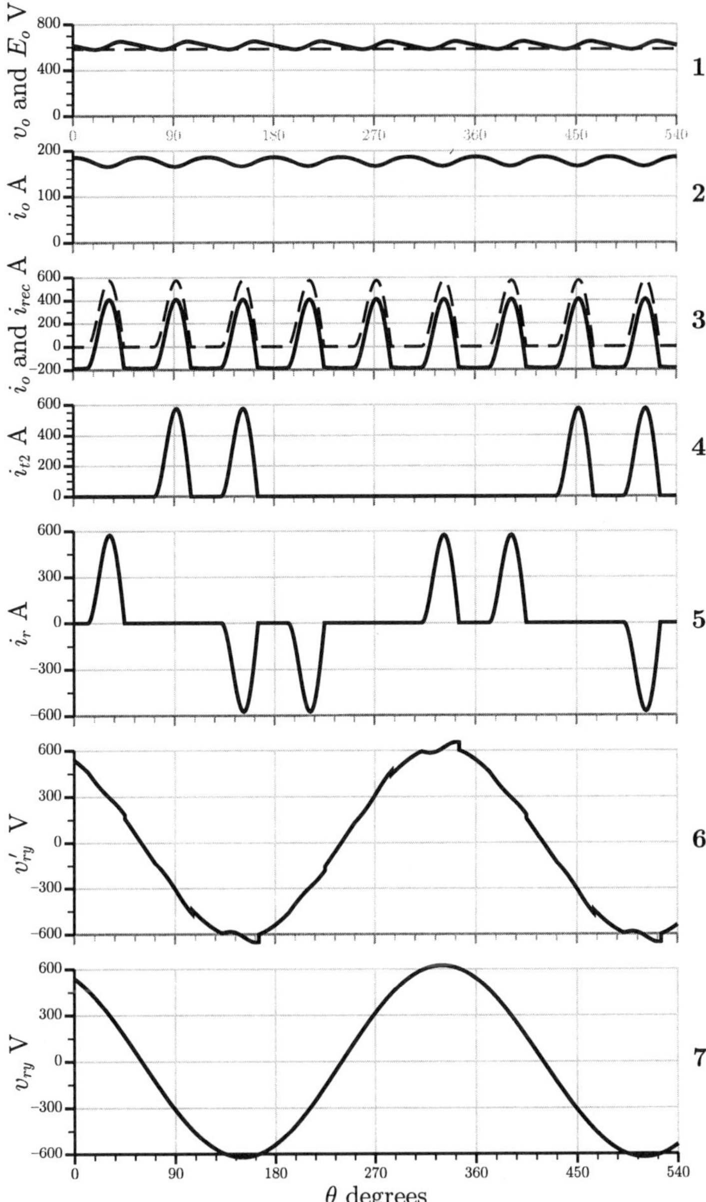

FIG. 7.9. Waveforms of, graphs 1 and 2, d.c. side quantities, graphs 3 and 4, rectifier quantities, and, graphs 5, 6, and 7, a.c. side quantities.

current in the other thyristors is similar but with the appropriate phase shift. Thus the current in thyristor T6 has a similar waveform but with a phase lag of 300°.

Graph 5 shows the current in the red a.c. line. This has four pulses per supply cycle, two negative and two positive. Each pulse is identical to a rectifier output current pulse. The positive pulses correspond to the conduction periods of thyristor T1 and the negative pulses to the conduction periods of thyristor T4. The waveforms of the yellow and blue line currents are similar but with appropriate phase shifts—120° lag for the yellow line and 240° lag for the blue line.

Graph 6 shows the waveform of the voltage between the red and yellow input terminals to the rectifier. The waveforms of the other two line to line voltages are similar but with appropriate phase shifts. During the discharging periods this voltage is identical to the busbar voltage. During the rectifier charging periods there is distortion due to the voltage drops in the line impedances. Four periods of small distortion and two periods of larger distortion occur during each supply cycle. The large distortions occur when current flows in both the red and yellow lines, i.e. when either thyristors T1 and T5 or T2 and T4 conduct. The smaller distortions occur when current flows in only one of the two lines, i.e. when either thyristor T3 or T6 conducts.

7.13 Power, r.m.s., and mean values in the steady state

The steady state expressions for the various currents and voltages permit the calculation of the mean and r.m.s. values which summarize rectifier operation.

The mean values of the d.c. side quantities and of the thyristor current are of interest. The mean value of a d.c. side quantity is obtained by integrating it over a rectifier segment and dividing the result by the width of the segment, $\pi/3$. For the output current and voltage and the capacitor current, this involves integration of the discharging mode expressions as well as the charging mode expression. Only the charging mode integral is required for the rectifier current since it is zero during the discharging modes.

There is a rectifier current pulse during each rectifier segment and each thyristor carries two of the six pulses which occur during each supply cycle. Thus the mean thyristor current is one-third of the mean rectifier current.

The integrals required for the determination of all the above mean values are particular cases of the integral of the basic capacitor smoothing expression which is given in eqn B.8.

The r.m.s. value of a quantity is obtained in four steps. First, the expression for the quantity is squared. Second, the resulting expression is integrated over the appropriate range. Third, the mean square is obtained by averaging the sum of the integrals. Finally the square root of this number is the required r.m.s. value. This process has much in common with the derivation of power.

The mean d.c. output power, which is a quantity of major interest in a rectifier, is the product of the mean output current and the mean output voltage. The mean total output power is the mean value of the instantaneous power which is the product of the output voltage expression and the output current expression. The mean total output power includes the power associated with the output harmonics. This harmonic component is usually useless to the load, it simply increases the load heating and reduces efficiency.

The rectifier instantaneous input power is the sum of the instantaneous phase powers, $v'_r i_r + v'_y i_y + v'_b i_b$, which since the a.c. system is three wire for which $i_b = -(i_r + i_y)$, can alternatively be expressed as $v'_{ry} i_r + v'_{yb} i_y$. The mean value of this quantity includes the total d.c. output, the losses in the thyristors and the losses in the output choke.

The kVA input as conventionally measured is $\sqrt{3}$ times the r.m.s. value of an a.c. line current times the r.m.s. value of an a.c. line to line voltage divided by 1000. The input power factor as conventionally measured is the ratio of the mean input power in kilowatts to the input kVA as conventionally measured.

The product of two basic capacitor smoothing expressions is essential to many of these calculations. If an r.m.s. value is being determined, the two expressions which are multiplied are the same. If power is being determined, one expression is a voltage and the other is a current. In view of the complexity of the basic capacitor smoothing expression, the product of two such expressions is extremely complex and is given, together with its integral, in eqns B.10 and B.32.

Example 7.11

Determine the output and input powers and the mean and r.m.s. values of all the rectifier currents and voltages for the case illustrated in Fig. 7.9 and discussed in Example 7.10.

The mean value of the output current is 176.944 A and its r.m.s. value is 177.089 A so that the r.m.s. ripple content of this current, which is the square root of the difference between the squares of the r.m.s. and mean values, is 7.175 A or 3.99% of the rated load current.

The mean value of the output voltage is 613.086 V and the r.m.s. value is 613.519 V. The r.m.s. ripple content of this voltage is 23.054 V or 4.15% of the rated load voltage.

A check on these results is provided by computing the load e.m.f. as the difference between the mean output voltage and the voltage drop in the load resistance due to the mean load current, i.e. $613.086 - 0.17 \times 176.944 = 583.006$ V, which is acceptably close to the actual value of 583 V.

The d.c. component of the output power, which is the product of the mean output voltage and mean output current, is 108.482 kW, 103.158 kW of which is useful load power and the remainder, 5.324 kW, is d.c. loss in the load resistance, a figure in agreement with the product of the square of the mean load current and the load resistance.

The average value of the total power delivered to the load, i.e. the mean value of the product of the instantaneous values of output voltage and current, is 108.490 kW indicating that the power associated with the output ripple is only 8.7 W, less than 0.01% of the rated load power. The very small value of the ripple power is due to the fact that the output current ripple is almost 90° out of phase with the voltage ripple, a matter discussed in more detail in Chapter 8.

The mean capacitor current is −0.0003 A, another satisfactory check on the accuracy of the results since it is acceptably close to the zero value which it should have because the loss of capacitor charge during the discharging periods is balanced by the gain during the charging periods.

The r.m.s. capacitor current is 222.832 A. Since this current has zero mean value, its r.m.s. value is comprised entirely from harmonics. The first non-zero harmonic, whose frequency is six times that of the a.c. supply, has the largest amplitude and makes the largest contribution to this r.m.s. value.

The mean rectifier current is 176.944 A, in agreement with the mean value of the output current. The r.m.s. value of the rectifier current is much larger at 279.258 A, indicating a large harmonic content. The harmonic content is in fact 216.047 A and is at six times the supply frequency.

The mean thyristor current is one-third of the mean rectifier current, 58.981 A, and the r.m.s. value is equal to 161.230 A, the r.m.s. value of the rectifier current divided by $\sqrt{3}$. The thyristor's current rating will be somewhere between these two values.

On the a.c. side only r.m.s. values are of interest, all mean values over a supply cycle being zero. The line to line voltage at the busbars is 440 V. The line to line voltage at the rectifier terminals is 439.707 V. The a.c. line current is 228.012 A.

The mean value of the power delivered from the busbars is 109.369 kW and the mean power delivered to the rectifier input terminals is 109.056 kW. The kVA at the rectifier input terminals as conventionally measured is $\sqrt{3}$ times the product of the r.m.s. line to line voltage and the r.m.s. line current, i.e. $\sqrt{3} \times 439.707 \times 228.012 \div 1000 = 173.653$ so that the input power factor, as conventionally measured, is 0.6280. Since the rectifier is operating at zero delay, the fundamental power factor is almost 1.0 and the relatively low value of the conventional power factor is indicative of the high harmonic content of the a.c. line currents.

The values of power provide a check on the accuracy of the results. The loss in the three a.c. lines is the I^2R loss in the line resistances, i.e. $3 \times 228.012^2 \times 0.002 = 311.938$ W. The difference between the power delivered by the busbars and the input power to the rectifier indicates a loss in the a.c. lines of 312.164 W which is in agreement with the above figure within the limits of computational accuracy. With the assumption of a constant voltage drop in a conducting thyristor, the power lost in the thyristors is six times the product of the mean thyristor current and the thyristor forward voltage drop, i.e. $6 \times 58.981 \times 1.6 = 566.219$ W. The difference between the input to the rectifier and the output to the load is 566.406 W which is also in acceptably close agreement to the directly calculated thyristor loss.

****

7.14 The load characteristics

Rectifier performance is often summarized by load characteristics which show the mean output voltage as a function of the mean output current for a given load and angle of delay. Although capacitor smoothing is normally employed with uncontrolled rectifiers, a comprehensive set of characteristics for the low supply impedance, low smoothing capacitance case is shown in Fig. 7.10. The starting point of a characteristic on the zero current axis is indicated by a black dot.

The characteristics cover the delay angle range of zero through 150° to emphasize the fact that a capacitor smoothed rectifier can operate in the negative output region provided the smoothing capacitor can withstand negative voltage, an ability not possessed by the electrolytic capacitor which is usually used.

Because conduction is discontinuous, the characteristics are nonlinear, most of them exhibiting a considerable reduction in output voltage with increase in load current, the greater part of the reduction occurring in the light load region, from zero load to about 20% of rated load. At zero load, the smoothing capacitor charges to the rectifier cutoff voltage and at higher loads it stabilizes at some lower value where the loss in charge during the discharging periods is just balanced by the gain during the charging periods.

For delay angles less than 30° the rectifier cutoff voltage is the peak value of the a.c. line to line voltage, in the example $\sqrt{2} \times 440 = 622.3$ V, so that the $\alpha = 0$ and the $\alpha = 30°$ characteristics start from the same no load point but with the 30° characteristic having a larger regulation,

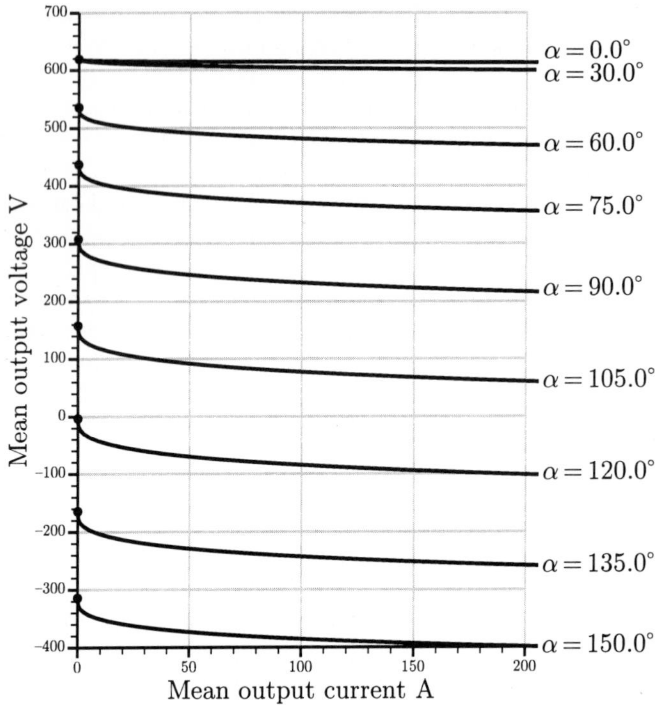

FIG. 7.10. Load characteristics for the low supply impedance case.

622.3 V down to 599.7 V at 200 A as compared to the zero delay angle case for which the corresponding voltage drop is from 622.3 V to 612.8 V. Thus in contrast to the larger delay angle cases, the regulation of a capacitor smoothed uncontrolled rectifier is low. In the case of the example it is about 1%.

7.15 Conclusion

The load characteristics of Fig. 7.10 show that, from the point of view of the load, capacitor smoothing is excellent when the delay angle is less than 30°, providing an output voltage which is almost equal to the peak value of the rectifier input line to line voltage at all loads. Taking the regulation as the drop in mean output voltage from zero load to the rated load of 180 A expressed as a percentage of the rated load voltage of 555.5 V, the regulation of the example at zero delay is 1.08% increasing to 3.36% at 30° delay. In contrast, at 60° delay the regulation has increased to 11.41% and reaches a maximum of 17.20% at 120° delay. Because of these large values of regulation, the capacitor smoothed rectifier is unattractive compared to the inductor smoothed rectifier for large delay angles and this is one of the reasons why it is generally used with an uncontrolled rectifier with diodes in place of thyristors so that the delay angle is zero. The remainder of this discussion is confined to such rectifiers.

The output voltage of an uncontrolled rectifier is always positive, a feature which permits the use of a physically smaller, lighter, and less expensive electrolytic capacitor in place of the bipolar capacitor which would be required for a controlled rectifier. This choice prevents the flow of generated load power back to the a.c. supply and therefore prohibits the use of regenerative braking if the load is a d.c. motor. A partial solution to this problem is provided by sensing the capacitor voltage and, if this rises above the cutoff voltage, $\sqrt{2}$ times the r.m.s. line to line input voltage to the rectifier, temporarily connecting a resistor in parallel with the capacitor to absorb the generated energy. To distinguish it from regenerative braking, this process is called *dynamic braking*. In many instances the regenerated energy is a small part of the total used by the process and its waste as heat in the dynamic braking resistor is a small price to pay for the much simpler diode rectifier.

Capacitor smoothing provides a very good d.c. output whose voltage is $\pi/3 = 1.047$ times greater than that available with inductor smoothing, $\sqrt{2}V_s$ as against $(3\sqrt{2}/\pi)V_s$, and which is almost independent of variations in load. The output ripple can be reduced to any level which is economically justified, by the installation of more smoothing capacitance. At zero delay and rated load current the r.m.s. values of the output voltage and output current ripples of the example are respectively 4.22% and 4.09% of the rated values. Increasing the smoothing capacitor by a factor of ten to 40 000 μF almost eliminates the output ripple, reducing it for both current and voltage to about 0.25%. Surprisingly the regulation, while still small, is increased to 3.07%. This unexpected result is associated with the lower natural frequency during the charging mode. The natural frequency during the charging mode is 141.9 Hz as against 449.8 Hz for the low capacitance case. At 141.9 Hz there is not time to fully recharge the capacitor during a rectifier segment whose duration is 1/360 s and the output voltage must settle to a somewhat lower level before equilibrium is attained.

The price paid for this excellent output performance is seen when the rectifier and a.c. supply quantities are considered. Current is delivered to the parallel combination of the load and capacitor in brief pulses which results in high r.m.s. currents, high harmonic content, and poor power factor. For the low capacitance case of Examples 7.10 and 7.11, the mean rectifier current is equal to the mean output current both being 176.9 A and yet, because it is delivered in pulses, the r.m.s. value of the rectifier current is almost 60% greater at 279.2 A. The r.m.s. value of the capacitor current is almost as large, 222.8 A, and is at the rectifier ripple frequency, six times the supply frequency or 360 Hz. This large, high frequency ripple and the high d.c. voltage capability of 622 V are very demanding requirements for an electrolytic capacitor.

Each diode must conduct one-third of the rectifier current pulses so that the mean diode current is 59.0 A while the r.m.s. value is nearly three times as big at 161.2 A.

The peak diode current of 573.0 A is also a significant feature in the selection of appropriate devices. The net result is that the diodes must have a significantly higher current rating than those used for an equivalent inductor smoothed rectifier.

An a.c. line current consists of two-thirds of the rectifier current pulses, one-third positive and one-third negative. The r.m.s. line current is 228.0 A and, with the r.m.s. line to line voltage of 439.7 V, gives an input kVA as conventionally measured of 173.6. Since the mean input power is 109.0 kW, the input power factor is 0.6280 in contrast to that of an inductor smoothed rectifier which would have a power factor of about 0.95 under similar circumstances. The rectifier is operating uncontrolled at zero delay, with a fundamental power factor of almost unity and the low conventional power factor is indicative of the high harmonic content of the a.c. line currents.

In discussing both the a.c. line currents and the capacitor rating, the potentially very large initial capacitor charging current must be borne in mind. This was discussed in connection with Fig. 7.8 in Section 7.9.

In summary, for a load which does not require either voltage control or regeneration, capacitor smoothing provides a d.c. output of high quality. However, the high harmonic content of the rectifier and a.c. supply currents and the poor input power factor restrict application to relatively low power levels where these matters are of little concern.

8

FOURIER ANALYSIS

8.1 Introduction

The repetitive non-sinusoidal waveforms associated with controlled rectifiers are obvious subjects for Fourier analysis. Any of the waveforms which have been determined in the preceding chapters can be analyzed into Fourier components and the spectra so derived provide insight into the operation of the system and highlight many factors of prime importance. The first part of this chapter illustrates the technique and the conclusions which can be drawn from its application by determining the Fourier components of the current and voltage waveforms already found in Chapter 5.

However, while the method is of considerable value in analyzing waveforms already derived, it has merit as a basic analytical tool replacing time domain analysis. Thus, instead of being involved with the solution of differential equations, the problem is transformed to the application of steady state a.c. circuit theory. Strictly speaking there are very few, if any, such situations where time domain solution can be avoided. However, by making appropriate and generally unimportant simplifications, many power electronic problems can be placed in this class. In the present context, controlled rectifiers, the crucial assumptions are that conduction is continuous and that waveforms associated with commutation can be omitted although the effects of commutation on the mean values of output current and voltage are still modeled, at least approximately, by the use of the a.c. equivalent rectifier circuit shown in Fig. 5.8. With these assumptions the waveform of the rectifier Thévenin voltage is known and can be analyzed into its spectral components and the effect of each of these on the load determined separately. The sum of all of these load current components is then the total load current.

An analysis of this type is started in the second half of the chapter, Section 8.13, but having determined the output current, Section 8.17, the problem of working back through the rectifier to determine the thyristor input currents immediately arises. This difficulty is overcome in Section 8.19 by the multiplication of two Fourier series.

The necessary tools, the essentials of Fourier analysis, and the product of Fourier series, are given in Appendices C and D.

The first part of the chapter, Sections 8.2 through 8.12, is devoted to the analysis of waveforms which have already been derived by time domain analysis. The second part, Sections 8.13 through 8.24, applies the direct approach to a semi-controlled rectifier. A first reading could concentrate on this latter part and in particular Sections 8.13 through 8.18, omitting the sections dealing with the multiplication of Fourier series.

8.2 Analysis of waveforms already derived

When a solution to the controlled rectifier problem has already been obtained by time domain analysis and spectral information is to be derived from it, the crucial waveform to be analyzed is that of the current in a thyristor since the spectra of all other currents and voltages can be derived from it.

The rectifier circuit which was analyzed in Chapter 5, and which is defined in Example 5.1, will be used as the illustrative example for Sections 8.2 through 8.12. In Example 5.9 it was determined that, when the delay angle is 45° and the load e.m.f. is 370 V, the current in thyristor T2 is for the range $0° < \theta < 105°$

$$i_{t2} = 0 \tag{8.1}$$

For the range $105° < \theta < 107.44°$ commutation from T1 to T2 occurs, and

$$i_{t2} = 9121.41\cos(\theta - 3.9501) - 1872.57 + 2293.48\exp(-\frac{\theta}{4.2122}) + 7031.32\exp(-\frac{\theta}{5.8141}) \tag{8.2}$$

For the range $107.44° < \theta < 165.0°$, T2 and T6 conduct, and

$$i_{t2} = 1386.75\cos(\theta - 2.9109) - 3639.82 + 4995.91\exp(-\frac{\theta}{4.2572}) \tag{8.3}$$

For the range $165.0° < \theta < 167.44°$, commutation from T6 to T4 occurs and

$$i_{t2} = 1248.22\cos(\theta - 3.4321) - 3745.13 + 5881.60\exp(-\frac{\theta}{4.2122}) \tag{8.4}$$

For the range $167.44° < \theta < 225°$, T2 and T4 conduct and the current in T2 is given by

$$i_{t2} = 1386.75\cos(\theta - 3.9581) - 3639.82 + 6389.16\exp(-\frac{\theta}{4.2572}) \tag{8.5}$$

For the range $225° < \theta < 227.44°$, during which period there is a commutation from thyristor T2 to T3, the decrease to zero of the current in T2 is described by

$$i_{t2} = 9199.52\cos(\theta - 3.0390) - 1872.57 + 3770.82\exp(-\frac{\theta}{4.2122}) - 10080.48\exp(-\frac{\theta}{5.8141}) \tag{8.6}$$

For the balance of the supply cycle, $227.44° < \theta < 360.0°$, thyristor T2 does not conduct and

$$i_{t2} = 0 \tag{8.7}$$

Equations 8.1 through 8.7 have the form of the generalized power electronics expression discussed in Appendix A and the integrals which are required for the Fourier analysis of the current waveform are given in equations A.33 through A.48. The application of these expressions is illustrated by Example 8.1.

Example 8.1

For the rectifier circuit described in Example 5.1, determine the amplitudes and phase angles of the first 21 Fourier components of the current in thyristor T2. Use the cosine form for the Fourier components.

Determine also the total r.m.s. value of the higher order Fourier components which have not been computed.

The procedure is straightforward. The amplitudes A_m and B_m of the cosine and sine components of the Fourier component of order m are found using the equations C.8, C.9, and C.10. The integrals required for each of the seven segments making up the current wave, equations 8.1 through 8.7, are given in equations A.33 through A.48.

The amplitudes and phase angles for the cosine form, eqn C.3, of the components are given in Table 8.1.

The average value of the current is the amplitude of the zero order Fourier component, 122.0 A, the value of which was obtained in Example 5.13.

The r.m.s. value of the current derived from the Fourier series by the use of eqn C.24 is 211.68 A, very little different from the true value of 212.21 A obtained directly from the time domain expressions in Example 5.13. The r.m.s. value of the remaining Fourier components, i.e. all whose order is greater than 20, is 14.93 A, the square root of the difference of the squares of these two values, i.e. $\sqrt{212.21^2 - 211.68^2}$.

The frequency spectrum and waveform of the thyristor current are shown in Fig. 8.1. The upper graph shows the frequency spectrum and the lower graph compares the actual waveform, solid line, with that derived from the Fourier components of Table 8.1, dashed line. The differences are small.

****

Table 8.1 *Fourier components of the current in thyristor T2. The order, frequency, amplitude, and phase of the first 21 Fourier components are given, for the cosine form*

Component number	Order	Frequency Hz	Amplitude A	Phase degrees
1	0	0.0	122.00	0.00
2	1	60.0	202.72	−166.42
3	2	120.0	102.94	27.12
4	3	180.0	0.00	137.70
5	4	240.0	56.87	−126.11
6	5	300.0	54.75	66.65
7	6	360.0	18.79	−104.30
8	7	420.0	12.90	−76.23
9	8	480.0	17.20	112.56
10	9	540.0	0.00	91.80
11	10	600.0	17.10	−41.60
12	11	660.0	17.96	151.49
13	12	720.0	4.45	−18.39
14	13	780.0	8.89	1.41
15	14	840.0	10.01	−165.94
16	15	900.0	0.00	168.53
17	16	960.0	9.99	40.95
18	17	1020.0	10.39	−125.64
19	18	1080.0	1.88	65.26
20	19	1140.0	6.55	82.99
21	20	1200.0	7.02	−83.72

8.2.1 *The odd triplens*

In three phase systems Fourier components whose order is an integral multiple of three times the supply frequency have special properties and in view of this they are given a special designation, the *triplens*. Thus the frequency of a triplen harmonic is $k \times 3 \times f_s$ where f_s is the supply frequency and k is any positive, nonzero integer. Reference to Table 8.1 shows that the amplitudes of the odd triplen components, i.e. the third, ninth, etc., are all zero to an accuracy of two decimal places. The computations in fact give extremely small values for these components and it is the very small cosine and sine components which have given rise to the phase angles. The amplitudes of the odd triplens are in fact truly zero, the very small

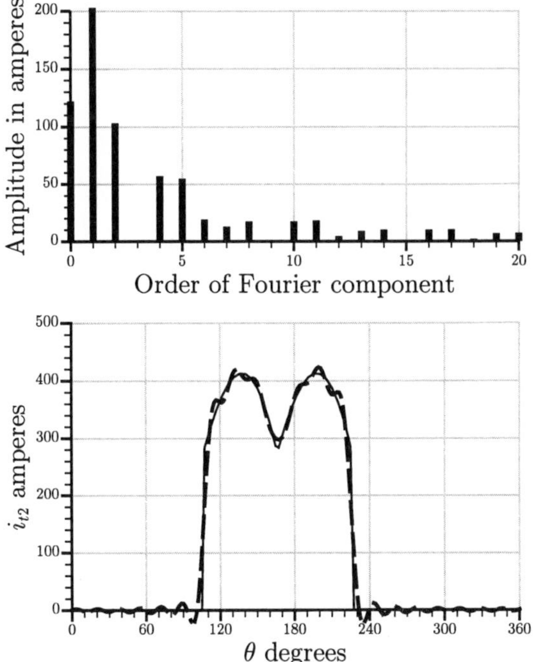

FIG. 8.1. The frequency spectrum and waveform of the current in thyristor T2. The actual waveform is shown in the lower graph by a solid line and that derived from the Fourier components by a dashed line.

computed values being due to the accumulation of roundoff errors. The proof of this statement is relatively simple and is illustrated by Fig. 8.2.

The waveform of the current in the thyristor can be regarded as the sum of the two waves in Fig. 8.2, the first being shown by a solid line and the second by a dashed line.

The first wave is composed of the first three segments of i_{t2}, eqns 8.1 through 8.3, plus the current in thyristor T6 during commutation from T6 to T4. This section is defined by eqn 8.6 phase advanced by 60°. Commutation from T6 to T4 occurs during the period $165° < \theta < 167.44°$, when the current in T6 is given by

$$i_{t6} = 9199.52\cos(\theta - 1.9918) - 1872.57 + 2940.61\exp(-\frac{\theta}{4.2122}) - 8418.97\exp(-\frac{\theta}{5.8141}) \qquad (8.8)$$

The dashed wave is composed of the current in thyristor T4 during commutation from T6 to T4 plus the last three sections of i_{t2}, eqns 8.5 through

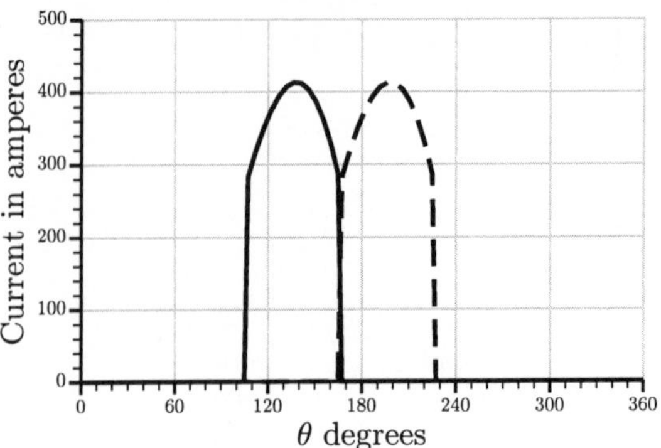

FIG. 8.2. Two pulses of identical waveform whose sum is the thyristor current.

8.7. The first section, for the range $165° < \theta < 167.44°$ segment is defined by eqn 8.2 phase retarded by 60°, i.e.

$$\begin{aligned} i_{t4} = & \; 9121.41\cos(\theta - 4.9973) - 1872.57 + \\ & 2940.80\exp(-\frac{\theta}{4.2122}) - 8418.97\exp(-\frac{\theta}{5.8141}) \end{aligned} \tag{8.9}$$

Each of the two waves of Fig. 8.2 can be analyzed into Fourier components, the two series being identical except for the phase difference of 60°, θ in the first series being replaced by $\theta - \pi/3$ in the second. Thus if i_f is the first current and i_s the second,

$$i_f = \sum_{m=0,\infty} I_{fm}\cos(m\theta + \phi_{cfm}) \tag{8.10}$$

and

$$i_s = \sum_{m=0,\infty} I_{fm}\cos(m(\theta - \frac{\pi}{3}) + \phi_{cfm}) \tag{8.11}$$

where I_{fm} is the amplitude of the mth order Fourier component of the current i_f and ϕ_{cfm} is its phase.

The Fourier series for the current i_{t2} is obtained by summing the Fourier components of these two series and for the mth order components the sum is

$$\begin{aligned} & I_{fm}\cos(m\theta + \phi_{cfm}) + I_{fm}\cos(m(\theta - \pi/3) + \phi_{cfm}) = \\ & I_{fm}\cos(m\theta + \phi_{cfm})\big[1 + \cos(\frac{m\pi}{3}) + \sin(\frac{m\pi}{3})\big] \end{aligned} \tag{8.12}$$

When m is an odd multiple of 3, $\cos(m\pi/3) = -1$ and $\sin(m\pi/3) = 0$ and the right-hand side of eqn 8.12 is zero. Thus all odd triplen Fourier components of a thyristor current are indeed zero.

8.3 The remaining currents and voltages

The determination of the spectra of the other currents and voltages of interest could proceed on similar lines. The expressions obtained for the variable by time domain analysis are converted to Fourier components by the methods used for the thyristor current with the aid of the integrals of the basic rectifier expression given in Appendix A. If one were writing a comprehensive computer program for a rectifier which would not only perform the time domain analysis but would also follow it with a frequency domain analysis, this would be an obvious and sensible way to proceed. However, it is instructive to take a different route by deriving the remaining spectra from the one already determined for the thyristor current. A thyristor current is the only variable for which this can be done and it is for this reason that the analysis commenced with this particular current.

8.4 The output current

The output current is the sum of the currents in the three thyristors in one side of the bridge. The waveforms of the thyristor currents are identical in shape, differing only in phase angle. Thus if the current in T2 is represented by $f(\theta)$, the current in T1, which leads i_{t2} by 120°, is represented by $f(\theta + 2\pi/3)$ and the current in T3, which lags i_{t2} by 120° is represented by $f(\theta - 2\pi/3)$. The output current, i_o, is the sum of these three currents,

$$i_o = f(\theta + \frac{2\pi}{3}) + f(\theta) + f(\theta - \frac{2\pi}{3}) \tag{8.13}$$

Considering the Fourier component of order m,

$$\begin{aligned} i_{om} &= I_{t2m}\big[\cos(m(\theta + \frac{2\pi}{3}) + \phi_{ct2m}) + \cos(m\theta + \phi_{ct2m}) + \\ &\qquad \cos(m(\theta - \frac{2\pi}{3}) + \phi_{ct2m})\big] \end{aligned} \tag{8.14}$$

Expanding the cosine terms so as to isolate the $2\pi/3$ components yields

$$i_{om} = I_{t2m}\cos(m\theta + \phi_{ct2m})\big[1 + 2\cos(m\frac{2\pi}{3})\big] \tag{8.15}$$

Considering k to be any positive nonzero integer, the Fourier component order can be considered to be either

$$m = 3k - 3 \quad \text{i.e. } 0, 3, 6, 9, 12 \ldots \tag{8.16}$$

or

$$m = 3k - 2 \quad \text{i.e. } 1, 4, 7, 10, 13 \ldots \tag{8.17}$$

or

$$m = 3k - 1 \quad \text{i.e. } 2, 5, 8, 11, 14 \ldots \tag{8.18}$$

Sorting the possible values of m into these three groups permits simplification of the term $2\cos(m2\pi/3)$ in eqn 8.15. Expressed in terms of k, the angle $m2\pi/3$ becomes $k2\pi - 2\pi$ in the case of eqn 8.16, $k2\pi - 4\pi/3$ in the case of eqn 8.17, and $k2\pi - 2\pi/3$ in the case of eqn 8.18. Now $k2\pi$ is an integral multiple of 2π and makes no difference to the circular functions of the angle so that it can be omitted from the $\cos(m2\pi/3)$ term of eqn 8.15. This leaves the remainders $\cos(0) = 1$ in the case of eqn 8.16, $\cos(-4\pi/3) = -0.5$ in the case of eqn 8.17, and $\cos(-2\pi/3) = -0.5$ in the case of eqn 8.18. Because of the implications of these angles in three phase systems, the series of eqn 8.16 is known as a *zero sequence series*, the series of eqn 8.17 as a *positive sequence* series and the series of eqn 8.18 as a *negative sequence series.*

Applying these conclusions, the amplitude of the Fourier component of order m of the output current in eqn 8.15 reduces to zero for all values of m other than multiples of 3. In the latter case its value is $3I_{t2m}$. It has already been seen that the odd triplens are zero so that the output current is made up of only the even triplens, 0, 6, 12, 18, etc. Hence

$$i_o = \sum_{m=0,\infty,6} 3I_{t2m} \cos(m\theta + \phi_{ct2m}) \tag{8.19}$$

where $m = 0, \infty, 6$ means that m takes values from zero to infinity in increments of 6, i.e. 0, 6, 12, 18 ...

Since it is already known that the output current waveform repeats every 60°, its Fourier components must be at multiples of six times the supply frequency. However, in addition to providing this obvious piece of information, eqn 8.19 relates the Fourier components to those already derived for a thyristor current.

Example 8.2

Determine the Fourier expression for the output current of the example, giving the first four terms of the series. Determine the mean value of the output current, estimate its r.m.s. value from these four terms, and determine the r.m.s. value of the remaining terms of the Fourier series.

Table 8.1 lists the first 21 Fourier components of the Fourier series for the thyristor current. It therefore defines the first four components of the series for the output current, the zero, sixth, twelfth, and eighteenth order components. Selecting the amplitudes and phase angles of these components and multiplying the amplitudes by 3 gives

$$i_o = 366.01 + 56.36\cos(6\theta - 1.8203) + 13.34\cos(12\theta - 0.3210) + 5.65\cos(18\theta + 1.1391) + \ldots$$

The mean value is the amplitude of the zero order component, 366.01 A, in exact agreement with the value derived from the time domain expressions in Example 5.13.

The r.m.s. value is the square root of the sum of the squares of the r.m.s. values of the Fourier components, see eqn C.24. The first four components listed above give an r.m.s. value of 368.32 A which is very close to the true value of 368.33 A obtained from the time domain expression derived in Example 5.13. The r.m.s. value of the remaining harmonics is the square root of the difference between the squares of the true r.m.s. value and the estimate and in this case is 2.04 A, a very small amount indicative of the rapid convergence of the Fourier series.

The spectrum of the output current and its waveform are shown in Fig. 8.3. The lower graph shows the actual waveform by the solid line and the approximation derived from the Fourier series by the dashed line. It is barely possible to distinguish the difference between the two, another indication of the very small contribution of Fourier components above the fourth.

****

8.5 The transformer secondary current

Referring to Fig. 5.9, the current in a transformer secondary winding is the difference between the current in a thyristor in the upper half of the bridge and that in its complementary thyristor in the lower half. Thus, the current in secondary winding number 2, i_2, is the difference, $i_{t2} - i_{t5}$. The waveform of the current i_{t5} is identical with that of i_{t2} except for a phase shift of 180°. Hence if

$$i_{t2} = \sum_m I_{t2m}\cos(m\theta + \phi_{ct2m}) \tag{8.20}$$

then

$$i_{t5} = \sum_m I_{t2m}\cos(m(\theta - \pi) + \phi_{ct2m}) \tag{8.21}$$

and

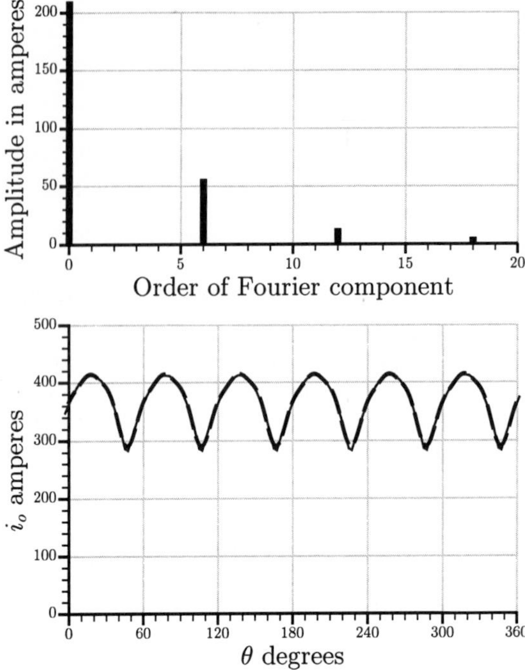

FIG. 8.3. The frequency spectrum and waveform of the output current. The actual waveform is shown in the lower graph by a solid line and the waveform derived from the Fourier series by a dashed line.

$$i_2 = \sum_m I_{t2m}\left[\cos(m\theta + \phi_{ct2m}) - \cos(m(\theta - \pi) + \phi_{ct2m})\right] \tag{8.22}$$

Expanding the second trigonometric term in eqn 8.22 and noting that $\sin(m\pi)$ is zero, the expression for the secondary current becomes

$$i_2 = \sum_m I_{t2m}\cos(m\theta + \phi_{ct2m})[1 - \cos(m\pi)] \tag{8.23}$$

Now $\cos(m\pi)$ has the value 1 if m is even and the value -1 if m is odd so that the frequency spectrum of the secondary current is comprised of only odd Fourier components with the same phase angle and twice the amplitude of the corresponding component in the thyristor in the upper half of the bridge which is supplied by the winding. Thus, the final expression for the secondary current is

$$i_2 = \sum_{m=1,\infty,2} 2I_{t2m}\cos(m\theta + \phi_{ct2m}) \tag{8.24}$$

Table 8.2 *The current in secondary winding number 2. The order, frequency, amplitude, and phase of the first 10 Fourier components are listed in columns 2 through 5*

Component number	Order	Frequency Hz	Amplitude A	Phase degrees
1	1	60.0	405.46	–166.43
2	3	180.0	0.00	137.70
3	5	300.0	109.49	66.65
4	7	420.0	25.79	–76.23
5	9	540.0	0.00	91.80
6	11	660.0	35.93	151.49
7	13	780.0	17.78	1.41
8	15	900.0	0.00	168.53
9	17	1020.0	20.78	–125.64
10	19	1140.0	13.11	82.99

This corresponds with expectations since the shape of the negative half of the waveform of the secondary current is identical with that of the positive half except for the change in sign and therefore there can be no even Fourier components, as discussed in Section C.5.

Example 8.3

Determine the first ten Fourier components of the current in the transformer secondary winding number 2 of the example and from them estimate the r.m.s. value of the current.

Using eqn 8.24, the required components are derived directly from those of Table 8.1 by selecting only the components of odd order, doubling their amplitudes, and keeping the same phase angles. This produces Table 8.2 for which it is again noted that the amplitudes of the triplens are zero.

The r.m.s. value of the secondary current computed from these ten Fourier components is 299.38 A, which compares favorably with the value of 300.12 A obtained directly from the time domain solution in Example 5.13. The r.m.s. value of the remaining Fourier components is 21.13 A.

The frequency spectrum and waveform of the secondary current are shown in Fig. 8.4. The rapid convergence of the Fourier series and the absence of even and triplen components is most evident in the spectrum. The waveform obtained from the ten Fourier components is shown by a dashed line for comparison with the actual waveform shown by a solid line.

****

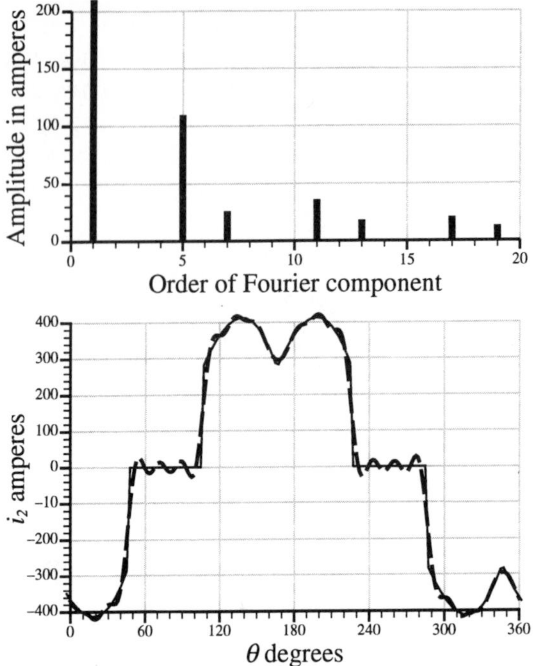

FIG. 8.4. The frequency spectrum and waveform of the current in transformer secondary winding number 2. The actual waveform is shown by a solid line and that derived from the Fourier series by a dashed line.

8.6 The current in a transformer primary winding

The current in a transformer primary winding is the sum of the reflection of the current in the corresponding secondary winding and the magnetizing current. Thus the current, i_{yb}, in the primary winding connected between the yellow and blue lines, is the current in secondary winding number 2 divided by the ratio of primary to secondary turns plus the magnetizing current, i_{myb}

$$i_{yb} = \frac{i_2}{n_{tps}} + i_{myb} \tag{8.25}$$

or, in terms of Fourier components,

$$i_{yb} = \sum_{m=1,\infty,2} \frac{I_{2m}}{n_{tps}} \cos(m\theta + \phi_{c2m}) + \sqrt{2} I_{mag} \cos(\theta + \phi_{cmyb}) \tag{8.26}$$

where n_{tps} is the number of primary turns per phase divided by the number of secondary turns per phase, I_{mag} is the r.m.s. value of the magnetizing current and ϕ_{cmyb} is its phase angle for the YB primary winding.

Table 8.3 *The current in the YB primary winding. The order, frequency, amplitude, and phase of the first 10 Fourier components are given in columns 2 through 5*

Component number	Order	Frequency Hz	Amplitude A	Phase degrees
1	1	60.0	35.60	-171.63
2	3	180.0	0.00	115.51
3	5	300.0	8.42	66.65
4	7	420.0	1.98	-76.23
5	9	540.0	0.00	-124.40
6	11	660.0	2.76	151.49
7	13	780.0	1.37	1.41
8	15	900.0	0.00	28.53
9	17	1020.0	1.60	-125.64
10	19	1140.0	1.01	82.99

The magnetizing component is combined with the first Fourier component, $m = 1$.

Example 8.4

Determine the first ten components of the Fourier series representing the current in the transformer primary winding connected between the yellow and blue terminals of the example.

The ratio of primary to secondary turns per phase is 13.0 and the first task is to divide the amplitudes of Table 8.2 by that number.

From the specification in Example 5.1, the magnetizing impedance at the supply frequency is $872.49\underline{/83.617^\circ}\ \Omega$. Since the open circuit voltage of the first secondary winding is the reference datum, the phase angle of the yellow–blue voltage is $-120°$. Thus the magnetizing current in the yellow–blue primary winding is

$$i_{myb} = \sqrt{2}\frac{3300}{872.49}\cos(\theta - 2.09440 - 1.45939) = 5.349\cos(\theta - 3.5538)$$

This added to the Fourier component of order 1 reflected from the secondary winding gives the primary Fourier component as $35.60\underline{/-171.63^\circ}$.

The first ten Fourier components are listed in Table 8.3 and give an r.m.s. value of 26.02 A, very close to the true value of 26.08 A. The r.m.s. value of the remaining components is 1.63 A.

The frequency spectrum and waveform of the primary current are shown in Fig. 8.5. The rapid convergence of the Fourier series and the absence of

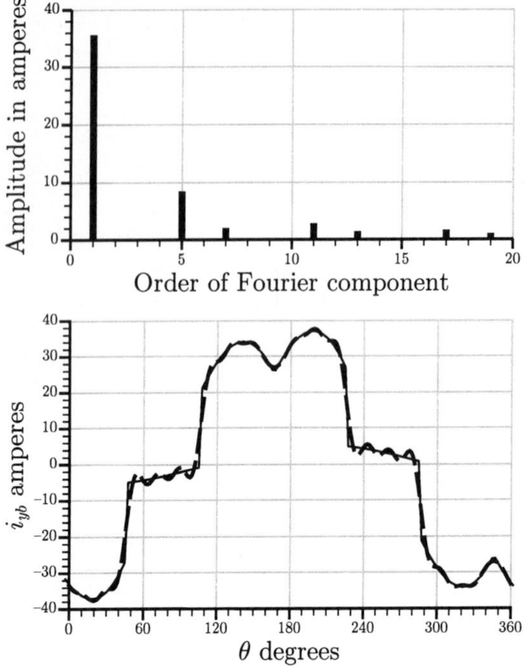

FIG. 8.5. The frequency spectrum and waveform of the current in the YB transformer primary winding. The actual waveform is shown by a solid line and that derived from the Fourier series by a dashed line.

even and triplen components is most evident in the spectrum in the first graph. The waveform obtained from the ten Fourier components is shown in the second graph by a dashed line for comparison with the actual waveform shown by a solid line. Even the small number of Fourier components used provide a good approximation to the actual waveform.

****

8.7 The a.c. supply line current

Referring to Fig. 5.9, the current in an a.c. supply line is the difference between the currents in the two primary windings connected to it. In particular the current in the red line is the difference between the current in the red–yellow primary and that in the blue–red primary.

$$i_r = i_{ry} - i_{br} \tag{8.27}$$

All three primary currents have the same waveform, the current in the red–yellow phase leading that in the yellow–blue phase by 120° and the current in the blue–red phase lagging that in the yellow–blue phase by 120°. Hence, the Fourier components of the current in the red line are obtained by generating the Fourier series for the currents in the red–yellow and blue–red primaries from the series already found for the yellow–blue primary, by introducing the appropriate phase lead or lag and then taking the difference of these two series. Considering the component of order m

$$i_{rm} = i_{rym} - i_{brm} \tag{8.28}$$

i.e.

$$i_{rm} = I_{ybm}\Big\{ \cos\big[m(\theta + \frac{2\pi}{3}) + \phi_{cybm}\big] - \cos\big[m(\theta - \frac{2\pi}{3}) + \phi_{cybm}\big]\Big\} \tag{8.29}$$

Expanding the cosine terms so as to isolate the $2\pi/3$ components yields

$$i_{rm} = -2I_{ybm}\sin(m\frac{2\pi}{3})\sin(m\theta + \phi_{cybm}) \tag{8.30}$$

Referring back to the distinctions which were made between various groupings of values of m in eqns 8.16, 8.17, and 8.18, if $m = 3k - 3$, i.e. the triplens, the value of $\sin(m2\pi/3)$ is zero. If $m = 3k - 2$, the value of $\sin(m2\pi/3)$ is $\sqrt{3}/2$ and if $m = 3k - 1$, the value of $\sin(m2\pi/3)$ is $-\sqrt{3}/2$. For convenience, since the cosine form has been used so far, the element $-\sin(m\theta + \phi_{cybm})$ can be converted to the cosine form by adding $\pi/2$ to its argument and the element $\sin(m\theta + \phi_{cybm})$ by subtracting $\pi/2$ from its argument. The result of these operations is

$$\begin{aligned} i_r = & \sum_{m=1,\infty,6} \sqrt{3}I_{ybm}\cos(m\theta + \phi_{cybm} + \frac{\pi}{2}) + \\ & \sum_{m=5,\infty,6} \sqrt{3}I_{ybm}\cos(m\theta + \phi_{cybm} - \frac{\pi}{2}) \end{aligned} \tag{8.31}$$

Since there were no triplens in the primary current, the amplitudes of all the primary current components are increased by a factor of $\sqrt{3}$ so that the r.m.s. value of the line current is $\sqrt{3}$ times the r.m.s. value of the primary phase current, an extension of the familiar result for sinusoidal three phase systems which is derived in every introductory course on power systems. However, the shapes of the primary current and line current waves are different because the phase of the negative sequence components is reversed relative to that of the positive sequence components, a fact illustrated in Example 8.5.

Table 8.4 *The current in the red a.c. supply line. The order, frequency, amplitude, and phase of the first 10 Fourier components are given in columns 2 through 5*

Component number	Order	Frequency Hz	Amplitude A	Phase degrees
1	1	60.0	61.66	–81.63
2	3	180.0	0.00	0.00
3	5	300.0	14.59	–23.35
4	7	420.0	3.44	13.77
5	9	540.0	0.00	0.00
6	11	660.0	4.79	61.49
7	13	780.0	2.37	91.41
8	15	900.0	0.00	0.00
9	17	1020.0	2.77	–215.64
10	19	1140.0	1.75	172.99

Example 8.5

Determine the first ten odd Fourier components of the current in the red a.c. supply line of the example and estimate the r.m.s. value of this current.

As an example of the technique consider the fifth order component. From Table 8.3 it is seen that in the yellow–blue primary phase this component has the value

$$i_{yb5} = 8.42\cos(5\theta + 1.1633)$$

The fifth order component is part of the second, negative sequence, group of terms in eqn 8.31 so that its value in the red line is

$$i_{r5} = \sqrt{3} \times 8.42\cos(5\theta + 1.1633 - \frac{\pi}{2}) = 14.59\cos(5\theta - 0.4075)$$

The frequencies, amplitudes, and phase angles of the first ten odd components, the even components being zero, are listed in Table 8.4. They give an r.m.s. value of 45.09 A, very close to the true value of 45.17 A derived directly from the time domain results in Example 5.13. These results indicate that the r.m.s. value of the remaining Fourier components is 2.82 A.

The frequency spectrum and waveform of this current are shown in Fig. 8.6. The spectrum is identical to that of the current in a transformer primary winding except for the scaling factor of $\sqrt{3}$. However, the waveform

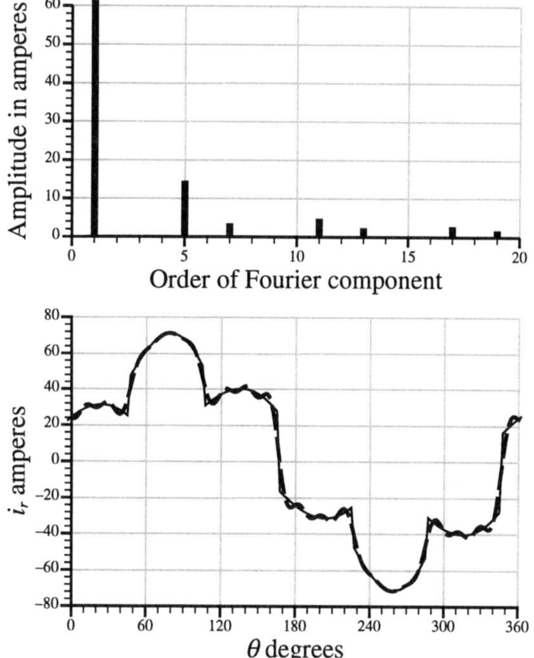

FIG. 8.6. The frequency spectrum and waveform of the current in the red a.c. supply line. The actual waveform is shown by the solid line and that derived from the Fourier components by the dashed line.

is quite different from that of the winding current due to the differences in phase angle, the $3k-2$ series components being phase advanced by 90° and the $3k-1$ components being retarded by 90°. Again the correct waveform is shown by the solid line and the waveform derived from the ten Fourier components by the dashed line.

****

8.8 The voltage waveforms

Having determined the current waveforms, the waveforms of all the voltages of interest can be obtained by making allowance for the voltages absorbed by the system impedances. For voltages on the a.c. side, the computation starts with the known sinusoidal bus bar voltages and works towards the transformer secondary terminals. For voltages on the d.c. side, the computation starts from the known load e.m.f. and works back towards the d.c. output terminals of the bridge. Derivation of the voltage waveforms does present a problem which has not been encountered with the current waveforms. The commutation notches are significant aspects of the voltage waveforms and accurate modeling of these very narrow features requires many Fourier components of high order. This requirement is illustrated by the examples.

8.9 Output voltage and voltage behind the choke

The output voltage is obtained by adding the voltage drop in the load impedance to the known load e.m.f. In the frequency domain this is a simple application of a.c. circuit theory. The load e.m.f. is the pure d.c. voltage, E_o, with no ripple. The Fourier series representing it therefore has only one term, the zero order component. The output voltage is then

$$v_o = E_o + \sum_{m=0,\infty,6} Z_{om} I_{om} \cos(m\theta + \phi_{com} + \zeta_{om}) \tag{8.32}$$

where Z_{om} is the magnitude of the load impedance at the frequency of the Fourier component of order m and ζ_{om} is its phase, I_{om} is the amplitude of the Fourier component of order m of the output current and ϕ_{com} is its phase. The expressions for the impedance and its phase are

$$Z_{om} = \sqrt{R_o^2 + m^2 X_o^2} \tag{8.33}$$

$$\zeta_{om} = \arctan(\frac{mX_o}{R_o}) \tag{8.34}$$

The voltage behind the choke is similarly obtained by adding the voltage drop in the smoothing choke

$$v_{bc} = E_o + \sum_{m=0,\infty,6} Z_{ocm} I_{om} \cos(m\theta + \phi_{com} + \zeta_{ocm}) \tag{8.35}$$

where Z_{ocm} is the magnitude of the combined load and choke impedance at the frequency of the Fourier component of order m and ζ_{ocm} is its phase. These quantities are given by

$$Z_{ocm} = \sqrt{(R_o + R_{ch})^2 + m^2 (X_o + X_{ch})^2} \tag{8.36}$$

$$\zeta_{ocm} = \arctan(\frac{m(X_o + X_{ch})}{R_o + R_{ch}}) \tag{8.37}$$

Example 8.6

Derive, for the example, the first four nonzero terms of the Fourier series representing the voltage behind the choke.

The combined resistance of the load and smoothing choke is 0.091 Ω and their combined reactance is 0.3695 Ω at the supply frequency, 60 Hz. The load e.m.f. is 370.0 V. The first four Fourier components of the output current have been found in Example 8.2. Substituting these values into eqn 8.35 yields the following components of v_{bc}

$$\begin{aligned}\mathbf{V}_{bc0} &= 370.0 + 0.091 \times 366.01 = 403.31\\ \mathbf{V}_{bc6} &= (0.091 + j2.2167) \times 56.36\underline{/-1.8203} = 125.04\underline{/-0.2907}\\ \mathbf{V}_{bc12} &= (0.091 + j4.4334) \times 13.34\underline{/-0.3210} = 59.15\underline{/1.2293}\\ \mathbf{V}_{bc18} &= (0.091 + j6.6501) \times 5.65\underline{/1.1391} = 37.59\underline{/2.6962}\end{aligned}$$

so that

$$\begin{aligned}v_{bc} = {} & 403.31 + 125.04\cos(6\theta - 0.2906) + 59.15\cos(12\theta + 1.2293)\\ & 37.59\cos(18\theta + 2.6962) + \ldots\end{aligned}$$

The r.m.s. value of the voltage obtained from this truncated series is 415.85 V compared to the true r.m.s. value of 417.08 V derived from the time domain solution. The square root of the difference between the squares of these two numbers is 31.96 V and is the r.m.s. value of all the remaining terms of the Fourier series.

The frequency spectrum of v_{bc} and its waveform are shown in Fig. 8.7. The actual waveform obtained from the time domain expressions derived in Chapter 5 is shown by the solid line and the waveform obtained from the first four terms of the Fourier series is shown by the dashed line.

****

Comparison with the results of Example 8.2 for the output current indicates that the Fourier approximation to the voltage waveform is, in general, far less precise and, in particular, shows no indication of the commutation notches. The estimate of the r.m.s. value of the voltage is less accurate than that of the current, indicating that the higher order Fourier terms which have been neglected are more significant. Comparison of the frequency spectra, Figs 8.3 and 8.7, shows that this is the case, the current series converging much more rapidly than the voltage series. The reason for this is illustrated by the computation of the voltage series. A voltage component is obtained as the product of a current component and an impedance. The dominant component in the impedance is the reactance, which increases linearly with the order of the component so that the voltage series converges correspondingly more slowly than the current series and far more terms are needed to produce an equally good approximation.

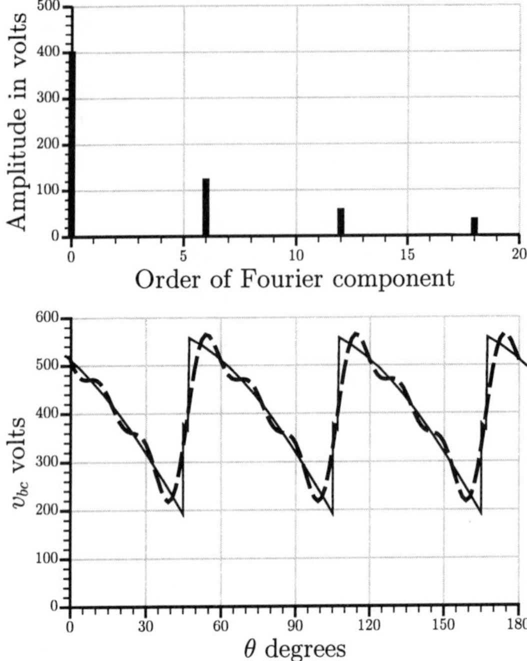

FIG. 8.7. The frequency spectrum and waveform of the voltage behind the choke. The actual waveform is shown in the lower graph by the solid line and the Fourier approximation by the dashed line.

An indication of the number of terms required is provided by focusing attention on the commutation notch. This feature is 2.437° wide and good modeling of it in the frequency domain requires that the wavelength of the highest Fourier component be much smaller than this width. *Much smaller* is not a very precise term but we will be on reasonably safe ground if it is taken to mean at least an order of magnitude smaller. Thus the highest Fourier component should have a wavelength less than 0.2437°, i.e. its order should be higher than $360/0.2437 = 1477.2$. Remembering that the order of the components of this series must be a multiple of six, this indicates that terms up to at least order 1482 should be included which means having at least 248 terms. To approximately model in this way a feature which can be precisely defined with far less effort by the time domain solution would be a waste of computing resources.

This is a general problem with the frequency domain approach. Parameters such as r.m.s. values can be estimated with reasonable accuracy by a few terms. A good approximation to features with continuous waveforms, such as the currents, is provided by a few terms. However, the accurate modeling of discontinuities, such as occur in the voltage waveforms, and of fine features, such as the commutation notches, requires an inordinately large number of terms.

8.10 The voltage at the transformer input terminals

The voltage between the red and yellow input terminals of the transformer, v'_{ry}, is obtained by subtracting the voltage drops in the impedances of the red and yellow lines from the voltage between the red and yellow busbars. Referring to Fig. 5.9, this is expressed as

$$v'_{ry} = v_{ry} - R_s i_r - X_s \frac{\mathrm{d}i_\mathrm{r}}{\mathrm{d}\theta} + R_s i_y + X_s \frac{\mathrm{d}i_\mathrm{y}}{\mathrm{d}\theta} \tag{8.38}$$

i.e.

$$v'_{ry} = \sqrt{2} V_s \cos(\theta) - (R_s + X_s \frac{\mathrm{d}}{\mathrm{d}\theta})(i_r - i_y) \tag{8.39}$$

where V_s is the r.m.s. value of the line to line voltage at the infinite bus.

Since the waveform of i_y is identical with that of i_r except for a phase lag of 120°, i.e. $i_y = i_r(\theta - 2\pi/3)$, the difference, $i_r - i_y$, is

$$i_r - i_y = \sum_{m=1,\infty,2} I_{rm} \Big\{ \cos(m\theta + \phi_{crm}) - \cos\big[m(\theta - 2\pi/3) + \phi_{crm}\big] \Big\} \tag{8.40}$$

The Fourier components must now be divided into positive, negative, and zero sequence terms as was done when deriving the expressions for the output current in eqns 8.16, 8.17, and 8.18. The series are of course modified by the knowledge that has been gained since that point. It is known for example that all triplens, i.e. all zero sequence components, are zero and that there are no even components on the a.c. side of the rectifier. Hence, only the positive and negative sequence series need be considered with the positive sequence series comprising terms of odd order 1, 7, 13, etc. i.e. $6k - 5$, and the negative sequence series terms of odd order, 5, 11, 17, etc. i.e. $6k - 1$, where k is any positive nonzero integer. With this information, eqn 8.40 reduces to

$$\begin{aligned} i_r - i_y = & \sum_{m=1,\infty,6} \sqrt{3} I_{rm} \cos(m\theta + \phi_{crm} + \frac{\pi}{6}) + \\ & \sum_{n=5,\infty,6} \sqrt{3} I_{rn} \cos(n\theta + \phi_{crn} - \frac{\pi}{6}) \end{aligned} \tag{8.41}$$

It is now a straightforward problem in a.c. circuit theory to substitute this expression for $i_r - i_y$ in eqn 8.39 to derive expressions for the transformer input voltage.

Example 8.7

Derive, for the example, the first ten odd Fourier components of the voltage between the red and yellow input terminals of the transformer.

Since the busbar voltage is sinusoidal, the first order component of the voltage between the red and yellow busbars is

$$v_{ry1} = \sqrt{3} \times 3300 \cos(\theta) = 4666.90 \cos(\theta)$$

and all other components are zero.

The first order component of i_r is, from Table 8.4,

$$i_{r1} = 61.66 \cos(\theta - 1.4248)$$

At this point a problem is encountered if the results are to be compared with those obtained by the time domain analysis of Chapter 5. There, in order not to further complicate an already complex problem, the transformer magnetizing impedance was moved to the system bus. In order to compute voltages on the same basis, the magnetizing current component must be removed from i_{r1}. After this is done the new current is designated i'_{r1}.

The magnetizing current in the yellow–blue winding is $5.349 \cos(\theta - 3.5538)$ from Example 8.4. This is converted to the magnetizing current component in the red line by multiplying the amplitude by $\sqrt{3}$ and advancing its phase by 90° to give

$$i_{rmag} = 9.2467 \cos(\theta - 1.9830)$$

Subtracting this from the line current yields

$$i'_{r1} = 61.66 \cos(\theta - 1.4248) - 9.2467 \cos(\theta - 1.9830)$$

i.e.

$$i'_{r1} = 54.03 \cos(\theta - 1.3339)$$

The expression for $i'_r - i'_y$ for the first order component is then, from eqn 8.41

$$i'_{r1} - i'_{y1} = 93.57 \cos(\theta - 0.8103)$$

and the line impedance to this component is $0.034 + j0.2262\ \Omega$. Substituting these expressions into eqn 8.39 yields the first order component of the transformer voltage

$$v'_{ry1} = 4649.40 \cos(\theta - 0.0026)$$

Table 8.5 *The voltage between the red and yellow input terminals of the rectifier transformer. The first 10 Fourier components are listed. Columns 2 through 5 give the order, frequency, amplitude, and phase of the component. Column 6 lists the mean three phase power associated with each component*

Component number	Order	Frequency Hz	Amplitude A	Phase degrees	Power kW
1	1	60.0	4649.40	–0.15	150.35
2	3	180.0	0.00	0.00	0.00
3	5	300.0	28.59	–145.07	–0.01
4	7	420.0	9.43	–47.46	0.00
5	9	540.0	0.00	0.00	0.00
6	11	660.0	20.63	–59.30	0.00
7	13	780.0	12.06	30.75	0.00
8	15	900.0	0.00	0.00	0.00
9	17	1020.0	18.44	23.86	0.00
10	19	1140.0	13.00	112.54	0.00

For the fifth order component, the busbar voltage is zero and

$$i_{r5} = 14.59\cos(5\theta - 0.4075)$$

so that

$$i_{r5} - i_{y5} = 25.27\cos(5\theta - 0.9311)$$

and the line impedance is $0.034 + j1.3095\ \Omega$. Substitution of these values into eqn 8.39 yields

$$v'_{ry5} = 28.59\cos(5\theta - 2.5319)$$

The values for the first ten components are listed in columns 4 and 5 of Table 8.5

****

8.11 Voltage between transformer secondary terminals

The voltage between the transformer secondary terminals is obtained in four steps.

- First, the voltage drop in the primary leakage impedance is subtracted from the voltage across a primary phase to give the primary winding e.m.f.

- Second, this e.m.f. is converted to the e.m.f. in the corresponding secondary winding by dividing it by the ratio of the number of turns on a primary phase to the number on a secondary phase.
- Third, the voltage drop in the secondary leakage impedance is subtracted from the secondary e.m.f. to obtain the voltage across a secondary phase.
- Fourth, the voltage in one secondary phase is subtracted from the voltage in another to obtain the line to line voltage between the phase ends.

This process readily translates into the frequency domain using the methods already described. Again reference is made to Fig. 5.9 to identify the voltages and currents.

8.11.1 *The primary winding e.m.f.*

Considering the red–yellow primary, its e.m.f., e_{ry}, is

$$e_{ry} = v'_{ry} - R_p i_{ry} - X_p \frac{\mathrm{d}i_{ry}}{\mathrm{d}\theta} \tag{8.42}$$

where R_p is the resistance of a primary winding and X_p is its leakage reactance at the supply frequency.

The current in the yellow–blue primary winding has already been found, eqn 8.26. The current in the red–yellow phase leads this by 120°, i.e.

$$i_{ry} = i_{yb}(\theta + \frac{2\pi}{3}) \tag{8.43}$$

In order to translate this time shift into the frequency domain, the positive and negative sequence components must be distinguished, the positive sequence ones having a phase lead of 120° and the negative ones a phase lag of the same amount

$$\begin{aligned} i_{ry} = & \sum_{m=1,\infty,6} I_{ybm} \cos(m\theta + \phi_{cybm} + \frac{2\pi}{3}) \\ & \sum_{n=5,\infty,6} I_{ybn} \cos(n\theta + \phi_{cybn} - \frac{2\pi}{3}) \end{aligned} \tag{8.44}$$

This expression, combined with eqn 8.42, yields phasor expressions for the Fourier components. For the positive sequence components, $m = 6k-5$,

$$\mathbf{E}_{rym} = \mathbf{V}_{rym} - (R_p + jmX_p)\, I_{ybm} \underline{/\phi_{cybm} + 2\pi/3} \tag{8.45}$$

and for the negative sequence components, $n = 6k - 1$,

$$\mathbf{E}_{ryn} = \mathbf{V}_{ryn} - (R_p + jnX_p)\, I_{ybn} \underline{/\phi_{cybn} - 2\pi/3} \tag{8.46}$$

8.11.2 *The secondary phase voltage*

In phasor terms, the component of order m of the voltage across secondary winding number 1 is

$$\mathbf{V}_{1m} = \frac{\mathbf{E}_{rym}}{n_{tps}} - (R_{sec} + jmX_{sec})\,\mathbf{I}_{1m} \tag{8.47}$$

The current in Phase 2 has already been found, eqn 8.24, and this must be advanced in time by $2\pi/3$ to obtain i_1. Translated to the frequency domain, this again involves separate treatment for the positive and negative sequence components, a phase lead of $2\pi/3$ for the positive sequence components and a phase lag of $2\pi/3$ for the negative sequence components.

For the positive sequence components, $m = 6k - 5$,

$$\mathbf{I}_{1m} = \mathbf{I}_{2m}\underline{/\phi_{c2m} + 2\pi/3} \tag{8.48}$$

and for the negative sequence components, $n = 6k - 1$,

$$\mathbf{I}_{1n} = \mathbf{I}_{2n}\underline{/\phi_{c2n} - 2\pi/3} \tag{8.49}$$

8.11.3 *The secondary line to line voltage*

The voltage, v_{12}, between secondary terminals 1 and 2 is the difference $v_1 - v_2$. Now v_2 has the same waveform as v_1 but with a delay of $2\pi/3$. In the frequency domain this translates into a phase lag of $2\pi/3$ for the positive sequence components and a phase lead of $2\pi/3$ for the negative sequence components. For the positive sequence components, $m = 6k - 5$,

$$\mathbf{V}_{12m} = \mathbf{V}_{1m} - \mathbf{V}_{2m} = \mathbf{V}_{1m}[1 - \cos(-\frac{2\pi}{3}) - j\sin(-\frac{2\pi}{3})]$$

i.e.

$$\mathbf{V}_{12m} = \sqrt{3}\,\mathbf{V}_{1m}\underline{/\pi/6} \tag{8.50}$$

and for the negative sequence components, $n = 6k - 1$,

$$\mathbf{V}_{12n} = \mathbf{V}_{1n} - \mathbf{V}_{2n} = \mathbf{V}_{1n}[1 - \cos(\frac{2\pi}{3}) - j\sin(\frac{2\pi}{3})]$$

i.e.

$$\mathbf{V}_{12n} = \sqrt{3}\,\mathbf{V}_{1n}\underline{/-\pi/6} \tag{8.51}$$

Example 8.8

Determine, for the example, the first ten Fourier components of the secondary line to line voltage, v_{12}.

The Fourier components of the voltage across the red–yellow primary winding are listed in Table 8.5. The Fourier components of the current in the yellow–blue primary are listed in Table 8.3. The first task is to translate the latter into components for the red–yellow phase by advancing the positive sequence components by $2\pi/3$ and retarding the negative sequence components by the same amount. The first order component is of positive phase sequence so that

$$\mathbf{I}_{ry1} = 35.60\underline{/-0.9012}$$

The next component, of third order, is of zero sequence and is zero. The following component is of fifth order and is of negative sequence so that

$$\mathbf{I}_{ry5} = 8.42\underline{/-0.9311}$$

The remaining Fourier components of i_{ry} are determined in a similar way.

The winding e.m.f. is now determined by the use of eqn 8.45 for the positive sequence terms and eqn 8.46 for the negative sequence terms. In regard to the first order term, the same problem is encountered as in Example 8.7. If the results are to be compared with those obtained in Chapter 5, the magnetizing current component must be removed from the current. When this is done

$$\mathbf{I}_{ry1} = 31.19\underline{/-0.8103}$$

Since the primary resistance is 0.45 Ω and the leakage reactance at the supply frequency of 60 Hz is 2.3750 Ω, the first order component of winding e.m.f. is

$$\mathbf{E}_{ry1} = 4649.40\underline{/-0.0026} - (0.45 + j2.3750) \times 31.19\underline{/-0.8103}$$

i.e.

$$\mathbf{E}_{ry1} = 4586.35\underline{/-0.0116}$$

The higher order magnetizing current components are zero so that the fifth order component is

$$\mathbf{E}_{ry5} = 28.59\underline{/-2.5319} - (0.45 + j11.875) \times 8.42\underline{/-0.9311}$$

i.e.

$$\mathbf{E}_{ry5} = 128.54\underline{/-2.4440}$$

The higher order components are obtained in a similar manner.

The e.m.f. in the secondary winding is obtained by dividing the primary components by 13, the ratio of primary to secondary turns per phase.

The resistance of a secondary phase is 0.0025 Ω and its leakage reactance at the supply frequency is 0.01546 Ω.

The components given in Table 8.2 for i_2 are now converted to Phase 1 using eqns 8.48 and 8.49. From $\mathbf{I}_{21}$ which is of positive sequence

$$\mathbf{I}_{11} = 405.46\underline{/-0.8103}$$

and from $\mathbf{I}_{25}$ which is of negative sequence

$$\mathbf{I}_{15} = 109.49\underline{/-0.9311}$$

with the remaining components being obtained in a similar way.

The voltage across Phase 1 is then, for the first order component

$$\mathbf{V}_{11} = 352.80\underline{/-0.0116} - (0.0025 + j0.1546) \times 405.46\underline{/-0.8103}$$

i.e.

$$\mathbf{V}_{11} = 347.58\underline{/-0.0103}$$

and for the fifth order component

$$\mathbf{V}_{15} = (0.0025 + j0.7330) \times 109.49\underline{/-0.9311}$$

i.e.

$$\mathbf{V}_{15} = 5.65\underline{/-1.0249}$$

Similar methods are employed for the remaining components

The line to line voltage components are now obtained using eqns 8.50 and 8.51 as appropriate so that for the first order component which is of positive sequence

$$\mathbf{V}_{121} = 602.02\underline{/0.5133}$$

and for the fifth order component which is of negative sequence

$$\mathbf{V}_{125} = 9.79\underline{/-1.5485}$$

All ten components, computed in this manner, are listed in Table 8.6.

The approximate r.m.s. value of the voltage determined from the ten components of Table 8.6 is 426.28 V, within 1% of the true value of 430.42 V determined from the time domain solution. The square root of the difference of the squares of these two numbers is 59.61 V and is the r.m.s. value of all the remaining terms of the Fourier series.

Table 8.6 *The voltage between transformer secondary winding terminals 1 and 2. Columns 2 through 5 list the order, frequency, amplitude, and phase of the first 10 Fourier components of the voltage. Column 6 lists the mean three phase power associated with each component*

Component number	Order	Frequency Hz	Amplitude A	Phase degrees	Power kW
1	1	60.0	602.09	28.73	149.08
2	3	180.0	0.00	0.00	0.00
3	5	300.0	31.81	–175.32	–0.10
4	7	420.0	10.49	–17.63	–0.01
5	9	540.0	0.00	0.00	0.00
6	11	660.0	22.95	–89.41	–0.01
7	13	780.0	13.42	60.66	0.00
8	15	900.0	0.00	0.00	0.00
9	17	1020.0	20.51	–6.21	0.00
10	19	1140.0	14.46	142.47	0.00

The frequency spectrum and waveform of this voltage are shown in Fig. 8.8. Again it is seen that the modeling of the waveform is significantly poorer than for the currents and that the commutation notches are barely indicated. Again it is concluded that this is due to the effect of the reactance increasing linearly with frequency and causing the Fourier series to converge more slowly.

****

8.12 Power

Since power is the product of voltage and current, the determination of instantaneous power involves the product of two Fourier series, a complex process whose consideration will be postponed to the second half of this chapter in Section 8.19. Fortunately, a knowledge of the average power over a supply cycle is sufficient for most needs and is a quantity which is easily obtained. The orthogonality of sinusoidal functions means that the average power associated with the product of a Fourier voltage component and a current component is zero except in the special case when they are of the same order. In this case the mean power is given by the familiar a.c. formula, the triple product of the r.m.s. value of the voltage, the r.m.s. value of the current, and the power factor, the cosine of their phase difference. Since the Fourier components are being specified by their amplitudes, this translates to eqn 8.52 for the d.c. side, zero order component, eqn 8.53 for

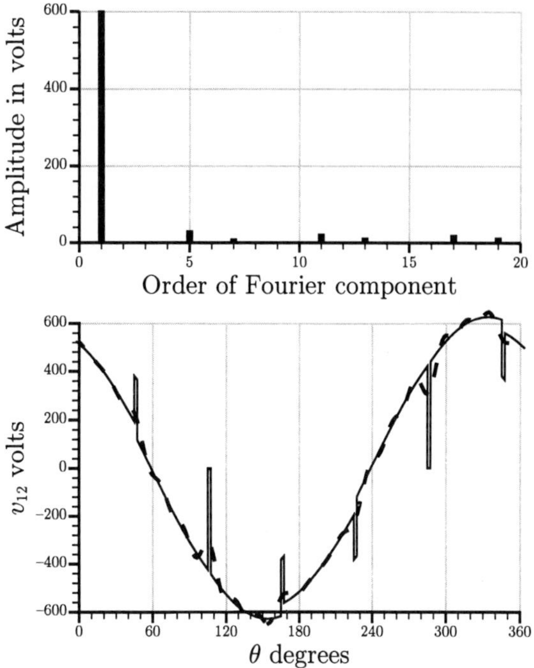

FIG. 8.8. The frequency spectrum and waveform of the voltage between terminals 1 and 2 of the transformer secondary. The actual waveform is shown in the lower graph by the solid line and the Fourier approximation by the dashed line.

the higher order d.c. side components, and eqn 8.54 for the three phase a.c. side

$$P_{dco} = V_o I_o \qquad m = 0 \tag{8.52}$$

$$P_{dcm} = \frac{1}{2} V_m I_m \cos(\phi_{vm} - \phi_{cm}) \qquad m \neq 0 \tag{8.53}$$

$$P_{acn} = \frac{3}{2} V_n I_n \cos(\phi_{vn} - \phi_{cn}) \tag{8.54}$$

where P_{dco} is the power associated with the zero order component on the d.c. side, P_{dcm} is the average power associated with the Fourier component of order $m \neq 0$ on the d.c. side, and P_{acn} is the total three phase power associated with the Fourier component of order n.

Example 8.9

Determine, for each Fourier component, the average power input to and output from the rectifier bridge.

The first order Fourier component of the voltage between lines 1 and 2 at the rectifier bridge input terminals taken from Table 8.6 is $602.02\underline{/0.5133}$ and the current in Line 2 taken from Table 8.2 is $405.44\underline{/-2.9046}$. This is a positive sequence component. Figure 8.9(a) shows the phasor relationships for positive sequence components between the line to line voltages, the line to neutral voltage, and the line currents. From this diagram it is seen that the voltage of Line 1 relative to the neutral is

$$\mathbf{V}_{11} = \frac{602.09}{\sqrt{3}}\underline{/0.5133 - \pi/6} = 347.62\underline{/-0.0221}$$

From Table 8.2, the corresponding current component in rectifier input Line 2 is $405.46\underline{/-2.9047}$. Inspection of Fig. 8.9(a) shows that the component in Line 1 leads this by 120°, i.e.

$$\mathbf{I}_{11} = 405.46\underline{/-2.9046 + 2\pi/3} = 405.46\underline{/-0.8103}$$

Thus the power factor of this first order component is

$$Pf_{11} = \cos(-0.0221 - (-0.8103)) = 0.7053$$

which is slightly less than the cosine of the angle of delay, 0.7071, because the commutation process introduces a small additional lag into the current. The mean three phase power associated with the component is

$$P_1 = \frac{3}{2}\frac{347.62 \times 405.46 \times 0.7053}{1000} = 149.11 \text{ kW}$$

Considering the fifth order Fourier component, from Table 8.6,

$$\mathbf{V}_{125} = 31.81\underline{/-3.0598}$$

so that the voltage between Line 1 and the transformer secondary neutral is, following the phasor relationships for negative sequence systems of Fig. 8.9(b),

$$\mathbf{V}_{15} = 18.37\underline{/-2.5362}$$

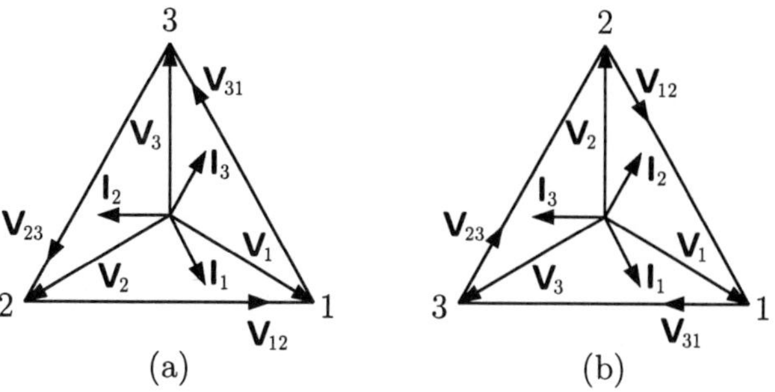

FIG. 8.9. Phasor relationships (a) for a positive sequence system and (b) for a negative sequence system.

From Table 8.2,

$$\mathbf{I}_{25} = 109.50\underline{/1.1633}$$

which, by inspection of the negative sequence phasor diagram, gives

$$\mathbf{I}_{15} = 109.50\underline{/-0.9311}$$

The power factor of this component is therefore

$$Pf_5 = \cos(-2.5362 - (-0.9311)) = -0.0343$$

The mean three phase power input associated with this component is then

$$P_5 = \frac{3}{2}\frac{18.37 \times 109.50 \times -0.0343}{1000} = -0.1035 \text{ kW}$$

The input power associated with each component is listed in the final column of Table 8.6.

The major output component is that of the zero order, d.c. term. From Example 8.6 the zero order voltage behind the choke is 403.31 V and from Example 8.2 the zero order component of the output current is 366.01 A. Since these are d.c. components, the amplitudes are also the r.m.s. values and the mean d.c. output power from the rectifier bridge is obtained from them as

$$P_{bco} = \frac{403.31 \times 366.01}{1000} = 147.61 \text{ kW}$$

From the same examples, the sixth order Fourier component of voltage is

$$\mathbf{V}_{bc6} = 125.04\underline{/-0.2906}$$

and for the current is

$$\mathbf{I}_{bc6} = 56.36\underline{/-1.8203}$$

so that the power factor of this component is

$$Pf_{bc6} = \cos(-0.2906 - (-1.8203)) = 0.0411$$

and the corresponding power is

$$P_{bc6} = \frac{1}{2}\frac{125.04 \times 56.36 \times 0.0411}{1000} = 0.1447 \text{ kW}$$

****

It will be noted that, on both the input and output sides, the power associated with the higher order Fourier components is very small relative to that associated with the first component, the first order component on the a.c. input side and the zero order component on the output d.c. side. This is only partially due to the fact that the amplitudes of the higher order components are relatively small. The very low power factor of these higher order components is a major factor leading to the very low harmonic power and is due to the inductive nature of the circuits which, at the higher frequencies of the higher order components, causes the current to lag almost 90° on the voltage.

The higher order components of output power are all positive and the higher order components of input power are all negative. In effect the rectifier bridge acts as a generator of high frequency power, taking some of the supply frequency power and converting it to higher frequencies some of which is delivered to the load and some of which is fed back to the a.c. network. The power associated with these higher order components does harm rather than good since it is converted to heat and reduces the efficiency. However, because of the very small size of the components, this is a relatively minor matter compared to other effects.

On the d.c. side the extra heat raises the temperature of the load and reduces its power handling capability, a factor which is amplified by the high efficiency of most loads. For example, consider that the load is a d.c. motor whose armature losses at rated load are 5% of its rated power, i.e. about 7500 W. The additional power loss associated with the higher Fourier components is 155 W, an increase of 2% which causes a corresponding reduction in the rated power from 150 kW to about 147 kW. Skin effect and iron losses which have not been taken into account will tend to increase the losses associated with the higher order components so that the circuit resistances will tend to increase with frequency. This will exacerbate this factor and the reduction in power handling capability may well be of the order of 5%.

The additional losses on the a.c. side are of little consequence compared to the potential effect on system protective devices and adjacent communications circuits, and the effect of the power factor of the first Fourier component. Considering the power factor first, conduction in the rectifier has been delayed by 45° in order to reduce the output voltage to 70% of its maximum value. This delay inevitably causes a corresponding lag in the current and the fundamental power factor is consequently 0.6967 lagging, slightly less than $\cos(\alpha)$ because of the additional lag associated with commutation. This situation deteriorates as the voltage is made smaller by further increase in delay angle. Some of the electric utility costs are associated with supplying voltage and current rather than with power, e.g. the size of the conductors to a factory and the losses in them are current related rather than power related. The utility is entitled to recover these costs and in some way or other the customer will be penalized for poor power factor and the cost per kilowatt hour will increase.

In the area of communications interference, consider the fifth order Fourier component. In effect a generator of 300 Hz power has been connected to the a.c. system and is injecting into it three phase currents whose amplitudes are 14.6 A and which are of negative phase sequence. This frequency is within the most important band for voice communication on the telephone network and can cause unacceptable interference. Most areas will have regulations giving the telephone company redress against such interference. The power system itself can experience problems since many of its protective relays are frequency and phase sequence sensitive and their operation can be affected in unforeseen ways. Other loads connected to the system may have power factor correction capacitors and these, having an impedance which is inversely proportional to the frequency, may act as a sink for the higher order components, again with potential unforeseen effects. These matters are gone into in greater detail in Chapter 9.

8.13 Fourier analysis as the prime analytical tool

By making two assumptions, that conduction is continuous and that the commutation process need not be modeled in the waveforms, Fourier analysis can be used as the prime analytical tool for the investigation of rectifier behavior thus avoiding completely the need for time domain analysis. It has already been seen that commutation is quickly accomplished, in the preceding example it occupied 2.77°, and the assumption that its effect on the waveforms need not be modeled is, in many cases, a relatively minor price to pay for the ease with which frequency domain analysis can solve complex rectifier problems. Ignoring the effect of commutation on the waveforms does not mean that its effect on the average values must also be ignored. A good approximation to this effect is obtained by retaining the voltage source V_{xst} which was included in the a.c. equivalent circuit in Fig. 5.8.

Provided the above two assumptions are acceptable, Fourier analysis may be directly applied to any rectifier problem. However, one of its most powerful virtues, the ability to deal with complex rectifier connections, is best seen by first applying it to single way rectifiers and then extending it to bridge rectifiers in an approach similar to that used in Chapter 3. This is the approach used here. The output voltage spectrum of a single way rectifier is first determined and then that result is extended to a semi-controlled bridge rectifier. For the examples it will be assumed that this rectifier has the same parameters as the fully controlled bridge used in the examples up to now and that it supplies the same load. To ensure that the mean load current remains at 366 A will require some increase of the delay angle of the bridge thyristors.

A time domain analysis would necessitate keeping track of the relative conducting periods of the diodes and thyristors. As was the case in Chapter 3, this would require considering the problem in two separate parts, for delays less than 60° and greater than 60°. With frequency domain analysis the problem becomes a straightforward one of two rectifiers with different delay angles connected in series, the analysis itself automatically keeping track of when the diodes and thyristors conduct.

The analysis begins by deriving the spectrum of the Thévenin voltage of a single way rectifier with p phases.

8.14 The Thévenin voltage of a p phase rectifier

The output voltage waveform of a simple, single way, p phase rectifier operating at a delay angle α, is shown in Fig. 8.10. The phase difference between adjacent phases is $2\pi/p$ and each conducting segment has a duration of $2\pi/p$. One segment and portions of the two flanking it are shown. The three segments are labelled A, B, and C and are respectively those produced by the phases whose voltages are $V_{pk}\cos(\theta + \phi_v + 2\pi/p)$, $V_{pk}\cos(\theta + \phi_v)$, and $V_{pk}\cos(\theta + \phi_v - 2\pi/p)$ where V_{pk} is the peak value of the open circuit secondary line to neutral voltage and ϕ_v is some arbitrary phase angle. The segment on which attention is focused is produced by $V_{pk}\cos(\theta + \phi_v)$. It starts at $\theta = \alpha - \phi_v - \pi/p$ and finishes at $\theta = \alpha - \phi_v + \pi/p$.

Since all segments are identical except for the appropriate phase shift, the Fourier series representing the rectifier Thévenin voltage consists only of components whose orders are integral multiples of the number of phases so that eqns C.21 and C.22 can be used to derive the Fourier components. Thus for $m = k \times p$, k being any positive integer including zero, the amplitude of the mth order cosine component, V_{rtcm} is

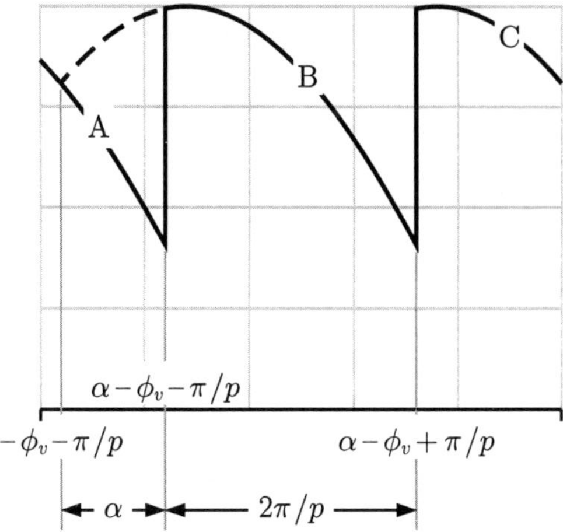

FIG. 8.10. The output voltage of a simple, p phase rectifier.

$$V_{rtcm} = \frac{p}{\pi}\int_{\alpha-\phi_v-\pi/p}^{\alpha-\phi_v+\pi/p} V_{pk}\cos(\theta+\phi_v)\cos(m\theta)d\theta \tag{8.55}$$

and the amplitude of the mth order sine component, V_{rtsm} is

$$V_{rtsm} = \frac{p}{\pi}\int_{\alpha-\phi_v-\pi/p}^{\alpha-\phi_v+\pi/p} V_{pk}\cos(\theta+\phi_v)\sin(m\theta)d\theta \tag{8.56}$$

Evaluation of these definite integrals gives

$$V_{rtcm} = \frac{p}{\pi}V_{pk}\left\{\frac{\sin[(m+1)\pi/p]\cos[(m+1)\alpha+m\phi_v]}{m+1}+\frac{\sin[(m-1)\pi/p]\cos[(m-1)\alpha-m\phi_v]}{m-1}\right\} \tag{8.57}$$

and

$$V_{rtsm} = \frac{p}{\pi}V_{pk}\left\{\frac{\sin[(m+1)\pi/p]\sin[(m+1)\alpha+m\phi_v]}{m+1}+\frac{\sin[(m-1)\pi/p]\sin[(m-1)\alpha-m\phi_v]}{m-1}\right\} \tag{8.58}$$

In applying these expressions, it must be remembered that the zero order, d.c., component is obtained by taking half the value of eqn 8.57 with $m=0$ as explained in Section C.2.

Since the minimum value of p is 2, the special case of $m = 1$ does not arise, see Appendix A, eqn A.40.

Remembering that m/p is the integer k and that $\sin(k\pi) = 0$ and $\cos(k\pi) = (-1)^k$,

$$\sin[(m+1)\frac{\pi}{p}] = \cos(m\frac{\pi}{p})\sin(\frac{\pi}{p}) = (-1)^{m/p}\sin(\frac{\pi}{p}) \tag{8.59}$$

and

$$\sin[(m-1)\frac{\pi}{p}] = -\cos(m\frac{\pi}{p})\sin(\frac{\pi}{p}) = -(-1)^{m/p}\sin(\frac{\pi}{p}) \tag{8.60}$$

With these substitutions eqns 8.57 and 8.58 can be written

$$V_{rtcm} = (-1)^{m/p}\frac{p}{\pi}V_{pk}\sin(\frac{\pi}{p})\left\{\frac{\cos[(m+1)\alpha + m\phi_v]}{m+1} - \frac{\cos[(m-1)\alpha + m\phi_v]}{m-1}\right\} \tag{8.61}$$

and

$$V_{rtsm} = (-1)^{m/p}\frac{p}{\pi}V_{pk}\sin(\frac{\pi}{p})\left\{\frac{\sin[(m+1)\alpha + m\phi_v]}{m+1} - \frac{\sin[(m-1)\alpha - m\phi_v]}{m-1}\right\} \tag{8.62}$$

The d.c. component is, by substituting $m = k = 0$ in eqn 8.61 and dividing the result by 2,

$$V_{rto} = \frac{p}{\pi}V_{pk}\sin(\frac{\pi}{p})\cos(\alpha) \tag{8.63}$$

and the maximum value of the d.c. component is

$$V_{rtmo} = \frac{p}{\pi}V_{pk}\sin(\frac{\pi}{p}) \tag{8.64}$$

Substituting this expression in eqns 8.61 and 8.62 yields

$$V_{rtcm} = (-1)^{m/p}V_{rtmo}\left\{\frac{\cos[(m+1)\alpha + m\phi_v]}{m+1} - \frac{\cos[(m-1)\alpha - m\phi_v]}{m-1}\right\} \tag{8.65}$$

and

$$V_{rtsm} = (-1)^{m/p} V_{rtmo} \left\{ \frac{\sin[(m+1)\alpha + m\phi_v]}{m+1} - \frac{\sin[(m-1)\alpha - m\phi_v]}{m-1} \right\} \tag{8.66}$$

Converting these cosine and sine terms to amplitude and phase gives

$$V_{rtm} = V_{rtmo} \sqrt{\frac{1}{(m+1)^2} + \frac{1}{(m-1)^2} - \frac{2\cos(2\alpha)}{(m+1)(m-1)}} \tag{8.67}$$

and for the cosine form

$$\phi_{vrtm} = \arctan\left(\frac{-\frac{\sin[(m+1)\alpha+m\phi_v]}{m+1} + \frac{\sin[(m-1)\alpha+m\phi_v]}{m-1}}{\frac{\cos[(m+1)\alpha+m\phi_v]}{m+1} - \frac{\cos[(m-1)\alpha+m\phi_v]}{m-1}} \right) \tag{8.68}$$

or for the sine form

$$\phi_{vrtm} = \arctan\left(\frac{\frac{\cos[(m+1)\alpha+m\phi_v]}{m+1} - \frac{\cos[(m-1)\alpha+m\phi_v]}{m-1}}{\frac{\sin[(m+1)\alpha+m\phi_v]}{m+1} - \frac{\sin[(m-1)\alpha+m\phi_v]}{m-1}} \right) \tag{8.69}$$

Again it is noted that eqn 8.67 applies only for $m > 0$. For $m = 0$ it must be divided by 2 to give the amplitude of the d.c. component.

A noteworthy feature of eqns 8.67 and 8.68 is that they do not explicitly contain the parameter p, the number of phases. This means that the amplitude and phase of a given Fourier component are independent of the number of phases except insofar as p restricts the possible values of m to $k \times p$. Thus, for a given delay angle, rectifiers having 2, 3, 4, 6, 8, 12, and 24 phases all have the same 24th order Fourier component relative to their mean rectifier Thévenin voltage.

Considering a delay angle α_2 which is the complement of a delay angle α_1, i.e. $\alpha_2 = \pi - \alpha_1$, then $\cos(2\alpha_2) = \cos(2\alpha_1)$ and the amplitudes of the Fourier components are the same in the two cases. Further, since $\cos(\alpha_2) = -\cos(\alpha_1)$ and $\sin(\alpha_2) = \sin(\alpha_1)$, the phase angles complement each other, i.e. ϕ_{vm} for the second rectifier equals π less ϕ_{vm} for the first rectifier.

8.15 Series connected rectifiers

The results just obtained for the frequency spectrum of the Thévenin voltage of a simple, single way rectifier are readily applied to the determination of the spectrum of the Thévenin voltage of any combination of series connected rectifiers. The Fourier series representing the Thévenin voltage waveform of a set of series connected rectifiers is the sum of the Fourier series for the individual rectifiers. Obvious candidates for this approach are the three phase bridge rectifiers, both fully and semi-controlled, and the various series connections discussed in Chapter 9. While consideration here is restricted to the three phase bridges, there is no restriction whatever on the individual rectifier voltages and delay angles, the analysis only being limited by the two initial assumptions, that conduction is continuous and that waveforms during commutation can be dispensed with.

The summation of the Fourier series is most easily accomplished with the Cartesian form of the series in which case for n rectifiers

$$v_{scm} = \left\{ \sum_{r=1,n} V_{rcm} \right\} \cos(m\theta) \tag{8.70}$$

$$v_{ssm} = \left\{ \sum_{r=1,n} V_{rsm} \right\} \sin(m\theta) \tag{8.71}$$

where V_{rcm} and V_{rsm} are the amplitudes of the mth order cosine and sine components for the rth rectifier, and v_{scm} and v_{ssm} are the amplitudes of the cosine and sine components of the mth order component of the combined output voltage.

8.15.1 *The fully controlled bridge rectifier*

The fully controlled three phase bridge rectifier is by far the most important example of the above principle. The Thévenin voltages for the two halves are equal and the delay angles are the same. However, the Thévenin voltage for the thyristors in the lower half of the bridge is phase shifted relative to that of the upper half as illustrated by Fig. 3.2. Considering the voltage v_r of that figure, it is seen that thyristor 4 in the bottom half of the bridge conducts half a cycle after thyristor 1 in the top half. Applying this to eqns 8.65 and 8.66 yields the following result for the Thévenin voltage

$$\begin{aligned} V_{rtcm} = (-1)^{\frac{m}{3}} V_{rtmo} \Bigg\{ & \frac{\cos[(m+1)\alpha + m\phi_v]}{m+1} - \\ & \frac{\cos[(m-1)\alpha - m\phi_v]}{m-1} + \\ & \frac{\cos[(m+1)\alpha + m(\phi_v - \pi)]}{m+1} - \\ & \frac{\cos[(m-1)\alpha - m(\phi_v - \pi)]}{m-1} \Bigg\} \end{aligned} \tag{8.72}$$

where the number of phases, p, has been set to 3 and m is an integral multiple of 3.

After collecting together related terms, eqn 8.72 reduces to

$$V_{rtcm} = (-1)^{\frac{m}{3}} V_{rtmo} \left\{ \frac{\cos[(m+1)\alpha + m\phi_v] + \cos[(m+1)\alpha + m(\phi_v - \pi)]}{m+1} \right. \tag{8.73}$$

$$\left. - \frac{\cos[(m-1)\alpha - m\phi_v] + \cos[(m-1)\alpha - m(\phi_v - \pi)]}{m-1} \right\}$$

Subtracting an odd multiple of π from an angle reverses the sign of its sine and cosine functions while subtracting an even multiple of π does not affect these functions. Thus, in eqn 8.73 the cosine sums will yield zero for the odd order Fourier components and twice the cosine for the even order components.

A similar analysis can be applied to the sine components with similar results. Remembering that m must be an integral multiple of the number of phases, in this case 3, it is seen that the Fourier series for the fully controlled bridge rectifier

- Is comprised only of triplens.
- That the order of these triplens is even, i.e. 0, 6, 12, 18...
- That the amplitude of these components is twice the amplitude of the corresponding components for the simple three phase rectifiers making up each half of the bridge.

The fact that this rather obvious result is produced provides a check on the validity of the analysis.

8.15.2 *The semi-controlled bridge rectifier*

The semi-controlled three phase bridge rectifier is a more challenging problem for Fourier analysis, being significantly more complex than the fully controlled bridge since the two halves have nothing in common except the three phase supply.

Putting $p = 3$ in eqn 8.64 yields the maximum value of the d.c. voltage for one of the three phase halves as

$$V_{rtmo} = \frac{3}{\pi}\frac{\sqrt{3}}{2}V_{pk} \tag{8.74}$$

which is in agreement with the value derived in Chapter 2, eqn 2.10, since here V_{pk} is the peak value of the line to neutral voltage and there V_s is the r.m.s. value of this same voltage.

The semi-controlled bridge is used for all succeeding examples and in order that the results be comparable with those already derived for a fully controlled bridge it will be assumed that the circuit conditions are identical except that the three thyristors in the bottom half of the bridge are replaced by diodes. So as to have the same time origin as used in Chapter 5 and in the first part of this chapter, the phase angle ϕ_v is set to zero. The mth order Fourier components of the thyristor rectifier in the top half of the three phase bridge are then

$$V_{rtcm} = (-1)^{\frac{m}{3}} V_{rtmo} \left\{ \frac{\cos[(m+1)\alpha]}{m+1} - \frac{\cos[(m-1)\alpha]}{m-1} \right\} \cos(m\theta) \quad (8.75)$$

and

$$V_{rtsm} = (-1)^{\frac{m}{3}} V_{rtmo} \left\{ \frac{\sin[(m+1)\alpha]}{m+1} - \frac{\sin[(m-1)\alpha]}{m-1} \right\} \sin(m\theta) \quad (8.76)$$

The components for the bottom half of the bridge are derived from these top half components by introducing the two factors which differentiate between the two halves. First, the delay angle for the diode half is set to zero. Second, the waveform of the lower half is delayed by 180°. Applying these conditions to eqns 8.5 and 8.6 yields

$$v_{dcm} = (-1)^{\frac{m}{3}} V_{rtmo} \left\{ \frac{1}{m+1} - \frac{1}{m-1} \right\} \cos[m(\theta - \pi)] \quad (8.77)$$

and

$$v_{rsm} = 0 \quad (8.78)$$

It is comforting to find that, after so much analysis, the sine component is zero, as it should be, since the rectifier Thévenin voltage of the diode half is an even function.

The term $\cos[m(\theta - \pi)]$ in eqn 8.77 expands to $\cos(m\pi)\cos(m\theta) + \sin(m\pi)\sin(m\theta)$ which, since $\sin(m\pi) = 0$, reduces to $\cos(m\pi)\cos(m\theta)$. The term $\cos(m\pi)$ is $(-1)^m$ and the product of this and the $(-1)^{m/3}$ term gives $(-1)^{4m/3}$. Since $m/3$ is an integer, it follows that $4m/3$ is even, so that the value of $(-1)^{4m/3}$ is always $+1$ no matter what the value of m. Thus eqn 8.77 can be rewritten as

$$v_{dcm} = V_{rtmo} \left[\frac{1}{m+1} - \frac{1}{m-1} \right] \cos(m\theta) \quad (8.79)$$

The solution for the bridge output voltage is now straightforward. The cosine components are obtained by summing eqns 8.75 and 8.79 and the sine component is given by eqn 8.76.

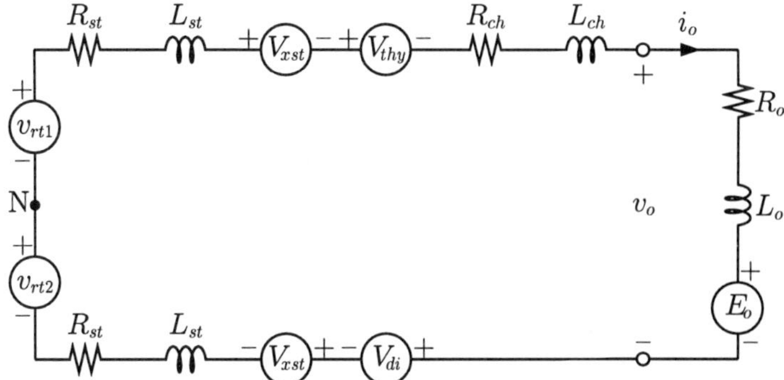

FIG. 8.11. The semi-controlled rectifier equivalent circuit.

The application of this procedure is illustrated in Example 8.10, but before that can be performed, the equivalent circuit of the rectifier must be established.

8.16 The equivalent circuit

The equivalent circuit of the semi-controlled bridge is shown in Fig. 8.11. It is essentially the a.c. equivalent circuit developed in Chapter 5 and illustrated in Fig. 5.8. However, the circuit has been modified to reflect the present viewpoint of two series connected rectifiers.

At the left of the circuit, two sources representing the Thévenin voltages of the two series connected rectifiers are shown joined at the system neutral N. Again it must be noted that this neutral is a notional one put there to aid in the visualization of the nature of the circuit. In reality it may or may not exist. Since each half of the bridge is a single way rectifier, the source Thévenin resistance, R_{st}, and inductance, L_{st}, are placed in the upper and lower lines rather than being combined in a single resistance and inductance of twice the size as was done in Fig. 5.8. The voltage source equivalent to commutation is treated similarly, being equally divided between the upper and lower lines. This division means that each voltage source has half the value of that shown in Fig. 5.8, i.e. for Fig. 8.11

$$V_{xst} = \frac{3X_{st}}{2\pi} I_o \tag{8.80}$$

Instead of having $2V_{thy}$ in the upper line as in Fig. 5.8, a voltage source V_{thy} is placed in the upper line and V_{di}, the voltage drop in a conducting diode, in the lower line. The remainder of the circuit is unchanged, representing the impedance of the smoothing choke and the impedance and e.m.f. of the load.

Example 8.10

If the rectifier of the previous example is replaced by a semi-controlled bridge, determine the delay angle which will produce the same mean load current for the same load e.m.f. Assume that the voltage drop in a conducting diode is the same as that in a thyristor—1.6 V. For this value of delay, determine the amplitude and phase of each nonzero Fourier component of the rectifier Thévenin voltage up to and including the 21st.

Using the same circuit data as used for Examples 8.1 through 8.9, which are specified in Example 5.1, the parameters of the equivalent circuit of Fig. 8.10 are

$$\begin{aligned}
R_{st} &= 0.005766\ \Omega \\
X_{st} &= 0.033525\ \Omega \\
V_{xst} &= 5.8588\ \text{V} \\
V_{thy} &= 1.6\ \text{V} \\
V_{di} &= 1.6\ \text{V} \\
R_{ch} &= 0.029\ \Omega \\
X_{ch} &= 0.248814\ \Omega \\
R_o &= 0.062\ \Omega \\
X_o &= 0.120637\ \Omega \\
E_o &= 370.0\ \text{V} \\
V_{rtmo} &= 296.884\ \text{V}
\end{aligned}$$

In Example 8.2 the mean value of the output current was found to be 366.01 A. The mean value of the rectifier Thévenin voltage which is required to drive this current through the equivalent circuit of Fig. 8.10 with the above parameter values is 422.45 V. The delay angle is found from this value using eqn 3.27 of Chapter 3.

$$422.45 = 296.88[1.0 + \cos(\alpha)]$$

from which the delay angle of the semi-controlled bridge is found to be 64.98°

Derivation of the Fourier components is now simply a matter of substituting these values into eqns 8.75, 8.76, and 8.79 and combining the results. Taking the third order component as an example of this process

$$\begin{aligned}
v_{tc3} &= -82.350\cos(3\theta) \\
v_{ts3} &= 186.855\sin(3\theta)
\end{aligned}$$

Table 8.7 *The first eight nonzero Fourier components of the semi-controlled rectifier Thévenin voltage*

Component number	Order	Frequency Hz	Amplitude V	Phase degrees
1	0	0.0	422.45	0.00
2	3	180.0	243.78	–129.96
3	6	360.0	103.04	–132.14
4	9	540.0	66.30	–142.10
5	12	720.0	48.89	–154.28
6	15	900.0	38.63	–167.41
7	18	1080.0	31.83	178.93
8	21	1260.0	26.99	164.92

$$v_{dcm} = -74.221\cos(3\theta)$$

which combine to give

$$v_{rt3} = -156.571\cos(3\theta) + 186.855\sin(3\theta) = 243.781\cos(3\theta - 2.26824)$$

The nonzero Fourier components of order zero through 21 are listed in Table 8.7 and the spectrum and waveform are shown in Fig. 8.12. The presence of the odd triplens, the third, ninth, etc., should be contrasted with their absence in the corresponding spectrum for the fully controlled bridge, Fig. 8.7. The large size of the third order component should be particularly noted.

The approximation to the rectifier Thévenin waveform provided by the first eight Fourier components is shown by the dashed line in the lower graph of Fig. 8.12. Fortunately in this case it is easy to obtain the true time domain expression and this, shown by the solid line in Fig. 8.12, provides a check on the accuracy of the frequency domain results.

A rectifier output voltage segment occupies 120° and, since the delay angle exceeds 60°, there will be a freewheeling period during each segment. Taking the segment from zero to 120°, there is a freewheeling period from zero to 6.37° when thyristor T3 and diode D6 conduct followed by an active period from 6.37° to 120° when thyristor T1 and diode D6 conduct. During the freewheeling period the rectifier Thévenin voltage is zero and during the active period it is the open circuit voltage of transformer secondary Terminal 1 relative to Terminal 2. Referring to the phasor diagram of Fig. 8.9(a), the rectifier voltage during the active subsegment is, in phasor terms,

$$\mathbf{V}_{13} = \mathbf{V}_1 - \mathbf{V}_3 = \sqrt{3}\,\mathbf{V}_1 \underline{/-\pi/6} = \sqrt{3}\,\frac{\mathbf{V}_{ry}}{n_{tps}} \underline{/-\pi/6}$$

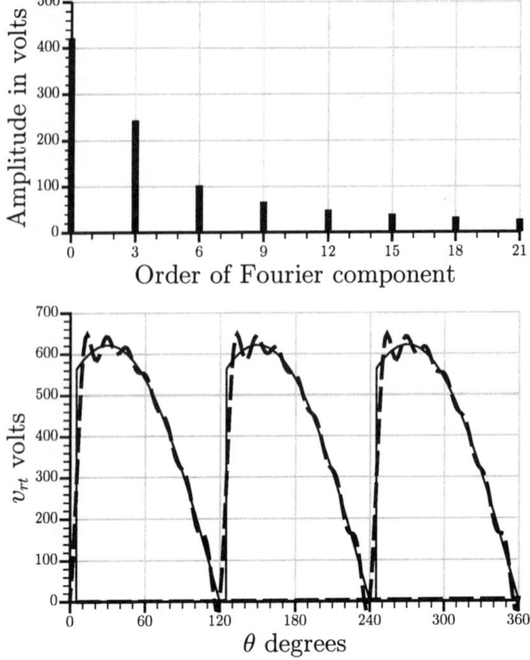

FIG. 8.12. The frequency spectrum and waveform of the rectifier Thévenin voltage. The actual waveform is shown in the lower graph by the solid line and the Fourier approximation by the dashed line.

Thus, the time domain expressions for the rectifier Thévenin voltage are for the range $0 < \theta < 6.37°$

$$v_{rt} = 0.0$$

and for the range $6.37° < \theta < 120°$

$$v_{rt} = \sqrt{3} \times \sqrt{2} \times \frac{3300}{13} \cos(\theta - \frac{\pi}{6}) = 621.8 \cos(\theta - \frac{\pi}{6})$$

The waveform derived from these expressions is shown in the lower graph of Fig. 8.12 by the solid line. Even with so few Fourier components, the curve derived from the truncated Fourier series is a reasonable approximation to the true waveform. Again it is noted that most of the error is due to the discontinuity which occurs at the firing of each thyristor and that accurate modeling of this feature would require a very large number of Fourier components.

That relatively few components can successfully model the gross characteristics is confirmed by considering the mean and r.m.s. values of the wave. The value of the zero order Fourier component is the same as that of true mean value derived from the time domain expressions, another indication that the results are probably correct. The r.m.s. value derived from the Fourier components is 467.43 V, 0.49% smaller than the true value, 469.75 V. The square root of the difference between the squares of these numbers is the r.m.s. value of the remaining components and is 46.59 V.

****

8.17 The output current

Derivation of the Fourier series representing the output current from that representing the rectifier Thévenin voltage is a simple application of d.c. and a.c. circuit theory. The d.c. resistance of the equivalent circuit is $2\,R_{st}+R_{ch}+R_o$ and the d.c. voltage opposing the zero order Fourier component is $2\,V_{xt}+V_{thy}+V_{di}+E_o$ so that the zero order current component is

$$I_{oo} = \frac{V_{rto} - 3X_{st}I_o/\pi + V_{thy} + V_{di} + E_o}{2R_{st} + R_{ch} + R_o} \tag{8.81}$$

Since there are no sources of a.c. in the equivalent circuit other than the Thévenin voltage, the higher order Fourier components of the output current are obtained in phasor form by dividing the voltage component by the circuit impedance at the appropriate frequency

$$\mathbf{I}_{om} = \frac{\mathbf{V}_{rtm}}{\mathbf{Z}_{tm}} \tag{8.82}$$

where

$$\mathbf{Z}_{tm} = (2R_{st} + R_{ch} + R_o) + jm(2X_{st} + X_{ch} + X_o) \tag{8.83}$$

Example 8.11

Determine, from the Fourier series representing the rectifier Thévenin voltage, the first eight nonzero components for the output current.

The d.c. component is given by

$$I_{oo} = \frac{422.45 - 384.92}{0.1025} = 366.01 \text{ A}$$

It is comforting that this is indeed the value of the mean load current.

The impedance to the third order component is

Table 8.8 *The first eight nonzero Fourier components of the rectifier output current*

Component number	Order	Frequency Hz	Amplitude A	Phase degrees
1	0	0.0	366.01	0.00
2	3	180.0	185.59	–215.48
3	6	360.0	39.31	–219.90
4	9	540.0	16.87	–230.61
5	12	720.0	9.33	–243.16
6	15	900.0	5.90	–256.51
7	18	1080.0	4.05	89.68
8	21	1260.0	2.94	75.56

$$\mathbf{Z}_{t3} = 0.1015 + j3 \times 0.4365 = 1.3135\underline{/1.4927}$$

so that the third order component of the output current is

$$\mathbf{I}_{o3} = \frac{243.78\underline{/-2.2682}}{1.3135\underline{/1.4927}} = 185.59\underline{/-3.7609}$$

The higher order components are obtained in a similar manner and are listed in Table 8.8.

The frequency spectrum of the output current is shown in the upper graph of Fig. 8.13 and the approximation to the waveform provided by the Fourier series is shown in the lower graph. The large size of the third order component is evident in both graphs but, as is generally the case for currents, the series converges rapidly and the wave is a good approximation to the true shape with little evidence of the ripple which is so prominent in the approximation to the voltage waveform.

The rectifier is close to the discontinuous conduction point. A small increase in load e.m.f. sufficient to decrease the mean value of the output current by about 100 A would send it into discontinuous conduction and Fourier analysis would no longer be applicable as a basic analytical tool. Time domain analysis as described in Chapter 6 would be required. In fact the current ripple is unacceptably large, much greater than for the fully controlled bridge, and extra inductance would have to be added to the circuit to provide greater smoothing. This would increase the size, weight, and cost of the smoothing choke to obtain an output waveform as good as that provided by the fully controlled rectifier. However, in order to provide comparable results, the unmodified circuit will be retained for the succeeding examples.

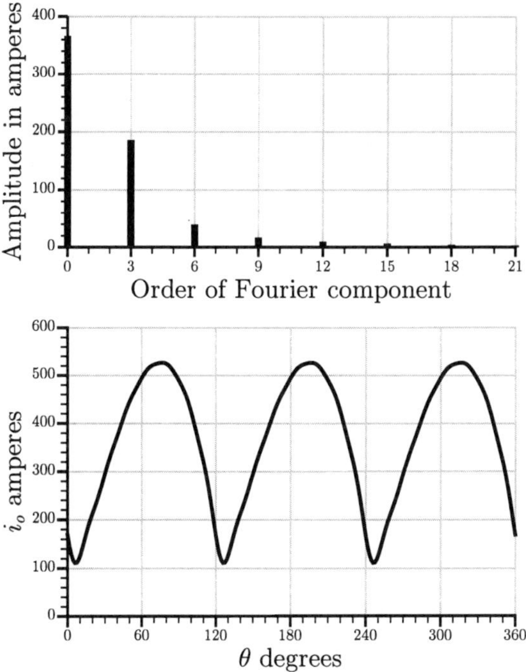

FIG. 8.13. The frequency spectrum and waveform of the output current. The frequency spectrum is shown in the upper graph and the waveform derived from it in the lower graph.

The partial consideration of commutation by the introduction of the d.c. voltage V_{xst} is based on the assumption that the mean current is being commutated. The large ripple makes the current relatively small at the commutation points, about 100 A in both thyristors and diodes, and makes commutation correspondingly easier. Under these circumstances the accuracy of the solution could be improved by omitting the commutation voltage. However, straining after accuracy in such minor matters seems inappropriate in view of the approximations inherent in the need to truncate the Fourier series after a reasonable number of terms

The approximation to the r.m.s. value of the output current is 390.10 A. The large difference between this and the mean value is due mainly to the large third order component and gives an r.m.s. ripple component of about 135 A. Although there is no simple means of obtaining the true time domain waveform it can be expected that, on the evidence of the rectifier Thévenin voltage, the r.m.s. value will be well within 0.5% of the true value.

****

8.18 The output voltage, voltage behind the choke, and the output power

The output voltage and voltage behind the choke can now be obtained by adding the voltages absorbed by the load and choke impedances to the load e.m.f. This method is described in Section 8.9 and the relevant equations are 8.32 through 8.37.

The power associated with each Fourier component is obtained as described in Section 8.12, the relevant equations being 8.52 through 8.54.

Example 8.12

Determine the first eight nonzero terms of the Fourier series representing the output voltage and voltage behind the choke, and the power associated with each component.

The component of zero order of the output voltage is obtained by adding the voltage absorbed by the load resistance to the load e.m.f., i.e.

$$V_{oo} = 370.0 + 0.062 \times 366.01 = 392.69 \text{ V}$$

The component of zero order of the voltage behind the choke is obtained by adding the voltage absorbed by the choke resistance to V_{oo}, i.e.

$$V_{bco} = 392.69 + 0.029 \times 366.01 = 403.31 \text{ V}$$

The next component is the one of third order. From Table 8.8, the third order component of the output current is $185.594\underline{/-215.484^\circ}$. The impedance of the load to this component is $0.367184\underline{/80.2789^\circ}$. The third order component of output voltage is the product of this current and impedance and is $68.1473\underline{/-135.205^\circ}$ V. The other components are obtained in a similar manner and are listed in Table 8.9.

The power output from the rectifier bridge which is associated with the third order component is the triple product of the r.m.s. values of the third order components of the voltage behind the choke, the output current, and the cosine of the phase angle between the two, i.e.

$$P_{bc3} = \frac{206.397}{\sqrt{2}}\frac{185.594}{\sqrt{2}}\cos[-130.177^\circ - (-215.484^\circ)] = 1.5673 \text{ kW}$$

The components of the power delivered to the load and the power output from the bridge are listed in columns 6 and 7 of Table 8.9.

The frequency spectrum and waveform of the output voltage are shown in Fig. 8.14. The remarks already made with regard to voltage waveforms

Table 8.9 *The Fourier components of the output voltage and voltage behind the choke and the powers associated with these components*

Order	V_{om} V	ϕ_{vom} degrees	V_{bcm} V	ϕ_{vbcm} degrees	P_{om} kW	P_{bcm} kW
0	392.69	0.00	403.31	0.00	143.73	147.62
3	68.15	−135.20	206.40	−130.18	1.07	1.57
6	28.56	−134.80	87.22	−132.25	0.05	0.07
9	18.35	−143.88	56.12	−142.18	0.01	0.01
12	13.52	−155.61	41.38	−154.33	0.00	0.00
15	10.68	−168.47	32.70	−167.45	0.00	0.00
18	8.80	178.05	26.94	178.90	0.00	0.00
21	7.46	164.16	22.84	164.89	0.00	0.00

apply with equal weight. The series converges more slowly than the current series due to the increase in circuit reactance with frequency so that the ripple in the waveform is more apparent. The discontinuities in the waveform at the commutation points are poorly modeled and good modeling would require an inordinately large number of Fourier components.

****

8.19 The diode and thyristor currents

In the first half of this chapter it was seen that the key to deriving the a.c. side waveforms is the currents in the bridge elements. There, only the current in a single thyristor was required. Here, because the top and bottom halves of the bridge operate at different angles of delay, the currents in a thyristor in the top half of the bridge and in a diode in the bottom half are required. Further progress with the present problem requires the Fourier series describing these currents. The potential complexity of this task is illustrated by comparing the series for the thyristor current developed in Example 8.1 with that for the output current developed in Example 8.2. The latter series has far fewer terms than the former and, while it is obvious how combining terms for the series representing the currents in three thyristors can result in the elimination of components, the process for creating components from a series where they do not exist is not obvious.

The procedure for accomplishing this task is in fact theoretically simple but computationally complex. It is the multiplication of Fourier series, the details of which are discussed in Appendix D.

With the chosen datum, it is known that thyristor T1 conducts the output current during the 120° segment from $-60° + \alpha$ to $60° + \alpha$ and that the other thyristors conduct during successive segments. In the lower half

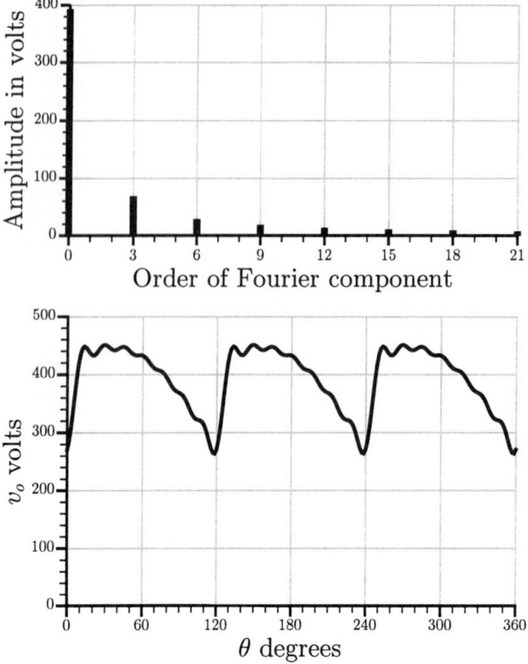

FIG. 8.14. The frequency spectrum and waveform of the output voltage.

of the bridge, diode D4 conducts the output current during the segment from 120° to 240° with the other diodes conducting during successive segments. Thus if the output current is multiplied by 1 during the appropriate segment and by zero everywhere else during a supply cycle, the waveforms of the thyristor and diode current will be generated. This trivial task in the time domain assumes greater complexity in the frequency domain and the first requirement is the creation of a function which will accomplish the multiplication. Because it effectively switches the output current function *on* when multiplying by 1 and *off* when multiplying by zero, it is known as a *switching function.*

8.19.1 *The switching function*

Figure 8.15 shows the required switching function which will be denoted by $S(\theta)$. It is repetitive with the same period as the a.c. supply and has the value 1 from δ to $\delta + 120°$ and is zero elsewhere throughout the cycle. The Fourier series representing the switching function is

$$S(\theta) = \sum_{n=0,\infty,1} S_{cn}\cos(n\theta) + S_{sn}\sin(n\theta) \tag{8.84}$$

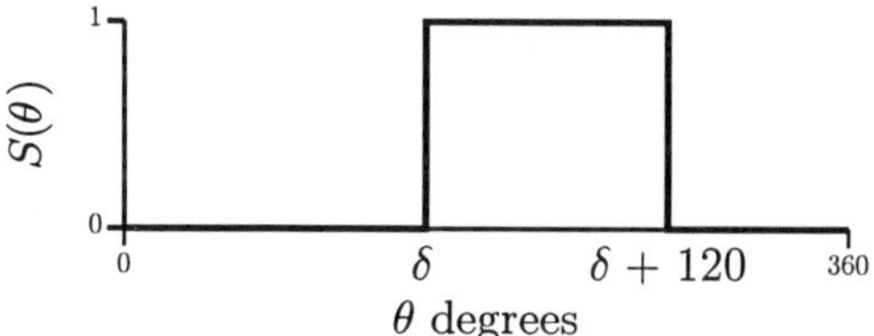

FIG. 8.15. The switching function for a three phase rectifier.

where, using eqns C.8, C.9, and C.10,

$$\text{for } n = 0 \qquad S_{co} = \frac{1}{3} \tag{8.85}$$

$$\text{for } n \neq 0 \qquad S_{cn} = \frac{1}{n}\frac{2}{\pi}\sin(\frac{n\pi}{3})\cos[n(\delta + \frac{\pi}{3})] \tag{8.86}$$

$$\text{and for } n \neq 0 \qquad S_{sn} = \frac{1}{n}\frac{2}{\pi}\sin(\frac{n\pi}{3})\sin[n(\delta + \frac{\pi}{3})] \tag{8.87}$$

For the present purpose there is little point in making a detailed examination of these functions but it is worth noting that, because of the $\sin(n\pi/3)$ term, there are no triplens, a consequence of the fact that this switching function is *on* for one-third of a cycle and *off* for two-thirds.

Example 8.13

Determine the first 23 Fourier components for the switching functions which are required to find the currents in thyristor T2 and the diode D5.

For thyristor T2 the angle δ is $\alpha + 60° = 124.98° = 2.1813$ rad. The values of the components computed using eqns 8.85 through 8.87 are listed in columns 3 and 4 of Table 8.10.

For diode D5 the angle δ is 120° or 2.0944 rad. The values of the components are determined using the same equations but with $\delta = 240°$ and are listed in columns 5 and 6 of Table 8.10.

The Cartesian form of the components is listed rather than the polar form employed in the other tables since this is the form most appropriate to computations involving the switching functions.

The table illustrates the point made above that the triplens are zero.

Figure 8.16 shows the approximations to the switching functions for thyristor T2 and diode D5 derived from these truncated Fourier series.

****

Table 8.10 *The first 23 Fourier components of the switching functions for the currents in thyristor T2 and diode D5. The amplitudes of the cosine and sine components are given in contrast to the other tables in which the polar form is displayed*

Component number	Order	S_{t2c}	S_{t2s}	S_{d5c}	S_{d5s}
1	0	0.3333	0.0000	0.3333	0.0000
2	1	-0.5492	-0.0479	0.2757	-0.4775
3	2	0.2715	0.0477	-0.1378	-0.2387
4	3	0.0000	0.0000	0.0000	0.0000
5	4	-0.1296	-0.0470	0.0689	-0.1194
6	5	0.1000	0.0464	-0.0551	-0.0955
7	6	0.0000	0.0000	0.0000	0.0000
8	7	-0.0646	-0.0450	0.0394	-0.0682
9	8	0.0529	0.0442	-0.0345	-0.0597
10	9	0.0000	0.0000	0.0000	0.0000
11	10	-0.0356	-0.0421	0.0276	-0.0477
12	11	0.0289	0.0409	-0.0251	-0.0434
13	12	0.0000	0.0000	0.0000	0.0000
14	13	-0.0181	-0.0384	0.0212	-0.0367
15	14	0.0136	0.0369	-0.0197	-0.0341
16	15	0.0000	0.0000	0.0000	0.0000
17	16	-0.0062	-0.0339	0.0172	-0.0298
18	17	0.0030	0.0323	-0.0162	-0.0281
19	18	0.0000	0.0000	0.0000	0.0000
20	19	0.0023	-0.0289	0.0145	-0.0251
21	20	-0.0046	0.0272	-0.0138	-0.0239
22	21	0.0000	0.0000	0.0000	0.0000
23	22	0.0084	-0.0236	0.0125	-0.0217

8.19.2 *The currents*

The Fourier series representing the current in a thyristor or diode is the product of the Fourier series for the output current and that for the appropriate switching function. Error is introduced at this point because both series must be truncated to a reasonable number of terms. However, as the example in Appendix D shows, the error will be small for components whose order is less than the lesser of the orders at which the two series are truncated. In the example the series have been truncated at the 21st order component so that terms in the product series whose order is less than or

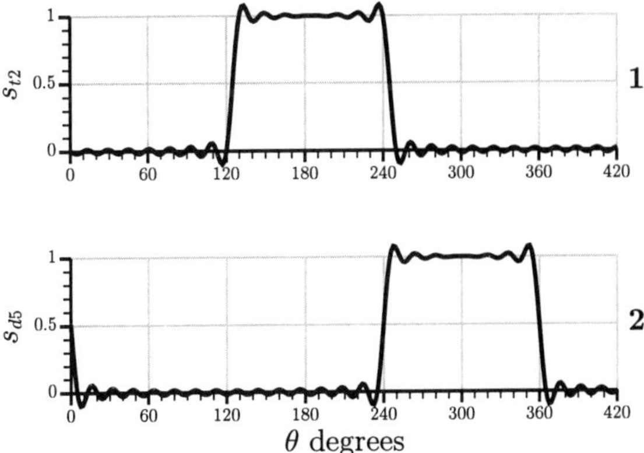

FIG. 8.16. The approximations to the switching functions, in Graph 1 for thyristor T2 and in Graph 2 for diode D5, provided by the truncated Fourier series of Table 8.10.

equal to 21 can be expected to be of reasonable accuracy.

A particularly simple situation arises when the load inductance is so large that the ripple in the output current is negligible. Then the Fourier series representing the output current has only one term, the zero order component, and the thyristor or diode current is simply the appropriate switching function series scaled by this constant. Clearly this is not the case in the example.

Example 8.14

Determine the Fourier series representing the currents in thyristor T2 and diode D5 up to the 21st order component.

The method follows that given in Appendix D and in particular eqn D.5. In this equation F_{cm} and F_{sm} are replaced by the cosine and sine components of i_o and G_{cn} and G_{sn} by the cosine and sine components of the appropriate switching function, S_{t2c} and S_{t2s} for i_{t2} and S_{d5c} and S_{d5s} for i_{d5}. The cosine and sine components of i_o are derived from the polar form given in Table 8.8 as

$$I_{ocm} = I_{om}\cos(\phi_{ocm})$$

and

$$I_{osm} = -I_{om}\sin(\phi_{ocm})$$

The calculation is long, involving many hundreds of individual computations and manual solution is out of the question. To give some feel for

Table 8.11 *Computation of the first order component of i_{t2} from the Fourier series for i_o and the switching function for thyristor T2. Component 1 is obtained when m + n = 1. Components 2 through 8 are obtained when m − n = 1 and 9 through 16 are obtained when m − n = −1*

Component number	m	n	I_{t2c}	I_{t2s}
1	0	1	−100.52	−8.76
2	3	2	−23.08	−11.02
3	6	5	−2.09	−0.56
4	9	8	−0.57	−0.11
5	12	11	−0.23	−0.03
6	15	14	−0.12	−0.01
7	18	17	−0.07	−0.01
8	21	20	−0.04	0.00
9	0	1	−100.52	−8.76
10	3	4	12.32	−3.43
11	6	7	1.54	−0.14
12	9	10	0.47	−0.01
13	12	13	0.20	0.01
14	15	16	0.10	0.01
15	18	19	0.06	0.00
16	21	22	0.04	0.00

it, Table 8.11 lists the computations required for the determination of the first order Fourier component of i_{t2}. If m is the order of a component in the series for i_o and n is the corresponding number for S_{t2}, m is an integral multiple of 3 and n takes every integer from zero upwards. The current series has been truncated at the eighth term, $m = 21$, and the switching function series has been truncated at the 23rd term, $n = 22$. Contributions to the first order component of i_{t2} arise in three ways, when $m + n = 1$, when $m - n = 1$, and when $m - n = -1$. There is only one $m + n$ term which is obtained when $m = 0$ and $n = 1$. This is component 1 in Table 8.11. There are seven $m - n = 1$ terms, $m = 3$ through 21 by threes and $n = 2$ through 20 by threes. These are components 2 through 8 in the table. There are eight $m - n = -1$ terms, $m = 0$ through 21 by threes and $n = 1$ through 22 by threes. These are components 9 through 16 the table. There are therefore 32 contributions to I_{t21}, 16 cosine and 16 sine components. These are listed in columns 4 and 5 of Table 8.11.

Table 8.12 *The first 21 Fourier components for i_{t2}, i_{d5}, and i_2*

Order	I_{t2} A	ϕ_{t2} degrees	I_{d5} A	ϕ_{t2} degrees	I_2 A	ϕ_{t2} degrees
0	122.00	0.00	122.00	0.00	0.00	0.00
1	215.03	171.22	214.39	52.96	368.58	−157.96
2	142.72	−18.99	141.19	103.60	249.02	−47.52
3	61.86	144.52	61.86	144.52	0.00	−90.00
4	14.11	−113.31	20.54	131.26	29.50	−74.34
5	21.84	−13.05	22.56	123.97	41.32	−34.91
6	13.10	140.10	13.10	140.10	0.00	−116.57
7	5.25	−163.19	10.36	109.20	11.42	−98.17
8	10.70	−27.88	11.48	126.64	21.63	−41.08
9	5.62	129.39	5.62	129.39	0.00	−131.19
10	3.67	161.11	7.66	99.62	6.73	−109.01
11	6.96	−44.25	7.73	129.39	14.66	−47.59
12	3.11	116.84	3.11	116.84	0.00	−133.36
13	3.05	135.70	6.41	94.84	4.56	−111.07
14	5.25	−60.59	6.06	133.06	11.23	−53.28
15	1.97	103.49	1.97	103.49	0.00	−133.07
16	2.73	114.92	5.75	92.09	3.40	−106.10
17	4.55	−76.34	5.46	137.62	9.58	−57.75
18	1.35	89.68	1.35	89.68	0.00	−135.02
19	2.74	96.90	5.61	90.51	2.91	−95.51
20	5.84	−87.02	7.01	144.86	11.56	−58.54
21	0.98	75.56	0.98	75.56	0.00	−131.66

The sum of all contributions to the cosine component, Table 8.11 column 4, is -212.51 A and the sum of all contributions to the sine component, column 5, is -32.82 A. The contribution of the components for which m is greater than 6 is very small, adding to the confidence that the result is a good approximation to the true series for I_{t2}.

Combining the cosine and sine components gives an amplitude of 215.03 A and a phase angle of 171.22°.

The first 21 terms of the Fourier series for I_{t2} and I_{d5} are listed in polar form in columns 2 through 5 of Table 8.12. The approximate waveforms of these currents derived from the Fourier series are shown in graph 1 of Fig. 8.17.

The truncated Fourier series indicate that the r.m.s. values of the thyristor and diode currents are each about 225 A. As expected the mean value of each current is one-third of the mean value of the output current.

****

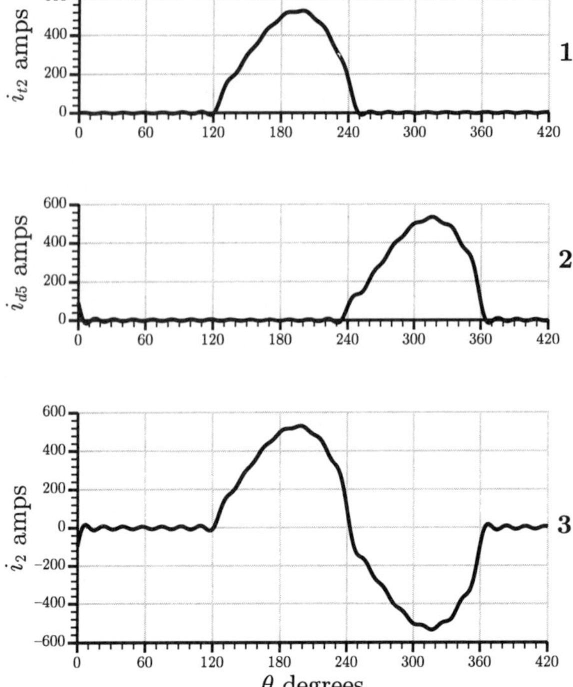

FIG. 8.17. The waveforms of the currents i_{t2}, i_{d5}, and i_2 determined using the truncated Fourier series given in Table 8.12.

8.20 The transformer secondary currents

The current in a transformer secondary is the difference between the currents in the thyristor and diode connected to it. For winding 2

$$i_2 = i_{t2} - i_{d5} \tag{8.88}$$

In the frequency domain the Fourier series representing i_{d5} is subtracted from that representing i_{t2} to obtain the series for i_2.

Example 8.15

Determine from the results of Example 8.14 the Fourier series representing the current in transformer secondary winding number 2.

From Table 8.12 it is seen that the zero order current in the diode is equal to that in the thyristor so that the zero order component of the transformer current is zero, as expected.

The triplens in the diode are equal to the triplens in the thyristor so that they also cancel and there are no triplens in the transformer. This also is as expected since triplens cannot flow in a three phase, three wire system.

The first order component of the thyristor current is $215.03\underline{/171.22^\circ}$ or, in Cartesian terms, $-212.51 + j32.82$. The corresponding component of the diode current is $214.39\underline{/52.96^\circ}$ or $129.13 + j171.14$. $\mathbf{I}_{21}$ is the difference between these two, $-341.64 - j138.31$ which, when transformed to polar form, is $368.58\underline{/-157.96^\circ}$.

The components of i_2 are given in columns 6 and 7 of Table 8.12. They indicate that the r.m.s. value of this current is about 318 A.

The waveform of i_2 is shown in graph 3 of Fig. 8.17.

****

8.21 The transformer primary currents

The transformer primary currents are the reflections of the corresponding secondary currents plus the magnetizing current. The process has been described in detail in Section 8.6 and is illustrated by Example 8.16.

Example 8.16

Determine the Fourier components up to the 20th order for the current in the yellow–blue transformer primary winding.

The secondary winding on the same limb of the transformer as the yellow–blue primary is number 2 and the reflected portion of i_{ry} is obtained by dividing the amplitudes of the components of i_2 obtained in Example 8.15 by the turns ratio, n_{tps}.

The magnetizing current has already been found in Example 8.4 to be $5.349 \cos(\theta - 3.5538)$. The first order component of i_{yb} is then, in phasor terms,

$$\mathbf{I}_{yb1} = \frac{368.58}{13.0}\underline{/-157.96^\circ} + 5.349\underline{/-203.62^\circ} = 32.320\underline{/-164.75^\circ}$$

There is no magnetizing current associated with the other components so that it is only necessary to divide the amplitudes of the components of i_2 by the turns ratio. Thus the second order component is

$$\mathbf{I}_{yb2} = \frac{249.02}{13.0}\underline{/-47.52^\circ} = 19.156\underline{/-47.52^\circ}$$

The components are listed in magnitude and phase in columns 2 and 3 of Table 8.13.

Table 8.13 *The first 20 Fourier components of the currents in the yellow–blue primary phase of the rectifier transformer and in the red line of the a.c. supply*

Order	I_{ybm} A	ϕ_{cybm} degrees	I_{rm} A	ϕ_{crm} degrees
1	32.32	–164.75	55.98	–74.75
2	19.16	–47.52	33.18	–137.52
3	0.00	–90.00	0.00	–90.00
4	2.27	–74.34	3.93	15.66
5	3.18	–34.91	5.51	–124.91
6	0.00	–116.57	0.00	–116.57
7	0.88	–98.17	1.52	–8.17
8	1.66	–41.08	2.88	–131.08
9	0.00	–131.19	0.00	–131.19
10	0.52	–109.01	0.90	–19.01
11	1.13	–47.59	1.95	–137.59
12	0.00	–133.36	0.00	–133.36
13	0.35	–111.07	0.61	–21.07
14	0.86	–53.28	1.50	–143.28
15	0.00	–133.07	0.00	–133.07
16	0.26	–106.10	0.45	–16.10
17	0.74	–57.75	1.28	–147.75
18	0.00	–135.02	0.00	–135.02
19	0.22	–95.51	0.39	–5.51
20	0.89	–58.54	1.54	–148.54

The frequency spectrum and waveform of i_{yb} are shown in Fig. 8.18. The absence of triplens and the large second order Fourier component are evident.

The approximation to the r.m.s. value of the current provided by the truncated Fourier series is 26.8 A.

****

8.22 The a.c. line currents

The Fourier components of the a.c. line currents are found by the method already described in Section 8.7. The phase sequence of the component must be taken into account and now, because there are components of even order as well as odd, there are more of each type. The zero sequence components are the triplens which are all zero so need not be considered further. The

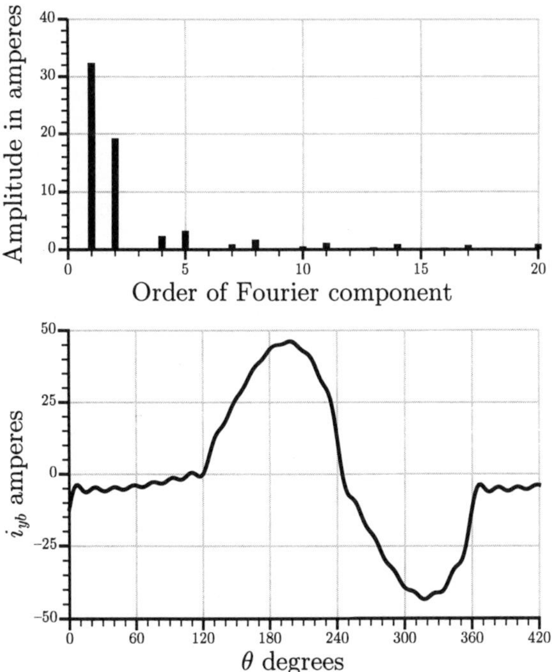

FIG. 8.18. The frequency spectrum and waveform of i_{yb}. The waveform derived from the truncated Fourier series is shown in the lower graph.

positive sequence components have orders which are integral multiples of three plus one, i.e. $m = 3k + 1$. The negative sequence components have orders which are an integral multiple of three minus one, i.e. $m = 3k - 1$. The computation is illustrated by Example 8.17.

Example 8.17

Determine, for the semi-controlled rectifier, the first 20 Fourier components of the current in the red a.c. line.

The first order component in the yellow–blue transformer primary winding taken from Table 8.13 is $32.32\underline{/-164.75°}$ A. This is a positive sequence component so that the first portion of eqn 8.31 is used to obtain the component for i_r as

$$\mathbf{I}_{r1} = 32.32 \times \sqrt{3}\underline{/-164.75 + 90.0} = 55.98\underline{/-74.75°}$$

The second order component of i_{yb} is $19.16\underline{/-47.52°}$ and is of negative sequence so that

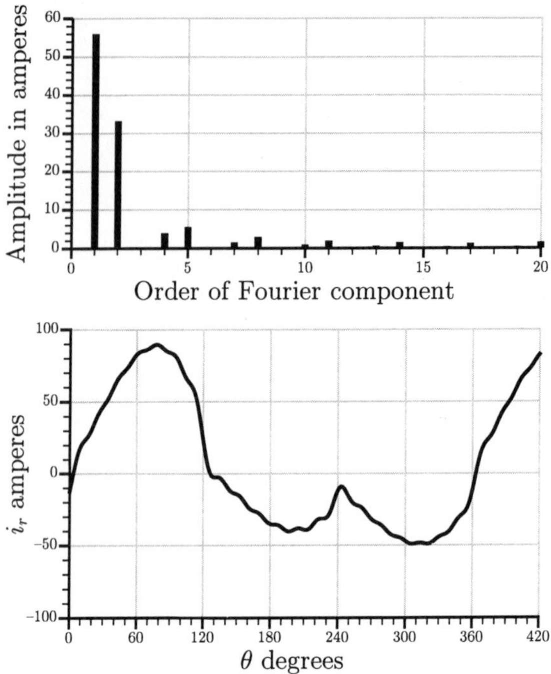

FIG. 8.19. The frequency spectrum and waveform of i_r. The waveform derived from the truncated Fourier series is shown in the lower graph.

$$\mathbf{I}_{r2} = 19.16 \times \sqrt{3}\underline{/-47.52 - 90.0} = 33.18\underline{/-137.52^\circ}$$

The Fourier components of i_r are listed in magnitude and phase in columns 4 and 5 of Table 8.13.

The frequency spectrum is shown in the upper graph of Fig. 8.19. It is identical to the spectrum of i_{yb} except for the scaling factor of $\sqrt{3}$. However, the waveform, shown in the second graph, is very different from that of the transformer primary current because of the different phase shifts experienced by the positive and negative sequence components.

The r.m.s. value of the line current is unaffected by the phase shift and is therefore $\sqrt{3}$ times the value for a primary current, 46.4 A.

****

8.23 The voltages on the a.c. side

The derivation of the Fourier series representing the a.c. side voltages follows the procedure already described in Section 8.11. In brief, taking due account of the phase sequence of the components, the voltage drops in the various system impedances caused by the Fourier components of the current are subtracted from the known sinusoidal busbar voltages. The process is illustrated in Example 8.18.

Example 8.18

Determine the first 20 components of the Fourier series representing the voltage between terminals 1 and 2 of the transformer secondary in the semi-controlled rectifier problem.

The busbar voltage is sinusoidal with an r.m.s. line to line value of 3300 V. The voltage between the red and yellow busbars is the datum phasor so that, using the peak value, $\mathbf{V}_{ry}$ is $4666.90\underline{/0.0}$ V.

The line impedance to the first order component is $0.034 + j0.2262\ \Omega$.

From Table 8.13, the first order component of the current in the red line is $55.98\underline{/-74.75°}$ A and is of positive phase sequence. Hence the first order component of the current in the yellow line is $55.98\underline{/-194.75°}$ A.

The first order component of the voltage across the red–yellow phase of the transformer primary is, allowing for the drops in the red and yellow lines,

$$\mathbf{V}'_{ry1} = 4666.90\underline{/0.0} - (0.034 + j0.2262) \times 55.98(\underline{/74.75°} - \underline{/-194.75°})$$

which, following the phasor diagram of Fig. 8.9(a), can be written

$$\mathbf{V}'_{ry1} = 4666.90\underline{/0.0} - (0.034 + j0.2262) \times 55.98 \times \sqrt{3}\underline{/-75.90° + 30.0°}$$

so that

$$\mathbf{V}'_{ry1} = 4649.14\underline{/-0.16°}$$

The first order component of current in the yellow–blue primary taken from columns 2 and 3 of Table 8.13 is $32.32\underline{/-164.75°}$ A so that the component in the red–yellow phase is $32.32\underline{/-44.75°}$ A. The leakage impedance of a primary phase to the first order component is $0.45 + j2.3750\ \Omega$ so that the first order component of the e.m.f. in the red–yellow primary is

$$\mathbf{E}_{ry1} = 4649.14\underline{/-0.16} - (0.45 + j2.3750) \times 32.32\underline{/-44.75°}$$

i.e.

$$\mathbf{E}_{ry1} = 4585.12\underline{/-0.72°}$$

Dividing this voltage by the turns ratio, 13, gives the e.m.f. in secondary winding number 1

$$\mathbf{E}_{11} = 352.70\underline{/-0.72°}$$

The first order component of current in secondary winding number 2 taken from columns 6 and 7 of Table 8.12 is $368.58\underline{/-157.396°}$. The component in phase number 1 is therefore $368.58\underline{/-39.31^\circ}$. The leakage impedance of a transformer secondary to a first order component is $0.0025 + j0.01546\ \Omega$. Hence the first order component of the voltage across secondary number 1 is

$$\mathbf{V}_{11} = 352.70\underline{/-0.72^\circ} - (0.0025 + j0.01546) \times 368.58\underline{/-39.31^\circ}$$

i.e.

$$\mathbf{V}_{11} = 348.54\underline{/-1.37^\circ}$$

Again following the relationships displayed in the phasor diagram of Fig. 8.9(a), the first order component of the voltage between secondary terminals 1 and 2 is obtained by multiplying the amplitude of $\mathbf{V}_{11}$ by $\sqrt{3}$ and advancing the phase by 30°. Thus

$$\mathbf{V}_{121} = 603.69\underline{/28.63^\circ}$$

The second order, and all higher order, components of busbar voltage are zero.

The line impedance to the second order component is $0.034 + j0.4524\ \Omega$. The second order component of the red line current is $33.18\underline{/-137.52^\circ}$ A and is of negative phase sequence. Hence the second order component of the current in the yellow line is $33.18\underline{/-17.52^\circ}$ A.

The second order component of the voltage across the red–yellow phase of the transformer primary is therefore

$$\mathbf{V}'_{ry2} = -(0.034 + j0.4524) \times 33.18(\underline{/-137.52^\circ} - \underline{/-17.52^\circ})$$

which, following the phasor diagram of Fig. 8.9(a), can be written

$$\mathbf{V}'_{ry2} = -(0.034 + j0.4524) \times 33.18 \times \sqrt{3}\underline{/-137.52^\circ - 30.0^\circ}$$

so that

$$\mathbf{V}'_{ry2} = 26.07\underline{/98.18^\circ}$$

The second order component of current in the yellow–blue primary taken from columns 2 and 3 of Table 8.13 is $19.16\underline{/-47.52^\circ}$ A so that the component in the red–yellow phase is $19.16\underline{/-167.52^\circ}$ A. The leakage impedance of a primary phase to the Fourier component of second order is

Table 8.14 *The first 20 Fourier components of the voltage between transformer secondary terminals 1 and 2 and the mean three phase power associated with these components*

Component number	Order	V_{12m} V	ϕ_{v12m} degrees	P_m kW
1	1	603.69	28.63	154.73
2	2	29.03	-292.44	-0.54
3	3	0.00	0.00	0.00
4	4	6.86	-16.80	-0.01
5	5	12.00	-276.88	-0.01
6	6	0.00	0.00	0.00
7	7	4.64	-39.58	0.00
8	8	10.05	-282.31	0.00
9	9	0.00	0.00	0.00
10	10	3.91	-49.99	0.00
11	11	9.37	-288.49	0.00
12	12	0.00	0.00	0.00
13	13	3.44	-51.82	0.00
14	14	9.13	-293.98	0.00
15	15	0.00	0.00	0.00
16	16	3.16	-46.71	0.00
17	17	9.45	-298.33	0.00
18	18	0.00	0.00	0.00
19	19	3.21	-36.03	0.00
20	20	13.42	-299.03	0.00

$0.45 + j4.7501\ \Omega$ so that the second order component of the e.m.f. in the red–yellow primary is

$$\mathbf{E}_{ry2} = 26.07\underline{/98.18^\circ} - (0.45 + j4.7501) \times 19.16\underline{/-167.52^\circ}$$

i.e.

$$\mathbf{E}_{ry2} = 117.47\underline{/97.31^\circ}$$

Dividing this voltage by the turns ratio, 13, gives the e.m.f. in secondary winding number 1

$$\mathbf{E}_{12} = 9.036\underline{/97.31^\circ}$$

The second order component of current in secondary winding number 2 taken from columns 6 and 7 of Table 8.12 is $249.02\underline{/-47.52^\circ}$. The component in phase number 1 is therefore $249.02\underline{/-167.52^\circ}$. The leakage impedance of a transformer secondary to a second order component

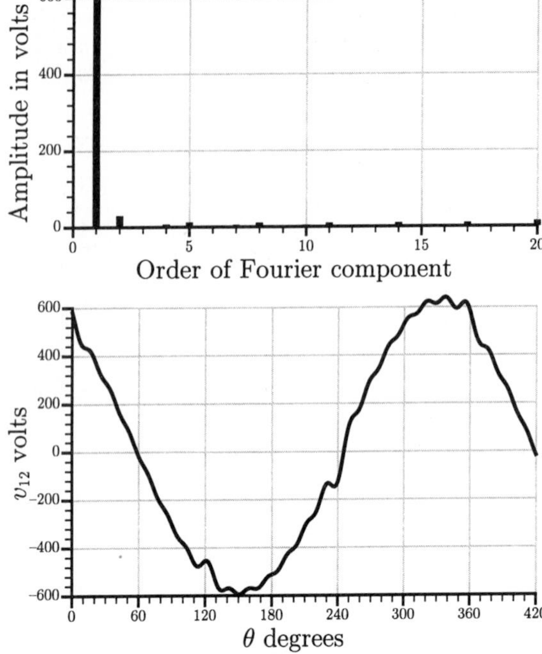

FIG. 8.20. The frequency spectrum and waveform of the voltage between terminals 1 and 2 of the transformer secondary.

is $0.0025 + j0.03091\ \Omega$. Hence the second order component of the voltage across secondary number 1 is

$$\mathbf{V}_{12} = 9.036\underline{/97.31^\circ} - +j0.03091) \times 249.02\underline{/-167.52^\circ}$$

i.e.

$$\mathbf{V}_{12} = 16.76\underline{/97.56^\circ}$$

Again following the relationships displayed in the phasor diagram of Fig. 8.9(b), the second order component of the voltage between secondary terminals 1 and 2 is obtained by multiplying the amplitude of $\mathbf{V}_{12}$ by $\sqrt{3}$ and retarding the phase by 30°. Thus

$$\mathbf{V}_{122} = 29.027\underline{/67.56^\circ}$$

The first 20 Fourier components of v_{12} are listed in columns 3 and 4 of Table 8.14

The frequency spectrum of v_{12} and the approximation to its waveform provided by the truncated Fourier series are shown in Fig. 8.20. The ripple is due to the limited number of terms in the series and to the consequent limited accuracy of the multiplication process.

****

8.24 Power

The mean power associated with each Fourier component entering the rectifier from the transformer secondary terminals is three times the triple product of the r.m.s. value of the phase voltage component, the r.m.s. value of the corresponding current component in the same phase, and the cosine of their phase difference. The powers are listed in column 5 of Table 8.14.

If the power delivered by the rectifier bridge to the combination of load and smoothing choke, Table 8.9, column 7, is compared with the power entering the bridge from the transformer, it is seen that the supply delivers 154.73 kW of 60 Hz, three phase power to the rectifier. Of this, 143.73 kW is delivered to the load as d.c. power and 1.13 kW of a.c. power is delivered to the load, mainly at 180 Hz. The rectifier bridge also sends back into the a.c. system 0.56 kW of a.c. power at frequencies other than the supply frequency, mainly at 120 Hz. As in the fully controlled bridge, the rectifier bridge, as well as playing its major role as a converter of system frequency a.c. into d.c., is also acting as a generator of high frequency power some of which it delivers to the load and some of which it injects into the a.c. power system.

8.25 Comparison of the two bridges

The cases of a fully controlled and a semi-controlled bridge supplying the same load from the same supply through the same transformer and smoothing choke have been analyzed. It is therefore appropriate to compare them and, although the comparison applies strictly to this particular case, it is nevertheless generally applicable.

The fully controlled bridge delivers a much smoother current to the load. The first Fourier component is of sixth order and has an amplitude of 56 A. The total r.m.s. value of the a.c. components of the load current is 41 A. The total a.c. power delivered to the combination of load and choke is 0.16 kW. The first a.c. component of the load current delivered by the semi-controlled bridge is of second order with an amplitude of 206 A. The total r.m.s. value of all the a.c. components of the load current is about 150 A. The a.c. power delivered to the combination of load and choke is 2.05 kW.

The situation on the a.c. side is similar, with the fully controlled bridge pumping far less high frequency energy back into the system. The first high frequency component of the a.c. system line current to the fully controlled bridge is of fifth order, 300 Hz, with an amplitude of 14.6 A. The total r.m.s. value of the high frequency components is 11.8 A. The power associated with these high frequency components is negligible, a few watts. The semi-controlled bridge injects into the a.c. supply a second order component of line current, 120 Hz, whose amplitude is 34.3 A. The total r.m.s. value of the high frequency components is about 25 A. The power associated with the high frequency components is about 0.6 kW.

Considering also that the semi-controlled bridge cannot invert, i.e. convert d.c. energy into a.c. energy, its disadvantages begin to loom large in comparison with its merits—a slightly better power factor and a substantially lower kVA input at reduced voltage. No doubt these considerations are responsible for the decreasing popularity of the semi-controlled bridge.

8.26 Frequency domain analysis as an analytic tool

Frequency domain analysis is a valuable supplement to time domain analysis, providing insights into the generation of unwanted frequency components of voltage and current, its major disadvantage being the very large number of Fourier components which are needed to accurately model fine scale features such as commutation.

As the prime analytical tool, frequency domain analysis has some significant drawbacks. It cannot model commutation and it cannot deal with discontinuous conduction. However, these are relatively minor matters on the output side of the rectifier where this method is simple and straightforward, and yields unambiguous results. Transferring these results to the a.c. side is another matter, involving the product of Fourier series with all the complexity that that entails. Of course complexity need not be a problem. Once a subroutine has been created which multiplies two series with a simple call statement, the complexity is overcome. However, there does remain a fundamental uncertainty. The series must be truncated after a reasonable number of terms and this always leaves doubt as to the size of the contribution made by the higher order components which have been missed.

A middle of the road approach uses Fourier analysis as the prime analytical tool only for d.c. side computations. If more is required, a full time domain analysis, with Fourier analysis of the results if desired, seems more appropriate since, just as subroutines can remove the labor involved in multiplying Fourier series, they can remove much of the complexity of time domain analysis.

9

WAVEFORM AND POWER FACTOR IMPROVEMENT

9.1 Introduction

In Chapter 2 it was seen how increasing the number of phases of a rectifier led to improvements in the waveforms of its output voltage and input current. The higher order Fourier components of the output voltage diminish and move to higher frequencies and the voltage approaches closer to the desired pure d.c. More and more secondary phases contribute to the primary input line current with increasing cancellation of harmonics. The price of this improvement is a very substantial increase in the size and complexity of the input transformer.

It was also seen that, while the controlled rectifier is an excellent device for varying the output voltage, there is a price to be paid for this feature; any reduction in output voltage by delay angle control is accompanied by a corresponding diminution of power factor.

In Chapter 3, Three Phase Bridge Rectifiers, two techniques for amelioration of this situation were introduced. Series connection of two simple rectifiers to form the fully controlled bridge, and sequential operation when simple controlled and uncontrolled rectifiers are connected in series to form the semi-controlled bridge.

These two techniques, series connection, and sequential operation, together with a third, parallel operation, will now be examined in more detail from the viewpoints of waveform and power factor improvement. The ultimate defense when these palliatives are insufficient, the use of static filters, will also be considered.

Series and parallel operation yield waveform improvements while minimizing the cost in increased equipment size but do nothing to improve the power factor. Sequential operation lowers the power factor penalty of reduced voltage operation at the cost of a deterioration in waveform quality. Static filters reduce individual or groups of unwanted Fourier components but create highly complex circuit conditions which can have undesirable side effects.

Throughout this chapter the rectifier is assumed to be ideal, fed from a three phase voltage source and feeding a d.c. current source, i.e. the non-ideal phenomena, which were the topics of Chapters 5 and 6, will be ignored.

9.2 Series connection

Series connection of two simple, three phase, single way, controlled rectifiers to form the fully controlled three phase bridge provides six phase operation from the viewpoint of the unwanted Fourier components and three phase operation from the viewpoint of kVA requirements. The connection is particularly simple since the same three phase supply feeds both simple rectifiers, the necessary 180° phase difference between the two being provided by connecting one in the common cathode and the other in the common anode arrangement.

The process can be extended with similar desirable consequences and one undesirable one—a transformer or transformers become necessary. However, this disadvantage is more apparent than real since the size of a system which would necessitate the additional concern about waveform would also most likely require the voltage adjustment and isolating capabilities of a transformer. Probably the commonest arrangement after the three phase bridge itself is the series connection of two three phase bridges as shown in Fig. 9.1. In this circuit the upper bridge will be identified as A and the lower bridge as B.

On the output side, the two bridges are placed in series by a connection from the negative terminal, A_a, of the first to the positive terminal, K_b, of the second. Each bridge is operated at the same delay angle, i.e.

$$\alpha_a = \alpha_b = \alpha \tag{9.1}$$

and the mean output voltage is twice the mean output voltage of a single unit, i.e.

$$V_o = V_{om}\cos(\alpha) \tag{9.2}$$

and

$$V_{om} = 2V_{aom} = 2V_{bom} \tag{9.3}$$

Each unit carries the same current, the output current I_o, which is assumed to be free of ripple.

9.2.1 *Output voltage waveform*

The two, three phase, supplies must have a phase difference of 30°. At the ripple frequency, which is six times the supply frequency, this corresponds to a 180° phase displacement so that much of the ripple is canceled and the ripple content of the combined output voltage corresponds to that of a twelve phase rectifier. This is illustrated by Fig. 9.2 which shows the waveforms of the output voltage of each bridge, v_{ao} and v_{bo}, and their sum, v_o. The figure has been drawn for a delay angle of 45°.

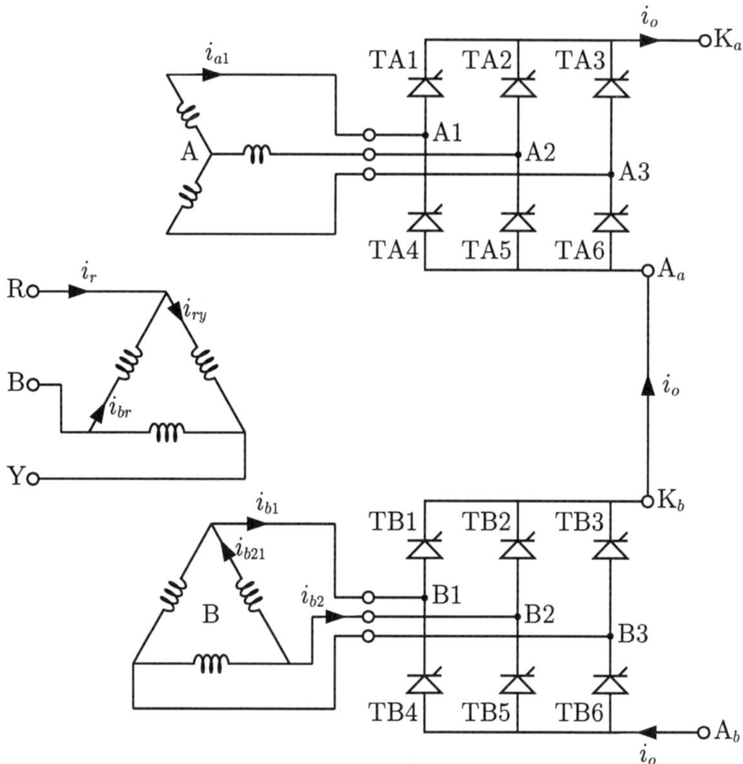

FIG. 9.1. Two three phase bridge rectifiers connected in series. The three phase supplies to the bridges are mutually displaced in phase by 30°

The Fourier series representing the output voltage of rectifier A is

$$v_{ao} = V_{om0}\cos(\alpha) + \sum_{m=6,\infty,6} V_{omm}\cos(m\theta + \phi_{am}) \tag{9.4}$$

where V_{om0} is the maximum value of the rectifier mean output voltage, i.e. the Fourier component of zero order, and V_{omm} is the amplitude and ϕ_{am} is the phase of the mth order Fourier component of the output voltage.

Since the a.c. voltages applied to rectifier B lag 30° on those applied to rectifier A, the Fourier series representing its output voltage can be derived from eqn 9.4 as

$$v_{bo} = V_{om0}\cos(\alpha) + \sum_{m=6,\infty,6} V_{omm}\cos[m(\theta - \frac{\pi}{6}) + \phi_{am}] \tag{9.5}$$

Putting the Fourier order $m = 6k$, where k is any positive nonzero integer, and expanding the cosine term of eqn 9.5 yields

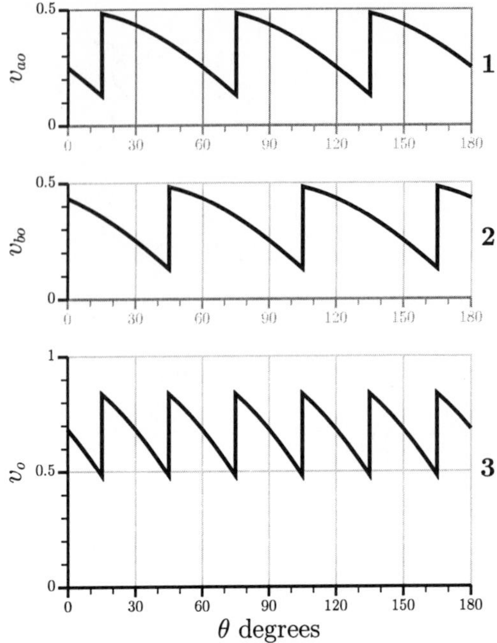

FIG. 9.2. Waveforms of the output voltages from the two bridges, v_{ao} and v_{bo}, and their sum, v_o.

$$v_{bo} = V_{om0}\cos(\alpha) + \sum_{m=6,\infty,6} V_{omm}\big[\cos(m\theta + \phi_{am})\cos(k\pi) + \sin(m\theta + \phi_{am})\sin(k\pi)\big]$$

Now $\cos(k\pi) = (-1)^k$ and $\sin(k\pi) = 0$ so that the output voltage v_o, which is the sum of v_{ao} and v_{bo}, is given by

$$v_o = 2V_{om0}\cos(\alpha) + \sum_{m=6,\infty,6} V_{omm}\cos(m\theta + \phi_{am})[1 + (-1)^k]$$

Since the value of $1+(-1)^k$ is zero for odd values of k and 2 for even values, it is seen that the series connection eliminates the Fourier components which are odd multiples of six so that

$$v_o = 2V_{om0}\cos(\alpha) + \sum_{m=12,\infty,12} 2V_{omm}\cos(m\theta + \phi_{am}) \tag{9.6}$$

i.e. the output is indeed that of a twelve phase rectifier.

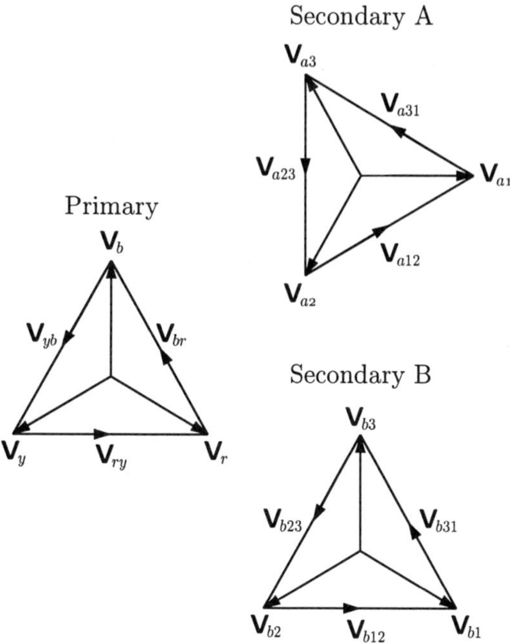

FIG. 9.3. Phasor diagrams for the primary and secondary voltages of the transformer supplying the two series connected three phase bridges.

9.2.2 *The transformer*

The necessary 30° phase shift between the supplies to the two bridges can be obtained by any appropriate means. Figure 9.1 shows the commonest of these, a three phase transformer with two separate secondary windings, one wye connected and the other delta connected. To obtain the same line to line input voltage to the rectifiers, the number of turns on a delta connected secondary, N_{tsb}, must be $\sqrt{3}$ times the number on a wye connected secondary, N_{tsa}. The separate secondary windings provide the necessary electrical isolation between the two bridges. In Fig. 9.1 the transformer primary is shown as connected in delta. Since there are no zero sequence currents flowing in the circuit, it could alternatively be connected in wye.

Figure 9.1 has been drawn so that windings which are parallel are on the same leg of the transformer and therefore the voltages induced in them have the same phase. Taking the usual datum, the voltage of the secondary A1 terminal relative to the system neutral, leads to the phasor diagrams of Fig. 9.3. It will be seen that the voltages applied to the second bridge lag by the desired 30° on the voltages applied to the first bridge.

9.2.3 *The circuit currents*

The waveforms of the various currents in the circuit are shown in Fig. 9.4. These have been drawn on the assumption that ripple in the output current is negligible and have been normalized to the output current, I_o, or its value referred to the primary via the first secondary winding A, $I'_o = I_o N_{tsap}$, as appropriate. The distance between ticks on the vertical axes is constant for all graphs and represents either the current I_o or I'_o as appropriate. The figure is drawn for a delay angle of 45°. It applies to any delay angle, α, provided the origin of the θ scale is moved to $45° - \alpha°$.

Graphs 1 through 3 show the waveforms of the currents in lines A1, A2, and A3, each space on the vertical axis representing I_o amperes. Each wave consists of two rectangular pulses of current, amplitude I_o, duration 120°, one positive, one negative, and separated by 60° intervals of zero current.

Graphs 4 through 6 show the waveforms of the currents in lines B1, B2, and B3. Their shapes and amplitudes are similar to those of the currents in lines A1, A2, and A3 but they lag by 30°.

Graphs 7 through 9 show the waveforms of the currents in the windings of transformer secondary B, i_{b21} flowing from B2 to B1, i_{b32} flowing from B3 to B2, and i_{b13} flowing from B1 to B3. These current have been derived from the line currents using the relationships of eqn 9.7 which apply provided the zero sequence currents are zero, a requirement fulfilled in this case

$$\begin{bmatrix} i_{b21} \\ i_{b32} \\ i_{b13} \end{bmatrix} = \frac{1}{3} \begin{bmatrix} 1 & -1 & \\ & 1 & -1 \\ -1 & & 1 \end{bmatrix} \begin{bmatrix} i_{b1} \\ i_{b2} \\ i_{b3} \end{bmatrix} \tag{9.7}$$

These currents have a stepped waveform, each step being of 60° duration and the height of the steps being $I_o/3$ and $2I_o/3$.

Graphs 10 through 12 show the waveforms of the currents in the primary windings, i_{ry} flowing from the red to the yellow terminal, i_{yb} from the yellow to the blue terminal and i_{br} from the blue to the red terminal. These waves are derived on the basis of ampere–turn balance

$$\begin{aligned} N_{tp} i_{ry} &= N_{tsa} i_{a1} + N_{tsb} i_{b21} \\ N_{tp} i_{yb} &= N_{tsa} i_{a2} + N_{tsb} i_{b32} \\ N_{tp} i_{br} &= N_{tsa} i_{a3} + N_{tsb} i_{b13} \end{aligned}$$

which, remembering that $N_{tsb} = \sqrt{3}\, N_{tsa}$, can be written as

$$i_{ry} = (i_{a1} + \sqrt{3} i_{b21}) N_{tsap} \tag{9.8}$$

$$i_{yb} = (i_{a2} + \sqrt{3} i_{b32}) N_{tsap} \tag{9.9}$$

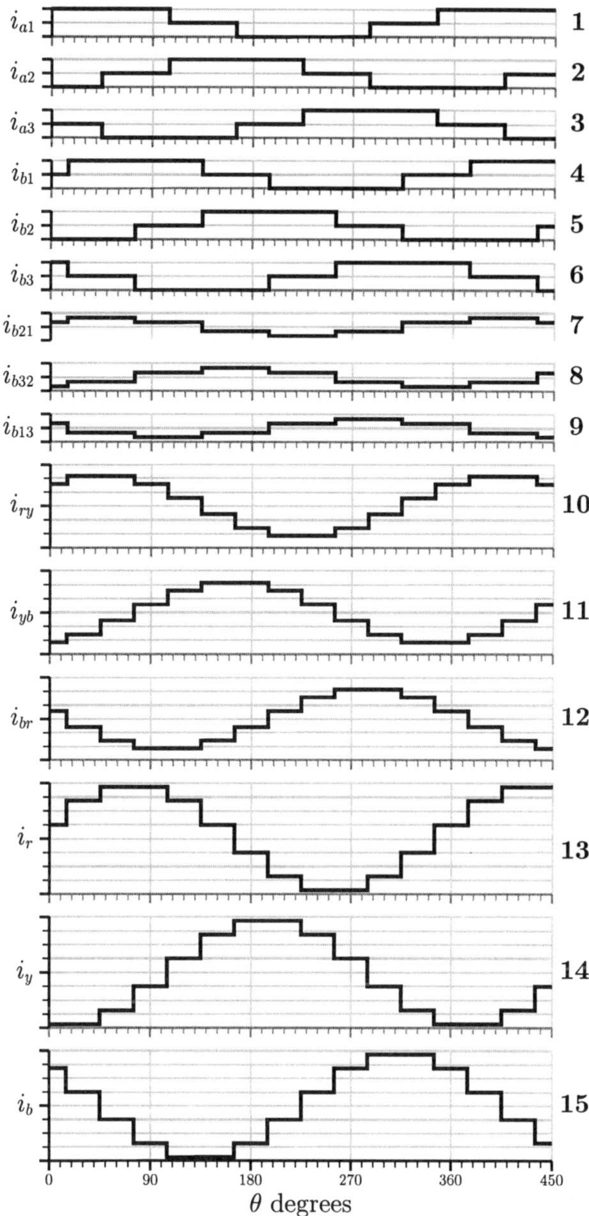

FIG. 9.4. Waveforms of the currents in the a.c. input lines to the rectifiers, the windings of secondary B, the primary windings, and the a.c. supply lines. The delay angle is 45°.

$$i_{br} = (i_{a3} + \sqrt{3}i_{b13})N_{tsap} \tag{9.10}$$

The waveforms have ten steps per cycle, most of 30° duration but two of 60° duration. The height of the steps is $N_{tsap}I_o/\sqrt{3}$, $(1 + 1/\sqrt{3})N_{tsap}I_o$, and $(1 + 2/\sqrt{3})N_{tsap}I_o$.

Graphs 13 through 15 show the waveforms of the a.c. line currents, i_r in the red line, i_y in the yellow line, and i_b in the blue line. The waveforms are similar to those of the primary winding currents in that there are ten steps per cycle with most of the steps being of 30° duration and two of 60° duration. The step heights are those of the primary phase currents multiplied by $\sqrt{3}$, i.e. $N_{tsap}I_o$, $(\sqrt{3}+1)N_{tsap}I_o$, and $(\sqrt{3}+2)N_{tsap}I_o$. The waves lag those of the primary phase currents by 30°.

9.2.4 *The r.m.s. currents*

The r.m.s. currents in the first secondary, I_{sa}, in the second secondary, I_{sb}, in the primary, I_p, and in an a.c. supply line, I_s, are

$$I_{sa} = \sqrt{\frac{2}{3}}I_o = 0.8165I_o \tag{9.11}$$

$$I_{sb} = \frac{\sqrt{2}}{3}I_o = 0.4714I_o \tag{9.12}$$

$$I_p = N_{tsap}\sqrt{\frac{4}{3} + \frac{2}{\sqrt{3}}}\, I_o = 1.5774\, N_{tsap}I_o \tag{9.13}$$

$$I_s = N_{tsap}\sqrt{4 + 2\sqrt{3}}\, I_o = 2.7321\, N_{tsap}I_o \tag{9.14}$$

9.2.5 *The transformer ratings*

For a given maximum d.c. output voltage, V_{om}, the r.m.s. secondary winding voltages must be

$$V_{sa} = \frac{\pi}{6\sqrt{6}}V_{om} = 0.2138V_{om} \tag{9.15}$$

$$V_{sa} = \frac{\pi}{6\sqrt{2}}V_{om} = 0.3702V_{om} \tag{9.16}$$

$$V_p = V_s = N_{tsap}V_{sa} \tag{9.17}$$

where V_s is the r.m.s. value of the supply line to line voltage.

Since, as expected, the currents in the secondary B are $1/\sqrt{3}$ times the currents in the secondary A and the voltages across the secondary B are $\sqrt{3}$ times those across the secondary A, the volt-ampere ratings of the secondaries are equal, each being $\pi/6$ times the d.c. output power rating, for a combined secondary rating of

$$kVA_{sec} = \frac{\pi}{3}\frac{V_{om}I_{orat}}{1000} = 1.0472\frac{V_{om}I_{orat}}{1000} \tag{9.18}$$

where I_{orat} is the rated value of the output current.

The primary rating is

$$kVA_p = 3\frac{V_pI_{prat}}{1000} = \pi\sqrt{\frac{1}{18}+\frac{1}{12\sqrt{3}}}\frac{V_{om}I_{orat}}{1000} = 1.0115\frac{V_{om}I_{orat}}{1000} \tag{9.19}$$

where I_{prat} is the rated value of the primary current.

These ratings are only marginally more than the rated d.c. output power indicating the success of the system in reducing higher order Fourier components of the transformer currents.

9.2.6 *The input kVA and power factor*

The input kVA as conventionally measured, kVA_s, is

$$kVA_s = \sqrt{3}\frac{V_sI_s}{1000} = \sqrt{\frac{1}{18}+\frac{1}{12\sqrt{3}}}\frac{V_{om}I_o}{1000} = 1.0115\frac{V_{om}I_o}{1000} \tag{9.20}$$

the same as the primary rating.

Since the circuit has been assumed to be ideal, without losses, the input power is equal to the output power and the power factor as conventionally measured is

$$Pf_s = \frac{P_o}{kVA_s} = \frac{\cos(\alpha)}{1.0115} = 0.9886\cos(\theta) \tag{9.21}$$

The fact that the power factor is only marginally less than $\cos(\alpha)$ is another indication of the low harmonic content of the input current.

Example 9.1

A d.c. load rated at 10 MW, 800 V, is to be fed from a 12 kV, 60 Hz, three phase supply via series connected bridge rectifiers. Determine the salient quantities in this system. The calculations can assume ideal components but the rectifier should be designed to provide 110% of the rated load voltage to allow for internal voltage drops and low supply voltage.

The salient values for a suitable transformer–rectifier combination are

$$\text{Rated load current} = \frac{10\times10^6}{800} = 12\,500\text{ A}$$

$$\text{Maximum output voltage} = 1.1 \times 800 = 880 \text{ V d.c.}$$
$$\text{Delay angle at rated d.c. voltage} = \arccos(\frac{800}{880}) = 24.62°$$
$$\text{Secondary line to line voltage} = \frac{\pi}{6\sqrt{2}} \times 880 = 325.8 \text{ V}$$
$$\text{Turns ratio, } N_{tp}/N_{tsa} = \frac{12\,000}{325.8/\sqrt{3}} = 63.79$$
$$\text{Turns ratio, } N_{tp}/N_{tsb} = \frac{12\,000}{325.8} = 36.83$$
$$\text{Rated mean thyristor current} = \frac{12\,500}{3} = 4167 \text{ A}$$
$$\text{Rated r.m.s. thyristor current} = \frac{12\,500}{\sqrt{3}} = 7217 \text{ A}$$
$$\text{Secondary current, } I_a = 12\,500\sqrt{\frac{2}{3}} = 10206 \text{ A}$$
$$\text{Secondary current, } I_b = 12\,500\frac{\sqrt{2}}{3} = 5893 \text{ A}$$
$$\text{Primary current, } I_p = \frac{12\,500}{63.79} \times 1.5774 = 309.1 \text{ A}$$
$$\text{AC line current, } I_s = \frac{12\,500}{63.79} \times 2.7321 = 535.3 \text{ A}$$
$$\text{Secondary rating, } kVA_{sec} = \frac{1.047 \times 880 \times 12\,500}{1000} = 11520 \text{ kVA}$$
$$\text{Primary rating, } kVA_p = \frac{1.0115 \times 880 \times 12\,500}{1000} = 11130 \text{ kVA}$$

Parallel connected thyristors would be required to carry the current.

****

9.2.7 *Spectral content of the currents*

In drawing the waveforms and calculating the r.m.s. values of the currents, it has been assumed that the output current ripple and commutation time are both negligible. These assumptions could be continued in order to determine the spectral content of the resulting rectangular current waveforms depicted in Fig. 9.4. However, it is as easy to take the more general approach of assuming that the waveform of the current in a thyristor has been found by a time domain analysis like that used in Chapter 5 and has been analyzed into its Fourier components. The Fourier series for a secondary line currents is then derived in the same manner as in Section 8.6.

Let the mth order component of the current in secondary line A1 be i_{a1m} where

$$i_{a1m} = I_{am}\cos(m\theta + \phi_{am}) \tag{9.22}$$

The waveforms of the thyristor currents are identical except for the appropriate phase shift so that the mth order currents in the other two lines of secondary A are

$$i_{a2m} = I_{am}\cos(m(\theta - 2\pi/3) + \phi_{am}) \tag{9.23}$$
$$i_{a3m} = I_{am}\cos(m(\theta + 2\pi/3) + \phi_{am}) \tag{9.24}$$

For secondary B, because its voltages lag by 30° on those of A,

$$i_{b1m} = I_{am}\cos(m(\theta - \frac{\pi}{6}) + \phi_{am}) \tag{9.25}$$
$$i_{b2m} = I_{am}\cos(m(\theta - \frac{5\pi}{6}) + \phi_{am}) \tag{9.26}$$
$$i_{b3m} = I_{am}\cos(m(\theta + \frac{\pi}{2}) + \phi_{am}) \tag{9.27}$$

In Chapter 8 it was established that m is odd, is not a multiple of 3 and can be divided into two series, one of positive sequence components and the other of negative sequence components. Letting k be any positive integer including zero, the positive sequence series is

$$m = 1 + 6k \tag{9.28}$$

and contains components of order 1, 7, 13, etc.

The negative sequence series is

$$m = 5 + 6k \tag{9.29}$$

and contains components of order 5, 11, 17, etc.

From eqn 9.7, the mth Fourier component of the current flowing from terminal 2 to terminal 1 in the transformer B secondary winding is

$$i_{b21m} = \frac{i_{b1m} - i_{b2m}}{3}$$

which, after substitution of eqns 9.25 and 9.26, becomes

$$i_{b21m} = \frac{I_{am}}{3}\left[\cos(m\theta - m\frac{\pi}{6} + \phi_{am}) - \cos(m\theta - m\frac{5\pi}{6} + \phi_{am})\right]$$

By combining the two cosine terms this becomes

$$i_{b21m} = -\frac{2I_{am}}{3}\sin(m\frac{\pi}{3})\sin(m\theta - m\frac{\pi}{2} + \phi_{am})$$

The second sine term can be expanded to yield

$$i_{b21m} = -\frac{2I_{am}}{3}\sin(m\frac{\pi}{3})\left[\sin(m\theta+\phi_{am})\cos(m\frac{\pi}{2})-\cos(m\theta+\phi_{am})\sin(m\frac{\pi}{2})\right]$$

Since m is odd, $\cos(m\pi/2)$ is zero so that this last equation reduces to

$$i_{b21m} = \frac{2I_{am}}{3}\sin(m\frac{\pi}{2})\sin(m\frac{\pi}{3})\cos(m\theta+\phi_{am}) \tag{9.30}$$

The total m.m.f., $\mathcal{F}_{s1m}$, created by the secondary current components i_{a1m} and i_{b21m} is

$$\mathcal{F}_{s1m} = N_{tsa}i_{a1m} + N_{tsb}i_{b21m}$$

which, since $N_{tsb} = \sqrt{3}\ N_{tsa}$, and by substituting eqns 9.22 and 9.30, becomes

$$\mathcal{F}_{s1m} = N_{tsa}I_{am}\cos(m\theta+\phi_{am})\left[1+\frac{2}{\sqrt{3}}\sin(m\frac{\pi}{2})\sin(m\frac{\pi}{3})\right] \tag{9.31}$$

When dealing with a three phase bridge rectifier it has been convenient to divide the values of m into three groups, the positive sequence group for which $m = 1 + 6k$, the zero sequence group for which $m = 3 + 6k$, and the negative sequence group for which $m = 5 + 6k$, where k is any integer including zero. It has been seen that the amplitudes of terms belonging to the zero sequence group are zero and so need no further consideration. In the present case the two series connected bridges appear to the transformer primary as a twelve phase rectifier and further progress with the analysis is helped by extending the subdivision of the Fourier components. The positive sequence components are split into two groups, one with Fourier components of order $1+12k$ and the other of order $7+12k$ and the negative sequence components are divided into two groups, the first with components of order $5 + 12k$ and the second with components of order $11 + 12k$.

These additional subdivisions of the positive and negative sequence series permit the following conclusions to be made

- When $m = 1 + 12k$, the value of $\mathcal{F}_{s1m}$ is $2\ N_{tsa}I_{am}\cos(m\theta + \phi_{am})$.
- When m is $5 + 12k$, the value of $\mathcal{F}_{s1m}$ is zero.
- When m is $7 + 12k$, the value of $\mathcal{F}_{s1m}$ is zero.
- When m is $11 + 12k$, the value of $\mathcal{F}_{s1m}$ is $2\ N_{tsa}I_{am}\cos(m\theta + \phi_{am})$.

Thus, Fourier components of orders $5 + 12k$ and $7 + 12k$ are eliminated from the transformer secondary m.m.f., leaving positive sequence components of order $1 + 12k$, i.e. $1, 13, 25\ldots$, and negative sequence components of order $11 + 12k$, i.e. $11, 23, 35\ldots$, to be balanced by the primary m.m.f. The expression for the current in the red–yellow transformer primary must balance the combined m.m.f. of the two secondary windings, A1 and B12, and is therefore

$$i_{ry} = \sum_{m=1,\infty,12} 2N_{tsap}I_{am}\cos(m\theta + \phi_{am}) + \sum_{n=11,\infty,12} 2N_{tsap}I_{an}\cos(n\theta + \phi_{an}) \tag{9.32}$$

It was found in Chapter 8 that the amplitudes of the line current components are $\sqrt{3}$ times those of the transformer primary components with the positive sequence components being retarded and the negative sequence components advanced in phase by 30°. Applying this to eqn 9.32 yields the current in the red line as

$$i_r = \sum_{m=1,\infty,m} 2\sqrt{3}\,N_{tsap}I_{am}\cos(m\theta - \frac{\pi}{6} + \phi_{am}) + \sum_{n=11,\infty,12} 2\sqrt{3}\,N_{tsap}I_{an}\cos(n\theta + \frac{\pi}{6} + \phi_{an}) \tag{9.33}$$

Hence, the rectifier appears as a twelve phase system from both the a.c. and d.c. sides while internally retaining the simplicity and effective utilization of material of the three phase bridge device.

Example 9.2

Determine the amplitudes and phase angles of the nonzero Fourier components up to order 61, of the currents i_{a1}, i_{b1}, i_{b21}, i_{ry}, and i_r in the rectifier of Example 9.1 for the case when the delay angle is 45° and the output current is 12 500 A. Assume that ripple in the d.c. output current is negligible and that commutation is instantaneous.

As indicated in Fig. 9.4, the waveform of the current in an a.c. line feeding a rectifier, e.g. i_{a1} or i_{b1}, has half wave symmetry with positive and negative rectangular pulses of current of amplitude I_o and duration 120°. A wave with this shape is shown as a function of the angle ψ in Fig. 9.5. Because of the half wave symmetry, there are no even order components in the series representing the current and, because of the 120° pulse duration, there are no triplens. The Fourier series can then be divided into positive sequence components of order $m = 1 + 6k$ and negative sequence

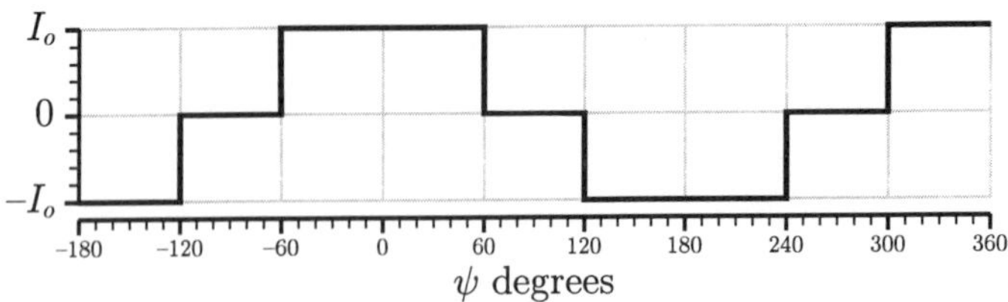

FIG. 9.5. Waveform of the current in a rectifier bridge input line.

components of order $m = 5 + 6k$, k being any positive integer including zero. The amplitude, I_m, of the mth order component, whether of positive or negative phase sequence, is

$$I_m = \frac{2\sqrt{3}}{\pi m} I_o \tag{9.34}$$

The wave of Fig. 9.5 is an even function and consequently has no sine Fourier components. The phase angle depends upon the phase sequence and for the positive phase sequence components is zero and for the negative sequence components is π. Thus, the Fourier series representing the wave shown in Fig. 9.5 is

$$i = \sum_{m=1,\infty,6} I_m \cos(m\psi) + \sum_{n=5,\infty,12} I_n \cos(m\psi + \pi) \tag{9.35}$$

The current i_{a1} lags the datum by α so that ψ is replace by $\theta - \alpha$ to obtain

$$i_{a1} = \sum_{m=1,\infty,6} I_m \cos[m(\theta - \alpha)] + \sum_{n=5,\infty,6} \cos[n(\theta - \alpha) + \pi] \tag{9.36}$$

The current i_{b1} lags the datum by $\alpha + \pi/6$ so that ψ is replace by $\theta - \alpha - \pi/6$ to obtain

$$i_{b1} = \sum_{m=1,\infty,6} I_m \cos[m(\theta - \alpha - \frac{\pi}{6})] + \sum_{n=5,\infty,n} I_n \cos[m(\theta - \alpha - \frac{\pi}{6}) + \pi] \tag{9.37}$$

For this example, $I_o = 12\,500$ A and $\alpha = 45° = \pi/4$ rad. Considering the 11th order component, this has an amplitude of 1253.02 A and, since it is of negative phase sequence, has a phase angle of 45° in the series for

Table 9.1 *Amplitudes and phase angles of the Fourier components of the secondary a.c. line currents i_{a1} and i_{b1} and the transformer B secondary current i_{b21} when the output current is 12 500 A and the delay angle is 45°*

Order m	I_{a1m} A	ϕ_{a1m} degrees	I_{b1m} A	ϕ_{b1m} degrees	I_{b21m} A	ϕ_{b21m} degrees
1	13783.2	–45.0	13783.2	–75.0	7957.7	–45.0
5	2756.6	–45.0	2756.6	165.0	1591.5	135.0
7	1969.0	45.0	1969.0	–165.0	1136.8	–135.0
11	1253.0	45.0	1253.0	75.0	723.4	45.0
13	1060.2	135.0	1060.2	105.0	612.1	135.0
17	810.8	135.0	810.8	–15.0	468.1	–45.0
19	725.4	–135.0	725.4	15.0	418.8	45.0
23	599.3	–135.0	599.3	–105.0	346.0	–135.0
25	551.3	–45.0	551.3	–75.0	318.3	–45.0
29	475.3	–45.0	475.3	165.0	274.4	135.0
31	444.6	45.0	444.6	–165.0	256.7	–135.0
35	393.8	45.0	393.8	75.0	227.4	45.0
37	372.5	135.0	372.5	105.0	215.1	135.0
41	336.2	135.0	336.2	–15.0	194.1	–45.0
43	320.5	–135.0	320.5	15.0	185.1	45.0
47	293.3	–135.0	293.3	–105.0	169.3	–135.0
49	281.3	–45.0	281.3	–75.0	162.4	–45.0
53	260.1	–45.0	260.1	165.0	150.1	135.0
55	250.6	45.0	250.6	–165.0	144.7	–135.0
59	233.6	45.0	233.6	75.0	134.9	45.0
61	226.0	135.0	226.0	105.0	130.5	135.0

i_{a1} and 75° in the series for i_{b1}. After normalizing the phase angles to the range −180 to +180°

$$I_{a111} = 1253.02\cos(11\theta + \frac{\pi}{4})$$

and

$$I_{b111} = 1253.02\cos(11\theta + \frac{5\pi}{12})$$

Since the 11th is a negative sequence component, in the series for i_{b2} it leads by 120° and is

$$i_{b211} = 1253.02\cos(11\theta + \frac{13\pi}{12})$$

Table 9.2 *Amplitudes and phase angles of the Fourier components of the transformer primary current i_{ry} and the a.c. supply line current i_r when the d.c. output current is 12 500 A and the delay angle is 45°*

Order m	I_{rym} A	ϕ_{rym} degrees	I_{rm} A	ϕ_{rm} degrees
1	432.12	–45.0	748.45	–75.0
11	39.28	45.0	68.04	75.0
13	33.24	135.0	57.57	105.0
23	18.79	–135.0	32.54	–105.0
25	17.28	–45.0	29.94	–75.0
35	12.35	45.0	21.38	75.0
37	11.68	135.0	20.23	105.0
47	9.19	–135.0	15.92	–105.0
49	8.82	–45.0	15.27	–75.0
59	7.32	45.0	12.69	75.0
61	7.08	135.0	12.27	105.0

The current i_{b21} is, from eqn 9.7, one-third of the difference between i_{b1} and i_{b2} so that its 11th order Fourier component is

$$i_{b2111} = 723.43\cos(11\theta + \frac{\pi}{4})$$

The transformer ampere–turns balance requires that the 11th order Fourier component of the primary current i_{ry} equal the sum of i_{a111} multiplied by the turns ratio N_{tsap} and i_{b2111} multiplied by the turns ratio N_{tsbp}, i.e.

$$i_{ry11} = 39.28\cos(11\theta + \frac{\pi}{4})$$

Since this is a negative sequence component, the corresponding component in the blue–red primary lags it by 120° and is

$$i_{br11} = 39.28\cos(11\theta - \frac{5\pi}{12})$$

The current in the red line is the difference between i_{ry} and i_{br} so that its 11th order component is

$$i_{r11} = 68.04\cos(11\theta + \frac{5\pi}{12})$$

The other components are obtained in a similar manner and are listed in Tables 9.1 and 9.2.

In Example 9.1 it was found that the r.m.s. input line current is 535.33 A. From Table 9.2 it is seen that the fundamental component of this current has an amplitude of 748.45 A, i.e. an r.m.s. value of 529.24 A. Thus, the r.m.s. value of all the harmonic components of the line current is $\sqrt{535.33^2 - 529.24^2} = 80.5$ A. Thus, despite the fact that this is a twelve phase system, this represents a significant amount of harmonic injection into the a.c. supply system. Later in the chapter it will be shown how the harmonics can be reduced by filters.

****

9.3 Parallel connection

The series connection just discussed is extremely effective at the voltage level used for the example. However, if the load were such as to require a much lower voltage, the thyristor losses, while still a small proportion of the total power flow, would become so great as to make parallel connection attractive. As an example, if the load voltage of the previous example were 200 V rather than 800, the rated load current would increase to 50 000 A. Since four thyristors are always conducting in series, and taking the voltage drop in each one as about 1.5 V, such a current entails a total thyristor power loss of the order of 300 kW!

This problem can be alleviated by connecting the two bridges in parallel, thus halving the thyristor loss to 150 kW. Alternatively four single way rectifiers can be connected in parallel, reducing the loss by another factor of 2–to 75 kW.

The output voltages of the two bridges are shown for a delay angle of 45° in Fig. 9.6, graph 1, the output of bridge A by a black line, and the output of bridge B by a gray line. Although the mean output voltages of the two bridges are equal, their instantaneous voltages differ substantially because of the 30° phase difference between their supplies. The difference voltage is shown in graph 2 of Fig. 9.6. Its mean value is zero and it varies at six times the supply frequency. In order to place the bridges in parallel this difference must be absorbed, a task performed by inserting an *interphase transformer* between the bridges as shown in Fig. 9.7 where the bridges are represented by a thyristor symbol within a box. The interphase transformer ensures that the output voltage is at all instants the average of the two bridge output voltages, i.e. that $v_o = (v_{ao} + v_{bo})/2$. The waveform of this output voltage is shown in graph 3 of Fig. 9.6. One cycle of ripple occupies 30° so that the output voltage is that of a twelve phase rectifier.

In its construction the interphase transformer is a normal transformer with identical primary and secondary windings. The two windings are connected in series aiding, as indicated in Fig. 9.7, so as to provide three

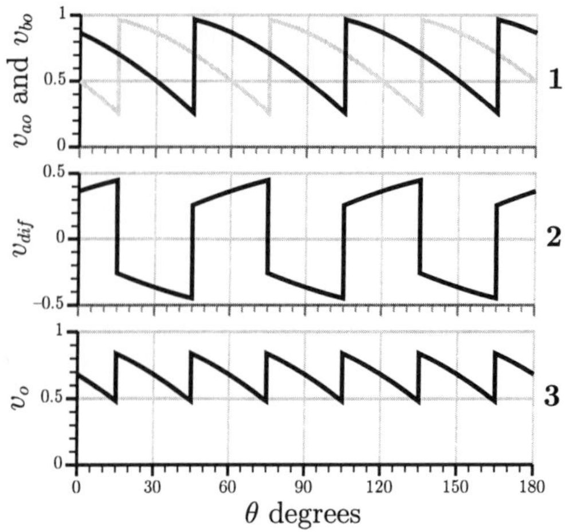

FIG. 9.6. Voltage waveforms for parallel connected bridge rectifiers. Graph 1 shows the output voltages of bridge A by a black line and of bridge B by a gray line. Graph 2 shows their difference. Graph 3 shows their average

terminals, two end connections, and a center tap. The end connections are joined to the two rectifiers and the center tap provides one of the d.c. output terminals.

The behavior of the interphase transformer is complex but the peculiarities of rectifier behavior which arise from it occur at light loads. Over the majority of the operating range a good approximation to the performance is obtained by treating the transformer as though it were ideal with negligible resistance and perfect coupling between the two windings. Here this approximation is adopted. Divergences from this ideal are discussed in Section 12.7.

In an ideal transformer, the number of volts per turn is the same for both primary and secondary and the ampere–turns of the two windings are equal. For the interphase transformer, whose two windings are identical, this means that the voltages across the two windings, v_{ita} and v_{itb}, are equal and the currents through the two windings, i_{ao} and i_{bo}, are equal. Now, since $v_{ita} = v_{ao} - v_o$ and $v_{itb} = v_o - v_{bo}$ it follows that $v_{ao} - v_o = v_o - v_{bo}$ so that the output voltage is the mean of the two rectifier outputs

$$v_o = \frac{1}{2}(v_{ao} + v_{bo}) \tag{9.38}$$

and that each rectifier supplies half the load current, i.e.

$$i_{ao} = i_{bo} = \frac{i_o}{2} \tag{9.39}$$

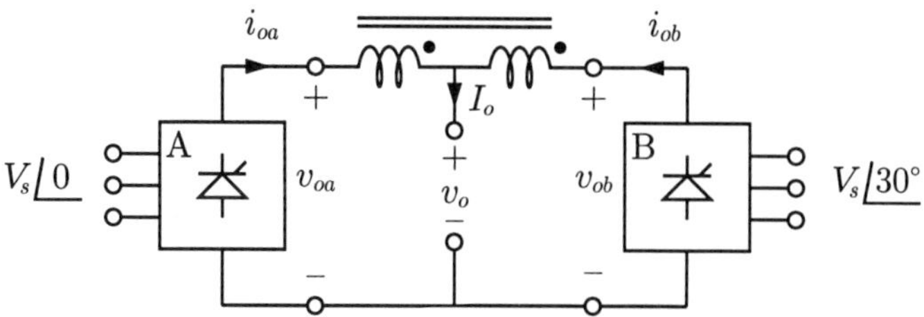

FIG. 9.7. Two rectifiers connected in parallel with the aid of an interphase transformer.

The output voltage, shown in graph 3 of Fig. 9.6, has a mean value slightly lower than the mean voltage of each rectifier due to the small voltage drops in the interphase transformer winding resistances. Its ripple amplitude is much smaller and its ripple frequency is doubled to twelve times the supply frequency.

The analysis of this system follows similar lines to that for the series connection but with minor, self evident, differences. The spectral content of the output voltage is similar to that of the series connection but the amplitudes of the components are halved, being equal to the corresponding components in a bridge rather than the sum for the two bridges. The waveform, and therefore the spectral content, of the transformer winding currents are similar but each secondary is carrying half the output current rather than the full amount.

As previously noted, interphase transformers can be used to parallel single way rectifiers if the load voltage is low enough to warrant this. Two single way, three phase rectifiers with a 60° phase difference between their supplies can be forced to operate in parallel by an interphase transformer operating at three times the supply frequency, thus providing six phase operation. A pair of such systems with 30° phase difference between their supplies can be forced to operate in parallel by the use of an interphase transformer operating at six times the supply frequency to provide a twelve phase output. This arrangement, illustrated schematically in Fig. 9.8, would have four simple, three phase, rectifiers supplied by two transformers, T1 and T2, each transformer having two secondary windings connected in antiphase, T1S1 and T1S2, and T2S1 and T2S2. The primary, T2P, of T1 is delta connected and the primary of T2 is wye connected so that there is a 30° phase difference between their secondary outputs. The number of turns on a primary winding of T1 is $\sqrt{3}$ times the number on a primary winding of T2 so that the secondary voltages are equal. The interphase transformer ITP1 absorbs the difference between the output voltages of the two rectifiers connected to it which is at three times the supply frequency. The interphase transformer ITP2 plays a similar role for the other

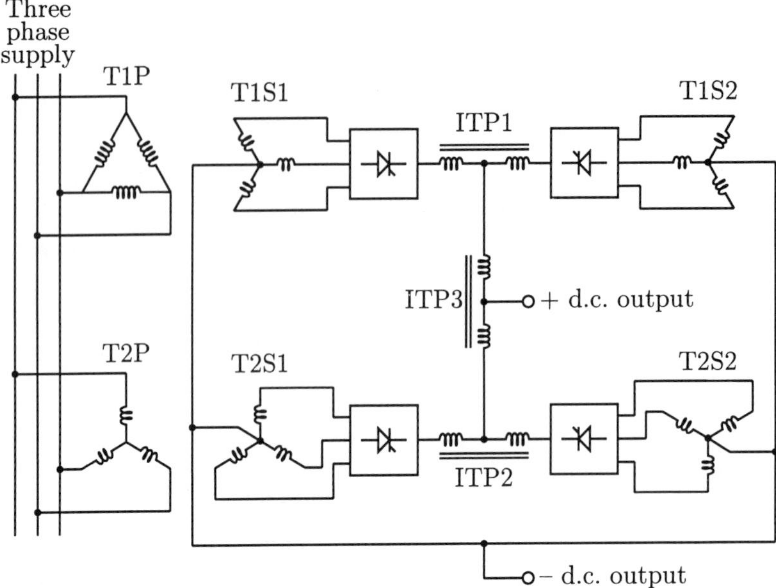

FIG. 9.8. Four simple, single way, three phase rectifiers connected in parallel with the aid of three interphase transformers to provide a twelve phase rectifier.

two rectifiers. The interphase transformer ITP3 absorbs the difference between the output voltage of ITP1 and ITP2 which is at six times the supply frequency. The result is a d.c. output voltage whose ripple is at twelve times the supply frequency. It is evident that the savings in thyristor losses may be substantially outweighed by the additional transformer losses and the increased capital cost of such a complex system.

9.4 More than twelve phases

The measures just described can be extended to increase the phase number beyond twelve. For example four bridges could be paralleled, by the use of three interphase transformers in an arrangement like that shown in Fig. 9.8, to provide 24 phase operation. However, there are few individual loads which justify such measures. More usual is a very substantial total load made up of individual loads each of which might justify a twelve phase connection. Such a situation is encountered in the electrochemical industry in the production of aluminum, sodium hydroxide, chlorine, etc. There are usually a number of similar electrolytic lines each operating more or less independently but at roughly the same load most of the time. Under these

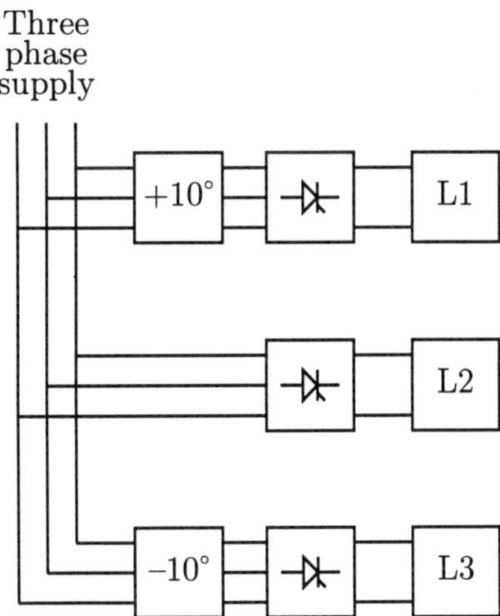

FIG. 9.9. Three loads, each supplied by a twelve phase rectifier, one fed directly from the three phase supply and the others fed through phase shifting transformers giving ±10° phase shift.

circumstances the a.c. supplies to the individual loads can be appropriately phase shifted so as to present to the a.c. system a higher phase number than twelve. As an example, if there are three separate loads, L1, L2, and L3, each supplied by a twelve phase rectifier as shown in Fig. 9.9, the supply to the first rectifier can be phase advanced relative to the a.c. system by 10°, the second rectifier can be operated at the a.c. system phase, and the supply to the third rectifier can be given a phase lag of 10° relative to the a.c. system. If the three loads are identical, this gives 36 phase operation, the first pair of supply side harmonics being the 35th and 37th. When the loads are not identical, residual harmonics derived from the 12 phase systems will exist, e.g. the 11th, 13th, 23rd, and 25th but they will be reduced in magnitude.

Figure 9.10 illustrates one of the many ways in which the necessary phase shift can be created. A three phase transformer with N_{tp} primary turns per phase has its primary windings wye connected. It has secondary windings with 0.201 N_{tp} turns. The primary windings are tapped at 0.885 N_{tp} from the star point and the secondary windings are connected to these taps as indicated, the blue secondary to the red primary, the red secondary to the yellow primary, and the yellow secondary to the blue primary. The voltages at the output terminals, R′, Y′, and B′, then lag the input voltages at the input terminals, R, Y, and B, by 10° for a positive sequence system and lead by 10° for a negative sequence system. Phase shifting transformers are discussed in greater detail in Section 12.5.

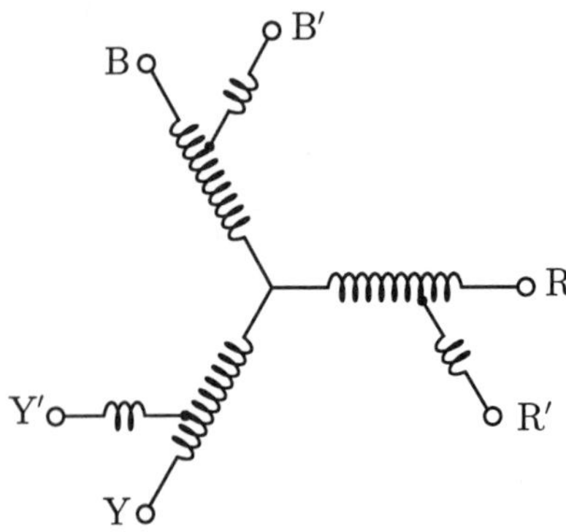

FIG. 9.10. A phase shifting transformer. The input terminals are marked R, Y, and B. The output terminals at which the phase shifted voltages appear are marked R′, Y′, and B′.

Example 9.3

Two similar loads, each rated at 800 V 12 500 A, are each supplied by a twelve phase rectifier–transformer combination identical with that specified in Examples 9.1 and 9.2. The three phase supply to one rectifier transformer is shifted by 15° relative to that of the other by a phase shifting transformer so that, when the two loads are equal, the system appears to the three phase system to have 24 phases. The rectifiers employ a combination of tap changing and delay angle variation to accommodate load variation.

First, determine the spectrum and waveform of the input a.c. line current when the d.c. output current of each rectifier is 12 500 A, the tap changer settings are such that the turns ratio, N_{tsap}, of each is 1/63.79, and the delay angle of each is 25°.

Second, determine the spectrum and waveform of the input a.c. line current when the first rectifier conditions are unchanged but the second rectifier is supplying a slightly reduced load and as a consequence has a turns ratio of 1/70.17 and a delay angle of 30° but is still supplying a d.c. output current of 12 500 A.

Third, compare the two cases with particular reference to simple twelve phase operation without the use of a phase shifter.

An expression for the current in the red a.c. line to a rectifier is given in eqn 9.33. Expressions for the currents in a.c. line 1 supplying rectifier A are given in eqns 9.34 and 9.36. To extend these expressions to this problem the rectifier which is directly connected to the a.c. system is designated D and the rectifier connected via the phase shifting transformer is E. The turns ratios N_{tsap} is denoted by N_{td} or N_{te} as appropriate. The phase advance provided by the phase shifting transformer supplying rectifier E is denoted by δ and the d.c. output currents of each rectifier are denoted by I_{do} and I_{eo}.

The current in the red a.c. line due to rectifier D can be written, with the aid of eqns 9.33, 9.34, and 9.36, as

$$i_{rd} = \sum_{m=1,\infty,12} I_{rdm}\cos(m\theta + \phi_{cdm}) + \sum_{n=11,\infty,12} I_{rdn}\cos(n\theta + \phi_{cdn}) \quad (9.40)$$

where

$$I_{dm} = \frac{12}{\pi}\frac{1}{m} I_{do} N_{td} \quad (9.41)$$

$$\phi_{cdm} = -m\alpha_d - \frac{\pi}{6} \quad (9.42)$$

$$I_{dn} = \frac{12}{\pi}\frac{1}{n} I_{do} N_{td} \quad (9.43)$$

$$\phi_{cdn} = -n\alpha_d - \frac{5\pi}{6} \quad (9.44)$$

The voltage supplying rectifier E is advanced in phase by δ and so θ in eqn 9.40 must be replaced by $\theta + \delta$. However, when the input currents to this rectifier are reflected into the a.c. system through the phase shifting transformer, the positive sequence components are phase retarded by δ and the negative sequence components are advanced by δ so that eqns 9.40 through 9.44 become

$$i_{re} = \sum_{m=1,\infty,12} I_{em}\cos(m\theta + \phi_{cem}) + \sum_{n=11,\infty,12} I_{en}\cos(n\theta + \phi_{cen}) \quad (9.45)$$

where

$$I_{em} = \frac{12}{\pi}\frac{1}{m} I_{eo} N_{te} \quad (9.46)$$

$$\phi_{cem} = -m\alpha_e - \frac{\pi}{6} + (m-1)\delta \quad (9.47)$$

$$I_{en} = \frac{12}{\pi}\frac{1}{n} I_{eo} N_{te} \quad (9.48)$$

$$\phi_{cen} = -n\alpha_e - \frac{5\pi}{6} + (n+1)\delta \quad (9.49)$$

Consider the 11th order Fourier component for the first case of equal loads on the two rectifiers. This is a negative sequence component so that for rectifier D, from eqns 9.43 and 9.44, its amplitude is 68.04 A and its phase is $-425°$ which, when normalized to the range $-180°$ to $+180°$, becomes $-65°$ or -1.1345 rad. Thus

$$i_{rd11} = 68.04\cos(11\theta - 1.1345)$$

For rectifier E, eqns 9.48 and 9.49 are used to obtain the same amplitude and a phase angle of $-605°$, which when normalized becomes 115° or 2.0071 rad. Thus

$$i_{re11} = 68.04\cos(11\theta + 2.0071)$$

Since these two components are in antiphase, their sum in the a.c. input line is zero.

Consider now the same component for the second case when rectifier E is on reduced load: the current due to rectifier D is unchanged at the value just found but the amplitude of the current due to rectifier E is reduced, because of the changed turns ratio, to 61.86 A and, because of the increased delay angle, its phase is decreased to $-660°$, which after normalization, becomes 60° or 1.0472 rad. Thus

$$i_{re11} = 61.86\cos(11\theta + 1.0472)$$

The resultant 11th order component of current in the red line is then

$$i_{r11} = 68.04\cos(11\theta - 1.1345) + 61.86\cos(11\theta + 1.0472)$$

i.e.

$$i_{r11} = 60.23\cos(11\theta - 0.1348)$$

The other components are determined in a similar manner and are listed up to the 97th order in Table 9.3.

The spectra and waveforms of the currents for the balanced case are shown in Fig. 9.11 and for the unbalanced case are shown in Fig. 9.12. In the waveforms, the oscillations and non-vertical transitions from one level to the next are the result of truncating the Fourier series at the 97th order component.

When the loads on the two rectifiers are equal and their operating conditions are the same, it will be seen from columns 2 and 3 of Table 9.3 that the two loads appear to the a.c. system as a single load having 24 phases. The harmonic pairs of the line currents are centered on even multiples of twelve times the supply frequency, e.g. $24 - 1 = 23$ and $24 + 1 = 25$ Hz. This is confirmed by the frequency spectrum and line current waveform which are shown in Fig. 9.11.

When the loads on the two rectifiers and their operating conditions differ slightly there is some deterioration in the line current waveform and,

Table 9.3 *Amplitudes and phase angles of the Fourier components of the a.c. supply line current, i_r, for two 12 phase rectifiers supplying similar loads. Columns 2 and 3 define the spectrum when the loads are identical. Columns 4 and 5 define the spectrum when the loads differ slightly, as defined in part 2 of Example 9.3*

	Equal loads		Unequal loads	
Order m	I_{rm} A	ϕ_{crm} degrees	I_{rm} A	ϕ_{crm} degrees
1	1496.9	55.00	1427.5	57.38
11	0.0	–63.43	60.2	7.72
13	0.0	–101.31	59.2	–58.23
23	65.1	5.00	33.5	58.23
25	59.9	–65.00	26.5	–7.72
35	0.0	180.00	40.8	–57.38
37	0.0	180.00	38.6	–122.62
47	31.8	–115.00	14.1	–172.28
49	30.5	175.00	15.7	121.77
59	0.0	90.00	13.0	–121.77
61	0.0	68.55	10.9	172.28
71	21.1	125.00	20.1	122.62
73	20.5	55.00	19.6	57.38
83	0.0	–43.60	8.0	7.72
85	0.0	90.00	9.1	–58.23
95	15.8	5.00	8.1	58.23
97	15.4	–65.00	6.8	–7.72

in particular, spectral components centered on odd multiples of twelve times the supply frequency appear. As an example, from column 4 of Table 9.3 it is seen that an eleventh order harmonic with an amplitude of 60.2 A and a thirteenth order harmonic with an amplitude of 59.2 A appear where there were none in the balanced case. However, this is still a substantial improvement on the situation in which no phase shifter is employed. Then, by running the program which created the data for Table 9.3 with zero phase shift between the two rectifiers, the amplitude of the eleventh order harmonic is found to be 136.1 A, and the amplitude of the thirteenth order harmonic is found to be 115.1 A.

Of course the inclusion of the phase shifter involves additional capital outlay and additional losses but these are potentially offset by a reduction

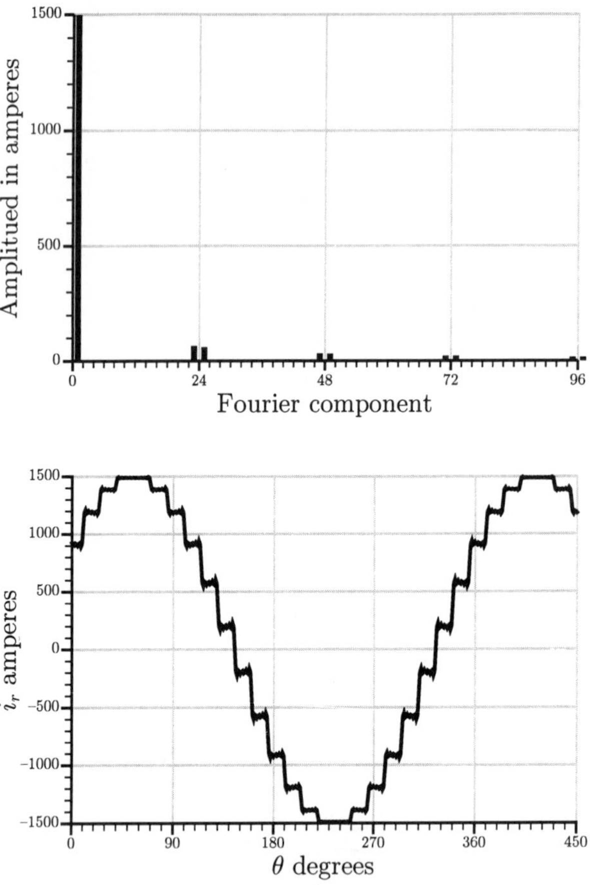

FIG. 9.11. Frequency spectrum and waveform of the a.c. line current when the loads and operating conditions of the two rectifiers are identical

in input current filtering, which the electric utility is likely to demand of a rectifier installation of this size.

****

9.5 Power factor improvement

The measures described so far yield improved operation as far as harmonics are concerned but do not alleviate the other major problem of a controlled rectifier—poor power factor at reduced voltage. The fundamental power factor of an ideal, fully controlled rectifier is $\cos(\alpha)$ since the fundamental component of the input current lags the voltage by the delay angle. In

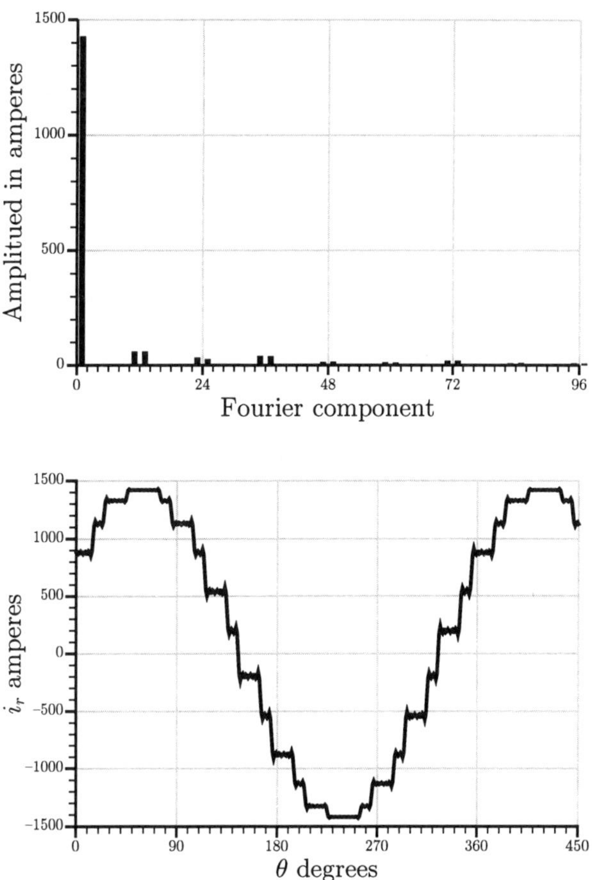

FIG. 9.12. Frequency spectrum and waveform of the a.c. line current when the loads and operating conditions of the two rectifiers differ slightly.

reality it is slightly less than $\cos(\alpha)$ because of the commutation overlap caused by a.c. side inductance as discussed in Chapter 5. Although reactive kVA is of no economic value to the customer, it entails capital and operating costs for the supplier and electricity rates reflect this fact. The customer is thereby encouraged, by economic pressure, to maintain a high power factor, frequently 0.9 or better. The operation of a large, fully controlled rectifier at large delay angles then becomes a costly luxury which it is in the owner's interest to avoid if possible.

There are three approaches to the problem of poor power factor at reduced voltage and they can be used separately or in combination depending on the load characteristics and the economics. The three methods are tap changing transformers, sequential operation of rectifiers, and the use of power factor improvement capacitors and they are reviewed in that order.

9.6 Ratio changing transformers

When the characteristics of the load are suitable, phasing back the rectifier can be completely or largely avoided by the use of ratio changing transformers. Loads such as battery charging and electrolysis are well suited to this approach since changes happen slowly and can be followed by the voltage changing mechanism of the transformer. Further, the range of voltage variation in these applications is usually modest, of the order of $\pm 10\%$ of the rated value.

Transformers for this duty can be divided into two types, continuously variable and discretely variable. The continuously variable transformers may be either of the variable ratio type where a carbon brush moves over an exposed portion of the winding, or of the variable position type whereby the coupling between the primary and secondary windings is changed by moving one relative to the other. Discrete ratio changes are achieved by tapping the secondary winding at suitable points and moving the output terminal from one tap to the next. Tap changing can be done off-load, when no current is flowing, or on-load. In the latter case, twice as many voltage steps are available as there are taps since the mechanism can interpolate between taps.

The continuously variable and on-load tap changing systems can be used to adjust the load current, for example with a battery charger to produce a desired current–time variation despite the increase in battery voltage as it gains charge. Off-load tap changing can be employed when the rectifier is called upon to charge batteries with different voltages or to feed an electrolytic line whose cell number varies in response to production and maintenance requirements.

These systems permit the use of a diode rectifier and thus maintain a high fundamental power factor. The fundamental power factor is not quite unity because the commutation overlap introduces some lag into the current and the conventional power factor is somewhat lower still because of the input current harmonics.

9.7 Hybrid arrangements

When the fluctuations in load are too rapid to be followed by the mechanical equipment, a hybrid arrangement can be employed combining ratio changing for the slower changes and delay angle variation for the faster ones. As an example, an on-load tap changer giving 5% step changes in voltage can be combined with delay angle variations in the range $0° < \alpha < 20°$ to give continuous and rapid variation in output voltage over any voltage range which can be accommodated by the tap changer. Because the delay angle variation is small, the power factor will not fall below 0.9.

9.8 Sequential operation

When the demands of the load require very fast variation of rectifier output voltage over a wide range, as for example with a reversing drive powered by a d.c. motor, there is no alternative to operation at large delay angles. Even under these circumstances it is possible to alleviate the penalties of poor power factor by sequential operation. A simple example of this has already been studied in the form of the semi-controlled bridge rectifier. This connects a simple controlled rectifier in series with a simple uncontrolled rectifier to provide a voltage variation of 100%, from the maximum value down to zero, by delay angle variation in the controlled half over the range zero to 180°. This process can be extended to the series connection of other rectifiers in an infinite variety of ways to satisfy almost any load demands.

The principle of sequential operation is illustrated by Fig. 9.13, which shows three of a chain of N_r rectifiers with their d.c. outputs connected in series. The rectifier inputs are all derived from the same three phase source and the electrical isolation which is necessary to permit series operation on the d.c. side and parallel operation on the a.c. side is provided by transformers labelled T_{i-1}, T_i, T_{i+1}, etc.

In sequential operation, only one rectifier is controlled at a time, the other rectifiers in the chain being operated at either zero delay or the maximum delay so as to minimize the reactive volt-ampere demand on the a.c. system, thus maximizing the power factor.

The mean d.c. output voltage, V_o, of the chain of rectifiers is the sum of the mean output voltages of the individual rectifiers, i.e.

$$V_o = \sum_{i=1,N_r} V_{oi} = \sum_{i=1,N_r} V_{oim} \cos(\alpha_i) \tag{9.50}$$

where V_{oi} is the mean output voltage of the ith rectifier whose delay is α_i and whose maximum output voltage is V_{oim}

The maximum output voltage, V_{om}, is obtained by setting all the delay angles to zero and is

$$V_{om} = \sum_{i=1,N_r} V_{oim} \tag{9.51}$$

All the rectifiers carry the same d.c. output current, I_o, and the total output power, P_{ot}, is the sum of the individual rectifier outputs, i.e.

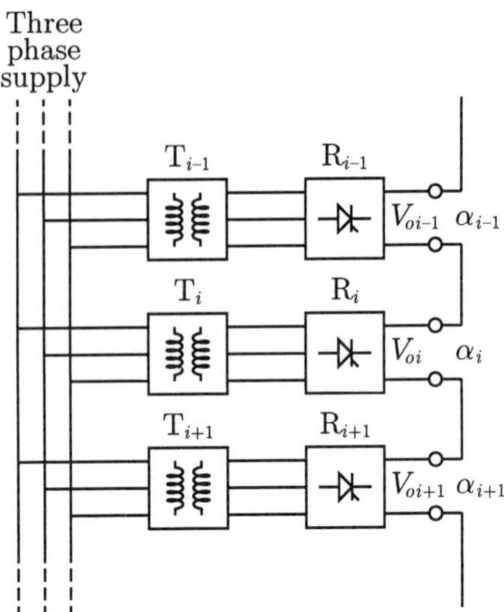

FIG. 9.13. Part of a chain of rectifiers which have their outputs connected in series and their transformer inputs in parallel across the same a.c. supply.

$$P_{ot} = \sum_{i=1,N_r} V_{oi} I_o = \sum_{i=1,N_r} V_{im} I_o \cos(\alpha_i) \tag{9.52}$$

Within the context of this chapter, where losses within the rectifier are being ignored, this total output power is also the total input power, P_t, so that

$$P_t = \sum_{i=1,N_r} V_{im} I_o \cos(\alpha_i) \tag{9.53}$$

The fundamental component of the input current to a rectifier lags the a.c. voltage by the delay angle so that the fundamental reactive volt-amperes of the ith rectifier, VAR_i, is given by

$$VAR_i = P_{oi} \tan(\alpha_i) = V_{oim} I_o \sin(\alpha_i) \tag{9.54}$$

The total reactive volt-ampere input, VAR_t, drawn from the a.c. system is the sum of these individual inputs, i.e.

$$VAR_t = \sum_{i=1,N_r} V_{im} I_o \sin(\alpha_i) \tag{9.55}$$

The total volt-ampere input, VA_t, is

$$VA_t = \sqrt{P_t^2 + VAR_t^2} \tag{9.56}$$

and the fundamental power factor, Pf_1, is

$$Pf_1 = \frac{P_t}{VA_t} \tag{9.57}$$

The application of these general expressions is illustrated by Example 9.4, which compares the operation of a single fully controlled rectifier with that of a chain of two, three, and four sequentially controlled rectifiers, the comparison being illustrated by Fig. 9.14.

Example 9.4

Compare the operation of a single, fully controlled rectifier with that of two, three, and four series connected rectifiers with sequential control. Make the comparison on the basis that all the rectifiers in a chain are similar, that the output current is constant at the rated value, I_o, and that the maximum delay angle is 160°.

The three rectifier case will be considered in detail, the analysis of the two and four rectifier cases being similar.

The maximum output voltage of each of the three rectifiers is one-third of the maximum output voltage, i.e.

$$V_{o1m} = V_{o2m} = V_{o3m} = 0.3333V_{om}$$

During the first stage of operation the delay angle of the first rectifier is increased from zero to 160° while the delays of the other two are held at zero.

At the start of the first stage when $\alpha_1 = 0$

- the output voltage is V_{om}
- the input power is $V_{om}I_o$
- the input reactive volt-amperes is zero
- the input volt-amperes is $V_{om}I_o$
- the fundamental power factor is 1.0

About half-way through the first stage, when $\alpha_1 = 90°$

$$\begin{aligned} V_o &= 0.3333V_{om}[\cos(90°) + \cos(0) + \cos(0)] = 0.6667V_{om} \\ P_t &= 0.6667V_{om}I_o \\ VAR_t &= 0.3333V_{om}I_o[\sin(90°) + \sin(0) + \sin(0)] = 0.3333V_{om}I_o \end{aligned}$$

$$VA_t = V_{om}I_o\sqrt{0.6667^2 + 0.3333^2} = 0.7454V_{om}I_o$$
$$Pf_1 = 0.6667/0.7454 = 0.8945 \text{ lagging}$$

At the end of the first stage when $\alpha_1 = 160°$

$$V_o = 0.3333V_{om}[\cos(160°) + \cos(0) + \cos(0)] = 0.3534V_{om}$$
$$P_t = 0.3534V_{om}I_o$$
$$VAR_t = 0.3333V_{om}I_o[\sin(160°) + \sin(0) + \sin(0)] = 0.1140V_{om}I_o$$
$$VA_t = V_{om}I_o\sqrt{0.3534^2 + 0.1140^2} = 0.3713V_{om}I_o$$
$$Pf_1 = 0.3534/0.3713 = 0.9517 \text{ lagging}$$

The second stage now begins with α_1 held constant at 160°, α_2 increasing from zero to 160°, and α_3 held constant at zero.

At the start of the second stage when $\alpha_2 = 0$, the values of V_o, P_t, VAR_t, VA_t, and Pf are those just given for the end of the first stage.

About half-way through the second stage when $\alpha_2 = 90°$

$$V_o = 0.3333V_{om}[\cos(160°) + \cos(90°) + \cos(0)] = 0.0201V_{om}$$
$$P_t = 0.0201V_{om}I_o$$
$$VAR_t = 0.3333V_{om}I_o[\sin(160°) + \sin(90°) + \sin(0)] = 0.4473V_{om}I_o$$
$$VA_t = V_{om}I_o\sqrt{0.0201^2 + 0.4473^2} = 0.4478V_{om}I_o$$
$$Pf = 0.0201/0.4478 = 0.0449 \text{ lagging}$$

At the end of the second stage, when $\alpha_2 = 160°$

$$V_o = 0.3333V_{om}[\cos(160°) + \cos(160°) + \cos(0)] = -0.2931V_{om}$$
$$P_t = -0.2931V_{om}I_o$$
$$VAR_t = 0.3333V_{om}I_o[\sin(160°) + \sin(160°) + \sin(0)] = 0.2280V_{om}I_o$$
$$VA_t = V_{om}I_o\sqrt{0.2931^2 + 0.2280^2} = 0.3713V_{om}I_o$$
$$Pf = -0.2931/0.3713 = -0.7892 \text{ lagging}$$

The third and final stage now begins with α_1 and α_2 held constant at 160° and α_3 increasing from zero to 160°.

At the start of this stage when $\alpha_3 = 0$, the values of V_o, P_t, VAR_t, VA_t, and Pf are those just given for the end of the second stage.

About half-way through the third stage when $\alpha_3 = 90°$

$$V_o = 0.3333V_{om}[\cos(160°) + \cos(160°) + \cos(90°)] = -0.6265V_{om}$$
$$P_t = -0.6265V_{om}I_o$$

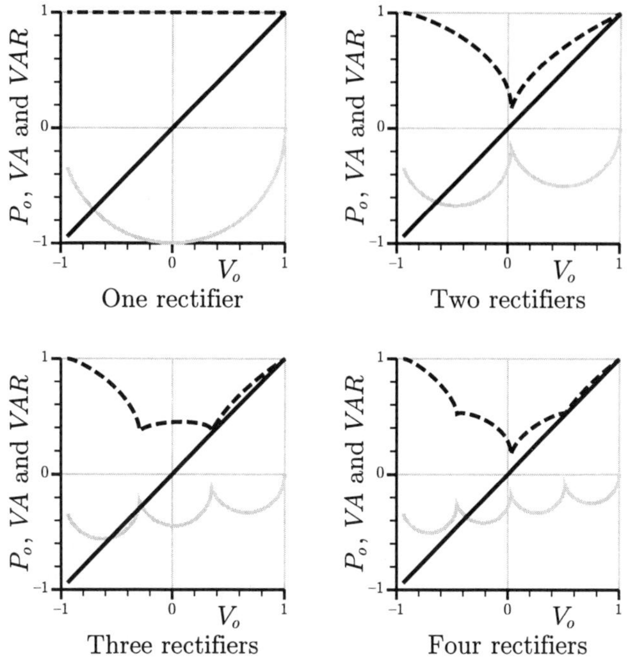

FIG. 9.14. Input characteristics of sequentially controlled rectifiers. Power is shown by solid black lines, VA by dashed black lines and VAR by gray lines.

$$\begin{aligned} VAR_t &= 0.3333V_{om}I_o[\sin(160°) + \sin(160°) + \sin(90°)] = 0.5613V_{om}I_o \\ VA_t &= V_{om}I_o\sqrt{0.6265^2 + 0.5613^2} = 0.8412V_{om}I_o \\ Pf &= -0.6265/0.8412 = -0.7448 \text{ lagging} \end{aligned}$$

At the end of the third stage when $\alpha_3 = 160°$

$$\begin{aligned} V_o &= 0.3333V_{om}[\cos(160°) + \cos(160°) + \cos(160°)] = -0.9397V_{om} \\ P_t &= -0.9397V_{om}I_o \\ VAR_t &= 0.3333V_{om}I_o[\sin(160°) + \sin(160°) + \sin(160°)] = 0.3420V_{om}I_o \\ VA_t &= V_{om}I_o\sqrt{0.9397^2 + 0.3420^2} = V_{om}I_o \\ Pf &= -0.9397 \text{ lagging} \end{aligned}$$

Similar calculations for the four alternative arrangements are summarized in Fig. 9.14, which shows for each alternative the input power by a solid black line, the fundamental input volt-amperes by a dashed black line, and the fundamental input reactive volt-amperes by a gray line, all as functions of the output voltage. The scales have been normalized to V_{om} in the case of the voltage axis and to $V_{om}I_o$ in the case of the power and volt-amperes.

The scales of the axes have been chosen so that one normalized unit always occupies the same length. This has been done so that the diagrams can also serve as the complex plane on which the normalized input current phasor can be drawn, a topic which is discussed next.

The VARs have been taken as negative in order to separate the curves and to provide, to another scale, a path which is the locus of the end of the fundamental input current phasor.

****

9.8.1 *The VAR locus and the input current*

Before discussing the conclusions which can be drawn from Fig. 9.14 it is appropriate to consider some of the figure's geometrical properties. The curves representing the reactive volt-amperes as a function of mean output voltage appear to be composed of circular arcs. This is in fact the case when the coordinates are expressed in the normalized form.

Considering a point P on the VAR trajectory when the ith rectifier is being controlled, $i-1$ rectifiers are at their maximum delay and $N_r - i$ rectifiers are still at zero delay. The output voltage is

$$V_o = (i-1)\frac{V_{om}}{N_r}\cos(\alpha_m) + \frac{V_{om}}{N_r}\cos(\alpha_i) + (N_r - i)\frac{V_{om}}{N_r} \qquad (9.58)$$

The x coordinate of P, x_p, is this voltage expressed as a fraction of V_{om}, i.e.

$$x_p = C_{xp} + R_p\cos(\alpha_i) \qquad (9.59)$$

where

$$C_{xp} = \frac{(i-1)\cos(\alpha_m) + N_r - i}{N_r} \qquad (9.60)$$

and

$$R_p = 1/N_r \qquad (9.61)$$

The VARs which are drawn from the a.c. supply are composed of the sum of $-(V_{om}/N_r)I_o\sin(\alpha_m)$ for each of the $i-1$ rectifiers which are at maximum delay, $-(V_{om}/N_r)\times\ I_o\sin(\alpha_i)$ for the rectifier which is being controlled, and zero for each of the $N_r - i$ rectifiers which are still at zero delay. The y coordinate of P, y_p, is the total VARs expressed as a fraction of $V_{om}I_o$ and is therefore

$$y_p = C_{yp} - R_p \sin(\alpha_i) \tag{9.62}$$

where

$$C_{yp} = -\frac{(N_r - i)\sin(\alpha_m)}{N_r} \tag{9.63}$$

and R_p has already been defined by eqn 9.61.

Equations 9.59 and 9.62 define a circle of radius $R_p = 1/N_r$ whose center is at $\{[(i-1)\cos(\alpha_m) + (N_r - i)]/N_r \ , \ -(N_r - 1)\sin(\alpha_m)/N_r\}$.

Further, the input power is $3V_sI_s\cos(\phi_s)$, where V_s is the a.c. supply line to neutral voltage, I_s is the a.c. line current, and $\cos(\phi_s)$ is the angle by which I_s lags V_s. Multiplying eqn 9.58 by I_o gives the output power which, because losses are being neglected, is also the input power. Thus, x_p represents the in-phase component of the a.c. line current to a scale $V_{om}I_o/(3V_s)$. Similarly y_p represents the quadrature component of the a.c. line current to the same scale. Hence, if the positive x axis of Fig. 9.14 is the direction of the phasor $\mathbf{V}_3$, then the line joining the origin to the point P represents the phasor $\mathbf{I}_s$.

9.8.2 *Reduction of the VAR demand*

The most significant feature of Fig. 9.14 is the substantial reduction in reactive volt-amperes achieved by sequential operation. The very large VAR demand of the single rectifier, $V_{om}I_o$, which occurs when the angle of delay is 90° and the mean output voltage is zero, is reduced to a maximum of

$$0.5V_{om}I_o[\sin(160°) + \sin(90°)] = 0.6710V_{om}I_o$$

in the two rectifier case,

$$0.3333V_{om}I_o[\sin(160°) + \sin(160°) + \sin(90°)] = 0.5613V_{om}I_o$$

in the three rectifier case, and

$$0.25V_{om}I_o[\sin(160°) + \sin(160°) + \sin(160°) + \sin(90°)] = 0.5065V_{om}I_o$$

in the four rectifier case.

Ultimately, if an infinite number of rectifiers could be used, the maximum VARs would be reduced to

$$V_{om}I_o \sin(160°) = 0.3420V_{om}I_o$$

With the reduction in VARs goes a corresponding reduction in the VA demand and an improvement in power factor.

The improvement of the input VAR and VA demands is purchased at the cost of more components and increased complexity in the rectifier and its controller. As more rectifiers are connected in series the output current must flow through more diodes and thyristors with additional losses and reduced efficiency. Input transformers must be employed whereas the single rectifier might be able to operate directly from the a.c. supply. The increased complexity could alternatively be directed to increasing the number of phases so decreasing the harmonic contents of both the output and input. These considerations make it unlikely that more than two rectifiers would be used. Most of the potential advantage is thereby realized and the consequent disadvantages are minimized.

The three phase bridge rectifier presents something of a special case in that the upper and lower halves can be operated sequentially without the need for transformer isolation and it thus gives the advantages without the need for an extra transformer. However, a price is still paid in that the output and inputs are those of a three phase rectifier rather than the six phase rectifier of the fully controlled bridge.

9.8.3 *The partially controlled rectifier*

In creating Fig. 9.14 it has been assumed that all the series connected rectifiers are controlled in sequence. This need not be the case. Some may be controlled and others not, the semi-controlled bridge being an example. In Fig. 9.14 it is represented by the first part of the two rectifier graph covering the voltage range from V_{om} down to zero. The semi-controlled bridge may be employed when the required output voltage covers this restricted range. Similar situations occur from time to time when the output voltage must be varied within a limited range, say V_{omax}–V_{omin}. Examples are (1) battery charging where the rectifier output must be increased a modest amount to compensate for the increase in battery voltage with charge and (2) the need to vary the speed of an inverter fed motor over a modest speed range while maintaining constant torque capability. As an example, an output voltage range from V_{om} to $0.5V_{om}$ can be provided by the series connection of a diode rectifier supplying $0.75V_{om}$ and a thyristor rectifier supplying $0.25V_{om}$. This is the situation described in Fig. 9.14 by the first circular arc of the VAR curve of the four rectifier graph. Comparison with the single rectifier graph indicates a very substantial reduction in VAR demand and almost unitary power factor over the whole voltage range. This situation is illustrated by Example 9.5.

Example 9.5

A load whose rated current is 120 A is supplied from a 440 volt, 60 Hz, three phase supply through a pair of rectifiers whose outputs are connected in series. The first rectifier is a diode bridge supplied via a transformer in which a secondary phase has 75% of the turns on a primary phase. The second rectifier is a thyristor bridge supplied via a transformer in which a secondary winding has 25% of the number of turns on a primary phase. Both transformers are wye–wye connected.

Assuming ideal components, a smooth output current and a maximum delay in the thyristor bridge of 160°, determine the delay angle, a.c. line current, fundamental power factor, total power factor, input power, input kVAR and input kVA as functions of the d.c. output voltage over the whole possible voltage range when the output current has the rated value.

The mean output voltage of the diode rectifier is, from eqn 3.1, $1.3505 \times 440.0 \times 0.75 = 445.67$ V. The mean output voltage from the thyristor rectifier is similarly $1.3505 \times 440.0 \times 0.25 \cos(\alpha) = 148.56 \cos(\alpha)$ V. The mean voltage delivered to the load is the sum of these voltages, $445.67 + 148.56 \cos(\alpha)$ V. When the delay angle is 90° the voltage is 445.67 V.

The mean power delivered to the load is the product of the output voltage and current and, for an output current of 120 A it is $[445.67 + 148.56 \cos(\alpha)] \times 120.0 \div 1000$ kW, i.e. $53.48 + 17.83 \cos(\alpha)$ kW. Since the output current is assumed to be ripple free, this is the total power delivered to the load. At a delay angle of 90° the second term of the foregoing expression for mean power is zero and the output power is 53.48 kW.

The currents in the a.c. lines connecting the transformers to the rectifiers have a rectangular waveform, amplitude 120 A, with an on time of 120° and an off time of 60° as shown in graphs 1 through 6 of Fig. 9.4. However, whereas the group B currents in that figure lag the group A currents by 30°, in this case the currents supplying the thyristor bridge lag those supplying the diode bridge by the delay angle.

The amplitude of the fundamental component of the current in an a.c. line from the transformer secondary to the diode rectifier is $[4 \cos(\pi/6)/\pi] \times 120 = 134.32$ A.

The amplitude of the fundamental component of the corresponding current in the thyristor rectifier is the same, 132.32 A, but lags that for the diode rectifier by the delay angle.

The waveforms of the transformer primary input currents are similar to those of the secondary output currents but their amplitudes are in proportion to the turns ratios. Thus, the amplitude of the rectangular waveform input current to the diode rectifier transformer is $120.0 \times 0.75 = 90$ A and that of the current supplying the thyristor rectifier is $120.0 \times 0.25 = 30$ A. Their fundamental components have amplitudes of 99.24 and 33.08 A. These waves are shown, for a delay angle of 90°, in graphs 1 and 2 of Fig. 9.15, the actual waveforms being shown by the black lines and the fundamental components by the gray lines.

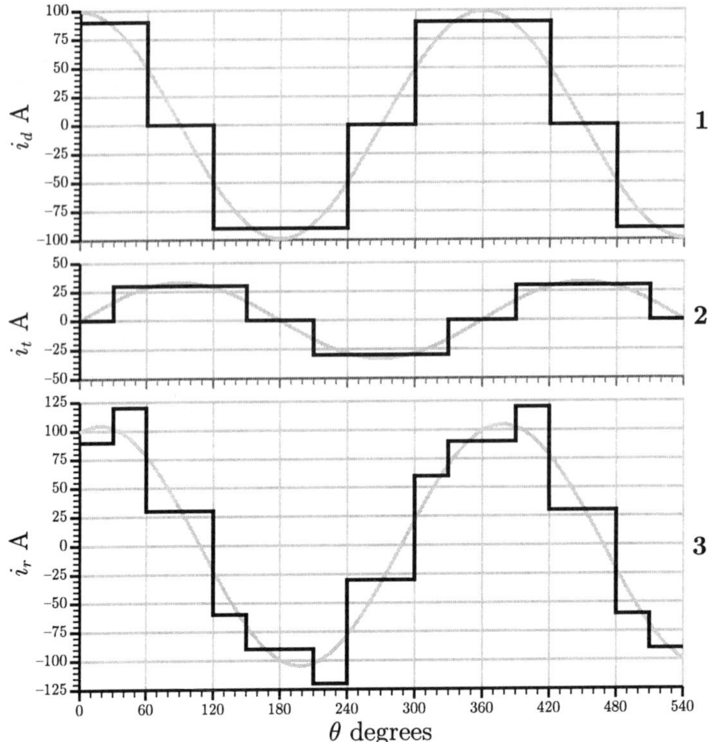

FIG. 9.15. The waveforms of the a.c. input line currents supplying the diode rectifier transformer, the thyristor rectifier transformer, and their sum, the a.c. line current. The delay angle is 90° and the d.c. output is assumed to be free of ripple.

The total current drawn from the supply is the sum of the currents feeding the diode rectifier and thyristor rectifier transformers and is shown in graph 3 of Fig. 9.15. The r.m.s. value of the supply current for this case, $\alpha = 90°$, is

$$\sqrt{\frac{60^2 \times 30° + 90^2 \times 60° + 120^2 \times 30° + 30^2 \times 60°}{180°}} = 77.46 \text{ A}$$

The amplitude and phase angle of the fundamental component are given by the phasor sum

$$99.24\underline{/0.0°} + 33.08\underline{/-90.0°} = 104.61\underline{/-18.43°}$$

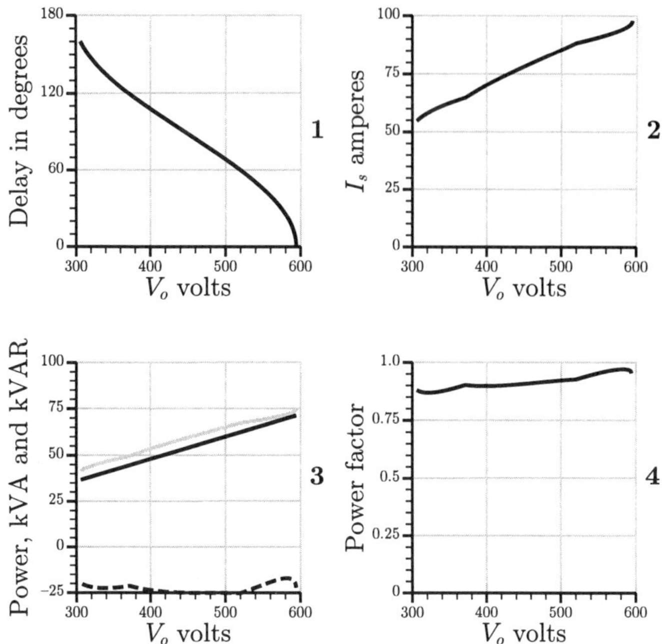

FIG. 9.16. The input characteristics of a partially controlled rectifier. A diode rectifier produces 75% of the maximum output voltage and a fully controlled rectifier contributes a voltage within the range ±25% so that the output voltage can be varied within the range 50% to 100%.

Since losses are neglected, the input power from the a.c. system is equal to the total power delivered to the d.c. load, $53.48 + 17.83\cos(\alpha)$ kW. At a delay of 90° the input power is 53.48 kW.

The input volt-amperes as conventionally measured is $\sqrt{3} \times 440.0$ multiplied by the r.m.s. value of the a.c. line current. When the delay angle is 90° the input kVA is 59.03.

The input volt-amperes reactive as conventionally measured is the value of the expression $\sqrt{kVA^2 - P_s{}^2}$ and at a delay of 90° it is 24.99 kVAR lagging.

The power factor as conventionally measured is the ratio of the input power to the input kVA as conventionally measured. At a delay of 90° it is 0.9060.

The fundamental power factor is the cosine of the angle of lag of the input current and for a delay of 90° it is $\cos(18.43°) = 0.9487$.

The characteristics of this rectifier are shown as functions of the mean output voltage over the full delay angle range of zero to 160° in Fig. 9.16. Graph 1 shows the delay angle. Graph 2 shows the r.m.s. current in the a.c. supply lines. Graph 3 shows the input power by a solid black line, the input kVA as conventionally measured by a gray line, and the input kVAR as conventionally measured by a dashed line. Graph 4 shows the power factor as conventionally measured.

The a.c. line current decreases as the voltage is lowered, the kVA is very little greater than the power, and the power factor remains high over the whole voltage range. This should be contrasted with the behavior of a single fully controlled rectifier over the same voltage range. This is illustrated by graph 1 of Fig. 9.14 over the output voltage range 1.0–0.5. The input current remains constant at 97.98 A (eqn 3.11), the input kVA remains constant at 74.67 (eqn 3.12), and the power factor as conventionally measured falls more or less linearly from 0.9549 at maximum output voltage to 0.4775 at one-half the maximum output voltage.

Thus, at the cost of some additional complication, the a.c. characteristics of the rectifier are substantially improved.

****

9.9 Static power factor improvement and filtering

The methods already described improve rectifier operation but at the best they leave a residual problem of harmonic currents and less than optimum power factor. At the present time, 1993, only one possibility remains for further improvement in high powered industrial rectifiers— static filtering. If the improvements in cost and performance of switching devices continue, it may become economically feasible to alleviate these problems by using active filtering and/or a different method of rectifier operation using forced commutation and pulse width modulation briefly described in Section 12.

Because of the high powers which must be handled, the only filter circuit components which can be used are the energy storage elements, capacitors and inductors, which can be manufactured with extremely low resistance losses. Properly employed these are most effective but their low losses introduce the possibility of resonances which can have the opposite of the desired effect, amplification rather than attenuation of the harmonics.

Figure 9.17 illustrates the basic structure of a rectifier. The rectifier is fed from an infinite bus by voltages of constant amplitude and sinusoidal waveform. A transformer may be interposed between the supply and the rectifier to provide a suitable voltage level and electrical isolation. The d.c. output of the rectifier supplies the load.

Smoothing of current on the d.c. output side is usually accomplished by adding inductance in series with, or capacitors in parallel with, the load. The two may be combined to provide a low pass filter. If the d.c. output is unipolar, electrolytic capacitors can be employed at great savings in size, weight, and cost.

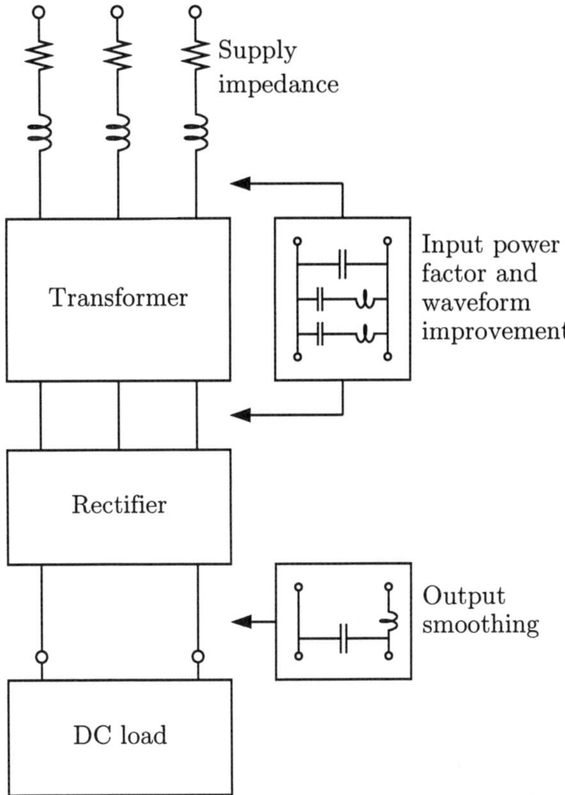

FIG. 9.17. The basic structure of a rectifier showing the possible locations of the power factor improvement capacitor, the a.c. waveform improvement filters, and the d.c. smoothing filter.

On the a.c. input side, power factor and waveform improvement can be accomplished by connecting appropriate LC networks in parallel with the rectifier. Power factor improvement is usually the most important consideration, although the growth of the power electronic load is making the electric utilities ever more conscious of the need to control harmonic currents injected into their systems by users. Power factor improvement is accomplished by connecting a three phase capacitor bank across the a.c. lines. When the harmonics must be reduced, tuned LC filters are employed for the lower harmonics. The filters have a capacitive impedance at the supply frequency and thus aid in power factor improvement. The combination of the power factor improvement capacitors, the supply inductance, and these filters limits the amplitudes of the higher harmonics relative to that of the fundamental.

9.9.1 *Output filtering*

The majority of d.c. loads, such as motor armatures, batteries, and electrolytic cells, are more sensitive to current harmonics than to voltage harmonics and the inclusion of a smoothing choke in the output of the rectifier is an effective and economical method for reducing these currents. The effects of the load and choke inductance in reducing current ripple have been investigated in great detail in Chapter 5 for those cases, the majority, when conduction is continuous, and in Chapter 6, when conduction in the rectifier is discontinuous.

When the load power is not too great, say less than 100 kW, the use of a capacitor in parallel with the load is an effective and widely used solution. This has been covered in considerable detail in Chapter 7. The value of capacitance required to make this method effective is normally so large as to be only economically achieved by the use of electrolytic capacitors, which unfortunately limits the output voltage to unipolar d.c. and thus prevents reversal of power flow from the d.c. to the a.c. side. It is the essence of capacitor smoothing that conduction in the rectifier is discontinuous, a condition which mandates time domain analysis. While concentrating on this major aspect of the problem, the analysis of Chapter 7 also permits solution in those rarer cases when there is sufficient inductance between the rectifier and the a.c. supply to ensure that conduction is continuous.

Frequency domain analysis can be used as the prime analytical tool, as described in the latter half of Chapter 8, for any problem where conduction is continuous and the effects of commutation can be ignored. As such it is an extremely useful tool for the analysis of the majority of inductor smoothing problems and the minority of capacitor smoothing problems. The techniques have been fully described in Chapter 8.

9.10 Power factor improvement using capacitors

The problems of power factor and waveform improvement of the a.c. side input current are interrelated and will be treated in sequence. As will be seen, they are closely related to the impedance of the a.c. supply and any attempt at improvement must take into account the particular supply situation.

The presence of the power factor improvement capacitor further complicates the already complex examples of the rectifier analysis studied in Chapters 5, 6, and 7. It is made amenable to a more comprehensible analysis by regarding the rectifier as a consumer of a sinusoidal current at the supply frequency and a generator of harmonic currents at multiples of the supply frequency. It is then modeled by a set of parallel connected current sources, one for each Fourier component as is done at the right-hand side of

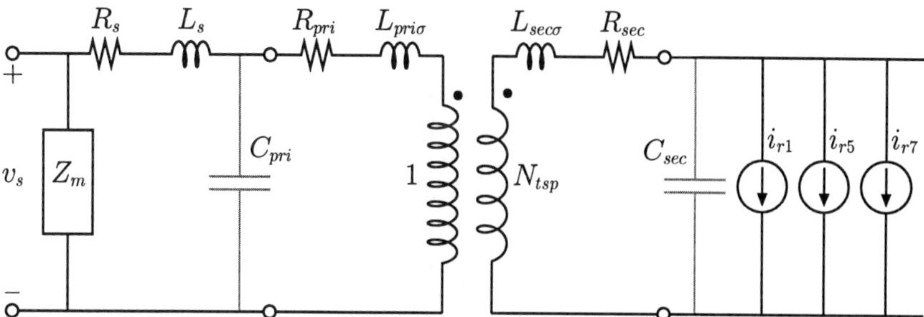

FIG. 9.18. Single phase equivalent circuit of a rectifier from the equivalent infinite supply bus up to the a.c. input terminals of the rectifier. The rectifier and load are replaced by current sources at the supply frequency and its harmonics.

Fig. 9.18, which is a single phase equivalent of the complete rectifier from the supply infinite bus up to the a.c. input terminals of the rectifying element itself. This seemingly cavalier treatment of the rectifier is justified by the fact that the currents drawn by it are essentially determined by the d.c. load and are only affected to a minor degree by the a.c. side impedances.

In Fig. 9.18, v_s is the supply voltage of sinusoidal waveform and Z_s, the series combination of R_s and L_s, is the a.c. system Thévenin impedance. $Z_{pri\sigma}$, the series combination of R_{pri} and $X_{pri\sigma}$, is the leakage impedance of a transformer primary phase. An ideal transformer, ratio $1 : N_{tsp}$, interposed between the primary and secondary sides, is followed by the secondary leakage impedance, $Z_{sec\sigma}$, R_{sec} in series with $X_{sec\sigma}$, which leads to the transformer output terminal and the rectifier. The rectifier is modeled by a set of current sources, i_{r1} for the fundamental at the same frequency as the supply voltage, i_{r5} for the fifth harmonic, i_{r7} for the seventh harmonic, etc. The magnetizing impedance, Z_m has been moved to the infinite bus as described in Chapter 5. Because of this it has no part in the analysis and is not shown in subsequent circuits. The power factor improvement capacitor is shown in gray in both its alternative locations, on the primary or on the secondary side of the transformer.

The first task of any computation is the referral of all impedances to one side or the other of the transformer. Here, everything is referred to the secondary side, a process illustrated by Example 9.6. If referral to the primary side is preferred, only minor changes are necessary.

It is assumed that there are no zero sequence components of current in any part of the a.c. side of the rectifier circuit, an assumption valid for the overwhelming majority of three phase rectifiers, the only exception being the extremely rare, three phase, single way rectifier which was described in Chapter 2.

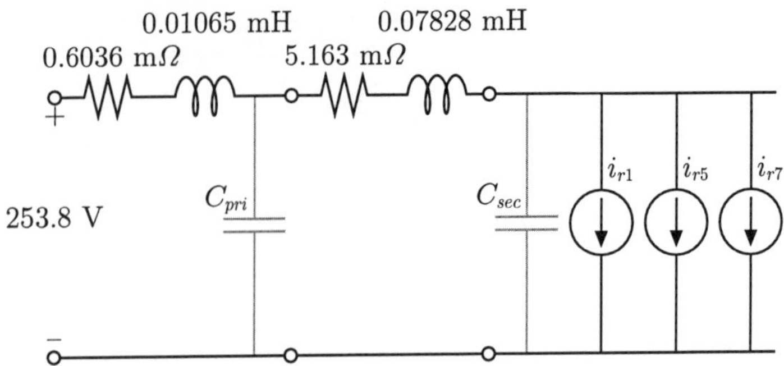

FIG. 9.19. The equivalent circuit for Example 9.6

Example 9.6

Determine the impedances of the equivalent circuit of Fig. 9.18 for the rectifier specified in Example 5.1. Refer the circuit to the secondary side of the transformer.

From Example 5.1, the line to line voltage at the infinite bus is 3300 V r.m.s. at a frequency of 60 Hz. The supply impedance is modeled by a resistance of 0.034 Ω in series with an inductance of 0.6 mH per input line. The transformer primary is delta connected with a leakage impedance of 0.45 Ω resistance in series with an inductance of 6.3 mH per phase. The secondary is wye connected with a leakage impedance of 0.0025 Ω in series with an inductance of 0.041 mH per phase. The number of primary turns per phase is 13 times the number of secondary turns per phase.

Since the transformer primary is delta connected, the first task is the conversion of the a.c. line impedance to an equivalent delta value so that it can be transferred into the delta. This is accomplished by multiplying its value by 3 to give a delta equivalent of 0.102 Ω resistance in series with 1.8 mH. This is transferred to the secondary by dividing by the square of the transformation ratio, i.e. by 13^2, to give 0.0006036 Ω in series with 0.01065 mH. Starting from the a.c. input terminals, this is the first impedance shown in Fig. 9.19.

The primary leakage impedance is referred to the secondary by division by the square of the transformation ratio to give 0.002663 Ω in series with 0.03728 mH. This is combined with the secondary leakage impedance to give 0.005163 Ω in series with 0.07828 mH. This is the second impedance of Fig. 9.19.

Since the primary is delta connected, the input voltage to the equivalent circuit is obtained by dividing the line to line voltage at the infinite bus, 3300 V r.m.s., by the transformation ratio to give 253.8 V r.m.s., the input voltage shown in Fig. 9.19.

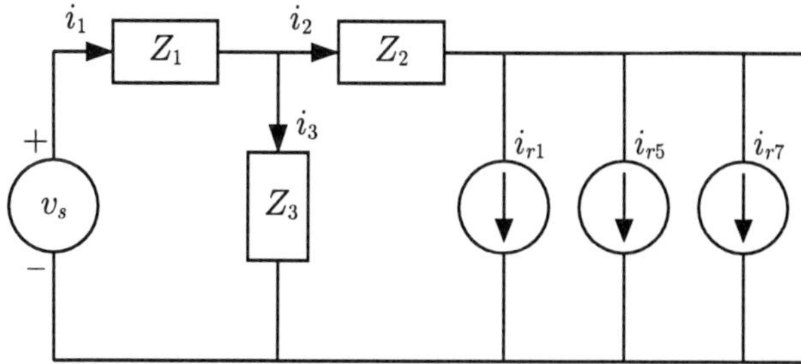

FIG. 9.20. Generalized rectifier a.c. equivalent circuit when a power factor improvement capacitor is included.

Figure 9.19 also shows in gray the power factor improvement capacitor in both its alternative locations, the one labelled C_{pri} being the actual value on the primary multiplied by the square of the transformation ratio.

****

9.10.1 *Circuit analysis*

The circuit of Fig. 9.20 is a generalization of that of Fig. 9.19. The power factor improvement capacitor is represented by the impedance Z_3 and the impedances Z_1 and Z_2 depend on the location of the capacitor. If it is connected on the supply side of the transformer, the impedance Z_1 is the equivalent supply system impedance, 0.0006036 Ω in series with 0.01065 mH in the case of Example 9.6, and the impedance Z_2 is the transformer leakage impedance, 0.05164 Ω in series with 0.07828 mH in the case of Example 9.6. If the capacitor is connected on the rectifier side, the impedance Z_1 is the sum of the equivalent supply system impedance and the equivalent leakage impedance of the transformer, i.e. in the case of Example 9.6 it is 0.05766 Ω resistance in series with 0.08893 mH inductance, and the impedance Z_2 is zero.

The circuit is analyzed by applying the superposition theorem, the effect of each source acting alone being determined. The actual system response is the sum of these individual responses. When not active, a source is replaced by its internal impedance, that for the bus bar voltage being zero and for the current sources infinite.

In the case of the fundamental, the effects of two sources, the voltage v_s and the current i_{r1} must be considered. In the case of the harmonics only the single current source representing the harmonic is active, the voltage source being replaced by a short circuit and the remaining current sources by open circuits.

The current at the fundamental frequency lags the bus bar voltage by an angle which is principally dependent on the delay angle and is secondarily dependent on the overlap angle. The quadrature component of this current is detrimental to the rectifier user in increasing the kVA demand beyond what is strictly necessary and in increasing the system voltage regulation. It is detrimental to the utility in increasing unnecessarily the system losses and regulation. These factors are typically translated into economic incentives to reduce the quadrature component, a reduction which is accomplished by connecting capacitors across the a.c. system at or close to the rectifier terminals.

The degree of compensation depends on economic considerations, essentially the capital cost of the capacitor versus the operating cost associated with the unwanted quadrature component. For a given capacitor the degree of compensation will vary with the load on the rectifier and its delay angle. It may well be desirable to stabilize the compensation by switching capacitor banks into and out of service or by having a parallel inductor controlled by an a.c. voltage regulator such as is described in Chapter 14.

In phasor terms the current component $\mathbf{I}_{1vs}$ produced by the voltage source and flowing in the impedance $\mathbf{Z}_1$ is given by

$$\mathbf{I}_{1vs} = \frac{\mathbf{V}_s}{\mathbf{Z}_1 + \mathbf{Z}_3} \qquad (9.64)$$

The current component $\mathbf{I}_{1r1}$ produced by the fundamental current source $\mathbf{I}_{r1}$ and flowing in the impedance $\mathbf{Z}_1$ is given by

$$\mathbf{I}_{1r1} = \frac{\mathbf{Z}_3}{\mathbf{Z}_1 + \mathbf{Z}_3}\mathbf{I}_{r1} \qquad (9.65)$$

The fundamental component of input current, i.e. the current $\mathbf{I}_{11}$ flowing in impedance $\mathbf{Z}_1$, is the sum of these two currents.

$$\mathbf{I}_{11} = \mathbf{I}_{1vs} + \mathbf{I}_{1r1} \qquad (9.66)$$

Substituting eqns 9.64 and 9.65 in eqn 9.66 yields

$$\mathbf{I}_{11} = \frac{\mathbf{V}_s}{\mathbf{Z}_1 + \mathbf{Z}_3} + \frac{\mathbf{Z}_3}{\mathbf{Z}_1 + \mathbf{Z}_3}\mathbf{I}_{r1} \qquad (9.67)$$

In eqn 9.67 it is usually the capacitor impedance, $\mathbf{Z}_3$ which is the unknown whose value to produce a given input power factor must be found. The required input power factor defines the relationship between the real and imaginary components of the input current so that the next task is the separation of the right-hand side of eqn 9.67 into real and imaginary components. This task is simplified by treating the input voltage $\mathbf{V}_s$ as the datum phasor so that its real part is V_s and its imaginary part is zero and by recognizing that the capacitor impedance can be written

$$\mathbf{Z}_3 = jX_3 \tag{9.68}$$

where $X_3 = -1/(\omega_s C)$ is the reactance of the capacitor at the supply frequency.

The ratio of the imaginary to the real part of the input current, K_{ir}, is obtained from the specified fundamental input power factor, P_{fs1}, as either

$$K_{ir} = \tan[\arccos(P_{fs1})] \tag{9.69}$$

if the power factor is leading or

$$K_{ir} = -\tan[\arccos(P_{fs1})] \tag{9.70}$$

if the power factor is lagging.

Applying these conditions to eqn 9.67, rationalizing the right-hand side and equating the imaginary part to the known fraction, K_{ir}, of the real part yields a quadratic equation in the capacitor reactance

$$K_2X_3^2 + K_1X_3 + K_0 \tag{9.71}$$

where

$$K_0 = (X_1 + K_{ir}R_1)V_s \tag{9.72}$$
$$K_1 = V_s - (R_1 - K_{ir}X_1)I_{r1r} - (X_1 + K_{ir}R_1)I_{r1i} \tag{9.73}$$
$$K_2 = K_{ir}I_{r1r} - I_{r1i} \tag{9.74}$$

and I_{r1r} is the real part and I_{r1i} is the imaginary part of $\mathbf{I}_{r1}$.

Solution of eqn 9.71 provides two real values for X_3, both negative, the magnitude of one being much larger than that of the other. The negative sign indicates a capacitive reactance and the root with the largest magnitude is the one required, the other root corresponding to a very large value of capacitance.

Example 9.7

For the controlled rectifier of Example 9.6, determine the size of capacitor which will improve the fundamental power factor to 0.95 lagging when the delay angle is 45° and the currents, $\mathbf{I}_{rm}$, are as listed in Table 9.4, column 3.

Table 9.4 *The peak amplitudes in amperes and phase angles in degrees of the currents in the red–yellow primary and referred to the secondary for the nonzero Fourier components of orders 1 through 19. The phase angles are relative to the red–yellow bus bar voltage*

Order m	$\mathbf{I}_{rym}$	$\mathbf{I}_{rm}$
1	$35.60 \angle -51.63$	$462.80 \angle -51.63$
5	$8.42 \angle -53.35$	$109.46 \angle -53.35$
7	$1.98 \angle 43.77$	$25.74 \angle 43.77$
11	$2.76 \angle 31.39$	$35.88 \angle 31.39$
13	$1.37 \angle 121.41$	$17.81 \angle 121.41$
17	$1.60 \angle 114.36$	$20.80 \angle 114.36$
19	$1.01 \angle -157.01$	$13.13 \angle -157.01$

The necessary capacitance is found by solving eqn 9.71 with the circuit parameters and driving voltage shown in Fig. 9.19.

The factor K_{ir} in eqn 9.71 to attain the desired fundamental power factor is $\tan[\arccos(0.95)] = -0.3287$, the value being negative because the required power factor is specified as 0.95 *lagging.* The calculation will be based on the red–yellow transformer primary winding and the corresponding secondary winding, number 1.

The next requirements are the real and imaginary current components, I_{r1r} and I_{rli}, the fundamental component of the rectifier input current is required to obtain these. Any values can be specified but it is convenient to relate the solution to a condition which has already been determined. The Fourier analysis performed on this rectifier in Chapter 8 is a convenient source of this current data. The transformer primary currents listed in Table 8.3 will be used as the basis of the computation. These data are the peak amplitudes and phase angles of the Fourier components of the current in the yellow–blue primary up to the nineteenth order. The data are converted to the red–yellow primary by advancing the phase of the positive sequence components by 120° and retarding the phase of the negative sequence components by 120°. From Chapter 8, the positive sequence Fourier components are the first, seventh, thirteenth, etc. and the negative sequence components are the fifth, ninth, seventeenth, etc. The corresponding currents referred to the secondary are obtained by multiplying the amplitudes by the transformation ratio, 13:1. Table 9.4 lists, in column 2, the peak amplitude and phase angle of the first seven nonzero Fourier components of the red–yellow primary current and, in column 3, this current referred to the secondary. Note that the phase angles have been normalized to the range $\pm 180°$.

From Table 9.4, the r.m.s. magnitude of the fundamental component of the current in secondary winding number 1 is $462.80/\sqrt{2} = 327.2$ A lagging the reference voltage by 51.63°. This current can therefore be expressed as the phasor, $327.2\cos(51.63°) - j327.2\sin(51.63°) = 203.1 - j256.6$ A r.m.s.

The impedances $\mathbf{Z}_1$ and $\mathbf{Z}_2$ have been found in Example 9.6. They depend on the location of the power factor improvement capacitor. If it is on the primary side of the transformer

$$\mathbf{Z}_1 = 0.0006036 + j0.004015$$

and

$$\mathbf{Z}_2 = 0.005163 + j0.02951$$

With these values eqn 9.71 becomes

$$189.9X_3^2 + 254.4X_3 + 0.9689 = 0$$

The two values of X_3 obtained from this equation are -0.003819 and $-1.336\ \Omega$. The larger, $-1.336\ \Omega$, is the required value and yields a capacitance of $10^6/(1.336 \times 377.0) = 1985\ \mu$F. However, this is the value referred to the secondary side on a line to neutral, i.e. wye connected, basis. It is referred to the primary side by dividing by the square of the transformation ratio to give 11.75 μF which, since the primary is delta connected, is the capacitance on a delta connected basis. If the capacitors are wye connected, the capacitance must be increased by a factor of 3 to give 35.24 μF per phase. At the supply frequency these capacitors present reactances of 225.8 and 75.28 Ω per phase. The voltage per phase when wye connected is 1905 V and the resulting capacitor current is 25.31 A leading to a total capacitor kVA rating for all three phases of $3 \times 1905 \times 25.31 \times 10^{-3} = 144.7$ kVA. In the delta connected case, the volts per phase is 3300 V resulting in a capacitor current of 14.61 A and a total capacitor kVA rating for all three phases of $3 \times 3300 \times 14.61 \times 10^{-3} = 144.7$ the same as when wye connected. Since the volume, weight, and cost of a capacitor are principally dependent on the kVA rating, it is essentially immaterial whether the connection is wye or delta although, for a variety of reasons the wye connection is usually preferred.

If the capacitor is on the secondary side of the transformer

$$\mathbf{Z}_1 = 0.005766 + j0.03353$$

$\mathbf{Z}_2$ is zero and eqn 9.71 becomes

$$189.9X_3^2 + 258.6X_3 + 8.029 = 0$$

The two values of X_3 obtained from this equation are -0.3180 and $-1.330\ \Omega$. The larger, $-1.330\ \Omega$, is the required value and yields a capacitance of

$10^6/(1.330 \times 377.0) = 1994\ \mu$F, almost identical with the value obtained in the previous case. Since the capacitor is now on the secondary side, this is the required capacitance per phase on a wye connected basis. If delta connected, the required capacitance per phase would be 664.8 μF. Again the kVA rating of the capacitor is independent of the connection, the total for all three phases being 145.3, almost the same as if it were on the primary side. However, as will be seen in the next example, the location can be highly important in respect to the response to the rectifier harmonics.

****

9.10.2 *Frequency response*

The harmonic current sources of Fig. 9.18 drive the parallel combination of the system impedance and power factor improvement capacitor. For the mth Fourier component, $\mathbf{I}_{rm}$, the currents, $\mathbf{I}_{1m}$, which flows in Z_1, and $\mathbf{I}_{3m}$, which flows in the capacitor are given by

$$\mathbf{I}_{1m} = \mathbf{I}_{rm} \frac{-jX_3/m}{R_1 + j(mX_1 - X_3/m)} \tag{9.75}$$

$$\mathbf{I}_{3m} = \mathbf{I}_{rm} \frac{R_1 + jmX_1}{R_1 + j(mX_1 - X_3/m)} \tag{9.76}$$

where X_1 and X_3 are the reactances at the supply frequency.

Division of both numerator and denominator by $-jX_3/m$ yields the transfer functions

$$\frac{\mathbf{I}_{1m}}{\mathbf{I}_{rm}} = \frac{1}{(1 - m^2X_1/X_3) + jmR_1/X_3} \tag{9.77}$$

and

$$\frac{\mathbf{I}_{3m}}{\mathbf{I}_{rm}} = \frac{-m^2X_1/X_3 + jmR_1/X_3}{(1 - m^2X_1/X_3) + jmR_1/X_3} \tag{9.78}$$

These are the familiar parallel resonance equations in a somewhat unfamiliar guise suited to the present problem. If the exciting frequency is such that $m^2X_1/X_3 \approx 1$, the denominators of both eqns 9.77 and 9.78 become small and it is possible for the supply and capacitor currents to be considerably larger than the driving current. This is inevitable when power factor improvement capacitors are added to a rectifier as Example 9.8 illustrates.

Example 9.8

Determine the resonant frequencies of the circuit with improved power factor found in Example 9.7, the transfer function defined by eqn 9.77, and the values of this transfer function at the frequencies of the Fourier components of the supply currents.

If the power factor improvement capacitor is on the primary side, the transfer function is

$$\frac{\mathbf{I}_{1m}}{\mathbf{I}_{rm}} = \frac{1}{1 - 0.003005m^2 + j0.0004516m}$$

At resonance $m = 18.24$ corresponding to a frequency of 1094 Hz, very close to the resonant frequency of the combination of the power factor improvement capacitance and the line inductance. Substitution of this value for m into eqn 9.77 establishes that the gain at resonance is 121.4. In view of this large value and the closeness of the resonance to the 17th and 19th Fourier components, problems can be anticipated with these components.

The values of the transfer function at the harmonic frequencies are given in column 2 of Table 9.5

If the power factor improvement capacitor is on the secondary side, the transfer function is

$$\frac{\mathbf{I}_{1m}}{\mathbf{I}_{rm}} = \frac{1}{1 - 0.02521m^2 + j0.004335m}$$

At resonance $m = 6.297$, i.e. at a frequency of 378.1 Hz, very close to the resonant frequency of the combination of the capacitor and the supply plus the transformer leakage impedances. At resonance the gain is 36.6. Thus, problems can be anticipated with the 5th and 7th order components of the rectifier input current.

The values of the transfer function for the Fourier components are listed in column 3 of Table 9.5.

Table 9.5 clearly shows the effects of the resonance. When the capacitor is on the primary side of the transformer, all the listed harmonics are amplified. In particular the 17th order component is increased by a factor of 7.583 and the 19th order component by a factor of 11.75. When the capacitor is on the secondary side, the 5th order component is increased by a factor of 2.699 and the 7th order component by a factor of 4.219, the higher harmonics being attenuated.

The effects of resonance are more dramatically illustrated by Fig. 9.21 which shows the Bode amplitude plots for the transfer functions, Fig. 9.21(a) being for the primary side case and Fig. 9.21(b) for the secondary side case.

Table 9.5 *The transfer function $\mathbf{I}_{sm}/\mathbf{I}_m$ when (a) the power factor improvement capacitor is on the primary side of the transformer and (b) it is on the secondary side. Phase angles are in degrees*

Order of the Fourier component	Transfer function: Primary capacitor	Transfer function: Secondary capacitor
5	$1.081 \angle -0.13$	$2.699 \angle -3.35$
7	$1.172 \angle -0.21$	$4.219 \angle -172.64$
11	$1.571 \angle -0.45$	$0.488 \angle -178.67$
13	$2.031 \angle -0.68$	$0.307 \angle -179.01$
17	$7.582 \angle -3.34$	$0.159 \angle -179.33$
19	$11.751 \angle -174.21$	$0.123 \angle -179.42$

In practice the circuit losses will increase with frequency because of effects not considered in the model and the amplification will not be as great as the values obtained in this calculation. Even so, it would be prudent to anticipate supply harmonic problems when connecting power factor improvement capacitors to a rectifier.

****

9.10.3 *The system currents*

The final task in the analysis is the determination of the actual currents from the equivalent circuit currents. The relationship between these currents is illustrated in Fig. 9.22.

The circuits of Fig. 9.22 each have four terminals labelled P, Q, S, and T. Figure 9.22(a) shows the case when the power factor improvement capacitor is on the primary side of the transformer and Fig. 9.22(b) shows the case when it is on the secondary. For each case the relationships between the equivalent circuit currents, i_1, i_2, and i_3 and the actual circuit currents are shown, the actual circuit currents being the transformer primary current, i_{pri}, the transformer secondary current, i_{sec}, and the capacitor current, i_{cap}. The current entering terminal P, i_p, and the current leaving terminal S, i_s, are also shown.

If the transformer primary is wye connected, the current in the a.c. supply line is the current i_p. If the primary is delta connected, the situation is more complicated since the Fourier components of i_p and whether they are of positive or negative phase sequence must then be considered. In either event the magnitude of the current component is increased $\sqrt{3}$ times. If it is of positive phase sequence, i.e. the fundamental, seventh, thirteenth, etc.

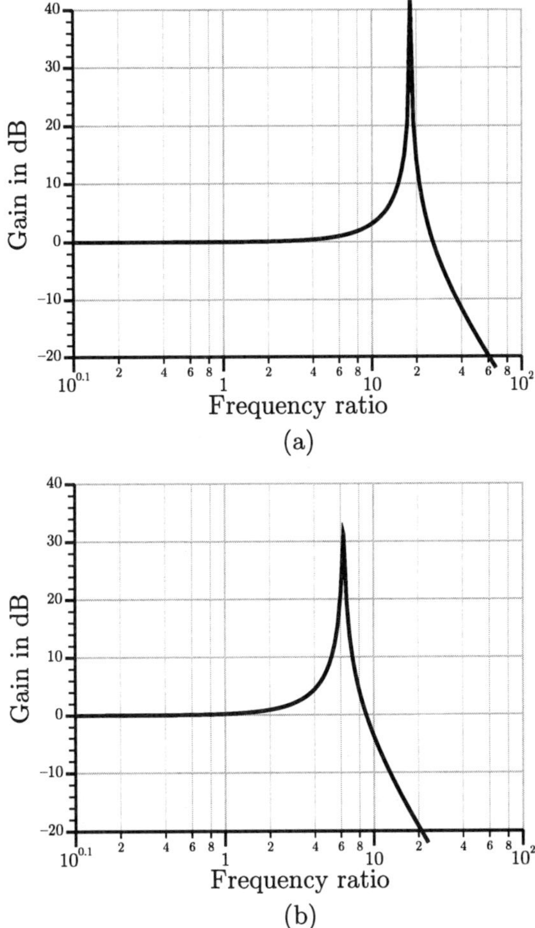

FIG. 9.21. Bode amplitude plots of the transfer function $\mathbf{I}_{1m} / \mathbf{I}_{rm}$ when (a) the power factor improvement capacitor is placed on the primary side of the transformer and (b) when it is placed on the secondary side of the transformer

components, its phase is retarded by 30°. If it is of negative phase sequence, i.e. the fifth, eleventh, seventeenth, etc. components, its phase is advanced by 30°

The situation on the secondary is similar. If the secondary is wye connected, then the a.c. line carrying current from the transformer secondary terminal S to the rectifier is i_s. If the secondary is delta connected, a Fourier current component in that a.c. line is obtained by increasing the

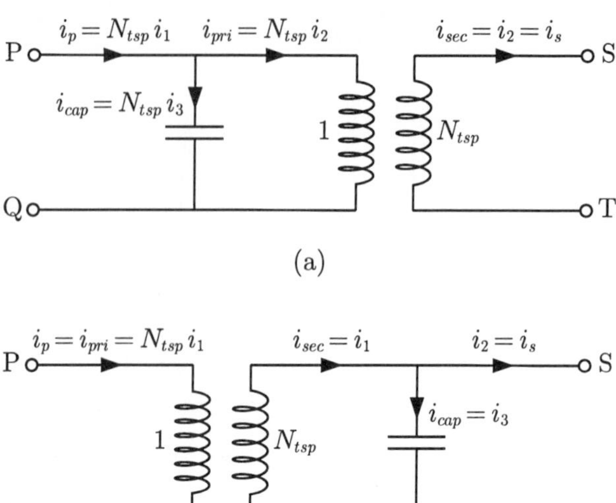

FIG. 9.22. Relationship between the equivalent circuit and the actual circuit currents. In (a) the capacitor is on the primary side of the transformer and in (b) it is on the secondary side.

corresponding component leaving terminal S by the factor $\sqrt{3}$ and retarding its phase by 30° if it is of positive phase sequence and advancing its phase by 30° if it is of negative phase sequence.

These points are covered in Section 8.7 and are illustrated in Example 9.9.

Example 9.9

For the rectifier of Example 9.7, determine the waveforms of the currents in an a.c. system input line when the capacitor is on the primary side of the transformer and compare this waveform with that obtained without power factor improvement.

The equivalent circuit impedances have been found in Examples 9.6 and 9.7. When the power factor improvement capacitor is on the primary side of the transformer, their values in ohms at 60 Hz are

$$\mathbf{Z}_1 = 0.0006036 + j0.004015$$
$$\mathbf{Z}_2 = 0.005163 + j0.02951$$
$$\mathbf{Z}_3 = -j1.336$$

With these values and the 60 Hz driving voltage given in Fig. 9.19 of 253.8 V r.m.s., the equivalent circuit harmonic currents are as listed in

Table 9.6 *Equivalent circuit currents when the capacitor is on the primary side of the transformer. Peak amplitudes in amperes and phase angles in degrees are shown*

Harmonic order m	$\mathbf{I}_{1m}$	$\mathbf{I}_{2m}$	$\mathbf{I}_{3m}$
1	$303.2\ \underline{/-18.19}$	$462.8\ \underline{/-51.64}$	$268.2\ \underline{/89.82}$
5	$118.3\ \underline{/-53.49}$	$109.5\ \underline{/-53.35}$	$8.894\ \underline{/-55.21}$
7	$30.18\ \underline{/43.56}$	$25.74\ \underline{/43.77}$	$4.445\ \underline{/42.33}$
11	$56.37\ \underline{/31.04}$	$35.88\ \underline{/31.49}$	$20.50\ \underline{/30.26}$
13	$36.18\ \underline{/120.7}$	$17.81\ \underline{/121.41}$	$18.37\ \underline{/120.1}$
17	$157.7\ \underline{/111.0}$	$20.80\ \underline{/114.4}$	$137.0\ \underline{/110.5}$
19	$154.3\ \underline{/28.78}$	$13.13\ \underline{/-157.0}$	$167.3\ \underline{/28.32}$

Table 9.6, which gives peak amplitudes in amperes and phase angles in degrees.

Comparison of the amplitudes of column 2 of Table 9.6 with those of column 3, which are the currents defined in column 3 of Table 9.4, shows the power factor improvement and corresponding amplitude reduction of the fundamental and the amplification of all the harmonics in accordance with Fig. 9.21(a). Thus, the phase angle of the fundamental is improved from 51.64° lagging to 18.19° lagging and its amplitude is reduced from 462.8 to 303.2 A. Taking one of the harmonics, the thirteenth, its amplitude is increased from 17.81 to 36.18 A.

The equivalent circuit currents are translated into actual circuit currents according to the procedures indicated in Fig. 9.22(a). As an example consider the fundamental components. Table 9.6 gives the current $\mathbf{I}_1$ as $303.2\underline{/-18.19°}$. Multiplying this by the transformation ratio, $N_{tsp} = 1/13$ gives the current $\mathbf{I}_p$ of Fig. 9.22(a), i.e. the current entering the parallel combination of the transformer primary and the capacitor, as $23.32\underline{/-18.19°}$. However, the transformer primary is delta connected so that the a.c. line current is obtained by multiplying the amplitude of this current by $\sqrt{3}$ and, since the fundamental is a positive sequence component, retarding its phase by 30°. Thus, the current in the red a.c. line is $40.40\underline{/-48.19°}$.

Since the capacitor is on the primary side of the transformer, the current in it is obtained by multiplying the current $\mathbf{I}_3$ by N_{tsp} to give $20.63\underline{/89.82°}$. This is the current in the red–yellow phase of the capacitor assuming the capacitor is delta connected like the transformer. If the capacitor is wye connected, the current in the red to neutral phase is obtained by multiplying the amplitude by $\sqrt{3}$ and retarding the phase by 30° to give $35.74\underline{/59.82°}$.

The current in the red–yellow transformer primary is, from Fig. 9.22(a), N_{tsp} times $\mathbf{I}_2$, i.e. $35.60\underline{/-51.64}$. The current in the red to neutral

Table 9.7 *Primary side circuit currents when the capacitor is on the primary side of the transformer. Peak amplitudes in amperes and phase angles in degrees are shown*

Harmonic order m	$\mathbf{I}_{sl}$	$\mathbf{I}_{pri}$	$\mathbf{I}_{cap}$
1	$40.40\underline{/-48.19}$	$35.60\underline{/-51.64}$	$20.63\underline{/89.82}$
5	$15.77\underline{/-23.49}$	$8.420\underline{/-53.35}$	$0.68\underline{/-55.21}$
7	$4.02\underline{/13.56}$	$1.980\underline{/43.77}$	$0.34\underline{/42.33}$
11	$7.51\underline{/61.04}$	$2.760\underline{/31.49}$	$1.58\underline{/30.26}$
13	$4.82\underline{/90.73}$	$1.370\underline{/121.41}$	$1.41\underline{/120.06}$
17	$21.01\underline{/141.02}$	$1.600\underline{/114.36}$	$10.54\underline{/110.52}$
19	$20.55\underline{/-1.22}$	$1.010\underline{/-157.01}$	$12.87\underline{/28.32}$

transformer secondary is the current $\mathbf{I}_2$, i.e. $462.8\underline{/-51.54°}$. Since the secondary is wye connected, this is also the current being delivered to the rectifier by the red secondary a.c. line. Had the secondary been delta connected, the current in this line would have been obtained from the secondary winding current by multiplying the amplitude by $\sqrt{3}$ and retarding the phase by 30°.

The harmonics are treated in a similar manner except for the phase shift. The phase of the positive sequence components, the seventh, thirteenth, and nineteenth, is retarded by 30° and that of the negative sequence components, the fifth, eleventh, and seventeenth, is advanced by the same amount.

A complete listing of the primary side values is given in Table 9.7 and of the secondary side values in Table 9.8. All magnitudes in these tables are peak amperes and all phase angles are in degrees. The column headings of these tables list $\mathbf{I}_{sl}$, the current in the red a.c. supply line, $\mathbf{I}_{pri}$, the current in the red–yellow transformer primary, $\mathbf{I}_{cap}$, the current in the red–yellow power factor improvement capacitor, $\mathbf{I}_{sec}$, the current in the transformer secondary number 1, and $\mathbf{I}_{rec}$, the current in the a.c. line connecting the transformer secondary terminal number 1 to an actual rectifier input terminal.

The Fourier components of column 2 of Table 9.8 have been used to create the waveform of the red a.c. input line current shown in Fig. 9.23(a) where it is shown in black. For comparison, the waveform without power factor improvement is shown in gray. It is evident that the black wave leads the gray wave, a consequence of the improvement in power factor. However, even more striking is the large harmonic content which is also a consequence of the introduction of the power factor improvement capacitors.

The waveform of the red a.c. input line current when the capacitor is

Table 9.8 *Actual secondary side circuit currents when the capacitor is on the primary side of the transformer. Magnitudes are peak amplitudes in amperes and phase angles are in degrees*

Harmonic order m	$\mathbf{I}_{sec}$	$\mathbf{I}_{rec}$
1	$462.8 \angle -51.64$	$462.8 \angle -51.64$
5	$109.5 \angle -53.35$	$109.5 \angle -53.35$
7	$25.7 \angle 43.77$	$25.7 \angle 43.77$
11	$35.9 \angle 31.49$	$35.9 \angle 31.49$
13	$17.8 \angle 121.41$	$17.8 \angle 121.41$
17	$20.8 \angle 114.36$	$20.8 \angle 114.36$
19	$13.1 \angle -157.01$	$13.1 \angle -157.01$

on the secondary side has been determined following a similar procedure and is shown in Fig. 9.23(b). The phase advance is evident but the most striking feature of the waveform is again the large harmonic content, this time at a lower frequency because of the lower resonant frequency, as shown in Fig. 9.21.

Considering Fig. 9.23, one is tempted to surmise that more harm than good has been done by the introduction of the capacitor. This is indeed so since the power factor as conventionally measured is still low despite the improvement in fundamental power factor. The r.m.s. current in an a.c. line is the square root of the sum of the squares of the Fourier component amplitudes of column 2 of Table 9.7 divided by $\sqrt{2}$. It is 37.69 A. The input kVA as conventionally measured is $\sqrt{3} \times 3300 \times 37.69 \times 10^{-3} = 215.4$. The r.m.s. value of the fundamental component of line current is $40.40 \div \sqrt{2} = 28.56$ A and the fundamental power factor is 0.95 lagging so that the input power is $\sqrt{3} \times 3300 \times 28.56 \times 0.95 \times 10^{-3} = 155.1$ kW. Thus, the power factor as conventionally measured is $155.1 \div 214.4 = 0.7194$, the decrease from 0.95 being a reflection of the high harmonic content. When the capacitor is on the secondary side, the conventional power factor is even lower at 0.6649.

****

Clearly the application of power factor improvement capacitors to a rectifier can give rise to new and serious problems which require thorough analytical exploration before practical implementation. These problems can be alleviated by filtering although, as will be seen, this too is far from a straightforward panacea.

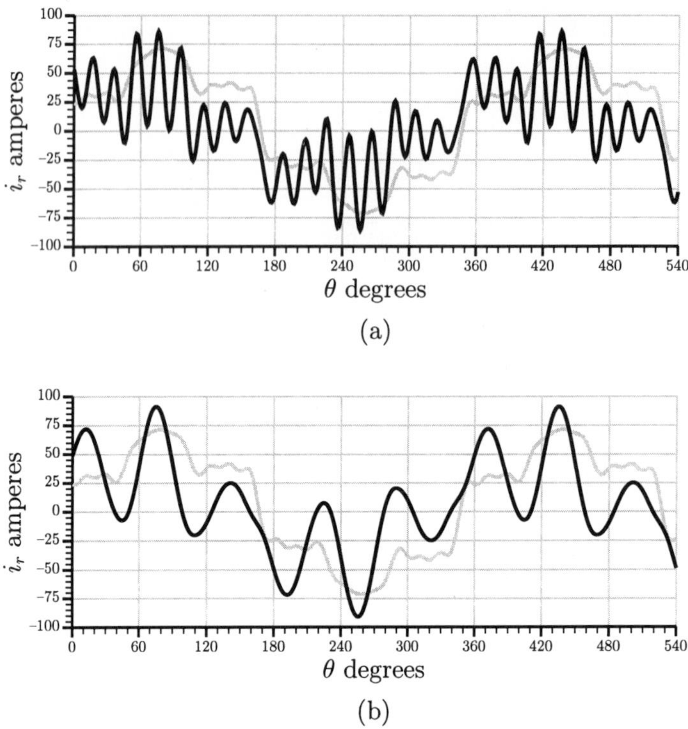

FIG. 9.23. Waveforms of the current in the red a.c. input line. The waveform before power factor improvement is shown by gray lines and the waveform after improvement by black lines. For part (a) the power factor improvement capacitor is connected on the primary side of the transformer and for part (b) it is connected on the secondary side

9.11 Filtering

The single phase equivalent circuit of Fig. 9.24 is an extension of the circuit shown in Fig. 9.20 by incorporating the filter circuits intended to improve the a.c. current waveforms. Like Fig. 9.20, the circuit is a single phase equivalent of the a.c. side of the rectifier from the equivalent supply system infinite bus up to the a.c. input terminals of the rectifier itself. The diagram shows the impedances Z_1, Z_2, and Z_3 of Fig. 9.20 plus impedances Z_4, Z_5, Z_6, and Z_7 representing four series resonant filters, each one tuned to a specific rectifier input current harmonic. Each filter provides at its resonant frequency a low impedance path into which a portion of a particular current harmonic is diverted away from the a.c. supply, thereby reducing the amount which the rectifier installation injects into the a.c. system.

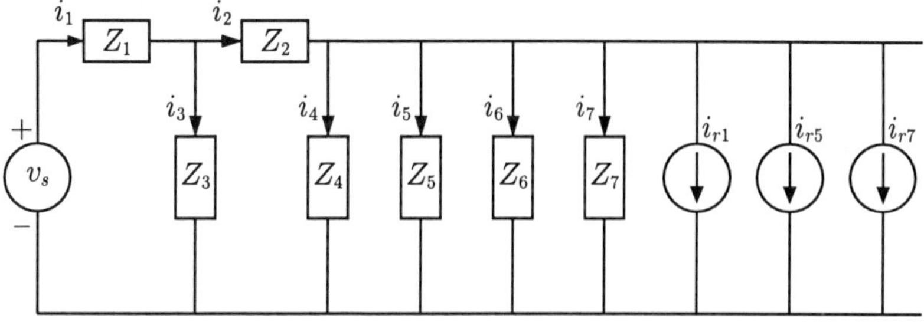

FIG. 9.24. The single phase equivalent circuit diagram for power factor improvement and filtering.

A filter impedance is represented by the series combination of a resistance, an inductance, and a capacitance with the resistance being as small as possible so that as much as possible of the harmonic current to whose frequency the filter is tuned will be diverted through it and will not pass into the a.c. supply system. Thus, the impedance Z_4 is, at a frequency ω, $\mathbf{Z}_4 = R_4 + j[\omega L_4 - 1/(\omega C_4)]$ and its resonant frequency $\omega_{res} = \sqrt{1/(L_4 C_4)}$. At the resonant frequency the filter then presents a purely resistive impedance whose value is R_4.

The analysis is carried out on the basis of the four filter circuits shown in Fig. 9.24 and it is assumed that all quantities have been referred to the secondary side of the transformer. It is a straightforward task to transfer all quantities to the primary side if that is more convenient and to increase the number of filters.

As was seen when considering power factor improvement, the values of the two impedances Z_1 and Z_2 depend upon the locations of the power factor improvement capacitors and the filters. If both are on the primary side of the transformer, then the impedance Z_1 is the supply impedance and the impedance Z_2 is zero. If the power factor improvement capacitors are on the primary side and the filters are on the secondary side, the impedance Z_1 is the supply impedance and the impedance Z_2 is the leakage impedance of the transformer. If the power factor improvement capacitors and the filters are on the secondary side of the transformer, the impedance Z_1 is the sum of the supply impedance and the transformer leakage impedance and impedance Z_2 is zero. The one case not covered by Fig. 9.24—power factor improvement capacitors on the secondary side and filters on the primary side, is unlikely since it is desirable to have as much impedance as is practicable between the filters and the infinite bus in order to maximize the diversion of harmonic currents from the a.c. supply through the filters. However, should this case arise its analysis would proceed in a way similar to that which follows.

Since many of the elements of Fig. 9.24 are paralleled, it is convenient to work with admittances rather than impedances, $\mathbf{Y}_1$ being the reciprocal of $\mathbf{Z}_1$, etc.

9.11.1 *Assumptions*

The general problem is fearsomely complex and it is appropriate to briefly review the assumptions which make it reasonably manageable. These assumptions are

- The supply system can be modeled by the voltage v_s, whose waveform is sinusoidal and whose radian frequency is ω_s, behind the inductive impedance Z_1.
- The resistances are independent of frequency.
- The transformer magnetizing impedance can be removed to the infinite bus with negligible error.
- The rectifier load can be represented by a set of parallel connected current sources, one for each Fourier component of the input current.

The first and second of these assumptions have the greatest significance for this particular problem. The Thévenin impedance of the supply may be far more complex than a simple series combination of inductance and resistance. In particular there may be power factor correction capacitors at various points within the system. These may resonate with the rectifier circuit inductances and give rise to unexpectedly large currents in the vicinity of the resonant frequencies.

In regard to the second assumption, significant harmonic currents are generated at up to 50 or more times the system frequency, i.e. up to 3000 Hz, and the resistances may vary substantially over such a wide frequency range. These are high powered circuits and the conductor cross sections may be large enough to create significant skin and proximity effects. There may well be solid metallic structures in close proximity to the conductors. Eddy currents induced in these contribute to the resistance and are strongly frequency dependent. Iron cored inductors are an especially important example of this effect with the rectifier transformer leakage impedance being the most significant circuit element in this regard.

The increase in resistance with frequency is generally benign in that it increases the damping of the harmonics relative to the fundamental. However, this is not the case with the filter inductors whose resistance must be kept to an absolute minimum. To this end these inductors are usually air cored and precautions are taken to keep solid metallic objects, e.g. the structure housing the filter, as far as possible from their magnetic fields and to use stranded conductors which will not exhibit significant skin and proximity effects.

The third assumption, removal of the transformer magnetizing impedance to the infinite bus, will affect waveforms throughout the circuit but the impedance is so large relative to the other circuit impedances that any error thus introduced will be minor. The errors can be minimized, but not eliminated, by increasing the magnetizing impedance by the sum of the transformer primary leakage impedance and the supply line impedance as was done in Chapter 5.

The significance of the fourth assumption has already been discussed in the previous section. In brief, this assumption is based on the fact that the rectifier input current depends principally on the load and only to a minor extent on the impedances of the supply and transformer.

9.11.2 *Analysis*

The determination of its frequency response is the principal aim of analysis of the equivalent circuit. The superposition theorem tells us that the response is the sum of the responses to the individual sources taken one at a time, all other sources being replaced by their internal impedances. Thus, in the case of Fig. 9.24 the response to five individual sources, v_s, i_1, i_5, i_7, i_{11}, and i_{13} must be determined and summed. The internal impedance of a current source is infinite so that the current sources are replaced by open circuits during the determination of the response to v_s, the equivalent circuit being as shown in Fig. 9.25(a). The internal impedance of a voltage source is zero so that it is replaced by a short circuit during the determination of the response to a current source, the equivalent circuit being as shown in Fig. 9.25(b).

The subscripts on the impedances have been augmented by a 1 in Fig. 9.25(a) to indicate that they are evaluated at the fundamental frequency and by an m in Fig. 9.25(b) to indicated that they are evaluated at the mth harmonic frequency. The subscripts on the impedance currents i_1 through i_7 have been augmented by vs in Fig. 9.25(a) indicating components due to v_s and by rm in Fig. 9.25(b) indicating components due to i_{rm}.

The analysis is performed by successive reduction of the circuit, until a solution can be obtained for the input current. Retracing this process then yields the currents in all the impedances. Since all the driving functions have sinusoidal waveforms, the analysis is performed using phasors.

9.11.2.1 *Response to* $\mathbf{I}_{rm}$ Figure 9.26 shows the four stages in circuit reduction required for the response to the driving function $\mathbf{I}_{rm}$, each stage being identified by a number to its right. Stage 1 is similar to the circuit of Fig. 9.25(b) but with three changes. First, all quantities are expressed

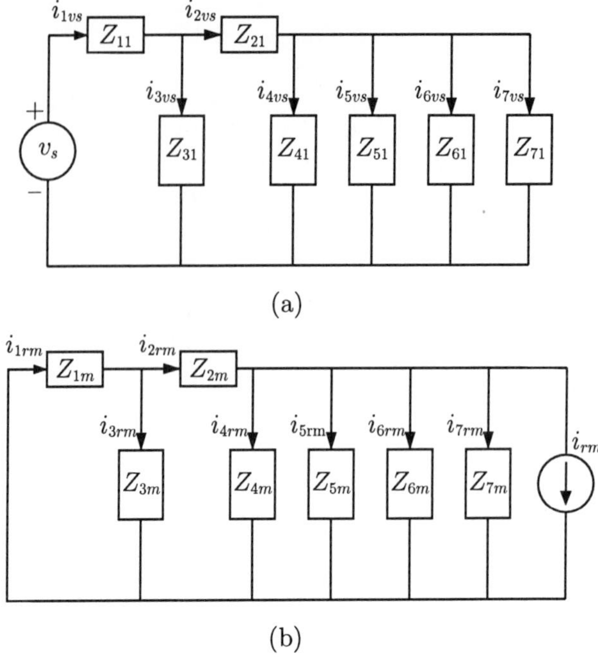

FIG. 9.25. The equivalent circuit during application of the superposition theorem. In (a) is shown its configuration when determining the response to v_s and in (b) its configuration when determining its response to the current source representing the Fourier component of order m.

as phasors. Second, the impedances have been replaced by the equivalent admittances. Third, the four filter admittances, $\mathbf{Y}_{4m}$ through $\mathbf{Y}_{7m}$, have been combined into a single admittance $\mathbf{Y}_{9m}$ given by

$$\mathbf{Y}_{9m} = \mathbf{Y}_{4m} + \mathbf{Y}_{5m} + \mathbf{Y}_{6m} + \mathbf{Y}_{7m} \tag{9.79}$$

At stage 2, the admittances $\mathbf{Y}_{1m}$ and $\mathbf{Y}_{3m}$ are combined into a single admittance $\mathbf{Y}_{8m}$ given by

$$\mathbf{Y}_{8m} = \mathbf{Y}_{1m} + \mathbf{Y}_{3m} \tag{9.80}$$

At stage 3, the admittances $\mathbf{Y}_{2m}$ and $\mathbf{Y}_{8m}$ are combined into a single admittance $\mathbf{Y}_{10m}$ given by

$$\mathbf{Y}_{10m} = \frac{\mathbf{Y}_{2m}\,\mathbf{Y}_{8m}}{\mathbf{Y}_{2m} + \mathbf{Y}_{8m}} \tag{9.81}$$

In the fourth and final stage, the admittances $\mathbf{Y}_{9m}$ and $\mathbf{Y}_{10m}$ are combined into a single admittance $\mathbf{Y}_{11m}$ given by

$$\mathbf{Y}_{11m} = \mathbf{Y}_{9m} + \mathbf{Y}_{10m} \tag{9.82}$$

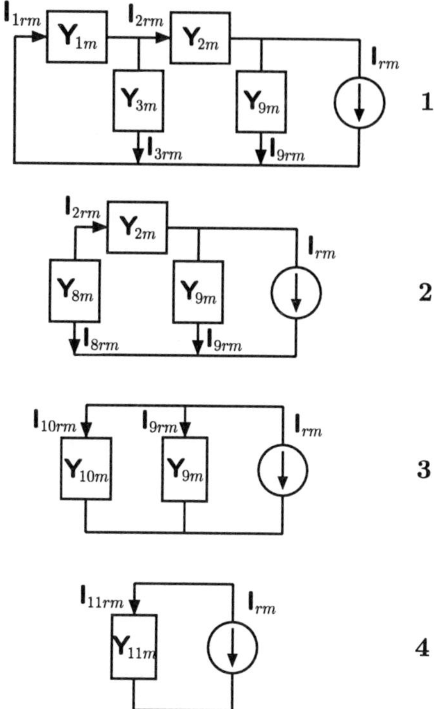

FIG. 9.26. Equivalent circuit reduction when determining the response to the mth order Fourier component of rectifier input current.

The various currents are now obtained by working back from reduction circuit 4 through to circuit 1. Thus, observation of the final reduction circuit indicates that

$$\mathbf{I}_{11rm} = -\mathbf{I}_{rm} \tag{9.83}$$

From circuit 3

$$\mathbf{I}_{10rm} = \frac{\mathbf{Y}_{10m}}{\mathbf{Y}_{11m}} \mathbf{I}_{11rm} \tag{9.84}$$

and

$$\mathbf{I}_{9rm} = \frac{\mathbf{Y}_{9m}}{\mathbf{Y}_{11m}} \mathbf{I}_{11rm} \tag{9.85}$$

From circuit 2

$$\mathbf{I}_{8rm} = \mathbf{I}_{10rm} \tag{9.86}$$

and

$$\mathbf{I}_{2rm} = -\mathbf{I}_{10rm} \tag{9.87}$$

From circuit 1

$$\mathbf{I}_{3rm} = \frac{\mathbf{Y}_{3m}}{\mathbf{Y}_{8m}} \mathbf{I}_{8rm} \tag{9.88}$$

and

$$\mathbf{I}_{1rm} = -\frac{\mathbf{Y}_{1m}}{\mathbf{Y}_{8m}} \mathbf{I}_{8rm} \tag{9.89}$$

Finally the current in any filter branch is the current $\mathbf{I}_{9rm}$ multiplied by the admittance of that branch and divided by the combined admittance of all the filter branches. Thus, the current in the seventh order filter whose admittance is $\mathbf{Y}_{5m}$ is given by

$$\mathbf{I}_{5rm} = \frac{\mathbf{Y}_{5m}}{\mathbf{Y}_{9m}} \mathbf{I}_{9rm} \tag{9.90}$$

9.11.2.2 *Response to* $\mathbf{V}_s$ Figure 9.27 shows the various circuit reduction stages required to determine the response to the supply voltage. Each stage is identified by a number to its right. Since the frequency of the voltage is the fundamental frequency, the subscript m used in determining the response to $\mathbf{I}_{rm}$ is replaced by 1.

Expressions for the admittances are found by working forwards through the circuits from 1 to 4. The admittance $\mathbf{Y}_{91}$ is, as in the Fig. 9.26, the parallel combination of the filter admittances, but for this calculation it is evaluated at the fundamental frequency.

The admittance $\mathbf{Y}_{121}$ is the series combination of $\mathbf{Y}_{21}$ and $\mathbf{Y}_{91}$, i.e.

$$\mathbf{Y}_{121} = \frac{\mathbf{Y}_{21}\mathbf{Y}_{91}}{\mathbf{Y}_{21} + \mathbf{Y}_{91}} \tag{9.91}$$

The admittance $\mathbf{Y}_{131}$ is the parallel combination of $\mathbf{Y}_{31}$ and $\mathbf{Y}_{121}$, i.e.

$$\mathbf{Y}_{131} = \mathbf{Y}_{31} + \mathbf{Y}_{121} \tag{9.92}$$

The admittance $\mathbf{Y}_{141}$ is the series combination of $\mathbf{Y}_{11}$ and $\mathbf{Y}_{141}$, i.e.

$$\mathbf{Y}_{141} = \frac{\mathbf{Y}_{11}\mathbf{Y}_{131}}{\mathbf{Y}_{31} + \mathbf{Y}_{131}} \tag{9.93}$$

The currents are now found by reversing the process, working through the circuits from 4 to 1. Thus, from Circuit 4

$$\mathbf{I}_{14vs} = \mathbf{V}_s\mathbf{Y}_{141} \tag{9.94}$$

From circuit 3

$$\mathbf{I}_{13vs} = \mathbf{I}_{14vs} \tag{9.95}$$

and

$$\mathbf{I}_{1vs} = \mathbf{I}_{14vs} \tag{9.96}$$

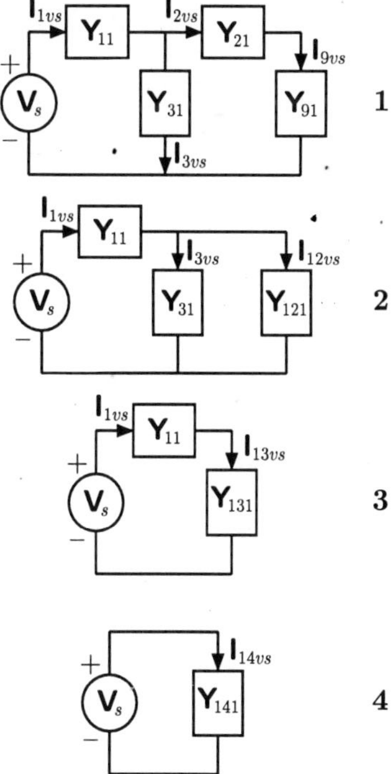

FIG. 9.27. Equivalent circuit reduction when determining the response to the supply voltage.

From circuit 2

$$\mathbf{I}_{12vs} = \frac{\mathbf{Y}_{121}}{\mathbf{Y}_{131}}\mathbf{I}_{13vs} \tag{9.97}$$

and

$$\mathbf{I}_{3vs} = \frac{\mathbf{Y}_{31}}{\mathbf{Y}_{131}}\mathbf{I}_{13vs} \tag{9.98}$$

From circuit 1

$$\mathbf{I}_{9vs} = \mathbf{I}_{12vs} \tag{9.99}$$

and

$$\mathbf{I}_{2vs} = \mathbf{I}_{12vs} \tag{9.100}$$

The current in a filter branch is the total filter current multiplied by the admittance of the branch and divided by the total filter admittance. Thus, the current in the fifth harmonic filter whose admittance is $\mathbf{Y}_{41}$ is given by

$$\mathbf{I}_{4vs} = \frac{\mathbf{Y}_{41}}{\mathbf{Y}_{91}}\mathbf{I}_{9vs} \tag{9.101}$$

9.11.2.3 *Branch currents* The current at the frequency of the Fourier component of order m in a branch of the circuit is produced by the current source at that frequency except for the case of the fundamental, $m = 1$. For the fundamental, the current is the phasor sum of that due to the voltage source, $\mathbf{V}_s$ and that due to the current source $\mathbf{I}_{r1}$. Thus, the current in the admittance $\mathbf{Y}_1$ has a component at the fundamental frequency represented by the phasor $\mathbf{I}_{1vs} + \mathbf{I}_{1r1}$, a component at the fifth harmonic frequency represented by the phasor $\mathbf{I}_{1r5}$, a component at the seventh harmonic frequency represented by the phasor $\mathbf{I}_{1r7}$, etc.

Since the majority of this book has been devoted to analysis, this would seem to be the appropriate point for an illustrative example but in this case a consideration of the filter design process provides additional insight into the complexities of filter behavior.

9.11.3 *Design*

Filter performance can be specified in numerous ways. For the present purpose it will be assumed that, for a specific d.c. load, the value of the required input fundamental power factor and the required attenuation for various harmonics have been given. While it is theoretically possible to obtain a closed form solution for the filter elements, the expressions involved are extremely complex and unwieldy. A simpler iterative approach will be adopted in which it is assumed that all elements other than the one being designed are known.

The iterative process starts with the filter admittances set to zero and determines the size of the power factor improvement capacitor to meet the specified fundamental input power factor. The filter admittances are then determined one by one to meet the harmonic attenuation specifications and using for the circuit admittances the values already determined. The iteration ends when the admittances of all the unknown branches, $\mathbf{Y}_3$ through $\mathbf{Y}_7$ have been found. Unfortunately these values are incorrect because they are based on a false assumption. For example $\mathbf{Y}_3$ was found on the basis that $\mathbf{Y}_4$ through $\mathbf{Y}_7$ were all zero. A second iteration is therefore performed exactly as the first but using the admittance values found during the first iteration. This yields improved values and the computation is repeated as many times as necessary to reach an acceptable level of accuracy. Fortunately, the process converges rapidly and four or five iterations should provide a satisfactory result. The process is clarified in Example 9.10 but some essential relationships must be derived before that can be performed.

9.11.3.1 *The capacitor* The determination of $\mathbf{Y}_3$, the admittance of the power factor improvement capacitor, requires first the determination of the input current, i.e. the current in admittance $\mathbf{Y}_1$ due to the voltage source $\mathbf{V}_s$ and the current source $\mathbf{I}_{1rm}$.

The input current component, $\mathbf{I}_{1vs}$, due to $\mathbf{V}_s$ is given by

$$\mathbf{I}_{1vs} = \frac{\mathbf{Y}_1(\mathbf{Y}_2\mathbf{Y}_3 + \mathbf{Y}_2\mathbf{Y}_9 + \mathbf{Y}_3\mathbf{Y}_9)}{\mathbf{Y}_1\mathbf{Y}_2 + \mathbf{Y}_2\mathbf{Y}_3 + \mathbf{Y}_1\mathbf{Y}_9 + \mathbf{Y}_2\mathbf{Y}_9 + \mathbf{Y}_3\mathbf{Y}_9}\mathbf{V}_s \tag{9.102}$$

The input current, $\mathbf{I}_{1r1}$, due to the current source $\mathbf{I}_{r1}$ is given by

$$\mathbf{I}_{1r1} = \frac{\mathbf{Y}_1\mathbf{Y}_2}{\mathbf{Y}_1\mathbf{Y}_2 + \mathbf{Y}_2\mathbf{Y}_3 + \mathbf{Y}_1\mathbf{Y}_9 + \mathbf{Y}_2\mathbf{Y}_9 + \mathbf{Y}_3\mathbf{Y}_9}\mathbf{I}_{r1} \tag{9.103}$$

The total fundamental component of the input current, $\mathbf{I}_{11}$, is the sum of these two currents and can be expressed as a function of $\mathbf{Y}_3$, i.e.

$$\mathbf{I}_{11} = \frac{\mathbf{N}_1\mathbf{Y}_3 + \mathbf{N}_0}{\mathbf{D}_1\mathbf{Y}_3 + \mathbf{D}_0} \tag{9.104}$$

where

$$\mathbf{N}_1 = \mathbf{V}_s\mathbf{Y}_1(\mathbf{Y}_2 + \mathbf{Y}_9) \tag{9.105}$$
$$\mathbf{N}_0 = \mathbf{V}_s\mathbf{Y}_1\mathbf{Y}_2\mathbf{Y}_9 + \mathbf{I}_{r1}\mathbf{Y}_1\mathbf{Y}_2 \tag{9.106}$$
$$\mathbf{D}_1 = \mathbf{Y}_2 + \mathbf{Y}_9 \tag{9.107}$$
$$\mathbf{D}_0 = \mathbf{Y}_1\mathbf{Y}_2 + \mathbf{Y}_1\mathbf{Y}_9 + \mathbf{Y}_2\mathbf{Y}_9 \tag{9.108}$$

The complex coefficients of this expression are now replaced by their Cartesian equivalents, i.e. $\mathbf{N}_0$ is replaced by $N_{0r} + jN_{0i}$, etc. and the capacitor admittance $\mathbf{Y}_1$ is replaced by j times its susceptance, i.e. $\mathbf{Y}_3 = jB_3 = j\omega_s C_3$ to give

$$\frac{N_{0r} - N_{0i}B_3 + j(N_{0i} + N_{1r}B_3)}{D_{0r} - D_{1i}B_3 + j(D_{0i} + D_{0r}B_3)} \tag{9.109}$$

This expression is rationalized and the denominator is discarded to leave the numerator

$$\begin{aligned}
&B_3^2(D_{1r}N_{1r} + D_{1i}N_{1i}) \\
&\quad +B_3(D_{1r}N_{0i} - D_{1i}N_{0r} - D_{0r}N_{1i} + D_{0i}N_{1r}) \\
&\qquad +D_{0r}N_{0r} + D_{0i}N_{0i} \\
&+j\left[B_3^2(D_{1r}N_{1i} - D_{1i}N_{1r})\right. \\
&\quad +B_3(D_{0r}N_{1r} + D_{0i}N_{1i} - D_{1r}N_{0r} - D_{1i}N_{0i}) \\
&\qquad \left.+D_{0r}N_{0i} - D_{0i}N_{0r}\right]
\end{aligned} \tag{9.110}$$

The specified fundamental input power factor, P_{fs1} defines the relationship between the real and imaginary parts of eqn 9.110. Defining K_{ir} as the ratio of the imaginary to the real part, then

$$K_{ir} = \tan[\arccos(P_{fs1})] \tag{9.111}$$

if the power factor is leading or

$$K_{ir} = -\tan[\arccos(P_{fs1})] \tag{9.112}$$

if the power factor is lagging. This known relationship yields a quadratic equation in the susceptance B_3

$$B_3^2 K_2 + B_3 K_1 + K_0 = 0 \tag{9.113}$$

where the coefficients K_0, K_1, and K_2 are

$$K_0 = D_{0r}N_{0i} - D_{0i}N_{0r} - K_{ir}(D_{0r}N_{0r} + D_{0i}N_{0i}) \tag{9.114}$$

$$\begin{aligned} K_1 = {} & D_{0r}N_{1r} + D_{0i}N_{1i} - D_{1r}N_{0r} - D_{1i}N_{0i} + \\ & K_{ir}(D_{0r}N_{1i} - D_{0i}N_{1r} + D_{1i}N_{0r} - D_{1r}N_{0i}) \end{aligned} \tag{9.115}$$

$$K_2 = D_{1r}N_{1i} - D_{1i}N_{1r} - K_{ir}(D_{1r}N_{1r} + D_{1i}N_{1i}) \tag{9.116}$$

While these appear to be formidable expressions, obtaining them in numerical form and solving for B_3 is a straightforward programming task. The solution yields two real positive values for B_3, one large and one small. The small value is the required one.

9.11.3.2 *The filter branches* The filter for the mth order Fourier component, $m > 1$, must produce the specified attenuation A_{sm} at its resonant frequency $\omega_m = m\omega_s$. At the resonant frequency the inductive and capacitive reactances of the filter cancel, leaving a purely resistive admittance whose conductance is G_{xm}

$$\mathbf{Y}_{xm} = G_{xm} \tag{9.117}$$

where x is the number of the impedance in Fig. 9.24 representing the filter, e.g. for the eleventh harmonic $x = 6$.

The complex attenuation $\mathbf{A}_m$ of the mth order Fourier component is

$$\mathbf{A}_m = \frac{\mathbf{I}_{rm}}{\mathbf{I}_{1rm}}$$

From eqns 9.83 through 9.89 the attenuation is given by

$$\mathbf{A}_m = \frac{\mathbf{Y}_{8m}\mathbf{Y}_{11m}}{\mathbf{Y}_{1m}\mathbf{Y}_{10m}} \tag{9.118}$$

From eqn 9.82, $\mathbf{Y}_{9m} + \mathbf{Y}_{10m}$ can be substituted for $\mathbf{Y}_{11m}$ to yield

$$\mathbf{A}_m = \frac{\mathbf{Y}_{8m}\mathbf{Y}_{9m} + \mathbf{Y}_{8m}\mathbf{Y}_{10m}}{\mathbf{Y}_{1m}\mathbf{Y}_{10m}} \tag{9.119}$$

The admittance $\mathbf{Y}_9$ is the sum of the admittance of the filter branch being designed, which from eqn 9.117 is G_{xm}, and the admittances of the balance of the filter branches, $\mathbf{Y}_{9mb}$. Making this substitution in eqn 9.119

$$\mathbf{A}_m = \frac{G_{xm}\mathbf{Y}_{8m} + \mathbf{Y}_{8m}\mathbf{Y}_{9mb} + \mathbf{Y}_{8m}\mathbf{Y}_{10m}}{\mathbf{Y}_{1m}\mathbf{Y}_{10m}} \tag{9.120}$$

The complex quantities in this equation can be replaced by their real and imaginary parts, e.g. $\mathbf{Y}_{8m}$ becomes $G_{8m} + jB_{8m}$, to give

$$\mathbf{A}_m = \frac{N_r + jN_i}{D_r + jD_i} \tag{9.121}$$

where

$$N_r = G_{xm}G_{8m} + G_{8m}G_{9mb} - B_{8m}B_{9mb} + G_{8m}G_{10m} - B_{8m}B_{10m} \tag{9.122}$$

$$N_i = G_{xm}B_{8m} + G_{8m}B_{9mb} + B_{8m}G_{9mb} + G_{8m}B_{10m} + B_{8m}G_{10m} \tag{9.123}$$

$$D_r = G_{1m}G_{10m} - B_{1m}B_{10m} \tag{9.124}$$

$$D_i = G_{1m}B_{10m} + B_{1m}G_{10m} \tag{9.125}$$

The magnitude of the attenuation must be the specified value, A_{sm}, i.e.

$$A_{sm}^2 = \frac{N_r^2 + N_i^2}{D_r^2 + D_i^2} \tag{9.126}$$

After expansion this yields the following quadratic equation in the unknown conductance G_{xm}

$$G_{xm}^2 K_2 + G_{xm}K_1 + K_0 = 0 \tag{9.127}$$

where the coefficients K_0, K_1 and K_2 are

$$\begin{aligned} K_0 = &-A_{sm}^2(G_{1m}^2G_{10m}^2 + B_{1m}^2G_{10m}^2 + B_{1m}^2B_{10m}^2) + \\ &G_{8m}^2G_{9mb}^2 + 2G_{8m}^2G_{9mb}G_{10m} + G_{8m}^2G_{10m}^2 + \\ &G_{8m}^2B_{9mb}^2 + 2G_{8m}^2B_{9mb}B_{10m} + G_{8m}^2B_{10m}^2 + \\ &G_{9mb}^2B_{8m}^2 + 2G_{9mb}G_{10m}B_{8m}^2 + G_{10m}^2B_{8m}^2 + \\ &B_{8m}^2B_{9mb}^2 + 2B_{8m}^2B_{9mb}B_{10m} + B_{8m}^2B_{10m}^2 \end{aligned} \tag{9.128}$$

$$K_1 = 2\,(G_{8m}^2G_{9mb} + G_{8m}^2G_{10m} + G_{9mb}B_{8m}^2 + G_{10m}B_{8m}^2) \tag{9.129}$$

$$K_2 = G_{8m}^2 + B_{8m}^2 \tag{9.130}$$

Table 9.9 *Filter attenuation factors and filter inductor quality factors*

Harmonic order	Attenuation factor	Quality factor
5	7	24
7	6	28
11	5	32
13	4	36

Solution of eqn 9.127 provides two values for the conductance, G_{xm}, of the filter branch, one positive and one negative. The positive value is the one required. The reciprocal of the conductance, R_{xm}, is the resistance of the filter inductor. Knowing the quality factor, Q_{xm} of the filter inductor at the filter resonant frequency, a value which can normally be obtained from the inductor manufacturer, the inductance L_x is obtained as

$$L_x = \frac{Q_{xm}}{m\omega_s G_{xm}} \tag{9.131}$$

At the resonant frequency the reactances of the filter inductor and capacitor are equal so that the filter capacitance is given by

$$C_x = \frac{1}{m^2\omega_s^2 L_x} \tag{9.132}$$

When all the filter branches have been designed, another iteration can be started to produce improved values. Once the values thus produced are acceptably stable, iteration can be terminated.

The utilization of this method is illustrated by Example 9.10.

Example 9.10

Design a power factor improvement and filtering network to meet the following specification for the rectifier of Examples 9.6 and 9.7 whose rectifier input current is defined by Table 9.4.

The fundamental input power factor is to be 0.95 lagging. The first four harmonics are to be attenuated by the factors listed in Table 9.9. The quality factors of the filter inductors at the various harmonic frequencies are listed in the same table. The power factor improvement capacitor is to be placed on the primary side of the transformer and the filters on the secondary side.

Since the power factor improvement capacitor is on the primary side of the transformer and the filters are on the secondary side, the equivalent circuit is that of Fig. 9.24 with $\mathbf{Z}_1$ equal to the line impedance and $\mathbf{Z}_2$ equal to the transformer leakage impedance. These impedances have been found in Examples 9.6 and 9.7 to be

$$\mathbf{Z}_1 = 0.0006036 + j0.004015\ \Omega \text{ at the supply frequency}$$

and

$$\mathbf{Z}_2 = 0.005163 + j0.02951\ \Omega \text{ at the supply frequency}$$

The voltage $\mathbf{V}_s$ is taken to be the datum phasor. It has been found in Example 9.6 to be

$$\mathbf{V}_s = 253.8 + j0.0 \text{ V r.m.s.}$$

These preliminaries completed, the first iteration can begin with the determination of the size of the power factor improvement capacitor using eqn 9.113 with all filter admittances set to zero. The quadratic equation is then

$$B_3^2 - 262.6B_3 + 195.9 = 0$$

It has two roots, $B_3 = 0.7483$ and $B_3 = 261.9$. The smaller of these is the required one and indicates an equivalent circuit capacitance of 1985 μF. At the supply frequency the impedance $\mathbf{Z}_3$ is

$$\mathbf{Z}_3 = -j1.336\ \Omega$$

The impedance of the first filter is now found following the procedure described above in Section 9.11.3.2 with the values of $\mathbf{Z}_1$, $\mathbf{Z}_2$, and $\mathbf{Z}_3$ known and the admittances of the remaining filters still zero. To attenuate the fifth harmonic current by a factor of 7, eqn 9.127 which defines the conductance of the filter at its resonant frequency, five times the supply frequency, is

$$G_4^2 + 0.4092G_4 - 1962 = 0$$

This has two roots, $G_4 = 44.09$ and $G_4 = -44.50$. The positive root is the one required. It fixes the filter resistance for this iteration at

$$R_4 = 0.02268\ \Omega$$

Using the Q factor of 24, this sets the reactances of the filter inductor and capacitor at $0.02268 \times 24 = 0.5443\ \Omega$ at 300 Hz. The filter impedance at the supply frequency is then

$$\mathbf{Z}_4 = 0.02268 - j2.613\ \Omega \text{ at 60 Hz}$$

Knowing the first four impedances, eqn 9.123 is determined for the second filter whose resonant frequency is seven times the supply frequency, 420 Hz.

To provide the required attenuation of the seventh harmonic by a factor of 6, the conductance is defined by

$$G_5^2 + 0.5332G_5 - 815.4 = 0$$

The required root is $G_5 = 28.29$ so that the resistance of the seventh order filter is

$$R_5 = 0.03535\ \Omega$$

Using the Q factor of 28, this sets the reactances of the filter inductor and capacitor at 0.9898 Ω at 420 Hz. The filter impedance at the supply frequency is then

$$\mathbf{Z}_5 = 0.03535 - j6.787\ \Omega \text{ at 60 Hz}$$

Similar calculations give values for the impedances of the eleventh and thirteen order filters as

$$\mathbf{Z}_6 = 0.05189 - j18.11\ \Omega \text{ at 60 Hz}$$

and

$$\mathbf{Z}_7 = 0.06472 - j30.11\ \Omega \text{ at 60 Hz}$$

The first iteration is now completed and the second is started with the impedance values already found. The filters all present a capacitive impedance at the supply frequency and the power factor reduction capacitor is reduced accordingly, the calculation yielding an equivalent circuit value of 355.5 μF.

Table 9.10 summarizes the computation by listing the values of the required root provided by the quadratic equations, either eqn 9.113 for the power factor improvement capacitor or eqn 9.127 for the filters. Convergence is rapid with the third significant figure accurate at the end of the fourth iteration and the fourth accurate at the end of the sixth iteration.

The final results are listed in Table 9.11. The values are for the equivalent circuit. Since the power factor improvement capacitor is on the primary side of the transformer, its value must be divided by the square of the transformation ratio, 13:1. With this transformation, the power factor improvement capacitor on the primary has a value of $521.40 \div 13^2 = 3.085$ μF. This is the value when connected in delta like the transformer primary winding. If the capacitor is wye connected, the capacitance must be three times this value, 9.256 μF per phase. These values are about one-quarter of those found in Example 9.7, the reduction being due to a substantial contribution to power factor improvement from the filter capacitors. The total capacitance required for the installation, 1951 μF per phase referred

Table 9.10 *The required roots of the quadratic equations at the end of iterations 1 through 7. The value in the second column is the susceptance of the power factor improvement capacitor in ohms at the supply frequency and the values in columns 3 through 6 are the filter conductances in mhos*

Iteration	B_3	G_4	G_5	G_6	G_7
1	0.7483	44.09	28.29	19.27	15.451
2	0.1340	41.76	25.23	13.77	8.447
3	0.2017	42.00	25.48	14.14	8.953
4	0.1961	41.98	25.46	14.11	8.910
5	0.1966	41.98	25.46	14.11	8.914
6	0.1966	41.98	25.46	14.11	8.913
7	0.1966	41.98	25.46	14.11	8.913

to the secondary, is about the same as the value found in Example 9.7 for power factor improvement without filtering.

****

The design process used in Example 9.10 produces a circuit which meets the specification as far as fundamental power factor and harmonic attenuation are concerned. As the results of Examples 9.8 and 9.9 have shown, this is no guarantee that the circuit will be satisfactory. Harmonics other than those to which the filters are specifically tuned may be increased and the final result may be far from the one anticipated. The next stage in ensuring a satisfactory outcome is the determination of the frequency response of the circuit.

Table 9.11 *Equivalent circuit parameters at the end of the seventh iteration*

Impedance number	Resistance Ω	Inductance μH	Capacitance μF
1	0.000604	10.65	0.00
2	0.005163	78.28	0.00
3	0.000000	0.00	521.40
4	0.023822	303.31	927.92
5	0.039272	416.69	344.61
6	0.070877	546.93	106.32
7	0.112191	824.11	50.52

9.11.4 *Frequency response*

The frequency responses of all sections of the circuit are worthy of consideration but the one of most importance and the one which will be considered is the relationship between the input and output currents of the equivalent circuit. Specifically we will consider the gain, G_m, which is the magnitude of the ratio of $\mathbf{I}_{1rm}$ to $\mathbf{I}_{rm}$ as a function of frequency. The reciprocal of the gain has been already found as the complex attenuation, $\mathbf{A}_m$, in eqn 9.118. From this the gain is

$$G_m = \left| \frac{\mathbf{Y}_{1m}\mathbf{Y}_{10m}}{\mathbf{Y}_{8m}\mathbf{Y}_{10m}} \right| \tag{9.133}$$

Unfortunately, when this expression is expanded to the level of the individual circuit resistances, inductances, and capacitances it becomes extremely unwieldy so that the most practical approach to it is via numerical analysis. The actual circuit parameters as found by the design process are used to determine values of the admittances at a specific frequency and from these the gain is determined. The process is repeated at different frequencies as many times as required to build up a picture of the gain as a function of frequency. Obviously minimum gains, i.e. maximum attenuation, will occur close to the filter resonant frequencies and it can be expected that maximum gains will also occur at intermediate frequencies due to the interaction between circuit elements. These minima and maxima are very sharp due to the low loss nature of the whole circuit and must be specifically searched for to obtain a valid frequency response. The process is illustrated by Example 9.11.

Example 9.11

Determine the gain of the filter circuit determined in Example 9.10 over the frequency ratio range 0.1–100, i.e. 6—6000 Hz, giving particular attention to the gain maxima and minima.

The gain must be determined at a sufficient number of values of the frequency ratio m to construct the frequency response without missing the minima and maxima. The values of the equivalent circuit admittances are essential to this process and the values at a frequency ratio of 4, i.e. a frequency of 240 Hz, are listed in Table 9.12.

Inserting the values of $\mathbf{Y}_1$, $\mathbf{Y}_8$, $\mathbf{Y}_{10}$, and $\mathbf{Y}_{11}$ from Table 9.12 into eqn 9.133 yields the gain

$$G_4 = \left| \frac{(2.3364 - \jmath 62.1741) \times (0.3199 - \jmath 7.4318)}{(2.3364 - \jmath 61.3878) \times (0.7033 - \jmath 2.5382)} \right| = 2.860$$

This gain corresponds to a logarithmic gain of 9.427 dB and, while it is not needed for this problem, the transfer function phase shift at this frequency is $-13.05°$.

Table 9.12 *The values of the fourteen equivalent circuit admittances at 240 Hz*

Admittance	Value in Ω^{-1}
$\mathbf{Y}_1$	$2.3364 - j62.1741$
$\mathbf{Y}_2$	$0.3698 - j8.4554$
$\mathbf{Y}_3$	$0.0000 + j0.7863$
$\mathbf{Y}_4$	$0.3568 + j3.8538$
$\mathbf{Y}_5$	$0.0234 + j0.7709$
$\mathbf{Y}_6$	$0.0024 + j0.1847$
$\mathbf{Y}_7$	$0.0008 + j0.0841$
$\mathbf{Y}_8$	$2.3364 - j61.3878$
$\mathbf{Y}_9$	$0.3834 + j4.8936$
$\mathbf{Y}_{10}$	$0.3199 - j7.4318$
$\mathbf{Y}_{11}$	$0.7033 - j2.5382$
$\mathbf{Y}_{12}$	$2.7445 + j11.0766$
$\mathbf{Y}_{13}$	$2.7445 + j11.8629$
$\mathbf{Y}_{14}$	$4.2904 + j14.3542$

The gain computation has been repeated a large number of times to produce the Bode amplitude plot of the gain of Fig. 9.28. This shows 0 dB gain at low frequencies, a high frequency roll off above 6000 Hz of 40dB per decade and at intermediate frequencies the four minima and five maxima listed in Table 9.13. The minima are close to but not quite at the resonant frequencies of the filters, e.g. there is one at a frequency ratio of 5.012 which is very close to the resonant frequency ratio of the fifth order filter at 5.0. The gains at these minima are slightly less than the gains at the resonant frequencies, e.g. the gain at the minimum at 11.044 Hz is 0.1935 which corresponds to an attenuation of 5.168, slightly greater than the attenuation of 5.0 at a frequency ratio of 11. The maxima occur at frequencies close to those of the minima, e.g. there is a maximum gain of 2.15 at a frequency ratio of 10.42, between the minima at frequency ratios of 7.030 and 11.045. The maxima at frequency ratios of 4.280 and 36.634 show particularly large gains of 16.29 dB and 41.30 dB respectively. Trouble with the 35th and 37th harmonics can be anticipated due to the maximum at a frequency ratio of 36.634.

****

9.11.5 *System currents*

The final task in the analysis is the determination of the circuit currents. The relationship between these currents and the equivalent circuit currents

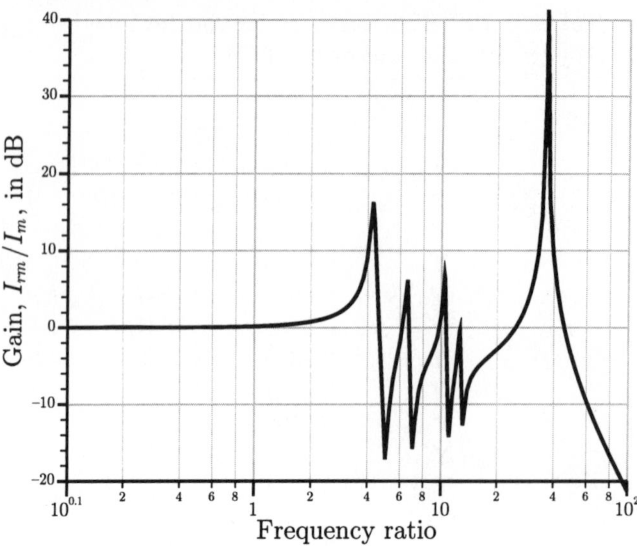

FIG. 9.28. Bode amplitude plot of the filter circuit transfer function

is illustrated by Fig. 9.29. Figure 9.29(a) shows the situation when the capacitor is on the primary side of the transformer and Fig. 9.29(b) when it is on the secondary side. In the figure the currents i_1, i_2, and i_3 are equivalent circuit currents, i_{pri} is the current in a primary winding, i_{sec} is the current in a secondary winding, i_{cap} is the current in one phase of the power factor improvement capacitor, i_9 is the combined current entering all the filter branches, i_p is the current entering terminal P, i_s is the current leaving terminal S, and N_{tsp} is the ratio of transformer secondary to primary turns per phase.

If the primary is wye connected, the a.c. line current supplying the transformer is i_p. If the primary is delta connected, the amplitude of a Fourier component of the a.c. line current is $\sqrt{3}$ times the amplitude of the corresponding component of i_p and its phase angle lags or leads by 30° depending on whether the phase sequence of the component is positive or negative.

If the secondary is wye connected, the current in an a.c. line supplying the rectifier is i_s. If the secondary is delta connected, the amplitude of the Fourier components of the current in the a.c. line supplying the rectifier is $\sqrt{3}$ times the amplitude of the corresponding component of i_s and its phase angle lags or leads by 30° depending on whether the phase sequence of the component is positive or negative.

The determination of the circuit currents is illustrated by Example 9.12.

Table 9.13 *Gain minima and maxima*

Type	Frequency Hz	Gain dB	Gain absolute
Maximum	4.280	16.294	6.526
Minimum	5.012	–16.962	0.142
Maximum	6.575	6.135	2.027
Minimum	7.030	–15.798	0.162
Maximum	10.423	6.673	2.156
Minimum	11.045	–14.266	0.194
Maximum	12.629	–0.278	0.969
Minimum	13.078	–12.759	0.230
Maximum	36.634	41.296	116.087

Example 9.12

Determine the Fourier components of the current in the red a.c. line to the rectifier transformer under the same load conditions as Example 9.11. From these determine the input power factor as conventionally measured and the waveform of the current.

The seven equivalent circuit impedances of Fig. 9.24 found in Example 9.10 are listed in Table 9.14. The table has four columns, the first of which gives the impedance symbol, the second gives its resistance, the third gives its inductive reactance at the supply frequency, and the fourth gives its capacitive reactance at the supply frequency. The impedance at any other frequency is obtained by multiplying the inductive reactance by the frequency ratio, the ratio of the frequency to the supply frequency, dividing the capacitive reactance by the frequency ratio, and subtracting the capacitive reactance from the inductive reactance. The fourteen admittances of Figs 9.26 and 9.27 are then found and the equivalent circuit currents are determined using eqns 9.83 through 9.90 and, when finding the fundamental, eqns 9.94 through 9.101. The values of $\mathbf{I}_2$ thus found are listed in column 2 of Table 9.15 where the current magnitudes are peak values and phase angles are in degrees.

Since the power factor improvement capacitor is on the primary side of the transformer, the current components in the red–yellow primary phase are found from the components of i_2 by multiplying the amplitude of the corresponding i_2 component by the transformation ratio $N_{tsp} = 1/13$, and using the same phase angle. The values of these components are listed in column 3 of Table 9.15.

The components of current in the red supply line are obtained from the corresponding components in the red–yellow primary by multiplying the amplitudes of the latter by $\sqrt{3}$ and by either retarding the phase angle

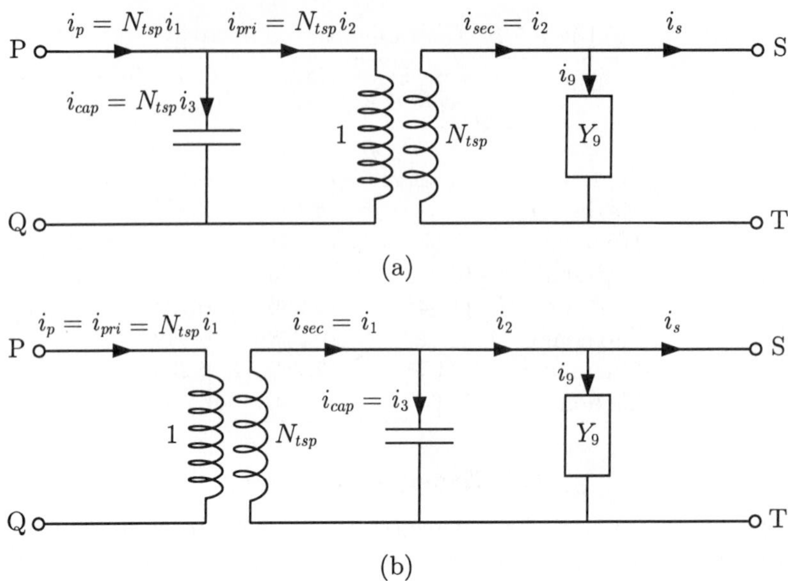

FIG. 9.29. Relationship between the equivalent circuit and actual circuit currents. In (a) the capacitor is on the primary and in (b) the capacitor is on the secondary side of the transformer

by 30° if the component is of positive phase sequence, the fundamental, seventh, thirteenth, and nineteenth, or advancing the phase by 30° if it is of negative phase sequence, the fifth, eleventh, and seventeenth. The values of these components are listed in column 4 of Table 9.15.

The r.m.s. value of the line current is obtained as the square root of half the sum of the squares of the peak amplitudes in column 4, i.e.

$$I_{line} = \sqrt{\frac{41.206^2 + 2.083^2 + 0.572^2 + 0.956^2 + 0.593^2 + 1.665^2 + 1.189^2}{2}}$$

i.e.

$$I_{line} = 29.22 \text{ r.m.s. A}$$

This value is an approximation to the r.m.s. value because of the truncation of the Fourier series at the nineteenth component. The actual current will be somewhat larger and may be significantly larger if some of the higher order components are amplified as noted in Example 9.11.

The voltage losses in the a.c. supply lines are small and the r.m.s. value of the line to line voltage at the transformer, 3299.83 V, is very little different from that at the infinite bus. The input kVA to the transformer as conventionally measured is $\sqrt{3} \times 3299.83 \times 29.22 \times 10^{-3} = 167.0$. The very

Table 9.14 *Equivalent circuit impedances at the supply frequency*

Impedance	Resistance Ω	Inductive reactance, Ω	Capacitive reactance, Ω
$\mathbf{Z}_1$	0.000604	0.004015	0.0000
$\mathbf{Z}_2$	0.005163	0.029511	0.0000
$\mathbf{Z}_3$	0.000000	0.000000	5.0870
$\mathbf{Z}_4$	0.023822	0.114345	2.8586
$\mathbf{Z}_5$	0.039272	0.157088	7.6973
$\mathbf{Z}_6$	0.070877	0.206187	24.9486
$\mathbf{Z}_7$	0.112189	0.310678	52.5046

small kVA absorbed by the line impedance will be ignored and this value will be taken as the input kVA to the transformer.

Because the waveform at the infinite bus is sinusoidal, power supplied by it is only associated with the fundamental component of current and is $\sqrt{3} \times 3300 \times (41.206/\sqrt{2}) \times \cos(-48.12°) \times 10^{-3} = 158.2$ kW.

The losses in the three a.c. lines are $3 \times 29.22^2 \times 0.034 = 0.087$ kW so that the power delivered to the rectifier transformer is $158.2 - 0.087 = 158.1$ kW and the power factor as conventionally measured is $158.1 \div 167.0 = 0.9467$. This is very little smaller than the fundamental power factor of 0.95 lagging, an indication that the harmonic content of the input current is small. This is confirmed by the waveform of the current shown in black in Fig. 9.30, which has been derived from the truncated Fourier series. For comparison, the waveform of the uncompensated current is shown in gray.

Table 9.15 *The Fourier components of the currents in equivalent circuit impedance* $\mathbf{Z}_2$*, the red–yellow transformer primary, and the red a.c. supply line. Peak amplitudes in amperes and phase angles in degrees are shown*

Order m	$\mathbf{I}_{2m}$	$\mathbf{I}_{rym}$	$\mathbf{I}_{rm}$
1	$337.780\underline{/-29.64}$	$25.983\underline{/-29.64}$	$41.206\underline{/-48.19}$
5	$15.329\underline{/-135.62}$	$1.179\underline{/-135.62}$	$2.083\underline{/-105.66}$
7	$4.124\underline{/-31.44}$	$0.317\underline{/-31.44}$	$0.572\underline{/-61.49}$
11	$6.491\underline{/-42.26}$	$0.499\underline{/-42.26}$	$0.956\underline{/-12.34}$
13	$3.859\underline{/61.13}$	$0.297\underline{/61.13}$	$0.593\underline{/31.03}$
17	$9.646\underline{/113.81}$	$0.742\underline{/113.81}$	$1.665\underline{/143.66}$
19	$6.382\underline{/-157.35}$	$0.491\underline{/-157.35}$	$1.189\underline{/-187.53}$

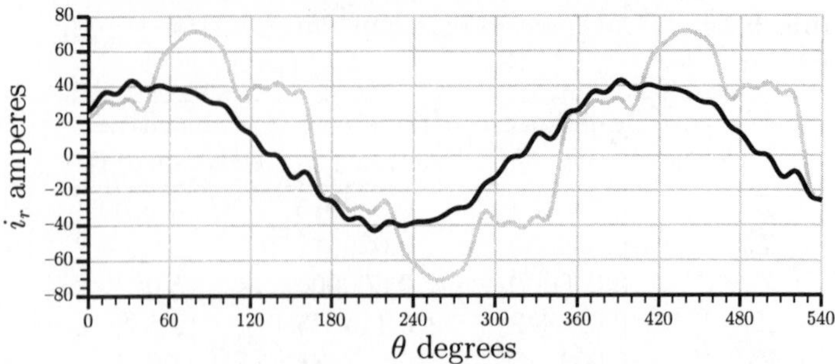

FIG. 9.30. Waveform of the current in an a.c. supply line shown by a black line. The waveform of the uncompensated current is shown for comparison by a gray line

From these numbers and the waveform it appears that the filter design has been successful. However, this optimistic conclusion should be tempered by the fact that it is based on a Fourier series truncated at the nineteenth order component. If higher order components had been included, problems may have been noted with the amplification of the 35th and 37th order components due to the resonance at 2198.1 Hz found in Example 9.11. It must also be kept in mind that, while these calculations are of necessity very precise, the data on which they are based, the various circuit impedances, are not so precise and slight shifts from the resonant frequencies found in Example 9.11 can profoundly alter the conclusions.

The value of the supply impedance is a particular cause for concern. It is very small, potentially very variable since the electric utility may change the system configuration at any time, and yet it is vital to the operation of the filter since it determines how much of a particular harmonic will be diverted into the filter. A frequent cause of trouble is the presence of capacitors elsewhere in the network. These may well be used by an adjacent customer and so are totally out of the control of the rectifier owner. They shunt part of the supply impedance and may resonate with one or more of the rectifier harmonics so that substantial harmonic currents may mysteriously appear in distant parts of the a.c. system. Obviously, any design for filtering and power factor improvement requires detailed analysis in close co-operation with the electric utility requiring it.

****

9.12 Pulse width modulation

None of the methods described so far for the improvement of power factor and waveform are ideal, they relieve some problems while leaving others

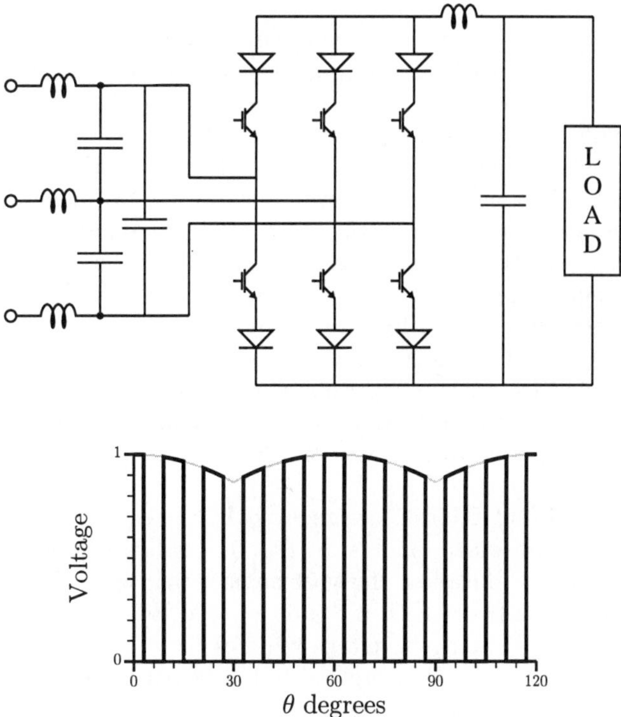

FIG. 9.31. A pulse width modulated rectifier.

unaffected or even worsened. One method, while not commercially practicable at the time of writing (1993), has the potential, in combination with simple filters, to greatly improve all aspects of rectifier operation. The method is pulse width modulation and relies upon rectifying devices which can switch off current as well as switching it on. Such devices are available today, they are the gate turn off thyristor and the power transistor in their various forms. Unfortunately they do not yet have the combination of power handling capability, switching speed, and cost which would make them commercially attractive in this situation. However, the development of improved and new devices proceeds rapidly and pulse width modulated rectifiers might well be the wave of the future.

The essence of pulse width modulated rectification is illustrated by Fig. 9.31, the upper part of which shows a three phase bridge rectifier with a power transistor and diode connected in series in each leg. The diodes prevent reverse conduction through the transistors. The bridge is fed from the three phase system via a low pass filter and feeds the d.c. load via another low pass filter. The lower part of the figure illustrates the control technique. The supply is rapidly connected to and disconnected from the load by the transistors so that voltage is delivered as a series of narrow pulses centered on the peak of each rectifier segment. The output voltage is reduced by making these pulses narrower and is increased by making them wider. Since the pulse pattern remains centered on the supply voltage segments, the fundamental power factor remains at unity. Substantial harmonics are generated by this process and the low pass filters control their amplitudes.

The operation of such a rectifier has much in common with that of the capacitor smoothed rectifier discussed in Chapter 7 and the analysis is correspondingly complex. The output capacitor is charged during a voltage pulse and then supplies the load current until the next pulse. Current is delivered to the capacitor in a series of narrow pulses and the rectifier input current reflects this, being a corresponding series of positive and negative current pulses. The input filters smooth this waveform so that the harmonic content of the currents drawn from the three phase system is acceptably low.

The difficulties of filtering are reduced by increasing the carrier frequency of the pulse width modulator. Fig. 9.23 shows five pulses per rectifier segment of 60° duration, the least number that would be useful. The carrier frequency is therefore $5 \times 6 \times 60 = 1800$ Hz for a 60 Hz system. A reasonable range of output voltage requires that the rectifier switching frequency capability be high compared to the carrier frequency, say ten or more times higher, e.g. at least 18 kHz. This is beyond the capabilities of today's high powered devices but may well become feasible as devices are improved.

The analysis of the system is even more complex than that of the capacitor smoothed rectifier because of the higher switching frequency, the input filter, and the many variables available for manipulation by the system designer. Much research remains to be done to discover optimum modulation and filtering methods.

10

THE RECTIFIER AND ITS LOAD

10.1 Introduction

In earlier chapters the load has been considered to be a constant e.m.f. behind an inductive impedance in order to investigate effects such as commutation, output current ripple, and discontinuous conduction. In this chapter the relationship between the rectifier and its load will be studied in greater depth with particular reference to one of the commonest and most demanding loads, the d.c. motor.

The commonest rectifier loads are lines of electrolytic cells in the electrochemical industry, battery chargers, d.c. motors, and inverters for supplying a.c. power at variable frequency and voltage. The electrolytic refining of metals such as copper and the production by electrolysis of materials such as aluminum, sodium hydroxide, and chlorine consume very large amounts of d.c. power derived from rectifiers but, being relatively static situations with a one way power flow, they are not of much interest in the present context.

The battery, also a significant user of rectifier derived power, is similarly of little interest in the present context. Power flow in it may reverse, requiring the rectifier to change to inverter operation but the load e.m.f. stays within a relatively narrow range. This is also true of pulse width modulated inverter loads which have constant d.c. voltage input and obtain a variable voltage output via pulse width modulation.

In contrast the d.c. motor presents a highly dynamic load whose e.m.f. quickly varies over a wide range and whose power flow can suddenly reverse and equally suddenly revert to the original direction. Feedback control and sophisticated controllers are essential to cope with these conditions and it is in the context of the d.c. motor drive that the majority of this chapter will be developed. A similar situation occurs when the load is a current source inverter or an inverter with a six step output voltage.

Inverter loads range from the very simple, requiring only unidirectional power flow through a diode rectifier with capacitor smoothing, to something as complex as an a.c. machine with rapidly varying, bidirectional power flow.

Considering a d.c. motor drive covers all these eventualities but, before commencing, the phrase *d.c. motor drive* requires some elaboration. By it is meant a separately excited d.c. motor with its armature supplied by one or two rectifiers and its field supplied by one or two separate rectifiers, the whole being monitored and regulated by an electronic controller. Another important class of drives is not considered, the d.c. series motor driving a vehicle and fed by a single phase rectifier which is supplied with single phase a.c. from an overhead wire or third rail. These very powerful single phase drives present particular problems which of themselves deserve book length treatment. Because of this and because they are outside the mainstream of industrial applications they are not considered.

The d.c. motor is typically operated in either one of two modes. In the first its speed is varied from zero to the base speed by variation of its armature voltage with its field current fixed at the normal value. In the second, speed increase above the base value is obtained by weakening the field, the armature voltage remaining fixed. The first range, zero to base speed, is often referred to as the *constant torque range* since the torque producing capability of the machine is constant at its normal value. The second range, above the base speed, is often referred to as the *constant power range* since the torque capability is diminished in proportion to the speed increase and the power capability is constant. The prime concern will be with the constant torque range.

10.2 The basic incompatibility

There is a basic incompatibility between a rectifier and its typical load. By its very nature a rectifier produces unidirectional current and obtains reversal of power flow by reversing its terminal voltage. The typical load, a battery, motor armature, or inverter, in contrast reverses power flow by reversing current flow, its terminal voltage remaining in the same direction. These conditions are usually expressed in terms of the terminal characteristics, mean current as a function of mean voltage. Figure 10.1 shows the axes of such a graph. They divide the plane of the graph into four quadrants as indicated. The rectifier operates in quadrants 1 and 2, positive current, positive or negative voltage. If the operating point is in quadrant 1, power flows through the rectifier from the a.c. to the d.c. side. If the operating point is in quadrant 2, the power flow is inverted from the d.c. to the a.c. side.

A battery operates in quadrants 1 and 4, positive voltage, positive or negative current. A machine armature, if rotation is unidirectional, also operates in quadrants 1 and 4. If the machine is required to rotate in either direction it operates in all four quadrants. It is from these considerations that the terms *two quadrant drive* and *four quadrant drive* arise.

Because of this incompatibility a rectifier and its load can normally operate only in the first quadrant with positive current, positive voltage,

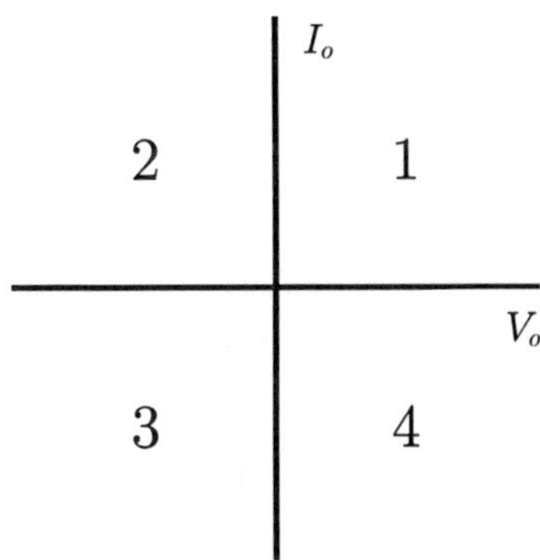

FIG. 10.1. The plane defined by the mean output voltage and mean output current showing the four operating quadrants. Power flow is positive in quadrants 1 and 3 and is negative in quadrants 2 and 4.

and positive power flow. Admittedly a rectifier supplying a d.c. machine armature can theoretically operate in quadrants 1 and 2 but this would need the armature rotation to reverse in order to reverse power flow, a highly unlikely situation.

In some instances the incompatibility may not be of concern. If the rectifier is only required to charge the battery or the machine always rotates in the same direction in the motoring mode there is no problem. This is a common situation for battery chargers but is far from common with d.c. machines. A d.c. drive is expensive and is only installed to meet special requirements such as wide speed variation with fast response. For such applications two or four quadrant operation is essential and means must be found for removing the incompatibility between the rectifier and motor.

10.2.1 *Reconciling the incompatibility*

The incompatibility is removed either by making the load fit the rectifier or the rectifier fit the load. There are two ways of making a d.c. machine armature fit the needs of the rectifier, field reversal and armature reversal as illustrated in Fig. 10.2(a) and 10.2(b). There is one way of making the rectifier fit the armature, the use of two rectifiers connected either as in Fig. 10.2(c) or 10.2(d).

Consider the machine motoring in the forward direction so that its field current is positive, its armature voltage is positive, current is flowing from the rectifier to the armature, and the system operates in the first quadrant. To reverse the power flow and go from motoring to generating, the field current can be reversed. In Fig. 10.2(a) this is shown accomplished by a reversing switch but more often it is done electronically, often by the dual

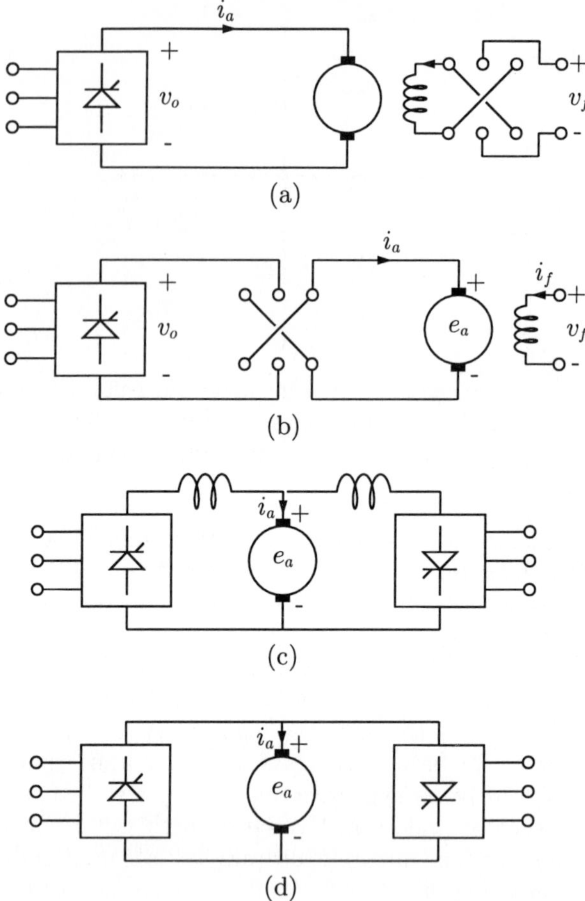

FIG. 10.2. Various ways in which the power flows of the rectifier and motor can be reconciled.

rectifier connection shown in Fig. 10.2(c). However it is accomplished, field reversal causes the armature e.m.f. to reverse and the net armature loop voltage $v_o - e_o$ goes from a small positive value to a large positive value. This would cause a very rapid rise in armature current to dangerously high levels were it not for an essential feature of d.c. drives, the current limit. The drive controller incorporates a current limit which prevents the armature current from significantly exceeding a preset value, typically two to four times the rated value. In this case the current limit automatically phases back the rectifier in concert with the e.m.f. so that there is a smooth transition to the second quadrant.

Field reversal is effective for many applications but is open to two objections. Because the field circuit is highly inductive with a great deal of energy stored in its magnetic field, reversal is comparatively slow, taking a time measured in seconds. Additionally there is a period, when the field current is passing through zero, when the torque capability of the machine is very low and control of the load is lost.

A similar but faster action can be obtained with armature reversal as shown in Fig. 10.2(b). This can be accomplished in times measured in tenths of seconds but it requires a large and expensive reversing contactor capable of handling several times the rated power of the drive. Also, while the time during which there is a loss of control over the load is reduced, it is not eliminated.

For the most demanding requirements dual rectifiers are necessary, Fig. 10.2(c) and 10.2(d). One rectifier supplies positive current to the armature and provides for operation in quadrants 1 and 2. The other supplies negative current to the armature and provides for operation in quadrants 3 and 4. The rectifiers can be operated in one of two modes, with or without circulating current. In the circulating current mode, Fig. 10.2(c), both rectifiers operate continuously, the nonlinearities associated with discontinuous conduction are eliminated and power flow reversal is accomplished without interruption. However, substantial inductors are required to limit the flow of ripple current between the two rectifiers which adds to the size and cost of the equipment and the continual circulation of d.c. current wastes a significant amount of energy.

A more sophisticated controller permits the elimination of the circulating current giving the suppressed circulating current mode illustrated in Fig. 10.2(d). Here only one rectifier operates at a time so that the ripple limiting inductors and the losses associated with the circulating current are eliminated. The price paid for this is a brief hiatus in the changeover from one rectifier to the other. This is analogous to that in the field reversal and armature reversal schemes but is so short, measured in milliseconds, that it is of little consequence. This system predominates today and the version with circulating current is limited to low power applications, such as field control for the field reversal scheme, where the benefits of a simpler controller and a continuous reversal of power flow outweigh the additional losses and the cost of the inductors.

The operation of the most complex of these schemes, the back to back rectifier circuits of Fig. 10.2(c) and 10.2(d), are reviewed in the remainder of this chapter. It has already been noted that a d.c. motor drive is costly, an expense incurred only when the special features of good controllability over a wide speed range and fast response are essential. Until the development of powerful and reliable force commutated inverters of wide frequency range, and competitive price, the d.c. motor drive was the only choice for such applications. Now, (1993), a.c. variable speed, variable frequency drives are competitive with high performance d.c. motor drives and may eventually displace them.

10.3 Closed loop control and current limiting

Closed loop control is necessary for high performance, with automatic current limiting as an essential feature. The d.c. motor is a low impedance device with the armature voltage and e.m.f. almost in exact balance. Only the small difference, of the order of a few per cent of the rated voltage, is necessary to drive the rated current through the armature. Thus, a sudden demand for full speed from rest, or even worse a sudden demand for speed reversal, could easily cause twenty to forty times rated armature current to flow. The machine cannot commutate currents of much more than a few times rated current and such large currents would result in a commutator flashover, follow-up arcing between brush arms, and the probable destruction of the armature. The automatic current limit prevents this from happening.

There are two types of current limit in use, the interventionist system which operates only when the current exceeds a preset value, and the current regulator where the demand for current is continuously regulated according to the demands placed on the drive but with a strict limit placed on the maximum demand. The interventionist system tends to be used for fractional and low horsepower drives while the regulating system tends to be used for drives of substantial power. For the time being, the study is carried out in the context of the conventional analog control system of the current regulator type shown in block diagram form in Fig. 10.3.

10.4 System description

The system will be analyzed in the constant torque mode with the field current held constant at its rated value, I_f, and speed controlled by variation of the rectifier output voltage, v_o.

A dual rectifier without circulating current supplies the armature. The two rectifiers are labelled G1A and G1B in Fig. 10.3 and the controller directs firing pulses to one or the other as required. The rectifiers are assumed to have biased cosine type control so that their static transfer functions in the continuous conduction mode are linear as shown in Fig. 10.4(a). Between the mean output voltage limits V_{omax} and V_{omin} the rectifier has a constant gain G_1. V_{omax} is the largest value of output voltage, given by a delay angle of zero and a control voltage, v_c, of V_{cmax}. V_{omin} is the smallest, i.e. the largest negative, value of output voltage given by the maximum prudent delay angle, α_{max}, corresponding to a control voltage V_{cmin}. For

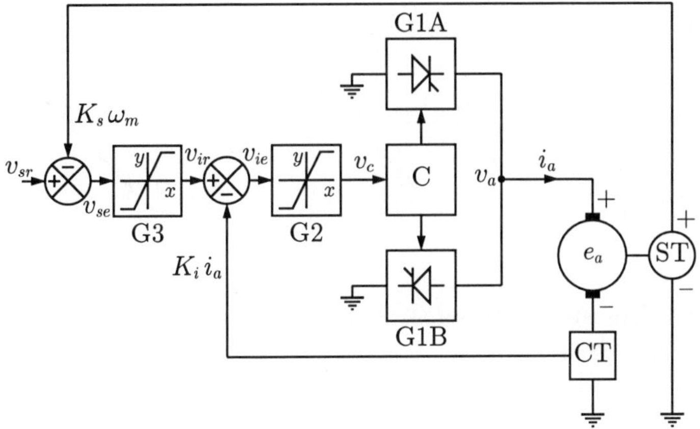

FIG. 10.3. Block diagram of the speed control system showing dual rectifiers, G1A and G1B, maintained in the circulating current free mode by a controller C. The armature current is controlled by the inner feedback loop and the speed is controlled by the outer loop. G2 amplifies the current error signal, v_{ie}, and G3 amplifies the speed error signal, v_{se}. ST is the speed transducer and CT is the current transducer.

control voltages exceeding V_{cmax} the delay angle remains at zero, the rectifier is fully on and the output voltage is constant at V_{omax}. For control voltages less than V_{cmin} the delay angle is held at α_{max} and the output voltage remains constant at V_{omin}. Rectifier dynamics is discussed in the following section and more details of the rectifier gate pulse controllers have been given in Chapter 4.

The rectifier control signal, v_c, is produced by a low power amplifier, G2, whose input is the armature current error signal, v_{ie}. The amplifier output is limited to the range $\pm KV_{cmax}$, K being somewhat greater than unity to ensure that the rectifier can be driven to its limits. The amplifier has the piecewise linear transfer characteristic shown in Fig. 10.4(b). Between the limits $\pm KV_{cmax}$ the gain is constant at G_2. The amplifier response is so fast compared to that of the motor and load that it can be considered infinite so that Fig. 10.4(b) represents the dynamic as well as the static transfer characteristic.

The current error voltage is the difference between the current reference voltage, v_{ir}, and the current feedback voltage, $K_i i_o$, where K_i is the gain in volts per ampere of the current transducer CT.

The current reference signal, v_{ir}, is provided by the amplifier G3 which has proportional plus integral gain with the transfer function $G_{3p}+G_{3i}/s$, s being the complex frequency. The integral component yields a system with

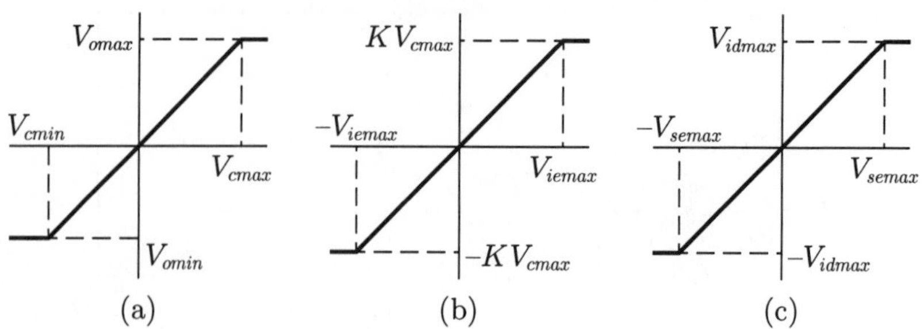

FIG. 10.4. Transfer characteristics of (a) a rectifier, (b) the amplifier G2, and (c) the amplifier G3.

zero steady state speed error, a matter of convenience for later discussions. The input to this amplifier is the speed error voltage, v_{se}, and the amplifier output is limited to the range $\pm V_{irmax}$.

The speed error voltage is the difference between the speed reference voltage, v_{sr}, and the speed feedback voltage, $K_s\omega_m$, where K_s is the gain, in volts per radian per second, of the speed transducer ST.

The a.c. supply feeding the rectifiers is taken to be a perfect three phase a.c. voltage source of sinusoidal waveform.

The armature e.m.f., e_a, is taken to be free of ripple and directly proportional to speed, the constant of proportionality being K_e volts per radian per second.

The electromagnetic torque, T_e, produced by the motor is proportional to the armature current, I_a, the constant of proportionality being K_t Newton meters per ampere. In the mks system of units the numerical values of K_e and K_t are equal.

The motor armature circuit impedance, including any impedance in the circuit which is external to the motor such as smoothing chokes and leads, is taken to be a resistance of R_o ohms in series with an inductance of L_o henries.

The motor is assumed to be directly coupled to the load by a rigid coupling. The total inertia of the complete rotating mass of motor and load combined is J kilogram meters^2. The coefficient of viscous friction is D Newton meters per radian per second, again for the combination of motor and load. The load torque is T_l Newton meters.

This is a simple system taking no account of potentially important matters such as the need to filter the current and speed feedback signals, elasticity of the coupling between the motor and load, and the introduction of differential feedback terms. However, it will well serve to illustrate the present needs.

10.5 Rectifier dynamics

The dynamic behavior of a rectifier is complex, exhibiting some characteristics of a time delay and some of a transportation lag depending on the operating conditions. Once a rectifier thyristor is fired, it continues to conduct until the a.c. supply voltages are such as to reverse bias it and allow it to recover its blocking capability. Under steady state conditions this occurs when the next thyristor is fired. Under transient conditions the situation is more complicated, depending on the change in delay angle and the characteristics of the load. To eliminate the latter effect it will be assumed that the load is an ideal current source which maintains a constant current circulating through the rectifier. This situation, although unlikely in practice, does have the merit of making a complex situation manageable and by the end of this chapter the reader should have no difficulty in determining rectifier behavior under any load conditions.

It is assumed that the width of the gate pulses is slightly greater than 60° so that the rectifier operates correctly when conduction is discontinuous.

Figure 10.5 illustrates the effect of sudden changes in delay angle on the rectifier output voltage. The six a.c. voltage waves corresponding to the rectifier input line to line voltages are shown in gray and are labelled about 30° past their negative peaks by the numbers of the conducting thyristors. Thus, when thyristors 2 and 6 conduct the output voltage segment is the appropriate portion of the wave labelled 26. The output voltage is shown by the heavy black line.

Fig. 10.5(a) shows the rectifier initially operating at a delay of 20°. This is followed by a period when the delay is 50° and finally the delay is reduced to 35°. Thyristor T2 is fired at $\theta = 20°$, this point being indicated by the Marker A above the graph, to effect a commutation from T1. The delay angle of the next thyristor, T4, can be increased to 50° at any time within the period from just after $\theta = 20°$ to just before $\theta = 20° + 60° = 80°$ but the effect of this change is not apparent until just after $\theta = 80°$. At this point, Marker B, T4 has not yet received a gate signal, T2 and T6 continue to conduct and the output voltage decreases along the wave v_{26} until T4 is fired at $\theta = 60° + 50° = 110°$, corresponding to the new delay. This point is indicated by the Marker C. Thus there is a transportation lag depending on when the change of delay angle is implemented. This is followed by a transition in output voltage level over the 30° period from $\theta = 80°$ when T4 would have been fired had the old delay been retained, Marker B, to $\theta = 110°$ corresponding to the new delay angle. The transportation lag varies from 60° if the decision to change the delay angle is implemented just after T2 is fired, Marker A, to zero if the decision is implemented just before the expected firing point of T4 at $\theta = 80°$, Marker B. Thus, the

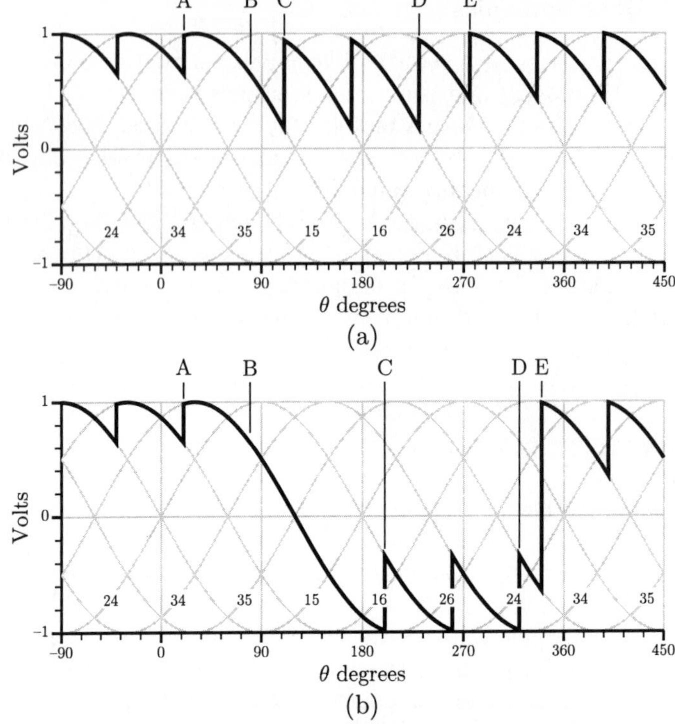

FIG. 10.5. The output voltage waveform of a rectifier when the delay angle is changed. Graph (a) shows a small change in delay angle, from 20° to 50° and then back to 35°. Graph (b) shows a large change in delay angle, from 20° to 140° and then to back to 40°.

duration of the transition varies from a maximum of 90° to a minimum of 30° depending on when the change is implemented. With a 60 Hz supply this corresponds to a total time delay of between 4.17 and 1.39 ms, up to 2.78 ms for the transportation lag and 1.39 ms for the actual change.

The change from a delay of 50° to 35° occurs between firing T5 and T1. T5 is fired at 230° corresponding to a delay of 50°, Marker D, and T1 is fired at $\theta = 275°$ corresponding to a delay of 35°, Marker E. The decision to change the delay can be implemented any time between these two points and takes effect immediately T1 is fired at $\theta = 275°$. Thus, there is a transportation lag which can be a maximum of 45° and a minimum of zero. There is no delay in the change in voltage level which takes effect immediately T1 is fired. Thus, at 60 Hz the time delay can range from 2.08 ms to zero.

These results are generally true for small changes in delay angle. If the delay is increased by $\Delta\alpha$ there is a transportation lag of between zero and 60° followed by a transition period $\Delta\alpha$. If the delay is decreased by $\Delta\alpha$ there is a transportation lag of between zero and $60° - \Delta\alpha$ and the duration of the transition period is zero.

The situation for large delay changes is more complicated and is illustrated by graph (b) of Fig. 10.5 which shows a transition from 20° delay to 140° and back to 40°. The change in delay is implemented some time after firing T2 at $\theta = 20°$, Marker A. The change becomes evident at the expected firing point of the next thyristor, T4, Marker B, and becomes effective when T4 is fired at $\theta = 200°$, Marker C. There is a transportation lag during the period when the delay increase is implemented, some time after firing T2, to the point at which a change becomes apparent, at the expected firing point of T4 at $\theta = 80°$. There is then a transition period from $\theta = 80°$, when the output voltage starts to decrease, to $\theta = 200°$ when T4 is fired and the change is fully implemented.

In contrast the change from 140° delay to 40° is effective immediately it is implemented. In the figure this is done at $\theta = 340°$, Marker E. With gate pulses slightly wider than 60°, gate signals will be sent to thyristors T1, T5, and T6. The largest driving voltage is between T1 and T6 so that it is this pair which picks up the load current, and transition to the new conditions is complete.

In practice large changes in delay will be complicated by the effect of the load as discussed later and the rule for small changes provides a good working basis for the study of this phenomenon. The time delays are small, ranging from zero to about 5 ms depending on the precise conditions under which the change is effected. In the majority of electromechanical and electrothermal systems, this delay is very short compared to other system delays and it is justifiable to neglect this complicated and variable phenomenon, considering the rectifier as a linear amplifier with negligible delay and positive and negative output voltage limits. This will be done in the balance of this chapter.

10.6 The system state equations

Block diagrams of the system under various operating conditions are shown in Fig. 10.6, the diagram for small signal speed control behavior being shown in (a), the behavior when limiting the armature current in (b) and the diagram when the rectifier is at its output voltage limit being shown in (c).

Starting at the right-hand end of Fig. 10.6(a) and working towards the left-hand end, the motor speed, ω_m, is the result of applying the difference between the electromagnetic torque, T_e, and the load torque, T_l, to the mechanical transfer function, G_m. This comprises the viscous friction and

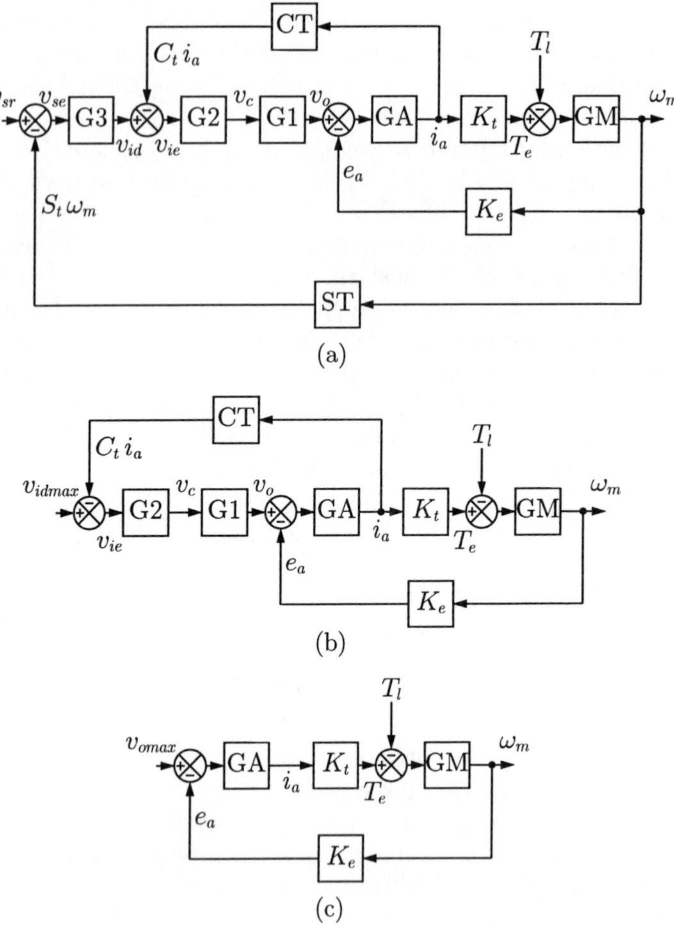

FIG. 10.6. Block diagrams for the speed control system, (a) for small signal operation, (b) when in current limit, and (c) when the rectifier output voltage is at its limit.

inertia and is $G_m = 1/(D+Js)$ where D is the coefficient of viscous friction and J is the inertia of the total rotating mass.

The electromagnetic torque is the product of the armature current, i_a, and the torque constant, K_t.

The armature current is the result of applying the difference between the rectifier output voltage, v_o, and the armature e.m.f., e_a, to the armature circuit transfer function, G_a. This is the series combination of the output circuit resistance, R_o, and inductance, L_o, and is $G_a = 1/(R_o + L_o s)$.

The armature e.m.f. is the product of the motor speed and the e.m.f.

constant, K_e.

The rectifier output voltage is derived from the current error signal, v_{ie}, amplified first by the gain G_2 of amplifier G2 to produce the rectifier control signal, v_c, and then by the rectifier gain, G_1.

The current error signal is the difference between the current demand signal, v_{id}, and the feedback signal, $C_t i_a$, produced by the current transducer.

The current demand signal is the speed error signal, v_{se}, multiplied by the gain, $G_{3p} + G_{3i}/s$, of amplifier G3.

The speed error signal is the difference between the speed reference signal, v_{sr}, and the signal, $S_t\omega_m$, produced by the speed transducer.

The system equations, starting at the output and progressing towards the input, are therefore

$$J\omega_m' + D\omega_m = T_e - T_l \tag{10.1}$$

$$L_o i_a' + R_o i_a = v_o - e_a \tag{10.2}$$

$$e_a = K_e\omega_m \tag{10.3}$$

$$v_o = G_1 G_2 v_{ie} \tag{10.4}$$

$$v_{ie} = v_{id} - C_t i_a \tag{10.5}$$

$$v_{id} = G_{3p} v_{se} + G_{3i}\int v_{se}\,\mathrm{dt} \tag{10.6}$$

$$v_{se} = v_{sr} - S_t\omega_m \tag{10.7}$$

where ω_m' and i_a' are the derivatives with respect to time of ω_m and i_a.

There are three state variables, the motor speed, ω_m, its armature position, θ_m, and its armature current, i_a, and two inputs, the speed reference, v_{sr}, and the load torque, T_l. The state equations are

$$\begin{bmatrix} R_o + G_1G_2K_i & G_1G_2G_{3p}K_s + K_e & G_1G_2G_{3i}K_s \\ -K_t & D & 0 \\ 0 & -1 & 0 \end{bmatrix} \begin{bmatrix} i_a \\ \omega_m \\ \theta_m \end{bmatrix} + \tag{10.8}$$

$$\begin{bmatrix} L_o & 0 & 0 \\ 0 & J & 0 \\ 0 & 0 & 1 \end{bmatrix} \begin{bmatrix} i_a' \\ \omega_m' \\ \theta_m' \end{bmatrix} = \begin{bmatrix} G_1G_2G_{3p} & 0 & 0 \\ 0 & -1 & 0 \\ 0 & 0 & G_1G_2G_{3i} \end{bmatrix} \begin{bmatrix} v_{sr} \\ T_l \\ \int v_{sr}\,\mathrm{dt} \end{bmatrix}$$

Example 10.1

For the case when the values of the system parameters are as listed below and for small signal operation, determine the numerical form of the state equations, the characteristic equation, the system natural frequencies, and expressions for the state variables in the steady state in terms of the inputs.

$$
\begin{aligned}
D &= 0.30 \text{ N m rad}^{-1}\text{ s}^{-1} \\
C_t &= 0.02 \text{ V A}^{-1} \\
G_1 &= 59.42 \\
G_2 &= 10.0 \\
G_{3p} &= 10.0 \\
G_{3i} &= 100.0 \\
J &= 4.40 \text{ kg m}^2 \\
K_e &= 3.24 \text{ V rad}^{-1}\text{ s}^{-1} \\
K_t &= 3.24 \text{ N m A}^{-1} \\
K_{vcmax} &= 12.0 \text{ V} \\
L_o &= 2.71 \text{ mH} \\
R_o &= 0.15\Omega \\
S_t &= 0.055 \text{ V rad}^{-1}\text{ s}^{-1} \\
V_{cmax} &= 10.0 \text{ V} \\
V_{cmin} &= -9.5 \text{ V} \\
V_{idmax} &= 10.0 \text{ V} \\
V_{iemax} &= 1.0 \text{ V} \\
V_{omax} &= 594.2 \text{ V} \\
V_{semax} &= 1.0 \text{ V}
\end{aligned}
$$

These parameters correspond more or less to a drive rated at 150 hp with a base speed of 1750 rpm and with fully controlled bridge rectifiers directly connected to a 440 V, 60 Hz, three phase supply. The rated torque of the motor is 610 N m and the rated current is 190 A. The load inertia is about equal to the motor inertia and the output circuit impedance is about double that of the armature alone.

Inserting the parameter values into the state equations, eqn 10.8, yields

$$
\begin{bmatrix} 12.034 & 330.05 & 3268.1 \\ -3.24 & 0.3 & 0 \\ 0 & -1 & 0 \end{bmatrix}
\begin{bmatrix} i_a \\ \omega_m \\ \theta_m \end{bmatrix}
+
\begin{bmatrix} 0.00271 & 0 & 0 \\ 4.4 & 0 & 0 \\ 0 & 0 & 1 \end{bmatrix}
\begin{bmatrix} i'_a \\ \omega'_m \\ \theta'_m \end{bmatrix}
=
$$

$$
\begin{bmatrix} 5942.0 & 0 & 0 \\ 0 & -1 & 0 \\ 0 & 0 & 59420.0 \end{bmatrix}
\begin{bmatrix} v_{sr} \\ T_l \\ \int v_{sr}\,\mathrm{dt} \end{bmatrix}
$$

The system is of third order with one real eigenvalue at -4420.3 and a complex conjugate pair at $-10.16 \pm j9.89$. The real root corresponds to a time constant of 0.23 ms. The frequency of the complex roots is 1.57 Hz and their time constant is 98.47 ms. Thus, the system is effectively of second order with an almost instantaneous initial response followed by a slower, but still rapid, and slightly oscillatory approach to the steady state.

For the steady state the state equations reduce to

$$\begin{aligned} I_o &= 1.6835\, V_{sr} + 0.3086\, T_l \\ \Omega_m &= 18.18\, V_{sr} \end{aligned}$$

The steady state speed in $\mathrm{rad\,s^{-1}}$ is 18.18 times the speed reference voltage and the application of load torque does not affect the steady state speed because of the integration component of amplifier G3. The steady state armature current is principally dependent on the load torque with a small component due to the speed reference voltage, because the friction is proportional to speed via the coefficient of viscous friction.

****

10.7 System dynamics

The system is of third order with three eigenvalues. While these could be all real, normally two of them form a complex conjugate pair as is the case in the example. Then the transient response, assuming a stable system, is a combination of an exponentially decaying sinusoidal oscillation and a monotonically decaying exponential, the amplitudes of these components depending on the boundary conditions and the system eigenvectors.

Requests for torque changes can generally be accommodated within the confines of the speed control system. In contrast a demand for a significant change in speed will usually result in saturation of one or more of the amplifiers, G1 through G3. Example 10.2 illustrates this point.

Example 10.2

Determine the system response when the speed reference voltage is suddenly increased from 6.00 to 6.02 V. Assume that the system was initially in the steady state.

Using the steady state equations derived in Example 10.1, the steady state armature current and speed corresponding to a speed reference voltage of 6.00 V and a load torque of 400 N m are 133.56 A and 109.09 $\mathrm{rad\,s^{-1}}$. Using eqns 10.1 through 10.7, the corresponding values of other system variables are

Speed error voltage	0.000
Current demand voltage	3.300
Current error voltage	0.629
Rectifier control voltage	6.286
Rectifier output voltage	373.5

The eigenvectors corresponding to the vector of eigenvalues

$$\{-4420.3 \quad -10.16 + j9.89 \quad -10.16 - j9.89\}$$

and the three state variables i_a, ω_m, and θ are

$$\begin{bmatrix} 1.000 & 1.000 & 1.000 \\ 1.67\,10^{-04} & -3.72\,10^{-02} - j3.65\,10^{-02} & -3.72\,10^{-02} + j3.65\,10^{-02} \\ 3.77\,10^{-08} & 8.62\,10^{-05} + j3.68\,10^{-03} & 8.62\,10^{-05} - j3.68\,10^{-03} \end{bmatrix}$$

After scaling these eigenvectors to fit the initial conditions, the time domain expressions for the state variables are

$$\begin{aligned} i_a &= 133.6 - 9.944\exp(-4420.4t) + \\ &\qquad 9.910\cos(9.887t + 0.01015)\exp(-10.16t) \\ \omega_m &= 109.5 + 0.001656\exp(-4420.4t) + \\ &\qquad 0.5167\cos(9.887t - 2.356)\exp(-10.16t) \\ \theta &= -4.841 \times 10^{-04} + 109.5t - 3.747 \times 10^{-07}\exp(-4420.4t) + \\ &\qquad 0.03645\cos(9.887t + 1.558)\exp(-10.16t) \end{aligned}$$

where the constant component of θ has been chosen to make θ zero at the time origin. All other variables of significance can be derived from these three expressions.

The response of the system is illustrated by Fig. 10.7 which shows the changes in nine variables over the 250 ms following the change. Each graph is identified by a number to its right. Graph 1 shows the step change in speed reference and graph 9 shows the effect on the drive speed which increases from the initial steady state value of 109.09 rad s^{-1} with a slight overshoot which peaks at 109.53 rad s^{-1}, 157.5 ms after the change and then settles back to the new steady state value of 109.46 rad s^{-1}.

Graph 2 shows the load torque unchanged at 400 N m and graph 3 shows the speed error signal which immediately increases from zero to 0.02 V, the change in speed reference voltage, and then gradually settles to zero as the drive speed responds to the change.

The current demand voltage, shown in graph 4, increases immediately by the change in speed reference signal multiplied by the proportional gain of G3, from 3.2997 to 3.4997 V and this has the effect of suddenly increasing

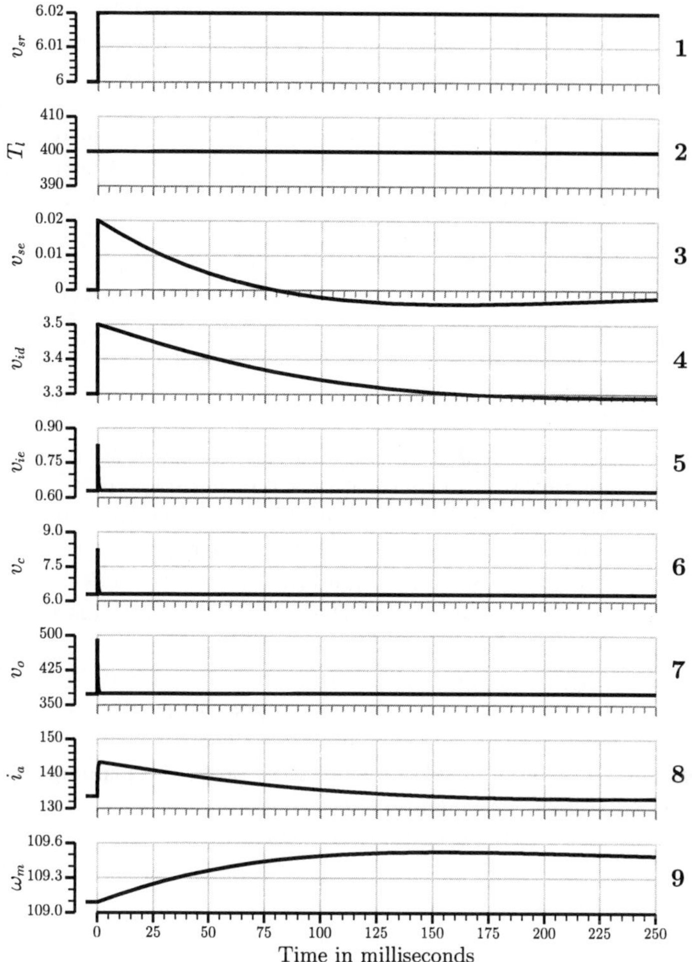

FIG. 10.7. The response of the drive to a sudden increase in speed reference voltage from 6.00 to 6.02 V. The graphs, which are numbered (1) through (9), show as functions of time (1) the speed reference voltage, (2) the load torque, (3) the speed error voltage, (4) the current demand voltage, (5) the current error voltage, (6) the rectifier control voltage, (7) the mean rectifier output voltage, (8) the armature current, and (9) the drive speed.

the current error voltage from 0.6286 to 0.8286 V. After amplification in G2, this results in a sudden increase in rectifier control voltage from 6.286 to 8.286 V, which changes the rectifier delay angle from 51.06° to 34.05° and results in the rectifier mean output voltage increasing from 373.5 to 492.3 V. The sudden increase of 118.8 V in armature driving voltage appears across the output circuit inductance and causes the armature current to increase at 43.85 $\mathrm{A\,ms^{-1}}$, a rate which quickly raises the output current to its maximum value of 143.3 A in 1.373 ms. The rapid increase in current results in an equally rapid increase in the current feedback signal and the current error voltage is thereby reduced to a small value. This very brief excursion of the current error signal and the corresponding excursions of the rectifier control voltage and output voltage are illustrated in graphs 5, 6, and 7.

As the drive accelerates to the new speed, the excess current which produces the acceleration gradually diminishes to the new steady state value of 133.6 A.

Although the accuracy of the results during the first few milliseconds of the transient is doubtful because of the discrete nature of rectifier control discussed earlier, the figure does illustrate the essentials of drive behavior. The inner current loop has a very small open loop time constant which is made even shorter by the very large amount of overdrive that can be produced by the rectifier. Because of this the system responds with great rapidity to even small changes. Any substantial change in speed demand will result first in the rectifier being driven to its voltage limit, producing a very rapid rise in current to the limiting value. This brings the rectifier back under control and the drive accelerates in the current limit mode until it is very close to the new speed. This behavior provides optimum response in that the acceleration is maintained throughout most of the change at the maximum value that the motor can sustain.

****

10.8 Large signal operation

Example 10.2 illustrates how a request for a very small change in speed takes the system close to its limits. In that case the request was for a speed increase of 0.3636 $\mathrm{rad\,s^{-1}}$, only about 0.2% of the base speed, and caused the mean output voltage of the rectifier to briefly reach 494 V. The majority of requests for speed alteration call for changes far greater than this minor adjustment and will result in the amplifiers reaching their ceiling voltages. Typically, the rectifier output voltage goes to its ceiling value and results in a very rapid increase in armature current. During this stage the system operates in the open loop mode shown in Fig. 10.6(c).

Operation in the open loop mode is brief since the rapidly rising armature current, fed back to the input of G2 via the current transducer CT, quickly reduces the current error signal, bringing amplifier G2 out of saturation and the rectifier under control. This puts the system in the current regulating mode of Fig. 10.6(b), and the drive accelerates with the armature current and electromagnetic torque more or less constant at the current limit setting.

The current regulating mode ends when the rising speed, fed back to the input of amplifier G3 via the speed transducer ST, is sufficiently close to the demand value to bring this amplifier out of saturation and the system reverts to the speed control mode, Fig. 10.6(a).

These three stages are illustrated by Example 10.3 in which the speed reference voltage is suddenly increased from 6 to 9 V corresponding to a speed increase from 60% to 90% of the base speed.

Example 10.3

Repeat the computations of Example 10.2 for the case when the speed reference voltage is suddenly increased from 6 to 9 V with the load torque remaining unchanged at 400 N m. Assume that the system is initially in the steady state.

The sudden increase in the speed reference voltage results in a sudden increase of the speed error voltage from zero to 3.0 V. When amplified by the proportional component of the speed error amplifier this would give, if the amplifier remained linear, a sudden increase in the current demand voltage of 30.0 V, from the value of 3.3 V at $t = 0-$ to 33.3 V at $t = 0+$. However, because the output of G3 is limited, the current demand voltage only rises to the limit value, 10.0 V. The inductance of the armature prevents a sudden change in armature current which is therefore still 133.56 A at $t = 0+$. Hence the current error voltage immediately increases from 0.6286 to 7.3288 V, sufficient to drive the output of amplifier G2 to its limit of 12 V. This reduces the rectifier delay angle to zero and produces the maximum mean rectifier output voltage, 594.2 V, placing the system in the open loop mode, shown in Fig. 10.6(c).

The sudden increase of 220.7 V in the load circuit voltage forces the armature current to increase rapidly, initially at $220.7 \div 2.71 = 81.44\ \mathrm{A\,ms^{-1}}$, and the build up of current results in an equally rapid reduction in current error voltage until, at $t = 4.389$ ms, it is down to 1.0 V, the rectifier delay angle is about to be increased, and the open loop mode terminates.

The state equations during the open loop mode are

$$\begin{bmatrix} 0.000271 & 0.0 \\ 0.0 & 0.3 \end{bmatrix} \begin{bmatrix} i'_a \\ \omega'_m \end{bmatrix} + \begin{bmatrix} 0.15 & 3.24 \\ -3.24 & 0.3 \end{bmatrix} \begin{bmatrix} i_a \\ \omega_m \end{bmatrix} = \begin{bmatrix} 594.2 \\ -400.0 \end{bmatrix}$$

This second order system is underdamped and has a pair of complex eigenvalues

$$\{-27.709 + j10.786 \quad -27.709 - j10.786\}$$

The corresponding matrix of eigenvectors is

$$\begin{bmatrix} 1.0 + j0.0 & 1.0 + j0.0 \\ -0.02312 - j0.009022 & -0.02312 + j0.009022 \end{bmatrix}$$

For continuity of current and speed at the time origin, the time domain solution is

$$\begin{aligned} i_a &= 139.84 + 7534.63\cos(10.786t - 1.5716)\exp(-27.709t) \\ \omega_m &= 176.92 + 186.99\cos(10.786t + 1.9420)\exp(-27.709t) \end{aligned}$$

The output voltage of the amplifier G2 becomes effective in controlling the rectifier when the mean armature current has risen to 450.0 A. Substituting this value in the armature current equation just given and solving for t puts the termination of the open loop mode at 4.389 ms after the change in speed reference voltage. The current regulating mode begins at this point.

The extent of the open loop mode is indicated on Fig. 10.8 by delimitation lines between graphs 8 and 9.

The state equations for the current regulating mode, whose circuit is shown in Fig. 10.6(b), are

$$\begin{bmatrix} 0.000271 & 0.0 \\ 0.0 & 4.4 \end{bmatrix} \begin{bmatrix} i'_a \\ \omega'_m \end{bmatrix} + \begin{bmatrix} 12.034 & 3.24 \\ -3.24 & 0.3 \end{bmatrix} \begin{bmatrix} i_a \\ \omega_m \end{bmatrix} = \begin{bmatrix} 594.2 \\ -400.0 \end{bmatrix}$$

This second order system has two real eigenvalues

$$\{-4440.39 \quad -0.26645\}$$

The corresponding matrix of eigenvectors is

$$\begin{bmatrix} 1.0 & 0.26925 \\ -1.6584\,10^{-4} & -1.0 \end{bmatrix}$$

For continuity of current and speed at the boundary between this mode and the open loop mode which precedes it, the time domain solution is

$$\begin{aligned} i_a &= 218.22 - 14.268\exp[-4440.39(t - 0.004389)] + \\ &\quad 246.05\exp[-0.26645(t - 0.004389)] \\ \omega_m &= 1023.4 + 0.002366\exp[-4440.39(t - 0.004389)] - \\ &\quad 913.82\exp(-0.26645(t - 0.004389)) \end{aligned}$$

The output of amplifier G3 is limited to 10.0 V and its proportional gain is 10.0 so that this amplifier is about to come out of its limit condition when the speed error voltage has fallen to 1.0 V, i.e. when the speed feedback voltage has risen to 8.0 V corresponding to a speed of 145.46 rad s^{-1}. Substitution of this value of speed into the speed equation just given and solving for t places the termination of the current limit mode and the commencement of the speed control mode at 154.519 ms.

The extent of the current regulating mode is indicated in Fig. 10.8 by delimitation lines between graphs 8 and 9.

The emergence of amplifier G3 from its limit condition puts the system back into the speed control mode whose circuit is shown in Fig. 10.6(a). The state equations, eigenvalues, and eigenvectors of this mode have been given in Example 10.2. Again, continuity of current and speed at the transition point define the initial conditions for this new mode and, as in Example 10.2, the initial rate of change of current rather than the initial value of armature position must be considered in solving for the initial conditions. The time domain equations for the state variables are then

$$\begin{aligned} i_a &= 138.61 - 0.3199\exp[-4420.35(t-0.1545)] + \\ &\quad 361.76\cos[9.8870(t-0.1545) - 0.5065]\exp[-10.156(t-0.1545)] \\ \omega_m &= 163.64 - 0.0000533\exp[-4420.35(t-0.1545)] + \\ &\quad 18.86\cos[9.8870(t-0.1545) - 0.5065]\exp[-10.156(t-0.1545)] \\ \theta &= -0.6726 + 163.64(t-0.1545) - 0.3199\exp[-4420.35(t-0.1545)] + \\ &\quad 361.76\cos[9.8870(t-0.1545) - 0.5065]\exp[-10.156(t-0.1545)] \end{aligned}$$

The response of the system is illustrated by Fig. 10.8 which shows the changes in nine variables over the 500 ms following the change. Each graph is identified by a number to its right. The extents of the three modes are indicated between graphs 8 and 9. Graph 1 shows the step change in speed reference and graph 9 shows the effect on the drive speed, which increases from the initial steady state value of 109.09 rad s^{-1} with a small overshoot that peaks at 165.20 rad s^{-1} at 360 ms and then settles back to the new steady state value of 163.63 rad s^{-1}.

Graph 2 shows the load torque unchanged at 400 N m and graph 3 shows the speed error signal which immediately increases from zero to 3.0 V, the change in speed reference voltage, and then gradually settles to zero as the drive speed responds to the change.

The current demand voltage, shown in graph 4, increases immediately from its initial steady state value of 3.3 V to its limit of 10.0 V and remains at this value until the termination of the current regulating mode at $t =$ 154.5 ms after which it steadily decreases as the speed approaches the new steady state value of 3.699 V.

The current error voltage, shown in graph 5, increases immediately from its initial steady state value of 0.629 to 7.329 V and then decreases rapidly

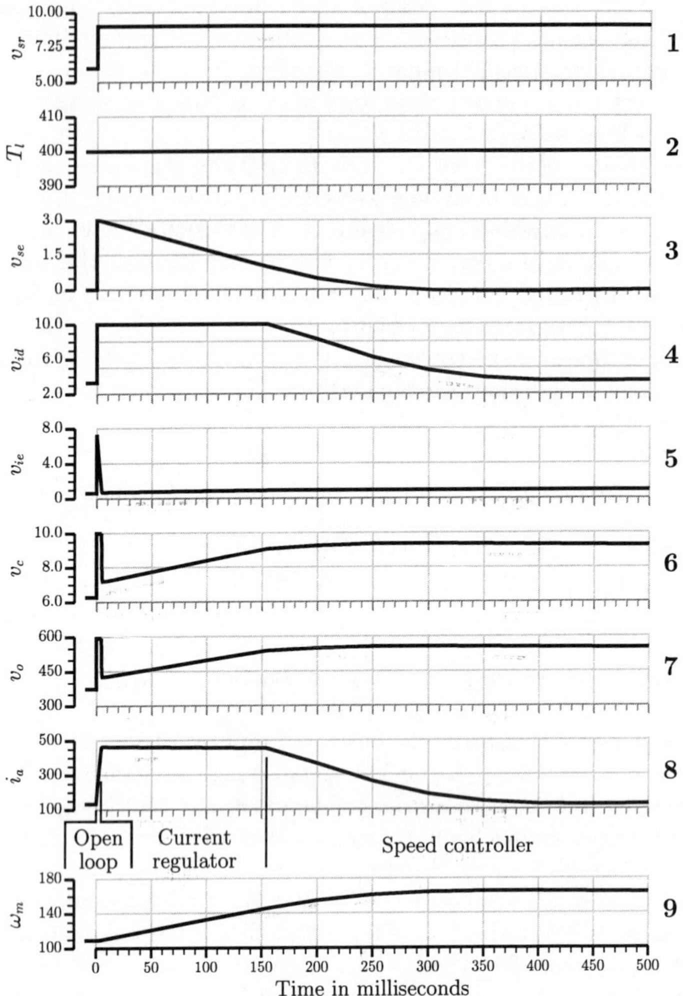

FIG. 10.8. The response of the drive to a sudden increase in speed reference voltage from 6.00 to 9.00 V. The graphs, which are numbered 1 through 9 show as functions of time (1) the speed reference voltage, (2) the load torque, (3) the speed error voltage, (4) the current demand voltage, (5) the current error voltage, (6) the rectifier control voltage, (7) the mean rectifier output voltage, (8) the armature current, and (9) the drive speed. The extent of the three modes which are operational during the transient is indicated between graphs 8 and 9.

as the armature current builds up during the open loop mode. At the end of this mode, when $t = 4.39$ ms, the current error voltage has decreased to 1.0 V. Further changes in this voltage are too small to be visible in graph 5 but during the current regulating mode it decreases to a minimum of 0.717 V at $t = 6.519$ ms and then increases to 0.908 V at the end of this mode. During the subsequent speed control mode v_{ie} increases to a maximum of 0.937 V at $t = 300$ ms and then decreases to its final steady state value of 0.927 V.

The rectifier control voltage, shown in graph 6, immediately rises to its maximum effective value, 10 V, and stays at this value until it comes back under control at the end of the open loop mode. Thereafter it mirrors the changes in the current error voltage multiplied by ten, the gain of amplifier G2. Thus, during the current regulating mode it decreases to a minimum of 7.17 V at 6.519 ms and then rises to 9.07 V at the end of the current regulating mode. In the speed control mode the voltage rises to a maximum of 9.37 V at 300 ms before settling back to its new steady state value of 9.27 V.

The mean rectifier output voltage, shown in graph 7, mirrors the rectifier control voltage multiplied by the rectifier gain, 59.42.

Initially the armature current, shown in graph 8, increases rapidly until the end of the open loop mode when it has reached 450.0 A. The current is held very close to this value throughout the current regulating mode, reaching a maximum of 464.1 A at 6.52 ms and then falling to 454.6 A at the end of this mode. This is of course the initial value for the speed control mode during which the current decreases steadily towards its final steady state value of 138.6 A.

Again the warning issued at this point in Example 10.2 is in order. Not much faith can be put in the precise results during the first few milliseconds because of the discrete nature of rectifier control discussed earlier. However, Fig. 10.8 does illustrate the essentials of drive large signal behavior.

****

10.9 The rectifier steady state transfer characteristic

The analysis given so far assumes continuous conduction and the illustrative examples have been chosen to provide this condition. However, if the response to a demand for speed reduction had been studied it would have been found that rectifier conduction becomes discontinuous, a condition in which the rectifier transfer function becomes nonlinear. A complete analysis including the effects of discontinuous conduction is too complex to be included here but the rectifier steady state transfer characteristic, the mean output voltage as a function of control voltage, is a useful tool for

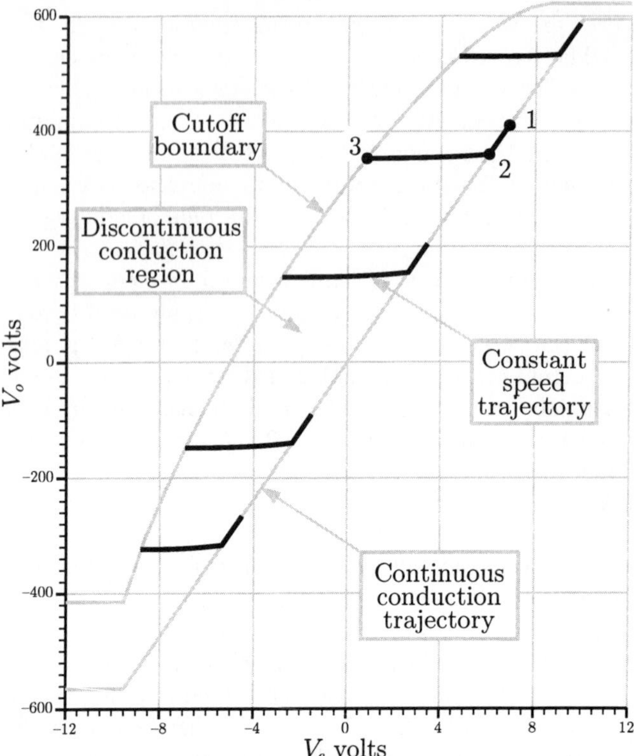

FIG. 10.9. The steady state transfer characteristics of a rectifier. The continuous conduction characteristic provides the right-hand bound. The cutoff characteristic provides the left-hand bound. Five constant speed trajectories are shown by black lines. The numerical values are for the example and the constant speed trajectories are for speed reference voltages of 9, 6, 2.5, –2.5, and –5.5 V.

obtaining an idea of the system behavior when conduction is discontinuous. The transfer characteristic for the rectifier of the example is shown in Fig. 10.9. Typically this characteristic shows three types of trajectory. The first is the continuous conduction characteristic, the three element, piecewise linear characteristic of Fig. 10.4(a) reproduced in Fig. 10.9.

The second type of trajectory is the cutoff characteristic. This shows the mean output voltage at cutoff as a function of control voltage. It is the curved line which lies above and to the left of the continuous conduction characteristic of Fig. 10.9. An operating point between the two characteristics corresponds to discontinuous conduction. The rectifier is cutoff for operating points on or above and to the left of the cutoff characteristic.

An operating point below and to the right of the continuous conduction characteristic is not possible except on a transient basis. Discontinuous conduction was studied extensively in Chapter 6.

The third type of path normally shown on the transfer characteristic is a constant speed trajectory. This shows the relationship between mean output voltage and control voltage for constant speed operation with load varying from a value at which continuous conduction is assured, down to zero. Such a characteristic illustrates the steady state behavior of the speed control system with proportional plus integral control of Examples 10.1, 10.2, and 10.3 for a fixed value of speed reference voltage. It approximates the behavior of a system with only proportional control. The numerical values of Fig. 10.9 are for the example. Five constant speed trajectories are shown by black lines. They are for speed reference voltages of 9.0, 6.0, 2.5, –2.5, and –5.5 V.

A constant speed trajectory starts on the continuous conduction characteristic at some value of rectifier mean output current in the continuous conduction range and progresses down the continuous conduction characteristic until the critical point is reached and conduction is just about to become discontinuous. Further diminution in load takes the trajectory from the continuous conduction characteristic to the cutoff characteristic where it terminates at zero load. Thus, crucial points on the trajectory are the first continuous conduction point, the critical point, and the cutoff point. For Fig. 10.9 the first point corresponds to twice the rated current, 380 A. The computations for the discontinuous conduction portion have been described in Chapter 6.

10.9.1 *The rated load point*

A continuous conduction point is easily found. At the specified speed, ω_m, the motor e.m.f. is

$$E_o = K_e \, \omega_m \tag{10.9}$$

and when the armature current is I_a the rectifier output voltage is

$$V_o = E_o + R_o I_o \tag{10.10}$$

The control voltage is

$$V_c = \frac{V_o}{G_1} \tag{10.11}$$

The electromagnetic torque is

$$T_e = K_t I_o \tag{10.12}$$

and the load torque is

$$T_l = T_e - D \, \omega_m \tag{10.13}$$

10.9.2 *The cutoff point*

The cutoff point is almost as easily defined. Equations 6.27 and 6.28 define, for a three phase bridge rectifier, the relationship at cutoff between the mean rectifier terminal voltage and the delay angle. At cutoff the rectifier terminal voltage is the armature e.m.f. and, since the speed is constant, this is given by eqn 10.9. Substituting this in eqn 6.28 and rearranging defines the delay angle at cutoff when the delay exceeds 30°

$$\alpha_{co} = \arccos(\frac{E_o}{\sqrt{2}V_{st}}) + \frac{\pi}{6} \tag{10.14}$$

where V_{st} is the r.m.s. value of the rectifier line to line input voltage. The corresponding control voltage is then given by

$$V_c = V_{cmax}\cos(\alpha_{co}) \tag{10.15}$$

10.9.3 *The critical point*

The critical point marking the boundary between continuous and discontinuous conduction is a little more difficult to determine. The problem is slightly different from that in Chapter 6 where it was assumed that the delay angle was known and that the critical e.m.f. was to be found. Here, the e.m.f. is known and the critical delay angle is required.

Starting with eqn 6.1, the values of I_8, ϕ_{c8}, and I_9 are known and the values of α and I_{10} are not known. These are found using the facts that, at the critical point, i_o is zero at $\theta = \alpha + \pi/3$ and at $\theta = \alpha + 2\pi/3$. Thus

$$I_8\cos(\alpha + \frac{\pi}{3} + \phi_{c8}) - I_9 + I_{10}\exp[-\frac{\alpha + \pi/3}{\tan(\zeta_3)}] = 0 \tag{10.16}$$

and

$$I_8\cos(\alpha + \frac{2\pi}{3} + \phi_{c8}) - I_9 + I_{10}\exp[-\frac{\alpha + 2\pi/3}{\tan(\zeta_3)}] = 0 \tag{10.17}$$

From eqn 10.17

$$I_{10}\exp[-\frac{\alpha + \pi/3}{\tan(\zeta_3)}] = -[I_8\cos(\alpha + \frac{2\pi}{3} + \phi_{c8}) - I_9]E_{xp} \tag{10.18}$$

where

$$E_{xp} = \exp[\frac{\pi/3}{\tan(\zeta_3)}]$$

Substituting this expression in eqn 10.16 yields

$$I_8[E_{xp}\cos(\alpha + \frac{2\pi}{3} + \phi_{c8}) - \cos(\alpha + \frac{\pi}{3} + \phi_{c8})] = I_9(E_{xp} - 1) \qquad (10.19)$$

This can be written as

$$A\cos(\alpha) - B\sin(\alpha) = C \qquad (10.20)$$

where

$$A = E_{xp}\cos(\frac{2\pi}{3} + \phi_{c8}) - \cos(\frac{\pi}{3} + \phi_{c8}) \qquad (10.21)$$

$$B = E_{xp}\sin(\frac{2\pi}{3} + \phi_{c8}) - \sin(\frac{\pi}{3} + \phi_{c8}) \qquad (10.22)$$

$$C = \frac{I_9(E_{xp} - 1)}{I_8} \qquad (10.23)$$

Thus

$$\alpha = \arccos(\frac{C}{\sqrt{A^2 + B^2}}) - \arctan(\frac{B}{A}) \qquad (10.24)$$

The critical value of control voltage is then

$$V_{ccrit} = V_{cmax}\cos(\alpha) \qquad (10.25)$$

The mean output is

$$V_{ocrit} = V_{ccrit}G_1 \qquad (10.26)$$

and the mean value of the output current is given by

$$I_{ocrit} = \frac{V_{ocrit} - E_o}{R_o} \qquad (10.27)$$

Intermediate points between the critical and the cutoff points are obtained using the techniques of Chapter 6.

Example 10.4

Determine the steady state transfer functions for the example, showing constant speed trajectories corresponding to speed reference voltages of 9.0, 6.0, 2.5, –2.5, and –5.5 V. Start the constant speed trajectories at twice the rated armature current, 380 A.

The complete transfer function is shown in Fig. 10.9. The continuous conduction characteristic is defined by four points. The rightmost point is for a control voltage of 12 V at which the mean rectifier output voltage has its maximum value of 594.2 V. The next point is that at which the rectifier just begins to come under control, $V_c = 10.0$ V and $V_o = 594.2$ V.

The third point is for the minimum controllable output voltage, $V_c = -9.5$ V, $V_o = -564.5$ V. The final point is when the control voltage has been decreased to its negative limit at $V_c = -12.0$ V at which point the mean output voltage is $V_o = -564.5$ V.

On the cutoff characteristic the output voltage remains at its maximum value of $\sqrt{2} \times 440 = 622.3$ V for control voltages in the range 12.0 down to 8.66 V, the latter being the value at which the delay angle is 30°. As the control voltage is reduced below 8.66 V the output voltage follows a path defined by eqns 10.14 and 10.15 until the delay angle reaches its maximum limit value of 161.8° corresponding to $V_c = -9.5$ V and $V_o = -414.8$ V. Further decrease in V_c to the minimum permitted value of –12.0 V leaves the delay angle and output voltage unchanged.

Computations for the five constant speed trajectories are similar so that only one need be considered here, the one for a speed reference voltage of 6.0 V corresponding to a steady state speed of 109.1 $\mathrm{rad\,s^{-1}}$ and a load e.m.f. of 353.5 V. Significant points on this characteristic are indicated and labelled 1, 2, and 3 in Fig. 10.9. Point 1 is determined using eqns 10.9 through 10.13. At this point the mean output current is 380 A, $V_c = 6.91$ V, $\alpha = 46.31°$, and $V_o = 410.5$ V.

Point 2 is the critical point and is determined using eqns 10.16 through 10.27. At this point $V_c = 6.057$ V, $\alpha = 52.72°$, $V_o = 359.9$ V, and $I_o = 43.12$ A.

Point 3 is the cutoff point determined using eqns 10.14 and 10.15. At this point the mean output current is zero and $V_c = 0.804$, $\alpha = 85.39°$, and $V_o = 353.5$ V.

****

Example 10.4 shows that, for a given change in load, far greater changes in V_c are required to control the rectifier in the discontinuous conduction mode than in the continuous conduction mode. For the 109.09 $\mathrm{rad\,s^{-1}}$ constant speed trajectory of the example, moving from location 3 to location 2, which corresponds to a modest increase of mean armature current from zero to 43.12 A, requires an increase of 5.25 V in control voltage, from 0.804 to 6.057 V, while the change in armature current of 336.9 A, from 43.12 A at location 2 to 380 A at location 1, only requires an increase in V_c of 0.851 V, from 6.057 to 6.908 V. In effect the rectifier becomes much harder to control when conduction is discontinuous. Detailed analysis of the dynamics of this problem requires extensive numerical analysis which is inappropriate here. However, some idea of its effect on system dynamics can be obtained by considering system perturbations which are small enough to keep the system within the discontinuous conduction mode. The system can then be considered to be linear about the operating point.

10.10 Small signal linearization

For small perturbations about a given operating point, the rectifier gain can be considered to be the slope, $\mathrm{d}V_o/\mathrm{d}V_c$, of the constant speed trajectory. In the continuous conduction region this is constant at 59.42. In the discontinuous region it is greatly diminished and is variable. The required slope is obtained through the intermediary of a series of other parameters. Thus, from eqn 10.15

$$\frac{\mathrm{d}V_c}{\mathrm{d}\alpha} = -V_{cmax}\sin(\alpha) \tag{10.28}$$

From eqn 10.10, since E_o is constant for a given characteristic,

$$\frac{\mathrm{d}V_o}{\mathrm{d}I_o} = R_o \tag{10.29}$$

Assuming discontinuous conduction of the Mode 2 type, the mean output current I_o is given by eqn 6.57 with the initiation angle δ set to $\alpha + \pi/3$. In this expression I_8 and I_{13} are independent of α, the only variables dependant on α being γ, δ, and I_{14}. Differentiation with respect to α therefore yields

$$\begin{aligned}\frac{\mathrm{d}I_o}{\mathrm{d}\alpha} = {} & \frac{3}{\pi}\left[I_8\cos(\gamma+\phi_{c8}) - I_{13} + I_{14}\exp(-\frac{\gamma}{\tan(\zeta_3)})\right]\frac{\mathrm{d}\gamma}{\mathrm{d}\alpha} - \\ & \frac{3}{\pi}\left[I_8\cos(\delta+\phi_{c8}) - I_{13} + I_{14}\exp(-\frac{\delta}{\tan(\zeta_3)})\right] - \\ & \frac{3}{\pi}\tan(\zeta_3)\left[\exp(-\frac{\gamma}{\tan(\zeta_3)}) - \exp(-\frac{\delta}{\tan(\zeta_3)})\right]\frac{\mathrm{d}I_{14}}{\mathrm{d}\alpha}\end{aligned}$$

The value of the first square bracketed term is the output current at the extinction angle, a point at which the current is zero. The value of the second square bracketed term is the output current at the initiation angle, a point at which the current is also zero. Thus, the derivative is given by the third term alone, i.e.

$$\frac{\mathrm{d}I_o}{\mathrm{d}\alpha} = -\frac{3}{\pi}\tan(\zeta_3)\left[\exp(-\frac{\gamma}{\tan(\zeta_3)}) - \exp(-\frac{\delta}{\tan(\zeta_3)})\right]\frac{\mathrm{d}I_{14}}{\mathrm{d}\alpha} \tag{10.30}$$

From eqn 6.39

$$\frac{\mathrm{d}I_{14}}{\mathrm{d}\alpha} = \left\{I_8\left[\sin(\delta+\phi_{c8}) - \frac{\cos(\delta+\phi_{c8})}{\tan(\zeta_3)}\right] + \frac{I_{13}}{\tan(\zeta_3)}\right\}\exp(\frac{\delta}{\tan(\zeta_3)}) \tag{10.31}$$

Thus, $\mathrm{d}V_o/\mathrm{d}\alpha$ is given by the product of eqns 10.29, 10.30, and 10.31. This quantity is then divided by eqn 10.28 to obtain the required incremental gain, $\mathrm{d}V_o/\mathrm{d}V_c$.

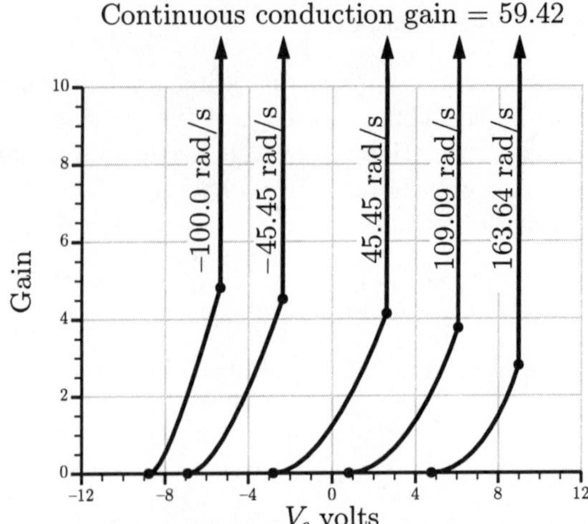

FIG. 10.10. The small perturbation gain of the rectifier as a function of its control voltage for the five constant speed trajectories of Fig. 10.9.

The incremental gain for the five constant speed trajectories of Fig. 10.9 is shown as a function of V_c in Fig. 10.10. The incremental gain is zero at the cutoff point, marked with a dot, and increases to a value of about 4 at the critical point. At this point, also marked with a dot, as the system passes from discontinuous to continuous conduction the incremental gain suddenly increases by a factor of about 15 to the continuous conduction value of 59.42, a value which is far above the upper edge of the figure. The order of magnitude of the change in gain, tenfold, is typical of all rectifiers.

The effect of this large reduction in gain is not as great as might be expected because of the closed loop control, a point illustrated by Example 10.5.

Example 10.5

Determine the eigenvalues of the speed control system for small perturbations about the following operating point, which is in the discontinuous conduction region about midway between points 2 and 3 of Fig. 10.9.

$$\omega_m = 109.09 \text{ rad s}^{-1} \qquad I_o = 6.377 \text{ A}$$
$$E_o = 353.5 \text{ V} \qquad V_c = 3.575 \text{ V}$$
$$V_o = 354.4 \text{ V} \qquad \alpha = 69.05°$$
$$\delta = 137.13° \qquad \gamma = 174.03°$$

At this point the incremental gain, $\mathrm{d}V_o/\mathrm{d}V_c$, is 0.956 and for small perturbations about it the eigenvalues are -105.3 and $-10.37 \pm j5.319$. The real eigenvalue has decreased substantially from the continuous conduction value of 4420.4 but still represents a rapid response, time constant 9.5 ms. The complex eigenvalues represent an oscillatory component with a frequency of 5.319 rad s^{-1} decaying with a time constant of 96.4 ms, not very much different from the continuous conduction values. The response to a small perturbation will be very similar to that shown in Fig. 10.7 but with the initial very rapid rise of current slowed to take about 20 ms. The steady state equations remain unchanged because of the integrator in the amplifier G3.

****

10.11 Large signal behavior

Example 10.2 illustrates that any speed change of significance will take the drive into a large signal mode, where some of the amplifiers will reach their output limits and the system configuration will change from the speed control to the current regulating or open loop modes. The study of the system is complicated not only by this mode switching but also by the ripple in the rectifier output voltage and current, and by discontinuous conduction. A complete study including all these matters is too complex to cover here. Instead the problem is approached via the steady state behavior which, while quantitatively in error under rapidly changing conditions, nevertheless provides a qualitative understanding of the processes involved and of the system performance.

10.11.1 *Speed increase*

Initially the drive is assumed to be operating at location 1 on the steady state transfer characteristic of Fig. 10.11 when the speed reference voltage is 6.0 V and the load torque is 400 N m. Under these conditions the speed is 109.1 rad s^{-1} and conduction is continuous with a mean armature current of 133.6 A. The speed error voltage is zero, the current demand voltage is 3.300 V, the current error voltage is 0.6286 V, the rectifier control voltage is 6.286 V, and the rectifier mean output voltage is 373.5 V.

Now suppose that the speed reference signal is suddenly increased to 9.0 V with the load torque unchanged. The new steady state operating point is location 2 in Fig. 10.11 at a speed of 163.6 rad s^{-1} and mean armature current of 138.6 A. The speed error voltage is zero, the current demand voltage is 3.699 V, the current error voltage is 0.9273 V, the rectifier control voltage is 9.273 V and the mean rectifier output voltage is 551.0 V. The transition from condition 1 to condition 2 has been studied in Example 10.3 and illustrated in Fig. 10.8. Conduction is continuous throughout the

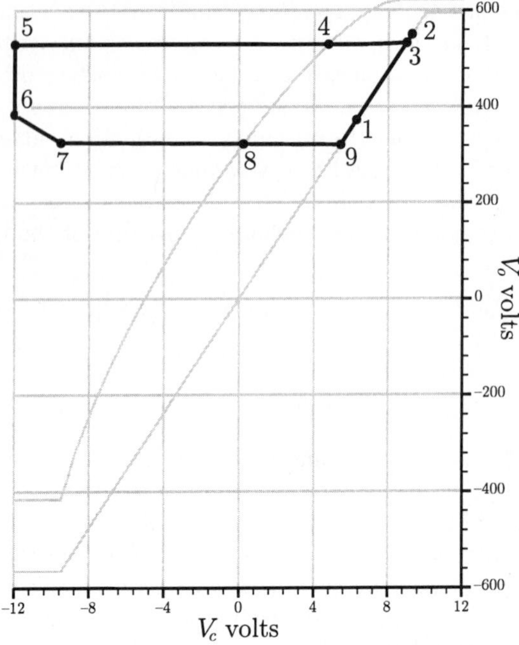

FIG. 10.11. The steady state transfer function showing, by heavy black lines, trajectories associated with a speed increase from point 1 to 2 and a corresponding speed decrease from point 2 back to 1 and passing via points 3 through 9.

transition. For a very brief initial period all amplifiers are at their positive limits, the rectifier is fully on, and the armature current increases very rapidly until it comes under the control of the current limit 4.4 ms after the step change at a value of 457.0 A. The drive accelerates in the current regulating mode for 150.1 ms with the current almost constant at about 450 A. Transition to the speed control mode occurs 154.5 ms after the initiation of the speed change when amplifier G3 comes off its 10.0 V positive limit and the speed is within 0.5% of its ultimate value 322 ms after the initiation of the change. The system moves up the continuous conduction characteristic from point 1 to 2 as indicated by the thick black line of Fig. 10.11.

An alternative simplified view of the change assumes that the system is in the current limit mode throughout the change with the current at 457 A producing an electromagnetic torque of 1481 N m. The mean speed is 136.4 rad s^{-1} so that the mean frictional torque is 40.9 N m. Combining the electromagnetic accelerating torque with the retarding load and friction torques gives a net accelerating torque of 1040 N m which, when divided by the drive inertia of 4.4 kg m^2, provides a mean acceleration of 236 rad s^{-2} for a transition time of 231 ms. In this time the drive speed has actually reached 160.0 rad s^{-1}, within 3.6 rad s^{-1} of its ultimate value.

Any substantial speed increase follows this pattern. There is a very brief period when the current builds up to the limit value, a much longer period of acceleration under current limit to a speed close to the target value, and a settling period in the speed control mode as the speed attains the target value and the current diminishes from the limit value to the value necessary to sustain the new speed.

10.11.2 *Speed decrease*

Deceleration is a very different process because of the unidirectional conductivity of the rectifier. The deceleration sequence will be followed by tracing the path on the steady state control characteristic as the system moves from point 2 back to point 1 following the trajectory marked by locations 3 through 9. Again the change in speed demand is assumed to be sudden, this time by instantaneously reducing the speed reference voltage from 9.0 to 6.0 V.

Since the speed cannot suddenly change, the instantaneous decrease in speed reference voltage immediately results in a speed error of –3.0 V. This places the output of amplifier G3 at its negative limit of –10 V. Like the speed, the current cannot suddenly change so the current error becomes $-10 - 138.6 \times 0.02 = -12.8$ V. This input puts the output of amplifier G2 at its negative limit of –12 V and fully phases back rectifier G1 to the maximum delay angle of 161.8° which places the rectifier in the cutoff region. All thyristors except the conducting pair are reverse biased and the current in the conducting pair is rapidly forced to zero as the a.c. voltage supplying this pair falls to zero and becomes negative. The situation when this has occurred is shown by location 5. The transition follows, more or less, the trajectory shown by the thick black line through locations 3 and 4 since, apart from the rapid current decay just noted, the transition is essentially instantaneous. At location 5 the rectifier control voltage is –12 V and the mean rectifier output voltage is the motor e.m.f. which, as the speed is still very nearly 163.6 rad s^{-1}, is very nearly 530.2 V. At location 5 with the rectifier cut off, the drive experiences a retarding torque of 449.1 N m, the combined effect of the load of 400 N m and the viscous friction torque of 163.6×0.3 N m, which produces a deceleration of $449.1/4.4 = 102.1$ rad s^{-2}. Since the friction torque is speed dependent the drive speed during cutoff is given by

$$\omega_m = -1333.3 + 1497.0 \exp(-\frac{t}{T}) \qquad (10.32)$$

where the time t is in seconds from the initiation of the change and T is the mechanical time constant, 14.67 s in this case.

As the drive slows, the e.m.f. decreases and the operating point follows the vertical path from location 5 to 6. Location 6 is reached when the output of amplifier G3 is about to come off its negative limit. During the transition, amplifier G3 is kept at its negative limit by the combined effects of its proportional and integral components. Were it purely proportional, it would come off its limit of –10.0 V, after 360.7 ms but the effect of the integrator continues to maintain it at the limit until 444.0 ms when the speed has fallen to 119.0 rad s^{-1} and the speed error voltage is –0.545 V. The rectifier control voltage is still at its negative limit of –12 V and the mean rectifier output voltage is the motor e.m.f. at 119.0 rad s^{-1}, 385.5 V.

After location 6, when G3 is again operational, the current demand voltage is given by

$$v_{id} = -119\,888 + 7\,933t + 119\,932\exp(-\frac{t}{14.67}) \tag{10.33}$$

This expression is ill conditioned since its value is the very small difference of very large quantities. For computational purposes it is better to substitute the series representing the exponential decay to obtain the computationally improved expression

$$v_{id} = 43.87 - 243.86t + 119932f(t) \tag{10.34}$$

where $f(t)$ is the exponential series less the first two terms which have been combined with the first two terms of eqn 10.33 to obtain eqn 10.34.

$$f(t) = \sum_{n=2,\infty} \frac{1}{!n}\left(-\frac{t}{T}\right)^n \tag{10.35}$$

a series which converges very rapidly.

A further drop in speed takes the system to location 7 at which point the rectifier control voltage has increased to its smallest effective value, 9.5 V, and the rectifier is about to come once again under control. This occurs 627.4 ms after the initiation of the change when the speed is 101.0 rad s^{-1} and the mean rectifier output voltage is the motor e.m.f. at this speed, 327.1 V. The speed error voltage is 0.4475 V, the current demand voltage is –0.95 V, and, since the armature current is still zero, the current error voltage is also –0.95 V.

The rectifier is still cut off but, now that it is once again under control, subsequent changes are very rapid and location 8 on the cutoff characteristic is reached after a further 9.6 ms at $t = 637.0$ ms. At location 8 the speed has fallen to 100.0 rad s^{-1}, significantly below the target value, 109.1 rad s^{-1}. The speed error voltage is 0.499 V, the current demand voltage is 0.0241 V, the current error voltage is 0.0241 V, the rectifier control voltage is 0.241 V and the mean rectifier output voltage is the motor e.m.f., 324.0 V.

The rectifier starts to conduct after location 8, the discontinuous conduction zone from location 8 to 9 being traversed very quickly. During this period the pulses of armature current create pulses of driving torque and an exact analysis of system behavior is complicated. However, this period is brief and the electromagnetic torque is small so that little error is introduced by ignoring the driving torque and assuming that the drive continues to decelerate under the influence of the load and friction torques alone, as described by eqn 10.32. With this assumption, point 9 is reached after 4.9 ms at t = 641.9 ms when the speed is 99.5 rad s^{-1}, the speed error voltage is 0.526 V, the current demand voltage is 0.543 V, the rectifier control voltage is 5.427 V and the mean rectifier output voltage is 322.5 V.

The system is now on the continuous conduction characteristic and accelerates to the target speed in a manner similar to that described in Example 10.3. Assuming that this occurs at the current limit, about 457 A, the target speed is reached after a further 40 ms at t = 683 ms. Including the settling time, the transition takes about 700 ms, more than twice the time for the corresponding speed increasing transition.

The speed undershoot is considerable, the minimum speed being about 99.5 rad s^{-1}, 10 rad s^{-1} below the target value, reached about 650 ms after the initiation of the reduction.

Deceleration is under the control of the load and friction torques and is thus directly dependent on the values of these very variable parameters. For example, if the load torque were zero, the retarding torque would be solely due to friction and would average about 40 N m for an average deceleration of about 9 rad s^{-2} and a deceleration time of about 6 s, ten times greater than the value found above. Of course the load and friction torques have some effect on the acceleration time but the effect is much smaller since acceleration is predominantly determined by the current limit.

In a high performance system the slowness and variability of deceleration and the significant undershoot would be unacceptable and steps would be taken to reduce if not eliminate the problem. This requires that the system be able to accept reverse current flow and can be accomplished either by field reversal, armature reversal, or dual rectifiers. The dual rectifier solution is illustrated.

10.12 The dual rectifier solution

This consideration of a dual rectifier system avoids a practical problem associated with the linear models of Fig. 10.6, the ripple in the rectifier output current, which requires the addition of a low pass filter in the current feedback path. After traversing this filter the residual ripple component becomes part of the current error voltage and is amplified by G2 and passed on to the rectifier where it causes jitter in the delay angle, which produces modulations in drive speed and in the rectifier a.c. input. It is difficult for the drive designer to provide sufficient attenuation to make the jitter acceptably small without adversely affecting the drive dynamics. For many years the current feedback signal was filtered in this way and there are still many drives attesting to the skill of designers in finding a compromise solution. However, the advent of digital control has removed this constraint and action can be taken on the basis of the instantaneous value of the armature current rather than its mean value. This dual rectifier example utilizes one of these methods. As for the linear example, the dual rectifier example has problems which will be addressed as they become apparent and at the end of the analysis suggestions are given for their alleviation.

The system considered is illustrated in Fig. 10.12. Its front end is similar to that of the previous system, the speed error voltage being amplified in the proportional plus integral amplifier G3 to produce the current demand voltage. At this point the system configuration changes and instead of the linear current comparator and amplifier G2, the current demand voltage is passed to two comparator amplifiers, CMP1 and CMP2, which provide signals to the rectifier controller, CTRL. The controller provides firing signals to rectifiers RECP and RECN and receives information from them. Rectifier RECP handles positive armature current and rectifier RECN handles negative current.

The comparators have different roles and different inputs. Comparator CMP1 has a single input, the current demand voltage v_{id}. It determines the sign of this signal and if the sign is positive the comparator indicates to the controller that rectifier RECP should conduct. If the signal is negative the comparator indicates that rectifier RECN should conduct. A change in sign of the current demand signal results in the following sequence of events which assumes a change from positive to negative, the sequence for the opposite change being similar, with the roles of the two rectifiers reversed.

When the sign of the current demand voltage changes from positive to negative, the first action of the controller is to block firing pulses to rectifier RECP. This leaves one pair of thyristors conducting and further action by the controller must await the decay to zero of this current. The direct determination of the current zero is not easy and an excellent alternative is indirect determination via the thyristor voltages. There will be a voltage of about +2 V across a conducting thyristor and if the voltage across a thyristor is within the range of say ±10 V it is prudent to assume that the thyristor is conducting. By monitoring all six thyristors for the time

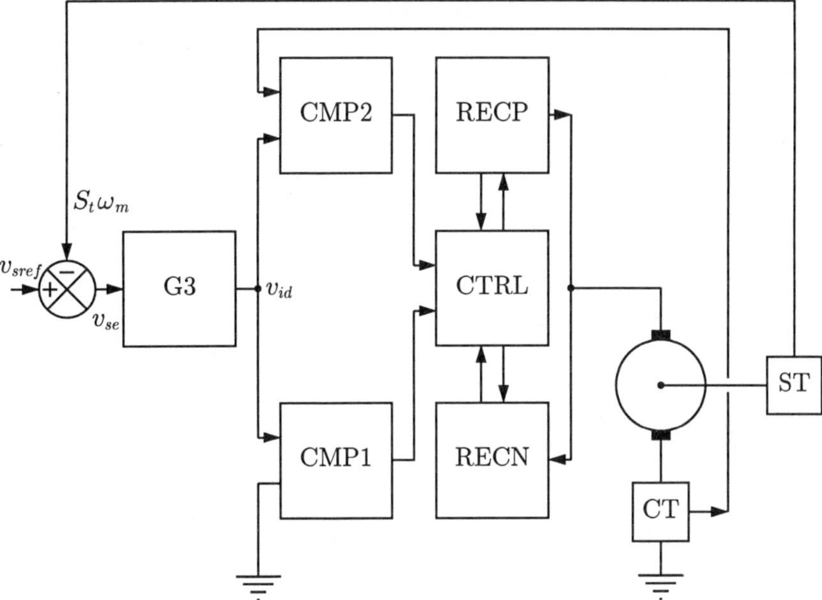

FIG. 10.12. Block diagram of a dual rectifier drive controlled by the value of the armature current at the switching instant. G3 is the speed error amplifier, CMP1 and CMP2 are comparators, CTRL is the controller, RECP is the rectifier supplying positive current, RECN is the rectifier supplying negative current, CT is the current transducer, and ST is the speed transducer.

when all their voltages are outside the forbidden band, the controller can determine when to initiate the next step. On receipt of the signal that rectifier RECP has ceased conduction the controller starts a brief delay of a few milliseconds duration, to ensure that all thyristors have fully recovered their blocking capability, before releasing firing pulses to rectifier RECN.

The location in time, i.e. the delay angle, of the firing pulses is determined by comparator CMP2. This compares the signal derived from the current transducer, CT, with the current demand voltage and informs the controller when the difference between the two signals is zero and in which direction it is passing through zero. Taking due account of the higher priority of the signal from comparator CMP1, the controller despatches firing pulses to one or other of the rectifiers. If the output voltage of the current transducer is passing from more than to less than the current demand signal, firing pulses are sent to the next thyristor pair of rectifier RECP. If it is passing from less than to more than the current demand signal, firing pulses are sent to the next thyristor pair of rectifier RECN. In this way the mean value of the armature current is kept at a level sufficient to maintain the drive speed at the desired value in a manner clarified by Fig. 10.13.

10.12.1 *Operation of the dual rectifier system*

The operation of this dual rectifier system is illustrated by Fig. 10.13, which shows, in three graphs identified by numbers at the right, the response to a sudden reduction in speed demand signal from 9 to 6 V. The figure shows in graph 1 the rectifier output voltage, in graph 2 the armature current, and in graph 3 the speed for the steady state just before the change and for about one-quarter of a second after the change. The system parameters, as far as they apply to the new system, are the same as those given in Example 10.1. Amplifier G3, the rectifiers, the motor, and the speed transducer are unchanged. The comparators and controller operate as described above. The current transducer provides an output signal proportional not only to the instantaneous value of the armature current but also to its rate of change. Without this latter component the system tends to go into subphase operation with the rectifier operating effectively with a three phase rather than the desired six phase output. The transfer function of the current transducer is taken as

$$v_{ia} = 0.02i_a - 0.000\,01\frac{\mathrm{d}i_a}{\mathrm{d}t} \tag{10.36}$$

The identification of thyristors in a rectifier and the phase of the voltages supplying it are those used throughout the book. The thyristor nomenclature is identified in Fig. 3.1 and the voltages are defined in Figs 3.2 and 3.3. In brief, in the steady state, thyristors T1 and T5 would conduct in the range $-60° + \alpha < \theta < 0° + \alpha$, thyristors T1 and T6 would conduct in the range $0° + \alpha < \theta < 60° + \alpha$, thyristors T2 and T6 would conduct in the range $60° + \alpha < \theta 120° + \alpha$, etc.

10.12.2 *Dynamic analysis*

It is not possible to control the system between firing pulses, so the dynamic analysis is based on the open loop system of Fig. 10.6(c). The state equation is similar to that given for the open loop mode in Example 10.3 but with the instantaneous rectifier output voltage replacing the mean voltage.

$$\begin{bmatrix} 0.000271 & 0.0 \\ 0.0 & 4.4 \end{bmatrix} \begin{bmatrix} i'_a \\ \omega'_m \end{bmatrix} + \begin{bmatrix} 0.15 & 3.24 \\ -3.24 & 0.3 \end{bmatrix} \begin{bmatrix} i_a \\ \omega_m \end{bmatrix} = \begin{bmatrix} v_o \\ -T_l \end{bmatrix} \tag{10.37}$$

where v_o is the rectifier output voltage and T_l is the load torque.

The output voltage is given by

$$v_o = V_{pk}\cos(\theta + \phi_{vs}) \tag{10.38}$$

In eqns 10.37 and 10.38, V_{pk} is the peak value of the supply voltage, $\theta = \omega_s t$, ϕ_{vs} is the phase angle of the supply voltage, and T_l is the load torque. The phase angle is $30° - n60°$, n being an integer depending on the particular pair of conducting thyristors. When T1 and T5 are conducting $n = 0$, when T1 and T6 are conducting $n = 1$, when T2 and T6 are conducting $n = 2$, etc. For the example $V_{pk} = 622.3$ V, $\omega_s = 377.0$ rad s^{-1} and $T_l = 400$ N m

The homogeneous equation is unchanged so that the eigenvectors of this system have the values already found in Example 10.3.

The steady state response to the two driving functions, the rectifier output voltage, and the load torque, is

$$i_a = 0.30732\,T_l + 0.9743\,V_{pk}\cos(\theta + \phi_{vs} - 1.424) \tag{10.39}$$

$$\omega_m = -0.01423\,T_l + 0.001903\,V_{pk}\cos(\theta + \phi_{vs} - 2.995) \tag{10.40}$$

Equations 10.39 and 10.40 illustrate the insensitivity of the speed to voltage ripple, the amplitude of the speed variation being about 500 times smaller than the variation in armature current, due essentially to the long mechanical time constant relative to the armature time constant.

10.12.3 *The initial steady state operation*

The initial steady state operation with a speed demand signal of 9.0 V is illustrated by the first complete rectifier segment of Fig. 10.13 which extends from 1.018 to 3.796 ms. Thyristors T1 and T6 are conducting. The delay angle is 21.99° and the expressions for the output voltage, armature current, and speed during this segment are

$$v_o = 622.3\cos(\theta - 0.5239) \tag{10.41}$$

$$\begin{aligned} i_a = {} & 122.9 + 606.3\cos(\theta - 1.9477) + \\ & 9708.8\cos(0.02861\,\theta + 1.5603)\exp(-0.07350\,\theta) \end{aligned} \tag{10.42}$$

$$\begin{aligned} \omega_m = {} & -5.6912 + 1.1842\cos(\theta + 2.7648) + \\ & 240.9\cos(0.02861\,\theta - 1.2093)\exp(-0.07350\,\theta) \end{aligned} \tag{10.43}$$

Despite the small scale, the current ripple is readily apparent in Fig. 10.13 and in contrast the speed appears to be constant. In fact the current varies from a low of 118.2 to a high of 150.2 A with a mean value of

138.61 A. The speed varies from a low of 163.632 to a high of 163.641 $\text{rad}\,\text{s}^{-1}$ with a mean value of 163.64 rad s^{-1}.

The current demand signal is 1.651 V and shows no variation during the segment. At the end of this rectifier segment $i_a = 118.2$ A and $\mathrm{d}i_a/\mathrm{d}t = -71\,286\ \text{A}\,\text{s}^{-1}$ so that the speed transducer output is $118.2 \times 0.02 - 71\,286 \times 0.000\,01 = 1.651$ V confirming that gate pulses are being emitted at the desired instants.

10.12.4 *The change in speed reference voltage*

The sudden reduction in speed reference voltage from 9.0 to 6.0 V occurs at some point in the next rectifier segment, i.e. after 3.796 and before 6.574 ms, this range being indicated by Markers A and B in graph 1 of Fig. 10.13. This causes a corresponding change in the speed error voltage from effectively zero (there is an extremely small variation in this voltage due to the very small variations in speed noted above) to –3.0 V which immediately puts the output of amplifier G3 at its negative limit, –10.0 V, thus indicating to the rectifier controller, via comparator CMP1, that maximum negative armature current is required. The controller blocks firing pulses to rectifier RECP. This leaves thyristors T2 and T6 conducting and the controller awaits the decay to zero of the current in this pair. This occurs at 7.738 ms and is indicated by Marker C in Fig. 10.13. At this point the speed is slightly reduced to 163.577 $\text{rad}\,\text{s}^{-1}$.

Although it has detected that the current in rectifier RECP is zero, the controller inserts a brief delay before releasing firing pulses to rectifier RECN in order to ensure that all thyristors in rectifier RECP have recovered their blocking capability. This delay has been set at 4 ms so that rectifier RECN receives its first firing pulses at 11.738 ms, indicated by Marker D in Fig. 10.13.

During this period the drive slows under the effect of the load and friction torques to 163.169 $\text{rad}\,\text{s}^{-1}$.

Consistent with the demand for maximum negative current, the delay angle of the incoming rectifier is set to zero. Thyristors T2 and T6 have the highest driving voltage at the instant of pulse release and therefore start to conduct. Because of the 4 ms delay, current pickup occurs slightly after the zero delay point, at an equivalent delay of 13.54° at which point the output voltage of rectifier RECN is –596.8 V, the negative sign being due to the reverse connection of this rectifier. This voltage and the motor e.m.f. of 544.9 V provide a total armature circuit driving voltage of –1141.6 V, which results in an initial rate of change of armature current of –421.3 $\text{A}\,\text{ms}^{-1}$ and rapidly drives the armature current to –1559 A as shown in graph 2 of Fig. 10.13. This current is, in an algebraic sense, far less than the current limit setting which is about –600 A for this rectifier and the

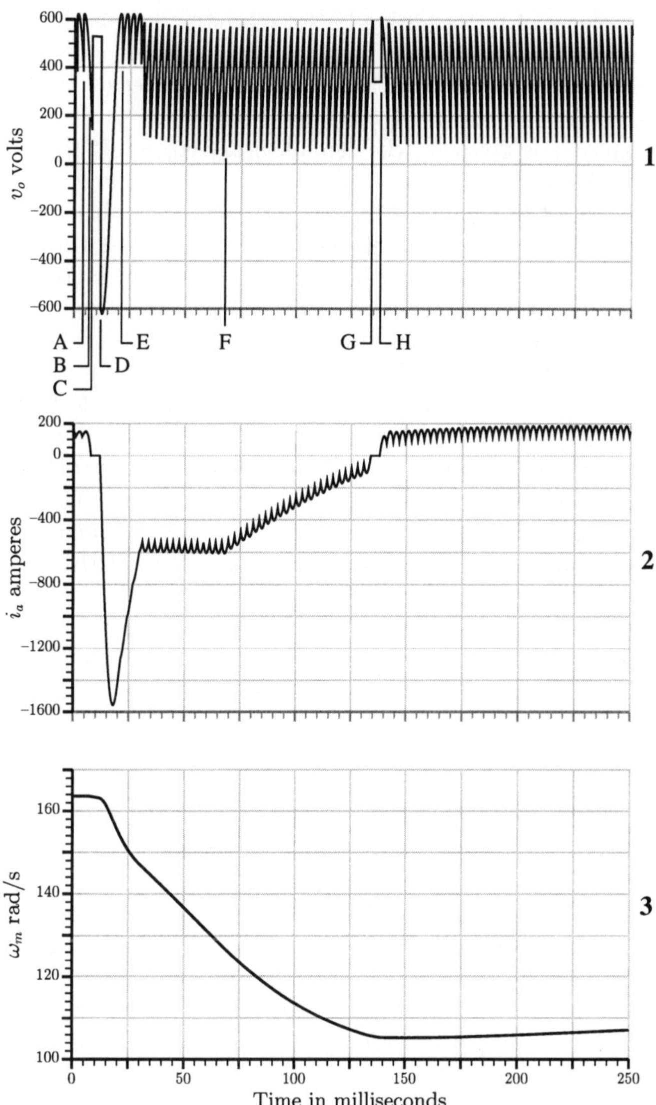

FIG. 10.13. The response of the drive to a sudden reduction of speed reference voltage from 9 to 6 V corresponding to a speed reduction from 163.6 to 109.1 rad s^{-1}. The graphs are identified by a number on the right, graph 1 showing the output voltage, graph 2 the armature current, and graph 3 the speed.

practical problems which it would cause are addressed at the end of this section. The release of further firing pulses is inhibited while the current is less than the limit value so that T2 and T3 continue to conduct until the maximum delay limit is reached. This limit has been set at 161.8° and is reached at 21.380 ms, indicated by Marker E in Fig. 10.13. At this instant the output voltage is 609.1 V, the armature current is –1261 A and the speed is 153.899 rad s^{-1}.

The despatch of firing pulses to the next thyristor pair occurs at the maximum delay angle at which point the output voltage is +414.8 V which, together with the motor e.m.f. of +498.6 V and the voltage absorbed by the circuit resistance, produces a rate of change of armature current of +56.19 A ms^{-1}, driving the current from its very large negative value towards the limit setting which is about –600 A.

The firing of successive thyristor pairs at the delay angle limit drives the current to its limit value at 31.054 ms at which point the current is –535.5 A and its rate of change is 71 076 A s^{-1} which, when transformed by the current transducer gives a current feedback voltage of $-535.5 \times 0.02 + 71\,076 \times 0.000\,01 = -10.0$ V, just sufficient to equal the current demand voltage which is still –10.0 V. The current is now held at the limit value until 68 ms when the speed has fallen sufficiently to bring the output of amplifier G3 off its negative limit. This point is indicated by Marker F in Fig. 10.13. At this instant the speed has fallen to 127.1 rad s^{-1}, still significantly above the target speed of 109.1 rad s^{-1}.

After 68 ms the current demand voltage is increasing in concert with the decrease in speed and at 132 ms, Marker G, it crosses zero and becomes positive. Comparator CMP1 detects this change and the controller blocks further firing pulses to rectifier RECN. The current in the last conducting thyristor pair reaches zero at 133.961 ms at which point the speed is 105.8 rad s^{-1}, about 4% below the target speed. The drive now coasts for the delay period of 4 ms and at 137.961 ms firing pulses are released to rectifier RECP, this point being indicated by Marker H. The current produced by rectifier RECP rapidly increases to a level slightly above that required to maintain the drive speed at the target value and the speed slowly rises to the target value, being 107.3 rad s^{-1} 250 ms after the change in speed demand voltage.

10.12.5 *Some practical considerations*

As well as indicating the operation of this type of control, the example illustrates some of the problems. The most obvious is the unacceptably high value of the armature current when rectifier RECN starts to conduct. Direct current machines will successfully commutate armature currents of about three times the rated value and the combination of the very high current and very high rate of change of current would cause a commutator flashover. Various measures are routinely applied to avoid this problem. One method initially sends firing pulses to the incoming rectifier at the complement of the delay angle of the outgoing rectifier. In this case this would bring rectifier RECN into operation at a delay of $180.0° - 21.99° = 158.0°$. Another technique starts the incoming rectifier at maximum delay and then phases it forward at a preset rate until the current limit setting is reached.

While the response of the system is rapid, it will settle to the target speed in about 1 s, it could be made more rapid by altering the characteristics of the speed error amplifier so as to maintain rectifier RECN at its current limit value for a longer period and to bring rectifier RECP back into operation at its current limit level rather than at the level required by the load. Both the proportional and integral gains could be increased and a differential component could be introduced to give the system some anticipation. Better still, nonlinear control techniques such as sliding mode control could be employed. Also worthy of consideration is a hybrid control system in which the current limit circuit comes into action only when the current exceeds a preset level, the system normally working as a simple speed controller without the inner current regulating loop.

10.13 Dual rectifiers with circulating current

Despite the popularity of the suppressed circulating current system, the dual rectifier with circulating current system as depicted in Fig. 10.2(c) is still employed in situations where its particular virtues are an advantage. The delay angles of the two rectifiers, α_a and α_b, are adjusted to produce a d.c. current circulating between the two rectifiers at a level sufficient to maintain continuous conduction at all times. This gives a linear transfer function under all operating conditions and provides for a smooth and continuous transition through zero load power without the brief hiatus which occurs with the suppressed circulating current system. The cost of these advantages lies in the additional expense of the inductors which must be placed in series with the rectifier outputs to limit the circulating current ripple and in the additional losses associated with the circulating current. The system tends to be employed in relatively low power applications, for example providing the field current for an electric machine.

The required d.c. level of circulating current is obtained by maintaining the delay angle of the second rectifier at sightly less than the complement of that of the first, i.e. $\alpha_b = 180° - \alpha_a - \delta$, δ being the *difference angle*. The difference angle can be preset to the value which maintains continuous conduction under the worst case conditions, i.e. when the ripple component of the circulating current is a maximum, a situation which occurs when the delay is about 90°. This strategy simplifies the control scheme at the cost of providing an excessive d.c. component at all other delay angles with the attendant losses. Alternatively δ can be continually adjusted to suit the operating conditions, minimizing energy loss at the expense of more complex control.

10.13.1 *The system equations*

The system will be studied in the context of the circuit of Fig. 10.14 where v_a is the voltage provided by rectifier A, v_b is the voltage provided by rectifier B, Z_c is the impedance of a circulating current limiting inductor, Z_o is the load impedance, and E_o is the d.c. e.m.f. produced by the load.

It is assumed that the supply and rectifier impedances are negligible so that internal voltage drops and commutation can be ignored. It is also assumed that the two smoothing reactors are identical and linear and that the load e.m.f. is pure d.c. with negligible ripple. Figure 10.14 shows the two rectifier currents, i_a and i_b, and the load current, i_o. The analysis will be in terms of these variables and the driving voltages, v_a and v_b.

Applying Kirchhoff's voltage law to each circuit mesh yields

$$v_a - E_o = (Z_c + Z_o)i_a - Z_o i_b \tag{10.44}$$

$$v_b + E_o = -Z_o i_a + (Z_c + Z_o)i_b \tag{10.45}$$

Over a given range of θ, the two voltages are represented by the sinusoids

$$v_a = \sqrt{2}V_s \cos(\theta_a + \phi_{va}) \tag{10.46}$$

and

$$v_b = \sqrt{2}V_s \cos(\theta_b + \phi_{vb}) \tag{10.47}$$

where V_s is the r.m.s. value of the line to line three phase input voltage and ϕ_{va} and ϕ_{vb} are phase angles.

Such equations have already been encountered on several occasions; they always lead to the same solution, the basic power electronics expression of Appendix 1. The solution for each current has four components, two exponentially decaying components associated with the eigenvalues of the homogeneous equations, a sinusoidally varying component produced by the a.c. driving voltage, and a d.c. component produced by the load e.m.f.

The homogeneous system has two eigenvalues

$$-\frac{2R_o + R_c}{2L_o + L_c} \quad \text{and} \quad -\frac{R_c}{L_c}$$

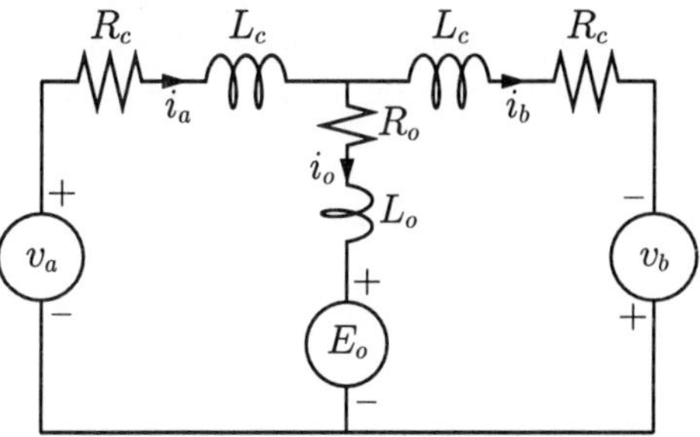

FIG. 10.14. Equivalent circuit for two rectifiers with circulating current supplying a d.c. active load.

These are the values appropriate to the use of time as the independent variable. If $\theta = \omega_s t$ is to be used instead of time as will be done here, the appropriate eigenvalues are then

$$s_1 = -\frac{2R_o + R_c}{2X_o + X_c} \tag{10.48}$$

and

$$s_2 = -\frac{R_c}{X_c} \tag{10.49}$$

where X_o and X_c are the reactances of the load and inductors at the a.c. supply frequency, ω_s.

The eigenvectors corresponding to these eigenvalues are $\{1 \quad -1\}$ and $\{1 \ 1\}$ clearly showing that the first corresponds to the parallel mode, current circulating between the load and the two rectifiers in parallel, and the second corresponds to the series mode, current circulating between the two rectifiers with none entering the load.

The third component of the currents, the steady state response to the a.c. driving voltages is obtained by applying a.c. circuit theory and is, in phasor terms,

$$\mathbf{I}_a = \mathbf{V}_a \frac{\mathbf{Z}_c + \mathbf{Z}_o}{\mathbf{Z}_c^2 + 2\mathbf{Z}_c\mathbf{Z}_o} + \mathbf{V}_b \frac{\mathbf{Z}_o}{\mathbf{Z}_c^2 + 2\mathbf{Z}_c\mathbf{Z}_o} \tag{10.50}$$

$$\mathbf{I}_a = \mathbf{V}_a \frac{\mathbf{Z}_o}{\mathbf{Z}_c^2 + 2\mathbf{Z}_c\mathbf{Z}_o} + \mathbf{V}_b \frac{\mathbf{Z}_c + \mathbf{Z}_o}{\mathbf{Z}_c^2 + 2\mathbf{Z}_c\mathbf{Z}_o} \tag{10.51}$$

where $\mathbf{I}_a$ and $\mathbf{I}_b$ are the phasors representing the rectifier output currents, $\mathbf{V}_a$ and $\mathbf{V}_b$ are the phasors representing the voltages v_a and v_b, and $\mathbf{Z}_a$ and $\mathbf{Z}_o$ are the complex impedances of the inductors and load at the supply frequency.

The fourth component of the currents, the steady state response to the d.c. e.m.f. of the load is obtained by applying d.c. circuit theory and is

$$i_a = -\frac{E_o}{R_c + 2R_o} \tag{10.52}$$

$$i_b = \frac{E_o}{R_c + 2R_o} \tag{10.53}$$

The load current is obtained by applying Kirchhoff's current law as

$$i_o = i_a - i_b \tag{10.54}$$

The load voltage is

$$v_o = E_o + R_o i_o + X_o \frac{\mathrm{d}i_o}{\mathrm{d}\theta} \tag{10.55}$$

The voltage across the inductor in series with rectifier A is

$$v_{la} = R_c i_a + X_c \frac{\mathrm{d}i_a}{\mathrm{d}\theta} \tag{10.56}$$

and across the inductor in series with rectifier B is

$$v_{lb} = R_c i_b + X_c \frac{\mathrm{d}i_b}{\mathrm{d}\theta} \tag{10.57}$$

10.13.2 *The driving voltages*

The next task is the determination of the driving voltages. This will be done in the context of a pair of identical, fully controlled, three phase, bridge rectifiers connected to the same three phase supply whose line to neutral and line to line voltages have been defined in Chapter 5 by eqns 5.19, 5.20, and 5.21. The waveforms of the line to line voltages are shown by grey lines in Fig. 10.15 and identified by the appropriate voltage symbol, v_{ry}, etc. The locations of three angles, α, β, and γ are indicated by vertical lines.

The output ripple of this system is at six times the a.c. supply frequency so that, for the steady state, only a 60° segment of the waveform need be considered to obtain a complete solution. The segment from $\theta = \alpha_a$ to $\theta = \gamma = \alpha_a + 60°$ will be considered. During this segment, using the nomenclature established in Chapter 5, thyristors T1 and T6 of rectifier A will be conducting. This segment is marked in Fig. 10.15, by a heavy black curve drawn for the case of $\alpha_a = 34°$. During this period it follows the path of v_{rb} so that

$$v_a = \sqrt{2}V_s \cos(\theta - \frac{\pi}{6}) \tag{10.58}$$

In general a complete 60° segment of rectifier A will coincide with parts of two segments of rectifier B so that the analysis must consider two sub-segments, one from α_a to β and the other from β to $\gamma = \alpha_a + 60°$ where β is

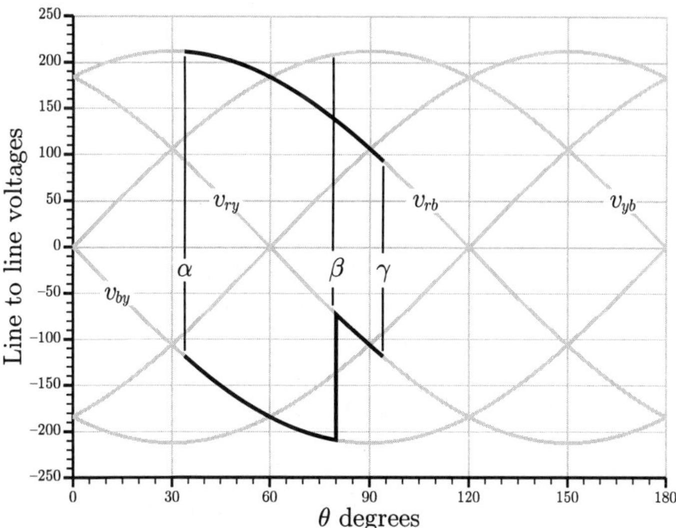

FIG. 10.15. Waveforms of the three phase rectifier input line to line voltages shown in gray and the subsegments when the rectifiers conduct shown in black.

the point at which there is a segment change in rectifier B. The waveform of v_b is also shown in Fig. 10.15 by heavy black lines for the two subsegments. The complement of α_a is $180° - 34° = 146°$ so that, if the difference angle is taken as 6°, the situation depicted in Fig. 10.15, $\alpha_b = 146° - 6° = 140°$. Thus, $\beta = -60° + 140° = 80°$ and the first subsegment extends from $\theta = 34°$ to $\theta = 80°$ during which period it follows the path of v_{by} so that

$$v_b = \sqrt{2}V_s \cos(\theta + \frac{\pi}{2}) \tag{10.59}$$

The second subsegment starts at $\theta = 80°$ and extends to $\theta = 94°$ during which period it follows the path of v_{ry} so that

$$v_b = \sqrt{2}V_s \cos(\theta + \frac{\pi}{6}) \tag{10.60}$$

These two subsegments have also been indicated in black in Fig. 10.15.

10.13.3 *Magnitudes of homogeneous solutions*

Once the a.c. voltages during the two subsegments are found, the resultant currents are determined and the expressions for the rectifier currents i_a and i_b are determined as follows.

For the first subsegment, $\alpha < \theta < \beta$

$$i_a = I_{11ab}\cos(\theta + \phi_{c11ab}) + I_{12} + I_{3ab}\exp(s_1\theta) + I_{4ab}\exp(s_2\theta) \quad (10.61)$$
$$i_b = I_{21ab}\cos(\theta + \phi_{c21ab}) + I_{22} - I_{3ab}\exp(s_1\theta) + I_{4ab}\exp(s_2\theta) \quad (10.62)$$

and for the second subsegment, $\beta < \theta < \gamma$

$$i_a = I_{11bc}\cos(\theta + \phi_{c11bc}) + I_{12} + I_{3bc}\exp(s_1\theta) + I_{4bc}\exp(s_2\theta) \quad (10.63)$$
$$i_b = I_{21bc}\cos(\theta + \phi_{c21bc}) + I_{22} - I_{3bc}\exp(s_1\theta) + I_{4bc}\exp(s_2\theta) \quad (10.64)$$

where all the parameters are now known except the four homogeneous component amplitudes, I_{3ab}, I_{4ab}, I_{3bc}, and I_{4bc}. These are found by applying the current continuity conditions. The circuit inductances ensure that all currents must be continuous across the boundaries between the first and second subsegment at $\theta = \beta$. In the steady state the values of the currents at the end of the segment, $\theta = \gamma$, and at the start of the segment, $\theta = \alpha$, must be equal. These two constraints define the unknown currents as the solution to the matrix equation 10.65

$$\begin{bmatrix} \exp(s_1\alpha) & \exp(s_2\alpha) & -\exp(s_1\gamma) & -\exp(s_2\gamma) \\ -\exp(s_1\alpha) & \exp(s_2\alpha) & \exp(s_1\gamma) & -\exp(s_2\gamma) \\ \exp(s_1\beta) & \exp(s_2\beta) & -\exp(s_1\beta) & -\exp(s_2\beta) \\ -\exp(s_1\beta) & \exp(s_2\beta) & \exp(s_1\beta) & -\exp(s_2\beta) \end{bmatrix} \begin{bmatrix} I_{3ab} \\ I_{4ab} \\ I_{3bc} \\ I_{4bc} \end{bmatrix} =$$
$$\begin{bmatrix} I_{11bc}\cos(\gamma + \phi_{c1bc}) - I_{11ab}\cos(\alpha + \phi_{c1ab}) \\ I_{21bc}\cos(\gamma + \phi_{c2bc}) - I_{21ab}\cos(\alpha + \phi_{c2ab}) \\ I_{11bc}\cos(\beta + \phi_{c1bc}) - I_{11ab}\cos(\beta + \phi_{c1ab}) \\ I_{21bc}\cos(\beta + \phi_{c1bc}) - I_{21ab}\cos(\beta + \phi_{c1ab}) \end{bmatrix} \quad (10.65)$$

This completes the solution for the rectifier currents and all other quantities, load current, load voltage, etc. can be obtained by elementary operations on these currents. The technique is illustrated by Example 10.6.

Example 10.6

Determine a complete solution for a pair of back to back connected rectifiers supplying current to the field winding of a d.c. machine when the delay angle of one rectifier is 34° and of the other is 140°, i.e. the situation depicted in Fig. 10.15. The rectifiers are fully controlled three phase bridges fed from the same three phase a.c. supply whose r.m.s. line to line voltage is 150 V and whose frequency is 60 Hz. The inductance and resistance of each ripple current limiting inductor are 7.2 mH and 0.25 Ω. The load is passive with a resistance of 4.0 Ω and an inductance of 100.0 mH.

The eigenvalues appropriate to the use of θ instead of t as independent variable are, from eqns 10.48 and 10.49,

$$s_1 = -0.1056 \quad \text{and} \quad s_2 = -0.09210$$

These values correspond to a time constant for the parallel mode of 25.11 ms and for the series mode of 28.80 ms, both long compared to the duration of a segment, 2.78 ms, so that the amplitude of the ripple currents should be modest.

The delay angles are those used for Fig. 10.15 where the 60° segment has been identified as occupying the period $34° < \theta < 94°$ with the first subsegment being of 46° duration and ending at 80° and with the second, of 14° duration, occupying the rest of the segment. Throughout the segment, for $34° < \theta < 94°$

$$v_a = 212.13\cos(\theta - \frac{\pi}{6})$$

During the first subsegment, for $34° < \theta < 80°$

$$v_b = 212.13\cos(\theta + \frac{\pi}{2})$$

During the second subsegment, for $80° < \theta < 94°$,

$$v_b = 212.13\cos(\theta + \frac{\pi}{6})$$

The current components produced by these voltages are, from eqns 10.50 and 10.51, during the first subsegment

$$i_a = 39.013\cos(\theta - 1.0153)$$

and

$$i_b = 38.951\cos(\theta - 0.8953)$$

and during the second subsegment

$$i_a = 67.428\cos(\theta - 1.4990)$$

and

$$i_b = 67.392\cos(\theta - 1.4589)$$

The d.c. current components are zero because the load e.m.f. is zero.

Substitution of these values into eqn 10.65 and solving yields the following values for the homogeneous components for the first subsegment

$$I_{3ab} = 21.611 \quad \text{and} \quad I_{4ab} = -13.063$$

and for the second subsegment

$$I_{3bc} = 22.991 \quad \text{and} \quad I_{4bc} = -49.428$$

Thus, the expressions for the rectifier currents during the first subsegment are

$$i_a = 39.013\cos(\theta - 1.0153) + 21.611\exp(-0.1056\,\theta) - 13.063\exp(-0.09210\,\theta)$$

and

$$i_b = 38.951\cos(\theta - 0.8953) - 21.611\exp(-0.1056\,\theta) - 13.063\exp(-0.09210\,\theta)$$

and during the second subsegment

$$i_a = 67.428\cos(\theta - 1.4990) + 22.909\exp(-0.1056\,\theta) - 49.428\exp(-0.09210\,\theta)$$

and

$$i_b = 67.392\cos(\theta - 1.4589) - 22.909\exp(-0.1056\,\theta) - 49.428\exp(-0.09210\,\theta)$$

The load current is the difference between the two rectifier currents so that during the first subsegment it is

$$i_o = 4.679\cos(\theta - 2.5128) + 43.222\exp(-0.1056\,\theta)$$

and during the second subsegment is

$$i_o = 2.701\cos(\theta - 3.0364) - 45.818\exp(-0.1056\,\theta)$$

The load voltage is the sum of the load e.m.f., in this case zero, and the voltage drops in the load resistance and inductance and is, during the first subsegment,

$$v_o = 177.34\cos(\theta - 1.0477) + 0.7927\exp(-0.1056\,\theta)$$

and during the second subsegment it is

$$v_o = 183.80\cos(\theta - 0.0200) - 0.8403\exp(-0.1056\,\theta)$$

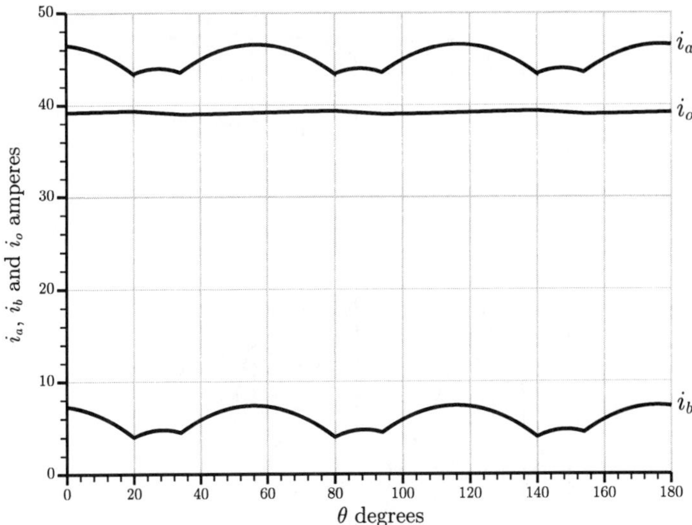

FIG. 10.16. Waveforms of the load and rectifier currents.

The voltages across the circulating current limiting inductors are the sums of the voltage drops in their resistances and inductances. For the first subsegment they are

$$v_{ca} = 106.34\cos(\theta + 0.4636) - 0.7927\exp(-0.1056\,\theta)$$

and

$$v_{cb} = 106.17\cos(\theta + 0.5837) + 0.7927\exp(-0.1056\,\theta)$$

and during the second subsegment they are

$$v_{ca} = 183.80\cos(\theta - 0.0200) - 0.8403\exp(-0.1056\,\theta)$$

and

$$v_{cb} = 183.70\cos(\theta + 0.0200) + 0.8403\exp(-0.1056\,\theta)$$

The waveforms of these currents are shown in Fig. 10.16. The load current has very little ripple, its mean and r.m.s. values both being 39.17 A with the r.m.s. value of the ripple being 0.11 A. Rectifier A is supplying the load and also sufficient current into rectifier B to maintain continuous conduction, the smallest value of current in rectifier B being 4.03 A at $\theta = 20°$, $80°$, etc. The mean value of i_a is 45.10 A and its r.m.s. value is 45.12 A so that its r.m.s. ripple is 1.09 A. The mean value of i_b is 5.93 A and its r.m.s. value is 6.04 A so that its r.m.s. ripple is 1.10 A, nearly but not quite the same as for i_a since some ripple current flows in the load.

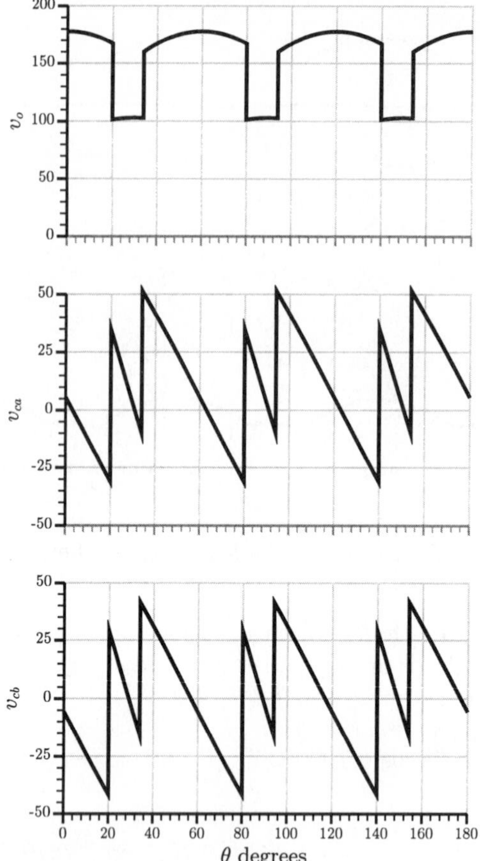

FIG. 10.17. The waveforms of the output voltage, and the voltages across the two inductors.

The waveforms of the load and inductor voltages are shown in Fig. 10.17. The inductor voltages have the familiar sawtoothed waveform but the waveform of the load voltage has a castellated form which has not previously been encountered. This is very typical of systems with circulating current. The circulating current limiting inductors cause the load voltage to be approximately the average of the two rectifier output voltages and the slopes of their saw toothed segments more or less cancel.

The mean power delivered to the load is 6.135 kW and the load losses due to harmonic currents are negligible. The mean power loss in the inductor in series with rectifier A is 0.509 kW and in the other inductor it is 9.1 W. The total input power, i.e. the power delivered by rectifier A less the power returned to the a.c. supply by rectifier B, is 6.653 kW so that the efficiency is 92.2%. Of course this figure would be reduced somewhat in practice because of the additional rectifier losses which have been neglected.

****

10.13.4 *The circulating current limiting inductors*

The size, weight, and cost of the circulating current limiting inductors is a significant item. Some idea of their size can be obtained from their equivalent 60 Hz kVA. The inductor in series with rectifier A is carrying an r.m.s. current of 45.12 A and is supporting an r.m.s. ripple voltage of about 24.72 V. Since the ripple is at 360 Hz, this corresponds to a 60 Hz voltage of about $24.72 \div 6 = 4.12$ V. Thus the 60 Hz kVA rating of an inductor is 0.186 and for the two inductors is twice this value, 0.372, about 6% of the load power, 6.135 kW, apparently not a serious matter. However, these rough calculations do not give a true picture. The system must be designed for the worst case conditions which, as far as ripple is concerned, occur at a delay of about 90°. The r.m.s. ripple is then 62.5 V leading to a much higher kVA rating for the two inductors combined of 0.94 kVA. The inductors must also be designed to carry the d.c. component of current, a substantial amount. To prevent this causing magnetic saturation in the inductor cores, the cores must have significant air gaps which will increase their size and weight far beyond the level expected for a purely a.c., gapless component such as a transformer.

10.14 Conclusion

An overview has been given of the interaction between a rectifier and its load and considerable time has been spent on the basics of two popular systems for overcoming the incompatibility between a rectifier and most loads. The operation of such systems is complex but is amenable to computer based analysis. However, it must be remembered that commercial systems, while following the broad generalities discussed here, may well adopt different specific methods for handling the peculiarities of rectifier–load behavior.

11

CYCLOCONVERTERS

11.1 Introduction

In the previous chapter it has been seen that, in dual rectifier systems, the switch from the positive to the negative current rectifier can be accomplished with great rapidity. Soon after the development of controlled rectifiers, in the early 1930's, it was realized that this provided the possibility of generating alternating currents of variable frequency directly from the fixed frequency a.c. supply, the positive rectifier supplying the positive half cycles of current and the negative rectifier the negative half cycles. At that early stage the name *cycloconverter* was applied to such a system and this proved to be so appropriate that it is still in use.

A cycloconverter is one or more pairs of back to back connected rectifiers, such as those discussed in the previous chapter, whose delay angles are modulated at some appropriate frequency so as to provide an a.c. output at the modulation frequency and at the desired voltage. Each pair of rectifiers produces one such output so that a three phase output, the usual requirement, requires three pairs, one for each phase. If each rectifier is a three phase bridge, this entails the use of at least 36 thyristors, each with its own gating and control circuit, a controller for each pair of rectifiers and a controller for the whole system. Such systems are therefore large and complicated and tend to be used only for large loads of say 1 MW and up.

11.2 Output frequency

A rectifier samples the input a.c. voltage at frequent intervals and directs the samples to its output. As an example, a three phase bridge rectifier connected to a 60 Hz system samples the system voltages 360 times per second and provides the opportunity to change conditions, via the delay angle, at this frequency. The 360 Hz can be regarded as a carrier which is modulated via the delay angle, at irregular and infrequent intervals in the case of a d.c. load but regularly and rapidly in the case of a cycloconverter. Modulation theory indicates that a system can be successfully modulated up to half the carrier frequency, however this is an ideal that is difficult to achieve and it is more reasonable to keep the modulation frequency below about one-tenth of the carrier frequency. This indicates the maximum reasonable output frequency for a cycloconverter—36 Hz for three phase bridges and 18 Hz for simple, single way, three phase rectifiers operating from a 60 Hz supply. Thus, the most successful applications combine high power and low frequency and the typical load is a large, low speed motor.

11.3 Cycloconverter circuits

There are many variations on the cycloconverter theme. The rectifiers can be simple, single way thus reducing complexity (18 thyristors for a three phase system) but limiting the output frequency to less than about 18 Hz. Alternatively bridges can be used, doubling the potential output frequency at the cost of twice as many thyristors. Each back to back connected pair of rectifiers can be operated in the circulating current mode or the circulating current free mode. The three phase load can be operated in the normal configuration, with isolated neutral, or with each phase independently connected to the pair of rectifiers feeding it.

Many different modulation strategies can be employed. A problem with cycloconverters is their poor input power factor and the amelioration of this problem leads away from the obvious sinusoidal modulation.

The study of cycloconverters is therefore very complex and whole books have been written on the topic. To limit this study to a single chapter, only one of the more popular configurations is reviewed. Initially this is done in the context of a pair of three phase bridges in the suppressed circulating current mode and this is followed by a consideration of three such pairs independently supplying the phases of a three phase load. This limited study illustrates all the major problems of analysis and the choices facing the system designer.

The system is shown symbolically in Fig. 11.1. Each rectifier is a six thyristor bridge, each being fed from the same three phase supply. The nomenclature established in Chapter 3 and used throughout the book is followed. Thus, the red line to neutral voltage is taken as datum and the thyristors are numbered 1 through 6 as in Fig. 3.1(c). The positive current rectifier is labelled P and the negative current rectifier is labelled N as indicated in Fig. 11.1 and subscripts p and n will be used where appropriate. With these conventions, the principal parameters are identified in Fig. 11.1.

The study commences with sinusoidal modulation and later explores the possibilities of non-sinusoidal modulation.

11.4 The line to line supply voltage

It will be convenient to work with the six line to line voltages which, taking the red line to neutral voltage as reference, are described in the time domain by eqns 11.1 through 11.6.

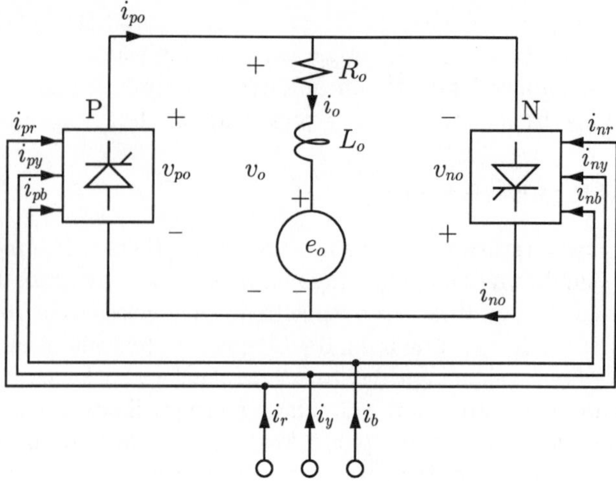

FIG. 11.1. A single phase, suppressed circulating current cycloconverter.

$$v_{br} = V_{pk}\cos(\theta + \frac{5\pi}{6}) \tag{11.1}$$

$$v_{by} = V_{pk}\cos(\theta + \frac{\pi}{2}) \tag{11.2}$$

$$v_{ry} = V_{pk}\cos(\theta + \frac{\pi}{6}) \tag{11.3}$$

$$v_{rb} = V_{pk}\cos(\theta - \frac{\pi}{6}) \tag{11.4}$$

$$v_{yb} = V_{pk}\cos(\theta - \frac{\pi}{2}) \tag{11.5}$$

$$v_{yr} = V_{pk}\cos(\theta - \frac{5\pi}{6}) \tag{11.6}$$

where V_{pk} is the peak value of the line to line supply voltage and $\theta = \omega_s t$, where ω_s is the radian frequency of the supply.

11.5 The firing and delay angles

Following the biased cosine method described in Chapter 4, obtaining the firing angle for a pair of conducting thyristors requires a reference voltage wave leading the appropriate line to line voltage by 30°. As an example the pair of thyristors T1 and T5 connect the red and yellow lines to the load so that, from eqn 11.3, their appropriate reference voltage is

$$v_{ryref} = V_{ref}\cos(\theta + \frac{\pi}{3}) \tag{11.7}$$

The firing angle is given by the intersections of this wave with the control signal, v_c, at which the difference, $v_{ref} - v_c$, is going through zero

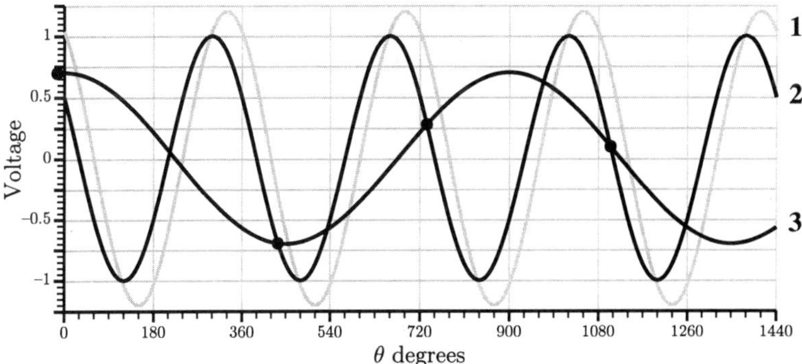

FIG. 11.2. Location of thyristor gating points. The supply line to line voltage is shown in gray and is labelled 1 to the right. Its reference voltage is the black curve at the same frequency and labelled 2. The control voltage is the black curve labelled 3. It has a modulation index of 0.7, a frequency ratio of 0.4, and a phase angle of zero. The gating points for thyristor T1 are indicated by black dots.

from positive to negative. Up to this point the control signal has always been a d.c. voltage. In the cycloconverter context it varies sinusoidally with a frequency f_r times the supply frequency, an amplitude M_i times the control voltage amplitude, and a phase angle ϕ_c. Thus, the control voltage for the thyristor pair T1 and T5 is

$$v_{cry} = M_i V_{ref} \cos(f_r\theta + \frac{\pi}{3} + \phi_{cry}) \tag{11.8}$$

M_i is known as the *modulation index* and f_r is known as the frequency ratio.

The situation is illustrated by Fig. 11.2 which shows the waveforms of a line to line voltage by a gray line, the reference voltage associated with it and the control voltage by solid black lines, and the thyristor gating points by black dots. The amplitudes are not important, that of the line to line voltage is set by the a.c. supply and the cycloconverter transformer, that of the reference voltage is set at a convenient value by the control system designer, and that of the control voltage is related to it via the modulation index. The figure has been drawn for the red to yellow line supply voltage which leads the datum by 30°, a frequency ratio of 0.4, a modulation index of 0.7, and a control voltage phase of 0°.

Of course, since we are dealing with bridge rectifiers, the firing angle so obtained is that for the incoming thyristor, in this case T1, the one whose anode is connected to the red supply line. The other thyristor in the pair, in this case T5 whose cathode is connected to the yellow line, will already be in the conducting mode.

11.5.1 *The delay angle*

The delay angle is the difference between the firing angle and the immediately preceding positive peak of the reference wave. In this case these peaks occur at $-\pi/3 + n2\pi$, n being any integer, positive or negative including zero.

11.5.2 *Large values of modulation index*

When the modulation index is large it is possible that the intersection will occur at a dangerously large value of delay angle. The controller must recognize this situation and limit the delay to the maximum safe value.

If the modulation index is greater than 1.0 it is possible that there will be no intersection between the reference and control waves within the permissible range of θ. In this event, if the control wave is greater than the reference, the delay angle is set to zero and alternatively, if the control wave is less than the reference, the delay angle is set to α_{max}.

Controller techniques for meeting these requirements are described in Chapter 4.

11.5.3 *The switching array*

For computations it is convenient to have all the rectifier switching information in a switching array. This defines each segment of the rectifier output voltage by the line to line voltage operative during the segment, its starting angle, and its finishing angle. The preparation and structure of such an array is illustrated by Example 11.1.

Example 11.1

Determine the switching arrays for a three phase bridge rectifier directly connected to a 440 V, 60 Hz, three phase supply when the control voltage has (i) a modulation index of 0.7, a frequency of 24 Hz, and a phase angle of zero and (ii) the same modulation index and frequency but a phase angle of 180°. The maximum permissible delay angle is 161.2°. The switching arrays should cover the range $0 < \theta < 1440.0°$

Consider first the switching instants associated with the red to yellow line to line voltage. From eqn 11.3 this voltage is given by

$$v_{ry} = 622.3\cos(\theta + 0.5236)$$

The associated reference voltage leads this by 30° and is

$$v_{ref} = V_{ref} \cos(\theta + 1.0472)$$

where V_{ref} is any value convenient for the control system designer.

Switching points can occur in the half cycles when the slope of this wave is negative, i.e. in the ranges $-60° < \theta < 101.2°$, $300° < \theta < 461.2°$, $660° < \theta < 821.2°$, etc.

For the first case, zero phase angle, the control voltage is

$$v_{cry} = 0.7\, V_{ref} \cos(0.4\theta)$$

so that the required gate instants are the zeros of the function

$$f(\theta) = \cos(\theta + 1.0472) - 0.7 \cos(0.4\theta)$$

One procedure for finding the zeros is to sample this function within a permissible range in six steps of $\alpha_{max} / 6$ until either a change in sign occurs or all six locations have been sampled. There are three possible outcomes. First, the value of every sample can be negative in which case the control wave is above the reference wave and the delay is zero. Second, the value of every sample can be positive in which case the control wave is below the reference wave and the delay is α_{max}. Third, there may be a sign change from positive to negative in which case there is a zero somewhere between the last two samples. At this point a linear interpolation can be made to obtain an approximate value of the intersection angle and then this can be refined using the Newton–Raphson technique.

Applying this method to the range $300° < \theta < 461.2°$, an intersection is located between 407.467° and 434.333°. At 407.467° the value of the function is 0.3692 and at 434.333° it is –0.0030. A linear interpolation locates the intersection close to 434.117°. A few Newton–Raphson iterations locate the intersection at 434.080°. Since zero delay corresponds to an intersection at 300°, the delay angle at the firing instant is 134.08°. Similar computations for the other segments of part 1 of the example and also for part 2 yield the switching arrays listed in Tables 11.1 and 11.2.

The meaning of the columns of Tables 11.1 and 11.2 is made clear by considering one row, for example that for segment 10 of Table 11.1. The first column provides an identifying number for the segment, in this case 10. The second column headed ID identifies the output voltage during that segment by the a.c. line letters. In this case the identification is YR indicating that the yellow line is connected to the positive d.c. bus via thyristor T2 and the red line is connected to the negative d.c. bus via thyristor T4. The third

Table 11.1 *Switching array when the modulation index is 0.7 and the phase angle of the control voltage is zero. Column 1 lists the segment number, column 2 identifies the voltage, column 3 gives the phase of the voltage, columns 4 and 5 give the segment start and finish, column 6 lists the delay angle, and column 7 gives the duration of the segment*

Seg. no.	ID	Phase degrees	Start degrees	Finish degrees	Delay degrees	Duration degrees
1	RY	30.0	–14.15	48.71	45.85	62.86
2	RB	–30.0	48.71	122.71	48.71	74.01
3	YB	–90.0	122.71	204.18	62.71	81.47
4	YR	–150.0	204.18	287.10	84.18	82.92
5	BR	150.0	287.10	365.59	107.10	78.48
6	BY	90.0	365.59	434.08	125.59	68.50
7	RY	30.0	434.08	492.07	134.08	57.99
8	RB	–30.0	492.07	543.74	132.07	51.67
9	YB	–90.0	543.74	592.43	123.74	48.69
10	YR	–150.0	592.43	639.80	112.43	47.37
11	BR	150.0	639.80	686.72	99.80	46.92
12	BY	90.0	686.72	733.78	86.72	47.06
13	RY	30.0	733.78	781.68	73.78	47.89
14	RB	–30.0	781.68	831.57	61.68	49.89
15	YB	–90.0	831.57	885.85	51.57	54.28
16	YR	–150.0	885.85	948.71	45.85	62.86
17	BR	150.0	948.71	1022.71	48.71	74.01
18	BY	90.0	1022.71	1104.18	62.71	81.47
19	RY	30.0	1104.18	1187.10	84.18	82.92
20	RB	–30.0	1187.10	1265.59	107.10	78.48
21	YB	–90.0	1265.59	1334.08	125.59	68.50
22	YR	–150.0	1334.08	1392.07	134.08	57.99
23	BR	150.0	1392.07	1443.74	132.07	51.67

column gives the phase angle of the a.c. line voltage, in this case $-150°$. Thus, the output voltage during segment 10 is

$$622.3\cos(\theta - \frac{5\pi}{6})\ \mathrm{V}$$

Columns 4 and 5 give the starting and finishing angles of the segment, column 6 gives the delay angle and column 7 the length of the segment. It is seen that segment 10 starts at $\theta = 592.4°$ and finishes at $639.8°$ giving a

Table 11.2 *Switching array when the modulation index is 0.7 and the phase angle of the control voltage is 180°. Column 1 lists the segment number, column 2 identifies the voltage, column 3 gives the phase of the voltage, columns 4 and 5 give the segment start and finish, column 6 lists the delay angle, and column 7 gives the duration of the segment*

Seg. no.	ID	Phase degrees	Start degrees	Finish degrees	Delay degrees	Duration degrees
1	BR	150.0	–48.71	14.15	131.29	62.86
2	BY	90.0	14.15	68.43	134.15	54.28
3	RY	30.0	68.43	118.32	128.43	49.89
4	RB	–30.0	118.32	166.22	118.32	47.89
5	YB	–90.0	166.22	213.28	106.22	47.06
6	YR	–150.0	213.28	260.20	93.28	46.92
7	BR	150.0	260.20	307.57	80.20	47.37
8	BY	90.0	307.57	356.26	67.57	48.69
9	RY	30.0	356.26	407.93	56.26	51.67
10	RB	–30.0	407.93	465.92	47.93	57.99
11	YB	–90.0	465.92	534.41	45.92	68.50
12	YR	–150.0	534.41	612.90	54.41	78.48
13	BR	150.0	612.90	695.82	72.90	82.92
14	BY	90.0	695.82	777.29	95.82	81.47
15	RY	30.0	777.29	851.29	117.29	74.01
16	RB	–30.0	851.29	914.15	131.29	62.86
17	YB	–90.0	914.15	968.43	134.15	54.28
18	YR	–150.0	968.43	1018.32	128.43	49.89
19	BR	150.0	1018.32	1066.22	118.32	47.89
20	BY	90.0	1066.22	1113.28	106.22	47.06
21	RY	30.0	1113.28	1160.20	93.28	46.92
22	RB	–30.0	1160.20	1207.57	80.20	47.37
23	YB	–90.0	1207.57	1256.26	67.57	48.69
24	YR	–150.0	1256.26	1307.93	56.26	51.67
25	BR	150.0	1307.93	1365.92	47.93	57.99
26	BY	90.0	1365.92	1434.41	45.92	68.50
27	RY	30.0	1434.41	1512.90	54.41	78.48

segment length of 47.4°. The zero delay point for this segment is at $\theta = 480°$ so that the delay angle is $592.4° - 480.0° = 112.4°$.

While the average length of a segment is 60° as in a rectifier, the continual variation of delay results in a corresponding variation in length, the duration of some segments being less than average and of others greater

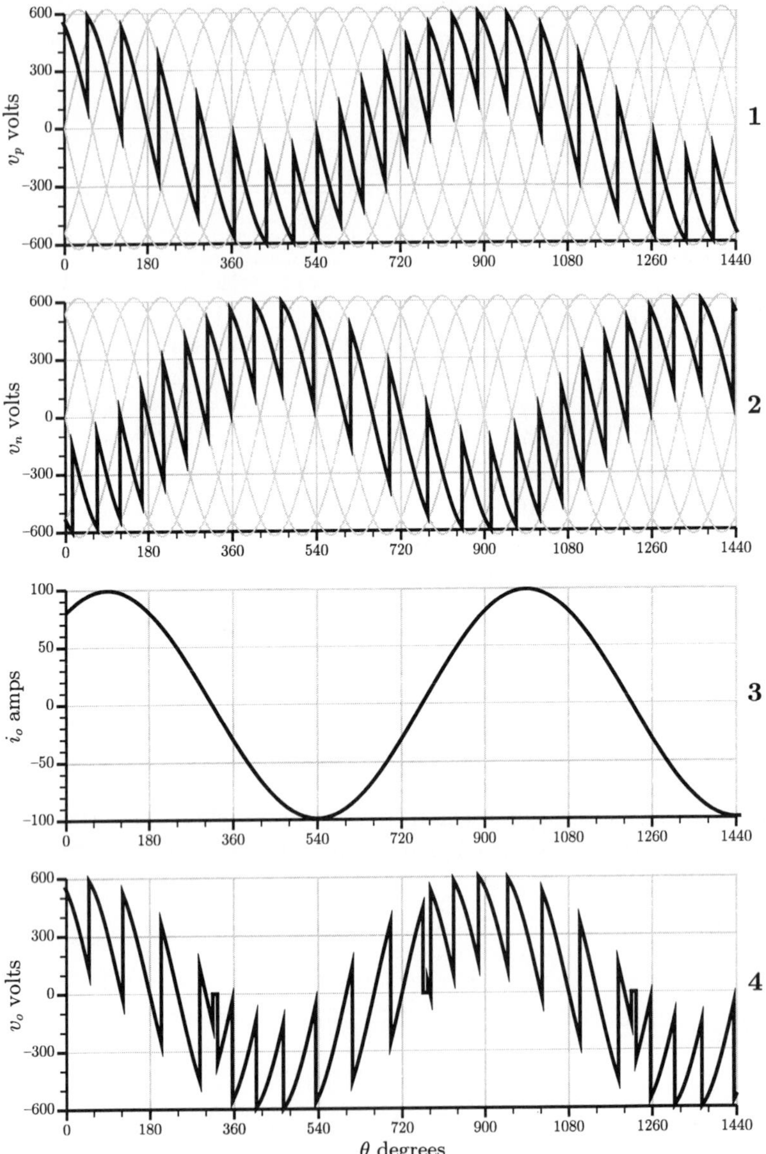

FIG. 11.3. Cycloconverter waveforms when the modulation index is 0.7 and the frequency ratio is 0.4. Each graph is identified by a number to its right. Graph 1 shows the output voltage of rectifier P and graph 2 shows that of rectifier N. Graph 3 shows the load current and graph 4 shows the load voltage.

than average. In Table 11.1, the longest segment is number 4 with a duration of 82.9° and the shortest is number 11 with a duration of 46.9°. This point is illustrated by graphs 1 and 2 of Fig. 11.3.

Figure 11.3 has been drawn for the modulation conditions of Example 11.1, it being assumed that the zero phase angle control signal is applied to rectifier P and the 180° phase angle one is applied to rectifier N. The figure covers 1.6 cycles of control voltage. Graph 1 shows the output voltage of rectifier P and graph 2 shows the output voltage of rectifier N *assuming they both conduct continuously* rather than alternately, each for half a control cycle, as is the actual situation. The six a.c. line to line voltages are shown in gray and the output voltage, defined by the segment data of Tables 11.1 and 11.2, is shown in black. It will be seen that the segments of short duration occur when the control voltage is increasing and those of long duration when it is decreasing. These groupings of long and short segments account for the difference in the number of segments between Tables 11.1 and 11.2. Since 1.6 control cycles are shown, there are two sets of long segments and one set of short segments listed in Table 11.1 and one set of long segments and two of short in Table 11.2.

****

11.6 The load current and voltage

Knowing the output voltage segments of each rectifier, it is possible to compute the load current and voltage waveforms. However, this is a complex task which will be postponed briefly so that the basics of cycloconverter operation can be seen unobscured by a cloud of detailed analysis. For the present purpose it is assumed that the load is a sinusoidal current source with an r.m.s. value of 70 A, a frequency of 24 Hz, and a phase angle of −36°. This would be approximated in real life if the load were highly inductive so that the ripple components of current were small compared to the 24 Hz component. The time domain expression for this current is

$$i_o = 99.0\cos(\theta - 0.6283) \tag{11.9}$$

Its waveform is shown in graph 3 of Fig. 11.3.

The positive half cycles of the load current are carried by rectifier P during which time rectifier N is off and the negative half cycles are carried by rectifier N during which time rectifier P is off. In the case of Fig. 11.3 the current is positive until 315°, is then negative for $180 \times 60 \div 24 = 450°$, until $\theta = 765°$, is positive from that point to 1215°, and is negative for the rest of the figure.

Thus, at the start of Fig. 11.3 rectifier P is conducting and, from Table 11.1, the output voltage is v_{ry}, the red line being connected to the d.c.

positive bus via thyristor T1 and the yellow line to the d.c. negative bus via T5. This situation lasts until 48.7° when the d.c. negative bus is connected to the blue line by firing thyristor T6. The output voltage is then v_{rb} until 122.7° when T2 is fired to transfer the positive bus to the yellow line making the output voltage v_{yb} for the next 81.5°, until 204.2°. This process continues while the output current is positive, until 315°. The waveform of the output voltage is shown in graph 4 of Fig. 11.3 and for the period $0 < \theta < 315°$ it is the same as that of rectifier P shown in graph 1.

At $\theta = 315°$ the cycloconverter controller detects that the current in rectifier P is zero, blocks its thyristors, and releases gate pulses to rectifier N. In drawing Fig. 11.3 it has been assumed that this changeover takes 10°, 0.453 ms, which is longer than would be expected in practice, so that it shows up clearly on the figure as the period when the output voltage is zero. After the changeover delay, at 325°, reference to Table 11.2 indicates that firing pulses are sent to thyristors T3 and T5 of rectifier N to connect the d.c. positive bus to the a.c. blue line and the d.c. negative bus to the yellow line. The d.c. output voltage is then v_{by} until 356.3° when thyristor T1 is fired to transfer the positive bus to the red line and make the output voltage v_{ry} for the next 51.7° until thyristor T6 is fired at $\theta = 407.9°$. This process continues throughout the negative current half cycle, until 765° when the output is transferred back to rectifier P. Thus, for the period $325° < \theta < 765°$ the output voltage is -1 times that of rectifier N, shown in graph 2, the -1 occurring because of the sign convention adopted in Fig. 11.1.

After the transfer delay, at 775°, rectifier P resumes conduction, a state which lasts until the next current reversal at $\theta = 1215°$. After the transfer delay, rectifier N conducts for the remainder of Fig. 11.3, from 1225° until 1440°.

Before further discussion of the output voltage waveform it is necessary to dispose of problems concerning the rectifier transfer delay. It has already been stated that this has been made longer than would normally be the case in order to display it clearly. There is a much more serious problem in that, in order not to have to bother with the problems of current ripple, it has been assumed that the load is a sinusoidal current source so that the current passes smoothly through zero from one half cycle to the next. The current must in fact be zero during the transfer period. During this period it has been assumed that the output voltage is zero, again so that the transfer delay will show clearly in Fig. 11.3. It will in fact be the load e.m.f. which would only be zero for a passive load. These problems are removed at the next stage of the analysis and in the meantime they are a small price to pay for the clarity of Fig. 11.3.

The output voltage waveform of graph 4 is made up of sections of the output waveforms of rectifiers P and N shown in graphs 1 and 2. The location of these sections depends on the phase of the load current relative to the control signal. In the case of Fig. 11.3 the current lags 36° on the control signal for rectifier P. Had it lagged say 136°, rectifier N would have conducted from zero to 115°, rectifier P from 125° to 565°, rectifier N from 575° to 1015°, and rectifier P from 1025° to the end of the figure. The waveforms of graphs 1 and 2 were drawn *assuming continuous conduction* in order that they could be employed for any load condition.

Since the output voltage waveform depends on the phase angle of the load current, exactness requires that a time domain analysis be made for every load condition of interest. This is developed in the next section.

11.7 Time domain analysis

In one sense, time domain analysis for the cycloconverter is very similar to that for a rectifier. The differential equation describing the load current in terms of the driving voltage, the load impedance, and the load e.m.f., is written down. The driving voltage is then selected from switching arrays such as those given in Tables 11.1 and 11.2 and the differential equation is solved for a given initial value of the load current. The value at the end of this segment is the initial value for the next segment. Other values such as the a.c. line currents and the input power are obtained in the manner described in Chapter 5. Unfortunately the initial value of current is not known and in Chapter 5 that problem was surmounted for the steady state by first making an intelligent estimate of the initial value and then iteratively refining it so that the current at the end of the segment equaled the initial value.

In the cycloconverter, the continuous variation of delay angle renders the rectifier approach invalid. In general, all that can be done is to start with an intelligent guess of the initial current and pursue the analysis segment by segment until the steady state is reached. This is easier said than done since the steady state may never be reached, a situation which occurs when the frequency ratio is irrational. Then every output cycle is unique. However, the differences between cycles are minor, occurring principally in the higher order components of the ripple, and after a few cycles one can be reasonably sure of being very close to the steady state.

If the frequency ratio is rational, the waveforms will eventually repeat and it becomes possible to find the steady state solution by iteration. This is the case for Fig. 11.3 for which the frequency ratio is 5 : 2 so that waveforms repeat regularly. The input waveforms, e.g. the a.c. line currents, repeat every five supply cycles during which time there are two cycles of output. For the output, the waveforms are even simpler since, being dependent on the 360 Hz carrier, they in fact repeat every output cycle, i.e. every 2.5 supply cycles. This can be clearly seen in Fig. 11.3 by observing the transfer points. Every cycle has the same waveform although the wave is not symmetrical, the negative half cycles being different from the positive half cycles since the number of carrier cycles per output cycle, 15, is odd.

11.7.1 *The output current equation*

Applying Kirchhoff's voltage law to the loop comprising the conducting rectifier and load, see Fig. 11.1, the differential equation defining the output current is

$$v_o = R_o i_o + X_o \frac{\mathrm{d}i_o}{\mathrm{d}\theta} + e_o \tag{11.10}$$

where the transformation from time variable t to the more convenient angle variable $\theta = \omega_s t$ has been made by replacing the load inductance by its reactance at the supply frequency and v_o depends on the conducting segment according to either Table 11.1 or 11.2 depending on whether the sign of the output current is positive or negative.

The driving voltage, v_o, has the form

$$v_o = V_{pk} \cos(\theta + \phi_{vo}) \tag{11.11}$$

where ϕ_{vo} is obtained from the switching arrays.

The load e.m.f., e_o, has the form

$$e_o = E_{pk} \cos(f_r\theta + \phi_{eo}) \tag{11.12}$$

where E_{pk} is its peak value and ϕ_{eo} is its phase angle.

The solution has three components, the steady state response to the driving voltage, v_o, the steady state response to the load e.m.f., and the transient solution. The steady state response to v_o is obtained using a.c. circuit theory as

$$I_{o1} \cos(\theta + \phi_{co1}) \tag{11.13}$$

where

$$I_{o1} = \frac{V_{pk}}{Z_{ofs}} \tag{11.14}$$

$$\phi_{co1} = \phi_{vo} - \zeta_{ofs} \tag{11.15}$$

$$Z_{ofs} = \sqrt{R_o^2 + X_o^2} \tag{11.16}$$

$$\zeta_{ofs} = \arctan(\frac{X_o}{R_o}) \tag{11.17}$$

and Z_{ofs} is the load impedance at the supply frequency and ζ_{ofs} is its phase angle.

The steady state response to e_o is similarly obtained as

$$-I_{o2}\cos(f_r\theta+\phi_{co2}) \tag{11.18}$$

where

$$I_{o2} = \frac{E_{pk}}{Z_{ofc}} \tag{11.19}$$

$$\phi_{co2} = \phi_{eo}-\zeta_{ofc} \tag{11.20}$$

$$Z_{ofc} = \sqrt{R_o{}^2+(f_rX_o)^2} \tag{11.21}$$

$$\zeta_{ofc} = \arctan(\frac{f_rX_o}{R_o}) \tag{11.22}$$

and Z_{ofc} is the load impedance at the control frequency and ζ_{ofc} is its phase angle.

The transient solution is

$$I_{o3}\exp[s_1(\theta-\phi_{co3})] \tag{11.23}$$

where

$$s_1 = -\frac{X_o}{R_o} \tag{11.24}$$

I_{o3} depends on the initial conditions and ϕ_{co3} is an angle whose value is constant for the segment. This angle is introduced to avoid excessively large values of I_{o3} when θ becomes large. It will be made equal to the value of θ at the start of the segment.

The complete solution is therefore

$$i_o = I_{o1}\cos(\theta+\phi_{co1})-I_{o2}\cos(f_r\theta+\phi_{co2})+I_{o3}\exp[s_1(\theta-\phi_{co3})] \tag{11.25}$$

This will be called the *basic cycloconverter expression*. Various functions of it are given in Appendix E.

If this segment starts at θ_{st} when the current is I_{ost} and ϕ_{co3} is set equal to θ_{st}, the amplitude of the transient component is given by

$$I_{o3} = I_{ost}-I_{o1}\cos(\theta_{st}+\phi_{co1})+I_{o2}\cos(f_r\theta_{st}+\phi_{co2}) \tag{11.26}$$

and the value of the output current at the end of the segment, I_{ofn}, which becomes the initial current for the next segment, is

$$I_{ofn} = I_{o1}\cos(\theta_{fn}+\phi_{co1})-I_{o2}\cos(f_r\theta_{fn}+\phi_{co2})+ \\ I_{o3}\exp[s_1(\theta_{fn}-\phi_{co3})] \tag{11.27}$$

Progressing in time, segment by segment, the computation eventually arrives at a segment where the sign of the finishing current is opposite to that of the starting current, indicating that the load current has passed through zero during the segment and that the cycloconverter controller will have changed rectifiers. The instant of crossing can be determined by first making a linear interpolation between the start and finish angles and then refining this using the Newton–Raphson technique.

The transfer delay is initiated at the current zero by blocking firing pulses to the outgoing rectifier. Firing pulses are released to the incoming rectifier at the end of this period. The output current is zero during the transfer period and the output voltage is the load e.m.f. The initial current for the first segment of the incoming rectifier is zero.

Incorporating these rules into a program which will iteratively solve for segment after segment is straightforward, although complicated. Enough segments are solved to ensure a reasonable approximation to the steady state and a good starting point ensures the efficiency of this process.

11.7.2 *The initial current*

An estimate of the initial current is found by considering the cycloconverter output frequency to be extremely low compared to the supply frequency. The mean output voltage is then $V_{omax}\cos(\alpha)$ where V_{omax} is the maximum d.c. voltage of which the rectifiers are capable. Under this assumption the cosine of the delay angle is the ratio of the control voltage to the amplitude of the reference voltage. From eqn 11.3 this is $M_i\cos(f_r\theta+\phi_c)$ so that the control signal frequency component of the output voltage is given by

$$v_o = M_i V_{omax}\cos(f_r\theta+\phi_c) \tag{11.28}$$

The difference between this and the load e.m.f. is the voltage

$$v_d = V_d\cos(f_r\theta+\delta) \tag{11.29}$$

which operates on the load impedance to produce a load current

$$i_d = I_d\cos(f_r\theta+\delta-\zeta_{ofc}) \tag{11.30}$$

When this expression is evaluated at the starting value of θ, it provides an estimate of the initial current.

The procedure is illustrated by Example 11.2.

Example 11.2

For the conditions of Example 11.1, i.e. a control frequency of 24 Hz, a modulation index of 0.7, zero phase angle for the positive rectifier control voltage, 180° phase angle for the negative rectifier control voltage, and a rectifier transfer delay of 10°, determine the waveforms of the load current and voltage if the load has a resistance of 0.34 Ω, an inductance of 5.6 mH, and an e.m.f. of sinusoidal waveform of r.m.s. value 254 V, frequency 24 Hz, and phase angle −5.6°.

The load impedance at the supply frequency is

$$0.34 + j2.1111 = 2.1384\underline{/80.8511^\circ}$$

The amplitude of the steady state response to the driving voltage is then either

$$\frac{622.25}{2.1384} = 290.997 \text{ A}$$

if the positive rectifier is conducting or

$$-\frac{622.25}{2.1384} = -290.997 \text{ A}$$

if the negative rectifier is conducting.
The phase angle of this current is

$$\phi_{vo} - 80.851^\circ$$

where ϕ_{vo} is selected from either Table 11.1 or 11.2 as appropriate, e.g. if the positive rectifier is conducting and $\theta = 483^\circ$, the value of ϕ_{vo} is obtained from Table 11.1, segment 7, as 30.0°.

The load impedance at the control frequency is

$$0.34 + j0.8445 = 0.9103\underline{/68.069^\circ}$$

so that the amplitude of the steady state response to the load e.m.f. is 394.591 A and its phase angle is

$$-5.6^\circ - 68.069^\circ = -73.669^\circ$$

The eigenvalue of the load circuit is

$$-0.34 \div 0.0056 = -60.71$$

when time in seconds is the independent variable and is

$$-0.34 \div 2.1111 = -0.1611$$

when θ is used as the independent variable, as is the case here.

The general expression for the load current is therefore

$$i_o = 291.00\cos(\theta - \phi_{vo} - 1.411) - 394.59\cos(0.4\theta - 1.2858) + I_{o3}\exp[-0.1611(\theta - \phi_{co3})]$$

The maximum d.c. voltage which can be produced by the rectifiers is 594.209 V so that the estimated fundamental component of the output voltage is

$$0.7 \times 594.21\cos(0.4\theta) = 415.94\cos(0.4\theta)$$

The load e.m.f. is $359.21\cos(0.4\theta - 0.09774)$ and the estimated fundamental net circuit driving voltage is the difference between the estimated fundamental component of output voltage and this e.m.f., i.e. 68.155 $\cos(0.4\theta + 0.5402)$ which, expressed as an r.m.s. phasor, is $48.193\underline{/30.951^\circ}$. The estimated fundamental component of output current in r.m.s. phasor terms is therefore

$$\frac{48.193\underline{/30.951^\circ}}{0.9103\underline{/68.069^\circ}} = 52.940\underline{/-37.118^\circ}$$

or, in the time domain, $74.868\cos(0.4\theta - 0.6478)$. The estimated value of the load current at $\theta = 0$ is therefore 59.7 A

With this initial value of current the problem is solved segment by segment using the current continuity condition to progress from one segment to the next, i.e. the final current for one segment is the initial current for the next. The waveforms of the output current and voltage are shown by black curves in Fig. 11.4, graphs 1 and 2. Graph 2 also shows the waveform of the load e.m.f. by a gray line. Each segment of the output current and voltage waves is identified by a number between delimiters located between the two graphs. Thus, the segment which extends from 534° to 613° is number 11. Note that these segment numbers are not the same as those used for the output voltage segments specified in Tables 11.1 and 11.2. Thus, the voltage segment corresponding to segment 11 of Fig. 11.4 is located in Table 11.1 as segment 12, v_{yr}, with a phase angle of −150.0°. The various parameters defining the output current and voltage segments are listed in Tables 11.3 and 11.4. From these tables it is found that segment 11 extends from 534.41° to 612.90° and that during it

$$v_o = -622.90\cos(\theta - 2.6180)$$

and

$$i_o = -291.00\cos(\theta - 4.0291) - 394.60\cos(0.4\theta - 1.2858) - 203.41\exp[-0.1610(\theta - 9.3273)]$$

9.3273 being the radian equivalent of 534.414°.

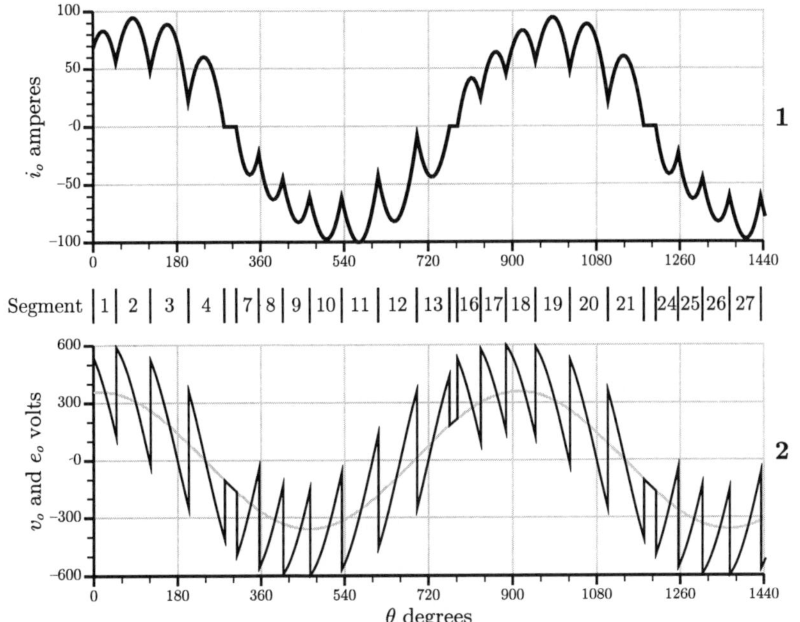

FIG. 11.4. Output current and voltage waveforms for the cycloconverter of Example 11.2.

Figure 11.4 will be considered segment by segment with the first few segments being skipped and returned to later because the graphs and Tables 11.3 and 11.4 reflect steady state conditions rather than the transient which would be obtained starting from the estimated initial current.

From Table 11.3, segment 4 terminates at 281.68° with the output current zero. The cycloconverter controller detects this current zero, blocks all further firing pulses to rectifier P and, after the rectifier transfer delay of 10°, releases firing pulses to rectifier N at 291.68°. The rectifier transfer segment from 281.68° to 291.68° is number 5. During it the output current is zero and the output voltage is the load e.m.f.

Reference to Table 11.2 reveals that when firing pulses are released to rectifier N at 291.68° it is in the middle of voltage segment 7 with supply voltage v_{br} connected to the output terminals via thyristors T3 and T4. The system driving voltage is then $-622.25\cos(\theta - 3.6652)$ and has the value -91.08 V which is aided, see the sign convention of Fig. 11.1, by the load e.m.f. of -128.926 V. The net circuit driving voltage is therefore -220.004 V which would, if it could, produce a negative rectifier output current. However, the rectifier can only conduct positive current and so no conduction occurs. The rectifier thyristors T3 and T4, although in the

Table 11.3 *Output segment data. Column 1 gives the number of the segment. Columns 2 and 3 list the start and finish of the segment. Columns 4, 5, and 6 list the output voltage amplitude, relative frequency, and phase angle*

Seg. no.	Start degrees	Finish degrees	V_{opk} V	f_r	ϕ_{vo} rad
1	0.00	48.71	622.25	1.00	0.5236
2	48.71	122.71	622.25	1.00	–0.5236
3	122.71	204.18	622.25	1.00	–1.5708
4	204.18	281.68	622.25	1.00	–2.6180
5	281.68	291.68	359.21	0.40	–0.0977
6	291.68	307.57	359.21	0.40	–0.0977
7	307.57	356.26	–622.25	1.00	1.5708
8	356.26	407.93	–622.25	1.00	0.5236
9	407.93	465.92	–622.25	1.00	–0.5236
10	465.92	534.41	–622.25	1.00	–1.5708
11	534.41	612.90	–622.25	1.00	–2.6180
12	612.90	695.82	–622.25	1.00	2.6180
13	695.82	765.03	–622.25	1.00	1.5708
14	765.03	775.03	359.21	0.40	–0.0977
15	775.03	781.68	359.21	0.40	–0.0977
16	781.68	831.57	622.25	1.00	–0.5236
17	831.57	885.85	622.25	1.00	–1.5708
18	885.85	948.71	622.25	1.00	–2.6180
19	948.71	1022.71	622.25	1.00	2.6180
20	1022.71	1104.18	622.25	1.00	1.5708
21	1104.18	1181.68	622.25	1.00	0.5236
22	1181.68	1191.68	359.21	0.40	–0.0977
23	1191.68	1207.57	359.21	0.40	–0.0977
24	1207.57	1256.26	–622.25	1.00	–1.5708
25	1256.26	1307.93	–622.25	1.00	–2.6180
26	1307.93	1365.92	–622.25	1.00	2.6180
27	1365.92	1434.41	–622.25	1.00	1.5708
28	1434.41	1512.90	–622.25	1.00	0.5236

conducting mode, are reverse biased by the combination of the a.c. voltage and the load e.m.f. This situation persists until the end of this voltage segment at 307.57°. The rectifier has been enabled too late in its conduction segment to pick up load current.

A simple controller such as the one described in Chapter 10 would

Table 11.4 *Parameters of eqn 11.25 defining the output current for the segments listed in the first column*

Seg. no.	I_{o1} A	ϕ_{co1} rad	I_{o2} A	f_r	ϕ_{co2} rad	I_{o3} A	s_1	ϕ_{co3} rad
1	291.0	−0.8875	394.6	0.4	−1.2858	−4.3	−0.1610	0.0000
2	291.0	−1.9347	394.6	0.4	−1.2858	151.1	−0.1610	0.8501
3	291.0	−2.9819	394.6	0.4	−1.2858	213.3	−0.1610	2.1417
4	291.0	−4.0291	394.6	0.4	−1.2858	152.7	−0.1610	3.5636
5	0.0	−4.0291	0.0	0.4	−1.2858	0.0	−0.1610	3.5636
6	0.0	1.2069	0.0	0.4	−1.2858	0.0	−0.1610	5.0909
7	−291.0	0.1597	394.6	0.4	−1.2858	468.9	−0.1610	5.3681
8	−291.0	−0.8875	394.6	0.4	−1.2858	287.8	−0.1610	6.2180
9	−291.0	−1.9347	394.6	0.4	−1.2858	90.7	−0.1610	7.1197
10	−291.0	−2.9819	394.6	0.4	−1.2858	−89.5	−0.1610	8.1318
11	−291.0	−4.0291	394.6	0.4	−1.2858	−203.4	−0.1610	9.3273
12	−291.0	1.2069	394.6	0.4	−1.2858	−203.4	−0.1610	10.6971
13	−291.0	0.1597	394.6	0.4	−1.2858	−86.0	−0.1610	12.1443
14	0.0	0.1597	0.0	0.4	−1.2858	0.0	−0.1610	12.1443
15	0.0	−0.8875	0.0	0.4	−1.2858	0.0	−0.1610	13.5268
16	291.0	−1.9347	394.6	0.4	−1.2858	−393.5	−0.1610	13.6428
17	291.0	−2.9819	394.6	0.4	−1.2858	−199.6	−0.1610	14.5136
18	291.0	−4.0291	394.6	0.4	−1.2858	−4.5	−0.1610	15.4609
19	291.0	1.2069	394.6	0.4	−1.2858	151.1	−0.1610	16.5580
20	291.0	0.1597	394.6	0.4	−1.2858	213.3	−0.1610	17.8497
21	291.0	−0.8875	394.6	0.4	−1.2858	152.7	−0.1610	19.2716
22	0.0	−0.8875	0.0	0.4	−1.2858	0.0	−0.1610	19.2716
23	0.0	−1.9347	0.0	0.4	−1.2858	0.0	−0.1610	20.7988
24	−291.0	−2.9819	394.6	0.4	−1.2858	468.9	−0.1610	21.0761
25	−291.0	−4.0291	394.6	0.4	−1.2858	287.8	−0.1610	21.9259
26	−291.0	1.2069	394.6	0.4	−1.2858	90.7	−0.1610	22.8277
27	−291.0	0.1597	394.6	0.4	−1.2858	−89.5	−0.1610	23.8398
28	−291.0	−0.8875	394.6	0.4	−1.2858	−203.4	−0.1610	25.0352

detect zero current in rectifier N and switch to rectifier P. However, in a cycloconverter this would have the unfortunate consequence of causing dither between the two rectifiers, which produces many possible types of undesirable behavior depending on the load parameters and the rectifier transfer delay. To avoid this, it is essential that the controller anticipate the possibility that the incoming rectifier will not pick up current on its first try and will allow it a second chance with the next segment. This has been done for Fig. 11.4 and rectifier N successfully starts conduction in segment 7 at 307.57° with thyristor T3 connecting the blue line to the d.c. positive bus and thyristor T5 connecting the negative d.c. bus to the yellow line.

The nonconducting period from 281.68° to 307.57° spans two output segments, numbers 5 and 6, number 5 being the rectifier transfer delay from 281.68° to 291.68° and number 6 being the period when rectifier N, although in the conducting mode, is reverse biased. Throughout these two segments the output voltage is the load e.m.f.

Rectifier N continues to supply negative current to the load during segments 7 through 13 but at the end of this last segment, at 765.03°, the current becomes zero and a transfer is made to rectifier P at 775.03° under similar conditions to those just described. From Table 11.1, voltage segment 13, the transfer connects v_{ry} to the load. However, as for the previous transfer, this is too late in the segment to initiate conduction, the rectifier being reverse biased. On the second try at 781.68°, voltage segment 14 of Table 11.1, rectifier P picks up the load current and a positive half cycle is initiated. The rectifier changeover hiatus covers segments 14 and 15, segment 14 being the rectifier transfer delay from 765.03° to 775.03° and segment 15, from 775.03° to 781.68° being the period when rectifier P, although in the conducting mode, is reverse biased.

It is now possible to see how the steady state condition was attained. Each half cycle of load current starts from zero current and is therefore influenced only by the timing of the current zero of the preceding half cycle. In general this can only be determined by solving the problem segment by segment until an acceptable approximation to the steady state is reached. However, an output frequency of 24 Hz was selected for this example so that there are $360 \div 24 = 15$ carrier cycles per output cycle. Thus, the output cycles are identical and the current one output cycle after the start, at $360 \div 0.4 = 900°$, is the correct value of the initial current. From Table 11.3, segment 18, the expression for the output current at this point is

$$291.00\cos(\theta - 4.0291) - 394.59\cos(0.4\theta - 1.2858) -$$
$$4.457\exp[-0.1610(\theta - 15.4609)]$$

which yields 68.48 A as the correct initial current. After finding this value the program should recommence with this initial value.

While the above technique is useful for demonstration purposes, it must be emphasized that it is only applicable for cases such as this where there is an integral number of carrier cycles per output cycle.

****

11.8 Cycloconverter controller characteristics

Example 11.2 has illustrated the requirements for a cycloconverter controller. It must be able to detect a current zero, block firing pulses to the outgoing rectifier, and have time to confirm that the outgoing rectifier really has stopped conducting before releasing firing pulses to the incoming rectifier. Further, since it is probable that the incoming rectifier will fail to pick up current at its first try, it must be allowed a second attempt before a further rectifier transfer is initiated.

In practice, two attempts will provide successful operation in the majority of circumstances. However, conditions can be envisaged when this would be inadequate. For example if conditions are such that there are several discontinuous conduction segments at crossover, a rectifier may require more than two attempts to successfully initiate a half cycle. The penalty for inadequate control is a wide variety of dithering behavior with neither rectifier being able to unequivocally gain ascendancy. Such a condition is hard to anticipate since it depends on the load characteristics, the control signal, and the rectifier change over delay and is best dealt with by a flexible controller tuned to specific circumstances.

11.9 Half cycle asymmetry

Careful inspection of Fig. 11.4 reveals that the output waves are asymmetrical. The negative half cycle contains seven nonzero current segments covering 457.5°, 183.0° at the control frequency, while a positive half cycle contains six nonzero segments covering 400.0°, 160.0° at the control frequency. It is evident from inspection that the output contains a d.c. component and computation confirms this, the d.c. component of current being −4.49 A. The presence of a d.c. current in an a.c. load is usually highly undesirable so that this aspect of cycloconverter behavior must be further investigated.

The asymmetry is affected by the phase angle of the control signal. All the waveforms shown so far have been for a control signal phase of zero for the positive rectifier and 180° for the negative rectifier. The output waveforms will of course repeat for every 60° of phase shift, the duration of a carrier cycle, but within this range the waveforms are affected by the control phase. Figure 11.5 illustrates this point by showing the waveform of the output current for control phase angles of zero, 15°, 30°, 45°, and 60°. The phase angle is shown slightly to the right and above the x axis of each graph.

To facilitate waveform comparisons, the graphs have been displaced in the x direction by the amount of the phase shift so that the waves as displayed appear exactly in phase. The graphs have been displaced in the

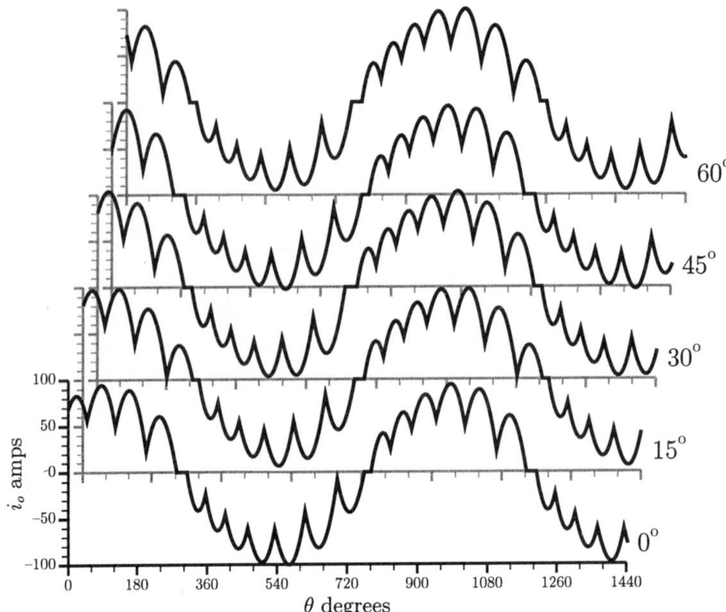

FIG. 11.5. Output current waveforms for various values of the control signal phase. The frequency and modulation index of the control signal are fixed at 24 Hz and 0.7. The phase angle of the control signal is given to the right of the graphs. The graph origins are shifted to facilitate waveform comparison.

y direction so that the x axis of one lies on the center line of the one below.

Careful inspection reveals the differences in waveforms. For example the rectifier crossover points differ from graph to graph. The lowest graph, the zero phase angle case, is the one already considered. It has six non-zero segments in a positive half cycle and seven in a negative half cycle. The topmost graph, the 60° phase angle case, is its inverse. This does not conform with the statement made above, that the output waveforms repeat every 60°. The apparent divergence from this contention reveals an interesting fact about the waveforms: either is possible depending on the initial value of output current used to start the computation. If in the zero phase case it had been chosen somewhat larger, the 0° waveform would have been like that of the 60° waveform and vice versa.

Tables 11.5 and 11.6 quantify the waveform differences by listing some basic parameters, mean, r.m.s., fundamental, and ripple values. These quantities are given for the output voltage in Table 11.5 and for the output current in Table 11.6. The ripple component listed in the final column of each table is the value for all Fourier components other than the

Table 11.5 *Parameters of the output voltages associated with the currents shown in Fig. 11.5. Column 1 gives the phase angle of the modulation signal and the remaining columns list output voltage parameters. Column 2 gives the r.m.s. value and column 3 the mean value. Columns 4 and 5 give the r.m.s. value and phase angle of the Fourier component at the control frequency and column 6 gives the r.m.s. value of all the other Fourier components combined*

ϕ_m degrees	V_{orms} V	V_{odc} V	V_{o1rms} V	ϕ_{vo1} degrees	V_h V
0.00	336.11	-1.527	299.42	1.37	152.72
15.00	335.71	1.476	298.84	16.03	152.95
30.00	336.49	1.632	299.98	32.09	152.44
45.00	336.38	-1.580	299.80	46.72	152.55
60.00	336.11	1.528	299.42	61.37	152.72

fundamental, i.e. it is the square root of the difference between the squares of the values given in columns 2 and 4.

Throughout this work the need to cross check complex computations whenever possible has been emphasized. These two tables provide such an opportunity. The d.c. voltage component of Table 11.5 divided by the circuit resistance should give the corresponding d.c. current in Table 11.6. Taking the 45° phase angle case, the d.c. voltage component is −1.580 V which, divided by the circuit resistance of 0.34 Ω, yields a d.c. current component of −4.647 A, the value in Table 11.6. At the reference frequency, 24 Hz, the difference between the fundamental component of voltage and the load e.m.f. divided by the load impedance should give the fundamental component of current. For the 45° case, Table 11.5 gives the fundamental component of voltage as $299.80\underline{/46.72^\circ}$. The load e.m.f. is $254.0\underline{/39.40^\circ}$ so that the difference voltage is $57.78\underline{/80.78^\circ}$. The circuit impedance at 24 Hz is $0.9103\underline{/68.07^\circ}$ so that the fundamental component of current is $57.78\underline{/80.78^\circ}$ divided by $0.9103\underline{/68.07^\circ}$, i.e. $63.48\underline{/12.72^\circ}$, the value given in Table 11.6.

11.10 Asynchronous operation

What might appear to be minor differences between the 24 Hz waveforms takes on a more serious aspect when it is considered that operation synchronized at exactly 0.4 times the supply frequency and at an exactly maintained phase angle is highly unlikely. Asynchronous operation is far more probable and the consequences of this can be seen by considering a control

Table 11.6 *Parameters of the output currents shown in Fig. 11.5. Column 1 gives the phase angle of the modulation signal and the remaining columns list output current parameters. Column 2 gives the r.m.s. value and column 3 the mean value. Columns 4 and 5 give the r.m.s. value and phase angle of the Fourier component at the control frequency and column 6 gives the r.m.s. value of all the other Fourier components combined*

ϕ_m degrees	I_{orms} A	I_{odc} A	I_{o1rms} A	ϕ_{co1} degrees	I_h A
0.00	63.52	–4.493	62.00	–33.62	13.82
15.00	61.95	4.340	60.43	–19.82	13.62
30.00	66.43	4.801	64.86	–0.81	14.34
45.00	65.02	–4.647	63.48	12.72	14.06
60.00	63.52	4.493	62.00	26.38	13.81

signal whose frequency is slightly displaced from 24 Hz. Such a signal can be interpreted as a 24 Hz signal whose phase angle systematically drifts over the full range of zero to 360° at the displacement frequency. The properties of the output voltage and current will correspondingly drift, repeating for every 60° change in phase angle. Thus, the d.c. level of the output current will vary slowly between about ±4.8 A creating a subharmonic at the difference frequency with an r.m.s. magnitude of about 3.4 A. Similarly, the r.m.s. amplitude of the fundamental component of current will vary between about 60.4 and 64.9 A at the same frequency.

In considering the subharmonic frequency, the starting point is the carrier frequency. If the rectifiers are fully controlled three phase bridges, this is six times the supply frequency. The period of the lowest order subharmonic is the shortest time which contains both an integral number of carrier cycles and an integral number of control cycles. In the example the carrier frequency is 360 Hz, the signal frequency is 24 Hz, and their ratio is 15 : 1—integral so that there are no subharmonics. For a ratio of 15.25 : 1, i.e. a control frequency of very nearly 23.61 Hz, the first integral ratio is 61 : 4 and there would be subharmonics at one-quarter, one-half, and three-quarters of the control frequency.

For steady input and load, all variations in the output are undesirable, the subharmonics being particularly so, and they are, unfortunately, an inevitable consequence of cycloconverter operation in which (unlike in their rival, inverters with a d.c. link) there is no buffer between the supply and the output. Fortunately the magnitude of the problem is reduced as the control frequency is reduced since there are then more carrier cycles per control cycle and variations between control cycles are correspondingly reduced. If the control frequency were close to 5 Hz, there would have been about 72 carrier cycles per control cycle. With the next integral ratios at 71 : 1 and 73 : 1 there is very little difference between successive control cycles. It is for this reason that cycloconverters are generally employed for the production of very low frequencies.

11.11 The input currents, power, and power factor

Since the model being considered is ideal with zero supply impedance, current transfer from one thyristor to the next is instantaneous and only two thyristors in a rectifier conduct simultaneously. Since circulating current is suppressed, only one rectifier conducts at a time. This greatly simplifies the search for the a.c. line currents. At any given time only two of the three lines carry current and that current is the output current entering the rectifier on one line and returning from it on the other. The entries in the second columns of Tables 11.1 and 11.2, the ID columns, aid in determining which lines are carrying current at a given time. However, in using them for this purpose it must be remembered that the segment numbers in Tables 11.1 and 11.2 are not the same as those in succeeding tables. The process is best illustrated by an example.

Example 11.3

Determine the waveforms of the a.c. input line currents for the case of the previous examples.

Consider segment 12 of Tables 11.3 and 11.4. From Table 11.3 this segment extends from 612.90° to 695.82° and from Fig. 11.4 it is seen that during it the negative rectifier is conducting. These facts direct attention to Table 11.2, segment 13. The ID for this segment is BR, indicating that the blue line is connected to the d.c. negative bus via thyristor T3 and the d.c. positive bus is connected to the red line via thyristor T4. The current in the blue line, i_b of Fig. 11.1, is then the same as the output current, the current in the red line is -1 times the output current, and the current in the yellow line is zero.

Progressing in this way through the output segments builds the a.c. line current waveforms shown in Fig. 11.6. The graphs are identified by a number to their right. Graph 1 shows the waveform of the output current for reference. Graphs 2, 3, and 4 show the waveforms of the three a.c. line currents. Graph 5 shows the waveform of the input power. For checking purposes, the waveform of the output power has also been drawn but only one graph is visible since the two waveforms are identical, as they should be since the idealized model has zero internal loss. The graphs also show,

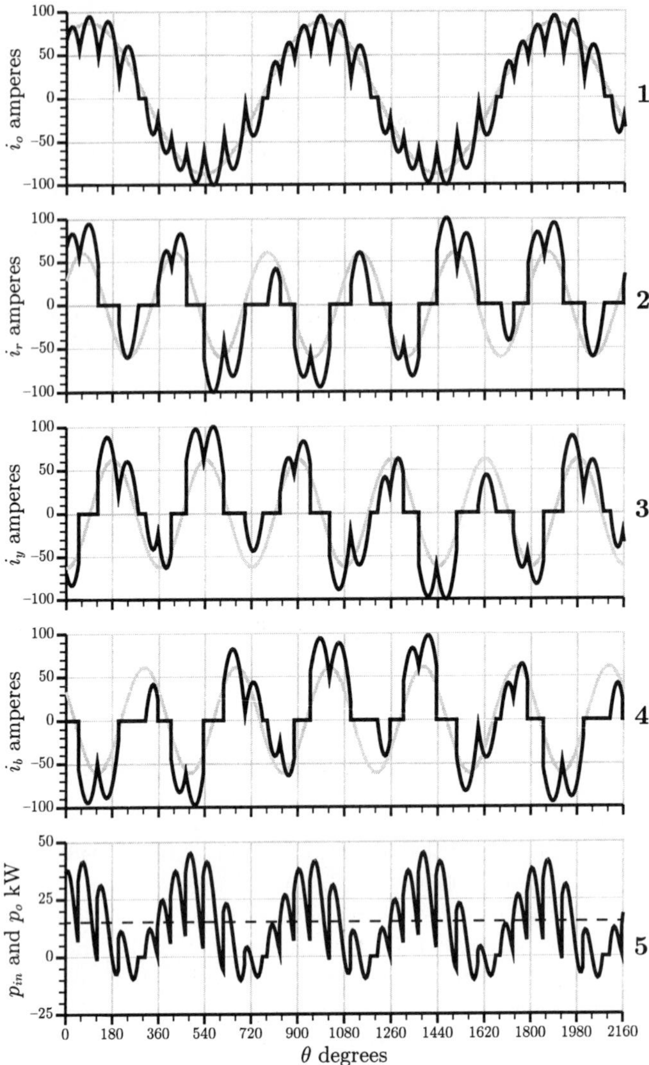

FIG. 11.6. Current and power waveforms for a control signal with a frequency of 24 Hz, a modulation index of 0.7, and zero phase. Graph 1 shows the load current, graphs 2, 3, and 4 show the currents in the red, yellow, and blue a.c. lines, and graph 5 shows the waveforms of the input and output powers which are identical. The control frequency Fourier component of the output current is shown in gray in graph 1. The supply frequency components of the a.c. line currents are shown in gray on graphs 2, 3, and 4. The mean value of the power is shown by a dashed line in graph 5.

in gray for graphs 1 through 4 and by a dashed black line for graph 5, appropriate Fourier components. For graph 1, this is the component at the output frequency, for graphs 2, 3, and 4, it is the component at the supply frequency, and for graph 5, it is the mean value.

Figure 11.6 covers a considerably greater range of θ than Figs 11.3, 11.4, and 11.5, 2160° as against 1440°. This is necessary to show complete cycles of input current waveform since these occupy five supply cycles. This is in contrast to the output waveforms which, since every output cycle contains an integral number (15) of carrier cycles, repeats every control cycle, i.e. every 2.5 supply cycles. A complete cycle of input therefore occupies two output cycles, five input cycles, 1800°. Inspection of Fig. 11.6, graphs 2, 3, and 4 reveals the correctness of this conclusion, the waveforms do repeat after 1800°.

From the a.c. line current waveforms, it is impossible to make other than the most sweeping conclusions regarding their relationships. Clearly they are not the same, clearly they are modulated at half the output frequency but it is very difficult to go beyond this without a full Fourier analysis, the subject of Section 11.12. However, before doing that some useful deductions can be made on the basis of a few key parameters. The currents in the red, yellow, and blue lines have r.m.s. values of 51.79, 51.98, and 51.82 A. Their mean values are well within the computational roundoff error and can be taken to be zero as expected. Their Fourier components at the supply frequency, the waves shown in gray, are respectively $42.82\underline{/-61.69^\circ}$, $44.24\underline{/-182.23^\circ}$, and $43.19\underline{/56.40^\circ}$, where the magnitudes are r.m.s. values. Clearly these are not balanced three phase components. In fact, based on the red line, there is a positive sequence component of $43.41\underline{/-62.52^\circ}$ and a negative sequence component of $0.86\underline{/71.91^\circ}$, the zero sequence component being zero since this is a three wire a.c. supply system.

****

11.11.1 *Power*

The waveforms of the total three phase input power and the output power for the example are shown in graph 5 of Fig. 11.6. Since the rectifiers are assumed to be ideal with zero internal loss, the input power equals at all times the output power and the two waveforms are identical. That this is the case in graph 5 is another indication of the correctness of these very complex computations.

Like the output current and voltage waveforms, the power waveform repeats every output cycle, i.e. every 900°. If the voltage and current waveforms were pure, ripple free sinusoids the power waveform would be at twice the a.c. frequency. This is nearly the case here, the slight variation at the output frequency being due to the d.c. component of the output current.

The mean power drawn from the a.c. system is 15.27 kW, the same as the mean power delivered to the load. The component at the output frequency delivered to the load is 15.21 kW and the difference, 65 W, is the power delivered to the load by all other Fourier components, a surprisingly small amount in view of the high harmonic content of all waveforms.

11.11.2 *Power factor*

In a situation such as this, power factor is, from a theoretical viewpoint, an elusive concept. One can think of the fundamental power factor of an input phase as the cosine of the phase difference between the fundamental component of line current and the corresponding line to neutral voltage. From this viewpoint the power factors of the red, yellow, and blue phases are 0.4743, 0.4659, and 0.4446. One could consider the cycloconverter power factor to be the mean of these three values, 0.4617. Alternatively one could think of the power factor as that of the positive sequence component of the fundamental current which is 0.4617, the power factors of the negative and zero sequence components being zero since the corresponding voltage components are zero.

While the foregoing approaches to power factor have theoretical attractions, their practical importance is overwhelmed by that of the conventional power factor, the ratio of the total input power to the conventional kVA, because of its importance in metering and billing. Problems are encountered even in such an apparently simple concept because of the phase imbalance. Here the input kVA has been taken as the product of the average of the three r.m.s. line currents and $\sqrt{3}$ times the r.m.s. line to line voltage divided by 1000. With this definition the kVA is 39.53 so that, with the total input power noted above at 15.27 kW, the conventional power factor is 0.3864.

Whichever power factor value is employed, it is impossible to avoid the conclusion that it is very low. The load has a fundamental power factor of 0.8193 and yet it appears to the three phase a.c. system as having a power factor of about half this. This approximate relationship, that the input power factor is about half the output power factor, applies generally to cycloconverters with sinusoidal modulation since so much of their time is spent in the low voltage, low power factor regions where the output voltage wave passes through zero. The poor power factor can be alleviated by the use of non-sinusoidal modulation, e.g. trapezoidal modulation, which reduces the relative time spent at low voltage levels, a matter discussed in Section 11.14.

Altogether, the combination of high harmonic content, imbalance, subharmonic modulation, and poor power factor makes the single phase cycloconverter, from the viewpoint of the a.c. system, a highly undesirable load. Fortunately cycloconverter loads of any appreciable power are invariably three phase and some of these undesirable aspects are thereby mitigated, as will be seen in Section 11.13.

11.12 Fourier analysis

Time domain analysis is essential for the determination of cycloconverter waveforms but Fourier analysis is a useful adjunct, giving insights into the structure of very complex waveforms. It has already been noted that in general, when the frequency ratio is irrational, the waveforms never repeat. They therefore have a continuous spectrum whose precise determination requires that the Fourier integral be performed over an infinite time. This is obviously not practical and recourse must be had to approximations, notably the fast Fourier transform applied over a finite time with an appropriate weighting function such as a Hanning window.

When the frequency ratio can be expressed as a rational fraction, as in the example where it is 5 : 2, the waveform does repeat and it can be described by a discrete Fourier series whose coefficients can be obtained by the standard techniques. The repeat time is the period of the wave multiplied by the appropriate portion of the rational fraction reduced to its lowest terms. Thus, in the example the ratio can be expressed as 20 : 8 or 10 : 4 but when reduced to its lowest terms, i.e. when the numerator and denominator do not have any common factor, it is 5 : 2. This means that a waveform cycle occupies 2 cycles of output frequency and 5 cycles of supply frequency. It is evident that, even when the frequency ratio is rational, the repeat time will generally be long. For example if the output frequency of the example were changed slightly from 24 to 23.7 Hz the ratio would be 60 : 23.7 which becomes the rational fraction 600 : 237. The numerator and denominator of this fraction have a common factor, 3, so that the ratio reduced to its lowest terms is 200 : 79. A waveform cycle would therefore occupy 79 output cycles or 200 supply cycles and would involve approximately 1200 distinct segments. While not infinite, this is a formidable amount of information. Fortunately most of the useful generalizations about cycloconverter waveforms can be deduced from the much more limited case of the example.

11.12.1 *Nomenclature*

Habitually, Fourier series are discussed in terms of a fundamental, the component of prime importance, and its harmonics. The utility of this nomenclature is very limited when applied to cycloconverter waveforms since there are then three frequencies of prime importance, the output frequency, the a.c. supply frequency, and the carrier frequency. Unless the signal frequency happens to be an integral submultiple of the supply frequency, a waveform cycle will occupy several output cycles and there will be Fourier components at corresponding submultiples of the output frequency. In the example, since a waveform cycle occupies two output cycles, i.e. five supply cycles, there will be Fourier components at one-half of the output frequency, i.e. one-fifth of the supply frequency. Such low frequency components are usually referred to as subharmonics.

While it is natural to think of the subharmonics in relationship to the output frequency, doing so takes a variable—the output frequency, as point of reference, entailing a shift in viewpoint every time the output frequency is changed. This is not convenient and it is therefore the standard practice to take the supply frequency as the reference.

Whichever reference frequency is employed, Fourier components of non-integral order are encountered. Thus, in the example and taking the supply frequency as reference, components of order 0, 0.2, 0.4, 0.6 ... can be expected, the component of order 0.4 being at the output frequency. The expressions for the Fourier components given in section C.2 are true, with some minor caveats, for fractional order. First, the integrals of eqns C.8, C.9, and C.10 are taken over a waveform cycle not one cycle of the unity order component. This means that in the example, where the angle θ is based on the supply frequency, the integral must be taken over the range β to $\beta + 2\pi \times 5$, β being any convenient angle. While any 10π range will do, it has been taken here as the range 281.68° to 2081.68°, i.e. for two cycles of output current from the start of the first negative half cycle. Second, where π occurs in the denominator of these expressions it is replaced by half the range of the integral. Thus, in the example the range of the integral is $2\pi \times 5$ so that π in the denominator is replaced by 5π.

Each waveform is made up of a series of segments each segment being defined by an appropriate basic cycloconverter expression, see eqn 11.25. Thus, the following integrals must be evaluated for every segment in the range and the results summed

$$\int_{\theta_{st}}^{\theta_{fn}} f(\theta)\cos(m\theta)\mathrm{d}\theta \quad \text{and} \quad \int_{\theta_{st}}^{\theta_{fn}} f(\theta)\sin(m\theta)\mathrm{d}\theta$$

where θ_{st} is the start and θ_{fn} is the finish of the segment.

The integrals are given in Sections E.6 and E.7 and the process is illustrated by Example 11.4

Example 11.4

For the example determine the Fourier components of the output voltage and current and the a.c. line currents up to and including order 9.2.

The frequency ratio is 5 : 2 so that a waveform cycle occupies five supply cycles. Thus, the analysis, starting from the zero order component, must progress in steps of 0.2 up to order 9.2, i.e. $1 + 5 \times 9.2 = 47$ Fourier components must be found.

The coefficients of the required basic cycloconverter expressions for the output voltage and current have been given in Tables 11.3 and 11.4. The method for deriving those for the a.c. line currents has been described in Example 11.3 and the coefficients for i_r are listed in Table 11.9. Considering segment 19, from Fig. 11.4 it is seen that the output current is positive during this segment so that the positive rectifier must be conducting. From Table 11.3 it is seen that this segment extends from 948.71° to 1022.71°. Turning now to Table 11.1, the switching array for the positive rectifier, it is found that the segment covering this range of θ is number 17 and that during it, the blue line is connected to the positive d.c. bus and the negative d.c. bus is connected to the red line. Hence, $i_r = -i_o$, i_y is zero and $i_b = i_o$. Turning now to Table 11.4, segment 19, the expressions for the line currents are

$$\begin{aligned} i_r &= -291.0\cos(\theta + 1.2069) + 394.6\cos(0.4\theta - 1.2858) - \\ &\qquad 151.1\exp[-0.1611(\theta - 16.5580)] \\ i_y &= 0.0 \\ i_b &= 291.0\cos(\theta + 1.2069) - 394.6\cos(0.4\theta - 1.2858) + \\ &\qquad 151.1\exp[-0.1611(\theta - 16.5580)] \end{aligned}$$

The contribution of the red line current during this segment to the Fourier component of order 3.4 is

$$\Delta I_{rc3.4} = \frac{1}{5\pi}\int_{16.5581}^{17.8497}\Big\{-291.0\cos(\theta + 1.2069) + 394.6\cos(0.4\theta - 1.2858) \\ -151.1\exp[-0.1611(\theta - 16.5580)]\Big\}\cos(3.4\theta)\,\mathrm{d}\theta$$

and

$$\Delta I_{rs3.4} = \frac{1}{5\pi}\int_{16.5581}^{17.8497}\Big\{-291.0\cos(\theta + 1.2069) + 394.6\cos(0.4\theta - 1.2858) \\ -151.1\exp[-0.1611(\theta - 16.5580)]\Big\}\sin(3.4\theta)\,\mathrm{d}\theta$$

where $\Delta I_{rc3.4}$ is the contribution to the cosine component and $\Delta I_{rs3.4}$ is the contribution to the sine component of the Fourier component of order 3.4 of the red line current and the limits of the integral have been expressed in radians.

These integrals are evaluated for every segment within a 10π range. This has been done for segments 5 through 38, and the results are summed to give the amplitude of the cosine and sine components. These are then converted to an amplitude and phase angle which are more convenient for both study and computation. When this is done the Fourier component of order 3.4 of the red line current is

$$i_{r3.4} = 13.44\cos(3.4\theta + 1.1317)$$

Repeating this process for each variable and each Fourier component yields the data of Tables 11.7 and 11.8 which list the peak amplitude and phase for each Fourier component. Where an amplitude is within the round-off error of the computation, it has been omitted from the tables. The frequency spectra of the waves are shown in Fig. 11.7 which has been extended up to the component of order 15. Each spectrum is identified by a number to its right, 1 being that for the output voltage, 2 for the output current, and 3, 4, and 5 for the a.c. line currents, i_r, i_y, and i_b.

****

11.12.2 *Frequency spectra*

Some conclusions regarding the specific spectra of Fig. 11.7 are readily drawn. It is seen that the orders of the components of the output voltage and current are multiples of 0.4. This is a consequence of the output waves repeating every 2.5 supply cycles, every 15 carrier cycles. The zero frequency component of voltage has an amplitude of 1.5283 V, too small to show in its spectrum. In contrast the zero order current component, amplitude 4.493 A, is readily seen in Spectrum 2. The ratio of these two values is the d.c. resistance of the load. The largest component of both voltage and current is the one at the output frequency, order 0.4. The phasor difference between this voltage component and the load e.m.f. divided by the current component is the load impedance at the output frequency. The higher order components are relatively small, the voltage components persisting to the limit of Fig. 11.7 and reflecting the discontinuities in the output voltage waveform. In contrast the output current components have almost disappeared by order 10, a consequence of the increase in load impedance with frequency.

The orders of the input current spectra are odd multiples of 0.2, a reflection of the fact that, relative to the waveform repetition frequency of $0.2\,f_s$, the line current waveforms are odd functions. The largest component is, as expected, the one at the supply frequency, order 1, but there are substantial components at frequency $0.2\,f_s$, a consequence of the modulation

Table 11.7 *Fourier components of the output voltage and current. Column 1 gives the order of the component, columns 2 and 3 the amplitude and phase of the voltage component, and columns 4 and 5 the amplitude and phase of the current component*

Order m	V_{om} V	ϕ_{vom} degrees	I_{om} A	ϕ_{com} degrees
0.00	1.53	–180.00	4.49	180.00
0.40	423.44	1.37	87.68	–33.62
0.80	4.72	–22.01	2.74	–100.53
1.20	12.42	161.90	4.86	79.52
1.60	8.28	105.58	2.44	21.34
2.00	12.18	–89.37	2.88	–174.74
2.40	13.57	–136.69	2.67	137.14
2.80	11.70	20.53	1.98	–66.20
3.20	29.02	–19.00	4.29	–106.11
3.60	10.87	131.69	1.43	44.27
4.00	72.74	149.04	8.61	61.35
4.40	9.60	–115.64	1.03	156.48
4.80	101.54	9.24	10.01	–78.85
5.20	7.81	–1.07	0.71	–89.31
5.60	62.79	–85.97	5.31	–174.33
6.00	5.43	116.90	0.43	28.37
6.40	71.99	–41.40	5.33	–129.95
6.80	2.35	–97.08	0.16	174.25
7.20	46.09	24.08	3.03	–64.62
7.60	8.94	159.25	0.56	70.47
8.00	23.41	100.64	1.39	11.78
8.40	32.50	–14.91	1.83	–103.81
8.80	11.41	–174.26	0.61	96.79
9.20	49.90	–169.74	2.57	101.26

by the load current. Other significant components can be seen at 1.8, 3.4, 4.2, and 5.0 f_s.

While not obvious, some general conclusions can be drawn. The output is derived from a 360 Hz carrier modulated by a 24 Hz signal and bands of components centered on the modulation frequency and on the carrier frequency and its multiples are therefore expected. With a little imagination, such bands centered on orders 6 and 12 can be distinguished in Spectra 1 and 2.

Table 11.8 *Fourier components of the a.c. line currents. Column 1 gives the order of the component, columns 2, 4, and 6 the amplitude, and columns 3, 5, and 7 the phase of the component in the red, the yellow, and the blue lines*

Order m	I_{rm} A	ϕ_{crm} degrees	I_{ym} A	ϕ_{cym} degrees	I_{bm} A	ϕ_{cbm} degrees
0.20	30.30	12.98	28.13	−106.15	29.65	137.04
0.60	8.19	143.52	4.76	−49.43	3.71	−19.79
1.00	60.55	−61.69	62.57	177.77	61.08	56.40
1.40	2.02	−156.97	4.05	6.66	2.19	171.58
1.80	9.40	−97.22	8.59	156.00	10.75	32.87
2.20	0.94	−9.33	1.97	102.87	1.83	−105.42
2.60	4.18	−64.25	2.18	−63.76	6.36	115.92
3.00	0.85	123.89	1.03	−158.26	1.47	−12.85
3.40	13.44	64.84	11.22	−173.20	12.12	−63.40
3.80	1.23	−101.94	1.10	−90.20	2.33	83.60
4.20	14.76	−69.74	15.66	54.13	14.34	175.36
4.60	1.87	109.13	2.17	27.94	3.08	−115.06
5.00	8.29	−89.84	7.07	28.41	7.95	141.67
5.40	0.32	91.23	1.67	−173.20	1.67	−4.16
5.80	5.12	−61.57	5.20	54.16	5.49	176.99
6.20	2.48	55.24	4.16	−71.44	3.34	145.13
6.60	2.16	−0.75	2.97	106.32	3.11	−115.17
7.00	1.65	21.06	0.15	160.71	1.54	−155.33
7.40	0.78	83.70	1.33	−175.85	1.42	−28.80
7.80	2.17	−100.89	1.74	−10.34	2.77	118.05
8.20	0.58	155.41	0.40	−21.44	0.18	−31.78
8.60	3.63	104.47	4.45	−120.64	3.19	5.64
9.00	0.73	−49.93	0.44	−133.91	0.89	100.81

In the case of a rectifier, the a.c. system line currents would have components at the supply frequency, and at odd multiples of this frequency excluding triplens, i.e. at order 1, 5, 7, 11 ... In the case of the cycloconverter, bands of Fourier components centered on these frequencies are therefore to be expected. Unfortunately, it is impossible to distinguish such bands in Spectra 3, 4, and 5 of Fig. 11.7 because the modulation frequency is so large that adjacent sidebands interfere. The bands could only be separated if the modulation frequency were much lower in relation to the supply frequency, e.g. at a frequency ratio greater than 100. Such a large ratio would enormously increase the amount of data to be processed and, as a consequence, the computation time.

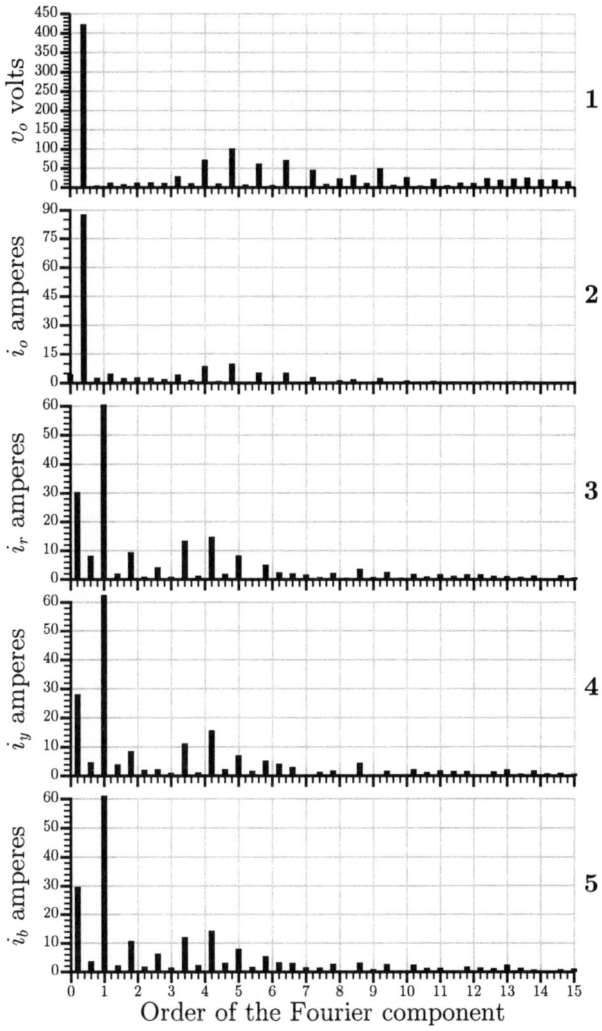

FIG. 11.7. Frequency spectra of (1) the output voltage, (2) the output current, (3), (4), and (5) the currents in the red, yellow, and blue a.c. input lines.

11.13 Three phase output

A cycloconverter would only rarely be used to supply a single phase load. By far the greatest number of practical applications is for large, low frequency, three phase loads. As an example, in recent years a number of large, low speed direct drives have been powered by three phase synchronous motors supplied by three phase cycloconverters, typical operating conditions being of the order of 5 000 kW at 5 Hz. The cycloconverter lends itself well to such an application, the low output frequency reducing many of the harmonic problems and the high power rating comfortably absorbing the cost of cycloconverter complexity. As will now be seen, the three phase output contributes greatly to the reduction in harmonic and subharmonic problems.

A typical load would have the three phases connected in wye with the neutral isolated. Unfortunately, while having considerable practical merit, such a system greatly increases the analytical complexity since the currents in the phases are mutually dependent, their sum at all times being forced to be zero by the isolated neutral. Fortunately the principal features of a three phase cycloconverter are equally well displayed by the analytically simpler case of a load with three independent phases and this is the route which will be taken.

When there is no coupling between the load phases, each one is analyzed separately by the methods illustrated by Examples 11.1 through 11.4 and the resulting a.c. input line currents are summed. The resultant currents have greatly reduced harmonic content and the power demand is much more uniform. The required analytical tools have already been developed and their application is illustrated by Example 11.5.

Example 11.5

Determine the a.c. input line currents for a three phase cycloconverter directly connected to a three phase, 440 V, 60 Hz system and supplying a three phase load, each load phase being identical to the one specified for Examples 11.1 through 11.4. Consider the load phases to be completely independent of each other so that there are three separate, independent systems, each like that shown in Fig. 11.1 and each connected to the same a.c. supply.

As in the previous examples, the modulation index is 0.7 and the modulation frequency is 24 Hz. The phase of the modulation signal is zero for the first load phase, $-120°$ for the second, and $-240°$ for the third.

The initial stage of the solution is the repetition of Examples 11.1 through 11.4 for each load phase. This produces a very large quantity of data, notably the basic cycloconverter expressions for the input line currents, i_r, i_y, and i_b, for each segment, for each output phase over one input cycle, i.e. five supply cycles. These data for the red line current are given in Tables 11.9, 11.10, and 11.11 and the waveforms of the three components of i_r defined by these tables are shown in Fig. 11.8, graphs 1, 2, and 3.

Table 11.9 *Coefficients of the basic cycloconverter expression for the current in the red a.c. input line when the phase angle of the modulation signal is zero, see eqn 11.25. For all segments the frequency ratio is 0.4 and the circuit eigenvalue is –0.1611*

Seg. no.	θ_{st} degrees	θ_{fn} degrees	I_{r1} A	ϕ_{cr1} rad	I_{r2} A	ϕ_{cr2} rad	I_{r3} A	ϕ_{cr3} rad
1	0.0	48.7	291.0	–0.888	394.6	–1.286	–4.3	0.000
2	48.7	122.7	291.0	–1.935	394.6	–1.286	151.1	0.850
3	122.7	204.2	0.0	–2.982	0.0	–1.286	0.0	2.142
4	204.2	281.7	–291.0	–4.029	–394.6	–1.286	–152.7	3.564
5	281.7	291.7	0.0	0.000	0.0	–1.286	0.0	3.564
6	291.7	307.6	0.0	1.207	0.0	–1.286	0.0	5.091
7	307.6	356.3	0.0	0.160	0.0	–1.286	0.0	5.368
8	356.3	407.9	291.0	–0.888	–394.6	–1.286	–287.8	6.218
9	407.9	465.9	291.0	–1.935	–394.6	–1.286	–90.7	7.120
10	465.9	534.4	0.0	–2.982	0.0	–1.286	0.0	8.132
11	534.4	612.9	–291.0	–4.029	394.6	–1.286	–203.4	9.327
12	612.9	695.8	–291.0	1.207	394.6	–1.286	–203.4	10.697
13	695.8	765.0	0.0	0.160	0.0	–1.286	0.0	12.144
14	765.0	775.0	0.0	0.000	0.0	–1.286	0.0	12.144
15	775.0	781.7	0.0	–0.888	0.0	–1.286	0.0	13.527
16	781.7	831.6	291.0	–1.935	394.6	–1.286	–393.5	13.643
17	831.6	885.8	0.0	–2.982	0.0	–1.286	0.0	14.514
18	885.8	948.7	–291.0	–4.029	–394.6	–1.286	4.5	15.461
19	948.7	1022.7	–291.0	1.207	–394.6	–1.286	–151.1	16.558
20	1022.7	1104.2	0.0	0.160	0.0	–1.286	0.0	17.850
21	1104.2	1181.7	291.0	–0.888	394.6	–1.286	152.7	19.272
22	1181.7	1191.7	0.0	0.000	0.0	–1.286	0.0	19.272
23	1191.7	1207.6	0.0	–1.935	0.0	–1.286	0.0	20.799
24	1207.6	1256.3	0.0	–2.982	0.0	–1.286	0.0	21.076
25	1256.3	1307.9	–291.0	–4.029	394.6	–1.286	287.8	21.926
26	1307.9	1365.9	–291.0	1.207	394.6	–1.286	90.7	22.828
27	1365.9	1434.4	0.0	0.160	0.0	–1.286	0.0	23.840
28	1434.4	1512.9	291.0	–0.888	–394.6	–1.286	203.4	25.035
29	1512.9	1595.8	291.0	–1.935	–394.6	–1.286	203.4	26.405
30	1595.8	1665.0	0.0	–2.982	0.0	–1.286	0.0	27.852
31	1665.0	1675.0	0.0	0.000	0.0	–1.286	0.0	27.852
32	1675.0	1681.7	0.0	–4.029	0.0	–1.286	0.0	29.235
33	1681.7	1731.6	–291.0	1.207	–394.6	–1.286	393.5	29.351
34	1731.6	1785.8	0.0	0.160	0.0	–1.286	0.0	30.222
35	1785.8	1848.7	291.0	–0.888	394.6	–1.286	–4.5	31.169
36	1848.7	1922.7	291.0	–1.935	394.6	–1.286	151.1	32.266

Table 11.10 *Coefficients of the basic cycloconverter expression for the current in the red a.c. input line when the phase angle of the modulation signal is $-120°$, see eqn 11.25. For all segments the frequency ratio is 0.4 and the circuit eigenvalue is –0.1611*

Seg. no.	θ_{st} degrees	θ_{fn} degrees	I_{r1} A	ϕ_{cr1} rad	I_{r2} A	ϕ_{cr2} rad	I_{r3} A	ϕ_{cr3} rad
1	0.0	12.9	0.0	0.160	0.0	–3.380	0.0	0.000
2	12.9	95.8	291.0	–0.888	–394.6	–3.380	203.4	0.225
3	95.8	165.0	291.0	–1.935	–394.6	–3.380	86.0	1.672
4	165.0	175.0	0.0	0.000	0.0	–3.380	0.0	1.672
5	175.0	181.7	0.0	–2.982	0.0	–3.380	0.0	3.055
6	181.7	231.6	–291.0	–4.029	–394.6	–3.380	393.5	3.171
7	231.6	285.8	–291.0	1.207	–394.6	–3.380	199.6	4.042
8	285.8	348.7	0.0	0.160	0.0	–3.380	0.0	4.989
9	348.7	422.7	291.0	–0.888	394.6	–3.380	151.1	6.086
10	422.7	504.2	291.0	–1.935	394.6	–3.380	213.3	7.378
11	504.2	581.7	0.0	–2.982	0.0	–3.380	0.0	8.800
12	581.7	591.7	0.0	0.000	0.0	–3.380	0.0	8.800
13	591.7	607.6	0.0	–4.029	0.0	–3.380	0.0	10.327
14	607.6	656.3	–291.0	1.207	394.6	–3.380	468.9	10.604
15	656.3	707.9	0.0	0.160	0.0	–3.380	0.0	11.454
16	707.9	765.9	291.0	–0.888	–394.6	–3.380	–90.7	12.356
17	765.9	834.4	291.0	–1.935	–394.6	–3.380	89.5	13.368
18	834.4	912.9	0.0	–2.982	0.0	–3.380	0.0	14.563
19	912.9	995.8	–291.0	–4.029	394.6	–3.380	–203.4	15.933
20	995.8	1065.0	–291.0	1.207	394.6	–3.380	–86.0	17.380
21	1065.0	1075.0	0.0	0.000	0.0	–3.380	0.0	17.380
22	1075.0	1081.7	0.0	0.160	0.0	–3.380	0.0	18.763
23	1081.7	1131.6	291.0	–0.888	394.6	–3.380	–393.5	18.879
24	1131.6	1185.8	291.0	–1.935	394.6	–3.380	–199.6	19.750
25	1185.8	1248.7	0.0	–2.982	0.0	–3.380	0.0	20.697
26	1248.7	1322.7	–291.0	–4.029	–394.6	–3.380	–151.1	21.794
27	1322.7	1404.2	–291.0	1.207	–394.6	–3.380	–213.3	23.086
28	1404.2	1481.7	0.0	0.160	0.0	–3.380	0.0	24.508
29	1481.7	1491.7	0.0	0.000	0.0	–3.380	0.0	24.508
30	1491.7	1507.6	0.0	–0.888	0.0	–3.380	0.0	26.035
31	1507.6	1556.3	291.0	–1.935	–394.6	–3.380	–468.9	26.312
32	1556.3	1607.9	0.0	–2.982	0.0	–3.380	0.0	27.162
33	1607.9	1665.9	–291.0	–4.029	394.6	–3.380	90.7	28.064
34	1665.9	1734.4	–291.0	1.207	394.6	–3.380	–89.5	29.076
35	1734.4	1812.9	0.0	0.160	0.0	–3.380	0.0	30.271
36	1812.9	1895.8	291.0	–0.888	–394.6	–3.380	203.4	31.641

Table 11.11 *Coefficients of the basic cycloconverter expression for the current in the red a.c. input line when the phase angle of the modulation signal is $-240°$, see eqn 11.25. For all segments the frequency ratio is 0.4 and the circuit eigenvalue is –0.1611*

Seg. no.	θ_{st} degrees	θ_{fn} degrees	I_{r1} A	ϕ_{cr1} rad	I_{r2} A	ϕ_{cr2} rad	I_{r3} A	ϕ_{cr3} rad
1	0.0	7.6	0.0	0.160	0.0	–5.475	0.0	0.000
2	7.6	56.3	291.0	–0.888	–394.6	–5.475	–468.9	0.132
3	56.3	107.9	291.0	–1.935	–394.6	–5.475	–287.8	0.982
4	107.9	165.9	0.0	–2.982	0.0	–5.475	0.0	1.884
5	165.9	234.4	–291.0	–4.029	394.6	–5.475	–89.5	2.896
6	234.4	312.9	–291.0	1.207	394.6	–5.475	–203.4	4.091
7	312.9	395.8	0.0	0.160	0.0	–5.475	0.0	5.461
8	395.8	465.0	291.0	–0.888	–394.6	–5.475	86.0	6.908
9	465.0	475.0	0.0	0.000	0.0	–5.475	0.0	6.908
10	475.0	481.7	0.0	–1.935	0.0	–5.475	0.0	8.291
11	481.7	531.6	0.0	–2.982	0.0	–5.475	0.0	8.407
12	531.6	585.8	–291.0	–4.029	–394.6	–5.475	199.6	9.278
13	585.8	648.7	–291.0	1.207	–394.6	–5.475	4.5	10.225
14	648.7	722.7	0.0	0.160	0.0	–5.475	0.0	11.322
15	722.7	804.2	291.0	–0.888	394.6	–5.475	213.3	12.614
16	804.2	881.7	291.0	–1.935	394.6	–5.475	152.7	14.036
17	881.7	891.7	0.0	0.000	0.0	–5.475	0.0	14.036
18	891.7	907.6	0.0	–2.982	0.0	–5.475	0.0	15.563
19	907.6	956.3	–291.0	–4.029	394.6	–5.475	468.9	15.840
20	956.3	1007.9	–291.0	1.207	394.6	–5.475	287.8	16.690
21	1007.9	1065.9	0.0	0.160	0.0	–5.475	0.0	17.592
22	1065.9	1134.4	291.0	–0.888	–394.6	–5.475	89.5	18.604
23	1134.4	1212.9	291.0	–1.935	–394.6	–5.475	203.4	19.799
24	1212.9	1295.8	0.0	–2.982	0.0	–5.475	0.0	21.169
25	1295.8	1365.0	–291.0	–4.029	394.6	–5.475	–86.0	22.616
26	1365.0	1375.0	0.0	0.000	0.0	–5.475	0.0	22.616
27	1375.0	1381.7	0.0	1.207	0.0	–5.475	0.0	23.999
28	1381.7	1431.6	0.0	0.160	0.0	–5.475	0.0	24.115
29	1431.6	1485.8	291.0	–0.888	394.6	–5.475	–199.6	24.986
30	1485.8	1548.7	291.0	–1.935	394.6	–5.475	–4.5	25.933
31	1548.7	1622.7	0.0	–2.982	0.0	–5.475	0.0	27.030
32	1622.7	1704.2	–291.0	–4.029	–394.6	–5.475	–213.3	28.322
33	1704.2	1781.7	–291.0	1.207	–394.6	–5.475	–152.7	29.744
34	1781.7	1791.7	0.0	0.000	0.0	–5.475	0.0	29.744
35	1791.7	1807.6	0.0	0.160	0.0	–5.475	0.0	31.271
36	1807.6	1856.3	291.0	–0.888	–394.6	–5.475	–468.9	31.548

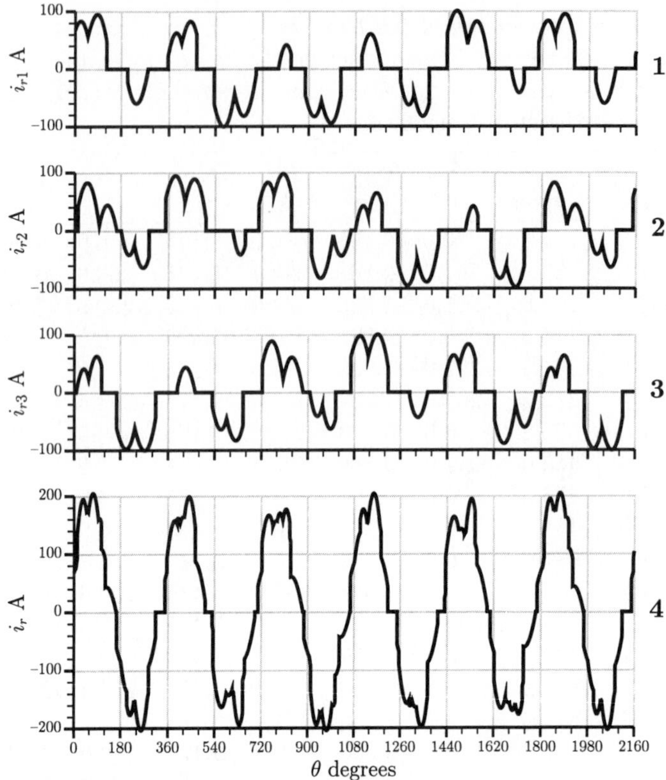

FIG. 11.8. Waveforms of the current in the red a.c. input line. Graphs 1, 2, and 3 show the components supplying output phases 1, 2, and 3. Graph 4 shows the actual line current which is the sum of the currents shown in graphs 1, 2, and 3.

The current in the red line is the sum of the three components and to obtain it the program must sort through the segments defined by the coefficients listed in Tables 11.9, 11.10, and 11.11 to create a new set of segments defining the actual line current. This process is illustrated by considering the interval occupied by segment 2 of Table 11.9. It covers the range $48.71° < \theta < 122.71°$ and defines the current drawn from the red a.c. line to supply the first load phase as

$$291.0\cos(\theta - 1.935) - 394.6\cos(0.4\theta - 1.286) + \\ 151.1\exp[-0.161(\theta - 0.650)]$$

This will be called segment 2.1, segment 2 of the component of the red line current supplying output phase 1.

From Table 11.10 it is found that during the above interval there are parts of two segments of the component of i_r supplying the second load phase. They are segments 2 and 3. Segment 2 extends from 12.90° to 95.82° during which time the current component is

$$291.0\cos(\theta - 0.888) + 394.6\cos(0.4\theta - 3.380) - \\ 203.4\exp[-0.161(\theta - 0.225)]$$

Segment 3 extends from 95.82° to 165.03° during which time the current component is

$$291.0\cos(\theta - 1.935) + 394.6\cos(0.4\theta - 3.380) - \\ 86.0\exp[-0.161(\theta - 1.673)]$$

These segments will be called 2.2 and 3.2, segments 2 and 3 of the component of the red line current supplying output phase 2.

Turning now to Table 11.11, it is found that three segments of the component of i_r supplying the third load phase are operative during segment 2.1. They are segments 2, 3, and 4. Segment 2 extends from 7.57° to 56.26° during which time the current component is

$$291.0\cos(\theta - 0.888) + 394.6cobs(0.4\theta - 5.475) + \\ 468.9\exp[-0.161(\theta - 0.132)]$$

Segment 3 extends from 56.26° to 107.93° and the current component is

$$291.0\cos(\theta - 1.935) + 394.6cobs(0.4\theta - 5.475) + \\ 287.8\exp[-0.161(\theta - 0.982)]$$

During segment 4, $107.93° < \theta < 165.92°$, the current is zero. These segments will be called 2.3, 3.3, and 4.3, segments 2, 3, and 4 of the component of the red phase current supplying output phase 3.

Combining this information, it is seen that there is a period in the middle of segment 2.1, from $\theta = 56.26°$ to $\theta = 95.82°$, when i_r is the sum of the currents defined by segments 2.1, 2.2, and 3.3 so that for this period i_r is given by the expression

$$769.9\cos(\theta - 1.601) - 789.1cobs(0.4\theta - 1.286) + \\ 40.2\exp[-0.161(\theta - 0.982)]$$

where, following the procedure used for all other segments, the final parameter, 0.982, has been set to the value of θ at the start of the segment, i.e. the radian equivalent of 56.26°.

Table 11.12 *Coefficients of the basic cycloconverter expression for the current in the red input line when supplying the three phase load. A complete cycle occupies 124 segments and data are given for only the first 15 of these. For all segments the frequency ratio is 0.4 and the circuit eigenvalue is* −0.1611

Seg. no.	θ degrees	θ degrees	I_{r1} A	ϕ_{cr1} rad	I_{r2} A	ϕ_{cr2} rad	I_{r3} A	ϕ_{cr3} rad
1	0.0	7.6	291.0	−0.888	394.6	−1.286	−4.3	0.000
2	7.6	12.9	582.0	−0.888	683.5	−1.809	−473.0	0.132
3	12.9	48.7	873.0	−0.888	789.2	−1.286	−262.6	0.225
4	48.7	56.3	769.9	−1.221	789.2	−1.286	−82.6	0.850
5	56.3	95.8	769.9	−1.601	789.2	−1.286	40.2	0.982
6	95.8	107.9	873.0	−1.935	789.2	−1.286	−39.2	1.672
7	107.9	122.7	582.0	−1.935	683.5	−0.762	211.0	1.884
8	122.7	165.0	291.0	−1.935	394.6	−0.239	79.7	2.142
9	165.0	165.9	0.0	0.000	0.0	0.000	0.0	2.880
10	165.9	175.0	291.0	−0.888	394.6	0.809	−89.5	2.896
11	175.0	181.7	291.0	−0.888	394.6	0.809	−87.3	3.055
12	181.7	204.2	582.0	−0.888	683.5	0.285	307.8	3.171
13	204.2	231.6	873.0	−0.888	789.2	0.809	136.2	3.564
14	231.6	234.4	769.9	−1.221	789.2	0.809	−16.2	4.042
15	234.4	281.7	769.9	−1.601	789.2	0.809	−145.7	4.091

Progressing in this way through the data of Tables 11.9, 11.10, and 11.11 yields a total of 124 distinct segments of i_r covering the range $0.0 < \theta < 2160°$, of which the segment just determined is number 5.

Tables 11.12, 11.13, and 11.14 define the first 15 segments for the red, yellow, and blue line currents which should be sufficient for readers to check the validity of their own computations. These 15 segments, plus the remainder of the 124 for each current, have been used to determine the waveforms of Fig. 11.9. This figure shows the waveforms of the three line currents, the corresponding line to neutral voltages in gray, and the total input power. Since it is assumed that the supply has zero internal impedance, the waveforms of the line to neutral voltages are sinusoidal with an amplitude of 359.3 V.

****

The waveforms of Fig. 11.9 are typical of three phase cycloconverters so that some general conclusions can be drawn. While the current waveforms clearly have appreciable harmonic content it is equally clear that they are a substantial improvement on those for the single phase case shown

Table 11.13 *Coefficients of the basic cycloconverter expression for the current in the yellow input line when supplying the three phase load. A complete cycle occupies 124 segments and data are given for only the first 15 of these. For all segments the frequency ratio is 0.4 and the circuit eigenvalue is* −0.1611

Seg. no.	θ degrees	θ degrees	I_{y1} A	ϕ_{cy1} rad	I_{y2} A	ϕ_{cy2} rad	I_{y3} A	ϕ_{cy3} rad
1	0.0	7.6	504.0	2.778	683.5	2.379	−164.9	0.000
2	7.6	12.9	769.9	2.588	789.2	1.856	307.4	0.132
3	12.9	48.7	873.0	2.254	789.2	1.856	262.6	0.225
4	48.7	56.3	582.0	2.254	394.6	1.856	233.7	0.850
5	56.3	95.8	291.0	2.254	394.6	2.903	−180.1	0.982
6	95.8	107.9	0.0	0.000	0.0	0.000	0.0	1.672
7	107.9	122.7	291.0	−2.982	394.6	−2.333	−90.7	1.884
8	122.7	165.0	582.0	−2.982	683.5	−1.809	126.3	2.142
9	165.0	165.9	582.0	−2.982	683.5	−1.809	112.1	2.880
10	165.9	175.0	504.0	2.778	683.5	−1.809	278.4	2.896
11	175.0	181.7	504.0	2.778	683.5	−1.809	271.4	3.055
12	181.7	204.2	769.9	2.588	789.2	−2.333	−127.1	3.171
13	204.2	231.6	873.0	2.254	789.2	−2.333	−136.2	3.564
14	231.6	234.4	582.0	2.254	683.5	−1.809	215.9	4.042
15	234.4	281.7	291.0	2.254	394.6	−1.286	140.3	4.091

in Fig. 11.6. It is particularly significant that the subharmonic content is greatly reduced.

The three current waveforms are now similar so that the load on the three input phases is essentially balanced. In fact for this case the three waveforms are identical except for the mutual phase displacement of 120°.

A linear three phase load draws steady, time invariant, power from a sinusoidal three phase supply. This is not the case with the cycloconverter, the input power waveform of Fig. 11.9 showing significant variation. Nevertheless, like the current waveforms, it is a substantial improvement on the single phase case.

11.13.1 *Some basic measures of the input*

A few measures of the waveforms provide a quantitative basis for the foregoing conclusions. Thus, the r.m.s. values of the line currents and the r.m.s. values of their fundamental components can be determined. The r.m.s. value of the harmonic content is the square root of the difference of the squares of these values. Since the line to neutral voltages have sinusoidal

Table 11.14 *Coefficients of the basic cycloconverter expression for the current in the blue input line when supplying the three phase load. A complete cycle occupies 124 segments and data are given for only the first 15 of these. For all segments the frequency ratio is 0.4 and the circuit eigenvalue is −0.1611*

Seg. no.	θ degrees	θ degrees	I_{b1} A	ϕ_{cb1} rad	I_{b2} A	ϕ_{cb2} rad	I_{b3} A	ϕ_{cb3} rad
1	0.0	7.6	291.0	0.160	394.6	-0.239	169.2	0.000
2	7.6	12.9	291.0	0.160	394.6	-0.239	165.6	0.132
3	12.9	48.7	0.0	0.000	0.0	0.000	0.0	0.225
4	48.7	56.3	291.0	1.207	394.6	1.856	-151.1	0.850
5	56.3	95.8	582.0	1.207	683.5	1.332	139.9	0.982
6	95.8	107.9	873.0	1.207	789.2	1.856	39.2	1.672
7	107.9	122.7	769.9	0.873	789.2	1.856	-120.3	1.884
8	122.7	165.0	769.9	0.493	789.2	1.856	-206.0	2.142
9	165.0	165.9	582.0	0.160	683.5	1.332	-112.1	2.880
10	165.9	175.0	291.0	0.160	394.6	1.856	-188.9	2.896
11	175.0	181.7	291.0	0.160	394.6	1.856	-184.1	3.055
12	181.7	204.2	291.0	0.160	394.6	1.856	-180.7	3.171
13	204.2	231.6	0.0	0.000	0.0	0.000	0.0	3.564
14	231.6	234.4	291.0	1.207	394.6	2.903	-199.6	4.042
15	234.4	281.7	582.0	1.207	683.5	-2.857	5.4	4.091

waveforms, the mean input power is the sum of the three phase powers $3V_{ln}I_{l1}\cos(\phi_1)$ where V_{ln} is the r.m.s. value of the a.c. supply line to neutral voltage, I_{l1} is the r.m.s. value of the fundamental component of the line current and ϕ_1 is their phase difference. The input kVA is $\sqrt{3}V_{ll}I_l \div 1000$ where V_{ll} is the r.m.s. value of the line to line voltage and I_l is the r.m.s. value of the line current. The fundamental power factor is $\cos(\phi_1)$ and the power factor as conventionally measured is the ratio of the mean power in kW to the kVA as conventionally measured.

Determination of the r.m.s. value of the line current and of the mean power requires that the integral of the product of two basic cycloconverter expressions be taken over the duration of a segment. In the case of the current it is the product of the basic expression with itself. In the case of the power, it is the product of the basic expression for the current with that for the voltage. The required integral is given in Section E.5.

Determination of the fundamental component requires that the integral of the product of the basic expression and both $\cos(\theta)$ and $\sin(\theta)$ be determined. These integrals are also given in Sections E.6 and E.7.

Application of these procedures is illustrated by Example 11.6.

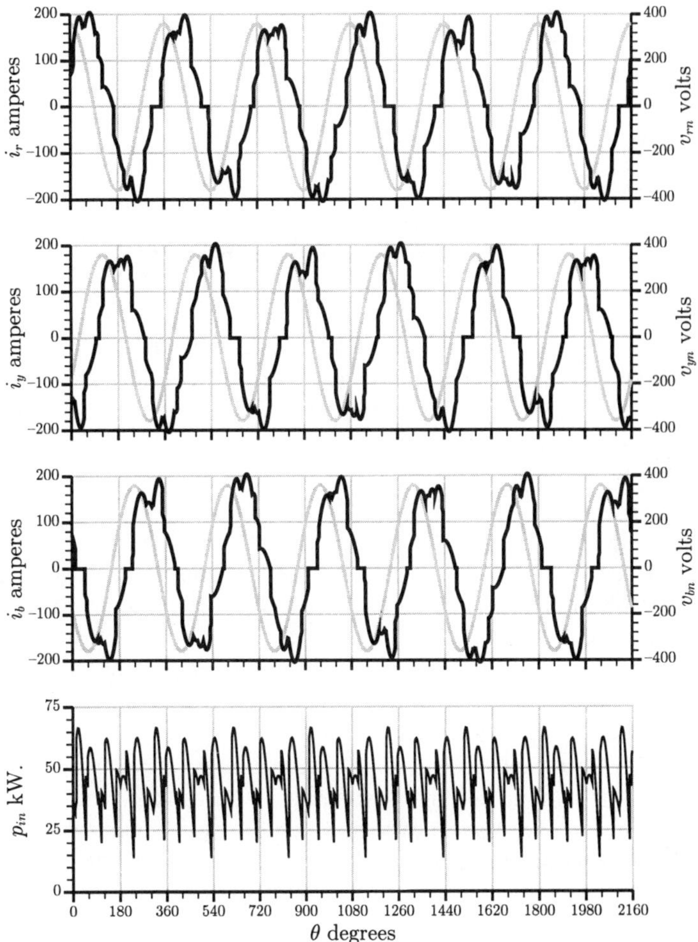

FIG. 11.9. The waveforms of the a.c. line to neutral voltages in gray, the a.c. line currents, and the total, three phase input power.

Example 11.6

For the three phase example, determine the r.m.s. value of the input line currents, the r.m.s. value of their fundamental component and from these the r.m.s. value of their total harmonic content. Also determine the mean input power and the input kVA and power factor as conventionally measured.

The waveforms of the three line currents are identical, differing only in their phase angle, so that they have the same r.m.s. value, the same fundamental amplitude, and the same harmonic content. The r.m.s. value

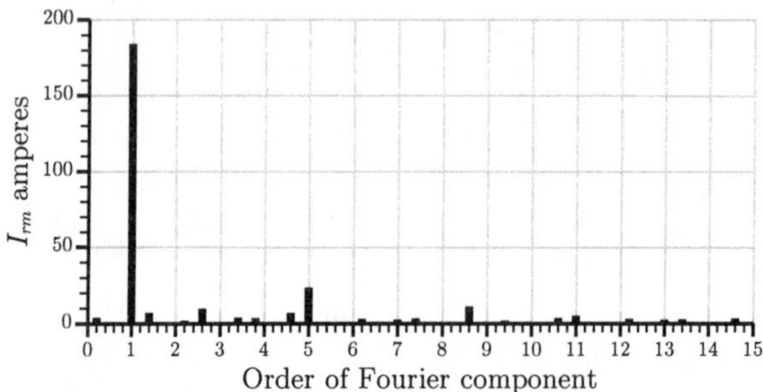

FIG. 11.10. Frequency spectrum of the current in the red a.c. line. The spectra of the other line currents are identical. A Fourier component of order 1 is at the supply frequency and a component of order 0.4 is at the control signal frequency.

of an input line current is 132.37 A and the r.m.s. value of its fundamental component is 130.24 A. The r.m.s. value of the non-fundamental content, including subharmonics, is the square root of the difference of the squares of these two values, i.e. $\sqrt{132.37^2 - 130.24^2} = 23.64$ A. This is 18% of the fundamental, a most notable improvement on the single phase case where the ratio depends somewhat on the phase angle but is greater than 60% for all the cases considered.

Some simplifications are possible in the computation of mean power. The fact that the input voltage waveforms are sinusoidal can be exploited to avoid having to express the voltage as a basic cycloconverter expression. Additional computational time can be saved by using the computational equivalent of the two wattmeter method for measuring three phase, three wire power. This method gives the input power as the sum of two products, the product of the red line current and the red to yellow line voltage and the product of the blue line current and the blue to yellow line voltage rather than the triple sum of the products of the line currents and line to neutral voltages.

The mean input power is 45.82 kW. The input kVA as conventionally measured is the triple product of $\sqrt{3}$, the r.m.s. line to line voltage, and the r.m.s. line current divided by 1000, i.e. $\sqrt{3} \times 440 \times 132.47 \div 1000 = 100.88$ kVA. The power factor as conventionally measured is the ratio of the mean input power to the kVA as conventionally measured, i.e. $45.82 \div 100.88 = 0.4542$. Thus, the power factor is still low but is an improvement on the value for the single phase case which, as found in section 11.11.2, depends slightly on the modulation phase angle and is about 0.4.

****

11.13.2 *Frequency spectra*

Since the waveforms of the a.c. line currents are identical except for a shift in phase, the amplitude spectrum of the red line current shown in Fig. 11.10 applies to all three. The amplitudes and phase angles of the components of all three line currents are listed in Table 11.15. Comparison with the spectra for the single phase case, Fig. 11.7 and Table 11.8, shows the beneficial effects of three phase operation. The subharmonic component at 0.2 times the supply frequency has been reduced from about 50% of the supply frequency component to 2.1% and the component at 0.6 times the supply frequency has been eliminated. The supply frequency component of the red line current is $184.19\underline{/-62.51^\circ}$ and forms part of a positive sequence set. The higher order components are small with the exception of the one of order 5 which has an amplitude of 23.27 A, 12.6% of the fundamental, and forms a negative sequence set.

The fundamental power factor is $\cos(-62.51^\circ) = 0.4616$, slightly higher than the power factor as conventionally measured.

11.14 Power factor improvement

The input power factor of a sinusoidally modulated cycloconverter is poor, about half that of the load, primarily due to the lag of the fundamental component of a.c. line current with a small portion due to the non-fundamental components. The problem arises because the cycloconverter must spend a considerable portion of its time producing low voltages in the regions where the sine wave output crosses the zero axis and, as was seen in Chapters 2 and 3, low voltage operation means low power factor operation.

An obvious remedy is to use square wave modulation with each rectifier being operated at zero delay when rectifying and α_{max} when inverting. The input power factor would then be high, of the order of 0.9. However, square wave operation tends to produce excessive harmonic content in the load current so recourse is had to trapezoidal modulation in which the output voltage is held at the positive and negative maxima for substantial portions of a cycle and is quickly ramped between these levels. Adjustment of the ratio of dwell time to ramp time can result in an improved input power factor with an acceptable level of output harmonics. Of course, the proportions of the modulation trapezoid fix the output voltage and any attempt to control the latter by operation away from zero delay and maximum delay during the dwell periods immediately destroys the merits of the remedy. This is therefore a suitable technique for a load which spends much of its time at maximum voltage, an example being a low speed drive in which the motor spends most of its time at full speed but which must be accelerated from standstill to that speed in a controlled manner.

Table 11.15 *Fourier components of the a.c. line currents supplying the three phase load. Column 1 gives the order of the component, columns 2, 4, and 6 give the amplitude of the component, and column 3, 5, and 7 give the phase of the component*

Order m	I_{rm} A	ϕ_{crm} degrees	I_{ym} A	ϕ_{cym} degrees	I_{bm} A	ϕ_{cbm} degrees
0.20	3.84	–28.09	3.84	91.91	3.84	–148.10
1.00	184.19	–62.51	184.19	177.49	184.19	57.49
1.40	7.07	–112.12	7.07	7.88	7.07	127.88
2.20	1.81	–174.14	1.81	65.86	1.81	–54.14
2.60	9.72	–113.81	9.72	6.19	9.72	126.19
3.40	3.91	85.07	3.91	–34.93	3.91	–154.93
3.80	3.70	–154.64	3.70	–34.64	3.70	85.36
4.60	6.88	127.75	6.88	7.75	6.88	–112.25
5.00	23.27	–93.26	23.27	26.74	23.27	146.74
5.80	0.68	67.91	0.68	–52.07	0.68	–172.09
6.20	2.99	–171.09	2.99	–51.08	2.98	68.92
7.00	2.62	49.16	2.62	–70.84	2.62	169.16
7.40	3.45	79.38	3.45	–160.62	3.45	–40.62
8.20	0.92	143.18	0.92	23.16	0.92	–96.84
8.60	11.14	116.36	11.14	–123.64	11.14	–3.64
9.40	1.81	89.43	1.81	–30.58	1.81	–150.58
9.80	0.40	–105.08	0.40	14.88	0.40	134.88
10.60	3.61	–167.06	3.61	72.94	3.61	–47.06
11.00	4.92	–6.65	4.92	113.36	4.92	–126.65
11.80	0.35	51.75	0.35	–68.29	0.35	171.75
12.20	2.98	64.04	2.98	–175.96	2.98	–55.96
13.00	2.47	91.53	2.46	–28.47	2.47	–148.46
13.40	2.65	100.70	2.65	–139.30	2.65	–19.30
14.20	1.09	66.56	1.09	–53.44	1.09	–173.44
14.60	3.22	147.32	3.22	–92.67	3.22	27.32

For loads requiring more flexible control, a combination of sinusoidal modulation and overmodulation has merit. The phase back and phase forward limits built into the controller maintain the delay angle between the safe limits of zero and α_{max} so that operation with a modulation index greater than unity is feasible. Operation at a modulation index of the order of 2.0 provides a close approximation to trapezoidal modulation and operation in the modulation index range below this value provides smooth control of output voltage albeit with the inevitable attendant reduction in power factor.

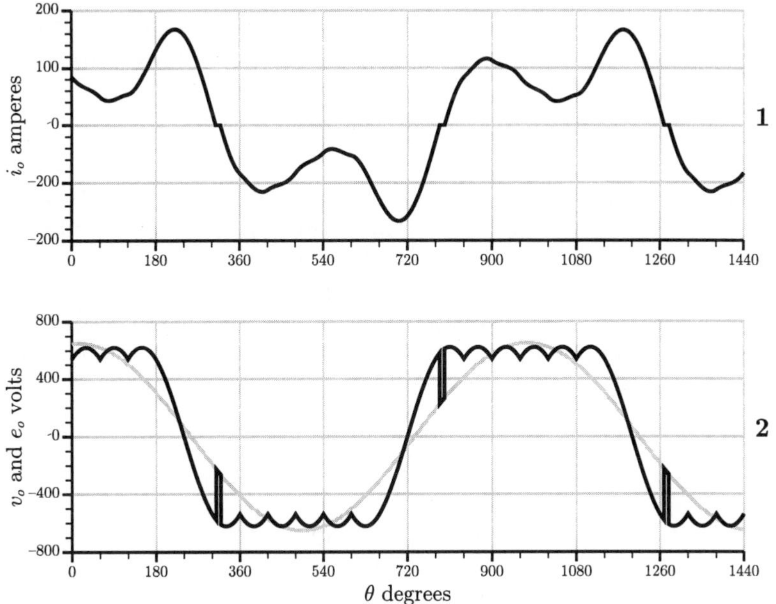

FIG. 11.11. Output waveforms when the modulation index is 2.0 and the output frequency is 22.5 Hz. Graph 1 shows the waveform of the output current and graph 2 shows that of the output voltage by a black line and of the load e.m.f. by a gray line.

Figure 11.11 shows, for the cycloconverter of the example, the waveforms of the output voltage and current for a modulation index of 2.0 and an output frequency of 22.5 Hz, i.e. a frequency ratio of 8 : 3. The essentially trapezoidal nature of the output voltage waveform, graph 2, is evident. Graph 2 also shows, by a gray line, the waveform of the sinusoidal load e.m.f. which has an r.m.s. value of 460 V and lags the modulation signal by 5.6°. The current waveform, graph 1, shows a pronounced third harmonic arising from the third harmonic content of the output voltage. This component would, of course, be eliminated if the three phase load were wye connected with isolated neutral.

11.15 An approximate theory

Cycloconverter theory is so complicated that the basics of behavior tend to be overwhelmed in a mass of detail. An approximate theory which concentrates on the basics while riding rough shod over the details has some merit in this situation. The theory presented below requires some rather sweeping assumptions and is as a consequence of limited value for quantitative calculations. However, it does have the merit of providing some insight into the actual cycloconverter waveforms which are encountered in practice.

11.15.1 *The output current and voltage*

Figure 11.12 illustrates the assumptions of the approximate theory and the progression of the argument. Graph 1 shows the waveforms of the fundamental components of the output voltage and current of a sinusoidally modulated cycloconverter, the voltage being the gray line and the current the black line. The approximate analysis is made on the basis of these components, ignoring all other components. The fundamental component of output current is represented by the expression

$$i_o = I_{o1}\cos(f_r\theta + \phi_{co1}) \tag{11.31}$$

where I_{o1} is its amplitude and ϕ_{co1} is its phase.

Zero crossings occur at $\theta = [(n - 0.5)\pi - \phi_{co1}] \div f_r$ radians where n is any integer, positive or negative including zero. When n is even, the zero crossing has a positive slope and when it is odd the slope is negative. The output current wave of Fig. 11.12 has been drawn with an amplitude of 0.7, a phase angle of $-60°$ and a frequency ratio of $12 : 1$. The range of the graph covers 1.5 output cycles and zero crossings occur in it at 1800°, 3960°, and 6120° corresponding to $n = 1$, 2, and 3.

The fundamental component of output voltage is similarly

$$v_o = V_{o1}\cos(f_r\theta + \phi_{vo1}) \tag{11.32}$$

Fig. 11.12 has been drawn with $V_{o1} = 0.7$ and $\phi_{vo1} = 0.0$.

11.15.2 *Output as seen from input terminals*

The positive rectifier is on during the positive half cycles of current and the negative rectifier during the negative half cycles. Thus, from the point of view of the a.c. supply, the load appears to be as illustrated in Fig. 11.12, graph 2 with the negative current half cycles and the associated voltage reversed in sign. For graph 2, within the ranges $-360° < \theta < 1800°$ and $3960° < \theta < 6120°$ the positive rectifier conducts and the current wave is represented by $0.7\cos(\theta/12 - \pi/3)$ and the voltage wave is represented by $0.7\cos(\theta/12)$. Within the ranges $1800° < \theta < 3960°$ and $6120° < \theta < 8280°$ the current wave is represented by $-0.7\cos(\theta/12 - \pi/3)$ and the voltage wave by $-0.7\cos(\theta/12)$.

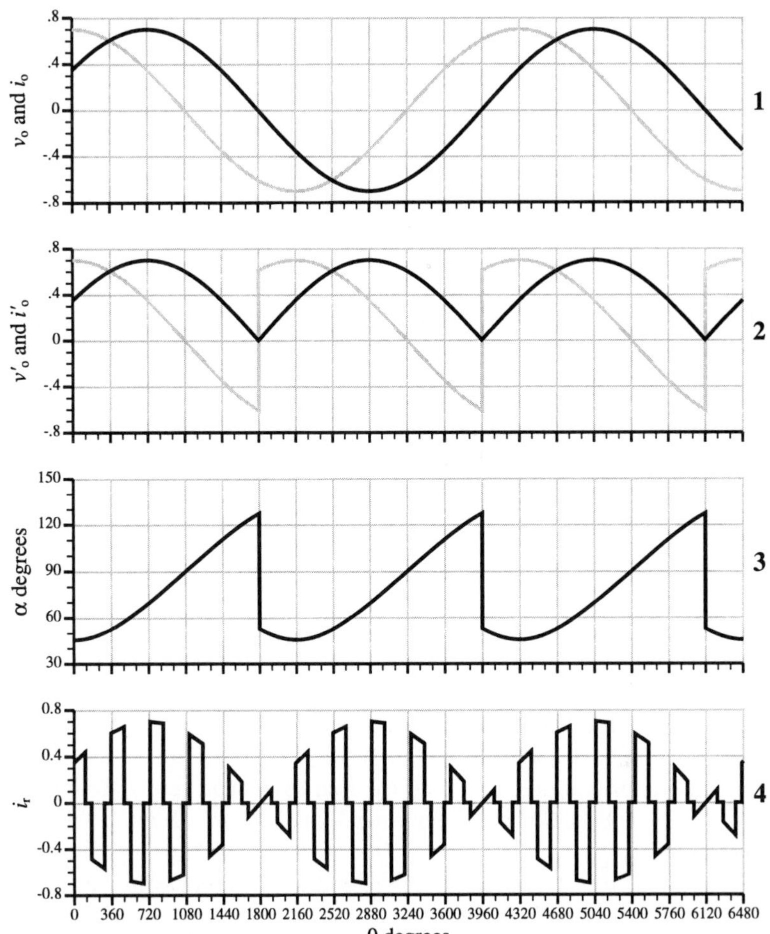

FIG. 11.12. Cycloconverter waveforms for the approximate theory for a modulation index of 0.7 and a frequency ratio of 12 : 1. Graph 1 shows the output voltage in gray and the output current in black. Graph 2 shows the same two waveforms as seen from the a.c. input. Graph 3 shows the delay angle and graph 4 the current in the red a.c. line.

11.15.3 *The delay angle*

If it is now assumed that the output frequency is very low compared to the supply frequency, a snapshot of the system taken at θ will appear to show the system in the steady state at a delay angle defined, for a positive current half cycle, by the expression

$$v_o = V_{omax} \cos(\alpha) \tag{11.33}$$

where v_o is the value of the output voltage at θ and V_{omax} is the maximum d.c. output voltage which a rectifier can produce.

The output voltage at any instant, θ, is given by the expression

$$v_o = M_i V_{omax} \cos(f_r\theta + \phi_{vo}) \tag{11.34}$$

Combining this with eqn 11.33 yields an expression for the delay angle,

$$\alpha = \arccos\left[M_i \cos(f_r\theta + \phi_{vo})\right] \tag{11.35}$$

The variation of delay angle in the negative rectifier is identical to this and the two delay angles are shown in graph 3 of Fig. 11.12.

11.15.4 *The basic a.c. input line current*

The waveform of the current in the red input line to a three phase bridge rectifier supplying a steady, ripple free output current I_o is shown in graph 3 of Fig. 3.4. The current is at the supply frequency and has a rectangular waveform whose amplitude alternates between $\pm I_o$, spending one-third of an input cycle at each extreme value, these extremes being separated by one-sixth cycle periods during which the line current is zero. Further, this current waveform lags the corresponding line to neutral voltage wave by the delay angle.

The *basic a.c. line current* will be defined as the current in the red a.c. input line when its amplitude is normalized to 1.0 and the rectifier delay angle is set to α. Its symbol will be i_{lbas}. Since the voltage between the red line and system neutral provides the datum phase, the basic a.c. line current is represented by the Fourier series

$$i_{lbas} = \sum \frac{4}{\pi m} \sin(m\frac{\pi}{2}) \cos(m\frac{\pi}{6}) \cos[m(\theta - \alpha)] \tag{11.36}$$

where the component order, m, is odd and excludes triplens.

Alternatively, expressed numerically, the series is

$$\begin{aligned} i_{lbas} = {} & 1.1027 \cos(\theta - \alpha) - 0.2205 \cos[5(\theta - \alpha)] + \\ & 0.1575 \cos[7(\theta - \alpha)] - 0.1002 \cos[11(\theta - \alpha)] \cdots \end{aligned} \tag{11.37}$$

11.15.5 *The cycloconverter a.c. input line current*

In a cycloconverter, the amplitude of the basic a.c. input line current is modulated according to the output current as seen from the input terminals and its phase is modulated according to the delay angle as seen from the output terminals.

The output current as seen from the input terminals is represented by a Fourier series of the type

$$\sum I_{am} \cos[2mf_r(\theta + \phi_{co1})] \tag{11.38}$$

where I_{am} is the amplitude of the component and m is any positive integer including zero. For the current wave of Fig. 11.12, graph 2, the series is

$$\begin{aligned} &0.4456 + 02971\cos(0.1667\theta - 2.0944) - \\ &0.0594\cos(0.3333\theta - 4.1889)\cdots \end{aligned} \tag{11.39}$$

The delay angle as seen from the input terminals is represented by a Fourier series of the type

$$\sum A_n \cos(2nf_r\theta + \phi_{an}) \tag{11.40}$$

where A_n is the amplitude of the component, ϕ_{an} is its phase, and n is any positive integer including zero. For the delay angle wave of Fig. 11.12, graph 3, the series is

$$\begin{aligned} &1.3265 + 0.5727\cos(0.1667\theta + 2.3631) + \\ &0.2264\cos(0.3333\theta - 2.8065)\cdots \end{aligned} \tag{11.41}$$

Substitution of these expressions for the amplitude and phase into the basic a.c. input line current eqn 11.37, yields the input line current, a process which has been carried out using the time domain expressions to generate the input line current wave of Fig. 11.12, graph 4. This is repeated, in gray, in Fig. 11.13, graph 1.

The process can also be carried out in the frequency domain by multiplying the amplitudes of the basic a.c. input line current by the series representing the output current amplitude, as seen from the input terminals, and replacing the delay angle as seen from the input terminals by its Fourier series. In general this is a complex process generating many terms but the procedure can be followed by considering only the first term of the basic a.c. input line current series, the first and second terms of the output current amplitude, and the first and second terms of the delay angle series. Using the numerical values, eqns 11.39 and 11.41, this yields

$$\begin{gathered}\left[0.4456 + 0.2971\cos(0.1667\theta - 2.0944)\right] \times \\ 1.1027\cos\left[\theta - 1.3265 - 0.5727\cos(0.1667\theta + 2.3631)\right]\end{gathered} \tag{11.42}$$

After some simplification eqn 11.42 becomes

$$\begin{gathered}\left[0.4914 + 0.3276\cos(0.1667\theta - 2.0944)\right] \times \\ \cos\left[\theta - 1.3265 + 0.4078\cos(0.1667\theta) + 0.4022\sin(0.1667\theta)\right]\end{gathered} \tag{11.43}$$

The argument of the second cosine function of eqn 11.43 has four components, the first, θ, representing an oscillation at the supply frequency and the second, third, and fourth representing the phase modulation associated with the variation of delay angle throughout an output half cycle.

11.15.6 *Amplitude modulation*

For the moment, the substitution

$$\psi = \theta - 1.3265 + 0.4078\cos(0.1667\theta) + 0.4022\sin(0.1667\theta) \tag{11.44}$$

is made into eqn 11.43 so that it becomes

$$\left[0.4914 + 0.3276\cos(0.1667\theta - 2.0944)\right]\cos(\psi) \tag{11.45}$$

This expression expands to the three component expression

$$\begin{gathered}0.4914\cos(\psi) + 0.1638\cos(\psi + 0.1667\theta - 2.0964) + \\ 0.1638\cos(\psi - 0.1667\theta + 2.0964)\end{gathered} \tag{11.46}$$

Ignoring for the moment the cosine components of ψ, eqn 11.46 is, from the viewpoint of modulation theory, composed of a carrier at the a.c. supply frequency, a lower sideband at the supply frequency less twice the output frequency, and an upper sideband at the supply frequency plus twice the output frequency. The two sidebands are symmetrical about the carrier, they have equal amplitudes, and are equally displaced from it in frequency by twice the output frequency. The sideband displacement is twice the output frequency because alternate half cycles of output are carried by the positive and negative rectifiers so that the load current as seen from the input terminals appears to vary at twice the output frequency, as illustrated in graph 2 of Fig. 11.12.

11.15.7 *Phase modulation*

Recognizing now that ψ is not a linear function of θ because of its cosine components, it is seen that each of the three terms of eqn 11.46 is phase modulated at twice the output frequency. Phase modulation is very similar to frequency modulation and produces bands of frequencies on either side of the carrier frequency. Since the three carriers of eqn 11.46 are so close together, at $f_s - 2f_o$, f_s and $f_s + 2f_o$, their sidebands will overlap and interfere with each other. Some idea of what happens can be obtained by considering in greater detail the first term of eqn 11.46.

First replacing ψ by eqn 11.44 yields

$$0.4914\cos\left[\theta - 1.3265 + 0.4078\cos(0.1667\theta) + 0.4022\sin(0.1667\theta)\right] \quad (11.47)$$

After a great deal of elementary but tedious trigonometric manipulation this expands to

$$\begin{aligned}
&-0.4768\sin(\theta)\sin[0.4078\cos(0.1667\theta)]\sin[0.4022\sin(0.1667\theta)] + \\
&\;\; 0.1189\cos(\theta)\sin[0.4078\cos(0.1667\theta)]\sin[0.4022\sin(0.1667\theta)] - \\
&\;\; 0.1189\sin(\theta)\cos[0.4078\cos(0.1667\theta)]\sin[0.4022\sin(0.1667\theta)] + \\
&\;\; 0.4768\cos(\theta)\cos[0.4078\cos(0.1667\theta)]\sin[0.4022\sin(0.1667\theta)] - \quad (11.48) \\
&\;\; 0.1189\sin(\theta)\sin[0.4078\cos(0.1667\theta)]\cos[0.4022\sin(0.1667\theta)] + \\
&\;\; 0.4768\cos(\theta)\sin[0.4078\cos(0.1667\theta)]\cos[0.4022\sin(0.1667\theta)] + \\
&\;\; 0.4768\sin(\theta)\cos[0.4078\cos(0.1667\theta)]\cos[0.4022\sin(0.1667\theta)] + \\
&\;\; 0.1189\cos(\theta)\cos[0.4078\cos(0.1667\theta)]\cos[0.4022\sin(0.1667\theta)]
\end{aligned}$$

This expression involves many terms of the type $\cos[X\cos(Y)]$ which can be replaced by series expansions involving Bessel functions of the first kind and integral order evaluated at X. Thus

$$\begin{aligned}
\cos[X\cos(Y)] &= J_0(X) - 2J_2(X)\cos(2Y) + 2J_4(X)\cos(4Y)\cdots \\
\cos[X\sin(Y)] &= J_0(X) + 2J_2(X)\cos(2Y) + 2J_4(X)\cos(4Y)\cdots \\
\sin[X\sin(Y)] &= 2J_1(X)\sin(Y) + 2J_3(X)\sin(3Y)\cdots \\
\sin[X\cos(Y)] &= 2J_1(X)\cos(Y) - 2J_3(X)\cos(3Y)\cdots
\end{aligned}$$

Within the range of X values of interest, the series converge rapidly and they will be truncated at the second order Bessel term for substitution into eqn 11.38. When this is done and is followed by a great deal more trigonometric simplification the final result is

$$0.0002\cos(1.6667\theta + 1.8151) + 0.0028\cos(1.5\theta + 1.0367) +$$
$$0.0193\cos(1.3333\theta + 0.2583) + 0.1350\cos(1.1667\theta - 0.5342) +$$
$$0.4519\cos(\theta - 1.3265) + 0.1350\cos(0.8333\theta + 1.0228) + \quad (11.49)$$
$$0.0193\cos(1.6667\theta - 2.9113) + 0.0028\cos(0.5\theta - 0.5481) +$$
$$0.0002\cos(0.3333\theta + 1.8151)$$

The effect of the phase modulation has been to split the single component at the carrier frequency into a band of nine components centered on the carrier frequency and displaced from it by integral multiples of twice the output frequency starting at $f_s - 8f_o$ and going, in increments of $2\,f_o$, to $f_s + 8\,f_o$.

The upper and lower sideband components of eqn 11.46 can be treated in a similar manner to produce similar results—numerous sideband components separated from the supply frequency by even multiples of the output frequency. When components of the same frequency are combined the final result for the spectral content of the current in the red input line is

$$0.0001\cos(1.8333\theta - 0.2813)(11.32) + 0.0007\cos(1.6667\theta - 0.9878) +$$
$$0.0039\cos(1.5\theta - 1.6559) + 0.0274\cos(1.3333\theta - 2.4705) +$$
$$0.0399\cos(1.1667\theta - 2.5464) + 0.4525\cos(\theta - 1.2765) + \quad (11.50)$$
$$0.2893\cos(0.8333\theta + 0.8968) + 0.0648\cos(0.6667\theta - 3.0835) +$$
$$0.0092\cos(0.5\theta - 0.7328) + 0.0011\cos(1.3333\theta + 1.5938) +$$
$$0.0001\cos(0.1667\theta - 2.3738)$$

This approximation to the frequency spectrum has been used to generate the waveform shown in black in Fig. 11.13, graph 1, superimposed on the actual waveform which has been transferred from graph 4 of Fig. 11.12 and is shown in gray. While the lack of higher spectral components in the approximation is evident, it does nevertheless illustrate quite well the major features of the actual wave—the amplitude and phase modulation at twice the output frequency.

The amplitudes of eqn 11.50 have been used to generate the frequency spectrum of Fig. 11.13, graph 2. Comparison of this with the spectrum shown in Fig. 11.7, graph 3, which has been derived from the waveform shown in Fig. 11.6, graph 2, illustrates the similarities even though the ratio of output to supply frequency is 2.5 : 1, far removed from the assumptions of the approximate analysis. The general forms of the spectra are similar although the true spectrum has small components displaced from the supply frequency by odd multiples of the output frequency. These arise from the small modulation of the output wave at one-half of the output frequency.

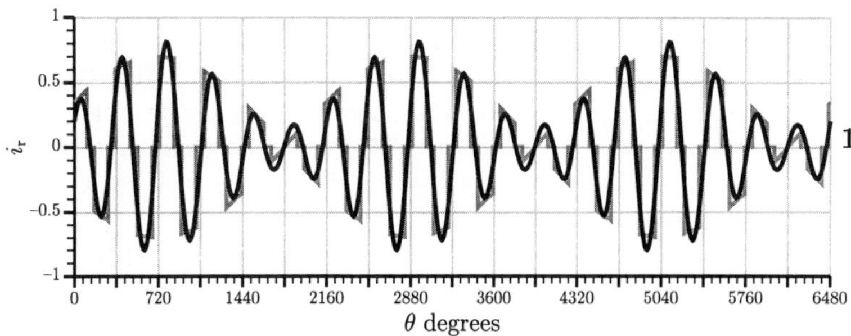

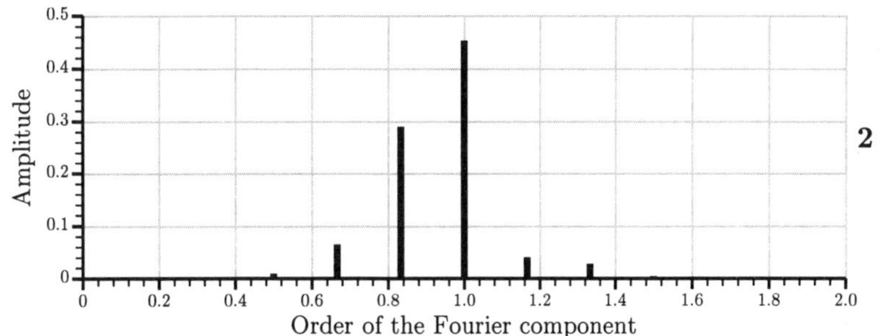

FIG. 11.13. Cycloconverter waveforms for the approximate theory for a frequency ratio of 12 : 1 and a modulation index of 0.7. Graph 1 shows the current in the red a.c. line in gray and the approximation in black. Graph 2 shows the spectrum of the line current determined by the approximate theory.

11.16 Conclusion

A three phase cycloconverter using full wave bridge rectifiers is a useful device for supplying power in the frequency range from zero up to about one-third of the a.c. supply frequency. Such a cycloconverter is a complicated device using a minimum of 36 thyristors and can normally only be justified for large loads above about 1000 kW. A somewhat simpler device with half the number of thyristors forming single way rectifiers has its maximum output frequency correspondingly halved.

Power flow in all cycloconverters is reversible, they operate comfortably with the output current in any phase relationship to the output voltage, leading or lagging, absorbing or generating power.

Harmonics and subharmonics are generally acceptable within the above frequency range. At higher frequencies subharmonic variations on both input and output sides may cause difficulties which become progressively worse as the output frequency is raised.

Poor power factor is a problem for all cycloconverters. This can be alleviated by the use of trapezoidal modulation at the cost of having a more or less fixed output voltage irrespective of frequency. When the output voltage must be varied with the frequency it is still possible to take advantage of trapezoidal modulation by accomplishing the output voltage variation by input voltage variation.

Single phase cycloconverters have severe subharmonic problems on the input side. These low frequency variations are generally unacceptable unless the capacity of the supply system is very large compared to the cycloconverter load.

12

TRANSFORMERS

12.1 Introduction

While it is possible to operate single way rectifiers directly from a four wire three phase supply and two way rectifiers, i.e. bridges, directly from a three wire a.c. supply, a transformer is frequently interposed between the a.c. system and the rectifier so as to attain one or more of the following goals

1. A d.c. output voltage unrelated to the a.c. input voltage.
2. Electrical isolation of the output from the input.
3. Reduction in the size of the commutation notches transferred from the d.c. side to the a.c. system.
4. Provision of a neutral point.

Such transformers have much in common with standard a.c. system transformers but also may have some aspects, e.g. different kVA ratings for the primary and secondary and more complex connections between phases, which set them apart as meriting special consideration. In addition, transformers belonging to the special group known as interphase transformers perform functions outside the range of normal transformer theory.

Rectifier circuits also make use of inductors which, while not transformers, have so many structural features in common with transformers that it is appropriate to consider them in the same context.

These matters are considered in this chapter and, while the reader can be expected to be acquainted with basic transformer construction and theory, there is an initial brief review of these topics in the light of matters particular to thyristor circuits.

12.2 Transformer structure and theory

A transformer comprises two or more closely coupled coils mounted on a laminated iron core, a structure shown symbolically in Fig. 12.1. Figure 12.1(a) shows a single phase core type transformer with the primary winding shown in gray and a secondary winding shown in black. Both primary and secondary are composed of two series connected coils. To obtain

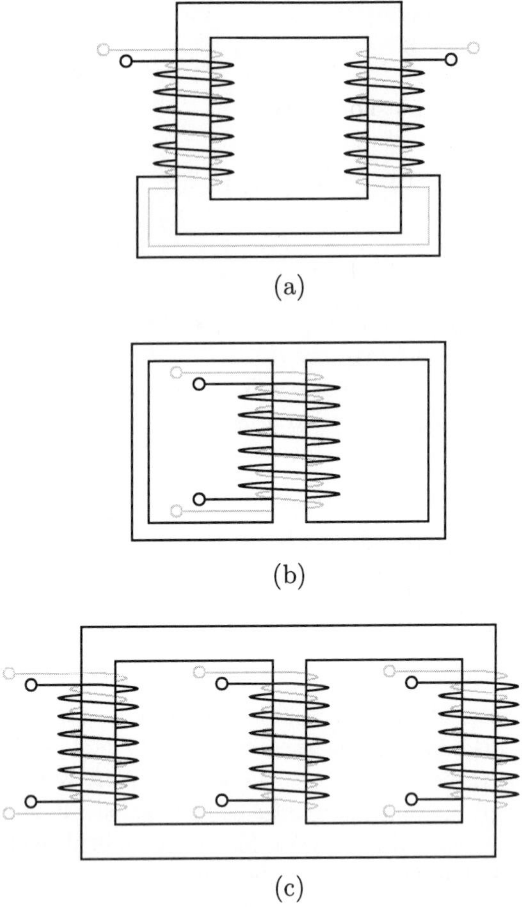

FIG. 12.1. Transformer construction. Part (a) shows a single phase core type transformer, part (b) shows a single phase transformer with a shell type core, and part (c) shows a three phase transformer.

close coupling between the primary and secondary, their coils are coaxial with the secondary coils outermost. There are many variants on this basic scheme, the primary might be the outer coil and the secondary the inner coil, the coils might be connected in parallel, there might be more than one secondary.

Single phase transformers can also use a shell type core as shown in Fig. 12.1(b). The windings are mounted on the center limb of a three limb core and the returning magnetic flux is carried by the two outer limbs which each have half the cross-sectional area of the center limb.

Three phase power can be handled by three single phase transformers but the three phase arrangement shown in Fig. 12.1(c) is usually preferred as being more economical. The core has three limbs like the shell type, single phase transformer but now the limbs are of equal cross-section, each carrying the primary and secondary windings of a phase. The windings can be either wye or delta connected.

While bearing the basic physical arrangements of Fig. 12.1 in mind, the course of the discussion will be clarified if the symbolic representations of Fig. 12.2 are used, Fig. 12.2(a) for the single phase transformer and Fig. 12.2(b) for the three phase transformer.

12.2.1 *Symbolic representation*

Magnetic flux conditions within the transformer are more easily displayed in the symbolic representations of Fig. 12.2. Here the primary and secondary windings have been separated so that they may be clearly distinguished. Figure 12.2(a) represents a single phase transformer with the primary winding, subscript p, and the secondary winding, subscript s, on separate limbs. The voltage v_p is applied to the primary and i_p is the current entering its dotted terminal. The voltage v_s is the output voltage from the secondary winding and i_s is the current leaving its dotted terminal. This convention, regarding the primary as the input side and the secondary as the output side, is inconsistent from a mathematical viewpoint but is the one generally adopted and will well serve the present study.

In Fig. 12.2 polarity dots indicate the positive ends of the windings. A wide variety of sign conventions relating currents and magnetic fluxes is possible. The one which will be used in this chapter has current entering a dotted terminal tending to create a magnetic flux which leaves the dotted end and enters the nondotted end of the core.

The magnetic flux set up by the primary and secondary currents follows extremely complex paths which can be studied using computational aids such as finite element analysis. However, for analytical purposes the flux can be regarded as having three components, a mutual flux, ϕ_m, which links both windings and two leakage components, $\phi_{p\sigma}$ which links the primary but not the secondary and $\phi_{s\sigma}$ which links the secondary but not the primary. There is a considerable disparity between these components. The mutual flux, being entirely within the iron core, is large while the leakage fluxes, taking a path which is predominantly nonferromagnetic, are small. The m.m.f. which produces the primary leakage flux is $n_p i_p$, that which produces the secondary leakage flux is $n_s i_s$, and their difference, $n_p i_p - n_s i_s$, produces the mutual flux, n_p and n_s being the number of primary and secondary turns. When the transformer is under significant load both the primary and secondary m.m.fs are large compared to their difference.

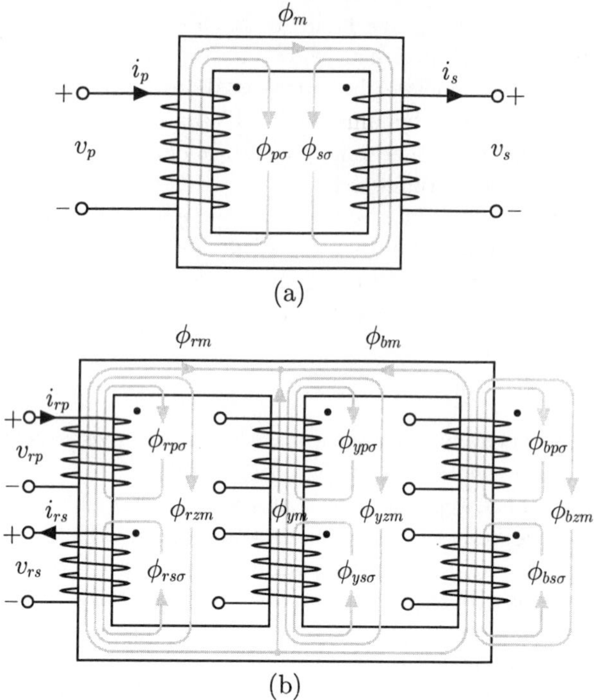

FIG. 12.2. Symbolic representations of (a) a single phase transformer and (b) a three phase transformer. For clarity the primary and secondary windings have been separated.

The e.m.f. produced in a winding by the mutual flux is equal to the product of the time rate of change of the flux and the number of winding turns. Thus if the primary e.m.f. is e_p, then the secondary e.m.f., e_s, is $e_p n_s / n_p$.

Conditions in a three phase transformer can be similarly analyzed but now the phase must be identified. This is done in Fig. 12.2(b) by using additional subscripts r, y, and b indicating the red, yellow, and blue phases, it being assumed that the red phase is on the left limb, the yellow phase on the center limb, and the blue phase on the right limb. The flux linking each coil is now shown in three parts. Considering the flux linking the red phase primary coil, there is a component ϕ_{rm}, a component ϕ_{rzm}, and a component $\phi_{rp\sigma}$.

The first flux component, ϕ_{rm}, also links the red phase secondary and combines in the top and bottom yokes of the transformer core with the corresponding fluxes for the yellow and blue phases, ϕ_{ym} and ϕ_{bm}, their sum always being zero. These components will be called the *ferromagnetic mutual fluxes.* The remaining portions of the mutual fluxes, ϕ_{rzm}, ϕ_{yzm}, and ϕ_{bzm}, must complete their path outside the transformer core, in nonferromagnetic material. These components will be called the *nonferromagnetic mutual fluxes.* The remaining components, $\phi_{rp\sigma}$ and the corresponding components for the remaining two primary coils and the three secondary coils, are *leakage fluxes* which link only the winding which creates them and no other. In considering the paths taken by the nonferromagnetic mutual fluxes and the leakage fluxes it must be remembered that the transformer is a three dimensional object so that, while the ferromagnetic mutual fluxes are confined to the magnetic core, the other components can occupy the totality of space surrounding the transformer.

When the transformer primary is connected to a three phase supply and the load is three phase and linear, the instantaneous sum of these mutual fluxes, both ferromagnetic and nonferromagnetic, is zero and the transformer behaves exactly as would a bank of three single phase transformers. However, under abnormal conditions, either supply system faults, load faults, or with certain rectifier connections, the sum of these fluxes is not zero. The excess, the nonferromagnetic component, must find its way back to the lower yoke of the transformer somehow and it is clear that much of this path, like that of the leakage fluxes, is in nonferromagnetic material. The flux is then small relative to the m.m.f. creating it. This behavior, known as *zero sequence behavior*, is very different from that of a bank of single phase transformers. For the time being, zero sequence behavior need not be considered. It will be taken up again in Section 12.6.

12.2.2 *Equivalent circuits*

The physical characteristics illustrated by Fig. 12.2 are incorporated in the transformer equivalent circuit shown in Fig. 12.3(a) which represents either a single phase transformer, or one phase of a three phase transformer, with input terminals on the left and output terminals on the right. In order to drive the current i_p into the positive input terminal, the input voltage, v_p, must first overcome the resistance, R_p, and the leakage inductance, $L_{p\sigma}$, of the primary winding. The resistance of a winding is somewhat greater than the d.c. value because of a.c. phenomena such as skin and proximity effects and iron losses associated with the leakage flux. The leakage inductance is the leakage flux linkage per ampere of winding current. Its computation is a complex matter of great concern to the transformer designer but it is sufficient for the present purpose to take it as $n_p\phi_{p\sigma}/i_p$ for the primary and $n_s\phi_{s\sigma}/i_s$ for the secondary. The series combination of the resistance and leakage inductance, $R_p + L_{p\sigma}\mathrm{d}/\mathrm{d}t$, will be referred to as the leakage impedance, $\mathbf{Z}_{p\sigma}$. More strictly this should be the operational leakage impedance

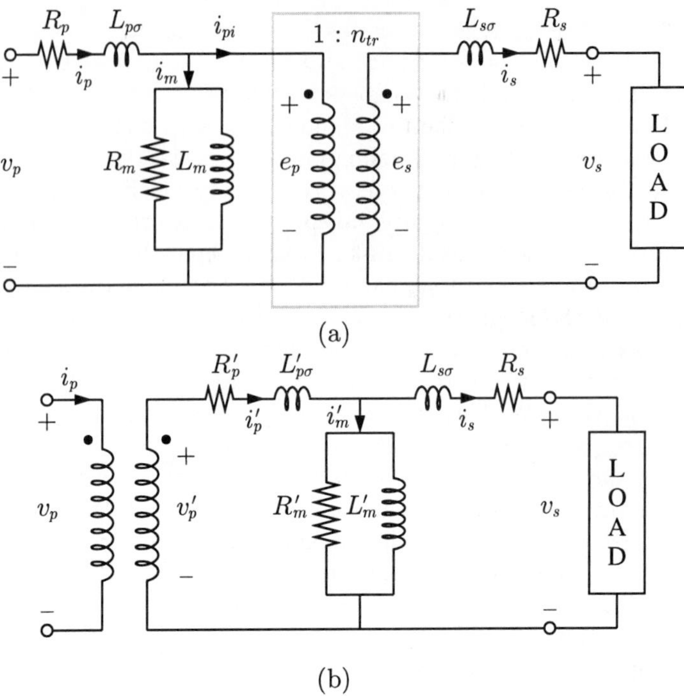

FIG. 12.3. Transformer equivalent circuits.

since it incorporates the operator d/dt rather than $j\omega$ but, since the context always makes clear whether the operational or the a.c. impedance is being used, the word *operational* has, in general, been omitted.

After passing through the leakage impedance, the input current divides, the part i_m going through the magnetizing impedance, $\mathbf{Z}_m$, and the remainder, i_{pi}, going through the primary of the ideal transformer. The portion i_m flowing through the magnetizing impedance is the *magnetizing current* and creates the mutual flux. It has two parts, the first flowing through the magnetizing inductance, L_m, and providing the m.m.f. required to produce the mutual flux. The second part flows through the magnetizing resistance, R_m, and supplies the iron losses associated with the mutual flux—hysteresis and eddy current losses in the laminations. The ferromagnetic core is easily magnetized so that the magnetizing current is small relative to the rated current of the transformer, typically less than 10%.

The ideal transformer, enclosed by the gray line, is a concept, rather than a reality, which makes the equivalent circuit possible. It is a transformer whose losses are zero and whose magnetic coupling is perfect. It has a transformation ratio n_{tr}, meaning that a primary input voltage is transformed into a secondary output voltage n_{tr} times as big and that a secondary output current is reflected as a primary input current n_{tr} times as big. The transformation ratio is very close to the turns ratio, n_s/n_p, and is often referred to as the turns ratio. Two e.m.fs are shown in Fig. 12.3(a), e_p on the primary side and e_s on the secondary side of the ideal transformer with $e_s = n_{tr}e_p$. The input current to the primary of the ideal transformer, i_{pi}, is similarly equal to $n_{tr}i_s$, i_s being the output current from the secondary. Thus, the input volt-amperes, e_pi_p, is at all instants equal to the output volt-amperes, e_si_s. The ideal transformer provides electrical isolation between the primary and secondary sides of the equivalent circuit.

Part of the secondary e.m.f. is absorbed in driving the secondary current, i_s, through the secondary leakage impedance, $R_s + L_{s\sigma}\mathrm{d}/\mathrm{d}t$, and the remainder, the secondary output voltage, v_s, overcomes the impedance presented by the load.

The equivalent circuit represents the behavior of one phase of a three phase transformer to positive or negative sequence voltages and currents which constitute the overwhelming majority of practical situations. In those rare situations when zero sequence currents flow, the impedance to such currents is similar in form but with a greatly reduced magnetizing impedance because the path of the zero sequence flux is predominantly in nonferromagnetic material. The magnetizing impedance during positive or negative sequence operation is large like that for a single phase transformer and is treated computationally in the same way. However, it is worth noting that it incorporates mutual couplings between phases, a phenomenon which does not occur in a three phase bank of single phase transformers.

The transformer losses, represented by the resistances of the equivalent circuit, are reduced as far as is economically justified. The same is not true of the leakage inductances which are often designed to have a specific value so as to limit the transformer short circuit current and, in the rectifier context, to reduce the severity of the commutation notches experienced by the a.c. supply.

12.2.3 *Transfer of impedance*

Most computations using the equivalent circuit are facilitated by the transference of impedances across the ideal transformer. Thus, the primary leakage and magnetizing impedances can be transferred to the secondary side of the transformer as shown in Fig. 12.3(b) provided allowance is made for the increase in voltage and decrease in current by the transformation ratio. This is done by increasing any impedance which is transferred from primary to secondary by the factor n_{tr}^2 and decreasing any impedance which is transferred from the secondary to the primary side by the same factor. That this process has been done will be indicated by attaching a prime to the affected component. Thus, in Fig. 12.3(b) the magnetizing inductance is shown as L'_m indicating that this inductance is n_{tr}^2 times the value shown in Fig. 12.3(a). Similarly the output voltage and current from the secondary of the ideal transformer are shown as v'_p and i'_p indicating that they are n_{tr} and $1/n_{tr}$ times the actual input values. Since there is very little possibility of confusion, the same nomenclature will be applied to transfers from the secondary to the primary side. Thus, if the ideal transformer were moved to the output terminals, the secondary leakage impedance would have been transferred from the secondary to the primary side, a process which would have required division by n_{tr}^2 and would have been indicated by writing the impedance as $R'_s + L'_s\mathrm{d}/\mathrm{d}t$.

Transfer of impedances across the ideal transformer is effective provided the secondary circuit is isolated from the primary, the normal situation. However, if the two sides of the transformer are deliberately interconnected, as in an autotransformer, this procedure must be applied with great care lest the move violate the real circuit electromagnetic topology.

12.2.4 *Nonlinearities*

The equivalent circuits are based on the assumption that a transformer is a linear device. In fact the ferromagnetic core constitutes a significant nonlinearity, the relationship between flux and magnetizing current having the shape of the familiar hysteresis loop. Such a loop is shown in graph 1 of Fig. 12.4. It is a dynamic loop, the word dynamic indicating that it has been taken at the excitation frequency and thus incorporates core eddy current losses as well as hysteresis loss.

Graphs 2 and 3 of Fig. 12.4 show the effects of creating a sinusoidally varying mutual flux. The black curve of graph 2 shows the sinusoidal waveform of the flux and the gray curve the cosinusoidal waveform of the winding e.m.f. The black curve of graph 3 is the magnetizing current required to create the flux. It has a peaky waveform because the ferromagnetic core is being taken some way into saturation. The current waveform can be analyzed into Fourier components and the two most important of these, the fundamental and the third harmonic, are shown in gray.

Figure 12.4 has been drawn on the assumption that the waveform of the winding e.m.f. is sinusoidal. In reality the harmonic components of the magnetizing current flowing in the primary leakage impedance create harmonic voltage drops which produce some distortion in the waveform of the e.m.f. even though the primary input voltage is sinusoidal. Fortunately, since the magnetizing current is small in comparison with the rated current, the distortion is small and is generally ignored, the magnetizing impedance being taken as that linear impedance which, under normal operating conditions, will conduct a sinusoidal current with the same r.m.s.

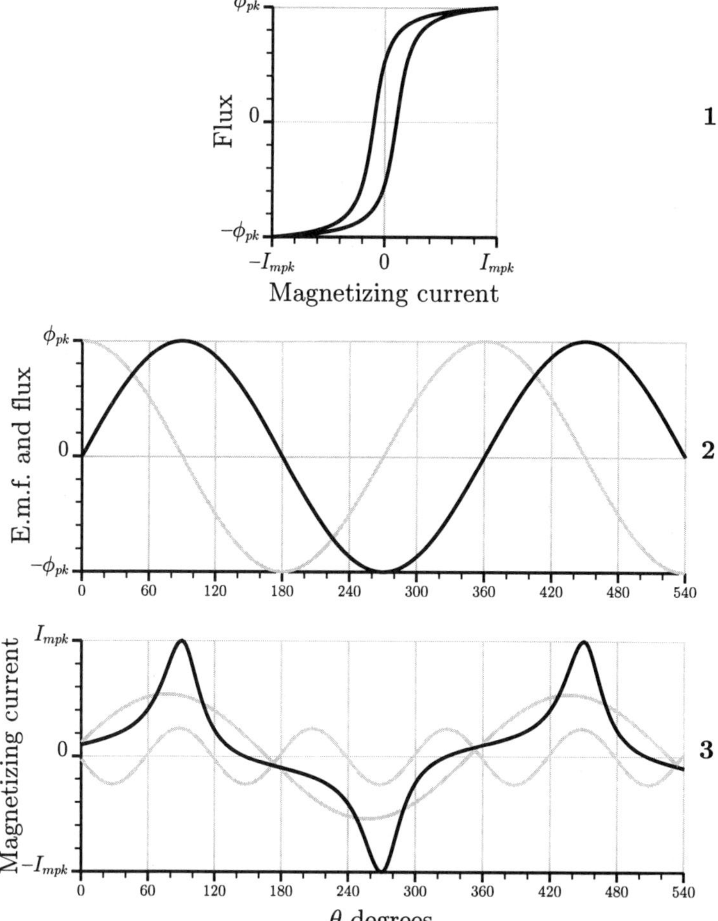

FIG. 12.4. The transformer magnetizing current. The graphs are identified by a number on the right. Graph 1 shows the dynamic hysteresis loop, graph 2 shows the winding e.m.f. in gray and the core flux in black, and graph 3 shows the magnetizing current in black and its fundamental and third harmonic components in gray.

value as the magnetizing current and will absorb the same amount of power as the ferromagnetic core.

Triplens, i.e. harmonics whose order is an integral multiple of three, in a three phase system form zero sequence systems. Since zero sequence currents cannot flow in a three wire, three phase system, the triplen component of the magnetizing current is suppressed in a three phase transformer with wye connected windings. This causes distortion of the winding e.m.fs

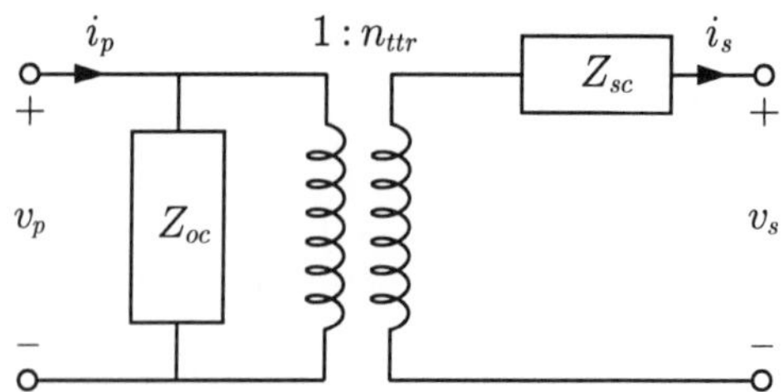

FIG. 12.5. The modified Thévenin equivalent circuit.

which appears as an oscillation of the neutral voltage about ground potential at three times the supply frequency. This is undesirable and every effort is made to have at least one of the windings delta connected so that the triplens can circulate round the closed mesh thus formed. When the wye–wye connection is unavoidable, an auxiliary delta connected winding is frequently added to carry the triplens.

12.2.5 *Equivalent circuit approximations*

Transformer computations using the equivalent circuits of Fig. 12.3 are complicated by the Tee configuration of the circuit, the magnetizing impedance branching from the junction of the primary and secondary leakage impedances. Various approximations are made to alleviate this situation. All these have as their basis the fact that in a practical transformer the leakage impedances are considerably smaller than the magnetizing impedance. The first and almost universal approximation, replacing the nonlinear magnetizing impedance by a linear equivalent, has already been discussed. Here, additional approximations are presented and are illustrated by Example 12.1.

Using Thévenin's theorem, the Tee equivalent circuit can be converted to the equivalent L, shown in Fig. 12.5, which is better suited to computation. The parameters of this circuit are directly measurable, the impedance $\mathbf{Z}_{oc}$ being the primary input impedance with the secondary open circuited, the Thévenin transformation ratio n_{ttr} being the ratio of the secondary voltage to the primary voltage with the secondary open circuited, and the impedance $\mathbf{Z}_{sc}$ being the secondary input impedance with the primary short circuited. The Thévenin transformation ratio is complex, having both an amplitude and phase, which introduces a new complication. Fortunately the phase is very small and it is usually ignored in power transformers, a practice which will be followed here. When this approximation is incorporated, the circuit will be called the *modified Thévenin equivalent circuit.* It is depicted in Fig. 12.5.

In terms of the parameters of the equivalent circuit of Fig. 12.3(a), the Thévenin circuit parameters are

$$\mathbf{Z}_{oc} = \mathbf{Z}_{p\sigma} + \mathbf{Z}_m \tag{12.1}$$

$$\mathbf{N}_{ttr} = \frac{\mathbf{Z}_m}{\mathbf{Z}_{p\sigma} + \mathbf{Z}_m} n_{tr} \tag{12.2}$$

$$\mathbf{Z}_{sc} = \mathbf{Z}_{s\sigma} + \frac{\mathbf{Z}'_{p\sigma}\mathbf{Z}'_m}{\mathbf{Z}'_{p\sigma} + \mathbf{Z}'_m} \tag{12.3}$$

The Thévenin transformation ratio, $\mathbf{N}_{ttr}$, is complex and can be expressed in terms of its modulus and argument as $n_{ttr}\underline{/\nu}$. The angle ν is very small, typically well under one degree, and it is the normal practice to ignore it, taking the Thévenin transformation ratio to be the real number n_{ttr}. This will be done from this point onwards.

In Chapter 5, when the effect of the transformer and supply impedances on commutation was being determined, a further approximation was made by moving the open circuit impedance to the equivalent infinite bus. This was done in order to avoid the problems posed by the non-sinusoidal transformer primary voltage and was justified by the small size of the supply impedance relative to the transformer impedance.

When studying supply systems, which contain numerous transformers with a variety of voltage levels and transformation ratios, it is usual to normalize everything to a single voltage level and use per unit values. This is not necessary in single transformer situations and ohmic values will be used here. However, it is worth remembering that, when normalized, the range of power transformer parameters is small, all transformers having much the same per unit impedances.

When making power system calculations, the contribution of the magnetizing currents to the overall solution is small and they are frequently ignored, the magnetizing impedances being replaced by open circuits.

Example 12.1 illustrates the effect of some of these approximations but before coming to it one other aspect of transformers must be considered, the phase shift in three phase transformers.

12.2.6 *Phase shift in three phase transformers*

In any transformer there is a phase difference between the secondary output voltage and the primary input voltage. This is produced by the voltages absorbed by the transformer internal impedances and is small, generally significantly less than 1°. The phase shift referred to in the phrase *phase shift in three phase transformers* is the much larger shift between the primary input line to line voltages and the secondary output line to line voltages generated as a consequence of the winding connections, wye or delta. Of course the small internal phase shift is still there and is added to the larger shift but it is not considered in the present context. This point is expanded upon in Example 12.1.

With the above proviso, there is no phase shift between the input voltage to a primary winding and the output voltage from a secondary winding. Thus, if the primary and secondary windings are connected in the same way, i.e. both wye or both delta, there is no phase shift. However, if one winding is wye connected and the other is delta connected there is a 30° phase difference between the line to line voltages of the input and output systems as illustrated in Fig. 12.6.

Figure 12.6 shows a transformer whose primary is delta connected and whose secondary is wye connected. The windings are shown symbolically in Fig. 12.6(a), parallel primary and secondary windings being on the same limb of the core. The windings are labelled 1, 2, and 3 and the a.c. lines are labelled R, Y, and B with subscripts p or s being added when necessary to distinguish between a primary and a secondary quantity. Figure 12.6(b) is a phasor diagram of the winding voltages, the primary and secondary voltages being in phase. Figure 12.6(c) is a phasor diagram showing the line to line voltages. On the primary, delta connected, side they are the same as the winding voltages. On the secondary, wye connected, side the winding voltages must be combined, the phasor voltage between the red and yellow output lines being the phasor difference between v_{1s} and v_{2s}. This produces a secondary system whose voltages lead those of the primary system by 30°. Exactly similar considerations show that a wye–delta connected transformer produces a secondary system whose voltages lag those of the primary system by 30°.

In power systems, where portions at different voltage levels may be interconnected by numerous transformers, care must always be taken to ensure that the phase shifts of the transformers are compatible. This particular aspect does not arise in the application of rectifier transformers but the phenomenon is exploited to excellent effect when a transformer with a single primary and dual secondary windings, one wye and one delta connected, is used to produce a 12 pulse output by connecting two rectifier bridges in series as illustrated in Fig. 9.1.

Example 12.1

A three phase transformer rated at 300 kVA delivers power from a 12 kV, 60 Hz system to a 440 V system. The transformer primary is delta connected and its secondary is wye connected. The primary leakage impedance at the supply frequency is $7.1 + j20.8\ \Omega$ per primary phase. The secondary leakage impedance is $0.0032 + j0.0093\ \Omega$ per secondary phase. The magnetizing impedance referred to a primary phase is represented by a resistance

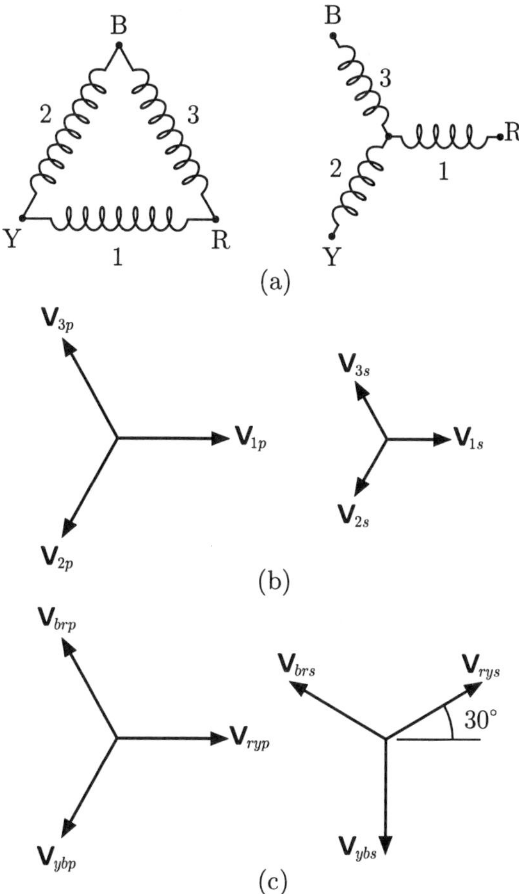

FIG. 12.6. Voltage phasors in a delta–wye connected three phase transformer. The transformer windings are shown in (a), the winding voltage phasors in (b), and the line to line voltage phasors in (c).

of 177 000 Ω in parallel with a reactance of 19 700 Ω. The transformation ratio, n_{tr}, is 1 : 0.021.

Compare results given by the modified Thévenin equivalent circuit with the true values when the primary voltage and input kVA have the rated values and the input power factor is 0.76 lagging.

The rated input voltage is 12 000 V r.m.s. line to line and, since the primary is delta connected, this is the rated primary phase voltage. The rated kVA is 300 so that the rated kVA per phase is 100. Thus, under the specified conditions, the primary input current is 100 000/12 000 = 8.3333

A r.m.s. The input power factor is 0.76 lagging so that this current lags the primary phase voltage by 40.4358°. The primary input voltage phasor to phase 1 will be taken as the datum phasor so that

$$\mathbf{V}_{1p} = 12\,000\underline{/0.0^\circ}$$

$$\mathbf{I}_{1p} = 8.3333\underline{/-40.4358^\circ}$$

The primary leakage impedance is 7.1 + j20.8 Ω so that the primary e.m.f., $\mathbf{E}_{1p}$, is

$$12\,000\underline{/0.0} - (7.1 + \text{j}20.8) \times 8.3333\underline{/-40.4358^\circ} = 11\,842.7\underline{/-0.4513^\circ}$$

The magnetizing impedance is 177 kΩ resistance in parallel with 19.7 kΩ reactance so that the magnetizing current, $\mathbf{I}_{1m}$, is $0.6049\underline{/-84.1004^\circ}$. Subtracting this from the input current yields the current phasor entering the primary winding of the ideal transformer, $\mathbf{I}_{1ip}$, as $7.9060\underline{/-37.5134^\circ}$.

Transforming the e.m.f. and current to secondary values yields the secondary e.m.f., $\mathbf{E}_{1s}$, as $248.698\underline{/-0.4513^\circ}$ and the secondary current, $\mathbf{I}_{1s}$, as $376.478\underline{/-37.5134^\circ}$

The secondary leakage impedance is 0.0032+j0.0093 and subtracting the voltage drop due to the secondary current flowing in this yields the output voltage, $\mathbf{V}_{1s}$, as $245.635\underline{/-0.9336^\circ}$. The output line to line voltage, $\mathbf{V}_{rys}$, has a magnitude $\sqrt{3}$ times that of this voltage and leads it by 30° so that it is $425.452\underline{/29.0664^\circ}$.

The primary copper loss is $3.0 \times 7.1 \times 8.3333^2 = 1479.2$ W. The secondary copper loss is $3.0 \times 0.0032 \times 376.478^2 = 1360.7$ W. The core loss is $3.0 \times 11843.8^2/177000.0 = 2377.1$ W. Thus, the total loss in the transformer is 5217.0 W and the efficiency is 97.712%.

At zero load, the output voltage is 251.724 so that the drop in output voltage from zero load to rated load is 3.026 V and, on the basis of the output voltage at rated load, the regulation is 2.479%.

For the modified Thévenin equivalent circuit of Fig. 12.5, the open circuit impedance is given by

$$\mathbf{Z}_{oc} = 2172.87 + \text{j}19\,479.75$$

the short circuit impedance is given by

$$\mathbf{Z}_{sc} = 0.006325 + \text{j}0.018464$$

and the transformation ratio is

$$\mathbf{N}_{ttr} = 0.020977\underline{/0.01390^\circ}$$

Since the modified circuit is being used, the very small phase shift is ignored and the scalar transformation ratio,

$$n_{ttr} = 0.020977$$

will be used.

The open circuit current is

$$\frac{12\,000}{2172.87 + j19\,479.75} = 0.6122\underline{/-83.6352^\circ}$$

Subtracting this from the primary input current and transforming the result by the Thévenin ratio gives the secondary output current phasor as $376.478\underline{/-37.4995^\circ}$.

Subtracting the voltage drop in the short circuit impedance from the Thévenin secondary e.m.f. yields the output voltage as

$$245.637\underline{/-0.9336^\circ}$$

Multiplying this by $\sqrt{3}$ and advancing its phase by 30° yields the output voltage between the red and yellow lines as $425.456\underline{/29.0664^\circ}$.

The open circuit loss is $3.0 \times 2172.87 \times 0.6122^2 = 2443.3$ W. The short circuit loss is $3.0 \times 0.006325 \times 376.478^2 = 2689.6$ W. The total loss is therefore 5123.9 W and the transformer efficiency is 97.75%.

The regulation, on the same basis as above, is 2.478%.

The differences between the true values and those obtained from the modified Thévenin equivalent circuit are, from a practical viewpoint, absolutely negligible.

****

12.3 Transformer ratings

The kVA rating of a transformer, all other factors being equal, is a reasonably good indicator of its size, weight, and cost and is therefore the most important single number characterizing it. The rating is half the sum of the winding kVA ratings, the latter being the product of the designed r.m.s. value of the winding voltage, V_w, and the designed r.m.s. value of its current, I_w, divided by 1000, i.e.

$$\text{kVA} = \frac{1}{2} \sum_{w=1,n_w} \frac{V_w I_w}{1000} \tag{12.4}$$

where n_w is the total number of windings.

The principal factors in the qualifying phrase used above, *all other factors being equal*, which affect the relationship between the kVA rating and the size, weight, and cost are the temperature withstand capability of the insulation, the saturation flux density of the magnetic core, and the duty cycle. Transformers fitting within a given class with regard to these factors, e.g. a 40°C temperature rise above ambient, a core of standard transformer laminations, and a continuous load, may legitimately be compared on the basis of their kVA rating.

The ratings of the primary and secondary windings of a standard a.c. transformer are the same, each being equal to the kVA of its designed load. This is not the case for a rectifier transformer, a fact demonstrated in Chapters 2 and 3. For a rectifier transformer, the rated load is the product of the rated d.c. load current and voltage and the transformer rating is always somewhat greater than this because of the non-sinusoidal current waveforms. Further, because some current harmonics which flow in the secondary may not flow in the primary, the rating of the primary may be somewhat less than that of the secondary. These matters have been considered in Chapters 2 and 3 and will now be expanded upon in the more complex case of the twelve phase system illustrated in Fig. 12.7.

12.3.1 *A twelve phase rectifier transformer*

Figure 12.7 shows a three phase transformer with a delta connected primary and two secondaries, one delta connected the other wye connected, each producing the same line to line voltage. Each secondary feeds a bridge rectifier and the two rectifiers are connected in series to feed the load. The arrangement was introduced in Section 9.2. Because of the 30° phase difference between the two secondaries, the output is twelve phase and the effect on the input is that of a twelve phase rectifier, although within the transformer the two secondaries operate six phase. The two secondaries are distinguished by the letter D for delta and W for wye and quantities pertaining to them by corresponding lower case subscripts d and w.

The arrangement of the transformer windings on the three phase core is illustrated by Fig. 12.8 where the primary windings are shown at the top, the delta secondary windings at the bottom, and the wye secondary windings in between. Each limb of the core carries three windings, one primary and two secondaries. Three of the six secondaries form the delta and the remaining three form the wye. The number of turns on a delta secondary is ideally $\sqrt{3}$ times the number on a wye secondary. However, since the number of turns must be a whole number, the ratio is in practice the nearest rational fraction, e.g. if a wye secondary has 100 turns, a delta secondary will have 173 turns giving a delta winding voltage 0.12% low, a negligible error.

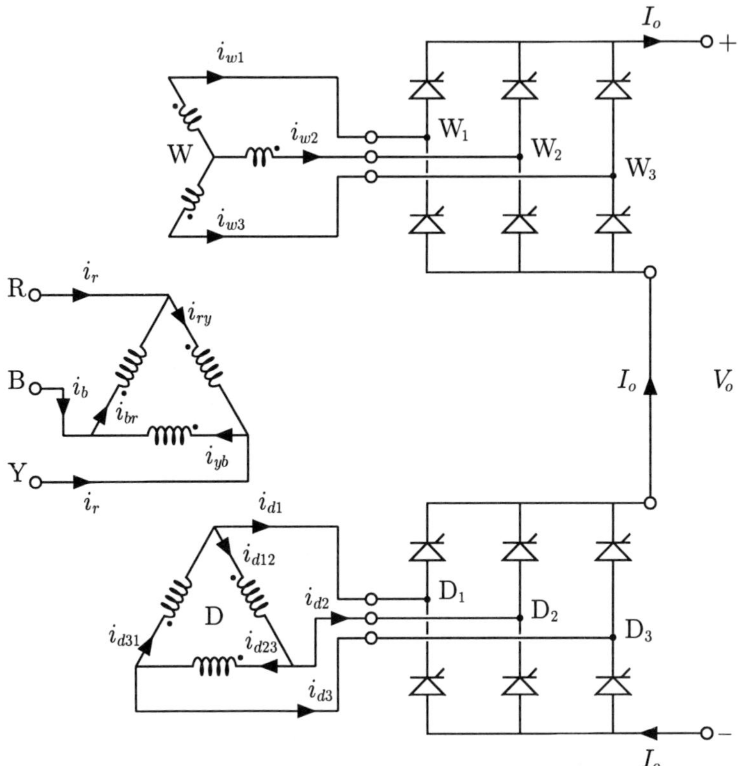

FIG. 12.7. A twelve phase rectifier using two series connected three phase bridge rectifiers.

The analysis of this system will be performed assuming that the load current is smooth, ripple free d.c. and will neglect the effect of transformer leakage impedance.

The value of the d.c. output voltage at zero delay, V_{om}, is twice that for either bridge rectifier unit and, using the result of eqn 3.15, is given by

$$V_{om} = 2\frac{3\sqrt{2}}{\pi}V_{aco} = 2.7009V_{aco}$$

where V_{aco} is the r.m.s. value of the a.c. line to line output voltage of either secondary. Inverting this expression yields

$$V_{aco} = \frac{\pi}{6\sqrt{2}}V_{om} = 0.3702V_{om}$$

This is the rated voltage of a delta secondary and is $\sqrt{3}$ times the rated voltage of a wye secondary.

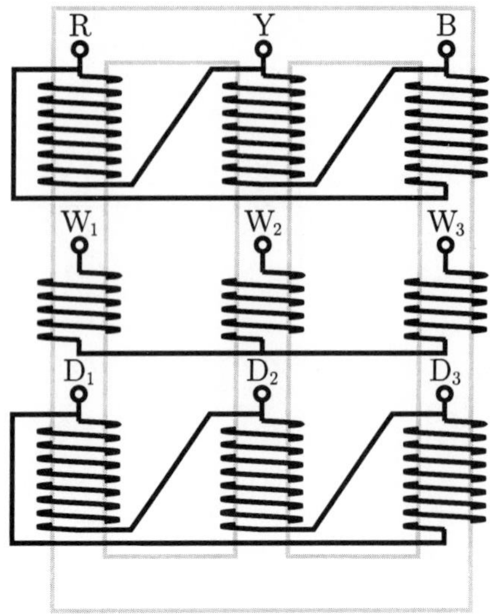

FIG. 12.8. The arrangement of the windings on the transformer core for the twelve phase rectifier. The outline of the core is shown in gray.

The voltage and current waveforms in the rectifier are illustrated in Fig. 12.9 in which the voltage of the red input line relative to the supply system neutral is taken as datum and the delay angle is assumed to be zero. The graphs are identified by numbers on the right. Graph 1 shows the three line to line input voltages, v_{ry}, v_{yb}, and v_{br}. The currents in the a.c. input lines to the rectifiers are shown in graphs 2, 3, and 4 for the wye rectifier and 5, 6, and 7 for the delta rectifier. As has just been seen in Fig. 12.6, there is a phase difference of 30° between the output currents from the delta secondary and those from the wye secondary. The a.c. line current to a rectifier has an amplitude of I_{dc} and is on for 120°, off for 60°, on in the reverse direction for 120°, and off for the remaining 60° of a cycle.

Figure 12.9 is drawn taking the voltage between the red supply line and the system neutral as datum and using the cosine form, i.e. $v_r = \sqrt{2}V_{rn}\cos(\theta)$. Voltages and currents are normalized using as base values the r.m.s. value of the rated input line to line voltage, the rated d.c. output current and the Thévenin transformation ratio, delta secondary open circuit voltage per phase/primary voltage per phase.

All a.c. input currents to the rectifiers have the same r.m.s. value, $\sqrt{2/3}\,I_{dc}$, and this is also the r.m.s. value of the current in a wye connected secondary.

The determination of the current in a delta connected secondary is complicated by the non-sinusoidal waveform of the output line current, so

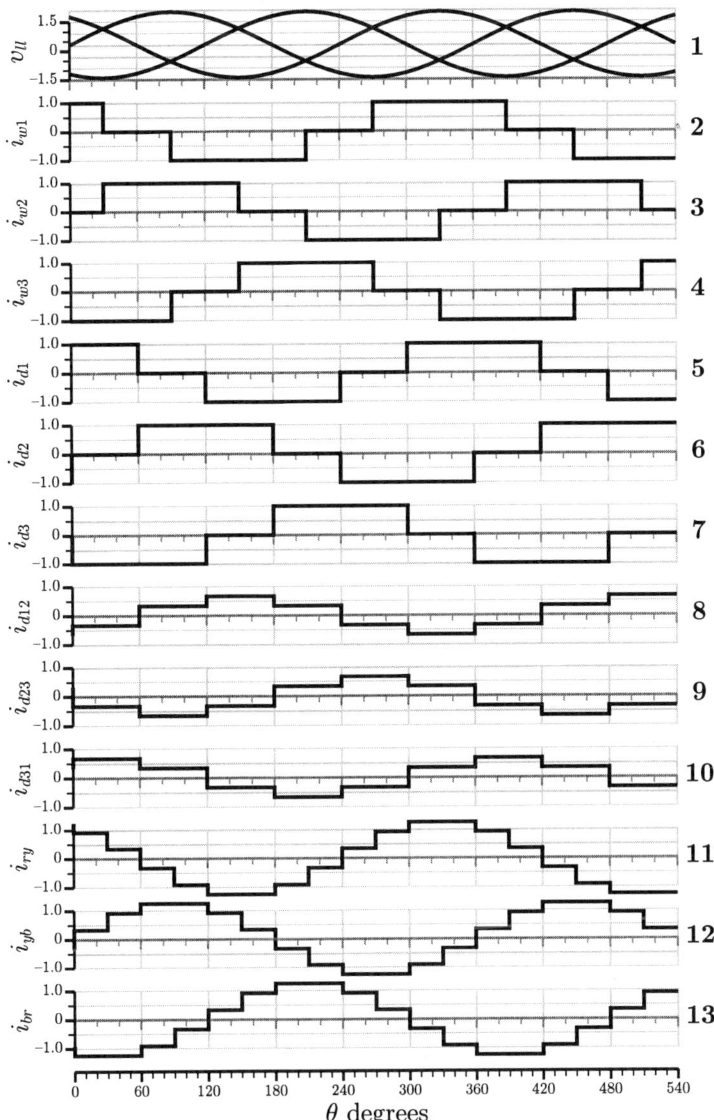

FIG. 12.9. Voltage and current waveforms in a twelve phase rectifier. The graphs have been drawn for a delay angle of zero and are normalized to the d.c. output current.

that it cannot be assumed that the r.m.s. value of the winding current is equal to that of the line current divided by $\sqrt{3}$ as is done when the waveforms are sinusoidal. Applying Kirchhoff's current law to each terminal of the delta secondary yields

$$\begin{bmatrix} i_{d1} \\ i_{d2} \\ i_{d3} \end{bmatrix} = \begin{bmatrix} -1 & 0 & 1 \\ 1 & -1 & 0 \\ 0 & 1 & -1 \end{bmatrix} \begin{bmatrix} i_{d12} \\ i_{d23} \\ i_{d31} \end{bmatrix} \tag{12.5}$$

Since there may be zero sequence currents circulating round the delta without appearing in the lines, the coefficient matrix is singular and cannot be inverted to yield an expression for the winding currents in terms of the line currents. In the present case it is known that there is no zero sequence voltage to produce such a current so that the current may be assumed to be zero. The sum of the winding currents is then zero and i_{d31} may be replaced by $-(i_{d12} + i_{d23})$ thus reducing eqn 12.5 to

$$\begin{bmatrix} i_{d1} \\ i_{d2} \end{bmatrix} = \begin{bmatrix} -2 & -1 \\ 1 & -1 \end{bmatrix} \begin{bmatrix} i_{d12} \\ i_{d23} \end{bmatrix} \tag{12.6}$$

Equation 12.6 can be inverted and, when this is done and the additional row for i_{d3} is inserted, yields eqn 12.7

$$\begin{bmatrix} i_{d12} \\ i_{d23} \\ i_{d31} \end{bmatrix} = \frac{1}{3} \begin{bmatrix} -1 & 1 & 0 \\ 0 & -1 & 1 \\ 1 & 0 & -1 \end{bmatrix} \begin{bmatrix} i_{d1} \\ i_{d2} \\ i_{d3} \end{bmatrix} \tag{12.7}$$

The waveforms of these three delta secondary winding currents are shown in graphs 8, 9, and 10 of Fig. 12.7. The waveforms are stepped rectangular with six steps per cycle. The steps are of equal duration, each lasting for one-sixth of a cycle, 60°. The step levels are at $I_{dc}/3$ and $2I_{dc}/3$. Hence, each delta winding current has an r.m.s. value of $\sqrt{2}I_{dc}/3$, less than that for a wye connected secondary by the usual factor, $\sqrt{3}$. This ratio, $1 : \sqrt{3}$, always applies when there is no zero sequence current.

Before the effect of these secondary currents on the primary can be determined, the transformation ratio for each winding must be found. The magnitude of the Thévenin ratio, the ratio of a secondary winding open circuit voltage to that of the associated primary, will be used. The voltage across a primary winding is the a.c. system line to line voltage, V_s. It was found above that the voltage across a delta secondary winding is $0.3702V_{dc}$ and that the voltage across a wye secondary winding is $0.2138V_{dc}$. The transformation ratio, n_{dttr}, for the delta secondary is therefore

$$n_{dttr} = 0.3702\frac{V_{dc}}{V_s} \tag{12.8}$$

and for the wye secondary is

$$n_{wttr} = 0.2138\frac{V_{dc}}{V_s} \tag{12.9}$$

The current in a primary winding has two functions, one being to balance the combined m.m.fs of its two secondaries and the other being to provide the magnetic flux linking the three windings. The magnetizing component forms a small part of the rated current and is normally ignored in kVA calculations. With the directions and winding polarities shown in Fig. 12.7, the m.m.f. balancing portion of a primary current is given by

$$i_p = -n_{dttr}i_d + n_{wttr}i_w \tag{12.10}$$

This computation has been performed to derive the primary current waveforms of graphs 11, 12, and 13 of Fig. 12.9. These are stepped rectangular waveforms with 12 steps per cycle. The steps are of equal duration, each lasting for one-twelfth of a cycle, 30°. Their Fourier spectra contain only the fundamental and harmonics whose orders are integral multiples of 12 because the m.m.f. components due to the secondary current harmonics whose order is an odd multiple of 6 cancel and are not reflected into the primary. The height of the steps is $I_{dc}n_{dttr}/3$, $I_{dc}n_{dttr}(1/3+1/\sqrt{3}) = 0.9107I_{dc}n_{dttr}$ and $I_{dc}n_{dttr}(2/3 + 1/\sqrt{3}) = 1.2440I_{dc}n_{dttr}$. Hence the r.m.s. value of a primary current is $0.9107I_{dc}n_{dttr}$.

Using these r.m.s. current values, the kVA rating of the wye secondary is

$$3\frac{\pi}{6\sqrt{6}}V_{om}\frac{\sqrt{2}}{\sqrt{3}}I_{dc} \times 10^{-3} = 0.5236V_{om}I_{dc} \times 10^{-3} \tag{12.11}$$

and the kVA rating of the delta secondary is

$$3\frac{\pi}{6\sqrt{2}}V_{om}\frac{\sqrt{2}}{3}I_{dc} \times 10^{-3} = 0.5236V_{om}I_{dc} \times 10^{-3} \tag{12.12}$$

so that, as might be expected, the ratings of the two secondaries are equal.

The total secondary rating is the sum of the two individual values and is $1.0472V_{om}I_{dc} \times 10^{-3}$, 4.7% greater than the rated load. The fundamental component of the current in a wye secondary has an r.m.s. value of $0.7797I_{dc}$ and of the current in a delta secondary has an r.m.s. value equal to this divided by $\sqrt{3}$, i.e. $0.4502I_{dc}$. On the basis of these fundamental values, the fundamental kVA ratings are each $0.5V_{om}I_{dc} \times 10^{-3}$ for a total of $V_{om}I_{dc} \times 10^{-3}$, establishing that the additional $0.0472V_{om}I_{dc} \times 10^{-3}$ is due to the harmonics of the secondary currents.

The primary rating is $3V_s \times 0.9107I_{dc}n_{dttr} \times 10^{-3}$ which, after substituting $0.3702V_{om}/n_{dttr}$ for V_s, yields the primary winding kVA rating as $1.0115V_{om}I_{dc} \times 10^{-3}$, a value only marginally above the ideal value of $V_{om}I_{dc}$. The fundamental component of a primary current has an r.m.s. value of $0.9003I_{dc}n_{dttr}$ so that the fundamental primary kVA is $3V_s \times 0.9003I_{dc}n_{dttr} \times 10^{-3} = V_{om}I_{dc} \times 10^{-3}$, again showing that the increase is due to current harmonics.

The difference between the primary and secondary kVA is $0.03568V_{om} \times I_{dc} \times 10^{-3}$. The r.m.s. values of the fifth and seventh harmonic components of a delta secondary winding current are $0.09003I_{dc}$ and $0.06431I_{dc}$. The values for a wye secondary being $\sqrt{3}$ times these values. The contribution of these harmonics to the total secondary kVA is

$$6 \times 0.3702V_{dc}(0.09003 + 0.06431)I_{dc} \times 10^{-3} = 0.3439V_{om}I_{dc} \times 10^{-3}$$

so that these two harmonics, which do not exist in the primary, account for nearly all the difference between the kVA ratings on the two sides of the transformer.

Similar computations will establish the winding ratings for any rectifier transformer.

12.4 The three phase, zigzag transformer

The first rectifier which was considered in Chapter 2 was the single way, three phase rectifier shown in Fig. 2.1. It was stated there that the standard three phase transformer shown in the figure would in practice be unsatisfactory because of the unbalanced magnetization of its core. The problem is due to the d.c. components of the secondary currents, see Fig. 2.8, graphs 2, 3, and 4, which cannot be reflected into the primary windings because of their zero frequency and which therefore produce an unbalanced d.c. m.m.f. which produces a d.c. magnetic flux in the transformer core. The replacement of mercury arc rectifiers in the 1960s by thyristor rectifiers enormously reduced the demand for the single way, three phase rectifier but, when it is required, the d.c. magnetization problem is overcome by using a transformer with a zigzag secondary winding as illustrated in Fig. 12.10.

The transformer has a normal primary winding which can be either delta or wye connected, the delta version being preferred. It is the one selected for Fig. 12.10. There are six identical secondary windings, two on each limb of the transformer core, which are interconnected to form the output winding shown in Fig. 12.10. The secondaries are designated 1*a* and 1*b*, on the same limb as the RY primary, 2*a* and 2*b*, on the same limb as the YB primary, and 3*a* and 3*b*, on the same limb as the BR primary. The arrangement of the windings on the core and their interconnections are shown in Fig. 12.11. The A group of secondary windings is indicated

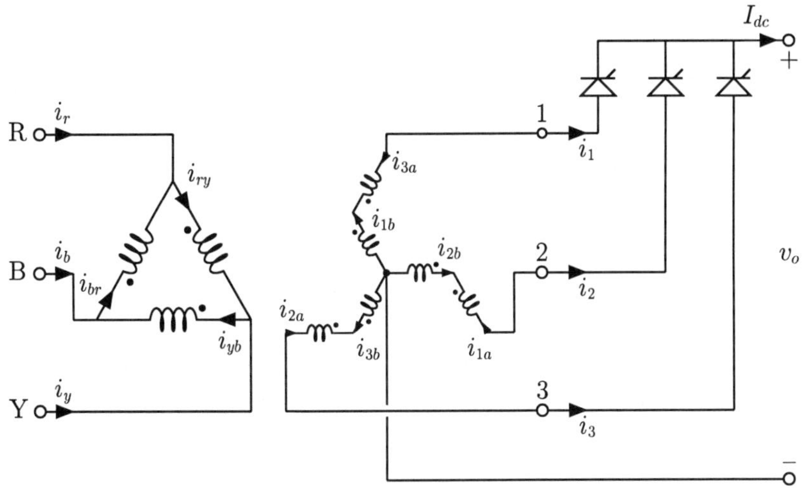

FIG. 12.10. A three phase, single way rectifier fed from a transformer with a zigzag secondary winding.

by an A to its right and the B group is indicated by a B to its right. The output terminals labelled 1, 2, and 3 correspond to the similarly numbered terminals in Fig. 12.10.

Again assuming a smooth, ripple free, d.c. output current, the current waveforms are shown in Fig. 12.12, the individual graphs being labelled by numbers on the right. The graphs are drawn for zero delay angle. Graph 1 shows the waveforms of the input line to line voltages, the voltage of the red line relative to the a.c. system neutral being taken as datum. Graphs 2 through 7 show the waveforms of the transformer secondary winding currents, i_{1a}, i_{2a}, i_{3a}, i_{1b}, i_{2b}, and i_{3b}. The currents in the A secondaries are negative since, as indicated on Fig. 12.10, it is assumed that a positive secondary current leaves a dotted terminal.

The secondary currents produce alternating core m.m.fs with no d.c. component which can therefore be completely balanced by a.c. currents in the primary windings. These currents are shown in graphs 8, 9, and 10.

While not necessary for the present needs, the a.c. line currents have been included in Fig. 12.12, graphs 11, 12, and 13.

As was done in Fig. 12.9, the waveforms of Fig. 12.12 are normalized taking the r.m.s. value of the a.c. line to line voltage as the voltage base, the rated d.c. output current as current base, and the Thévenin transformation ratio, open circuit voltage in a secondary winding/voltage across the corresponding primary winding, as base. This is designated by the symbol n_{ttr}.

Equation 2.8 shows that, in order to produce a given maximum d.c.

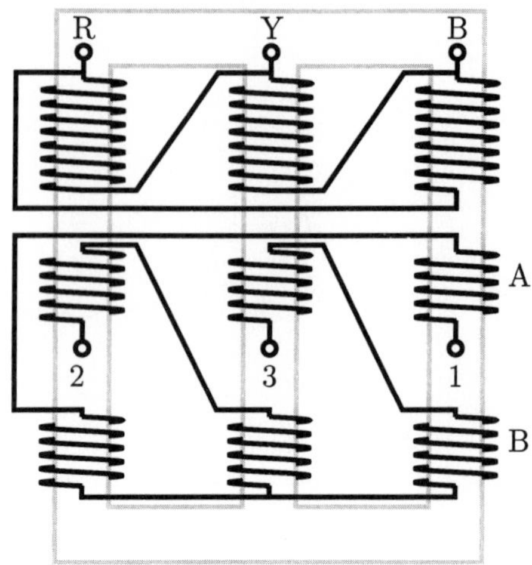

FIG. 12.11. The arrangement of the windings on the core of the zigzag transformer. The outline of the core is shown in gray.

output voltage, V_{om}, the output voltage from the transformer must have an r.m.s. line to neutral value of

$$\frac{\pi}{3}\sqrt{\frac{2}{3}}V_{om} = 0.8550V_{om}$$

Hence, allowing for the 30° phase shift between the voltages in the series connected secondary windings, the r.m.s. value of a secondary voltage is this value divided by $\sqrt{3}$, i.e.

$$V_{sec} = \frac{1}{\sqrt{3}}\frac{\pi}{3}\sqrt{\frac{2}{3}}V_{om} = 0.4937V_{om} \tag{12.13}$$

The magnitude of the Thévenin transformation ratio is therefore

$$n_{ttr} = \frac{V_s}{0.4937V_{om}} = 2.0255\frac{V_s}{V_{om}} \tag{12.14}$$

The r.m.s. value of the current in a secondary winding is $I_{dc}/\sqrt{3} = 0.5774I_{dc}$ and the r.m.s. value of the current in a primary winding is $I_{dc} \times \sqrt{2/3} = 0.8165I_{dc}$.

The kVA rating of all six secondary windings combined is therefore

$$kVA_{sec} = 6 \times 0.4937V_{om} \times 0.5774I_{dc} = 1.7104V_{om}I_{dc} \tag{12.15}$$

It was found in Chapter 2, eqn 2.27, that the kVA rating of a standard secondary is $1.4810V_{om}I_{dc}$ so that the solution of the transformer magnetization problem requires an increase in secondary size of $1.7104/1.4810 = 1.1549$, a 15.5% increase.

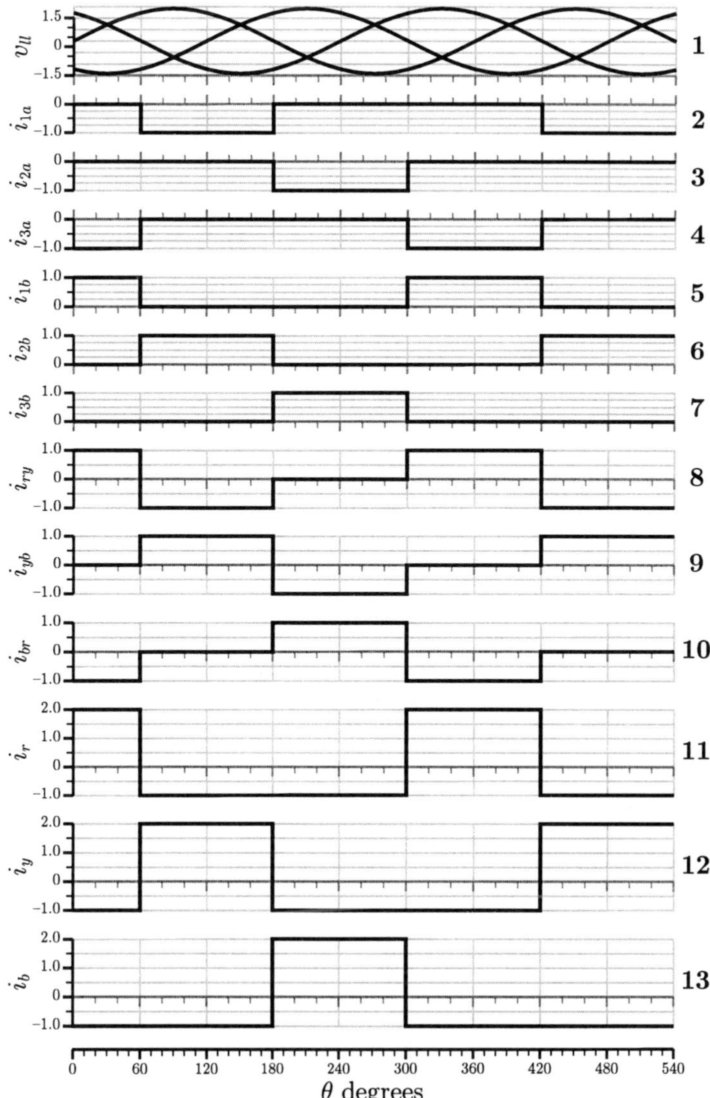

FIG. 12.12. Voltage and current waveforms in the three phase single way rectifier. Individual graphs are identified by a number on the right.

The kVA rating of the primary is

$$kVA_{prim} = 3V_s\sqrt{\frac{2}{3}}\frac{I_{dc}}{n_{ttr}} = 3\frac{n_{ttr}V_{om}}{2.0255}\sqrt{\frac{2}{3}}\frac{I_{dc}}{n_{ttr}} = 1.2093V_{om}I_{dc} \qquad (12.16)$$

The average kVA, what might be termed the kVA rating of the transformer, is $(1.7104+1.092)\div 2 = 1.4599$ times the kVA rating of the load, indicating that this transformer is about 50% larger than an a.c. transformer supplying an a.c. load with the same kVA as the d.c. load. A bridge rectifier would require a transformer with a kVA rating 4.7% greater than that of the load and would operate six phase on both input and output. Thus, when compared to an equivalent bridge rectifier, the increase in transformer size plus the increased transformer complexity and the higher harmonic content of both the output and input waveforms are a substantial price to pay for reducing the number of thyristors from six to three, a consideration which undoubtedly accounts for the rarity of this type of rectifier.

12.5 Phase shifting transformers

The use of phase shifting transformers in rectifier applications has been discussed in Section 9.4. In brief, some applications, notably in the electrochemical industries, consume vast quantities of d.c. energy in a number of more or less equal segments. For example an aluminum smelter will probably have several production lines each with several hundred series connected electrolytic cells. The lines are more or less identical, each consuming about the same amount of power, but for operational flexibility and reliability each line is operated independently with individual cells being taken out of service as necessary for maintenance. Under normal operating circumstances the lines are more or less identical, each consuming about the same amount of power. The a.c. supply waveforms can then be greatly improved by using phase shifting transformers to make small adjustments to the phase of the a.c. supply to the rectifiers for each electrolytic line. Although the phase shift can be made within the rectifier transformers it is generally preferable to have these identical since manufacturing costs are thereby reduced and a single spare transformer can back up the whole plant. The phase shifting is therefore normally done in special three phase transformers which, since the necessary electrical isolation is provided by the rectifier transformers, are usually autotransformers. The analysis is therefore given in this context.

Figure 12.13 shows a wye connected three phase autotransformer with input terminals R, Y, and B and terminals 1, 2, and 3 at which a phase advanced output is available. Polarity dots identify the coil ends. The same principles can be employed to produce a delta connected phase shifting

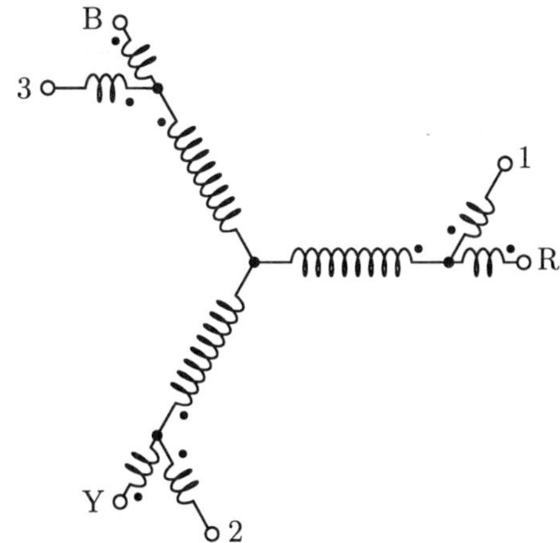

FIG. 12.13. A three phase autotransformer with three phase input at terminals R, Y and B and a phase advanced three phase output at terminals 1, 2, and 3.

autotransformer or an isolating phase shifting transformer and to produce phase lag if this is required. The transformer has three sets of windings whose locations and interconnections are indicated in Fig. 12.14 where the sets are identified by letters to the left as A, B, and C. The usual proviso is made, that the windings will in practice be placed one over the other as indicated in Fig. 12.1.

The lower ends of the A group are connected together to form the neutral point and the B group is connected in series aiding with the A group to provide the input terminals. The A and B components of a pair are on the same limb of the transformer core. A coil from the C group has one end connected to the junction between an A and a B group coil and its other end is connected to an output terminal. A coil from the C group is on a different limb from that of the A and B group coils to which it is connected, the precise location and connection of the coil depending on the amount of phase shift required. The two figures are drawn for a transformer producing a phase lead, with the dotted end of a C group coil connected to the junction of an A and B group coil and the coil itself being located on a limb so as to produce a voltage leading that of its A and B group coils by 240°. This latter point is illustrated by Fig. 12.15 which shows the coil connections for one of the three phases and the corresponding voltage phasor diagram.

Here the analysis assumes the transformer to be ideal, departures from the ideal being considered in Section 12.6 and Example 12.3. Following the phasor diagram of Fig. 12.15, the input voltage phasor to the phase

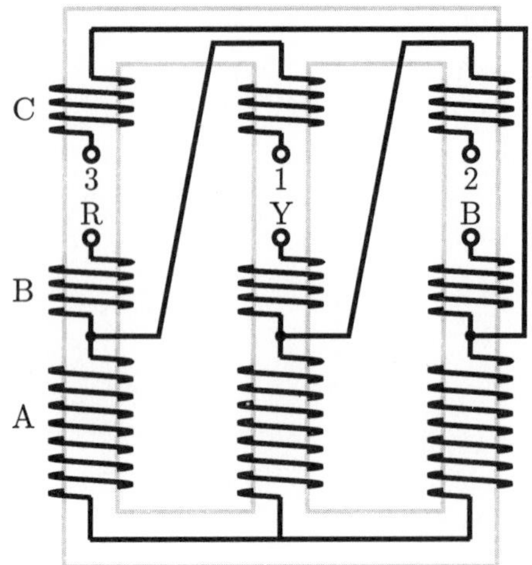

FIG. 12.14. Location of the windings on the transformer core. The core is outlined in gray. The coils are arranged in three groups identified by the letters A, B, and C to their left.

in question is $\mathbf{V}_1$ and is taken as the datum phasor. The voltage phasors for windings A and B are $\mathbf{V}_a$ and $\mathbf{V}_b$. They are in phase with $\mathbf{V}_1$ so that $V_a + V_b = V_1$. The voltage phasor for the C winding is $\mathbf{V}_c$ and it leads $\mathbf{V}_1$ by ϕ_{vc}. The desired output voltage phasor is $\mathbf{V}_o$ leading $\mathbf{V}_1$ by ϕ_{vo} and $\mathbf{V}_o = \mathbf{V}_a - \mathbf{V}_c$. The input and output voltage phasors are known and the first step in the analysis is the determination of the winding voltages, $\mathbf{V}_a$, $\mathbf{V}_b$, and $\mathbf{V}_c$. This can be done by resolving the phasors parallel to and perpendicular to $\mathbf{V}_1$ to obtain the three scalar equations

$$V_a + V_b = V_1 \tag{12.17}$$

$$V_a - V_c \cos(\phi_{vc}) = V_o \cos(\phi_{vo}) \tag{12.18}$$

$$-V_c \sin(\phi_{vc}) = V_o \sin(\phi_{vo}) \tag{12.19}$$

The known quantities in these equations are the magnitudes of the input and output voltages, the phase of the output voltage and the phase of the voltage across winding C. Equation 12.19 yields the magnitude of the voltage across winding C as

$$V_c = -V_o \frac{\sin(\phi_{vo})}{\sin(\phi_{vc})} \tag{12.20}$$

Substituting this value into eqn 12.18 yields the magnitude of the voltage across winding A as

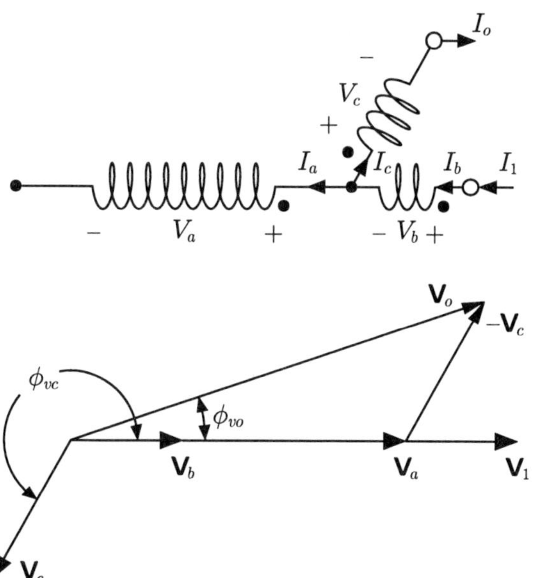

FIG. 12.15. One of the three phases of the phase shifting transformer. The upper diagram shows the winding connections and polarities and the directions of positive voltages and currents. The voltage phasor diagram is shown in the lower display.

$$V_a = V_o \cos(\phi_{vo}) + V_c \cos(\phi_{vc}) \tag{12.21}$$

Substitution of this value in eqn 12.17 yields the magnitude of the voltage across winding B as

$$V_b = V_1 - V_a \tag{12.22}$$

The currents must now be determined and to do this it is assumed that the magnitude and phase of the output current are known and that the winding currents and input current are required. Inspection of the circuit diagram of Fig. 12.15 immediately provides the current in the C winding and the input current in terms of the output current and the current in winding B. Thus

$$\mathbf{I}_c = \mathbf{I}_o \tag{12.23}$$

and

$$\mathbf{I}_1 = \mathbf{I}_b \tag{12.24}$$

Applying Kirchhoff's current law to the node at which the three windings join yields

$$\mathbf{I}_a - \mathbf{I}_b + \mathbf{I}_c = 0 \tag{12.25}$$

The fourth relationship which is necessary to determine the currents is provided by the m.m.f. balance for a leg of the transformer core. For the ideal transformer, the instantaneous sum of the m.m.fs on any limb must be zero. Considering the limb on which the A and B coils are located, this yields the phasor equation

$$n_a\mathbf{I}_a + n_b\mathbf{I}_b + n_c\mathbf{I}_x = 0 \tag{12.26}$$

where $\mathbf{I}_x$ is the current in the C group coil on the same limb as the A and B group coils and n_a, n_b, and n_c are the numbers of turns on the A, B, and C group coils.

Reference to Fig. 12.13 shows that $\mathbf{I}_x$ leads $\mathbf{I}_c$ by 120° so that eqn 12.26 can be rewritten

$$n_a\mathbf{I}_a + n_b\mathbf{I}_b + n_c\mathbf{I}_c\underline{/120^\circ} = 0 \tag{12.27}$$

Now, although the numbers of turns on the windings are not known, their ratios are known. They are $n_b/n_a = V_b/V_a$ and $n_c/n_a = V_c/V_a$. Dividing eqn 12.27 by n_a, substituting these voltage ratios for the turns ratios, and replacing $\mathbf{I}_c$ by $\mathbf{I}_o$ yields

$$\mathbf{I}_a + \frac{V_b}{V_a}\mathbf{I}_b + \frac{V_c}{V_a}\mathbf{I}_o\underline{/120^\circ} = 0 \tag{12.28}$$

Substituting $\mathbf{I}_o$ for $\mathbf{I}_c$ in eqn 12.25 yields

$$\mathbf{I}_a - \mathbf{I}_b + \mathbf{I}_o = 0 \tag{12.29}$$

These two phasor equations yield the following solutions for the phasors $\mathbf{I}_a$ and $\mathbf{I}_b$

$$\mathbf{I}_a = -\frac{1 + \frac{V_c}{V_b}\underline{/120^\circ}}{1 + \frac{V_a}{V_b}}\mathbf{I}_o \tag{12.30}$$

$$\mathbf{I}_b = \frac{1 - \frac{V_c}{V_a}\underline{/120^\circ}}{1 + \frac{V_b}{V_a}}\mathbf{I}_o \tag{12.31}$$

Now that the winding voltages and currents have been determined, the kVA ratings of the transformer windings can be found. The autotransformer connection will yield a transformer considerably smaller than the equivalent isolating transformer as is illustrated in Example 12.2.

Example 12.2

An electrochemical plant has three electrochemical lines, each one rated at 5 MW. Each line is fed by a twelve phase rectifier and, to improve the waveform of the current drawn from the 12 kV a.c. system supplying the plant, the a.c. input to the first rectifier transformer is phase advanced by 10°, the second rectifier transformer is directly connected to the supply, and the a.c. input to the third rectifier transformer is given a phase lag of 10°. Thus, when all three electrolytic lines are operating at the same power level, the plant appears to the a.c. system as a 36 phase rectifier.

The phase lag and phase lead are provided by autotransformers using the connection of Fig. 12.13, accepting 12 kV as the line to line input voltage and producing three phase, 12 kV, line to line output voltages, with the necessary phase shift.

For the phase advancing autotransformer determine, for the rated load of 5 MW, the voltages and currents in its windings and its kVA rating. Assume that the transformer is ideal and that the rectifier is operating at zero delay so that the power drawn from the phase shifting transformer can be assumed to be at unity power factor.

In eqns 12.17, 12.18, and 12.19, $V_1 = V_o = 12\,000/\sqrt{3} = 6928.2$ V, $\phi_{vo} = 10°$, and $\phi_{vc} = 240°$. Substituting these values in eqns 12.20, 12.21, and 12.22 yields the winding voltages

$$V_a = 6128.4 \qquad V_b = 799.9 \qquad V_c = 1389.2$$

At the rated load of 5 MW the magnitude of the output current is $5 \times 10^6/\sqrt{3} \times 12\,000 = 240.6$ A. At unity power factor and for this phase advancing transformer, this current will lead the datum phasor by 10°. Thus

$$\mathbf{I}_o = 240.6\underline{/10.0°}$$

from eqn 12.30

$$\mathbf{I}_a = 41.9\underline{/-85.0°}$$

from eqn 12.31

$$\mathbf{I}_b = 240.6\underline{/0.0°}$$

from eqn 12.23

$$\mathbf{I}_c = 240.6\underline{/10.0°}$$

and from eqn 12.24

$$\mathbf{I}_1 = 240.6\underline{/0.0°}$$

The kVA values for the three groups of windings are

$$\begin{aligned} kVA_a &= 3 \times 6128.4 \times 41.9 \times 10^{-3} &= 770.9 \\ kVA_b &= 3 \times 799.9 \times 240.6 \times 10^{-3} &= 577.2 \\ kVA_c &= 3 \times 1389.2 \times 240.6 \times 10^{-3} &= 1002.6 \end{aligned}$$

Thus, the total kVA for the transformer is

$$kVA_t = 770.9 + 577.2 + 1002.6 = 2350.7$$

An isolating transformer performing the same function would have a primary winding sustaining the full line to neutral input voltage of 6928.2 V and carrying the full input current of 240.6 A so that its rating would be

$$3 \times 6928.2 \times 240.6 \times 10^{-3} = 5000 \text{ kVA}$$

It would have a secondary sustaining an r.m.s. voltage of 6128.4 V, the same as the A winding of the autotransformer and another secondary sustaining 1389.2 V, the same as the C winding of the autotransformer. Both secondaries would carry the output current of 240.6 A so that the combined rating of the two secondaries would be

$$3 \times (6128.4 + 1389.2) \times 240.6 \times 10^{-3} = 5426 \text{ kVA}$$

The combined rating of the primary and two secondary windings would be 10 426 kVA, 4.43 times that of the autotransformer. The windings of such a transformer would cost and weigh more than four times as much as those of the equivalent autotransformer. This gives an indication of the considerable savings which can accrue from the use of autotransformers in those situations where electrical isolation between primary and secondary is not required and where there is not much difference between the input and output voltages.

****

12.6 Multiwinding transformers

So far in this chapter three examples of transformers with more than two windings per phase have been encountered, the twelve phase rectifier transformer, the three phase zigzag rectifier transformer, and the phase shifting autotransformer. Each of these has three windings per phase for a total of nine windings in a three phase transformer. Such transformers come under the general heading of multiwinding transformers and are encountered from time to time in rectifier circuits.

In Section 12.2, the analysis of a transformer with two windings per phase was developed via the equivalent circuit, an approach which relates directly to the physical processes within the transformer and which is easy to use. Classical transformer theory was developed well before computers were invented and the same equivalent circuit technique was successfully applied to multiwinding transformers. However, the analytical complexity of a transformer increases as the square of the number of windings per phase so that a transformer with three windings per phase is more than twice as complex to analyze as one with two windings per phase. This complexity combined with the ready availability of computers gives considerable attraction to the circuit theory approach in which the transformer is described as a series of mutually coupled coils. This approach has the merit of handling with ease the large number of couplings between coils but requires that special efforts be made to maintain contact with the reality of the physical system. The circuit theory approach is used here.

Kirchhoff's voltage law applied to a three phase multiwinding transformer with N windings per phase results in a $3N$ by $3N$ impedance matrix with the coil resistances and self inductances on the leading diagonal and the coil mutual inductances in all other locations. It can be argued that linear circuit theory cannot be applied to a transformer because of the high degree of nonlinearity exhibited by its ferromagnetic core. However, in practice in the vast majority of transformer problems this does not present a significant difficulty provided the relationship of the inductances to the physical reality is constantly born in mind.

The equations relating the coil voltages and currents can be arranged in any convenient order, that for the present purpose being the one which groups together the coils on a single limb. Designating these coils as belonging to the red, yellow, or blue group distinguished by subscripts r, y, and b, the equations are of the type

$$\begin{bmatrix} \mathbf{V}_r \\ \mathbf{V}_y \\ \mathbf{V}_b \end{bmatrix} = \begin{bmatrix} \mathbf{Z}_{rr} & \mathbf{Z}_{ry} & \mathbf{Z}_{rb} \\ \mathbf{Z}_{yr} & \mathbf{Z}_{yy} & \mathbf{Z}_{yb} \\ \mathbf{Z}_{br} & \mathbf{Z}_{by} & \mathbf{Z}_{bb} \end{bmatrix} \begin{bmatrix} \mathbf{I}_r \\ \mathbf{I}_y \\ \mathbf{I}_b \end{bmatrix} \tag{12.32}$$

where $\mathbf{V}_r$ is the N subvector of voltages applied to the red windings, $\mathbf{I}_r$ is the N subvector of currents in those windings, $\mathbf{Z}_{rr}$ is the N by N impedance submatrix for those windings, $\mathbf{Z}_{ry}$ is the N by N mutual impedance matrix coupling the red and yellow windings, and $\mathbf{Z}_{rb}$ is the N by N mutual impedance matrix coupling the red and blue windings. Similar definitions apply to the remaining symbols. It should be noted that upper case, bold face, sanserif symbols, which have hitherto been reserved for complex quantities, are being used here to represent matrices.

At this point a change in nomenclature with regard to the current vector must be noted. Hitherto in this chapter the usual two windings per phase transformer convention has been used. This considers as positive an input current to the primary and an output current from the secondary. If the mathematics is to keep track of the problem, a more consistent convention is desirable. For the purposes of this analysis a current entering the dotted terminal of a coil will be considered as positive irrespective of whether it is on the primary or secondary and the polarity dots will be positioned so that a positive current magnetizes the transformer limb on which the coil is mounted, in a specified direction. This arrangement of polarity dots, but not this arrangement of currents, is shown in Fig. 12.2.

Since the system is symmetrical, the phases being identical, the self impedances are the same, i.e. $\mathbf{Z}_{rr} = \mathbf{Z}_{yy} = \mathbf{Z}_{bb}$, and the mutual impedances are the same, i.e. $\mathbf{Z}_{rb} = \mathbf{Z}_{br} = \mathbf{Z}_{ry} = \mathbf{Z}_{yr} = \mathbf{Z}_{yb} = \mathbf{Z}_{by}$, so that eqn 12.32 can be simplified to

$$\begin{bmatrix} \mathbf{V}_r \\ \mathbf{V}_y \\ \mathbf{V}_b \end{bmatrix} = \begin{bmatrix} \mathbf{Z}_{rr} & \mathbf{Z}_{ry} & \mathbf{Z}_{ry} \\ \mathbf{Z}_{ry} & \mathbf{Z}_{rr} & \mathbf{Z}_{ry} \\ \mathbf{Z}_{ry} & \mathbf{Z}_{ry} & \mathbf{Z}_{rr} \end{bmatrix} \begin{bmatrix} \mathbf{I}_r \\ \mathbf{I}_y \\ \mathbf{I}_b \end{bmatrix} \tag{12.33}$$

In the majority of three phase problems there is no zero sequence current, i.e. $\mathbf{I}_r + \mathbf{I}_y + \mathbf{I}_b = 0$, and it is convenient to include this information in eqn 12.33 by dividing the current vectors into two parts, a zero sequence portion denoted by the subscript z, and a nonzero sequence portion denoted by the subscript pn. Thus

$$\begin{bmatrix} \mathbf{I}_r \\ \mathbf{I}_y \\ \mathbf{I}_b \end{bmatrix} = \begin{bmatrix} \mathbf{I}_{rpn} + \mathbf{I}_z \\ \mathbf{I}_{ypn} + \mathbf{I}_z \\ \mathbf{I}_{bpn} + \mathbf{I}_z \end{bmatrix} \tag{12.34}$$

where

$$\mathbf{I}_z = \frac{1}{3}\left[\mathbf{I}_r + \mathbf{I}_y + \mathbf{I}_b\right] \tag{12.35}$$

and

$$\mathbf{I}_{rpn} + \mathbf{I}_{ypn} + \mathbf{I}_{bpn} = 0 \tag{12.36}$$

The subscript pn has been used for the balance of a current vector after the zero sequence component has been removed since this is the sum of the positive and negative sequence components. For the present purposes there is no need to divide this into separate sequence components.

Substituting eqn 12.34 in 12.33 yields

$$\begin{bmatrix} \mathbf{V}_r \\ \mathbf{V}_y \\ \mathbf{V}_b \end{bmatrix} = \begin{bmatrix} \mathbf{Z}_{rr} & \mathbf{Z}_{ry} & \mathbf{Z}_{ry} \\ \mathbf{Z}_{ry} & \mathbf{Z}_{rr} & \mathbf{Z}_{ry} \\ \mathbf{Z}_{ry} & \mathbf{Z}_{ry} & \mathbf{Z}_{rr} \end{bmatrix} \begin{bmatrix} \mathbf{I}_{rpn} \\ \mathbf{I}_{ypn} \\ \mathbf{I}_{bpn} \end{bmatrix} + \begin{bmatrix} \mathbf{Z}_{rr} & \mathbf{Z}_{ry} & \mathbf{Z}_{ry} \\ \mathbf{Z}_{ry} & \mathbf{Z}_{rr} & \mathbf{Z}_{ry} \\ \mathbf{Z}_{ry} & \mathbf{Z}_{ry} & \mathbf{Z}_{rr} \end{bmatrix} \begin{bmatrix} \mathbf{I}_z \\ \mathbf{I}_z \\ \mathbf{I}_z \end{bmatrix} \tag{12.37}$$

which, after expansion, reduces to the three separate equations

$$\mathbf{V}_r = \left[\mathbf{Z}_{rr} - \mathbf{Z}_{ry}\right]\mathbf{I}_{rpn} + \left[\mathbf{Z}_{rr} + 2\mathbf{Z}_{ry}\right]\mathbf{I}_z \tag{12.38}$$

$$\mathbf{V}_y = \left[\mathbf{Z}_{rr} - \mathbf{Z}_{ry}\right]\mathbf{I}_{ypn} + \left[\mathbf{Z}_{rr} + 2\mathbf{Z}_{ry}\right]\mathbf{I}_z \tag{12.39}$$

$$\mathbf{V}_b = \left[\mathbf{Z}_{rr} - \mathbf{Z}_{ry}\right]\mathbf{I}_{bpn} + \left[\mathbf{Z}_{rr} + 2\mathbf{Z}_{ry}\right]\mathbf{I}_z \tag{12.40}$$

thus reducing the original computation of order $3N$ to three separate computations, each of order N, a much simpler task.

The analysis must now come to grips with the details of the problem and this will be done in the context of a transformer with three windings per phase. The analysis will be illustrated by application to two of the three such situations already encountered. In this context, the submatrices $\mathbf{Z}_{rr}$ and $\mathbf{Z}_{ry}$ are 3 by 3. The subscripts a, b, and c used to distinguish between them will continue to identify the coils on a transformer limb.

Since inductance is reciprocal, i.e. if L_{xy} is the mutual inductance between coil X and coil Y and L_{yx} is the mutual inductance between coil Y and coil X then $L_{yx} = L_{xy}$, the impedance matrix $\mathbf{Z}_{rr}$ is

$$\mathbf{Z}_{rr} = \begin{bmatrix} R_a + L_{rara}\frac{\mathrm{d}}{\mathrm{dt}} & L_{rarb}\frac{\mathrm{d}}{\mathrm{dt}} & L_{rarc}\frac{\mathrm{d}}{\mathrm{dt}} \\ L_{rarb}\frac{\mathrm{d}}{\mathrm{dt}} & R_b + L_{rbrb}\frac{\mathrm{d}}{\mathrm{dt}} & L_{rbrc}\frac{\mathrm{d}}{\mathrm{dt}} \\ L_{rarc}\frac{\mathrm{d}}{\mathrm{dt}} & L_{rbrc}\frac{\mathrm{d}}{\mathrm{dt}} & R_c + L_{rcrc}\frac{\mathrm{d}}{\mathrm{dt}} \end{bmatrix} \tag{12.41}$$

and the impedance matrix $\mathbf{Z}_{ry}$ is

$$\mathbf{Z}_{ry} = \begin{bmatrix} L_{raya}\frac{\mathrm{d}}{\mathrm{dt}} & L_{rayb}\frac{\mathrm{d}}{\mathrm{dt}} & L_{rayc}\frac{\mathrm{d}}{\mathrm{dt}} \\ L_{rayb}\frac{\mathrm{d}}{\mathrm{dt}} & L_{rbyb}\frac{\mathrm{d}}{\mathrm{dt}} & L_{rbyc}\frac{\mathrm{d}}{\mathrm{dt}} \\ L_{rayc}\frac{\mathrm{d}}{\mathrm{dt}} & L_{rbyc}\frac{\mathrm{d}}{\mathrm{dt}} & L_{rcyc}\frac{\mathrm{d}}{\mathrm{dt}} \end{bmatrix} \tag{12.42}$$

The resistances in these expressions are simply the resistances of the coils A, B, and C, the only complication being that they may have to be increased somewhat above the d.c. values to allow for a.c. phenomena such as skin and proximity effects. The inductances present a more complicated problem because of the mutual couplings between the windings. To determine them, first use the fact that inductance is flux linkage per ampere so that

$$L_{xy} = \frac{n_y \phi_{xy}}{i_x} \tag{12.43}$$

where n_y is the number of turns on coil Y and ϕ_{xy} is the portion of the flux created by current i_x in coil X which links with coil Y.

Now ϕ_{xy} is the ratio of the m.m.f. of coil X to the reluctance, $\mathcal{R}_{xy}$, of the path taken by the flux, i.e.

$$\phi_{xy} = \frac{n_x i_x}{\mathcal{R}_{xy}} \tag{12.44}$$

It is convenient to separate transformer flux into the three components shown in Fig. 12.2, the mutual flux between limbs, ϕ_m, the mutual flux for a single limb, ϕ_{zm}, and the leakage flux, ϕ_σ. Equation 12.44 then subdivides into

$$\phi_{xy} = \frac{n_x i_x}{\mathcal{R}_{xym}} + \frac{n_x i_x}{\mathcal{R}_{xyzm}} + \frac{n_x i_x}{\mathcal{R}_{xy\sigma}} \tag{12.45}$$

Combining eqns 12.43 and 12.45 yields the following expression for the mutual inductance between coil X and coil Y

$$L_{xy} = L_{xym} + L_{xyzm} + L_{xy\sigma} = \frac{n_x n_y}{\mathcal{R}_{xym}} + \frac{n_x n_y}{\mathcal{R}_{xyzm}} + \frac{n_x n_y}{\mathcal{R}_{xy\sigma}} \tag{12.46}$$

It is seen that each inductance is the sum of a ferromagnetic mutual component, a nonferromagnetic mutual component, and a nonferromagnetic leakage component. Depending on the circumstances, some of these components may be zero. When coil X and coil Y are on different limbs of the transformer core, only the ferromagnetic mutual component links them. When the coils are on the same limb, in addition to the ferromagnetic mutual component there is the linkage provided by the nonferromagnetic mutual component which is common to all coils on a limb and there may also be linkage provided by a component of the leakage flux which links with some but not all of the coils on that limb.

Applying this result to the three windings per limb transformer yields the following expressions for the twelve distinct inductances which appear in its inductance matrix

$$L_{rara} = n_a^2\left(\frac{1}{\mathcal{R}_m} + \frac{1}{\mathcal{R}_{zm}} + \frac{1}{\mathcal{R}_{aa\sigma}} + \frac{1}{\mathcal{R}_{ab\sigma}} + \frac{1}{\mathcal{R}_{ac\sigma}}\right) \tag{12.47a}$$

$$L_{rarb} = n_a n_b\left(\frac{1}{\mathcal{R}_m} + \frac{1}{\mathcal{R}_{zm}} + \frac{1}{\mathcal{R}_{ab\sigma}}\right) \tag{12.47b}$$

$$L_{rarc} = n_a n_c\left(\frac{1}{\mathcal{R}_m} + \frac{1}{\mathcal{R}_{zm}} + \frac{1}{\mathcal{R}_{ac\sigma}}\right) \tag{12.47c}$$

$$L_{rbrb} = n_b^2\left(\frac{1}{\mathcal{R}_m} + \frac{1}{\mathcal{R}_{zm}} + \frac{1}{\mathcal{R}_{bb\sigma}} + \frac{1}{\mathcal{R}_{ab\sigma}} + \frac{1}{\mathcal{R}_{bc\sigma}}\right) \tag{12.47d}$$

$$L_{rbrc} = n_b n_c\left(\frac{1}{\mathcal{R}_m} + \frac{1}{\mathcal{R}_{zm}} + \frac{1}{\mathcal{R}_{bc\sigma}}\right) \tag{12.47e}$$

$$L_{rcrc} = n_c^2\left(\frac{1}{\mathcal{R}_m} + \frac{1}{\mathcal{R}_{zm}} + \frac{1}{\mathcal{R}_{cc\sigma}} + \frac{1}{\mathcal{R}_{ac\sigma}} + \frac{1}{\mathcal{R}_{bc\sigma}}\right) \tag{12.47f}$$

$$L_{raya} = -n_a^2\frac{1}{2}\frac{1}{\mathcal{R}_m} \tag{12.47g}$$

$$L_{rayb} = -n_a n_b\frac{1}{2}\frac{1}{\mathcal{R}_m} \tag{12.47h}$$

$$L_{rayc} = -n_a n_c\frac{1}{2}\frac{1}{\mathcal{R}_m} \tag{12.47i}$$

$$L_{rbyb} = -n_b^2 \frac{1}{2} \frac{1}{\mathcal{R}_m} \quad (12.47j)$$

$$L_{rbyc} = -n_b n_c \frac{1}{2} \frac{1}{\mathcal{R}_m} \quad (12.47k)$$

$$L_{rcyc} = -n_c^2 \frac{1}{2} \frac{1}{\mathcal{R}_m} \quad (12.47l)$$

Replacing the reciprocal reluctances in these expressions by the corresponding inductances for a single turn coil yields

$$L_{rara} = n_a^2 \left(L_m + L_{zm} + L_{aa\sigma} + L_{ab\sigma} + L_{ac\sigma} \right) \quad (12.48a)$$

$$L_{rarb} = n_a n_b \left(L_m + L_{zm} + L_{ab\sigma} \right) \quad (12.48b)$$

$$L_{rarc} = n_a n_c \left(L_m + L_{zm} + L_{ac\sigma} \right) \quad (12.48c)$$

$$L_{rbrb} = n_b^2 \left(L_m + L_{zm} + L_{bb\sigma} + L_{ab\sigma} + L_{bc\sigma} \right) \quad (12.48d)$$

$$L_{rbrc} = n_b n_c \left(L_m + L_{zm} + L_{bc\sigma} \right) \quad (12.48e)$$

$$L_{rcrc} = n_c^2 \left(L_m + L_{zm} + L_{cc\sigma} + L_{ac\sigma} + L_{bc\sigma} \right) L_{raya} \quad (12.48f)$$

$$L_{raya} = -n_a^2 \frac{1}{2} L_m \quad (12.48g)$$

$$L_{rayb} = -n_a n_b \frac{1}{2} L_m \quad (12.48h)$$

$$L_{rayc} = -n_a n_c \frac{1}{2} L_m \quad (12.48i)$$

$$L_{rbyb} = -n_b^2 \frac{1}{2} L_m \quad (12.48j)$$

$$L_{rbyc} = -n_b n_c \frac{1}{2} L_m \quad (12.48k)$$

$$L_{rcyc} = -n_c^2 \frac{1}{2} L_m \quad (12.48l)$$

so that the positive–negative sequence inductance matrix is

$$\mathbf{L}_{pn} = \begin{bmatrix} L_{pn11} & L_{pn12} & L_{pn13} \\ L_{pn12} & L_{pn22} & L_{pn23} \\ L_{pn13} & L_{pn23} & L_{pn33} \end{bmatrix} \quad (12.49)$$

where

$$L_{pn11} = n_a^2 \left(\frac{3}{2} L_m + L_{zm} + L_{aa\sigma} + L_{ab\sigma} + L_{ac\sigma} \right) \quad (12.50a)$$

$$L_{pn12} = n_a n_b \left(\frac{3}{2} L_m + L_{zm} + L_{ab\sigma} \right) \quad (12.50b)$$

$$L_{pn13} = n_a n_c \left(\frac{3}{2} L_m + L_{zm} + L_{ac\sigma} \right) \quad (12.50c)$$

$$L_{pn22} = n_a^2\left(\frac{3}{2}L_m + L_{zm} + L_{aa\sigma} + L_{ab\sigma} + L_{ac\sigma}\right) \quad (12.50d)$$

$$L_{pn23} = n_b n_c\left(\frac{3}{2}L_m + L_{zm} + L_{bc\sigma}\right) \quad (12.50e)$$

$$L_{pn33} = n_a^2\left(\frac{3}{2}L_m + L_{zm} + L_{aa\sigma} + L_{ab\sigma} + L_{ac\sigma}\right) \quad (12.50f)$$

and the zero sequence inductance matrix is

$$\mathbf{L}_z = \begin{bmatrix} L_{z11} & L_{z12} & L_{z13} \\ L_{z12} & L_{z22} & L_{z23} \\ L_{z13} & L_{z23} & L_{z33} \end{bmatrix} \quad (12.51)$$

where

$$L_{z11} = n_a^2\left(L_{zm} + L_{aa\sigma} + L_{ab\sigma} + L_{ac\sigma}\right) \quad (12.52a)$$
$$L_{z12} = n_a n_b\left(L_{zm} + L_{ab\sigma}\right) \quad (12.52b)$$
$$L_{z13} = n_a n_c\left(L_{zm} + L_{ac\sigma}\right) \quad (12.52c)$$
$$L_{z22} = n_b^2\left(L_{zm} + L_{bb\sigma} + L_{ab\sigma} + L_{bc\sigma}\right) \quad (12.52d)$$
$$L_{z23} = n_b n_c\left(L_{zm} + L_{bc\sigma}\right) \quad (12.52e)$$
$$L_{z33} = n_c^2\left(L_{zm} + L_{cc\sigma} + L_{ac\sigma} + L_{bc\sigma}\right) \quad (12.52f)$$

The various inductances can be determined by either computation or measurement. Classical texts on transformer design give detailed descriptions of how this can be done. The tedium of the necessary computations has been largely removed by computerization but there is now a strong tendency to replace the classical techniques entirely by field computation methods based on the finite element technique.

No matter how the inductances are determined, one crucial factor must be borne in mind, the primacy of the nonferromagnetic inductances in determining the transformer characteristics of greatest interest. These inductances are very small compared to the ferromagnetic inductances, typically more than two orders of magnitude less, and they are constant for the overwhelming majority of operating conditions of interest. In contrast the ferromagnetic inductances vary widely depending on the magnetic state of the transformer core. The solution is to select a reasonable value for L_m, e.g. that value which will give the correct r.m.s. value of magnetizing current, and then add to this the appropriate nonferromagnetic components. This results in large inductances whose precise values are not important but whose small differences from the corresponding ferromagnetic components are vital.

This aspect of the analysis necessitates exploitation of the full computer accuracy. As an example, the ferromagnetic component may be about 1.0 H and the nonferromagnetic component may be 0.0083 H. For computations, the ferromagnetic component must be fixed at precisely 1.0 H (or 0.95 or 1.05 depending on which is the most appropriate for the core magnetic conditions) and then the nonferromagnetic component is added to give a total inductance of 1.0 + 0.0083 = 1.0083 H as the value presented to the computer. This may seem absurd when the error in determining the inductances may be ±1% or more but it is the 0.0083 H which is important in the computation and it is this value which must be preserved.

In making the calculation the computer is presented with an inductance matrix dominated by the large ferromagnetic mutual components but whose solution is dominated by the small nonferromagnetic components. The matrix is ill-conditioned and because of this the computer may have difficulty in obtaining a valid solution. For example if, using this approach, the decision is made to neglect the magnetizing current as is frequently done in the equivalent circuit method, then the ferromagnetic inductances will be infinite, the inductance matrix will be singular, and it will be impossible to obtain a solution. Provided one is certain that the problem lies in the inductances and not in the program, one remedy is to reduce the ferromagnetic inductances to a level at which a solution can be obtained. Another useful approach is to normalize the problem to a fixed number of turns. This is the equivalent of transferring all quantities in the equivalent circuit to one side of the transformer. The ferromagnetic components of all inductances are then equalized and the computation becomes more stable.

The technique is illustrated by the following two examples, the first of which involves the steady state response of a phase shifting transformer to three phase excitation with voltages of sinusoidal waveform.

Example 12.3

The windings of the phase shifting autotransformer of Example 12.2 have the following parameters

Winding A

$$\begin{aligned} \text{Resistance} &= 1.47\ \Omega \\ \text{Ferromagnetic inductance} &= 1.23\ \text{H} \\ \text{Nonferromagnetic inductance} &= 0.0144\ \text{H} \\ \text{Leakage inductance, } L_{aa\sigma} &= 0.0032\ \text{H} \\ \text{Leakage inductance, } L_{ab\sigma} &= 0.0023\ \text{H} \\ \text{Leakage inductance, } L_{ac\sigma} &= 0.0019\ \text{H} \end{aligned}$$

Winding B

$$\begin{aligned} \text{Resistance} &= 0.048\ \Omega \\ \text{Leakage inductance, } L_{bb\sigma} &= 0.000055\ \text{H} \\ \text{Leakage inductance, } L_{bc\sigma} &= 0.000095\ \text{H} \end{aligned}$$

Winding C

$$\begin{aligned} \text{Resistance} &= 0.058\ \Omega \\ \text{Leakage inductance, } L_{cc\sigma} &= 0.000164\ \text{H} \end{aligned}$$

$$\begin{aligned} \text{Turns ratio, } n_b/n_a &= 0.130516 \\ \text{Turns ratio, } n_c/n_a &= 0.226682 \end{aligned}$$

The transformer operates from a 12 kV, 60 Hz three phase supply. The nominal rated output conditions are 5 MVA at an r.m.s. line to line voltage of 12 kV leading the input voltage by 10°.

Determine the regulation and efficiency at rated load when the load power factor is 0.8 lagging.

The transformer is wye connected so that there are no zero sequence currents. Attention is therefore concentrated on the PN portions of eqns 12.38, 12.39, and 12.40.

Since this is a steady state a.c. problem, phasor currents and voltages and complex impedances will be used.

Since the a.c. supply and load are balanced three phase, only the solution for the red phase need be found, eqn 12.38, the solutions for the yellow and blue phases being the same except for negative and positive phase shifts of 120°.

Using eqns 12.50a to obtain the inductances of the PN impedance matrix and then converting these to reactances at 60 Hz, it is found that eqn 12.38 becomes

$$\begin{bmatrix} \mathbf{V}_{ra} \\ \mathbf{V}_{rb} \\ \mathbf{V}_{rc} \end{bmatrix} = \begin{bmatrix} 1.47 + \mathrm{j}703.677 & \mathrm{j}92.356 & \mathrm{j}159.615 \\ \mathrm{j}92.356 & 0.48 + \mathrm{j}12.864 & \mathrm{j}20.775 \\ \mathrm{j}159.615 & \mathrm{j}20.775 & 0.58 + \mathrm{j}36.834 \end{bmatrix} \begin{bmatrix} \mathbf{I}_{ra} \\ \mathbf{I}_{rb} \\ \mathbf{I}_{rc} \end{bmatrix} \tag{12.53}$$

The three phase symmetry of the problem indicates that $\mathbf{I}_{rc} = \mathbf{a}\mathbf{I}_{yc}$ where $\mathbf{a} = \exp(\mathrm{j}2\pi/3)$ so that

$$\begin{bmatrix} \mathbf{I}_{ra} \\ \mathbf{I}_{rb} \\ \mathbf{I}_{rc} \end{bmatrix} = \begin{bmatrix} 1 & 0 & 0 \\ 0 & 1 & 0 \\ 0 & 0 & \mathbf{a} \end{bmatrix} \begin{bmatrix} \mathbf{I}_{ra} \\ \mathbf{I}_{rb} \\ \mathbf{I}_{yc} \end{bmatrix} \tag{12.54}$$

From Fig. 12.15 it is noted that $\mathbf{I}_{rb} = \mathbf{I}_{r1}$ and $\mathbf{I}_{yc} = \mathbf{I}_{ro}$ and, by applying Kirchhoff's current law at the node common to the three windings, $\mathbf{I}_{ra} = \mathbf{I}_{rb} - \mathbf{I}_{yc} = \mathbf{I}_{r1} - \mathbf{I}_{ro}$. Expressing this in matrix form

$$\begin{bmatrix} \mathbf{I}_{ra} \\ \mathbf{I}_{rb} \\ \mathbf{I}_{yc} \end{bmatrix} = \begin{bmatrix} 1 & -1 \\ 1 & 0 \\ 0 & 1 \end{bmatrix} \begin{bmatrix} \mathbf{I}_{r1} \\ \mathbf{I}_{ro} \end{bmatrix} \tag{12.55}$$

Combining eqns 12.54 and 12.55 yields

$$\begin{bmatrix} \mathbf{I}_{ra} \\ \mathbf{I}_{rb} \\ \mathbf{I}_{rc} \end{bmatrix} = \begin{bmatrix} 1 & -1 \\ 1 & 0 \\ 0 & \mathbf{a} \end{bmatrix} \begin{bmatrix} \mathbf{I}_{r1} \\ \mathbf{I}_{ro} \end{bmatrix} \tag{12.56}$$

Having transformed the currents of the original problem to the input and output currents, the same must now be done for the voltages. From Fig. 12.15 it is noted that $\mathbf{V}_{r1} = \mathbf{V}_{ra} + \mathbf{V}_{rb}$ and that $\mathbf{V}_{ro} = \mathbf{V}_{ra} - \mathbf{V}_{yc}$. Again, using the three phase symmetry, $\mathbf{V}_{yc} = \mathbf{a}^2\mathbf{V}_{rc}$ so that this last equation can be rewritten $\mathbf{V}_{ro} = \mathbf{V}_{ra} - \mathbf{a}^2\mathbf{V}_{rc}$. Combining these relationships into a single matrix equation yields

$$\begin{bmatrix} \mathbf{V}_{r1} \\ \mathbf{V}_{ro} \end{bmatrix} = \begin{bmatrix} 1 & 1 & 0 \\ 1 & 0 & -\mathbf{a}^2 \end{bmatrix} \begin{bmatrix} \mathbf{V}_{ra} \\ \mathbf{V}_{rb} \\ \mathbf{V}_{rc} \end{bmatrix} \tag{12.57}$$

Premultiplication of eqn 12.53 by the transformation matrix of eqn 12.57 and replacement of the current vector by the right-hand side of eqn 12.56 yields eqn 12.58 relating the input and output voltages and currents.

$$\begin{bmatrix} \mathbf{V}_{r1} \\ \mathbf{V}_{ro} \end{bmatrix} = \begin{bmatrix} 1.518+\mathrm{j}901.343 & -157.692+\mathrm{j}886.318 \\ -157.752+\mathrm{j}886.318 & 1.52798+\mathrm{j}900.216 \end{bmatrix} \begin{bmatrix} \mathbf{I}_{r1} \\ \mathbf{I}_{ro} \end{bmatrix} \tag{12.58}$$

Equation 12.58 can be used directly to obtain the open circuit output voltage since under these conditions $\mathbf{I}_{ro} = 0.0 + \mathrm{j}0.0$. Taking $\mathbf{V}_{r1}$ as the datum phasor, $\mathbf{V}_{r1} = 12\,000/\sqrt{3} + \mathrm{j}0.0 = 6928.20 + \mathrm{j}0.0$ and, from eqn 12.58, the open circuit input current is $7.687\underline{/89.90^\circ}$ and the open circuit output voltage is

$$\mathbf{V}_{rooc} = 6915.77\underline{/10.0006^\circ} \tag{12.59}$$

In order to obtain a solution under load, the output voltage, $\mathbf{V}_{ro}$, in eqn 12.58 must be replaced by the load voltage drop, $\mathbf{Z}_o\mathbf{I}_{ro}$ where $\mathbf{Z}_o$ is the load impedance per phase. Under the given conditions, 5 MVA load at 0.8 power factor lagging, the nominal rated phase voltage is 6928.20 V and the nominal rated load current is 240.563 A so that the nominal rated load impedance is 28.8 Ω. At a power factor of 0.8 lagging this yields a complex load impedance of 23.04 + j17.28 Ω. Hence, $(23.04 + \mathrm{j}17.28)\mathbf{I}_{ro}$ may be substituted for $\mathbf{V}_{ro}$ in eqn 12.58 and, after transferring this to the right-hand side,

$$\begin{bmatrix} \mathbf{V}_{r1} \\ 0 \end{bmatrix} = \begin{bmatrix} 1.518+\mathrm{j}901.343 & -157.692+\mathrm{j}886.318 \\ -157.752+\mathrm{j}886.318 & -24.568+\mathrm{j}917.496 \end{bmatrix} \begin{bmatrix} \mathbf{I}_{r1} \\ \mathbf{I}_{ro} \end{bmatrix} \tag{12.60}$$

Solution of this equation yields the input and output currents as

$$\mathbf{I}_{r1} = 235.937\underline{/-40.5816^\circ}$$

and

$$\mathbf{I}_{ro} = 231.284\underline{/-29.1437^\circ}$$

from which it is deduced that

$$\begin{aligned}\mathbf{I}_{ra} &= 46.7878\underline{/119.184^\circ}\\ \mathbf{I}_{rb} &= 235.937\underline{/-40.5816^\circ}\\ \mathbf{I}_{rc} &= 231.284\underline{/90.8563^\circ}\\ \mathbf{I}_{yc} &= 231.284\underline{/-29.1437^\circ}\end{aligned}$$

Multiplying the load impedance by the output current yields the output voltage

$$\mathbf{V}_{ro} = 6660.99\underline{/7.72624^\circ}$$

so that the regulation, expressed as a percentage of the open circuit output voltage, is

$$\text{Regulation} = 3.68396\%$$

Inserting the winding currents into eqn 12.53 yields the winding voltages

$$\begin{aligned}\mathbf{V}_{ra} &= 5978.05\underline{/-1.13968^\circ}\\ \mathbf{V}_{rb} &= 958.733\underline{/7.12409^\circ}\\ \mathbf{V}_{rc} &= 1190.78\underline{/-1.58311^\circ}\\ \mathbf{V}_{yc} &= 1190.78\underline{/-121.583^\circ}\end{aligned}$$

From the input voltage and current the total three phase input power is given by

$$P_{in} = 3724.38 \text{ kW}$$

From the output voltage and current the total three phase output power is given by

$$P_{out} = 3697.40 \text{ kW}$$

The power lost in the transformer is the difference between these two quantities

$$P_{loss} = 26.98 \text{ kW}$$

and the efficiency is

$$\text{Efficiency} = 99.28\%$$

Some commentary on these results is necessary. First, the requirement for six figure accuracy in the impedances must be reiterated. Other numerical values can be rounded to engineering accuracies but, as noted earlier, the accuracy of the impedance values must be high enough to retain the correct values of leakage impedances even though these are swamped by the mutual components. Second, the phase shift under load is only 7.73° rather than the desired 10.0°. It is far more important that the 10° phase shift be achieved at rated load than at no load and the transformer would in fact be redesigned with a higher no load phase shift in order to produce the required value under load. Third, the output power at 3693 kW is considerably less than the expected value of $5000 \times 0.8 = 4000$ kW. This is due to the voltage drop in the transformer and to the phase shift caused by the transformer leakage inductances. Again redesign of the transformer is indicated. Finally, the efficiency value is high because no account has been taken of iron losses. These could be expected to be of the same order as the full load copper losses, i.e. about 27 kW and the efficiency would be reduced to about 98.5 %.

Iron loss can be taken into account in a steady state a.c. problem such as this by giving the ferromagnetic mutual impedance a small positive real component so that it matches the actual magnetizing impedance. In doing this it must be noted that the latter is usually expressed as the parallel combination of a reactance and a considerably larger resistance while the value required for this computation is the equivalent series combination of a reactance with a relatively small resistance.

****

The next example illustrates the determination of the Thévenin equivalent circuit of the three phase rectifier with zigzag transformer.

Example 12.4

Determine the Thévenin equivalent circuit of a three phase rectifier whose rated output power is 90 kW and whose rated output voltage is 60 V at zero delay. The rectifier is fed from a 440 V, three phase, 60 Hz supply via a zigzag transformer using a circuit similar to that shown in Fig. 12.10.

Taking winding group A to be the primary and groups B and C to be interconnected as shown in Fig. 12.11 to form the secondary, the transformer data is as follows, where all reactances are at 60 Hz and the magnetizing and zero sequence magnetizing impedances are referred to the primary.

$$\begin{aligned}
\text{Primary resistance } R_a &= 0.0271\ \Omega \\
\text{Secondary resistance } R_b &= 0.000241\ \Omega \\
\text{Secondary resistance } R_c &= 0.000249\ \Omega \\
\text{Magnetizing impedance } \mathbf{Z}_m &= 2.06 + \mathrm{j}23.42\ \Omega \\
\text{Zero sequence mutual impedance } \mathbf{Z}_{zm} &= \mathrm{j}0.0464\ \Omega \\
\text{Leakage impedance } \mathbf{Z}_{aa\sigma} &= \mathrm{j}0.0395\ \Omega
\end{aligned}$$

$$\begin{aligned}
\text{Leakage impedance } \mathbf{Z}_{ab\sigma} &= \text{j}0.00292\ \Omega\\
\text{Leakage impedance } \mathbf{Z}_{ac\sigma} &= \text{j}0.00221\ \Omega\\
\text{Leakage impedance } \mathbf{Z}_{bb\sigma} &= \text{j}0.000183\ \Omega\\
\text{Leakage impedance } \mathbf{Z}_{bc\sigma} &= \text{j}0.000115\ \Omega\\
\text{Leakage impedance } \mathbf{Z}_{cc\sigma} &= \text{j}0.000167\ \Omega\\
\text{Turns ratio } n_b/n_a &= 0.0682\\
\text{Turns ratio } n_c/n_a &= 0.0682
\end{aligned}$$

The voltage drop in a conducting thyristor can be taken as 1.8 V and it may be assumed that ripple in the d.c. output current is negligible.

Although the B and C secondary windings are nominally identical, a fact reflected by them having the same number of turns, they will normally be physically different. In the standard manufacturing process the A, B, and C windings of a phase will form three concentric cylinders so that, for example, the length of conductor in a C winding will be different from that in a B winding and as a consequence their resistances will be different. This is reflected in the above data in which the resistances and leakage reactances of the B and C windings are different.

The computation starts with the application of Kirchhoff's voltage law to the nine coils of the basic transformer to create the equation

$$\mathbf{V}_9 = \mathbf{Z}_{99}\mathbf{I}_9 \tag{12.61}$$

where $\mathbf{V}_9$ is the vector of the nine coil voltages, $\mathbf{Z}_{99}$ is the nine by nine impedance matrix, and $\mathbf{I}_9$ is the vector of the nine coil currents.

The first task is the determination of $\mathbf{Z}_{99}$. The tedium of doing this is appreciably reduced by using the symmetries illustrated by eqn 12.33. Since the impedances are given as complex numbers in ohms at 60 Hz, it will be convenient to continue in this way without converting to inductances and using operational impedances. Further, the magnetizing impedance has a real component indicating the extent of the transformer core iron losses. This will be included in the impedance matrix. Equations 12.48a through 12.48l indicate how the elements of the impedance matrix are constructed.

For the symmetrical submatrix $\mathbf{Z}_{rr}$

$$\begin{aligned}
\mathbf{Z}_{rr11} &= 0.0271 + 2.06 + \text{j}(23.42 + 0.0464 + 0.0395 + \frac{0.00292}{0.0682} + \frac{0.00221}{0.0682})\\
&= 2.0871 + \text{j}23.5811
\end{aligned}$$

$$\begin{aligned}
\mathbf{Z}_{rr12} &= 2.06 \times 0.0682 + \text{j}(23.42 \times 0.0682 + 0.0464 \times 0.0682 + 0.00292)\\
&= 0.140492 + \text{j}1.60333
\end{aligned}$$

$$\begin{aligned}\mathbf{Z}_{rr13} &= 2.06\times 0.0682 + \mathrm{j}(23.42\times 0.0682 + 0.0464\times 0.0682 + 0.00221)\\ &= 0.140492 + \mathrm{j}1.60262\end{aligned}$$

$$\begin{aligned}\mathbf{Z}_{rr22} &= 0.000241 + 2.06\times 0.0682^2 + \mathrm{j}(23.42\times 0.0682^2 + 0.0464\times 0.0682^2\\ &\qquad +0.00292\times 0.0682 + 0.000183 + 0.000115)\\ &= 0.00982255 + \mathrm{j}0.109645\end{aligned}$$

$$\begin{aligned}\mathbf{Z}_{rr23} &= 2.06\times 0.0682^2 + \mathrm{j}(23.42\times 0.0682^2 + 0.0464\times 0.0682^2\\ &\qquad +0.000115)\\ &= 0.00958155 + \mathrm{j}0.109263\end{aligned}$$

$$\begin{aligned}\mathbf{Z}_{rr33} &= 0.000249 + 2.06\times 0.0682^2 + \mathrm{j}(23.42\times 0.0682^2 + 0.0464\times 0.0682^2\\ &\qquad +0.00221\times 0.0682 + 0.000115 + 0.000167)\\ &= 0.00983055 + \mathrm{j}0.109581\end{aligned}$$

For the symmetrical submatrix $\mathbf{Z}_{ry}$

$$\mathbf{Z}_{ry11} = -\frac{1}{2}2.06 - \mathrm{j}\frac{1}{2}23.42 = -1.03 - \mathrm{j}11.71$$

$$\begin{aligned}\mathbf{Z}_{ry12} &= -\frac{1}{2}2.06\times 0.0682 - \mathrm{j}\frac{1}{2}23.42\times 0.0682\\ &= -0.070246 - \mathrm{j}0.798622\end{aligned}$$

$$\begin{aligned}\mathbf{Z}_{ry13} &= -\frac{1}{2}2.06\times 0.0682 - \mathrm{j}\frac{1}{2}23.42\times 0.0682\\ &= -0.070246 - \mathrm{j}0.798622\end{aligned}$$

$$\begin{aligned}\mathbf{Z}_{ry22} &= -\frac{1}{2}2.06\times 0.0682^2 - \mathrm{j}\frac{1}{2}23.42\times 0.0682^2\\ &= -0.00479078 - \mathrm{j}0.0544660\end{aligned}$$

$$\begin{aligned}\mathbf{Z}_{ry23} &= -\frac{1}{2}2.06\times 0.0682^2 - \mathrm{j}\frac{1}{2}23.42\times 0.0682^2\\ &= -0.00479078 - \mathrm{j}0.0544660\end{aligned}$$

$$\begin{aligned}\mathbf{Z}_{ry33} &= -\frac{1}{2}2.06\times 0.0682^2 - \mathrm{j}\frac{1}{2}23.42\times 0.0682^2\\ &= -0.00479078 - \mathrm{j}0.0544660\end{aligned}$$

The temptation to round these numbers to engineering accuracy must be resisted. As has been noted previously, accurate results depend on the small leakage impedance values being preserved within these large numbers. Six significant figures will usually provide sufficient accuracy in the computations but it is wise to maintain the maximum accuracy of which the computer is capable.

The basic impedance matrix, $\mathbf{Z}_{99}$, assembled using these elements and expressed column by column is

Column 1	Column 2
$+2.0871 \times 10^{+0} + j2.3581 \times 10^{+1}$	$+1.4049 \times 10^{-1} + j1.6033 \times 10^{+0}$
$+1.4049 \times 10^{-1} + j1.6033 \times 10^{+0}$	$+9.8226 \times 10^{-3} + j1.0965 \times 10^{-1}$
$+1.4049 \times 10^{-1} + j1.6026 \times 10^{+0}$	$+9.5816 \times 10^{-3} + j1.0926 \times 10^{-1}$
$-1.0300 \times 10^{+0} - j1.1710 \times 10^{+1}$	$-7.0246 \times 10^{-2} - j7.9862 \times 10^{-1}$
$-7.0246 \times 10^{-2} - j7.9862 \times 10^{-1}$	$-4.7908 \times 10^{-3} - j5.4466 \times 10^{-2}$
$-7.0246 \times 10^{-2} - j7.9862 \times 10^{-1}$	$-4.7908 \times 10^{-3} - j5.4466 \times 10^{-2}$
$-1.0300 \times 10^{+0} - j1.1710 \times 10^{+1}$	$-7.0246 \times 10^{-2} - j7.9862 \times 10^{-1}$
$-7.0246 \times 10^{-2} - j7.9862 \times 10^{-1}$	$-4.7908 \times 10^{-3} - j5.4466 \times 10^{-2}$
$-7.0246 \times 10^{-2} - j7.9862 \times 10^{-1}$	$-4.7908 \times 10^{-3} - j5.4466 \times 10^{-2}$

Column 3	Column 4
$+1.4049 \times 10^{-1} + j1.6026 \times 10^{+0}$	$-1.0300 \times 10^{+0} - j1.1710 \times 10^{+1}$
$+9.5816 \times 10^{-3} + j1.0926 \times 10^{-1}$	$-7.0246 \times 10^{-2} - j7.9862 \times 10^{-1}$
$+9.8306 \times 10^{-3} + j1.0958 \times 10^{-1}$	$-7.0246 \times 10^{-2} - j7.9862 \times 10^{-1}$
$-7.0246 \times 10^{-2} - j7.9862 \times 10^{-1}$	$+2.0871 \times 10^{+0} + j2.3581 \times 10^{+1}$
$-4.7908 \times 10^{-3} - j5.4466 \times 10^{-2}$	$+1.4049 \times 10^{-1} + j1.6033 \times 10^{+0}$
$-4.7908 \times 10^{-3} - j5.4466 \times 10^{-2}$	$+1.4049 \times 10^{-1} + j1.6026 \times 10^{+0}$
$-7.0246 \times 10^{-2} - j7.9862 \times 10^{-1}$	$-1.0300 \times 10^{+0} - j1.1710 \times 10^{+1}$
$-4.7908 \times 10^{-3} - j5.4466 \times 10^{-2}$	$-7.0246 \times 10^{-2} - j7.9862 \times 10^{-1}$
$-4.7908 \times 10^{-3} - j5.4466 \times 10^{-2}$	$-7.0246 \times 10^{-2} - j7.9862 \times 10^{-1}$

Column 5	Column 6
$-7.0246 \times 10^{-2} - j7.9862 \times 10^{-1}$	$-7.0246 \times 10^{-2} - j7.9862 \times 10^{-1}$
$-4.7908 \times 10^{-3} - j5.4466 \times 10^{-2}$	$-4.7908 \times 10^{-3} - j5.4466 \times 10^{-2}$
$-4.7908 \times 10^{-3} - j5.4466 \times 10^{-2}$	$-4.7908 \times 10^{-3} - j5.4466 \times 10^{-2}$
$+1.4049 \times 10^{-1} + j1.6033 \times 10^{+0}$	$+1.4049 \times 10^{-1} + j1.6026 \times 10^{+0}$
$+9.8226 \times 10^{-3} + j1.0965 \times 10^{-1}$	$+9.5816 \times 10^{-3} + j1.0926 \times 10^{-1}$
$+9.5816 \times 10^{-3} + j1.0926 \times 10^{-1}$	$+9.8306 \times 10^{-3} + j1.0958 \times 10^{-1}$
$-7.0246 \times 10^{-2} - j7.9862 \times 10^{-1}$	$-7.0246 \times 10^{-2} - j7.9862 \times 10^{-1}$
$-4.7908 \times 10^{-3} - j5.4466 \times 10^{-2}$	$-4.7908 \times 10^{-3} - j5.4466 \times 10^{-2}$
$-4.7908 \times 10^{-3} - j5.4466 \times 10^{-2}$	$-4.7908 \times 10^{-3} - j5.4466 \times 10^{-2}$

Column 7	Column 8
$-1.0300 \times 10^{+0} - j1.1710 \times 10^{+1}$	$-7.0246 \times 10^{-2} - j7.9862 \times 10^{-1}$
$-7.0246 \times 10^{-2} - j7.9862 \times 10^{-1}$	$-4.7908 \times 10^{-3} - j5.4466 \times 10^{-2}$
$-7.0246 \times 10^{-2} - j7.9862 \times 10^{-1}$	$-4.7908 \times 10^{-3} - j5.4466 \times 10^{-2}$
$-1.0300 \times 10^{+0} - j1.1710 \times 10^{+1}$	$-7.0246 \times 10^{-2} - j7.9862 \times 10^{-1}$
$-7.0246 \times 10^{-2} - j7.9862 \times 10^{-1}$	$-4.7908 \times 10^{-3} - j5.4466 \times 10^{-2}$
$-7.0246 \times 10^{-2} - j7.9862 \times 10^{-1}$	$-4.7908 \times 10^{-3} - j5.4466 \times 10^{-2}$
$+2.0871 \times 10^{+0} + j2.3581 \times 10^{+1}$	$+1.4049 \times 10^{-1} + j1.6033 \times 10^{+0}$
$+1.4049 \times 10^{-1} + j1.6033 \times 10^{+0}$	$+9.8225 \times 10^{-3} + j1.0965 \times 10^{-1}$
$+1.4049 \times 10^{-1} + j1.6026 \times 10^{+0}$	$+9.5815 \times 10^{-3} + j1.0926 \times 10^{-1}$

Column 9
$-7.0246 \times 10^{-2} - j7.9862 \times 10^{-1}$
$-4.7908 \times 10^{-3} - j5.4466 \times 10^{-2}$
$-4.7908 \times 10^{-3} - j5.4466 \times 10^{-2}$
$-7.0246 \times 10^{-2} - j7.9862 \times 10^{-1}$
$-4.7908 \times 10^{-3} - j5.4466 \times 10^{-2}$
$-4.7908 \times 10^{-3} - j5.4466 \times 10^{-2}$
$+1.4049 \times 10^{-1} + j1.6026 \times 10^{+0}$
$+9.5816 \times 10^{-3} + j1.0926 \times 10^{-1}$
$+9.8306 \times 10^{-3} + j1.0958 \times 10^{-1}$

The B and C secondary windings are connected in series opposing as indicated in Fig. 12.11 and the resulting zigzag transformer is represented by the equation

$$\mathbf{V}_6 = \mathbf{Z}_{66}\mathbf{I}_6 \tag{12.62}$$

where $\mathbf{V}_6$ is the six element voltage column vector, $\mathbf{Z}_{66}$ is a six by six impedance matrix, and $\mathbf{I}_6$ is the six element current column vector. $\mathbf{Z}_{66}$ is derived from $\mathbf{Z}_{99}$ by the following procedure.

From the circuit diagram, Fig. 12.10, the six voltages of the zigzag transformer are related to the nine winding voltages of the basic transformer by

$$\begin{bmatrix} v_{ra} \\ v_{1n} \\ v_{ya} \\ v_{2n} \\ v_{ba} \\ v_{3n} \end{bmatrix} = \begin{bmatrix} 1 & 0 & 0 & 0 & 0 & 0 & 0 & 0 & 0 \\ 0 & 0 & 1 & 0 & 0 & 0 & 0 & -1 & 0 \\ 0 & 0 & 0 & 1 & 0 & 0 & 0 & 0 & 0 \\ 0 & -1 & 0 & 0 & 0 & 1 & 0 & 0 & 0 \\ 0 & 0 & 0 & 0 & 0 & 0 & 1 & 0 & 0 \\ 0 & 0 & 0 & 0 & -1 & 0 & 0 & 0 & 1 \end{bmatrix} \begin{bmatrix} v_{ra} \\ v_{rb} \\ v_{rc} \\ v_{ya} \\ v_{yb} \\ v_{yc} \\ v_{ba} \\ v_{bb} \\ v_{bc} \end{bmatrix} \tag{12.63}$$

where v_{1n}, v_{2n}, and v_{3n} are the voltages between terminals 1, 2, and 3 of Fig. 12.10 and the secondary neutral.

Equation 12.63 can be written

$$\mathbf{V}_6 = \mathbf{C}_{v1}\mathbf{V}_9 \tag{12.64}$$

Inspection of Fig. 12.10 shows that $\mathbf{I}_6$ and $\mathbf{I}_9$ are related by

$$\begin{bmatrix} i_{ra} \\ i_{rb} \\ i_{rc} \\ i_{ya} \\ i_{yb} \\ i_{yc} \\ i_{ba} \\ i_{bb} \\ i_{bc} \end{bmatrix} = \begin{bmatrix} 1 & 0 & 0 & 0 & 0 & 0 \\ 0 & 0 & 0 & -1 & 0 & 0 \\ 0 & 1 & 0 & 0 & 0 & 0 \\ 0 & 0 & 1 & 0 & 0 & 0 \\ 0 & 0 & 0 & 0 & 0 & -1 \\ 0 & 0 & 0 & 1 & 0 & 0 \\ 0 & 0 & 0 & 0 & 1 & 0 \\ 0 & -1 & 0 & 0 & 0 & 0 \\ 0 & 0 & 0 & 0 & 0 & 1 \end{bmatrix} \begin{bmatrix} i_{ra} \\ i_1 \\ i_{ya} \\ i_2 \\ i_{ba} \\ i_3 \end{bmatrix} \tag{12.65}$$

where i_1, i_2, and i_3 are the input currents to terminals 1, 2, and 3 of Fig. 12.10

Equation 12.65 can be written

$$\mathbf{I}_9 = \mathbf{C}_{c1}\mathbf{I}_6 \tag{12.66}$$

Premultiplication of eqn 12.61 by $\mathbf{C}_{v1}$ and substitution of the right-hand side of eqn 12.66 for $\mathbf{I}_9$ yields

$$\mathbf{V}_6 = \mathbf{C}_{v1}\mathbf{V}_9 = \mathbf{C}_{v1}\mathbf{Z}_{99}\mathbf{I}_9 = \mathbf{C}_{v1}\mathbf{Z}_{99}\mathbf{C}_{c1}\mathbf{I}_6$$

which, by comparison with eqn 12.62, shows that

$$\mathbf{Z}_{66} = \mathbf{C}_{v1}\mathbf{Z}_{99}\mathbf{C}_{c1}$$

Evaluating this transformation yields $\mathbf{Z}_{66}$ as

Column 1	Column 2
$+2.0871 \times 10^{+0} + j2.3581 \times 10^{+1}$	$+2.1074 \times 10^{-1} + j2.4012 \times 10^{+0}$
$+2.1074 \times 10^{-1} + j2.4012 \times 10^{+0}$	$+2.9235 \times 10^{-2} + j3.2816 \times 10^{-1}$
$-1.0300 \times 10^{+0} - j1.1710 \times 10^{+1}$	$+0.0000 \times 10^{-1} + j0.0000 \times 10^{-1}$
$-2.1074 \times 10^{-1} - j2.4020 \times 10^{+0}$	$-1.4372 \times 10^{-2} - j1.6373 \times 10^{-1}$
$-1.0300 \times 10^{+0} - j1.1710 \times 10^{+1}$	$-2.1074 \times 10^{-1} - j2.4020 \times 10^{+0}$
$+0.0000 \times 10^{-1} + j0.0000 \times 10^{-1}$	$-1.4372 \times 10^{-2}, -j1.6373 \times 10^{-1}$

Column 3	Column 4
$-1.0300 \times 10^{+0} - j1.1710 \times 10^{+1}$	$-2.1074 \times 10^{-1} - j2.4020 \times 10^{+0}$
$+0.0000 \times 10^{-1} + j0.0000 \times 10^{-1}$	$-1.4372 \times 10^{-2} - j1.6373 \times 10^{-1}$
$+2.0871 \times 10^{+0} + j2.3581 \times 10^{+1}$	$+2.1074 \times 10^{-1} + j2.4012 \times 10^{+0}$
$+2.1074 \times 10^{-1} + j2.4012 \times 10^{+0}$	$+2.9235 \times 10^{-2} + j3.2816 \times 10^{-1}$
$-1.0300 \times 10^{+0} - j1.1710 \times 10^{+1}$	$+0.0000 \times 10^{-1} + j0.0000 \times 10^{-1}$
$-2.1074 \times 10^{-1} - j2.4020 \times 10^{+0}$	$-1.4372 \times 10^{-2}, -j1.6373 \times 10^{-1}$

Column 5	Column 6
$-1.0300 \times 10^{+0} - j1.1710 \times 10^{+1}$	$+0.0000 \times 10^{-1} + j0.0000 \times 10^{-1}$
$-2.1074 \times 10^{-1} - j2.4020 \times 10^{+0}$	$-1.4372 \times 10^{-2} - j1.6373 \times 10^{-1}$
$-1.0300 \times 10^{+0} - j1.1710 \times 10^{+1}$	$-2.1074 \times 10^{-1} - j2.4020 \times 10^{+0}$
$+0.0000 \times 10^{-1} + j0.0000 \times 10^{-1}$	$-1.4372 \times 10^{-2} - j1.6373 \times 10^{-1}$
$+2.0871 \times 10^{+0} + j2.3581 \times 10^{+1}$	$+2.1074 \times 10^{-1} + j2.4012 \times 10^{+0}$
$+2.1074 \times 10^{-1} + j2.4012 \times 10^{+0}$	$+2.9235 \times 10^{-2}, +j3.2816 \times 10^{-1}$

The first element of the rectifier equivalent circuit, its input voltage, can now be determined. This varies throughout a supply cycle depending on which thyristor is conducting. When T1 conducts it is the open circuit value of v_{1n}, when T2 conducts it is the open circuit value of v_{2n}, and when T3 conducts it is the open circuit value of v_{3n}. These voltages are obtained by noting that, when the transformer is open circuited, currents i_1, i_2, and i_3 are all zero. Hence, under these conditions rows 2, 4 and 6 of eqn 12.62 and columns 2, 4, and 6 of $\mathbf{Z}_{66}$ can be deleted to obtain three equations relating the primary winding currents and voltages. The voltages are the three phase supply line to line voltages and, taking the voltage of the red line to system neutral as reference, they are, in phasor form with r.m.s. amplitudes,

$$\begin{aligned} \mathbf{V}_{ra} &= 440.0\underline{/30.0^\circ} \\ \mathbf{V}_{ya} &= 440.0\underline{/-90.0^\circ} \\ \mathbf{V}_{ba} &= 440.0\underline{/150.0^\circ} \end{aligned} \tag{12.67}$$

The primary currents which flow in response to these voltages are

$$\begin{aligned} \mathbf{I}_{ra} &= 12.42\underline{/-54.95^\circ} \\ \mathbf{I}_{ya} &= 12.42\underline{/-174.95^\circ} \\ \mathbf{I}_{ba} &= 12.42\underline{/65.05^\circ} \end{aligned}$$

Substituting these currents into the current vector $\mathbf{I}_6$ and setting the secondary currents to zero yields the secondary open circuit voltages as

$$\begin{aligned} \mathbf{V}_{1n} &= 51.86\underline{/0.028^\circ} \\ \mathbf{V}_{2n} &= 51.86\underline{/-119.972^\circ} \\ \mathbf{V}_{3n} &= 51.86\underline{/120.028^\circ} \end{aligned}$$

Note the phase shift between these voltages and the primary voltages. If the transformer were ideal these voltages would lag the primary voltages by 30° and their amplitudes would be $2 \times \cos(30^\circ) \times n_b/n_a = 0.1181$ times the primary amplitudes. The actual values differ slightly from these ideal values because of the voltage drops in the primary leakage impedances associated with the magnetizing currents.

The Thévenin impedance

The second element of the rectifier equivalent circuit, the Thévenin impedance of the supply system and transformer as seen from the operative secondary terminal, all sources being replaced by their internal impedances, can now be found. In view of the three phase symmetry, no matter which secondary terminal is selected as the operative one, the Thévenin impedance will have the same value. Selecting terminal 1 as the active one, terminals 2 and 3 are open and, since zero supply system impedance is assumed, the primary terminals are short circuited. Applying these conditions to eqn 12.62 gives a voltage vector in which the voltage phasors $\mathbf{V}_{ra}$, $\mathbf{V}_{rb}$, and $\mathbf{V}_{rc}$ are zero and a current vector in which currents $\mathbf{I}_2$ and $\mathbf{I}_3$ are zero. The latter condition allows rows 4 and 6 to be deleted from the equation and columns 4 and 6 from $\mathbf{Z}_{66}$ to give a matrix equation with four rows. All elements of the voltage vector are zero except the second which, since the computer requires numbers, will be arbitrarily set equal to 1 V. The solution for the current $\mathbf{I}_1$ is then derived and the Thévenin impedance is obtained as 1 V divided by this current. The result is a Thévenin impedance of $0.0007413 + j0.001295\ \Omega$.

The Thévenin impedance just determined is a 60 Hz value and it must be remembered that the combination of the rectifier and its load results in currents with a wide range of frequencies in the transformer and supply. The Thévenin impedance varies with frequency although, in view of the predominantly inductive nature of the transformer and supply, this variation should not be great except at low frequencies. It is wise to check this expectation by determining the Thévenin resistance and inductance over a range of frequencies. The results of doing this at 6, 30, 60, and 120 Hz are shown in Table 12.1. It will be seen that the variation is very small, well within the present needs, and the 60 Hz values will be used, a resistance of $0.0007413\ \Omega$ in series with an inductance of $3.435\ \mu$H.

The effects of commutation on the output voltage must now be determined. The approximate procedure established in Chapter 5 will be followed, ignoring the effects of resistance during commutation and assuming that the load behaves as a constant current source during this period. Commutation from thyristor 1 to 2 will be considered. The first task is the determination of the output voltage assuming commutation is instantaneous, i.e. the voltage at terminal 2. The output voltage and the rate of change of current during commutation are then determined. The voltage lost during commutation is the difference between the two voltages and, with the rate of change of current, gives the resistive equivalent of commutation for the rectifier equivalent circuit.

Table 12.1 *Thévenin resistance and inductance as functions of frequency*

Frequency Hz	Resistance mΩ	Inductance μH
6	0.7406	3.453
30	0.7413	3.436
60	0.7413	3.435
120	0.7413	3.435

The basic equation for this approximate analysis is that relating voltage and rate of change of current when resistance is neglected,

$$\mathbf{V}_6 = \mathbf{L}_{66}\frac{\mathrm{d}\mathbf{I}_6}{\mathrm{d}t} \tag{12.68}$$

The inductance matrix, $\mathbf{L}_{66}$ is obtained by taking the imaginary part of the impedance matrix, $\mathbf{Z}_{66}$ and dividing it by $\omega_s = 377.0$, the radian frequency of the supply.

$$\mathbf{L}_{66} = \begin{bmatrix} +6.25509\times10^{-2} & +6.36949\times10^{-3} & -3.10617\times10^{-2} & -6.37137\times10^{-3} & -3.10617\times10^{-2} & +0.00000\times10^{-1} \\ +6.36949\times10^{-3} & +8.70465\times10^{-4} & +0.00000\times10^{-1} & -4.34304\times10^{-4} & -6.37137\times10^{-3} & -4.34304\times10^{-4} \\ -3.10617\times10^{-2} & +0.00000\times10^{-1} & +6.25509\times10^{-2} & +6.36949\times10^{-3} & -3.10617\times10^{-2} & -6.37137\times10^{-3} \\ -6.37137\times10^{-3} & -4.34304\times10^{-4} & +6.36949\times10^{-3} & +8.70465\times10^{-4} & +0.00000\times10^{-1} & -4.34304\times10^{-4} \\ -3.10617\times10^{-2} & -6.37137\times10^{-3} & -3.10617\times10^{-2} & +0.00000\times10^{-1} & +6.25509\times10^{-2} & +6.36949\times10^{-3} \\ +0.00000\times10^{-1} & -4.34304\times10^{-4} & -6.37137\times10^{-3} & -4.34304\times10^{-4} & +6.36949\times10^{-3} & +8.70465\times10^{-4} \end{bmatrix} \tag{12.69}$$

The output voltage when not commutating

During normal operation, i.e. when not commutating, the current in the secondary connected to the conducting thyristor is $-I_o$ and the currents in the other two secondaries are zero. In all cases the rates of change of the secondary currents are zero. The rates of change of the primary currents are therefore determined by first deleting rows 2, 4, and 6 of eqn 12.68 and columns 2, 4, and 6 of $\mathbf{L}_{66}$ to obtain

$$\begin{bmatrix} v_{ra} \\ v_{ya} \\ v_{ba} \end{bmatrix} = \begin{bmatrix} +6.2551 \times 10^{-2} & -3.1062 \times 10^{-2} & -3.1062 \times 10^{-2} \\ -3.1062 \times 10^{-2} & +6.2551 \times 10^{-2} & -3.1062 \times 10^{-2} \\ -3.1062 \times 10^{-2} & -3.1062 \times 10^{-2} & +6.2551 \times 10^{-2} \end{bmatrix} \begin{bmatrix} \frac{di_{ra}}{dt} \\ \frac{di_{ya}}{dt} \\ \frac{di_{ba}}{dt} \end{bmatrix} \tag{12.70}$$

Equation 12.70 can be inverted to yield

$$\begin{bmatrix} \frac{di_{ra}}{dt} \\ \frac{di_{ya}}{dt} \\ \frac{di_{ba}}{dt} \end{bmatrix} = \begin{bmatrix} +7.8706 \times 10^{+2} & +7.7638 \times 10^{+2} & +7.7638 \times 10^{+2} \\ +7.7638 \times 10^{+2} & +7.8706 \times 10^{+2} & +7.7638 \times 10^{+2} \\ +7.7638 \times 10^{+2} & +7.7638 \times 10^{+2} & +7.8706 \times 10^{+2} \end{bmatrix} \begin{bmatrix} v_{ra} \\ v_{ya} \\ v_{ba} \end{bmatrix} \tag{12.71}$$

The output voltages are obtained in terms of the rates of change of current by deleting rows 1, 3, and 5 from eqn 12.68 and columns 2, 4, and 6 from $\mathbf{L}_{66}$

$$\begin{bmatrix} v_{1n} \\ v_{2n} \\ v_{3n} \end{bmatrix} = \begin{bmatrix} +6.3695 \times 10^{-3} & +0.0000 \times 10^{-1} & -6.3714 \times 10^{-3} \\ -6.3714 \times 10^{-3} & +6.3695 \times 10^{-3} & +0.0000 \times 10^{-1} \\ +0.0000 \times 10^{-1} & -6.3714 \times 10^{-3} & +6.3695 \times 10^{-3} \end{bmatrix} \begin{bmatrix} \frac{di_{ra}}{dt} \\ \frac{di_{ya}}{dt} \\ \frac{di_{ba}}{dt} \end{bmatrix} \tag{12.72}$$

The output voltages are expressed in terms of the input voltages by replacing the current rate of change vector in eqn 12.72 by the right-hand side of eqn 12.71 to give

$$\begin{bmatrix} v_{1n} \\ v_{2n} \\ v_{3n} \end{bmatrix} = \begin{bmatrix} +6.65786 \times 10^{-2} & -1.46223 \times 10^{-3} & -6.95232 \times 10^{-2} \\ -6.95231 \times 10^{-2} & +6.65786 \times 10^{-2} & -1.46225 \times 10^{-3} \\ -1.46225 \times 10^{-3} & -6.95231 \times 10^{-2} & +6.65786 \times 10^{-2} \end{bmatrix} \begin{bmatrix} v_{ra} \\ v_{ya} \\ v_{ba} \end{bmatrix} \tag{12.73}$$

Using the input voltages already defined in eqn 12.67, the output voltages are

$$\mathbf{V}_{1n} = 51.86\underline{/0.0^\circ} \tag{12.74}$$

$$\mathbf{V}_{2n} = 51.86\underline{/-120.0^\circ} \tag{12.75}$$

$$\mathbf{V}_{3n} = 51.86\underline{/120.0^\circ} \tag{12.76}$$

The output voltage during commutation

During commutation from thyristor 1 to thyristor 2,

- The current i_3, and hence its rate of change, are zero
- Terminals 1 and 2 are connected together via the conducting thyristors 1 and 2 so that the two voltages v_{1n} and v_{2n} are the same
- The output current remains constant with the value I_o.

Since $i_1 + i_2 = -I_o$ it follows that $\mathrm{d}i_1/\mathrm{d}t + \mathrm{d}i_2/\mathrm{d}t = 0$, i.e. $\mathrm{d}i_1/\mathrm{d}t = -\mathrm{d}i_2/\mathrm{d}t$.

Since i_3 is zero during this commutation, row 6 can be deleted from eqn 12.68 and column 6 from $\mathbf{L}_{66}$ to produce eqn 12.77

$$\begin{bmatrix} v_{ra} \\ v_{1n} \\ v_{ya} \\ v_{2n} \\ v_{ba} \end{bmatrix} = \begin{bmatrix} +6.25509\times10^{-2} & +6.36949\times10^{-3} & -3.10617\times10^{-2} & -6.37137\times10^{-3} & -3.10617\times10^{-2} \\ +6.36949\times10^{-3} & +8.70465\times10^{-4} & +0.00000\times10^{-1} & -4.34304\times10^{-4} & -6.37137\times10^{-3} \\ -3.10617\times10^{-2} & +0.00000\times10^{-1} & +6.25509\times10^{-2} & +6.36949\times10^{-3} & -3.10617\times10^{-2} \\ -6.37137\times10^{-3} & -4.34304\times10^{-4} & +6.36949\times10^{-3} & +8.70465\times10^{-4} & +0.00000\times10^{-1} \\ -3.10617\times10^{-2} & -6.37137\times10^{-3} & -3.10617\times10^{-2} & +0.00000\times10^{-1} & +6.25509\times10^{-2} \end{bmatrix} \begin{bmatrix} \frac{\mathrm{d}i_{ra}}{\mathrm{d}t} \\ \frac{\mathrm{d}i_1}{\mathrm{d}t} \\ \frac{\mathrm{d}i_{ya}}{\mathrm{d}t} \\ \frac{\mathrm{d}i_2}{\mathrm{d}t} \\ \frac{\mathrm{d}i_{ba}}{\mathrm{d}t} \end{bmatrix} \tag{12.77}$$

Since the two unknown output voltages, v_{1n} and v_{2n} are equal, a new four element voltage vector in which all elements are known can be created using the transformation of eqn 12.78

$$\begin{bmatrix} v_{ra} \\ v_{ya} \\ 0 \\ v_{ba} \end{bmatrix} = \begin{bmatrix} 1 & 0 & 0 & 0 & 0 \\ 0 & 0 & 1 & 0 & 0 \\ 0 & -1 & 0 & 1 & 0 \\ 0 & 0 & 0 & 0 & 1 \end{bmatrix} \begin{bmatrix} v_{ra} \\ v_{1n} \\ v_{ya} \\ v_{2n} \\ v_{ba} \end{bmatrix} \tag{12.78}$$

Since $\mathrm{d}i_1/\mathrm{d}t = -\mathrm{d}i_2/\mathrm{d}t$, the five element rate of change of current vector of eqn 12.77 can be expressed in terms of a four element vector.

$$\begin{bmatrix} \frac{\mathrm{d}i_{ra}}{\mathrm{d}t} \\ \frac{\mathrm{d}i_1}{\mathrm{d}t} \\ \frac{\mathrm{d}i_{ya}}{\mathrm{d}t} \\ \frac{\mathrm{d}i_2}{\mathrm{d}t} \\ \frac{\mathrm{d}i_{ba}}{\mathrm{d}t} \end{bmatrix} = \begin{bmatrix} 1 & 0 & 0 & 0 \\ 0 & 0 & -1 & 0 \\ 0 & 1 & 0 & 0 \\ 0 & 0 & 1 & 0 \\ 0 & 0 & 0 & 1 \end{bmatrix} \begin{bmatrix} \frac{\mathrm{d}i_{ra}}{\mathrm{d}t} \\ \frac{\mathrm{d}i_{ya}}{\mathrm{d}t} \\ \frac{\mathrm{d}i_2}{\mathrm{d}t} \\ \frac{\mathrm{d}i_{ba}}{\mathrm{d}t} \end{bmatrix} \tag{12.79}$$

Premultiplication of eqn 12.77 by the transformation matrix of eqn 12.78 and replacement of the rate of change of current vector by the right-hand side of eqn 12.79 provides a relationship between the applied voltages and the rates of change of current, eqn 12.80

$$\begin{bmatrix} v_{ra} \\ v_{ya} \\ 0 \\ v_{ba} \end{bmatrix} = \begin{bmatrix} +6.25509\times10^{-2} & -3.10617\times10^{-2} & -1.27409\times10^{-2} & -3.10617\times10^{-2} \\ -3.10617\times10^{-2} & +6.25509\times10^{-2} & +6.36949\times10^{-3} & -3.10617\times10^{-2} \\ -1.27409\times10^{-2} & +6.36949\times10^{-3} & +2.60954\times10^{-3} & +6.37137\times10^{-3} \\ -3.10617\times10^{-2} & -3.10617\times10^{-2} & +6.37137\times10^{-3} & +6.25509\times10^{-2} \end{bmatrix} \begin{bmatrix} \frac{di_{ra}}{dt} \\ \frac{di_{ya}}{dt} \\ \frac{di_{2}}{dt} \\ \frac{di_{ba}}{dt} \end{bmatrix} \tag{12.80}$$

Inversion of the inductance matrix of eqn 12.80 yields the rates of change of current in terms of the applied voltages, eqn 12.81

$$\begin{bmatrix} \frac{di_{ra}}{dt} \\ \frac{di_{ya}}{dt} \\ \frac{di_{2}}{dt} \\ \frac{di_{ba}}{dt} \end{bmatrix} = \begin{bmatrix} +2.97782\times10^{+3} & -3.18846\times10^{+2} & +1.60965\times10^{+4} & -3.19170\times10^{+2} \\ -3.18844\times10^{+2} & +1.33459\times10^{+3} & -8.04707\times10^{+3} & +1.32407\times10^{+3} \\ +1.60965\times10^{+4} & -8.04708\times10^{+3} & +1.18268\times10^{+5} & -8.04946\times10^{+3} \\ -3.19168\times10^{+2} & +1.32407\times10^{+3} & -8.04945\times10^{+3} & +1.33491\times10^{+3} \end{bmatrix} \begin{bmatrix} v_{ra} \\ v_{ya} \\ 0 \\ v_{ba} \end{bmatrix} \tag{12.81}$$

Substitution of the primary input voltages defined by eqn 12.67, and remembering that the phasors are r.m.s. values, gives the rates of change of current as

$$\begin{bmatrix} \frac{di_{ra}}{dt} \\ \frac{di_{ya}}{dt} \\ \frac{di_{2}}{dt} \\ \frac{di_{ba}}{dt} \end{bmatrix} = \begin{bmatrix} \sqrt{2}\times1.45061\times10^{6}\cos(\theta+0.52352) \\ \sqrt{2}\times7.25207\times10^{5}\cos(\theta-2.61246) \\ \sqrt{2}\times1.06237\times10^{7}\cos(\theta+0.52351) \\ \sqrt{2}\times7.25422\times10^{5}\cos(\theta-2.62368) \end{bmatrix} \tag{12.82}$$

where $\theta = \omega_s t$.

Remembering that $di_1/dt = -di_2/dt$, these values can be substituted into the rate of change of current vector of eqn 12.77 and the fourth (or second) row evaluated to give the output voltage at terminals 1 and 2 as

$$v_2 = \sqrt{2} \times 25.9315 \cos(\theta - 1.04727) \tag{12.83}$$

Subtracting this voltage from the output voltage assuming instantaneous commutation, $\mathbf{V}_{2n}$ of eqn 12.75, gives the voltage loss due to commutation, v_δ.

$$v_\delta = \sqrt{2} \times 51.8617 \cos(\theta - 2.09440) - \sqrt{2} \times 25.9315 \cos(\theta - 1.04727)$$

i.e.

$$v_\delta = \sqrt{2} \times 44.9139 \cos(\theta - 2.61810) \tag{12.84}$$

The duration of commutation, t_c, is the time taken for the current in the incoming thyristor to rise from zero to the value of the output current, I_o. Because the sign convention makes all input currents to the transformer positive, the rate of change which is required is -1 times the rate of change of i_2 just found in eqn 12.82. It is convenient to obtain this sign reversal by subtracting π from the phase angle rather than by changing the sign of the amplitude so that

$$\frac{di_2}{dt} = \sqrt{2} \times 1.06237 \times 10^7 \cos(\theta - 2.61809) \tag{12.85}$$

It is also convenient to change this rate of change with respect to time to a rate of change with respect to θ by dividing the amplitude by the radian frequency of the supply, ω_s so that

$$\frac{di_2}{d\theta} = \sqrt{2} \times 28\,180.2 \cos(\theta - 2.61809) \tag{12.86}$$

The current in thyristor T2 is the integral of eqn 12.86 with respect to θ and with the constant of integration set by the fact that the current is zero when the thyristor is fired at $\theta = \theta_{st}$. Hence,

$$i_2 = \sqrt{2} \times 28\,180.2[\cos(\theta + 2.09431) - \cos(\theta_{st} + 2.09431)] \tag{12.87}$$

where the sine terms obtained directly from the integral have been converted to cosine terms by subtracting $\pi/2$ from the phase angle and normalizing the resulting phase angle to the range $\pm\pi$.

Commutation ends at $\theta = \theta_{fn}$ when the current has risen to the value of the output current, I_o. Hence, θ_{fn} is defined by

$$\sqrt{2} \times 28\,180.2[\cos(\theta_{fn} + 2.09431) - \cos(\theta_{st} + 2.09431)] = I_o \qquad (12.88)$$

which sets the value of $\cos(\theta_{fn} + 2.09431) - \cos(\theta_{st} + 2.09431)$ as

$$\cos(\theta_{fn} + 2.09431) - \cos(\theta_{st} - 2.09431) = \frac{I_o}{\sqrt{2} \times 28\,180.2} \qquad (12.89)$$

The average voltage lost during the commutation, V_δ, is the integral of v_δ, eqn 12.84, with respect to θ over the range θ_{st} to θ_{fn} averaged over one-third of a cycle, $2\pi/3$. Thus,

$$V_\delta = \sqrt{2} \times 21.4448[\cos(\theta_{fn} - 2.09429) - \cos(\theta_{st} - 2.09429)] \qquad (12.90)$$

where, as for the thyristor current, the sine functions produced by the integral have been converted to cosine functions by subtracting $\pi/2$ from the phase angle and normalizing the resulting angle to the range $\pm\pi$.

The phase angles in eqns 12.89 and 12.90 are almost identical, 2.09431 and 2.09429 rad. In fact they should be identical, the small difference between them being due to the accumulation of computer roundoff error during the many calculations. Hence, the value of the cosine difference given by eqn 12.89 can be substituted in eqn 12.90 to give

$$V_\delta = 0.000\,761\,0 \times I_o \qquad (12.91)$$

i.e. the average voltage loss is proportional to the mean output current, the constant of proportionality being 0.000 761 0. Hence, in the d.c. equivalent circuit, Fig. 5.6, the voltage loss due to commutation can be represented by a resistance of 0.7610 mΩ and in the a.c. equivalent circuit, Fig. 5.8, it can be represented by a voltage source whose value is $0.000\,761\,0 \times I_o$.

These data have been incorporated in the rectifier equivalent circuits shown in Fig. 12.16. Figure 12.16(a) shows the d.c. equivalent circuit which is suitable for the estimation of the mean output current, I_o. The mean rectifier Thévenin voltage, V_{rt} is obtained from eqn 2.8 and is $60.65\cos(\alpha)$ V, the rectifier Thévenin resistance is 0.000 741 3 Ω, the resistance equivalent to commutation is 0.000 761 0 Ω and the voltage absorbed by the conducting thyristor is 1.8 V. Figure 12.16(b) shows the a.c. equivalent circuit which is suitable for the estimation of the waveform of the output current. The rectifier Thévenin voltage, v_{rt}, applies during the 120° segment which extends from $60° + \alpha$ to $180° + \alpha$ when thyristor T2 conducts. This voltage must be appropriately phase shifted for other segments. The rectifier

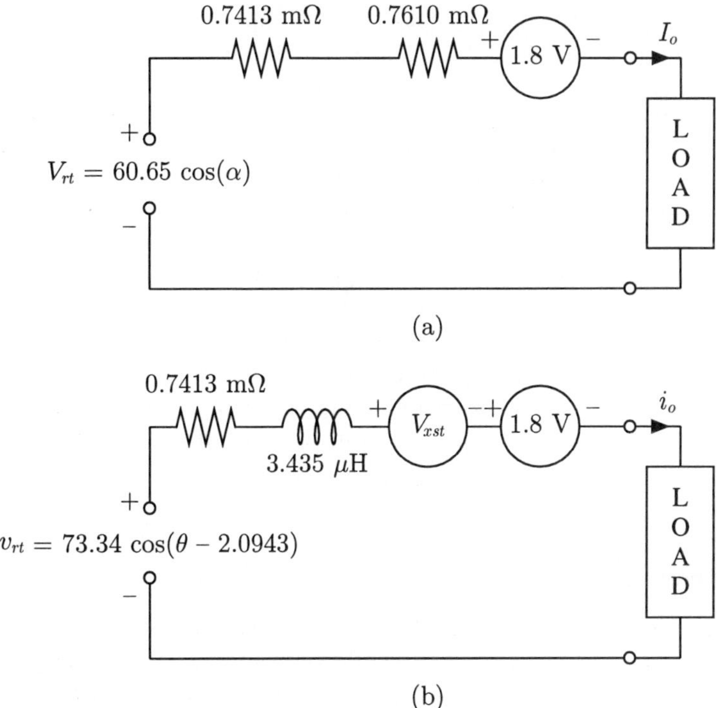

FIG. 12.16. The rectifier equivalent circuit, (a) for d.c. calculations and (b) for output current waveform calculations. The applied voltage is that which would be obtained if thyristor 2 were conducting. It applies for $60° + \alpha \leq \theta \leq 180° + \alpha$. The voltage loss due to commutation, V_{xst} in the a.c. circuit, is the voltage drop in the resistive equivalent of commutation in the d.c. equivalent circuit, i.e. $0.000\,761 \times I_o$.

Thévenin impedance is represented by a resistance of 0.7413 mΩ in series with an inductance of 3.435 μH. The voltage loss due to commutation is represented by the voltage source V_{xst} whose value is $0.000\,761\,3 \times I_o$. As was demonstrated in Chapter 5, the errors in the estimated mean output current and estimated output current waveform will normally be small.

The mean output voltage is the mean rectifier Thévenin voltage less the voltage absorbed by a conducting thyristor, 1.8 V, and the voltage absorbed by the equivalent circuit resistance, $(0.000\,741\,3 + 0.000\,761\,0) \times I_o$. At the rated output current, 1500 A, the voltage absorbed by the rectifier and transformer would be $1.8 + 2.253 = 4.053$ V and at zero delay the regulation would be 3.71% and the efficiency on the basis of the equivalent circuit would be 93.33%. In reality the efficiency would be somewhat lower since the equivalent circuit does not take into account some of the transformer input losses. If a three phase bridge rectifier were used, the thyristor voltage drop would be increased to 3.6 V and the efficiency would be reduced to 90.37%. The reduction in voltage drop and improvement in efficiency would be the major motivation for the use of a single way rectifier in a low voltage application such as this. The improvement would be correspondingly reduced at higher output voltages and increased at lower voltages.

The improvement in efficiency is bought at the price of the greater output ripple from the single way circuit. This could however be reduced to the bridge value if two single way three phase rectifiers with a phase shift of 30° were operated in parallel with the aid of an interphase transformer as described in Chapter 9 and in greater detail in the following section. The improvement would have to be such as to justify the extra complexity.

****

12.7 Interphase transformers

The function of an interphase transformer is to force nominally identical rectifiers to operate in parallel. The rectifiers are similar in every respect except that their a.c. input voltages differ in phase, the difference being 2π divided by the product of the rectifier phase number and the number of rectifiers. The interphase transformer absorbs most of the difference in the rectifier output ripple voltages and this results in a final output voltage whose ripple content is equivalent to that of a single rectifier whose phase number is the product of the phase number of the rectifiers and the number of rectifiers. As an example, consider an interphase transformer designed to force three similar three phase single way rectifiers to operate in parallel. The output voltage ripple from each rectifier is at three times the supply frequency and the a.c. supplies to the rectifiers would differ in phase by $360/(3 \times 3) = 40°$. The final output voltage would be equivalent to that from a nine phase rectifier.

Interphase transformers were commonly employed when mercury arc rectifiers performed the role now played by silicon diodes and thyristors. These devices absorbed about ten times the voltage of their solid state equivalents, about 20 V instead of 2 V, so that the incentive for parallel operation was considerably greater than today. Interphase transformers are uncommon today but there is still a niche for them, notably in low voltage applications where even the small voltage absorbed by a solid state rectifier is significant. The discussion will be kept within reasonable bounds by restricting it to the most common form of interphase transformer, the one which forces two rectifiers to operate in parallel. It will be considered in the context of a pair of three phase bridge rectifiers whose output voltage ripple is six phase. The phase difference between the a.c. supplies to the rectifiers is $360/(6 \times 2) = 30°$ and the final output is equivalent to that of a twelve phase rectifier. The circuit diagram is shown in Fig. 12.17 which identifies most of the quantities required for the analysis.

The interphase transformer is single phase with two identical windings which are connected in series aiding to form a center-tapped autotransformer. Such an interphase transformer in idealized form was briefly discussed in Section 9.3 and that treatment is sufficient for an understanding of the principal mode of action of the device. However, the real device at low values of load current loses its ability to force parallel operation. It is essential that the system designer be aware of this and, if necessary, take steps to neutralize it. This section provides that awareness. The treatment is simplified by considering the system to be ideal with the exception of the interphase transformer which is represented by a linear circuit model.

In the circuit diagram of Fig.12.17 the rectifiers are labelled A and B with A connected to a three phase a.c. supply whose r.m.s. line to line voltage is V_s and whose phase angle is zero. Rectifier B is connected to an a.c. supply of the same voltage but advanced in phase relative to rectifier A by 30° thus providing to the load the effect of a twelve phase rectifier. The supplies could be obtained from a three phase transformer with two independent secondary windings, one delta connected, the other wye connected, as illustrated in Fig. 9.1. It will be assumed that the output current ripple is negligible, an assumption recognized in Fig. 12.17 by representing the load as a current source I_o. The rectifier output voltages, v_{ao} and v_{bo}, are shown respectively by black and gray lines in graph 1 of Fig.12.18 which has been drawn for a delay angle of 45°.

12.7.1 *The circuit equations*

The equations defining the relationship between the interphase transformer currents and voltages are

$$v_{ai} = Z_s i_{ai} - Z_m i_{bi} \tag{12.92}$$

$$v_{bi} = Z_m i_{ai} - Z_s i_{bi} \tag{12.93}$$

where $Z_s = R_s + L_s \mathrm{d}/\mathrm{d}t$ and $Z_m = L_m \mathrm{d}/\mathrm{d}t$ are the operational self and mutual impedances of the interphase transformer windings. Applying Kirchhoff's voltage law to the two loops of Fig. 12.17 yields

$$v_{ai} = v_{ao} - v_o \tag{12.94}$$

$$v_{bi} = v_o - v_{bo} \tag{12.95}$$

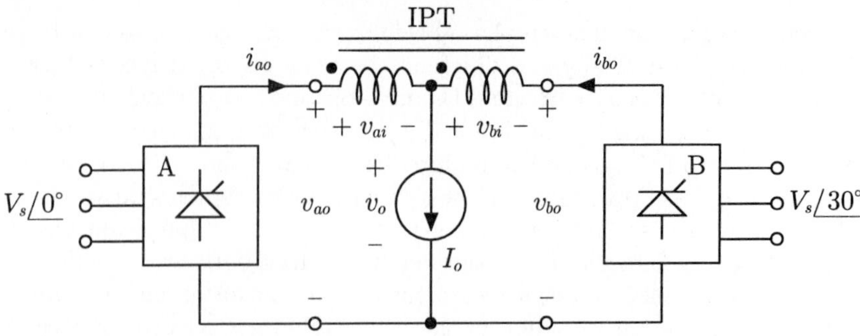

FIG. 12.17. Two rectifiers connected in parallel with the aid of an interphase transformer.

Finally applying Kirchhoff's current law to the center tap of the interphase transformer yields

$$i_{ai} + i_{bi} = I_o \tag{12.96}$$

In these five equations, the voltages v_{ao} and v_{bo} and the current I_o are known and solutions are required for v_{ai}, v_{bi}, i_{ao}, i_{bo}, and v_o. The method of solution uses eqn 12.96 to express i_{bo} in terms of i_{ao} and I_o and eqns 12.94 and 12.95 to eliminate v_{ai} and v_{bi} to obtain a pair of equations in i_{ao} and v_o which are readily reduced to an expression for v_o and a first order differential equation defining i_{ai}

$$v_o = \frac{v_{ao} + v_{bo}}{2} - R_s\frac{I_o}{2} \tag{12.97}$$

and

$$R_s i_{ao} + (L_s + L_m)\frac{\mathrm{d}i_{ao}}{\mathrm{d}t} = \frac{v_{ao} - v_{bo}}{2} + R_s\frac{I_o}{2} \tag{12.98}$$

As usual, it is convenient to change the independent variable from time, t, to angle, $\theta = \omega_s t$, the product of the radian frequency of the supply and time. This change brings in its train a change from inductance to reactance so that eqn 12.98 becomes

$$R_s i_{ao} + (X_s + X_m)\frac{\mathrm{d}i_{ao}}{\mathrm{d}\theta} = \frac{v_{ao} - v_{bo}}{2} + R_s\frac{I_o}{2} \tag{12.99}$$

where $X_s = \omega_s L_s$ and $X_m = \omega_s L_m$.

The rectifier output voltages, v_{ao} and v_{bo}, are made up of segments of sinusoids, each segment having a length of $2\pi/p$ where p is the phase number of the rectifier, and with a phase shift of π/p between the B rectifier and the A rectifier. Their difference $v_{ao} - v_{bo}$ is also sinusoidal. It will be designated as $v_\delta = V_\delta \cos(\theta + \delta)$ so that eqn 12.99 becomes

$$R_s i_{ao} + (X_s + X_m)\frac{di_{ao}}{dt} = \frac{1}{2}V_\delta \cos(\theta + \delta) + R_s\frac{I_o}{2} \qquad (12.100)$$

The solution to this equation is

$$i_{ao} = I_1 + I_2\cos(\theta + \phi_{c2}) + I_3\exp(-\frac{\theta}{\tan(\zeta_t)}) \qquad (12.101)$$

where I_1 is the steady state response to the d.c. driving function $R_s I_o/2$, $I_2\cos(\theta + \phi_{c2})$ is the steady state response to the a.c. driving function $V_\delta\cos(\theta+\delta)/2$, and $I_3\exp(-\theta/\tan(\zeta_t))$ is the solution to the homogeneous equation. With the exception of I_3, all parameters in this expression are known, with values given by eqns 12.102 through 12.106

$$I_1 = I_o/2 \qquad (12.102)$$

$$I_2 = V_\delta/(2Z_t) \qquad (12.103)$$

$$Z_t = \sqrt{R_s^2 + (X_s + X_m)^2} \qquad (12.104)$$

$$\tan(\zeta_t) = R_s/(X_s + X_m) \qquad (12.105)$$

$$\phi_{c2} = \delta - \zeta_t \qquad (12.106)$$

The amplitude of the homogeneous solution, I_3, depends on the boundary conditions.

12.7.2 *Continuous conduction*

When the load current is sufficiently large, the currents i_{ao} and i_{bo} flow continuously. The application of the circuit equations under these conditions is illustrated by Example 12.5

Example 12.5

Two similar three phase bridge rectifiers are operated in parallel with the aid of an interphase transformer. The a.c. supplies to the rectifiers have a frequency of 60 Hz and an r.m.s. line to line voltage of 440 V and the supply to rectifier B leads that to rectifier A by 30°. The rated load is 4730 kW at 550 V. At the supply frequency, the self impedance of an interphase transformer winding is represented by a resistance of 0.0009 Ω in series with an inductance of 0.62 mH. The mutual inductance between the windings is 0.60 mH.

Determine the waveforms of the interphase transformer voltages and currents when the load current is one-tenth the rated value, i.e. 860 A.

Make the calculations for a delay angle of 45°. Assume that the a.c. supplies are ideal, neglect voltage drops in the rectifiers and assume that ripple in the output current is negligible.

The red line to neutral voltage of the a.c. supply to rectifier A will be taken as datum and the red line to neutral voltage of rectifier B will be assumed to lead this by 30°. An output cycle of a rectifier occupies 60° so this is the period which must be studied to determine the behavior of the interphase transformer. The 60° period from α to $\alpha + 60°$ when thyristors T1 and T6 of rectifier A conduct will be studied. During this period

$$v_{ao} = \sqrt{2}V_s \cos(\theta - \pi/6) = 622.25 \cos(\theta - \pi/6)$$

During the first half of this period when thyristors T1 and T6 of the second rectifier conduct, $\alpha < \theta < \alpha + 30°$,

$$v_{bo} = 622.25 \cos(\theta)$$

and the difference voltage, $v_\delta = v_{ao} - v_{bo}$, is

$$v_\delta = 2\sqrt{2}V_s \sin(\pi/12) \cos(\theta - 7\pi/12) = 322.10 \cos(\theta - 7\pi/12)$$

During the second half when thyristors T2 and T6 of the second rectifier conduct, $\alpha + 30° < \theta < \alpha + 60°$,

$$v_{bo} = 622.25 \cos(\theta - \pi/3)$$

and the difference voltage is

$$v_\delta = 322.10 \cos(\theta + 3\pi/12)$$

These waveforms repeat during every 60° period, the waveforms of v_{ao} and v_{bo} being shown in black and gray in graph 1 of Fig. 12.18.

The load voltage, v_o, for the period $\alpha < \theta < \alpha + 30°$ is, from eqn 12.97,

$$v_o = -0.387 + 601.05 \cos(\theta - 0.2618)$$

and during the period $\alpha + 30° < \theta < \alpha + 60°$ it is

$$v_o = -0.387 + 601.05 \cos(\theta - 0.7854)$$

The waveform of the load voltage is shown in graph 3 of Fig. 12.18.

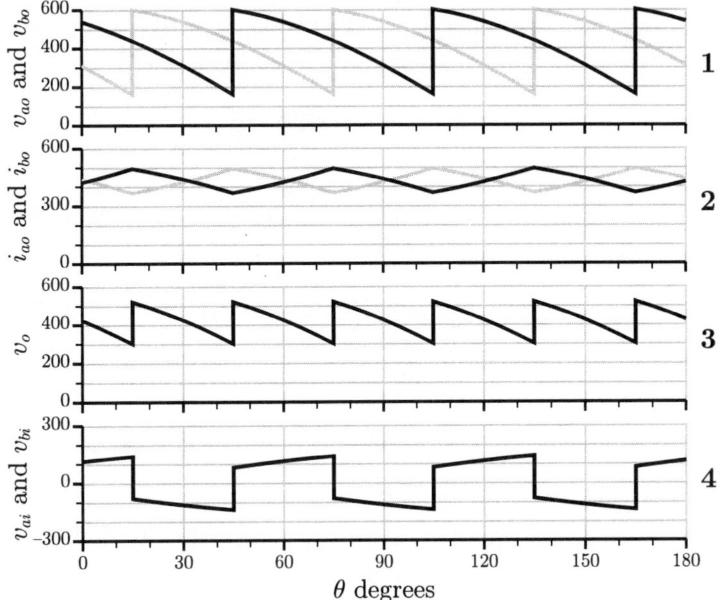

FIG. 12.18. Waveforms in the system of Fig. 12.17. The waveforms are drawn for a delay angle of 45° and conduction is continuous.

The interphase transformer winding voltages are now obtained using eqns 12.94 and 12.95 during the period $\alpha < \theta < \alpha + 30°$ as

$$\begin{aligned} v_{ai} &= 0.387 + 161.05\cos(\theta - 1.8326) \\ v_{bi} &= -0.387 + 161.05\cos(\theta - 1.8326) \end{aligned}$$

and during the period $\alpha + 30° < \theta < \alpha + 60°$ they are

$$\begin{aligned} v_{ai} &= 0.387 + 161.05\cos(\theta + 0.7854) \\ v_{bi} &= -0.387 + 161.05\cos(\theta + 0.7854) \end{aligned}$$

The waveforms of these voltages are shown in graph 4 of Fig. 12.18.

From eqn 12.101, i_{ao} during the period $\alpha < \theta < \alpha + 30°$ is

$$i_{ao} = 430.0 + 350.16\cos(\theta - 3.4014) + I_x\exp(-0.001957\theta)$$

and during the period $\alpha + 30° < \theta < \alpha + 60°$ it is

$$i_{ao} = 430.0 - 350.16\cos(\theta - 3.4014) + I_y\exp(-0.001957\theta)$$

where I_x and I_y depend on the boundary conditions.

Current continuity requires that these two expressions for i_{ao} yield the same value when $\theta = \alpha + 30°$. Setting $\alpha = 45°$, this yields the relationship

$$I_x - I_y = 478.62$$

In the steady state the value of i_{ao} when $\theta = \alpha + 60°$ obtained from the second expression must equal that obtained when $\theta = \alpha$ from the first expression. This yields the relationship

$$I_x - 0.99795 I_y = 478.13$$

From this and the previous equation it is found that $I_x = 239.19$ and $I_y = -239.44$.

Equation 12.96 can be used to obtain i_{bo} with the result that during the period $\alpha < \theta < \alpha + 30°$

$$\begin{aligned} i_{ao} &= 430.0 + 350.16\cos(\theta - 3.4014) + 239.19\exp(-0.001957\theta) \\ i_{bo} &= 430.0 - 350.16\cos(\theta - 3.4014) - 239.19\exp(-0.001957\theta) \end{aligned}$$

and during the period $\alpha + 30° < \theta < \alpha + 60°$

$$\begin{aligned} i_{ao} &= 430.0 + 350.16\cos(\theta - 0.7350) - 239.44\exp(-0.001957\theta) \\ i_{bo} &= 430.0 - 350.16\cos(\theta - 0.7350) + 239.44\exp(-0.001957\theta) \end{aligned}$$

The waveforms of these currents are shown in black and gray in graph 2 of Fig. 12.18.

****

Graphs 1 and 3 of Fig. 12.18 show how the two rectifier output voltages are averaged by the interphase transformer to produce an output voltage corresponding to that of a rectifier with twice the number of phases. Graph 4 shows the voltages appearing across the interphase transformer windings. They are nearly identical, differing only in the small voltage drop produced by the d.c. output current flowing in the transformer resistance, so that although they are drawn in black and gray they are indistinguishable on the scale of the graph.

The waveforms of the rectifier output currents, i_{ao} and i_{bo}, are shown in graph 2 of Fig. 12.18. Each has a mean value equal to one-half of the load current plus a ripple at the rectifier output ripple frequency. This ripple component provides the transformer magnetizing m.m.f., half being provided by one winding and half by the other. Clearly, as the load current is reduced there comes a point at which the ripple will cause the rectifier output current to try to become negative. This it cannot do and conduction in that rectifier ceases, the whole of the load current being carried by the other rectifier for a brief period. This discontinuous conduction mode of operation is similar in some ways to that in a single rectifier which was discussed at length in Chapter 6. The two phenomena are likely to occur simultaneously, producing very complex behavior. However, there is not room for a complete study here and so discontinuous conduction in the interphase transformer will be treated as a separate phenomenon by continuing to model the load by a d.c. current source. This will at least provide an understanding of the phenomenon and an appreciation of the steps which can be taken to alleviate it.

12.7.3 *Discontinuous conduction*

Despite the assumption of a ripple free load current, the currents in the interphase transformer windings contain some ripple associated with the necessity to magnetize the transformer core. When the load current is less than the peak to peak value of this ripple, conduction in the rectifiers becomes discontinuous. This situation is illustrated by Fig. 12.19 which shows the same waveforms as Fig. 12.18 at the same delay angle, 45°, but with the load current reduced to 70 A, at which level conduction is discontinuous. Graph 2 shows the current i_{ao} in black and i_{bo} in gray. Starting at $\theta = 45°$, when thyristor T6 of rectifier A is fired, rectifier B is carrying the whole of the load current, 70 A, and the current in rectifier A is zero. Looking now at graph 1 which shows the waveform of v_{ao} in black and that of v_{bo} in gray, it is seen that firing T6 puts rectifier A into conduction with a significantly higher output voltage than that of rectifier B. As a consequence the current i_{ao} increases and i_{bo} falls until i_{ao} reaches 70 A and i_{bo} reaches zero. This occurs when $\theta = 63.22°$. Since i_{bo} cannot become negative, conduction in rectifier B ceases at this point and does not resume until $\theta = 75°$ when the next thyristor in that rectifier is fired. For the period $63.22° < \theta < 75°$ rectifier A carries the whole of the load current. After rectifier B is enabled at 75°, the process repeats with the roles of the two rectifiers interchanged, i_{ao} decreasing from 70 A to zero at 93.22° and i_{bo} increasing from zero to 70 A in the same period. Between 93.22° and 105° rectifier A is off and rectifier B carries the whole of the load current.

A rectifier segment, e.g. the 60° period from 45° to 105°, repeats until the delay angle and/or the load current are changed. It is convenient to divide it into four subsegments, the first when both rectifiers conduct, the second when only rectifier A conducts, the third when both conduct, and the fourth when only rectifier B conducts. The extent of these subsegments will be defined by vectors of four starting angles, θ_{st1} through θ_{st4}, and

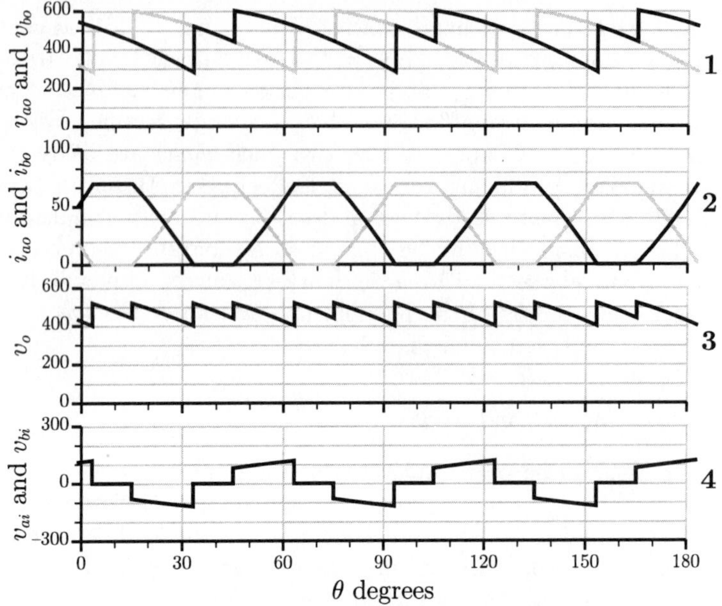

FIG. 12.19. Waveforms in the system of Fig. 12.17. The delay angle is 45°, the load current is 70 A and conduction is discontinuous.

four finishing angles, θ_{fn1} through θ_{fn4}, the starting angle θ_{st2} being equal to the finishing angle θ_{fn1}, etc.

12.7.3.1 *Subsegment 1* During subsegment 1, i_{ao} is defined by eqn 12.101 with I_3 set by the requirement that the current be zero at the start of the segment, θ_{st1}. This yields the following expression for I_3

$$I_3 = -[I_1 + I_2 \cos(\theta_{st1} + \phi_{c2})] \exp[\frac{\theta_{st1}}{\tan(\zeta_t)}] \tag{12.107}$$

The output voltage is defined by eqn 12.97 and the interphase transformer winding voltages are defined by eqns 12.94 and 12.95.

12.7.3.2 *Subsegment 2* During subsegment 2, only rectifier A conducts so that $i_{ao} = I_o$ and $i_{bo} = 0$. The output voltage is the output voltage of rectifier A less the drop caused by the output current flowing in interphase transformer winding A. Substituting the above values of i_{ao} and i_{bo} into eqns 12.92 and 12.93 and remembering that the rate of change of I_o is zero yields

$$v_{ai} = R_s I_o \tag{12.108}$$

$$v_o = v_{ao} - R_s I_o \tag{12.109}$$
$$v_{bi} = R_m I_o \tag{12.110}$$
$$v_{bo} = v_o - R_m I_o \tag{12.111}$$

12.7.3.3 *Subsegment 3* During subsegment 3 both rectifiers conduct and the expressions derived for continuous conduction apply with the exception that the initial value of i_{ao} is I_o and that of i_{bo} is zero. This yields the value of I_3 as

$$I_3 = [I_1 - I_2 \cos(\theta_{st3} + \phi_{c2} - \pi/6)] \exp[\frac{\theta_{st3}}{\tan(\zeta_t)}] \tag{12.112}$$

12.7.3.4 *Subsegment 4* During subsegment 4, only rectifier B conducts so that $i_{ao} = 0$ and $i_{bo} = I_o$. The output voltage is the output voltage of rectifier B less the drop caused by the output current flowing in interphase transformer winding B. Substituting the above values of i_{ao} and i_{bo} into eqns 12.92 and 12.93 and remembering that the rate of change of I_o is zero yields

$$v_{ai} = R_m I_o \tag{12.113}$$
$$v_o = v_{ao} - R_s I_o \tag{12.114}$$
$$v_{bi} = R_s I_o \tag{12.115}$$
$$v_{bo} = v_o - R_m I_o \tag{12.116}$$

The application of these expressions is illustrated by Example 12.6.

Example 12.6

Determine the waveforms of the voltages and currents in the rectifier with interphase transformer of Example 12.5 when the delay angle is 75° and the load current is 70 A. Assume that conduction is discontinuous.

The same segment as in Example 12.5 will be considered, i.e. that covering the range $\alpha < \theta < \alpha + \pi/3$. The four subsegments of which this is composed cover the ranges $\alpha < \theta < \beta$, $\beta < \theta < \alpha + \pi/6$, $\alpha + \pi/6 < \theta < \gamma$, and $\gamma < \theta < \alpha + \pi/3$ where β and γ have yet to be determined.

During the first subsegment, the current i_{ao} is given by eqn 12.101 with $I_1 = 35.0$, I_2 unchanged at 350.16, ϕ_{c3} unchanged at -3.4014, and ζ_t unchanged at 89.89°. From eqn 12.107 $I_3 = 268.32$. Thus, during this subsegment when both rectifiers conduct, the expressions for i_{ao} and i_{bo} are

$$i_{ao} = 35.0 + 350.16 \cos(\theta - 3.4014) + 268.32 \exp(-0.0019568\theta)$$
$$i_{bo} = 35.0 - 350.16 \cos(\theta - 3.4014) - 268.32 \exp(-0.0019568\theta)$$

At the start of the segment, $\theta = 45°$, these expressions yield $i_{ao} = 0.0$ and $i_{bo} = 70.0$ as indeed they should.

When $\theta = \beta$, the current in rectifier B has decreased to zero. There are many ways in which β can be found. An appropriate one is to make an initial search for its value by incrementing θ in 5° steps until i_{ao} changes sign. This locates β between 60° at which $i_{bo} = 14.348$ and 65° at which $i_{bo} = -8.169$. A linear interpolation between these points gives a first approximation for the extinction angle of $\beta = 63.186°$. The Newton–Raphson technique refines this to $\beta = 63.231°$ in a single iteration. Substituting this angle for θ in the above expressions for the rectifier output currents yields $i_{ao} = 70.0$ and $i_{bo} = 0.0$ confirming that the extinction point is correct.

During this first subsegment, from eqns 12.94, 12.95, and 12.97 the voltages are

$$\begin{aligned} v_{ao} &= 622.25\cos(\theta - 0.5236) \\ v_{bo} &= 622.25\cos(\theta) \\ v_o &= -0.0315 + 601.05\cos(\theta - 0.2618) \\ v_{ai} &= 0.0315 + 161.05\cos(\theta - 1.8326) \\ v_{bi} &= -0.0315 + 161.05\cos(\theta - 1.8326) \end{aligned}$$

The second subsegment extends from 63.231° to 75.0° and during it

$$\begin{aligned} i_{ao} &= 70.0 \\ i_{bo} &= 0.0 \\ v_{ao} &= 622.25\cos(\theta - 0.5236) \\ v_{bo} &= -0.0630 + 622.25\cos(\theta - 0.5236) \\ v_o &= -0.0630 + 622.25\cos(\theta - 0.5236) \\ v_{ai} &= 0.0630 \\ v_{bi} &= -0.0 \end{aligned}$$

At the start of the third subsegment, $\theta = 75°$, rectifier B again begins to conduct, its current increasing from the initial value of zero and the current in rectifier A correspondingly decreasing from its initial value of 70 A. The expressions for the two currents are

$$\begin{aligned} i_{ao} &= 35.0 + 350.16\cos(\theta - 0.7834) - 268.60\exp(-0.0019568\theta) \\ i_{bo} &= 35.0 - 350.16\cos(\theta - 0.7834) + 268.60\exp(-0.0019568\theta) \end{aligned}$$

Since the system is symmetrical, the conduction angle is equal to that already obtained for the first subsegment, 18.23°, so that the extinction angle, γ, is 93.23°. Substitution of this angle for θ in the above expressions for currents yields $i_{ao} = 0.0$ and $i_{bo} = 70.0$, confirming the correctness of the extinction angle.

The voltages during subsegment 3 are given by eqns 12.94, 12.95, and 12.97. They are

$$\begin{aligned} v_{ao} &= 622.25\cos(\theta - 0.5236) \\ v_{bo} &= 622.25\cos(\theta - 1.0472) \\ v_o &= -0.0315 + 601.05\cos(\theta - 0.7854) \\ v_{ai} &= 0.0315 + 161.05\cos(\theta - 0.7854) \\ v_{bi} &= -0.0315 + 161.05\cos(\theta - 0.7854) \end{aligned}$$

The fourth subsegment extends from 93.231° to 105.0° and during it

$$\begin{aligned} i_{ao} &= 0.0 \\ i_{bo} &= 70.0 \\ v_{ao} &= -0.0630 + 622.25\cos(\theta - 1.0472) \\ v_{bo} &= 622.25\cos(\theta - 1.0472) \\ v_o &= -0.0630 + 622.25\cos(\theta - 1.0472) \\ v_{ai} &= 0.0 \\ v_{bi} &= -0.0630 \end{aligned}$$

This completes the information required to draw Fig. 12.19 since the segment repeats until either the load current or delay angle are changed.

****

12.7.3.5 *The mean output voltage* Comparison of graphs 4 of Figs 12.18 and 12.19 shows that a smaller portion of a rectifier output voltage is absorbed by the interphase transformer when conduction is discontinuous within the interphase transformer. This results in an increase in output voltage similar to that found in Chapter 6 where discontinuous conduction within a rectifier was reviewed.

The mean value of the output voltage is obtained by considering the output voltage during subsegments 1 and 2. During subsegment 1 the output voltage is given by eqn 12.97 and during subsegment 2 it is given by eqn 12.109. The mean output voltage, V_o is then

$$V_o = \frac{6}{\pi}\left\{\int_{\theta_{st1}}^{\theta_{fn1}} \frac{v_{ao} + v_{bo}}{2} - R_s\frac{I_o}{2}\,\mathrm{d}\theta + \int_{\theta_{st2}}^{\theta_{fn2}} v_{ao} - R_s I_o\,\mathrm{d}\theta\right\} \tag{12.117}$$

Inserting numerical values appropriate to Example 12.6, this integral becomes

$$V_o = 1.9099 \int_{0.7854}^{1.1036} 601.05\cos(\theta - 0.2618) - 0.0009 \times 35\, d\theta +$$

$$3.8197 \int_{1.1036}^{1.3090} 622.25\cos(\theta - 0.5236) - 0.0009 \times 70\, d\theta$$

which when evaluated yields

$$V_o = 471.25 \text{ V}$$

Figure 12.20 shows, in black, load characteristics for the rectifier of the example at various values of delay angle, the cosine of the delay being shown to the right of each characteristic.

12.7.3.6 *The critical point* The point at which conduction changes from continuous to discontinuous is the critical point and the locus of such points for all values of delay angle is the critical boundary.

The critical point occurs when the extinction angle, β, defined in Example 12.5, is $\alpha + 30°$. To determine this point, I_3 in eqn 12.101 is set to the value corresponding to discontinuous conduction given in eqn 12.107. The value of θ is then set to $\alpha + \pi/6$ and the resulting value of i_{ao} is the critical value, i.e.

$$\begin{aligned} I_{ocrit} = {} & \tfrac{I_{ocrit}}{2}[1 - \exp(-\tfrac{\pi/6}{\tan(\zeta_t)})] + \\ & I_2[\cos(\alpha + \pi/6 + \phi_{c2}) - \cos(\alpha + \phi_{c2})\exp(-\tfrac{\pi/6}{\tan(\zeta_t)})] \end{aligned} \tag{12.118}$$

Solution of eqn 12.118 yields

$$I_{ocrit} = 2I_2 \frac{\cos(\alpha + \pi/6 + \phi_{c2})\exp(\frac{\pi/6}{\tan(\zeta_t)}) - \cos(\alpha + \phi_{c2})}{1 + \exp(\frac{\pi/6}{tan(\zeta_t)})} \tag{12.119}$$

The critical boundary for the rectifier of the example is shown in gray in Fig. 12.20. The critical current depends on the delay angle, being small as the delay angle approaches both zero and α_{max} and reaching its maximum value at a delay of 90°. Even under worst case conditions, $\alpha = 90°$, the critical current is small, in the case of the example 181.2 A, just over 2% of the rated load current. Nevertheless the rise in voltage at very low loads can be substantial with the worst case again occurring at a delay of 90°. From Fig. 12.20, in this worst case the voltage rises from zero at the critical point to 159.2 V at zero load, i.e. 29% of the rated value.

As mentioned earlier, the type of discontinuous conduction discussed here will in practice occur simultaneously with rectifier discontinuous conduction as discussed in Chapter 6. The remedy is the same for both phenomena, feedback control of the output voltage.

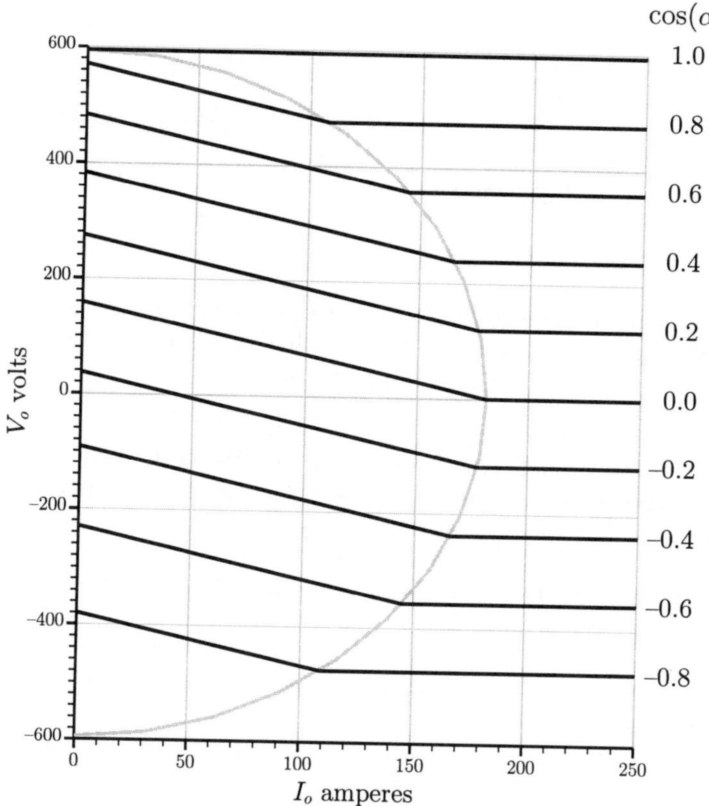

FIG. 12.20. Load characteristics of the rectifier system. Load characteristics are shown by black lines computed for fixed delay angles as identified by the values of $\cos(\alpha)$ at the right. The critical boundary is shown in gray, conduction being continuous to its right and discontinuous to its left.

12.7.3.7 *Design considerations* The output current at the critical point is kept as low as possible by minimizing the interphase transformer magnetizing current by keeping the peak flux density in its core well below the saturation level.

The high frequency of flux variation in the core, in the case of the example at six times the supply frequency, also indicates a lower than normal flux density to prevent excessive core loss.

The use of low peak flux density is also indicated by the possibility of unbalanced d.c. magnetization. The theory assumes a pair of perfectly matched rectifiers operating at the same delay angle. In practice it is probable that there will be small differences between rectifiers and between their delay angles which will result in small differences between their mean d.c. output voltages. In view of the low resistance of the interphase transformer windings, even a small difference in voltage will produce a substantial d.c. current flowing in the same direction in both windings. This will bias the core flux in one direction or the other and take the flux swings associated with the ripple voltage components absorbed by the transformer closer to saturation.

The above considerations lead to the use of a peak core flux density about half that used for a normal transformer. However, despite this it is wise to incorporate in the delay angle generators of the two rectifiers a feedback mechanism which will ensure that the difference between their mean output voltages is minimized at all times.

12.7.3.8 *Interphase transformer rating* Some idea of the size of the interphase transformer can be obtained by calculating its equivalent rating at the supply frequency. The equivalent current is easily found but the equivalent voltage requires more consideration.

Since the ripple in the winding currents is very small relative to the rated load current, the rated current of an interphase transformer winding may be taken to be half the rated load current.

The rated voltage depends on the core flux and, since this does not vary sinusoidally with time, the flux swing under operating conditions must be found. An expression for the voltage appearing across an interphase transformer winding is obtained by combining eqns 12.94 and 12.97

$$v_{ai} = \frac{v_{ao} - v_{bo}}{2} + \frac{R_s I_o}{2}$$

Since the voltage drop $R_s I_o/2$ is very small, it may be ignored for the present purpose so that

$$v_{ai} = \frac{v_{ao} - v_{bo}}{2} \tag{12.120}$$

From $\theta = \alpha$ to $\theta = \alpha + \pi/6$ this gives

$$v_{ai} = \sqrt{2} V_s \sin(\frac{\pi}{12}) \cos(\theta - \frac{7\pi}{12}) \tag{12.121}$$

If the analysis is restricted to delay angles greater than 15°, this voltage leads to a change in flux linkage, $\Delta\lambda$, of

$$\Delta\lambda = \int_{\alpha}^{\alpha+\pi/6} \sqrt{2} V_s \sin(\frac{\pi}{12}) \cos(\theta - \frac{7\pi}{12}) \frac{d\theta}{\omega_s}$$

which after some manipulation yields

$$\Delta\lambda = \frac{0.1895 V_s \sin(\alpha)}{\omega_s} \tag{12.122}$$

If the transformer winding voltages were of sinusoidal waveform at the supply frequency and had an r.m.s. value of V_{equiv}, i.e. if

$$v_{ai} = \sqrt{2} V_{equiv} \sin(\theta)$$

the core flux linkage swing would be

$$\Delta\lambda = \int_0^{\pi} \frac{\sqrt{2} V_{equiv} \sin(\theta)}{\omega_s} \, d\theta$$

which yields

$$\Delta\lambda = \frac{2\sqrt{2} V_{equiv}}{\omega_s} \tag{12.123}$$

Equating eqns 12.122 and 12.123 and solving for V_{equiv} yields

$$V_{equiv} = 0.06699 \, V_s \sin(\alpha) \tag{12.124}$$

The interphase transformer equivalent supply frequency rating is the product of half the rated load current and this equivalent voltage, i.e.

$$kVA_{ipt} = \frac{0.06699 \, V_s \sin(\alpha) I_o}{2000}$$

However, it must be remembered that an interphase transformer is operated at a lower flux density than a normal transformer, thus increasing its size. This is taken into account by doubling the above rating to give

$$kVA_{ipt} = \frac{0.06699 \, V_s \sin(\alpha) I_o}{1000}$$

For three phase bridge rectifiers, the maximum value of the load voltage, V_{omax} is $3\sqrt{2}V_s/\pi$ so that the above expression becomes

$$kVA_{ipt} = \frac{0.0496 \, V_{omax} I_o \sin(\alpha)}{1000} \tag{12.125}$$

Thus, under worst case conditions, i.e. a delay angle of 90°, the equivalent supply frequency rating of the interphase transformer is about 5% that of the load. At a delay of 15°, the smallest value for which the formula applies, the rating is about 1.3% that of the load.

Equation 12.124 indicates that when the delay angle is zero the rating of the interphase transformer is zero. This obviously incorrect result comes about because eqn 12.124 is only correct for delay angles equal to or greater than 15°. A similar but somewhat more complex analysis is required when the delay is less than 15°.

12.7.3.9 *Losses and efficiency* Obviously an interphase transformer would not be employed unless the benefits outweighed the cost. Some simple calculations provide the basis on which such a decision can be made.

For the example, assuming that each thyristor absorbs 2 V when conducting, the loss in the rectifiers would be $4 \times 8600 \div 1000 = 34.4$ kW at rated load. The loss in the interphase transformer due to the rated load current is $2 \times 0.0009 \times 4300^2 \div 1000 = 33.3$ kW. In comparison to this, its iron loss will be small and will be ignored. Thus, the combined rectifier and interphase transformer loss will be about 67.7 kW. The same two rectifiers could be reconnected in series to give the same twelve phase type of output without any change in the main transformer. The rectifier losses would then be $8 \times 8600 \div 1000 = 68.8$ kW. The reduction in loss brought about by the use of the interphase transformer is therefore of the order of 1.1 kW, hardly sufficient to justify the expenditure.

The losses in the interphase transformer can be reduced by reducing the resistance of its winding. If for example the resistance were reduced from 0.0009 to 0.0003 Ω, the power saving at rated load would be 23.3 kW. The small size of the interphase transformer relative to the equipment as a whole may make such a large reduction in resistance feasible.

More sophisticated calculations would be required before the decision to include an interphase transformer is made but these numbers give an indication of the considerations involved.

12.8 Inductors

Inductors are frequently encountered in power electronic circuits where they perform a number of functions. They are employed on the input a.c. side as filter elements, they are used within the rectifier to encourage equitable current sharing between parallel connected diodes and thyristors, and on the output side they function as filter elements to reduce current ripple to an acceptable level. In both appearance and construction inductors are usually very similar to a single phase transformer with a shell type laminated iron core as illustrated in Fig. 12.1(b), the facts that they only have a single coil and that the core in all likelihood contains a significant non-ferromagnetic gap not being obvious from the external appearance. Some inductors are constructed differently, some on the lines of a single phase core type transformer, Fig. 12.1(a), some with multiple limbs and coils on the lines of a three phase transformer, Fig. 12.1(c), and some do not contain any ferromagnetic material at all.

Several particular merits account for the predominance of the shell type construction. The iron core almost completely encases the copper coil and thereby affords significant protection from mechanical damage. The encompassing core channels the substantial magnetic fields produced by the inductor and minimizes and controls that portion which spreads into the surrounding equipment. The core provides a sound mechanical structure which minimizes the noise and vibration emitted by the inductor. When applicable, multiphase construction, several coils on a multilimbed core, can advantageously exploit mutual couplings between coils. Coils without ferromagnetic cores can be employed in situations where their wide spreading magnetic fields can be tolerated. Such situations are typically high voltage, high current applications with the equipment situated outdoors.

Iron cored inductors with a nonmagnetic gap in the core are frequently more briefly called *airgapped inductors* even though the gap is seldom of air, some solid, strong, nonconducting, nonmagnetic material being employed instead which permits the two portions of the iron core to be firmly clamped together to minimize noise and vibration. Similarly, but with more justification, inductors without a ferromagnetic core are frequently referred to as air cored inductors.

Airgapped inductors are considered in the following analysis in the context of the shell type construction. While some design considerations will be introduced, the discussion does not go into the details of design. These are dealt with in depth in texts devoted to transformers in particular and ferromagnetic apparatus in general.

12.8.1 *Airgapped inductors*

A d.c. side inductor intended to minimize the ripple component of the d.c. output current must carry a large d.c. current while presenting a high impedance to the flow of a.c. components of current. The latter aim can be met by using a ferromagnetic core with the highest permeability consistent with the economic requirements. However, this solution is incompatible with the need to carry a large d.c. component of current because of ferromagnetic hysteresis and saturation.

Hysteresis and saturation are nonlinear phenomena which lead to behavior which is hard to predict when an iron cored coil is magnetized by a current with both d.c. and a.c. components, as in an output filter inductor. This complex behavior is discussed at length in texts devoted to magnetic materials but for the present purpose a highly simplified review will suffice.

Figure 12.21 shows a typical hysteresis loop for an inductor with a complete iron core. The loop has been drawn by plotting the coil flux linkages as a function of the coil current and is equivalent to the standard loop, which shows the flux density B as a function of the magnetizing force H, except for horizontal and vertical scaling factors. The major loop shows the behavior when the current in the inductor swings symmetrically between a large negative and a large positive value. If the swing in current is stopped at the point marked A, and the current is diminished before being

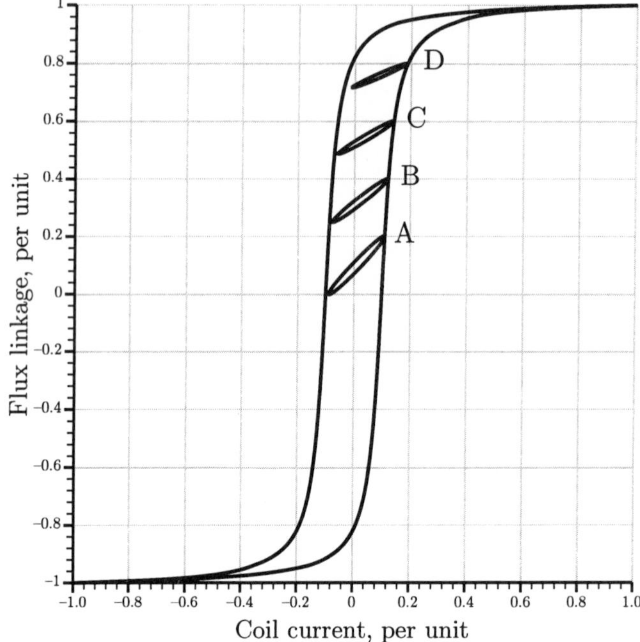

FIG. 12.21. Hysteresis loop for an inductor with a continuous iron core.

returned to A to resume its upward progress, the core material follows the minor hysteresis loop with one corner located at A. Repeating the process at points B, C, and D results in other minor loops being traced as indicated. Two general deductions can be made regarding the minor loops. First, their inclination to the current axis decreases as the flux density increases. Second, even at low flux densities the inclination is substantially less than the slope of the major loop at a similar flux density. As the flux density approaches the saturation level, the slope of the minor loops becomes very small, approaching that of an air cored coil.

Inductance is flux linkage per ampere and in the nonlinear situation represented by Fig. 12.21, it is an ambiguous quantity. The particular definition required here is the value relevant to small perturbations of current, the ripple component, about a quiescent operating point, the d.c. component. This is known as the *incremental inductance.* The incremental inductance is proportional to the slope of the minor hysteresis loop round the quiescent operating point. Figure 12.21 shows that it is very variable, depending on the d.c. current, and becomes very small as the flux density approaches the saturation level.

The introduction of an air gap of suitable length in the core resolves these difficulties. Figure 12.22 shows the effect of introducing a small gap

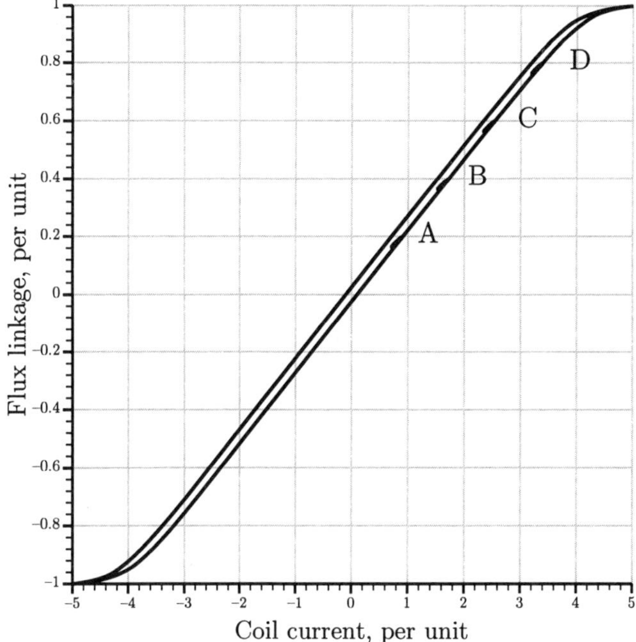

FIG. 12.22. Hysteresis loop for an inductor with an iron core containing a small nonmagnetic gap.

into the core whose hysteresis loop is shown in Fig. 12.21. Note that the horizontal scale is five times greater than for Fig. 12.21. Some rough calculations indicate the length of the gap. Assuming the relative permeability of the unsaturated iron to be of the order of 1000, a not unreasonable figure, and the mean magnetic length of the iron to be 50 cm, the equivalent nonmagnetic length of the iron is $500 \div 1000 = 0.5$ mm. Figure 12.22 has been drawn assuming an air gap about four times this value, i.e. 2 mm. In most applications the gap would be proportionally much greater, nevertheless the effect of this very short gap is dramatic in reducing the effects of nonlinearity and hysteresis.

The design of a d.c. side inductor aims to keep the core flux below the saturation level at all times and yet maintain the highest possible impedance to the flow of ripple components of current. The factors affecting this are illustrated by Example 12.7 in which the calculations are of a very preliminary nature, ignoring the effects of flux fringing in the air gap and core nonlinearity. Such a calculation would be refined by using either the classical design methods or one of the numerous analytical programs based on finite element analysis.

Example 12.7

Do a preliminary design study for an inductor which must carry a maximum d.c. current of 450 A. Assume shell type construction and a coil of 96 turns. The width and depth of the iron core through the coil are both 18 cm. Assume that the core reluctance is negligible in relation to that of the air gap, that the magnetic flux is confined to the core, and ignore flux fringing in the air gap. Aim for a mean core flux density of about 1.0 Wb/m.^{2}

The first task is the determination of the length of the air gap which must be sufficient to maintain the core flux density close to or below the target value at all times. The air gap length is obtained by first determining the reluctance of the magnetic circuit as the ratio of the coil m.m.f. to the flux passing through it. Since this is a preliminary study subject to later refinement, the work is simplified by a number of approximations. The iron core must be laminated so that part of its cross-section is occupied by the insulation between laminations. This will be ignored and the cross-sectional area of the iron will be taken to be $0.18 \times 0.18 = 0.0324\,\text{m}^2$ The reluctance of the air gap, $\mathcal{R}_{ag}$, which, since the reluctance of the iron is being ignored, is also the total reluctance of the magnetic circuit and is given by

$$\mathcal{R}_{ag} = \frac{a_g}{\mu_o A_g}$$

where μ_o is the permeability of free space $= 4\pi \times 10^{-7}$, a_g is the length of the air gap, and A_g is the effective cross-sectional area of the air gap. Since flux fringing in the air gap is being ignored, $A_g = 0.18 \times 0.18 = 0.0324\,\text{m}^2$ Thus

$$\mathcal{R}_{ag} = 24.56 \times 10^6 \times a_g$$

The maximum m.m.f. produced by the d.c. current is $450 \times 96 = 43\,200$ ampere–turns. The aim is to have this m.m.f. produce a flux of about $1.0 \times 0.0324 = 0.0324$ Wb. Hence the air gap length is given by

$$43\,200 = 0.0324 \times 24.56 \times 10^6 \times a_g$$

from which $a_g = 0.05429$ m $= 54.29$ mm which will be rounded to 54 mm. With this air gap length the reluctance is 1.326×10^6 H and the flux produced is $43\,200/1.326 \times 10^6 = 0.03258\,\text{Wb}$. This flux produces a linkage of $96 \times 0.03258 = 3.128$ Wb-turns with the coil.

The linkage of 3.128 Wb-turns produced by a coil current of 450 A yields the coil inductance as $3.128/450 = 0.006\,949$ H $= 6.949$ mH.

An idea of the dimensions of the inductor can be obtained by considering the size of the conductors, but before doing this attention is drawn to

Fig. 12.23 which shows a cross-section through the core parallel to the laminations. The conductor cross-sections are shown in black and the air gap in gray.

Assuming a current density of 180 A/cm^2, the cross-sectional area of the conductor is $450/180 = 2.5\,\text{cm}^2$ At this point a suitable conductor would be selected from a table of standard sizes. In the absence of such tables, a conductor of 22.4 mm by 11.2 mm giving an area of $2.51\,\text{cm}^2$ is selected.

The total area of copper which must pass through a core window is $2.51 \times 96 = 240.8\,\text{cm}^2$. The window must also provide room for the conductor insulation, the coil former, ventilation ducts, etc. Factors based on experience are employed to allow for these items. A factor of 2.5 will be used to obtain a window area of 602 cm^2. The shape of this area must accommodate a reasonable distribution of conductors and choosing to wind the coil in eight concentric layers, each having twelve turns, as shown in Fig. 12.23, leads to a window depth of 37.5 cm and a width of 16.4 cm.

Since the width of the center limb of the core is 18 cm the width of the yokes and outer limbs which carry half the flux are each 9 cm. This makes the total width of the core $9 + 16.4 + 18 + 16.4 + 9 = 68.8$ cm and the total height of the core $9 + 37.5 + 9 = 55.5$ cm. The depth of the core is 18 cm so that the volume of the core is 57 661 cm^3 and, taking the density of iron as 7.8 gr cm^{-3}, the core weighs about 450 kg.

The overall width and depth of the coil are both about $16.4 + 18 + 16.4 = 50.8$ cm The outer periphery of the coil is therefore $50.8 \times 4 = 203.2$ cm. The inner periphery is $18 \times 4 = 72$ cm The average length of a turn is therefore about $(203.2 + 72.0)/2 = 137.6$ cm. The total length of the conductor is $137.6 \times 96 = 13\,210$ cm. Taking the resistivity of copper to be 1.721×10^{-6} ohm-cm at 20° C, the resistance of the coil at this temperature is $1.721 \times 10^{-6} \times 13\,210/(2.24 \times 1.12) = 0.00906\ \Omega$. Taking the temperature coefficient of the resistance of copper to be 0.00393 per degree Celsius, the resistance at 80° C is $0.00906(1 + 0.00393 \times 60) = 0.0112\ \Omega$.

The volume of copper in the coil is about $2.24 \times 1.12 \times 13\,210 = 33\,141$ cm^3. Taking the density of copper as 8.89 gr cm^{-3}, the weight of copper is about 295 kg.

A cross-sectional view of the inductor parallel to the laminations is shown in Fig. 12.23. This is drawn to a scale of one-tenth of full size and shows the air gaps in gray and the conductors in black.

This inductor is a massive object. With about 450 kg of iron and about 295 kg of copper, its total weight including insulation and fittings will be of the order of 1 t. When hot, the d.c. voltage drop in the coil will be about $0.0112 \times 450 = 5.04$ V and the loss in it will be about $0.0112 \times 450^2/1000 = 2.27$ kW. Its reactance at 360 Hz is about 15.7 Ω. To give some measure of scale to these numbers, imagine that the inductor is part of a three phase bridge rectifier directly coupled to a 440 volt, 60 Hz supply. The

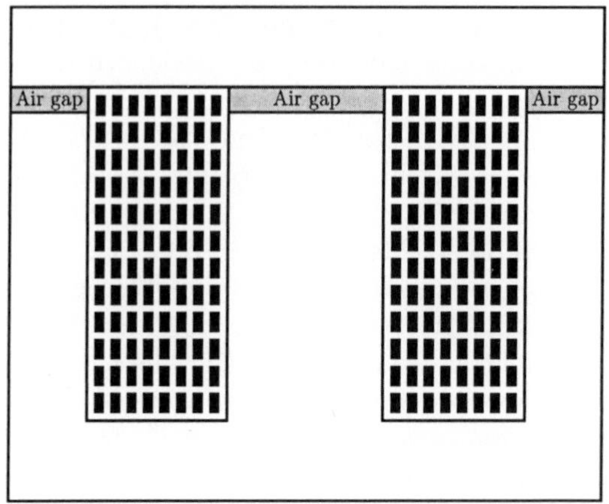

FIG. 12.23. Cross-section of the inductor parallel to the laminations. The drawing is one tenth of full size. The air gaps are shown in gray and the conductors in black.

maximum output voltage which the rectifier can produce is 594 V and its maximum power output is $594 \times 450 \div 1000 = 267.3$ kW. At maximum current, the inductor absorbs about 0.85% of the maximum output voltage and of the maximum output power. The effectiveness of the inductor can be judged from the ripple current. The maximum voltage ripple occurs when the delay angle is 90° when the lowest component, the one at 360 Hz, has a peak amplitude of 205 V. Assuming zero inductance in the load, the ripple current produced by this voltage would have an amplitude of $205/15.7 = 13.0$ A. This is 2.9% of the maximum d.c. current.

While these calculations are approximate and would be refined to take into account the standard available core and conductor sizes and the manufacturing facilities, and would be optimized to meet some criterion such as minimum weight, minimum cost, etc., they do give an indication of the major factors affecting the design of a d.c. side inductor.

****

13

SINGLE PHASE AC VOLTAGE CONTROLLERS

13.1 Introduction

AC voltage controllers convert a.c. at fixed frequency and voltage into a.c. of the same frequency but variable voltage. Control is accomplished by triacs or back to back connected thyristors in the supply lines as shown in Fig. 13.1. Figure 13.1(a) shows a triac controlling a single phase load and Fig. 13.1(b) shows six thyristors, a back to back pair in each line, controlling a three phase load. At the price of some asymmetry in operation, the thyristor pair can be replaced by a combination of thyristor and diode as shown in Fig. 13.1(c). The diode is less expensive than the thyristor it replaces and the number of thyristor firing circuits is halved.

These controllers are used to regulate the output of lamps, control the temperature of heating loads such as ovens and furnaces, start and, within limits, control the speed of induction motors, and regulate the power factor of a.c. loads. Triacs can be used for powers up to about 10 kW. Thyristors can control powers up to megawatt levels.

For three phase loads there are many possible thyristor arrangements which have been described in the literature. Most require access to the interconnections between the load phases and thus become special purpose designs. The general purpose arrangement of Fig. 13.1(b) which enables the controller to interface with any type of a.c. load appears to be preferred and is the one which is analyzed in Chapter 14.

Two types of control can be distinguished, integral cycle control and fractional cycle control.

13.2 Integral cycle control

Integral cycle control applies full power to the load for a certain number of a.c. cycles by continuously gating the thyristors. Power flow is then blocked for a period by removing the gate signal. This cycle of operation is repeated indefinitely resulting in the load voltage waveform shown in Fig. 13.2. This very simple arrangement is suitable for loads whose time of response to an input power change is long compared to the a.c. period. Thermal loads such as ovens and furnaces have this characteristic, often having time constants measured in minutes. To avoid very high rates of rise of current, oscillations

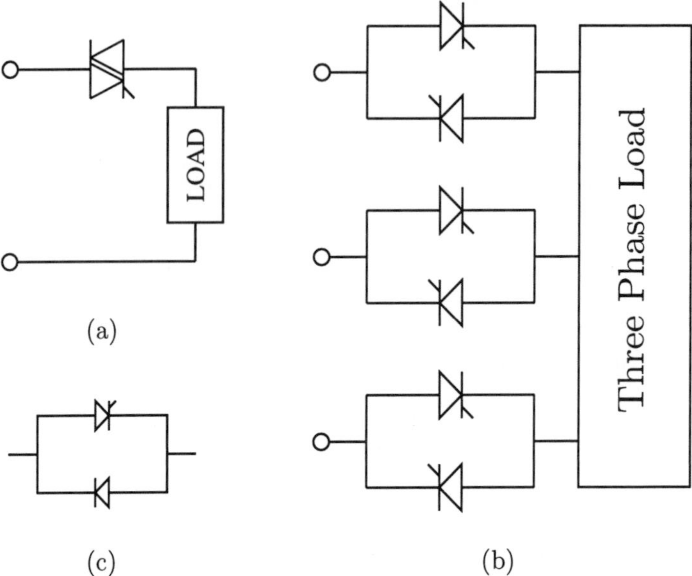

FIG. 13.1. AC load controllers. In (a) is shown a triac controlling a single phase load. In (b) six thyristors in three back to back connected pairs control a three phase load. In (c) is shown an alternative solid state switch comprising a thyristor and a diode connected back to back.

in parasitic inductances and capacitances, and radio frequency interference, gate signal is usually applied at a voltage zero. Gate signal can be removed at any time, the conducting thyristor or triac recovering at the next current zero.

The first few half cycles and the last half cycle of an on period will contain transient components. However these rapidly decay and are of minor consequence so that it can be said, with very little error, that the mean power, P_o, delivered to the load is related to the rated power, P_r, i.e. the power with full voltage continuously applied, by

$$P_o = P_r \frac{t_{on}}{t_{on} + t_{off}} \tag{13.1}$$

where t_{on} is the on time and t_{off} is the off time.

Since interruption of power can only occur at the end of a half cycle, control is discrete rather than continuous, an integral number of half cycles of supply voltage being applied to the load. However, with the standard 50 or 60 Hz supply a controller cycle time of 1 s permits control in steps of

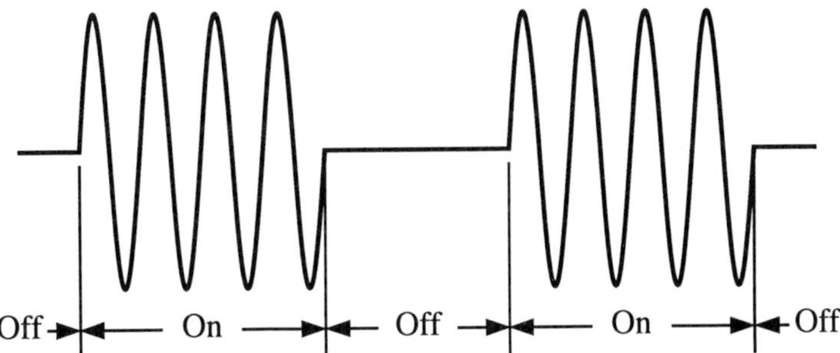

FIG. 13.2. The load current waveform when using an integral cycle controller.

about 1% so that, with appropriate loads, the discrete nature of the control does not present a problem.

This system will not be analyzed further since the techniques described for fractional cycle control can be applied with very little modification should investigation of the transient components be necessary.

13.3 Fractional cycle control

Fractional cycle control applies power during a selected portion of each half cycle of the a.c. supply. It must be used for loads whose response time to power changes is comparable with the period of the supply. Incandescent lamps and induction motors are examples of such loads and the familiar domestic lamp dimmer is an example of this type of controller.

Single phase loads are considered in this chapter and three phase loads in Chapter 14. For generality, the load will be assumed to be active, comprising a series combination of a resistance R_o, an inductance L_o, and an e.m.f. e_o of sinusoidal waveform at the supply frequency as shown in Fig. 13.3(a). Such a load fits reasonably with the majority of practical situations. A lamp or oven will be almost purely resistive and can be represented by setting the inductance and e.m.f. to zero. An induction motor can be modeled with reasonable accuracy by the full complement of resistance, inductance, and e.m.f. provided these quantities are appropriate to the machine's speed.

The time origin is chosen so that the supply voltage is

$$v_s = \sqrt{2}V_s \sin(\theta) \tag{13.2}$$

where V_s is the r.m.s. value of the supply voltage, $\theta = \omega_s t$, and ω_s is the radian frequency of the supply. This wave is shown by the gray line of Fig. 13.3(b).

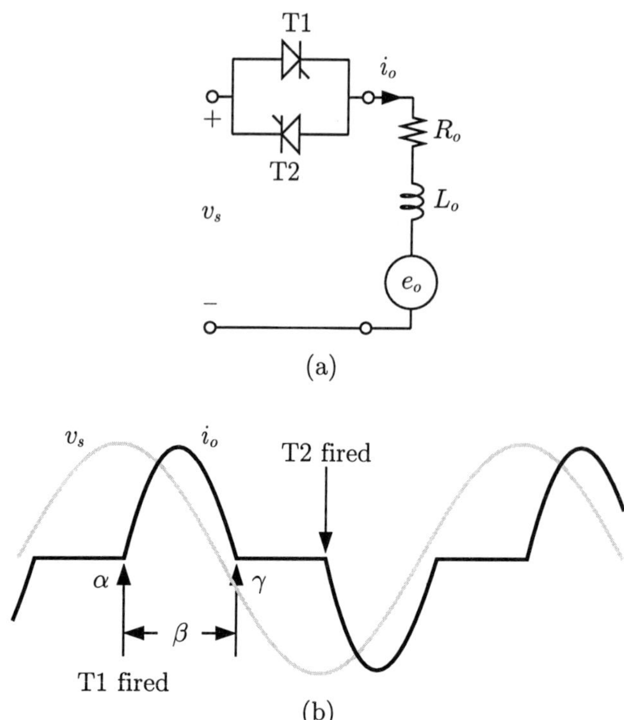

FIG. 13.3. A solid state switch controlling a single phase, active, inductive load. In (a) is shown the circuit diagram and in (b) the waveforms of the load current, the solid black line, and the input voltage, the gray line.

The waveform of the e.m.f. is assumed to be sinusoidal at the same frequency as the supply, with an r.m.s. value E_o and phase lead of ϵ on the supply voltage, i.e.

$$e_o = \sqrt{2}E_o \sin(\theta + \epsilon) \tag{13.3}$$

It is assumed that thyristor T1 is fired at $\theta = \alpha$ and that conduction continues for an angle β until $\theta = \gamma = \alpha + \beta$.

This sequence is then repeated during the next half cycle with T2 replacing T1 and then the whole cycle is repeated indefinitely.

The principal angles are given the following names

- α delay angle
- β conduction angle
- γ extinction angle

The waveform of the load current is shown by the black line of Fig. 13.3(b) and the locations of the delay angle, the extinction angle, and the extent of the conduction angle are indicated.

The aim of the analysis is to first determine the extinction angle as a function of the delay angle and then to determine all relevant values such as load voltage and current, thyristor voltage and current, etc. Thyristor voltage drops and are neglected and it is assumed that the supply impedance is negligible.

13.4 Uncontrolled operation

As the starting point for the analysis, the behavior of the circuit when uncontrolled, i.e. when gating signals are continuously applied to each thyristor, is determined. The problem is then one in steady state a.c. circuit theory which in phasor terms yields

$$\mathbf{I}_o = \frac{\mathbf{V}_d}{\mathbf{Z}} \tag{13.4}$$

where $\mathbf{V}_d$ is the difference between the supply voltage and the load e.m.f. and $\mathbf{Z}$ is the impedance of the circuit

$$\mathbf{V}_d = \mathbf{V}_s - \mathbf{E}_o \tag{13.5}$$

The real and imaginary components of $\mathbf{V}_d$ are given by

$$V_{dr} = V_s - E_o \cos(\epsilon) \tag{13.6}$$

$$V_{di} = -E_o \sin(\epsilon) \tag{13.7}$$

The situation described by eqns 13.4 through 13.7 is depicted in the phasor diagram of Fig. 13.4. This shows to scale the two cases of Example 13.1. Case (i), $E_o = 93.0\underline{/-5.6^\circ}$, is shown in (a) and case(ii), $E_o = 108.5\underline{/10.3^\circ}$, is shown in (b).

Depending on the magnitude and phase angle of the load e.m.f., the output current may lag or lead the supply voltage and power can flow from the supply to the load or from the load to the supply. The phase angle of the current relative to the supply voltage is denoted by ϕ_c.

Since the solid state controller can only delay conduction and cannot, under normal operating conditions, force it to start before the natural point, $\theta = \phi_c$, the phase of the uncontrolled current relative to the supply voltage is the minimum effective delay angle.

Under uncontrolled conditions the conduction angle is 180° and the extinction angle is the delay angle plus 180°. Under these conditions and

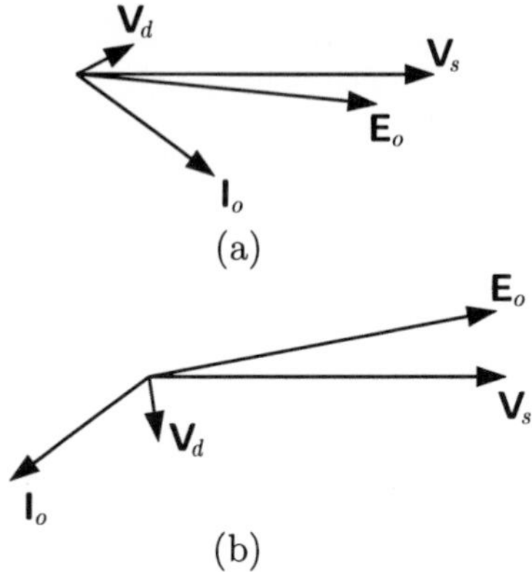

FIG. 13.4. Phasor diagrams showing the relationships between the supply voltage phasor, $\mathbf{V}_s$, the load e.m.f., $\mathbf{E}_o$, their difference, $\mathbf{V}_d$, and the resulting load current, $\mathbf{I}_o$. In (a) the load e.m.f. lags the supply voltage and in (b) it leads.

with the chosen time origin, the delay angle is the phase angle, ϕ_c, of the current.

Example 13.1

A single phase induction motor rated at one-half horse power is fed from a 110 V, 60 Hz supply. At rated load it can be represented by a series combination of a resistance of 1.68 Ω, an inductance of 8.9 mH and an e.m.f. of sinusoidal waveform.

Determine the minimum effective delay angle, output current, input power, and input power factor for the two cases (i) when the r.m.s. value of the load e.m.f. is 93.0 V lagging the supply voltage by 5.6° and (ii) when the r.m.s. value is 108.5 V leading the supply voltage by 10.3°.

Note that the load e.m.f. of case (i) is used for all succeeding examples in this chapter.

Using standard a.c. circuit theory, for the first case when $E_o = 93.0$ V and $\epsilon = -5.6°$

$$\begin{aligned} \alpha_{min} &= 35.92° \\ \mathbf{V}_d &= 19.66\underline{/27.49°} \\ I_o &= 5.240 \text{ A} \\ P_1 &= 466.8 \text{ W} \\ Pf &= 0.8099 \end{aligned}$$

For the second case when $E_o = 108.5$ V and $\epsilon = 10.3°$

$$\alpha_{min} = 143.9°$$
$$\mathbf{V}_d = 19.67\underline{/-80.49°}$$
$$I_o = 5.242 \text{ A}$$
$$P_1 = -465.9 \text{ W}$$
$$Pf = -0.8099$$

****

If the delay angle is made larger than the uncontrolled value, the extinction angle will be less than the uncontrolled value, the conduction angle will be less than 180°, and there will be a period of zero current flow when the conducting thyristor has recovered and the incoming thyristor has not yet been fired. This circuit behavior will be determined after thyristor T1 is fired.

13.5 The circuit equations

When a thyristor is conducting, Kirchhoff's voltage law for the load loop yields

$$v_d = v_s - e_o = R_o i_o + L_o \frac{\mathrm{d}i_o}{\mathrm{d}t}$$

It will be convenient to work with the angle $\theta = \omega_s t$ rather than time and this expression is converted to this end using the relationships $\mathrm{d}i_o/\mathrm{d}t = \omega_s \mathrm{d}i_o/\mathrm{d}\theta$ and substituting the load reactance at the supply frequency, X_o, for the product $\omega_s L_o$ to obtain

$$v_d = v_s - e_o = R_o i_o + X_o \frac{\mathrm{d}i_o}{\mathrm{d}\theta} \tag{13.8}$$

The input current equals the output current

$$i_1 = i_o \tag{13.9}$$

Since the small voltage drops in the conducting thyristors are being neglected, the output voltage equals the input voltage

$$v_o = v_s \tag{13.10}$$

and the voltage across the thyristors is zero

$$v_{th} = 0 \tag{13.11}$$

When both thyristors are blocking

$$i_o = 0 \tag{13.12}$$

$$v_o = e_o \tag{13.13}$$
$$i_1 = 0 \tag{13.14}$$
$$v_{th1} = v_s - e_o = v_d \tag{13.15}$$
$$v_{th2} = -(v_s - e_o) = -v_d \tag{13.16}$$

Equation 13.8 is a first order differential equation with a sinusoidally varying driving function, v_d. Its solution is the sum of the particular integral and the homogeneous solution.

In obtaining the particular integral, reference to the phasor diagram Fig. 13.4 shows that the difference voltage, v_d, can be written as

$$v_d = \sqrt{2}V_d \sin(\theta + \delta) \tag{13.17}$$

where

$$V_d = \sqrt{[V_s - E_o \ \cos(\epsilon)]^2 + [E_o \ \sin(\epsilon)]^2} \tag{13.18}$$

and

$$\delta = \arctan\left[\frac{-E_o \sin(\epsilon)}{V_s - E_o \cos(\epsilon)}\right] \tag{13.19}$$

The particular integral, which is the steady state response, i_{ss}, to this driving voltage, is then

$$i_{ss} = I_a \sin(\theta + \phi_c) \tag{13.20}$$

where

$$I_a = \sqrt{2}V_d/Z_o \tag{13.21}$$
$$Z_o = \sqrt{R_o{}^2 + X_o{}^2} \tag{13.22}$$
$$\phi_c = \delta - \zeta \tag{13.23}$$
$$\zeta = \arctan(X_o/R_o) \tag{13.24}$$

The homogeneous solution is the transient response, i_t,

$$i_t = I_b \exp(-\theta/\tan(\zeta)) \tag{13.25}$$

where the value of I_b depends upon the initial conditions.

Thus, the complete solution is

$$i_o = I_a \sin(\theta + \phi_c) + I_b \exp(-\theta/\tan(\zeta)) \tag{13.26}$$

This expression is the basic power electronics expression of eqn A.1 with

$$X_{11} = 0 \tag{13.27a}$$

$$X_{12} = I_a \tag{13.27b}$$
$$\phi_{12} = \phi_c \tag{13.27c}$$
$$X_{13} = I_b \tag{13.27d}$$
$$T_{13} = \tan(\zeta) \tag{13.27e}$$
$$X_{16} = 0 \tag{13.27f}$$
$$X_{17} = 0 \tag{13.27g}$$

Initially, at $\theta = \alpha$, the value of i_o is zero, a fact which determines I_b as

$$I_b = -I_a \sin(\alpha + \phi_c) \exp(\frac{\alpha}{\tan(\zeta)}) \tag{13.28}$$

Example 13.2

Determine for the example operating with an e.m.f. of $93.0\underline{/-5.6°}$ V, i.e. the same value as in part (i) of the Example 13.1, and with the delay angle set at 95°, the expression for the output current during the first half cycle.

From case (i) of Example 13.1

$$v_d = \sqrt{2}19.66 \sin(\theta + 0.4797)$$

and applying this to eqns 13.20 through 13.28

$$i_o = 7.4109\ \sin(\theta - 0.6269) - 14.58\ \exp(-\theta/1.997)$$

****

13.6 The extinction and conduction angles

The extinction angle, γ, is defined by the next current zero after α of eqn 13.26. Thus

$$I_a\ \sin(\gamma + \phi_c) + I_b\ \exp(-\frac{\gamma}{\tan(\zeta)}) = 0 \tag{13.29}$$

The solution of this transcendental equation must be obtained numerically by some method of successive approximation. Provided a good initial value for γ can be obtained, the Newton–Raphson method is extremely efficient. The uncontrolled extinction angle, $\gamma = \phi_c + \pi$, is a good initial approximation.

Table 13.1 *The Newton–Raphson approximation to the extinction angle*

Iteration number	γ degrees	$f(\gamma)$	$f'(\gamma)$
1	215.92	-2.210	-6.304
2	195.83	-0.089	-5.641
3	194.93	-0.000	-5.590

Using the Newton–Raphson method, if γ_1 is an approximation to the true value, $f(\gamma_1)$ denotes the value of eqn 13.26 with $\gamma = \gamma_1$ and $f'(\gamma_1)$ is the value of its derivative with respect to γ at γ_1, then

$$f(\gamma_1) = I_a \ \sin(\gamma_1 + \phi_c) + I_b \exp(-\frac{\gamma_1}{\tan(\zeta)}) \tag{13.30}$$

and

$$f'(\gamma_1) = I_a \ \cos(\gamma_1 + \phi_c) - \frac{I_b}{\tan(\zeta)} \exp(-\frac{\gamma_1}{\tan(\zeta)}) \tag{13.31}$$

and a better approximation, γ_2, is

$$\gamma_2 = \gamma_1 - \frac{f(\gamma_1)}{f'(\gamma_1)} \tag{13.32}$$

Example 13.3

For the example with $E_o = 93$ V, $\epsilon = -5.6°$ and $\alpha = 95°$, determine the extinction and conduction angles.

The angle of lag of the uncontrolled current is 0.6269 rad or 35.92° so that an initial approximation to γ is $35.92° + 180° = 215.92°$. Successive approximations to γ using the Newton–Raphson method starting with this value are given in Table 13.1.

The approximation to the extinction angle has an accuracy of better than 0.001° after only three iterations. Its value is 194.93° and the conduction angle is directly obtained as $\beta = \gamma - \alpha$, so that for this case

$$\beta = 99.93°$$

****

13.7 The r.m.s. output current

The r.m.s. value of the output current, I_{orms}, is given by

$$I_{orms} = \sqrt{\frac{1}{\pi} \int_{\alpha}^{\gamma} [I_a \ \sin(\theta + \phi_c) + I_b \ \exp(-\frac{\theta}{\tan(\zeta)})]^2 \ d\theta} \qquad (13.33)$$

This integral is given in eqn A.6, with the parameter values listed in eqns 13.27a through 13.27g. After some simplification

$$I_{orms} = \left\{ \frac{1}{2\pi} \left\{ I_a^2[\beta - \sin(\beta)\cos(\alpha + \gamma + 2\phi_c)] - \right.\right. \qquad (13.34)$$

$$4I_a I_b \sin(\zeta)[\sin(\gamma + \delta)\exp(\frac{-\gamma}{\tan(\zeta)}) - \sin(\alpha + \delta)\exp(\frac{-\alpha}{\tan(\zeta)})] -$$

$$\left.\left. I_b{}^2 \tan(\zeta)[\exp(\frac{-2\gamma}{\tan(\zeta)}) - \exp(\frac{-2\alpha}{\tan(\zeta)})] \right\} \right\}^{1/2}$$

Example 13.4

Determine the r.m.s. value of the output current for the example when $E_o = 93$ V, $\epsilon = -5.6°$, and $\alpha = 95°$.

Using eqn 13.34 with the values of extinction and conduction angles found in Example 13.3

$$I_{orms} = 1.565 \text{ A}$$

a value which should be compared with the uncontrolled current of 5.240 A.

****

13.8 The output voltage

The output voltage, v_o, equals the supply voltage when a thyristor is conducting and the load e.m.f. when both thyristors are blocking. Thus, for $\alpha + n\pi < \theta < \gamma + n\pi$

$$v_o = \sqrt{2}V_s \ \sin(\theta) \qquad (13.35)$$

and for $\gamma + n\pi < \theta < \alpha + (n+1)\pi$

$$v_o = \sqrt{2}E_o \ \sin(\theta + \epsilon) \qquad (13.36)$$

where n is any integer, positive or negative including zero.

The waveforms of the supply voltage, the load voltage, and the load current for the example are shown in graphs 1, 2, and 3 of Fig. 13.5.

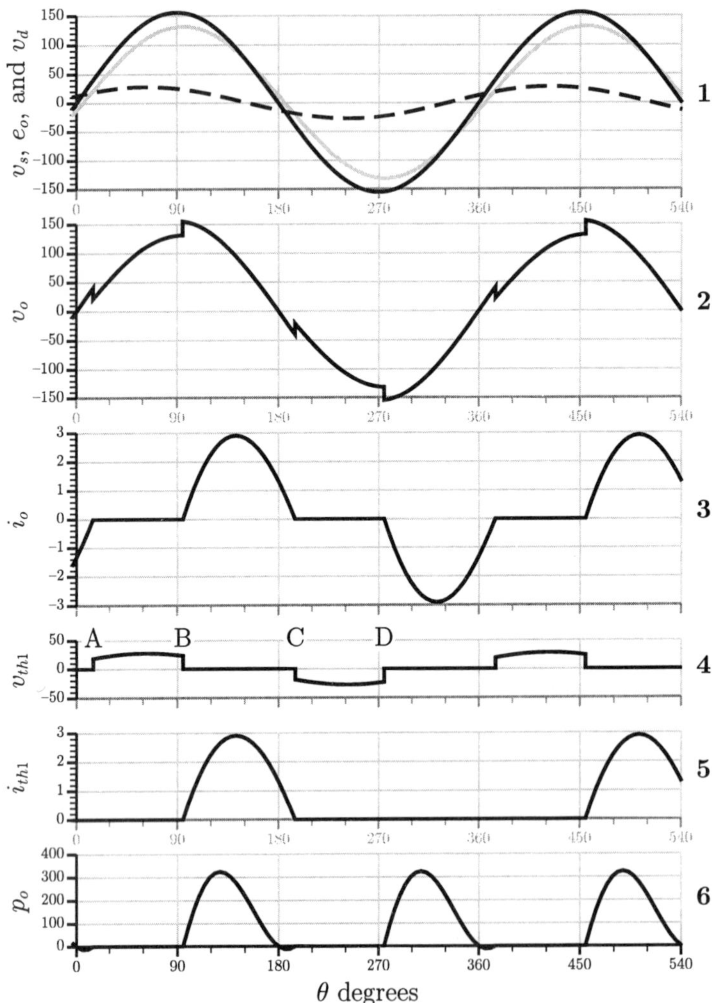

FIG. 13.5. Controller waveforms. The graphs are identified by a number on the right. Graph 1 shows the supply voltage by a thick black line, the load e.m.f. by a gray line, and their difference by a black, dashed line. Graphs 2 and 3 show the output voltage and current. Graphs 4 and 5 show the voltage across thyristor T1 and the current through it. Graph 6 shows the load power. All voltages are in volts, currents are in amperes and power is in watts.

13.9 The r.m.s. output voltage

The r.m.s. value of the output voltage, V_{orms}, is given by

$$V_{orms} = \left\{ \frac{1}{\pi} \left\{ \int_{\alpha}^{\gamma} \left[\sqrt{2}V_s \ \sin(\theta)\right]^2 \mathrm{d}\theta + \int_{\gamma}^{\alpha+\pi} \left[\sqrt{2}E_o \sin(\theta+\epsilon)\right]^2 \mathrm{d}\theta \right\} \right\}^{\frac{1}{2}}$$

i.e.

$$V_{orms} = \left\{ \frac{1}{\pi} \left\{ V_s^2 \left[\beta - \sin(\beta)\cos(\alpha+\gamma)\right] + E_o^2 \left[\pi - \beta + \sin(\beta)\cos(\alpha+\gamma+2\epsilon)\right] \right\} \right\}^{\frac{1}{2}} \tag{13.37}$$

Example 13.5

Determine the r.m.s. value of the output voltage for the example with $E_o = 93$ V, $\epsilon = -5.6°$, and $\alpha = 95°$.

It was found in Example 13.3 that, under these conditions, the extinction angle is 194.92° and the conduction angle is 99.92°. Substituting these values in eqn 13.37 yields the r.m.s. value as

$$V_{orms} = 98.70 \text{ V}$$

****

13.10 The thyristor voltage

Since the small on-state thyristor voltage drop is being neglected, the voltage across a thyristor is zero when either thyristor conducts and is equal to $v_s - e_o$ when both thyristors are blocking. For thyristor T1 this means that for $\alpha + n\pi < \theta < \gamma + n\pi$

$$V_{th1} = 0 \tag{13.38}$$

and for $\gamma + n\pi < \theta < \alpha + (n+1)\pi$

$$V_{th1} = \sqrt{2}V_s \ \sin(\theta) - \sqrt{2}E \ \sin(\theta+\epsilon) \tag{13.39}$$

The waveform of V_{th1} is shown in graph 4 of Fig. 13.5. The voltage is small because the load e.m.f. is almost equal and opposite to the supply voltage. With a passive load the maximum voltage to be withstood by the thyristor is the peak value of the supply voltage, $\sqrt{2}V_s$.

With an active load it is prudent to assume that the load e.m.f. might fall out of synchronism with the supply if the thyristors are off. The maximum voltage to be withstood by a thyristor is then $\sqrt{2}V_s + \sqrt{2}E_o$.

Thus, the thyristors of the example should be capable of withstanding about 310 V on a repetitive basis, which, with a factor of safety of two, would call for thyristors rated at about 650 V.

It is noted from Fig. 13.5 graph 4, that there are four discontinuities in the thyristor voltage waveform. They have been labelled A, B, C, and D. Discontinuity A corresponds to the recovery of T2. Discontinuity B corresponds to turn-on of T1. Discontinuity C corresponds to recovery of T1. Discontinuity D corresponds to turn-on of T2. Only discontinuity A applies both a positive voltage and a positive dv/dt to T1. Premature triggering of T1 at A due to excessive dv/dt can be prevented by use of a suitable snubber, an RLC circuit which reduces the stresses experienced by the thyristors. A single snubber protects both thyristors and the supply line inductance may be sufficient to eliminate the need for a special snubber inductance.

13.11 The thyristor current

Each thyristor carries the load current for half the time as shown for thyristor T1 in Fig. 13.5, graph 5. The r.m.s. value of a thyristor current is therefore $1/\sqrt{2}$ times the r.m.s. load current, i.e.

$$I_{thrms} = \frac{I_{orms}}{\sqrt{2}} \tag{13.40}$$

The mean thyristor current, I_{th}, is given by

$$I_{th} = \frac{1}{2\pi}\int_{\alpha}^{\gamma} i_o \, d\theta$$

i.e.

$$\begin{aligned} I_{th} = & \frac{1}{2\pi}\Big\{I_a[\cos(\alpha+\phi_c) - \cos(\gamma+\phi_c)] \\ & +I_b \tan(\zeta)[\exp\frac{-\alpha}{\tan(\zeta)}) - \exp(\frac{-\gamma}{\tan(\zeta)})]\Big\} \end{aligned} \tag{13.41}$$

Example 13.6

Determine the mean and r.m.s. values of a thyristor current for the example when $E_o = 93$ V, $\epsilon = -5.6°$, and $\alpha = 95°$. Also, for comparison, determine these values when the delay angle has the minimum effective value.

It was found in Example 13.3 that, under these conditions, the extinction angle is 194.92°. Substituting this value in eqn 13.41 yields the mean value as

$$I_{th} = 0.5302 \text{ A}$$

In Example 13.4 it was found that, under these conditions, the r.m.s. value of the output current is 1.565 A so that, from eqn 13.40

$$I_{thrms} = 1.107 \text{ A}$$

The largest value of thyristor current occurs when the delay angle has its minimum effective value so that there is no control of the load current. Under these conditions

$$I_{th} = \frac{I_a}{\pi} = 2.36 \text{ A}$$

and

$$I_{thrms} = \frac{I_a}{2} = 3.71 \text{ A}$$

****

13.12 The output power, volt-amperes, and power factor

During the interval $\alpha < \theta < \gamma$ the power delivered to the load is

$$p_o = v_o\, i_o$$

i.e.

$$p_o = \frac{V_s I_a}{\sqrt{2}}[\cos(\phi_c) - \cos(2\theta + \phi_c)] - \sqrt{2} V_s I_b\ \sin(\theta) \exp(\frac{-\theta}{\tan(\zeta)}) \quad (13.42)$$

The waveform of the output power is shown in graph 6 of Fig. 13.5.

During an off period the output power is zero so that the mean power, P_o, delivered to the load is

$$P_o = \frac{1}{\pi} \int_{\alpha}^{\gamma} p_o \mathrm{d}\theta$$

This integral is given in eqn A.6, and, with the values defined in eqns 13.27a through 13.27g, it reduces to

$$P_o = \frac{V_s I_a}{2\sqrt{2}\pi}\Big[2\beta\cos(\phi_c) - \sin(2\gamma+\phi_c) + \sin(2\alpha+\phi_c)\Big] - $$
$$-\frac{\sqrt{2}V_s I_b}{\pi}\sin(\zeta)\Big[\sin(\gamma+\zeta)\exp(\frac{-\gamma}{\tan(\zeta)}) - \sin(\alpha+\zeta)\exp(\frac{-\alpha}{\tan(\zeta)})\Big] \quad (13.43)$$

The output VA as conventionally measured is

$$VA_o = V_{orms} I_{orms} \quad (13.44)$$

The output power factor as conventionally measured is

$$Pf_o = \frac{P_o}{VA_o} \quad (13.45)$$

Example 13.7

Determine the mean value of the output power, the output volt-amperes, and the power factor as conventionally measured for the example when $E_o = 93$ V, $\epsilon = -5.6°$, and $\alpha = 95°$.

Using eqns 13.43 through 13.45 and the results derived in earlier examples it is found that

$$\text{Output power} = 90.20 \text{ W}$$
$$\text{Output VA} = 154.49 \text{ VA}$$
$$\text{Output power factor} = 0.5839$$

The power factor when uncontrolled, i.e. with the delay angle set to its minimum effective value, is 0.8099 and the reduction in power factor at the delay of 95° is due to two effects, the increased lag of the fundamental component of current and the harmonic content of both the output voltage and the output current waves.

****

13.13 The input power, volt-amperes, and power factor

Since losses in the thyristors are being neglected, the input power equals the output power at all times so that the mean input power, P_1, is

$$P_1 = P_o \quad (13.46)$$

The mean input VA, VA_1, as conventionally measured is

$$VA_1 = V_s I_{1rms} \tag{13.47}$$

and the input power factor as conventionally measured is

$$Pf_1 = \frac{P_1}{VA_1} \tag{13.48}$$

Example 13.8

Determine the values of the mean input power and the input volt-amperes and power factor as conventionally measured when $E_o = 93$ V, $\epsilon = -5.6°$, and $\alpha = 95°$.

From eqns 13.46 through 13.48

$$\begin{aligned} \text{Input power} &= 90.20 \text{ W} \\ \text{Input VA} &= 172.18 \text{ VA} \\ \text{Input power factor} &= 0.5239 \end{aligned}$$

The input power factor is substantially less than the normal uncontrolled 60 Hz power factor of 0.8099 for two reasons. First, the delay introduced by the controller increases the phase lag of the current. Second, the current harmonics introduce additional VA.

****

13.14 Fourier analysis

Since the shape of a negative half wave is identical with that of a positive half wave, the amplitudes of the even harmonics are zero and half range integrals can be employed as described in Section C.5.

13.14.1 *Fourier components of the output voltage*

The output voltage may be expressed either in the Cartesian form

$$v_o(\theta) = \sum_{m=1,\infty,2} V_{ocm} \cos(m\theta) + V_{osm} \sin(m\theta) \tag{13.49}$$

or in the polar form

$$v_o(\theta) = \sum_{m=1,\infty,2} V_m \sin(m\theta + \phi_{vm}) \tag{13.50}$$

The relationships between the amplitudes of the cosine and sine components, V_{ocm} and V_{osm}, and the amplitude and phase of the mth order component are

$$V_{om} = \sqrt{(V_{ocm}^2 + V_{osm}^2)} \tag{13.51}$$

and

$$\phi_{vm} = \arctan(\frac{V_{ocm}}{V_{osm}}) \tag{13.52}$$

From eqns C.8, C.9, and C.10, the amplitudes of the cosine and sine components are

$$\begin{aligned} V_{ocm} &= \frac{2}{\pi}\int_{\alpha}^{\gamma} \sqrt{2}V_s \sin(\theta)\cos(m\theta)\mathrm{d}\theta + \\ &\quad \frac{2}{\pi}\int_{\gamma}^{\alpha+\pi} \sqrt{2}E_o \sin(\theta+\epsilon)\cos(m\theta)\mathrm{d}\theta \end{aligned} \tag{13.53}$$

and

$$\begin{aligned} V_{osm} &= \frac{2}{\pi}\int_{\alpha}^{\gamma} \sqrt{2}V_s \sin(\theta)\sin(m\theta)\mathrm{d}\theta + \\ &\quad \frac{2}{\pi}\int_{\gamma}^{\alpha+\pi} \sqrt{2}E_o \sin(\theta+\epsilon)\sin(m\theta)\mathrm{d}\theta \end{aligned} \tag{13.54}$$

Evaluating the integrals of eqns 13.53 and 13.54 yields

$$\begin{aligned} V_{ocm} &= \frac{\sqrt{2}V_s}{\pi}\left\{\frac{\cos[(m+1)\alpha] - \cos[(m+1)\gamma]}{m+1} - \right. \\ &\quad \left. \frac{\cos[(m-1)\alpha] - \cos[(m-1)\gamma]}{m-1}\right\} - \\ &\quad \frac{\sqrt{2}E_o}{\pi}\left\{\frac{\cos[(m+1)(\alpha+\pi)+\epsilon] - \cos[(m+1)\gamma+\epsilon]}{m+1} - \right. \\ &\quad \left. \frac{\cos[(m-1)(\alpha+\pi)-\epsilon] - \cos[(m-1)\gamma-\epsilon]}{m-1}\right\} \end{aligned} \tag{13.55}$$

and

$$V_{osm} = \frac{\sqrt{2}V_s}{\pi}\left\{\frac{\sin[(m+1)\alpha] - \sin[(m+1)\gamma]}{m+1} - \frac{\sin[(m-1)\alpha] - \sin[(m-1)\gamma]}{m-1}\right\} - \frac{\sqrt{2}E_o}{\pi}\left\{\frac{\sin[(m+1)(\alpha+\pi)+\epsilon] - \sin[(m+1)\gamma+\epsilon]}{m+1} - \frac{\sin[(m-1)(\alpha+\pi)-\epsilon] - \sin[(m-1)\gamma-\epsilon]}{m-1}\right\} \tag{13.56}$$

Since m is odd, both $m+1$ and $m-1$ are even and, as the addition of an even multiple of π to an angle does not change the values of the circular functions of the angle, π can be omitted in the above expressions from the angle $(\alpha+\pi)$. The simplified expressions are then

$$V_{ocm} = \frac{\sqrt{2}V_s}{\pi}\left\{\frac{\cos[(m+1)\alpha] - \cos[(m+1)\gamma]}{m+1} - \frac{\cos[(m-1)\alpha] - \cos[(m-1)\gamma]}{m-1}\right\} - \frac{\sqrt{2}E_o}{\pi}\left\{\frac{\cos[(m+1)(\alpha)+\epsilon] - \cos[(m+1)\gamma+\epsilon]}{m+1} - \frac{\cos[(m-1)(\alpha)-\epsilon] - \cos[(m-1)\gamma-\epsilon]}{m-1}\right\} - \tag{13.57}$$

and

$$V_{osm} = \frac{\sqrt{2}V_s}{\pi}\left\{\frac{\sin[(m+1)\alpha] - \sin[(m+1)\gamma]}{m+1} - \frac{\sin[(m-1)\alpha] - \sin[(m-1)\gamma]}{m-1}\right\} - \frac{\sqrt{2}E_o}{\pi}\left\{\frac{\sin[(m+1)\alpha+\epsilon] - \sin[(m+1)\gamma+\epsilon]}{m+1} - \frac{\sin[(m-1)\alpha-\epsilon] - \sin[(m-1)\gamma-\epsilon]}{m-1}\right\} \tag{13.58}$$

13.14.2 *The fundamental, a special case*

The first order component or fundamental, $m = 1$, is a special case since the denominator term $m - 1$ is then zero. The correct value of the integral is obtained by the application of l'Hôpital's rule and is given in eqns A.40 and A.48.

$$V_{oc1} = \frac{\sqrt{2}V_s}{\pi}\left\{\frac{\cos(2\alpha) - \cos(2\gamma)}{2}\right\} - \frac{\sqrt{2}E_o}{\pi}\left\{\frac{\cos(2\alpha + \epsilon) - \cos(2\gamma + \epsilon)}{2} - (\pi - \beta)\sin(\epsilon)\right\} \tag{13.59}$$

and

$$V_{os1} = \frac{\sqrt{2}V_s}{\pi}\left\{\frac{\sin(2\alpha) - \sin(2\gamma)}{2} + \beta\right\} - \frac{\sqrt{2}E_o}{\pi}\left\{\frac{\sin(2\alpha + \epsilon) - \sin(2\gamma + \epsilon)}{2} - (\pi - \beta)\cos(\epsilon)\right\} \tag{13.60}$$

13.14.3 *Fourier components of the output current*

The Fourier components of the output current can be obtained either by analysis of the current waveform, eqn 13.26, or from the voltage components by a.c. circuit analysis. The latter method will be used.

The load reactance at the supply frequency has been denoted by the symbol X_o, so that the reactance at the frequency of the mth order Fourier component is mX_o and the impedance to this component is

$$Z_{om}/\underline{\zeta_{om}} = R_o + jmX_o \tag{13.61}$$

so that

$$Z_{om} = \sqrt{R_o^2 + m^2X_o^2} \tag{13.62}$$

and

$$\zeta_{om} = \arctan(mX_o/R_o) \tag{13.63}$$

The amplitude, I_m, of the mth current component is then, provided $m \neq 1$,

$$I_{om} = \frac{V_{om}}{Z_{om}} \tag{13.64}$$

and its phase, ϕ_{cm}, is

$$\phi_{cm} = \phi_{vm} - \zeta_{om} \tag{13.65}$$

Again $m = 1$ constitutes a special case since the load e.m.f. is sinusoidal at the supply frequency. This opposes the fundamental component of the driving voltage so that, in phasor terms,

$$\mathbf{I}_{o1}\underline{/\phi_{c1}} = \frac{\mathbf{V}_{o1}\underline{/\phi_{v1}} - \mathbf{E}_o\underline{/\epsilon}}{\mathbf{Z}_1} \tag{13.66}$$

13.14.4 *Power of the Fourier components*

The mean output power, P_{om}, associated with the mth order Fourier component is

$$P_{om} = \frac{1}{2}V_{om}I_{om}\cos(\phi_{vm} - \phi_{cm}) \tag{13.67}$$

Since losses in the thyristors are being neglected, this is also the input power associated with the mth Fourier component, P_{1m}, i.e.

$$P_{1m} = P_{om} \tag{13.68}$$

Example 13.9

Determine the frequencies, amplitudes, and phase angles of the first six nonzero Fourier components of the output voltage and output current for the example and also determine the mean power associated with each of these components when $E = 93$ V, $\epsilon = -5.8°$, and $\alpha = 95°$.

Equations 13.59 and 13.60 are used to determine the fundamental component of the output voltage and eqns 13.57 and 13.58 to determine the remaining components. Equations 13.61 through 13.63 are then used to determine the load impedance at the Fourier component frequencies, eqn 13.66 for the fundamental component of current, and eqns 13.64 and 13.65 for the remaining components. Equation 13.67 then gives the mean power associated with each component.

The data for all six components are listed in Table 13.2. It will be noted that the voltage and current series converge rapidly and that the power series converges extremely rapidly. The frequency spectrum of the output current and the waveform derived from this spectrum are shown in Fig. 13.6. This wave is almost indistinguishable from the time domain solution shown in Fig. 13.5, graph 3, another indication of the rapid convergence of the Fourier series.

The total input power associated with the six components of Table 13.2 is 90.20 W, the same as the value found by the time domain solution

Table 13.2 *Fourier components of the output voltage and current for Example 13.9 and the mean powers associated with these components*

Order	Freq.	Voltage		Current		Power
		Amplitude	Phase	Amplitude	Phase	
	Hz	V	degrees	A	degrees	W
1	60	138.66	-4.97	1.966	-53.85	89.34
3	180	10.18	99.77	0.997	19.24	0.84
5	300	1.18	-157.83	0.070	-242.12	0.00
7	420	3.94	63.08	0.167	-22.83	0.02
9	540	0.43	-96.55	0.014	-183.37	0.00
11	660	2.40	25.89	0.065	-61.58	0.00

in Example 13.7. Of the total, 89.34 W, 99.04%, is associated with the fundamental, 0.84 W, 0.93%, is associated with the third harmonic and the remainder, 0.02 W, 0.02%, is contributed by all other harmonics.

The smallness of the harmonic content is surprising in view of the wide divergence of the current waveform from sinusoidal. However it is typical of a.c. controllers, a fact which will also be noted for the three phase controller discussed in Chapter 14.

****

13.14.5 *The r.m.s. values*

Although analytic expressions have been found for the r.m.s. values of the output current and voltage, eqns 13.34 and 13.37, the Fourier series provide a simple alternative method for deriving these values as described in Section C.7. While these expressions are not in closed form, the series converge so rapidly that a result of acceptable accuracy is obtained by summing a few terms

$$V_{orms} = \left\{\frac{1}{2}\sum_{m} V_{om}^2\right\}^{\frac{1}{2}} \tag{13.69}$$

$$I_{orms} = \left\{\frac{1}{2}\sum_{m} I_{om}^2\right\}^{\frac{1}{2}} \tag{13.70}$$

Example 13.10

Compare the approximate r.m.s. values of the output voltage and current determined from the six terms of the Fourier series derived in Example 13.9 with the values derived analytically in Examples 13.4 and 13.5.

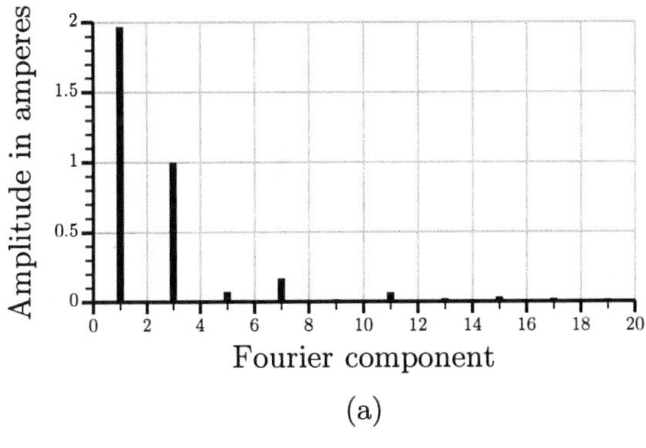

(a)

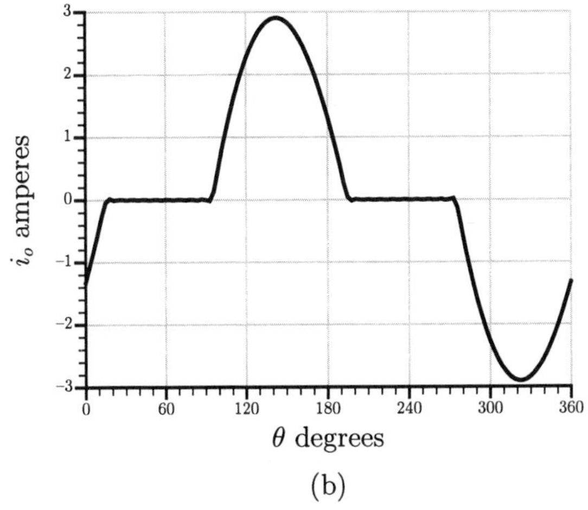

(b)

FIG. 13.6. The frequency spectrum of the output current and the waveform derived from that spectrum.

The true r.m.s. value of the output current found in Example 13.4 is 1.565 A. The r.m.s. value of the six Fourier components of Table 13.2 is

$$\sqrt{(1.966^2 + 0.997^2 + 0.070^2 + 0.167^2 + 0.014^2 + 0.065^2)/2} = 1.558 \text{ A}$$

The r.m.s. value of the fundamental is, from the first line of Table 13.2 $1.966/\sqrt{2} = 1.390$ A. Thus, the r.m.s. value of the harmonics is $\sqrt{1.565^2 - 1.390^2} = 0.719$ A. The r.m.s. value of the combination of the third through the eleventh harmonics is

$$\sqrt{(0.997^2 + 0.070^2 + 0.167^2 + 0.014^2 + 0.065^2)/2} = 0.718 \text{ A}$$

The r.m.s. value of the output voltage, as found in Example 13.5, is 98.40 V. The r.m.s. value of the six Fourier components of Table 13.2 is

$$\sqrt{(138.66^2 + 10.18^2 + 1.18^2 + 3.94^2 + 0.43^2 + 2.40^2)/2} = 98.37 \text{ V}$$

The r.m.s. value of the fundamental is, from the first line of Table 13.2, $138.66/\sqrt{2} = 98.05$ V. Thus, the r.m.s. value of the harmonics is $\sqrt{98.40^2 - 98.05^2} = 8.29$ V. The r.m.s. value of the combination of the third through the eleventh harmonics is

$$\sqrt{(10.18^2 + 1.18^2 + 3.94^2 + 0.43^2 + 2.40^2)/2} = 7.95 \text{ V}$$

The small differences between the values derived from the truncated Fourier series and the true values are another manifestation of the very small contribution of harmonics beyond the eleventh which was noted in the previous example.

****

13.15 The control characteristics

The designer of a control system using an a.c. voltage controller needs to know the shape of the control characteristic, the relationship between the quantity being controlled and the control variable. The parameters involved depend on the circumstances. The controlled variable might be the delay angle or some function of the delay angle such as its cosine, as is generally used in a controlled rectifier. The output parameter might be the output power in the case of a heating load or the fundamental component of the output current in the case of a motor controller or some other variable.

Figure 13.7 shows, for the example, the relationships between the r.m.s. values of the fundamental components of output voltage and current and the fundamental component of the mean output power and the delay angle. The delay angle range is from the minimum effective value of 35.92° to the maximum value of 152.51°. The relationships are typical of all a.c. controllers and exhibit considerable nonlinearity especially as the delay approaches its maximum value.

13.16 Gating considerations

Depending on the load e.m.f., the uncontrolled current can have any phase relative to the supply voltage. This would present a problem to the designer of a general purpose control scheme since the datum from which control would begin could be anywhere within a cycle. Fortunately the real problem is far more restricted. Either the load will be passive and essentially resistive, such as a lamp or oven, or it will be an inductor or induction motor in the motoring mode and therefore with the current lagging between

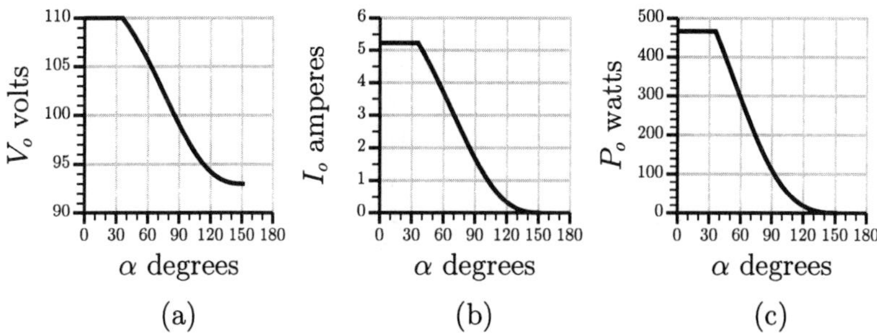

FIG. 13.7. Control characteristics for the example. The graphs show as functions of delay angle in (a) the r.m.s. value of the fundamental component of the output voltage, in (b) the r.m.s. value of the fundamental component of the output current, and in (c) the fundamental component of the output power. The effective range of delay angle is 35.92°–152.51°.

zero and 90° relative to the supply voltage. This latter situation will be considered as more general and incorporating all the salient requirements.

For the example operating in the natural uncontrolled state, the current would lag the supply voltage by 35.92° so that this is the minimum effective delay angle. For the example the net driving voltage, i.e. v_d, leads the supply voltage by 27.49° thus setting the maximum effective delay angle at $180° - 27.49° = 152.51°$. Thus, the effective delay angle range is $35.92° < \theta < 152.51°$. It is not practicable to limit the range of α so precisely and it therefore is necessary to apply wide firing pulses to the thyristors so that, even when the actual delay angle is less than the minimum, the thyristor will still fire properly. As an example, if the minimum delay angle were set at zero and the firing pulses were 90° wide, this would cover all uncontrolled current lag angles from zero to 90°. A controller of this type is assumed for Fig. 13.7 so that when the delay angle is reduced below the minimum value the system continues to operate at maximum voltage, current, and power.

14

THREE PHASE AC VOLTAGE CONTROLLERS

14.1 Introduction

Single phase a.c. controllers are limited to low power applications, typically below 1 kW, such as lamp dimmers and power factor control of fractional horsepower motors. Loads of substantial power, say above 10 kW will be three phase and will therefore require three phase controllers. With rare exceptions the analysis of a three phase load is significantly more complex than for a single phase load because the three phases interact. An essentially resistive load such as an oven may have the three phases wye connected. If the neutral point is isolated as is usually the case, this imposes the constraint that the sum of the phase currents must at all times be zero. An electromagnetic load such as an induction motor adds the further complication that the phases are coupled via their common magnetic field. Add to these load considerations the wide variety of possible controller configurations and the problem of analysis takes on a breadth and a complexity inappropriate for comprehensive treatment here. The problem will therefore be reviewed in the context of a general purpose controller with six thyristors, a back to back connected pair in each line. Such a controller can be manufactured as a separate entity to suit a wide variety of loads and can be connected between the supply and the load without any modification of the latter. Most other controller arrangements, whatever may be their other merits, have the very considerable demerit of requiring disturbance of the internal wiring of the load.

The load will be assumed to be balanced three phase, inductive and active with e.m.fs of sinusoidal waveform at the supply frequency and with isolated neutral. Such a load is reasonably representative of an induction motor and can represent a three phase inductor or oven by setting the e.m.fs to zero. This situation, which is typical of the majority of applications, will serve to illustrate all major points of the analysis.

14.2 Definition of the symbols

The situation is illustrated by Fig. 14.1 where the nomenclature is defined. The potential of the supply neutral is taken as the datum and all single suffix potentials are with reference to it. The supply lines are referred to as

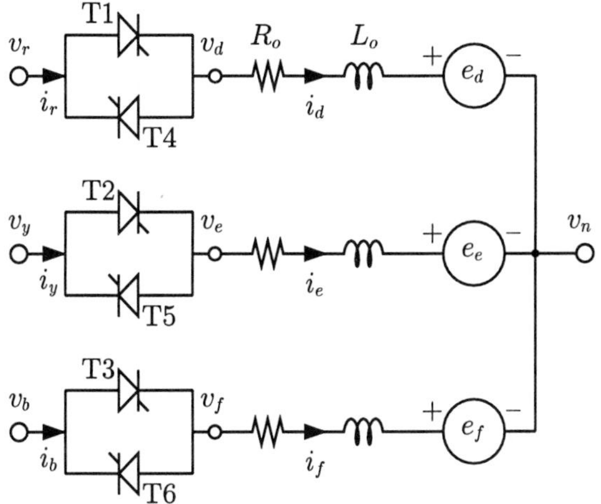

FIG. 14.1. A three phase active load connected to a three phase supply via a three phase, six thyristor controller.

red, yellow, and blue with corresponding lower case subscripts so that the potentials of the supply terminals are v_r, v_y and v_b. These potentials form a balanced three phase set of sinusoidal waveform and the time origin is chosen so that

$$v_r = \sqrt{2}V_s \ \sin(\theta) \tag{14.1}$$

$$v_y = \sqrt{2}V_s \ \sin(\theta - \frac{2\pi}{3}) \tag{14.2}$$

$$v_b = \sqrt{2}V_s \ \sin(\theta + \frac{2\pi}{3}) \tag{14.3}$$

where V_s is the r.m.s. value of the supply line to neutral voltage, $\theta = \omega_s t$, and ω_s is the radian frequency of the supply.

The supply line currents are i_r, i_y, and i_b and although their waveforms are not known, it is known, from the symmetry of the system, that they are similar except for the appropriate $2\pi/3$ phase shift so that if

$$i_r = f(\theta) \tag{14.4}$$

then

$$i_y = f(\theta - \frac{2\pi}{3}) \tag{14.5}$$

and

$$i_b = f(\theta + \frac{2\pi}{3}) \tag{14.6}$$

The supply impedance is assumed to be negligible.

The load phases are identified as D, E, and F so that the potentials of the load terminals are v_d, v_e, and v_f. Again, from the symmetry of the system, it can be said that if

$$v_d = g(\theta) \tag{14.7}$$

then

$$v_e = g(\theta - \frac{2\pi}{3}) \tag{14.8}$$

and

$$v_f = g(\theta + \frac{2\pi}{3}) \tag{14.9}$$

The load neutral is isolated and has a potential v_n relative to the supply neutral.

The load e.m.fs form a balanced three phase set of sinusoidal waveform, r.m.s. value E_o, having the same phase sequence as the supply and with a phase lead of ϵ so that

$$e_d = \sqrt{2}E_o \ \sin(\theta + \epsilon) \tag{14.10}$$

$$e_e = \sqrt{2}E_o \ \sin(\theta + \epsilon - \frac{2\pi}{3}) \tag{14.11}$$

$$e_f = \sqrt{2}E_o \ \sin(\theta + \epsilon + \frac{2\pi}{3}) \tag{14.12}$$

The currents in the load phases are i_d, i_e, and i_f and again, from the system symmetry, if

$$i_d = h(\theta) \tag{14.13}$$

then

$$i_e = h(\theta - \frac{2\pi}{3}) \tag{14.14}$$

and

$$i_f = h(\theta + \frac{2\pi}{3}) \tag{14.15}$$

The forward thyristors, T1, T2, and T3, are fired sequentially at 120° intervals. The reverse thyristors, T4, T5, and T6, are fired 180° after the corresponding forward thyristors, i.e. T4 is fired half a cycle after T1, T5 half a cycle after T2, and T6 half a cycle after T3.

Modeling a conducting thyristor by a small, constant voltage drop of the order of 2 V introduces a small d.c. term into the current expressions. This additional complication to an already complex problem is avoided by neglecting the voltage drop in a conducting thyristor so that if either T1 or T4 is conducting, $v_d = v_r$ etc. The error thus introduced is minimal and, if its consideration is essential for a given problem, the expressions for the currents and their functions are given in Appendix A.

The same concepts of delay angle, extinction angle, and conduction angle which were introduced in Chapter 13 are used. They will be measured from the time origin so that T1 conducts from α to γ, T2 conducts from $\alpha + 2\pi/3$ to $\gamma + 2\pi/3$, etc.

14.3 The three phase load

The phases of a normal three phase load are identical and are mutually coupled through a common magnetic field. Each phase has a resistance R_o, a self inductance L_{sp}, and a mutual inductance with the other phases. The symmetry of the system ensures that the value of this mutual inductance between phases, whose symbol will be M_{bp}, is the same between any pair of phases. Generally this mutual coupling is negative with a magnitude somewhat less than half the value of the self inductance.

Kirchhoff's voltage law applied to the three phases yields

$$v_r = e_d + R_o i_d + L_{sp}\frac{di_d}{dt} + M_{bp}\frac{di_e}{dt} + M_{bp}\frac{di_f}{dt} + v_n \tag{14.16}$$

$$v_y = e_e + R_o i_e + L_{sp}\frac{di_e}{dt} + M_{bp}\frac{di_f}{dt} + M_{bp}\frac{di_d}{dt} + v_n \tag{14.17}$$

$$v_b = e_f + R_o i_f + L_{sp}\frac{di_f}{dt} + M_{bp}\frac{di_d}{dt} + M_{bp}\frac{di_e}{dt} + v_n \tag{14.18}$$

When the load neutral is isolated, as is normally the case, the instantaneous sum of the three phase currents is zero.

$$i_d + i_e + i_f = 0 \tag{14.19}$$

so that

$$\frac{di_d}{dt} + \frac{di_e}{dt} + \frac{di_f}{dt} = 0 \tag{14.20}$$

Using this result, eqns 14.16 through 14.18 can be rewritten as

$$v_r = e_d + R_o i_d + L_o\frac{di_d}{dt} + v_n \tag{14.21}$$

$$v_y = e_e + R_o i_e + L_o\frac{di_e}{dt} + v_n \tag{14.22}$$

$$v_b = e_f + R_o i_f + L_o\frac{di_f}{dt} + v_n \tag{14.23}$$

where

$$L_o = L_{sp} - M_{bp} \tag{14.24}$$

Thus, any balanced three phase load with isolated neutral, the type shown in Fig. 14.1, can be analyzed as three separate single phase loads whose inductance, L_o, is the difference between the self inductance of a phase, L_{sp}, and the mutual inductance between phases, M_{bp}. This result will be applied during the remainder of this chapter, it being understood that, for a mutually coupled load, the inductance L_o is the difference between the self inductance of a phase and the mutual inductance between phases.

It will also be convenient to work in terms of the angle $\theta = \omega_s t$ rather than time and to use the symbol X_o, a reactance at the supply frequency, instead of the product $\omega_s L_o$.

In addition to this simplification, it will be useful to denote the difference between the voltage applied to a phase and the e.m.f. generated within that phase by the symbol v_δ, $v_{\delta r}$ representing the difference $v_r - e_d$, etc. Thus, there are three difference voltages

$$v_{\delta r} = v_r - e_d = \sqrt{2}V_\delta \sin(\theta + \delta) \tag{14.25}$$

$$v_{\delta y} = v_y - e_e = \sqrt{2}V_\delta \sin(\theta + \delta - \frac{2\pi}{3}) \tag{14.26}$$

$$v_{\delta b} = v_b - e_f = \sqrt{2}V_\delta \sin(\theta + \delta + \frac{2\pi}{3}) \tag{14.27}$$

With these substitutions, the form in which the three phase equations will be used is

$$v_r - e_d = v_{\delta r} = R_o i_d + X_o \frac{\mathrm{d}i_d}{\mathrm{d}\theta} + v_n \tag{14.28}$$

$$v_y - e_e = v_{\delta y} = R_o i_e + X_o \frac{\mathrm{d}i_e}{\mathrm{d}\theta} + v_n \tag{14.29}$$

$$v_b - e_f = v_{\delta b} = R_o i_f + X_o \frac{\mathrm{d}i_f}{\mathrm{d}\theta} + v_n \tag{14.30}$$

Summing these equations and using eqns 14.19 and 14.20 yields

$$v_n = \frac{v_r + v_y + v_b}{3} - \frac{e_d + e_e + e_f}{3}$$

which, since the voltages and e.m.fs form balanced three phase sets, shows that, when all three phases are connected to the a.c. system,

$$v_n = 0 \tag{14.31}$$

The application of these equations is illustrated by Example 14.1.

Example 14.1

The flow of power from a three phase supply to an active load is regulated by a three phase solid state controller whose circuit is illustrated in

Fig. 14.1. The line to line voltage of the supply is 440 V r.m.s., the supply frequency is 60 Hz and the supply impedance is negligible. The load is of balanced three phase construction with the three phases wye connected and the neutral isolated. Each load phase has a resistance of 0.07 Ω, a self inductance of 0.22 mH, and the mutual inductance between phases is −0.10 mH. The load e.m.fs are of sinusoidal waveform, have an r.m.s. value of 232 V, and lag the corresponding line to neutral supply voltage by 3.0°.

Determine the load phase currents when the system is operating uncontrolled, i.e. with the delay angle set at its minimum effective value so that full voltage is applied to the load throughout a cycle.

Note that this problem is used throughout this chapter and the results derived here are used in subsequent examples.

Since the delay angle is set to its minimum effective value, the voltages applied to the load terminals are the supply voltages and this is a problem in steady state a.c. circuit theory. The first task is the determination of the equivalent load impedance. The equivalent inductance is, from eqn 14.24,

$$L_o = 0.22 - (-0.10) = 0.32 \text{ mH}$$

The equivalent load reactance is

$$X_o = 377.0 \times 0.00032 = 0.1206 \ \Omega \text{ at } 60 \text{ Hz}$$

The equivalent load impedance is

$$\mathbf{Z}_o = 0.07 + \text{j}0.1206 = 0.1395 \ \underline{/\ 59.88^\circ}$$

The r.m.s. value of the line to neutral supply voltage is

$$V_s = 440 \div \sqrt{3} = 254.0 \text{ V}$$

The difference voltage for the red phase is

$$v_{\delta r} = \sqrt{2} \times 254.0 \sin(\theta) - \sqrt{2} \times 232 \ \sin(\theta - 0.05236)$$

i.e.

$$v_{\delta r} = \sqrt{2} \times 25.44 \sin(\theta + 0.4976)$$

where the lag of the load e.m.f. of 3.0° has been converted to 0.05236 rad and the lead of the difference voltage of 0.4976 rad when converted to degrees becomes 28.51°. The current in the red phase is this difference voltage divided by the phase impedance and has an r.m.s. value of

$$I_o = 25.44 \div 0.1395 = 182.4 \text{ A}$$

and a phase lead of

$$\phi_c = 28.51 - 59.88 = -31.36° = -0.5474 \text{ rad}$$

The power factor of the load is

$$Pf_o = \cos(-0.5474) = 0.8539$$

The power delivered to the load is

$$P_o = \sqrt{3} \times 440 \times 182.4 \times 0.8539 \div 1000 = 118.6 \text{ kW}$$

The load volt-amperes is

$$VA_o = \sqrt{3} \times 440 \times 182.4 \div 1000 = 139.0 \text{ kVA}$$

****

14.4 The operating modes

The conduction periods of all six thyristors are shown in Figs 14.2 and 14.3 by horizontal gray shaded boxes, the horizontal scale being in degrees and the six vertical levels applying to the six thyristors being as listed to the left of the vertical axis. Figure 14.2 has been drawn for a delay angle of 70° and an extinction angle of 209.6° for a conduction angle of 139.6°, values appropriate for one of the examples used in this chapter. At all times current is flowing between the supply and the load and there are periods when three thyristors conduct with all three phases connected to the supply and periods when two thyristors conduct so that only two phases are connected to the supply. For the case shown for $70° < \theta < 89.6°$ thyristors T1, T3, and T5 conduct and for $89.6° < \theta < 130°$ only T1 and T5 conduct. This mode of operation is called the 3-2 mode.

As the delay angle is increased and the extinction angle consequently diminishes, the overlap between conducting thyristors diminishes correspondingly and eventually becomes zero. Looking specifically at the conduction overlap between T3 and T1 which is the same as between T5 and T6, T1 and T2, T6 and T4, T2 and T3, and T4 and T5, the overlap starts when T1 is fired at $\theta = \alpha$ and ends when T3 is extinguished at $\theta = \gamma - 120°$. Thus,

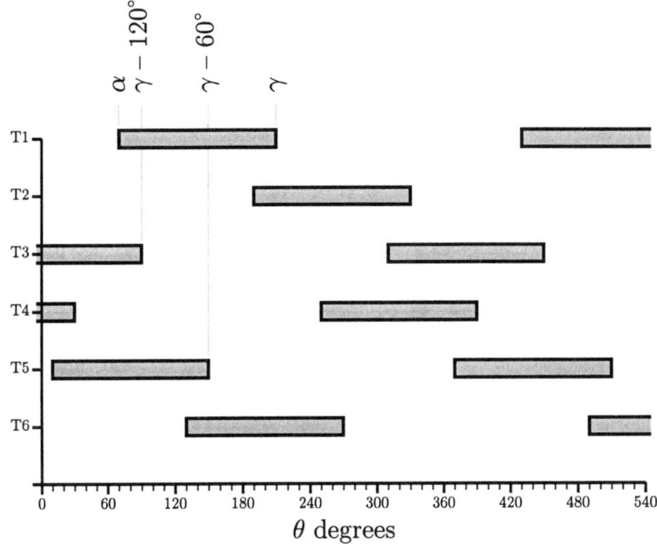

FIG. 14.2. Conduction periods of the thyristors when the system is in the 3-2 mode. The case shown has a delay angle of 70° and an extinction angle of 209.6°.

the overlap is $\gamma - 120° - \alpha = \beta - 120°$. This overlap becomes zero when the conduction angle is reduced to 120° by increasing the delay angle.

Figure 14.3 shows the thyristor conduction periods when the delay angle is 100° and the extinction angle is 200.2° for a conduction angle of 100.2°. There are now either two thyristors conducting or one conducting. However, in view of the isolated neutral, it is impossible for only one thyristor to conduct so that this is in reality the 2-0 mode rather than the 2–1 mode.

In the case shown in Fig. 14.3, T1 and T5 would conduct from α to $\gamma - 60°$ at which point the current becomes zero, T5 recovers and blocks further current flow until T6 is fired at $\alpha + 60°$. Clearly T1 must be in the conducting mode when this occurs, a condition which can be ensured by appropriate gating, a matter already discussed in Section 13.16.

Assuming a suitable gating procedure, the analysis is commenced in the 3-2 mode with small delay angles. The end of the 3-2 mode is then determined by finding when the conduction angle has fallen to 120° after which the ensuing 2-0 mode is analyzed.

Examination of Figs 14.2 and 14.3 shows that a supply cycle can be divided into six, 60° segments during which the current waveforms are identical except for an appropriate phase shift. For example, the segment from α to $\alpha + 60°$ sees T1, T3, and T5 conducting currents

$$i_{t1}(\theta) = f_1(\theta) \tag{14.32}$$

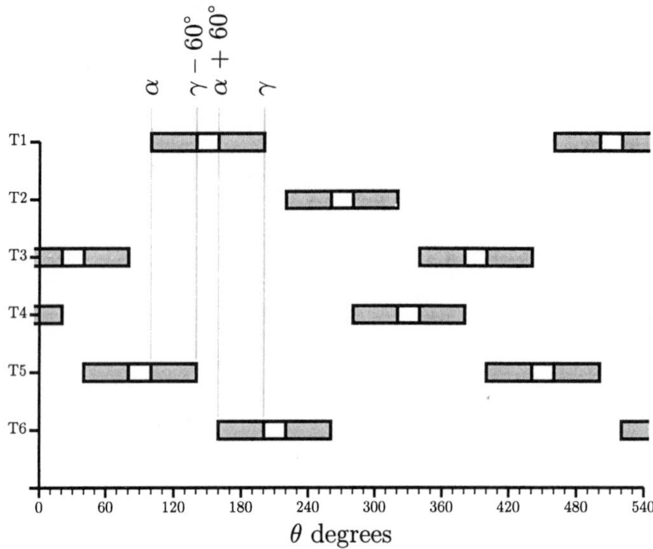

FIG. 14.3. Conduction periods of the thyristors when the system is in the 2-0 mode. The case shown has a delay angle of 100° and an extinction angle of 200.2°.

$$i_{t3}(\theta) = f_3(\theta) \tag{14.33}$$

$$i_{t5}(\theta) = f_5(\theta) \tag{14.34}$$

For the next 60° segment, T6 plays the role of T1, T5 plays the role of T3, and T1 plays the role of T5. Thus, for this segment

$$i_{t6}(\theta) = f_1(\theta - \frac{\pi}{3}) \tag{14.35}$$

$$i_{t5}(\theta) = f_3(\theta - \frac{\pi}{3}) \tag{14.36}$$

$$i_{t1}(\theta) = f_5(\theta - \frac{\pi}{3}) \tag{14.37}$$

Similar arguments allow the currents during any 60° segment to be expressed in terms of the currents during the first segment so that the solution for one 60° segment only need be obtained and the whole problem is solved.

14.5 The 3-2 mode

Consider the segment from α to $\alpha + 60°$. For $\alpha < \theta < \gamma - 120°$, T1, T3, and T5 are conducting and the circuit configuration is as shown in Fig. 14.4.

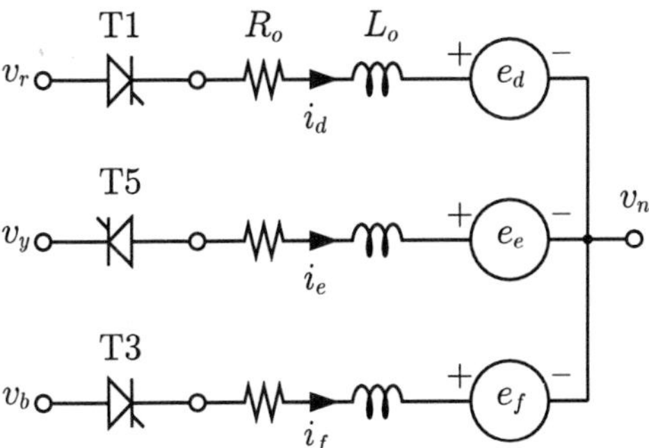

FIG. 14.4. The circuit configuration when the three thyristors T1, T3, and T5 are conducting.

Applying Kirchhoff's voltage law to the loops R–N–Y and B–N–Y and Kirchhoff's current law to the load neutral yields

$$v_r - v_y = e_d + R_o i_d + X_o \frac{\mathrm{d}i_d}{\mathrm{d}\theta} - e_e - R_o i_e - X_o \frac{\mathrm{d}i_e}{\mathrm{d}\theta} \tag{14.38}$$

$$v_b - v_y = e_f + R_o i_f + X_o \frac{\mathrm{d}i_f}{\mathrm{d}\theta} - e_e - R_o i_e - X_o \frac{\mathrm{d}i_e}{\mathrm{d}\theta} \tag{14.39}$$

and

$$i_d + i_e + i_f = 0 \tag{14.40}$$

where the equations are written in terms of the angle $\theta = \omega_s t$ as being more convenient than t.

From these three equations by successive elimination and introducing the difference voltages, $v_{\delta r} = v_r - e_d$, etc. (see eqns 14.25 through 14.30) it is found that

$$v_{\delta r} = R_o i_d + X_o \frac{\mathrm{d}i_d}{\mathrm{d}\theta} \tag{14.41}$$

$$v_{\delta y} = R_o i_e + X_o \frac{\mathrm{d}i_e}{\mathrm{d}\theta} \tag{14.42}$$

$$v_{\delta b} = R_o i_f + X_o \frac{\mathrm{d}i_f}{\mathrm{d}\theta} \tag{14.43}$$

$$v_n = 0 \tag{14.44}$$

where

$$V_\delta = \sqrt{[V_s - E_o\ \cos(\epsilon)]^2 + E_o \sin(\epsilon)^2} \tag{14.45}$$

and

$$\delta = \arctan(\frac{-E_o\ \sin(\epsilon)}{V_s - E_o\ \cos(\epsilon)}) \tag{14.46}$$

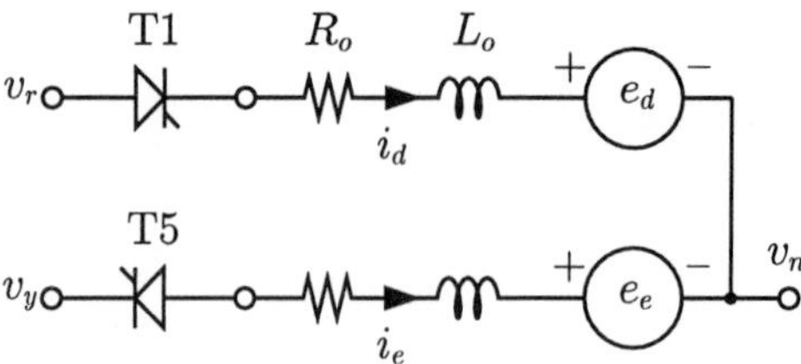

FIG. 14.5. The circuit configuration when the two thyristors T1 and T5 are conducting.

The solution to eqns 14.41 through 14.43 is

$$i_d = \sqrt{2}I_s \ \sin(\theta + \phi_c) + I_{bd} \ \exp(-\frac{\theta}{\tan(\zeta_o)}) \tag{14.47}$$

$$i_e = \sqrt{2}I_s \ \sin(\theta + \phi_c - \frac{2\pi}{3}) + I_{be} \ \exp(-\frac{\theta}{\tan(\zeta_o)}) \tag{14.48}$$

$$i_f = \sqrt{2}I_s \ \sin(\theta + \phi_c + \frac{2\pi}{3}) + I_{bf} \ \exp(-\frac{\theta}{\tan(\zeta_o)}) \tag{14.49}$$

where I_s is the r.m.s. value of the uncontrolled a.c. line current, i.e. the value when α has its minimum effective value, and is given by

$$I_s = \frac{V_\delta}{Z_o} \tag{14.50}$$

$$Z_o = \sqrt{{R_o}^2 + {X_o}^2} \tag{14.51}$$

$$\zeta_o = \arctan(\frac{X_o}{R_o}) \tag{14.52}$$

$$\phi_c = \delta - \zeta_o \tag{14.53}$$

The values of I_{bd}, I_{be}, and I_{bf} depend upon the boundary conditions and, in order that eqn 14.40 be satisfied

$$I_{bf} = -(I_{bd} + I_{be}) \tag{14.54}$$

For the second part of the 60° segment from $\gamma - 120°$ to $\alpha + 60°$, when T1 and T5 conduct, i_f has become zero at $\gamma - 120°$, the extinction angle for T3, and the circuit configuration becomes that of Fig. 14.5.

It is evident that

$$i_e = -i_d \tag{14.55}$$

so that the loop equation is

$$\frac{\sqrt{3}}{2}\sqrt{2}V_\delta \sin(\theta + \delta + \frac{\pi}{6}) = R_o i_d + X_o \frac{\mathrm{d}i_d}{\mathrm{d}\theta} \tag{14.56}$$

Further

$$v_n = v_r - e_d - R_o i_d - X_o \frac{di_d}{d\theta} \tag{14.57}$$

or alternatively

$$v_n = v_y - e_e + R_o i_d + X_o \frac{di_d}{d\theta} \tag{14.58}$$

Adding these two equations yields

$$v_n = \frac{v_r + v_y}{2} - \frac{e_d + e_e}{2} \tag{14.59}$$

i.e.

$$v_n = \frac{V_\delta}{\sqrt{2}} \sin(\theta + \delta - \frac{\pi}{3}) \tag{14.60}$$

The potential v_f during the nonconducting period is obtained by summing the neutral voltage and the phase e.m.f.

$$v_f = e_f + v_n$$

which, using the expression for v_n of eqn 14.59, becomes

$$v_f = \frac{v_r + v_y}{2} + e_f - \frac{e_d + e_e}{2}$$

By using the fact that these are three phase voltages (eqns 14.1 through 14.12) this equation becomes

$$v_f = \frac{3}{\sqrt{2}} E_o \sin(\theta + \epsilon + \frac{2\pi}{3}) - \frac{V_s}{\sqrt{2}} \sin(\theta + \frac{2\pi}{3}) \tag{14.61}$$

The solution to eqn 14.56 yields

$$i_d = \frac{\sqrt{3}}{2} \sqrt{2} I_s \sin(\theta + \phi_c + \frac{\pi}{6}) + I_{cd} \exp(-\frac{\theta}{\tan(\zeta_o)}) \tag{14.62}$$

where I_{cd} depends on the boundary conditions.

The two thyristor stage ends at $\theta = \alpha + 60°$ when T6 is fired.

14.5.1 *Current continuity conditions*

Since all the circuits are inductive, the three currents i_d, i_e, and i_f are continuous throughout a cycle. This continuity condition is applied at the three boundaries $\theta = \alpha$, $\theta = \gamma - 120°$ and $\theta = \alpha + 60°$ to determine the values of I_{bd}, I_{be} and I_{cd}.

At the first boundary when $\theta = \alpha$, $i_d = 0$, and $i_e = -i_f$. The second of these conditions is automatically satisfied by eqn 14.40. The first condition gives

$$I_{bd} = -\frac{\sqrt{2}I_s \ \sin(\alpha + \phi_c)}{E_{x1}} \tag{14.63}$$

where

$$E_{x1} = \exp(-\frac{\alpha}{\tan(\zeta_o)}) \tag{14.64}$$

a pair of equations which fully define I_{bd}.

At the second boundary when $\theta = \gamma - 120°$, $i_f = 0$, and $i_e = -i_d$. Again the second condition is automatically satisfied by eqn 14.40. The first condition yields

$$I_{bf} = -\frac{\sqrt{2}I_s \sin(\gamma + \phi_c)}{E_{x2}} \tag{14.65}$$

where

$$E_{x2} = \exp(-\frac{\gamma - \frac{2\pi}{3}}{\tan(\zeta_o)}) \tag{14.66}$$

Moving now to the second part of the segment when only T1 and T5 conduct, at $\theta = \gamma - 120°$, i_d defined by eqn 14.62 is equal to i_d defined by eqn 14.47, i.e.

$$\sqrt{2}I_s \ \sin(\gamma + \phi_c - \frac{2\pi}{3}) + I_{bd}E_{x2} = \frac{\sqrt{3}}{2}\sqrt{2}I_s \ \sin(\gamma + \phi_c - \frac{\pi}{2}) + I_{cd} \ E_{x2}$$

which, after some manipulation, reduces to

$$I_{cd} = I_{bd} - \frac{\sqrt{2}I_s \sin(\gamma + \phi_c)}{2E_{x2}} \tag{14.67}$$

Thus, knowing I_{bd} and γ, this equation defines I_{cd}.

Finally, at the third boundary, $\theta = \alpha + 60°$, at which point i_d is given by eqn 14.62 as

$$i_d(\alpha + \frac{\pi}{3}) = \frac{\sqrt{3}}{2}\sqrt{2}I_s \ \sin(\alpha + \phi_c + \frac{\pi}{2}) + I_{cd} \ \exp(-\frac{\alpha + \frac{\pi}{3}}{\tan(\zeta_o)})$$

Replacing $\sin(\alpha + \phi_c + \frac{\pi}{2})$ by $\cos(\alpha + \phi_c)$ and writing

$$E_{x3} = \exp(-\frac{\alpha + \frac{\pi}{3}}{\tan(\zeta_o)}) \tag{14.68}$$

so that

$$i_d(\alpha + \frac{\pi}{3}) = \frac{\sqrt{3}}{2}\sqrt{2}I_s \ \cos(\alpha + \phi_c) + I_{cd} \ E_{x3} \tag{14.69}$$

14.5.2 *Correlation of boundary conditions*

Since successive 60° segments are identical except for a cyclic change in thyristors, the value of i_d at $\theta = \alpha + 60°$, eqn 14.69, must equal the value of $-i_e$ at α. Thus, subtracting eqn 14.48 with $\theta = \alpha$ from eqn 14.69 yields

$$\frac{\sqrt{3}}{2}\sqrt{2}I_s\cos(\alpha+\phi_c)+I_{cd}E_{x3}-\sqrt{2}I_s\,\sin(\alpha+\phi_c-\frac{2\pi}{3})-I_{be}E_{x1}=0 \quad (14.70)$$

Substituting the right-hand side of eqn 14.67 for I_{cd}, combining the sinusoidal terms, and rearranging yields

$$I_{be} = \frac{E_{x3}}{E_{x1}}I_{bd} + \frac{\sqrt{2}I_s}{2E_{x1}}[\sin(\alpha + \phi_c) + \frac{E_{x3}}{E_{x2}}\sin(\gamma + \phi_c)] \quad (14.71)$$

Using eqn 14.54 to substitute $-(I_{bd} + I_{be})$ for I_{bf} in eqn 14.65 yields

$$I_{be} = -I_{bd} + \frac{\sqrt{2}I_s}{E_{x2}}\sin(\gamma + \phi_c) \quad (14.72)$$

Subtracting eqn 14.72 from 14.71, substituting the right side of eqn 14.66 for E_{x2} and, using eqns 14.63, yields a function of the extinction angle, $f(\gamma)$, where

$$f(\gamma) = (\frac{1}{2} - \frac{E_{x3}}{E_{x1}})\sin(\alpha + \phi_c) + (E_{x1} - \frac{E_{x3}}{2})\sin(\gamma + \phi_c)\exp(-\frac{\gamma - \frac{2\pi}{3}}{\tan(\zeta_o)}) \quad (14.73)$$

which, when γ has the correct value, is zero. This transcendental equation must be solved for γ by successive approximation. The Newton–Raphson method is excellent with $\gamma = \pi - \phi_c$ as the initial estimate. The derivative of $f(\gamma)$ with respect to γ is

$$f'(\gamma) = (E_{x1} - \frac{E_{x3}}{2})\cos(\gamma+\phi_c) - \frac{1}{\tan(\zeta_o)}\sin(\gamma+\phi_c)\exp(\frac{\gamma - \frac{2\pi}{3})}{\tan(\zeta_o)}) \quad (14.74)$$

and a better estimate γ_2 is obtained from an estimate γ_1 by

$$\gamma_2 = \gamma_1 - \frac{f(\gamma_1)}{f'(\gamma_1)} \quad (14.75)$$

Example 14.2

For the example, determine the extinction and conduction angles when the angle of delay is 70°.

Table 14.1 *The Newton–Raphson approximation to the extinction angle*

Iteration number	γ degrees	$f(\gamma)$ A	$f'(\gamma)$ A s^{-1}
1	211.364	-0.0279	-0.9035
2	209.597	-0.0005	-0.9029
3	209.565	-0.0000	-0.9029
4	209.564	-0.0000	-0.9029

The exponential coefficient E_{x1} and the current amplitude I_{bd} are first found from eqns 14.64 and 14.63.

$$E_{x1} = 0.4922$$
$$I_{bd} = -327.19A$$

The exponential coefficient E_{x3} is now found from eqn 14.68

$$E_{x3} = 0.2681$$

The extinction angle can now be found using the Newton–Raphson method, eqns 14.73, 14.74, and 14.75. Table 14.2 shows the progress of the iteration starting from $\gamma = \pi - \phi_c$. The extinction angle is found to be 209.56° to an accuracy of 0.001° in four iterations. The conduction angle is therefore 139.56° and the fact that it exceeds 120° establishes that the system is in mode 3-2.

****

14.5.3 *Summary of the solution for the 3-2 mode*

Since the solution for the 3-2 mode is rather complicated, it is appropriate to briefly summarize the steps involved.

1. Determine E_{x1} and I_{bd} from eqns 14.64 and 14.63.

2. Determine γ by the Newton–Raphson method, starting from $\gamma = \pi - \phi_c$, using eqn 14.73 for $f(\gamma)$, eqn 14.74 for $f'(\gamma)$, and eqn 14.75 for the improved value of γ. Determine the value β and, if it is equal to or greater than 120°, continue with the calculation. If it is less than 120°, the system is in the 2-0 mode and the analysis must be changed to suit this case.

3. Determine I_{be} from a combination of eqn 14.72 and 14.66.

4. Determine I_{bf} from eqn 14.54.
5. Determine I_{cd} from eqn 14.67.

This procedure yields the solution for the 60° segment extending from α to $\alpha + 60°$. For the first portion from α to $\gamma - 120°$

$$i_d = \sqrt{2}I_s \sin(\theta + \phi_c) + I_{bd} \exp(-\frac{\theta}{\tan(\zeta_o)}) \tag{14.76}$$

$$i_e = \sqrt{2}I_s \sin(\theta + \phi_c - \frac{2\pi}{3}) + I_{be} \exp(-\frac{\theta}{\tan(\zeta_o)}) \tag{14.77}$$

$$i_f = \sqrt{2}I_s \sin(\theta + \phi_c + \frac{2\pi}{3}) + I_{bf} \exp(-\frac{\theta}{\tan(\zeta_o)}) \tag{14.78}$$

For the second portion from $\gamma - 120°$ to $\alpha + 60°$

$$i_d = \frac{\sqrt{3}}{2}\sqrt{2}I_s \sin(\theta + \phi_c + \frac{\pi}{6}) + I_{cd} \exp(-\frac{\theta}{\tan(\zeta_o)}) \tag{14.79}$$

$$i_e = -i_d \tag{14.80}$$

$$i_f = 0 \tag{14.81}$$

Equations 14.76 through 14.81 define the current waveforms during the 60° interval from α to $\alpha + 60°$. The values during the rest of a cycle are derived from these by appropriate phase shifts.

Example 14.3

Determine expressions for the load currents during the 60° period from $\theta = 70°$ to 130° for the example when the delay angle is 70°.

In Example 14.1 it was found that the r.m.s. value of the uncontrolled output current was 182.4 A so that $\sqrt{2}I_s$, its peak value, is 257.92 A. The lead of this current on the datum voltage is ϕ_c which was found in the same example to be −31.36°.

In Example 14.2 it was found that the extinction angle is 209.56°.

Using these values and following the procedure summarized above, the values of the exponential coefficients are

$$E_{x1} = 0.4922$$
$$E_{x2} = 0.4037$$
$$E_{x3} = 0.2681$$

The transient current component amplitudes are

$$I_{bd} = -327.19 \text{ A}$$

$$I_{be} = 347.26 \text{ A}$$
$$I_{bf} = -20.07 \text{ A}$$

and

$$I_{cd} = -337.23 \text{ A}$$

so that the expressions for the load currents are, for the period when thyristors T1, T3, and T5 conduct, $70° < \theta < 89.56°$

$$\begin{aligned} i_d &= 257.92 \sin(\theta - 0.5474) - 327.19 \exp(-\frac{\theta}{1.7234}) \\ i_e &= 257.92 \sin(\theta - 2.6411) + 347.26 \exp(-\frac{\theta}{1.7234}) \\ i_f &= 257.92 \sin(\theta + 1.5470) - 20.07 \exp(-\frac{\theta}{1.7234}) \end{aligned}$$

and for the period when only T1 and T5 conduct, $89.56° < \theta < 130°$

$$\begin{aligned} i_d &= 223.36 \sin(\theta - 0.0238) - 337.22 \exp(-\frac{\theta}{1.7234}) \\ i_e &= -i_d \\ i_f &= 0 \end{aligned}$$

****

14.6 The modal boundary

As the delay angle is increased, the extinction angle decreases and eventually reaches a value at which the overlap between the conduction periods of thyristors T3 and T1, or any such corresponding pair, has decreased to zero. This is the boundary between the 3-2 mode and the 2-0 mode. The delay angle at this point will be called the *critical delay* and denoted by the symbol α_{crit}. The corresponding extinction angle is the *critical extinction angle*, denoted by γ_{crit}. Inspection of Fig. 14.2 shows that at the critical point

$$\gamma_{crit} - \frac{2\pi}{3} = \alpha_{crit} \tag{14.82}$$

Substituting this value into eqn 14.73 yields the following function in α_{crit}

$$\begin{aligned} g(\alpha_{crit}) = {} & (\frac{1}{2} - \frac{E_{x3}}{E_{x1}}) \sin(\alpha_{crit} + \phi_c) + \\ & (E_{x1} - \frac{E_{x3}}{2}) \sin(\alpha_{crit} + \phi_c + \frac{2\pi}{3}) \exp(\frac{\alpha_{crit}}{\tan(\zeta_o)}) \end{aligned}$$

Remembering that the values of the exponential coefficients are $E_{x1} = \exp(-\alpha/\tan(\zeta_o))$, eqn 14.64, and $E_{x3} = \exp(-(\alpha+\frac{\pi}{3})/\tan(\zeta_o))$, eqn 14.68, this expression reduces to

$$g(\alpha_{crit}) = -\frac{3}{4E_{x4}}\sin(\alpha_{crit}+\phi_c) + (1-\frac{1}{2E_{x4}})\sin(\frac{2\pi}{3})\cos(\alpha_{crit}+\phi_c) \tag{14.83}$$

where

$$E_{x4} = \exp(\frac{\pi/3}{\tan(\zeta_o)}) \tag{14.84}$$

At the critical delay, $g(\alpha_{crit})$ is zero so that

$$\alpha_{crit} = \arctan(\frac{2E_{x4}-1}{\sqrt{3}}) - \phi_c \tag{14.85}$$

Example 14.4

Determine the critical delay for the example.

From eqn 14.84, $E_{x4} = 1.8361$ and substituting this value into eqn 14.85 yields

$$\alpha_{crit} = 88.41° = 1.5431 \text{ rad}$$

****

14.7 The 2-0 mode

If the delay angle exceeds the critical value, the system is in the 2-0 mode. Considering the same 60° segment as for the 3-2 mode, i.e. the one from α to $\alpha + 60°$, reference to Fig. 14.3 shows that the segment is divided into two parts, the first from α to $\gamma - 60°$ when T1 and T5 conduct and the second from $\gamma - 60°$ to $\alpha + 60°$ when there is no conduction. Successive 60° segments are obtained by appropriately phase shifting the solution for this segment.

For the first part of the segment, $\alpha < \theta < \gamma - 60°$, the situation has already been described by Fig. 14.5 and eqns 14.55 through 14.62. However, under the conditions of mode 2-0 the value of eqn 14.62 is zero when $\theta = \alpha$ so that

$$I_{cd} = -\frac{\sqrt{3}}{2}\frac{\sqrt{2}I_s\sin(\alpha+\phi_c+\pi/6)}{E_{x1}} \tag{14.86}$$

The value of eqn 14.62 is again zero when $\theta = \gamma - 60°$, i.e.

$$\frac{\sqrt{3}}{2}\sqrt{2}I_s\sin(\gamma+\phi_c-\frac{\pi}{6})+I_{cd}E_{x4}\ \exp(-\frac{\gamma}{\tan(\zeta_o)}=0$$

which, by substituting the value of I_{cd} just found in eqn 14.86 and canceling the common factor $(\sqrt{3}/2)\sqrt{2}I_s$, yields the following equation defining the extinction angle

$$\sin(\gamma+\phi_c-\frac{\pi}{6})-\frac{E_{x4}}{E_{x1}}\sin(\alpha+\phi_c+\frac{\pi}{6})\exp(-\frac{\gamma}{\tan(\zeta_o)})=0 \qquad (14.87)$$

Again, this is a transcendental equation in the extinction angle and again the Newton–Raphson technique is an excellent method for arriving at a solution of acceptable accuracy with

$$f(\gamma)=\sin(\gamma+\phi_c-\frac{\pi}{6})-\frac{E_{x4}}{E_{x1}}\sin(\alpha+\phi_c+\frac{\pi}{6})\exp(-\frac{\gamma}{\tan(\zeta_o)}) \qquad (14.88)$$

and

$$f'(\gamma)=\cos(\gamma+\phi_c-\frac{\pi}{6})+\frac{E_{x4}}{E_{x1}\tan(\zeta_o)}\sin(\alpha+\phi_c+\frac{\pi}{6})\exp(-\frac{\gamma}{\tan(\zeta_o)}) \qquad (14.89)$$

Again $\gamma=\pi-\phi_c$ is a suitable starting value.

14.7.1 *The currents*

During the conduction period, $\alpha<\theta<\gamma-60°$, the expressions for the currents are

$$i_d=\frac{\sqrt{3}}{2}\sqrt{2}I_s\sin(\theta+\phi_c+\frac{\pi}{6})+I_{cd}\ \exp(-\frac{\gamma}{\tan(\zeta_o)}) \qquad (14.90)$$

$$i_e=-i_b \qquad (14.91)$$

$$i_f=0 \qquad (14.92)$$

During the nonconducting period, $\gamma-60°<\theta<\alpha+60°$, all currents are zero.

Example 14.5

Determine the extinction angle for the example when the delay angle is 100° and expressions for the currents in the load phases during the 60° segment from 100° to 160°.

Table 14.2 *The Newton–Raphson approximation to the extinction angle*

Iteration number	γ degrees	$f(\gamma)$ A	$f'(\gamma)$ A s^{-1}
1	211.364	-0.0877	-0.5250
2	201.796	-0.0105	-0.3952
3	200.277	-0.0003	-0.3722
4	200.231	0.0000	-0.3715
5	200.230	0.0000	-0.3715

Since the delay exceeds the critical value found in Example 14.4, the system is operating in mode 2-0 and eqn 14.90 is used for the current in load phase D. Equation 14.86 yields $I_{cd} = -607.98$ A so that

$$i_d = 223.36 \sin(\theta - 0.0238) - 607.98\ \exp(-\frac{\theta}{1.7234})$$

Equations 14.88 and 14.89 are used for the Newton–Raphson iteration with the values for E_{x4}, ϕ_c, and ζ_o already found and $E_{x1} = 0.3632$. Thus,

$$f(\gamma) = \sin(\gamma - 0.0238) - 4.9977 \exp(-\frac{\gamma}{1.7324})$$

The progress of the Newton–Raphson iteration is given in Table 14.2 which shows that the extinction angle is found to be 200.23° to an accuracy of better than 0.001° after five iterations.

During the first part of the 60° segment, $100.0° < \theta < 140.23°$, the load currents are

$$\begin{aligned} i_d &= 223.36 \sin(\theta - 0.0238) - 607.98 \exp(-\frac{\theta}{1.7234}) \\ i_e &= -i_d \\ i_f &= 0 \end{aligned}$$

During the second part of the segment, for $140.23° < \theta < 160°$, no thyristors conduct and all three load currents are zero

$$i_d = i_e = i_f = 0$$

This example and Example 14.2 illustrate how the extinction angle is relatively insensitive to changes in delay. An increase of the delay angle of 30°, from 70° to 100°, only results in a decrease in the extinction angle of 9.33° from 209.56° to 200.23°.

****

14.7.2 *The conduction angle in mode 2-0*

While the phrase *conduction angle* has been used in connection with mode 2-0 in exactly the same sense as originally defined, i.e. from the first firing of a thyristor during any one cycle to its final extinction during the same cycle, in the case of Example 14.5 a duration of 100.23°, it must be remembered that the thyristor does not conduct during the whole of this period. Considering specifically thyristor T1, this is first fired at $\theta = 100.0°$ and it and T5 conduct for a period of 40.23° until $\theta = 140.23°$. T5 has then reached its extinction point and stops conducting and with it T1. There is no current in T1 until T6 is fired at $\theta = 160.00°$. Provided T1 is still in a conducting state, by the use of wide firing pulses or a complementary pulse, conduction recommences and continues until the final extinction of T1 at $\theta = 200.23°$. Thus, T1 actually conducts for only about 70% of the conduction angle, from 100° to 140.23° and from 160° to 200.23°. This point is illustrated by Fig. 14.3, where the gray filling has been omitted from the conduction bars during the nonconducting portion, and by the load current waveform of Fig. 14.7, graph 2.

14.8 The cutoff point

When, at the trigger point, $\theta = \alpha$, the driving voltage round the loop containing the incoming thyristor is zero and going negative in the case of a forward conducting thyristor, T1, T2, or T3, or going positive in the case of a backward conducting thyristor, T4, T5, or T6, no conduction takes place and the cutoff point has been reached. The delay angle at this point will be called the *cutoff delay*, symbol α_{co}. When the system is approaching cutoff it will be in the 2-0 mode and for the T1–T5 thyristor pair the driving voltage is $v_{\delta d} - v_{\delta e} = \sqrt{6}V_\delta \sin(\theta + \delta + \pi/6)$. The appropriate zero occurs when $\alpha_{co} + \delta + \pi/6 = \pi$, i.e. when

$$\alpha_{co} = \frac{5\pi}{6} - \delta \tag{14.93}$$

Example 14.6

Determine for the example the delay angle at cutoff.

It was found in Example 14.1 that the phase angle of the difference voltage is 0.4976 rad or 28.51°. The delay angle at cutoff is therefore 2.1204 rad or 121.49°. The effective delay angle range for this system is therefore from a minimum value of 0.5474 rad = 31.36° to a maximum of 2.1204 rad = 121.49° with the system in the 3-2 mode for the range 31.36° to 88.41° and in the 2-0 mode from 88.41° to 121.49°.

****

14.9 The voltages

The voltages of interest are the potentials of the load phase ends, v_d, v_e, and v_f, the voltage of the load neutral, v_n, and the voltages across the thyristors, v_{t1}, etc. Expressions for these are derived for the 60° segment from α to $\alpha + 60°$.

14.9.1 *Three thyristors conducting*

In the 3-2 mode, when thyristors T1, T3, and T5 are conducting, the voltages of the load phase ends are the same as the supply line voltages

$$v_d = v_r \tag{14.94}$$

$$v_e = v_y \tag{14.95}$$

$$v_f = v_b \tag{14.96}$$

It has already been shown, eqn 14.31, that the load neutral voltage is zero in these circumstances.

The voltage across a thyristor is the difference between the voltages of the line and the load phase end to which it is connected. There is therefore zero voltage across all thyristors when three thyristors conduct and, in particular, the voltage across T1 is

$$v_{t1} = 0 \tag{14.97}$$

14.9.2 *Two thyristors conducting*

For the selected segment, $\alpha < \theta < \alpha + 60°$, the two thyristors which conduct, in either the second part of mode 3-2 or the first part of mode 2-0, are T1 and T5, so that

$$v_d = v_r \tag{14.98}$$

$$v_e = v_y \tag{14.99}$$

An expression for the potential of the load neutral has already been found (eqn 14.59) which, using the fact that this is a three phase, three wire system, can be reduced to

$$v_n = -\frac{v_b - e_f}{2} = -\frac{v_{\delta f}}{2} \tag{14.100}$$

The voltage of the end of the nonconducting phase can now be obtained as the sum of the load neutral voltage and the phase e.m.f., in this case

$$v_f = v_n + e_f \tag{14.101}$$

which by the use of eqn 14.100 becomes

$$v_f = \frac{3e_f - v_b}{2} \tag{14.102}$$

The voltages across the conducting thyristors and their complementary thyristors are zero and the voltage across the nonconducting thyristor, T3, is given by

$$v_{t3} = v_b - v_f \tag{14.103}$$

which by the use of eqns 14.102 and 14.100 becomes

$$v_{t3} = -3v_n = \frac{3}{2}v_{\delta f} \tag{14.104}$$

The voltage across its complementary thyristor, T6, is the negative of this value

$$v_{t6} = -v_{t3} = -\frac{3}{2}v_{\delta f} \tag{14.105}$$

14.9.3 *No thyristors conducting*

When no thyristors conduct, the potentials of the phase ends are the sums of the load neutral voltage and the phase e.m.f.

$$v_d = e_d + v_n \tag{14.106}$$

$$v_e = e_e + v_n \tag{14.107}$$

$$v_f = e_f + v_n \tag{14.108}$$

The thyristor voltages are typified by that for T1

$$v_{t1} = v_r - v_d = v_r - e_d - v_n \tag{14.109}$$

All these voltages depend on a knowledge of the load neutral potential. Unfortunately, because the load is now floating, this potential depends upon a variety of ill-defined quantities such as the leakage impedance of the thyristors and the stray capacitances of the phases to ground. On the assumption that these quantities are more or less balanced, the potential of the load neutral will be taken to be zero

$$v_n = 0 \tag{14.110}$$

Example 14.7

Determine expressions for the circuit voltages during the period $\alpha < \theta < \alpha + 60°$ for the cases of Examples 14.2 and 14.5.

In Example 14.2 the delay angle is 70° and the system is in the 3-2 mode with an extinction angle of 209.56°. For the period $70° < \alpha < 89.56°$ the three thyristors T1, T3, and T5 conduct and the potentials of the load phase ends are

$$v_d = v_r = 359.26 \ \sin(\theta)$$
$$v_e = v_y = 359.26 \ \sin(\theta - 2.0944)$$
$$v_f = v_b = 359.26 \ \sin(\theta + 2.0944)$$

The potential of the load neutral is zero and all thyristor voltages are zero.

During the period $89.56° < \theta < 130°$ only two thyristors, T1 and T5, conduct. The voltages of the phase ends are then

$$v_d = v_r = 359.26 \ \sin(\theta)$$
$$v_e = v_y = 359.26 \ \sin(\theta - 2.0944)$$

and, from eqn 14.102

$$v_f = 312.90 \sin(\theta + 2.0120)$$

The potential of the load neutral is obtained from eqn 14.100 as

$$v_n = 35.97 \ \sin(\theta + 2.5920)$$

The voltage across thyristor T3 is obtained from eqn 14.104 as

$$v_{t3} = 53.97 \ \sin(\theta + 2.5920)$$

In Example 14.5 the delay angle is 100° and the system is in the 2-0 mode with an extinction angle of 200.23°. For the period $100° < \alpha < 140.23°$ the two thyristors T1 and T5 conduct and the potentials of the load phase ends are

$$v_d = v_r = 359.26 \ \sin(\theta)$$
$$v_e = v_y = 359.26 \ \sin(\theta - 2.0944)$$

and, from eqn 14.102

$$v_f = 312.90 \quad \sin(\theta + 2.0120)$$

The potential of the load neutral is obtained from eqn 14.100 as

$$v_n = 17.99 \ \sin(\theta - 0.5496)$$

The voltage across thyristor T3 is obtained from eqn 14.104 as

$$v_{t3} = 53.97 \ \sin(\theta + 2.5920)$$

During the period $140.23° < \theta < 160°$ no thyristors conduct and, with the assumption that the potential of the load neutral is zero under these circumstances,

$$v_n = 0.0$$

$$v_d = e_d = 328.10 \ \sin(\theta - 0.0524)$$
$$v_e = e_e = 328.10 \ \sin(\theta - 2.1468)$$
$$v_f = e_f = 328.10 \ \sin(\theta + 2.0420)$$
$$v_{t3} = v_b - v_f = 35.97 \ \sin(\theta + 2.5920)$$

****

14.10 The current and voltage waveforms

The waveforms of interest are those of the currents in a load phase, in a supply line, and in a thyristor and of the voltages of a load phase end, the load neutral, and across a thyristor. The symmetry of the circuit ensures that all the waveforms can be derived from one example of each by appropriate phase shifting. The currents in phase D, the red line, and thyristor T1 and the voltages of terminal D, the load neutral, and across thyristor T1 will be determined.

All these currents and voltages, with the single exception of thyristor current, have half wave symmetry so that attention can be concentrated on the half cycle which extends from the instant of firing T1 at $\theta = \alpha$ to the instant of firing its complement, T4, at $\theta = \alpha + 180°$. The second half cycle covering the period $\alpha + 180° < \theta < \alpha + 360°$ can be obtained by inversion and phase shifting. The exception, the current in T1, is easily dealt with since it is zero throughout the second half cycle.

The various quantities are defined for the 3-2 mode in Table 14.3 in which the half cycle is divided into six subsegments, two for each 60° segment, like the one considered up to now. The subsegment number is given in the first row and its starting and finishing point in rows 2 and 3.

Rows 4 through 9 show the thyristor states by an X for conducting and 0 for nonconducting.

Row 10 gives the currents i_d and i_r which are the same. For brevity they are shown as functions of θ, $f_1(\theta)$, $f_2(\theta)$, $f_3(\theta)$, and $g_1(\theta)$. The functions f_1, f_2, and f_3 are defined by eqns 14.47 through 14.49 with the values of I_{bd}, I_{be}, and I_{bf} determined by the procedure defined in Subsection 14.5.3. The function g_1 is defined by eqn 14.62 with I_{cd} given by eqn 14.67. This current pattern is repeated in each 60° segment with appropriate changes in thyristors.

Row 11 gives the current in thyristor T1. This current is zero throughout the following half cycle.

Row 12 defines the voltage of the load neutral using eqn 14.31 when three thyristors conduct and eqn 14.100 when two thyristors conduct.

Row 13 gives the potential of the end of the D phase. This is the same as the potential of the red line while T1 is conducting and is given by eqn 14.102, appropriately phase shifted, when T1 is not conducting.

Table 14.3 *Expressions for the current and voltage waveforms for the 3-2 mode during the half cycle $\alpha < \theta < \alpha + 180°$*

Subseg.	1	2	3	4	5	6
Start	α	$\gamma - 120°$	$\alpha + 60°$	$\gamma - 60°$	$\alpha + 120°$	γ
End	$\gamma - 120°$	$\alpha + 60°$	$\gamma - 60°$	$\alpha + 120°$	γ	$\alpha + 180°$
T1	X	X	X	X	X	0
T2	0	0	0	0	X	X
T3	X	0	0	0	0	0
T4	0	0	0	0	0	0
T5	X	X	X	0	0	0
T6	0	0	X	X	X	X
$i_d = i_r$	$f_1(\theta)$	$g_1(\theta)$	$-f_2(\theta - 60)$	$g_1(\theta - 60)$	$f_3(\theta - 120)$	0
i_{t1}	i_d	i_d	i_d	i_d	i_d	0
v_n	0	$-v_{\delta f}/2$	0	$-v_{\delta e}/2$	0	$-v_{\delta d}/2$
v_d	v_r	v_r	v_r	v_r	v_r	$e_d + v_n$
v_{t1}	0	0	0	0	0	$v_r - v_d$

Row 14 defines the voltage across thyristor T1. It is zero while the thyristor is conducting and is given by eqn 14.105, appropriately phase shifted, when T1 is not conducting.

The various quantities are defined for the 2-0 mode in Table 14.4, whose layout and content are similar to those of Table 14.3.

Row 10 gives the currents i_d and i_r which are the same. In the function $g_1(\theta)$ which is defined by eqn 14.62, the value of I_{cd} is given by eqn 14.86.

Row 11 gives the current in T1. This current is zero throughout the following half cycle.

Row 12 defines the voltage of the load neutral, using eqn 14.100, when two thyristors conduct. It is zero, according to the assumption of balanced stray impedances, when no thyristors conduct.

Row 13 gives the potential of the end of the D phase. This is the same as the potential of the red line while T1 is conducting. It is given by eqn 14.106, appropriately phase shifted, when T1 is not conducting.

Row 14 defines the voltage across T1. It is zero while the thyristor is conducting and is given by eqn 14.104, appropriately phase shifted, when T1 is not conducting.

Example 14.8

Determine expressions for the current in load phase D, the potentials of the load neutral and the terminal of phase D, and the voltage across

Table 14.4 *Expressions for the current and voltage waveforms for the 2-0 mode during the half cycle $\alpha < \theta < \alpha + 180°$*

Subseg.	1	2	3	4	5	6
Start	α	$\gamma - 60°$	$\alpha + 60°$	γ	$\alpha + 120°$	$\gamma + 60°$
End	$\gamma - 60°$	$\alpha + 60°$	γ	$\alpha + 120°$	$\gamma + 60°$	$\alpha + 180°$
T1	X	0	X	0	0	0
T2	0	0	0	0	X	0
T3	0	0	0	0	0	0
T4	0	0	0	0	0	0
T5	X	0	0	0	0	0
T6	0	0	X	0	X	0
$i_d = i_r$	$g_1(\theta)$	0	$g_1(\theta - 60)$	0	0	0
i_{t1}	i_d	0	i_d	0	0	0
v_n	$-v_{\delta f}/2$	0	$-v_{\delta e}/2$	0	$-v_{\delta d}/2$	0
v_d	v_r	e_d	v_r	e_d	$e_d + v_n$	e_d
v_{t1}	0	$v_r - v_d$	0	$v_r - v_d$	$v_r - v_d$	$v_r - v_d$

thyristor T1 for the cases of Examples 14.2 and 14.5 over the half cycle $\alpha < \theta < \alpha + 180°$.

Plot the waveforms of these quantities over a complete cycle.

In the case of Example 14.2 the delay angle is 70°, the extinction angle is 209.56°, and the system is in the 3-2 mode. Table 14.3 therefore defines the waveforms. The reference potential for all quantities is the potential of the red line relative to the supply neutral which is

$$v_r = 359.26 \ \sin(\theta)$$

The first subsegment extends from 70° to 89.56° and during it, it was found in Example 14.3 that the current in load phase D, which is the same as the current in the red line and in thyristor T1, is

$$i_d = i_r = i_{t1} = 257.92 \ \sin(\theta - 0.5474) - 327.19 \ \exp(-\frac{\theta}{1.7234})$$

This is the function $f_1(\theta)$ of Table 14.3.

The potentials of the load neutral and load terminal D and the voltage across thyristor T1 are

$$v_n = 0.0$$
$$v_d = 359.26 \ \sin(\theta)$$

$$v_{t1} = 0.0$$

The second subsegment extends from 89.56° to 130.0° and during it, it was found in Example 14.3 that the current in load phase D, which is the same as the current in the red line and the current in thyristor T1, is

$$i_d = i_r = i_{t1} = 223.36\ \sin(\theta - 0.0238) - 337.23\ \exp(-\frac{\theta}{1.7234})$$

This is the function $g_1(\theta)$ of Table 14.3.

The potentials of the load neutral and load terminal D and the voltage across thyristor T1 are

$$\begin{aligned} v_n &= 17.99\ \sin(\theta - 0.5496) \\ v_d &= 359.26\ \sin(\theta) \\ v_{t1} &= 0.0 \end{aligned}$$

The third subsegment extends from 130.0° to 149.56°, 60° after the first subsegment. In Example 14.3 it was found that the current in load phase E during the first subsegment is

$$i_e = 257.92\ \sin(\theta - 2.6418) + 347.26\ \exp(-\frac{\theta}{1.7234})$$

This is the function $f_2(\theta)$ of Table 14.3 and when phase shifted by 60° and inverted, because this is the current in a reverse connected thyristor, gives the current in load phase D, which is the same as the current in the red line and the current in thyristor T1, as

$$i_d = i_r = i_{t1} = 257.92\ \sin(\theta - 0.5474) - 637.60\ \exp(-\frac{\theta}{1.7234})$$

The potentials of the load neutral and load terminal D and the voltage across thyristor T1 are

$$\begin{aligned} v_n &= 0.0 \\ v_d &= 359.26\ \sin(\theta) \\ v_{t1} &= 0.0 \end{aligned}$$

The fourth subsegment extends from 149.56° to 190.0°, 60° after the second subsegment when the current in load phase D is as given above. Delaying this expression by 60° yields

$$i_d = i_r = i_{t1} = 223.36\ \sin(\theta - 1.0710) - 619.18\ \exp(-\frac{\theta}{1.7234})$$

The potentials of the load neutral and load terminal D and the voltage across thyristor T1 are

$$v_n = 17.99\ \sin(\theta + 1.5448)$$

$$v_d = 359.26 \sin(\theta)$$
$$v_{t1} = 0.0$$

The fifth subsegment extends from 190.0° to 209.56°, 120° after the first subsegment. In Example 14.3 it was found that the current in load phase F during the first subsegment is

$$i_f = 257.92 \sin(\theta + 1.5470) - 20.07 \exp(-\frac{\theta}{1.7234})$$

This is the function $f_3(\theta)$ of Table 14.3 and when delayed by 120° it yields

$$i_d = i_r = i_{t1} = 257.93 \sin(\theta - 0.5474) - 67.66 \exp(-\frac{\theta}{1.7234})$$

The potentials of the load neutral and load terminal D and the voltage across thyristor T1 are

$$v_n = 0.0$$
$$v_d = 359.26 \sin(\theta)$$
$$v_{t1} = 0.0$$

During the sixth subsegment, which extends from 209.56° to 250.0°, thyristor T1 is blocking, load phase D is disconnected from the supply, and the currents in that phase, in the red line, and in T1 are zero.

$$i_d = i_r = i_{t1} = 0.0$$

The potentials of the load neutral and load terminal D and the voltage across thyristor T1 are

$$v_n = 17.99 \sin(\theta - 2.6440)$$
$$v_d = 312.90 \sin(\theta - 0.0824)$$
$$v_{t1} = 53.96 \sin(\theta + 0.4976)$$

With the exception of the thyristor current, the next half cycle is similar but with the signs reversed. The role played by thyristor T1 is now taken over by T4 and the current in T1 is zero throughout the half cycle.

The waveforms of these currents and voltages are shown in Fig. 14.6, each graph being identified by a number at the right-hand end of its horizontal axis. The waveform of the voltage of the red line is shown for reference in graph 1, the current in load phase D, which is also the current in the red line, in graph 2, and the potential of terminal D in graph 3. The

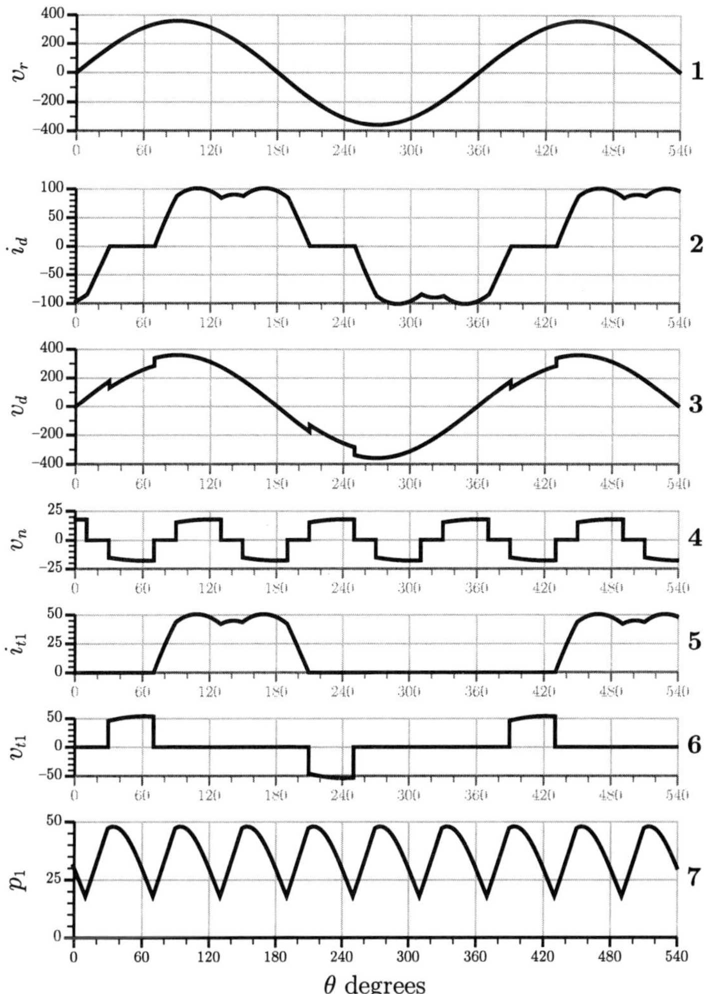

FIG. 14.6. Waveforms for a delay angle of 70° when the system operates in the 3-2 mode. Graph 1 shows the potential of the red line relative to the supply neutral. Graph 2 shows the current in load phase D. Graph 3 shows the potential of the terminal of load phase D. Graph 4 shows the potential of the load neutral. Graph 5 shows the current in thyristor T1 and graph 6 shows the voltage across that thyristor. Graph 7 shows the total instantaneous input power to the load.

potential of the load neutral is shown in graph 4, note that it is at three times the supply frequency. The current in thyristor T1 is shown in graph 5 and the voltage across thyristor T1 in graph 6. The thyristor voltage is small because the load e.m.f. is nearly equal and opposite to the supply voltage. However, as for the single phase case, it would be prudent to design the system for the possibility that the load might fall out of synchronism with the supply, in which case the thyristor voltage could be as high as $\sqrt{2}(V_s + E_o)$ which in this case is 687 V.

Continuing now with the second part of this example: in the case of Example 14.5 the delay angle is 100°, the extinction angle is 200.23°, and the system is in the 2-0 mode. Table 14.4 therefore defines the waveforms. As always, the reference potential for all quantities is the potential of the red line relative to the supply neutral which is

$$v_r = 359.26\ \sin(\theta)$$

The first subsegment extends from 100° to 140.23° and during it, it was found in Example 14.5 that the current in load phase D, which is the same as the current in the red line and in thyristor T1, is

$$i_d = i_r = i_{t1} = 223.36\ \sin(\theta - 0.0238) - 607.98\ \exp(-\frac{\theta}{1.7234})$$

This is the function $g_1(\theta)$ of Table 14.4.

The potentials of the load neutral and load terminal D and the voltage across thyristor T1 are

$$\begin{aligned} v_n &= 17.99 \sin(\theta - 0.5496) \\ v_d &= 359.26 \sin(\theta) \\ v_{t1} &= 0.0 \end{aligned}$$

The second subsegment extends from 140.23° to 160.0° and during it thyristor T5 is reverse biased so that, although T1 is nominally on, no current can flow through it, so that

$$i_d = i_r = i_{t1} = 0.0$$

According to the assumption that the leakage and stray impedances are balanced

$$v_n = 0.0$$

and the potential of load terminal D and the voltage across thyristor T1 are

$$v_d = 328.10\ \sin(\theta - 0.0524)$$

$$v_{t1} = 35.97 \ \sin(\theta + 0.4976)$$

The third subsegment extends from 160.0° to 200.23°, 60° after the first subsegment. Assuming that the control system ensures that T1 is in the conducting mode, the expression for the current in it and in the red line and load phase D is given by delaying by 60° the expression given above for the current during the first subsegment

$$i_d = i_r = i_{t1} = 223.36 \ \sin(\theta - 1.0710) - 1116.30 \ \exp(-\frac{\theta}{1.7234})$$

The potentials of the load neutral and load terminal D and the voltage across thyristor T1 are

$$\begin{aligned} v_n &= 17.99 \ \sin(\theta - 1.5448) \\ v_d &= 359.26 \ \sin(\theta) \\ v_{t1} &= 0.0 \end{aligned}$$

The fourth subsegment extends from 200.23° to 220.0°, and there is no current in load phase D

$$i_d = i_r = i_{t1} = 0.0$$

The potentials of the load neutral and load terminal D and the voltage across thyristor T1 are

$$\begin{aligned} v_n &= 0.0 \\ v_d &= 328.10 \ \sin(\theta - 0.0524) \\ v_{t1} &= 35.97 \ \sin(\theta + 0.4976) \end{aligned}$$

The fifth subsegment extends from 220.0° to 260.23°, and there is no current in load phase D

$$i_d = i_r = i_{t1} = 0.0$$

The potentials of the load neutral and load terminal D and the voltage across thyristor T1 are

$$\begin{aligned} v_n &= 17.99 \ \sin(\theta - 2.6440) \\ v_d &= 312.90 \ \sin(\theta - 0.0824) \end{aligned}$$

$$v_{t1} = 53.96 \ \sin(\theta + 0.4976)$$

During the sixth subsegment, which extends from 260.23° to 280.0°, there is no current in load phase D

$$i_d = i_r = i_{t1} = 0.0$$

The potentials of the load neutral and load terminal D and the voltage across thyristor T1 are

$$\begin{aligned} v_n &= 0.0 \\ v_d &= 328.10 \ \sin(\theta - 0.0524) \\ v_{t1} &= 35.97 \ \sin(\theta + 0.4976) \end{aligned}$$

With the exception of the thyristor current, the next half cycle is similar but with the signs reversed. The role played by thyristor T1 is now taken over by T4 and the current in T1 is zero throughout the half cycle.

The waveforms of these currents and voltages are shown in Fig. 14.7. The waveform of the voltage of the red line is shown for reference in graph 1, the current in load phase D, which is also the current in the red line, in graph 2, and the potential of load terminal D in graph 3. Graph 4 shows the potential of the load neutral, note that it is at three times the supply frequency. Graph 5 shows the current in thyristor T1 and graph 6 the voltage across it. Again it must be noted that, although the thyristor voltage is modest, prudence indicates that the thyristors should be capable of withstanding the combined peak values of the supply voltage and load e.m.f.

****

14.11 The input and output powers

Since thyristor losses are assumed to be zero, the input power, p_1, equals at all times the power delivered to the load, p_o. The latter is the sum for all three phases of the products of the phase voltages and currents, i.e.

$$p_o = (v_d - v_n)i_d + (v_e - v_n)i_e + (v_f - v_n)i_f$$

which expands to

$$p_o = v_d i_d + v_e i_e + v_f i_f - v_n(i_d + i_e + i_f)$$

The last term in this expression is zero because the isolated load neutral forces the sum of the three phase currents to be always zero. Hence, the input and output powers are given by

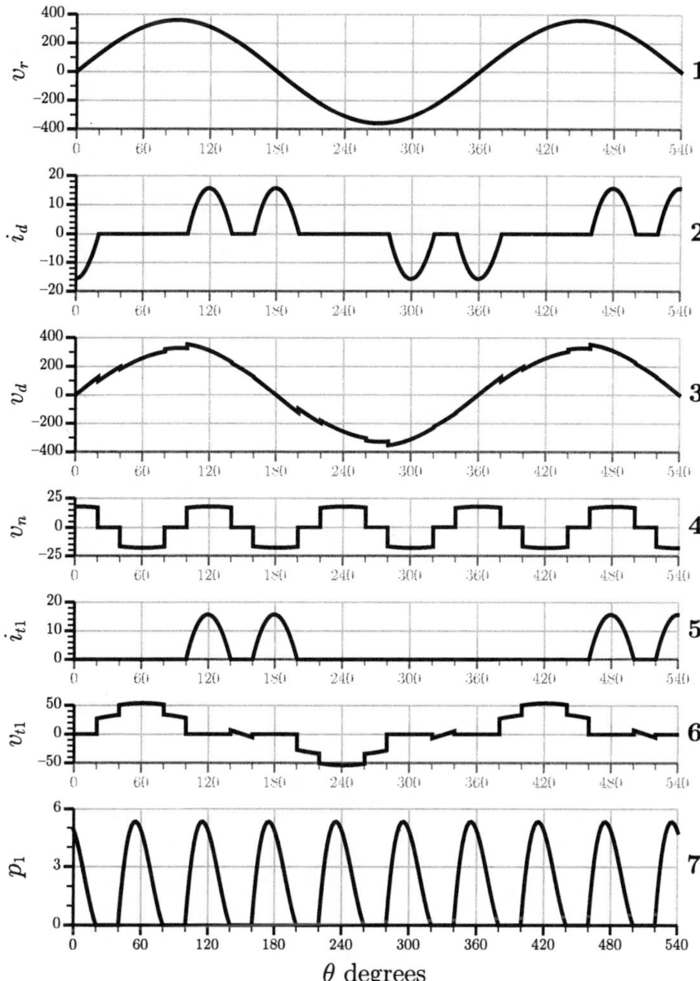

FIG. 14.7. Waveforms for a delay angle of 100° when the system operates in the 2-0 mode. Graph 1 shows the potential of the red line relative to the supply neutral. Graph 2 shows the current in load phase D. Graph 3 shows the potential of the terminal of load phase D. Graph 4 shows the potential of the load neutral. Graph 5 shows the current in thyristor T1 and graph 6 shows the voltage across this thyristor. Graph 7 shows the total instantaneous input power to the load.

$$p_1 = p_o = v_d i_d + v_e i_e + v_f i_f \tag{14.111}$$

Because every 60° segment is like every other except for a cyclic change in the conducting thyristors, expressions for the power are only required for the segment under study, the one from α to $\alpha + 60°$. During this segment, when three thyristors are conducting, the potentials of the load phase terminals equal the supply line potentials, $v_d = v_r$, $v_e = v_y$, and $v_f = v_b$, which are defined by eqns 14.1 through 14.3. The load phase currents are given by eqns 14.47 through 14.49 with I_{bd}, I_{be}, and I_{bf} determined by the procedure outlined in the Subsection 14.5.2. Thus, the power is the sum of the uncontrolled three phase power and the power associated with the transient components of the currents

$$\begin{aligned} p_i = p_o = {} & P_u + \sqrt{2}V_s I_{bd}\ \sin(\theta)\exp(-\frac{\theta}{\tan(\zeta)}) \\ & +\sqrt{2}V_s\ I_{be}\ \sin(\theta - \frac{2\pi}{3})\exp(-\frac{\theta}{\tan(\zeta)}) \\ & +\sqrt{2}V_s\ I_{bf}\ \sin(\theta + \frac{2\pi}{3})\exp(-\frac{\theta}{\tan(\zeta)}) \end{aligned} \tag{14.112}$$

where P_u is the uncontrolled three phase power given by

$$P_u = 3V_s I_s \cos(\phi_c) \tag{14.113}$$

Equation 14.112 could be further reduced by using the fact that $I_{bd} + I_{be} + I_{bf} = 0$ (eqn 14.54) but there is little advantage to be gained by doing this.

When two thyristors are conducting $v_d = v_r$ and i_d is given by eqn 14.62, $v_e = v_y$ and $i_e = -i_d$, and i_f is zero. Thus, the power is

$$p_1 = p_o = v_d i_d + v_e i_e = (v_d - v_e) i_d$$

i.e.

$$\begin{aligned} p_1 = p_o = {} & \frac{P_u}{2} - \frac{3\sqrt{2}}{4} V_s \sqrt{2} I_s \cos(2\theta + \phi_c \frac{\pi}{3}) + \\ & \sqrt{6} V_s I_{cd} \sin(\theta + \frac{\pi}{p} 6)\exp(-\frac{\theta}{\tan(\zeta)}) \end{aligned} \tag{14.114}$$

The power is of course zero when no thyristors are conducting.

Example 14.9

Determine expressions for the power delivered to the load for the example for the cases of Examples 14.2 and 14.5.

Considering first the case of Example 14.2, the delay angle is 70°, the extinction angle is 209.56°, and the system is in the 3-2 mode with three thyristors conducting for $70° < \theta < 89.56°$ and two thyristors conducting for $89.56° < \theta < 130°$. The supply voltage, V_s, is 254.03 V and, from Example 14.3, $\sqrt{2}I_s$ is 257.92 A, ϕ_c is –0.5474 rad, I_{bd} is –327.19 A, I_{be} is 347.26 A, and I_{bf} is –20.07 A so that, from eqn 14.112

$$\begin{aligned} p_1 = p_o &= 118.68 - 117.55\ \sin(\theta)\exp(-\frac{\theta}{1.7234}) - \\ &124.76\sin(\theta - \frac{2\pi}{3})\exp(-\frac{\theta}{1.7234}) - \\ &7.21\sin(\theta + \frac{2\pi}{3})\exp(-\frac{\theta}{1.7234})\ \text{kW} \end{aligned}$$

The last three terms can be usefully combined to give the simpler expression

$$p_1 = p_o = 118.68 + 210.12\ \sin(\theta - 2.5665)\exp(-\frac{\theta}{1.7234})\ \text{kW}$$

During the period $89.56° < \theta < 130°$ when two thyristors conduct, the power is given by eqn 14.114 with I_{cd} already obtained in Example 14.3 as –337.23 A.

$$\begin{aligned} p_1 = p_o &= 59.34 - 69.49\ \cos(2\theta + 0.4998) - \\ &209.84\sin(\theta + \frac{\pi}{6})\exp(-\frac{\theta}{1.7234})\ \text{kW} \end{aligned}$$

The waveform of the power flow is shown in graph 7 of Fig. 14.6. It has a significant variation at six times the supply frequency in contrast with the uncontrolled three phase power which is a steady, unvarying flow.

Moving now to the case of Example 14.5, the delay angle is 100°, the extinction angle is 200.23°, and the system is in mode 2-0 with two thyristors conducting from $100° < \theta < 140.23°$ and no conduction from $140.23° < \theta < 160°$. From Example 14.5, $I_{cd} = -607.98$ A so that, from eqn 14.114,

$$\begin{aligned} p_1 = p_o &= 59.34 - 69.49\ \cos(2\theta + 0.4998) - \\ &378.32\sin(\theta + \frac{\pi}{6})\exp(-\frac{\theta}{1.7234})\ \text{kW} \end{aligned}$$

During the period $140.23° < \theta < 160°$ when there is no conduction, the power is of course zero.

The waveform of the power is shown in graph 7 of Fig. 14.7. It is small, which is characteristic of mode 2-0, and is delivered to the load in a series of brief pulses, six per supply cycle.

****

14.12 The r.m.s. values of the currents and voltages

The r.m.s. values of significance are those of the currents in a load phase, an a.c. line, and a thyristor and of the voltages between a pair of a.c. lines, between a pair of load terminals, across a load phase, and between the load neutral and the supply neutral. These values are obtained in the usual way by squaring, integrating, averaging, and taking the square root. In its essentials this is exactly the same process as that used in Chapter 13 but in its details it is considerably more complex. In the single phase case there is only one mode of operation and only two subsegments per half cycle. Here there are two modes and six subsegments per half cycle. However, by tracing the expressions listed in Tables 14.3 and 14.4, the determination becomes a routine computational procedure whose broad outline will be given here and whose details are clarified by Example 14.10.

Consider first a current i_x during the nth subsegment which extends from θ_n to θ_{n+1}. This is defined by an expression of the type

$$i_x = I_{x1}\ \sin(\theta + \phi_{x1}) + I_{x2}\ \exp(-\frac{\theta}{\tan(\zeta_o)}) \tag{14.115}$$

where the values of θ_n, θ_{n+1}, I_{x1}, ϕ_{x1}, and I_{x2} are obtained by the procedures already described.

Denoting by Q_n the value of the integral of the square of this expression with respect to θ over the duration of the subsegment, then, using eqn A.5 with the appropriate coefficients in the basic power electronics expression set to zero

$$\begin{aligned} Q_n = {} & \frac{1}{2}I_{x1}^2\left\{(\theta_{n+1}-\theta_n) - \frac{1}{2}\Big[\sin(2(\theta_{n+1}+\phi_{x1})) - \sin(2(\theta_n+\phi_{x1}))\Big]\right\} \\ & -2I_{x1}I_{x2}\sin(\zeta_o)\Big[\sin(\theta_{n+1}+\phi_{x1}+\zeta_o)\exp(-\frac{\theta_{n+1}}{\tan(\zeta_o)}) \\ & -\sin(\theta_n+\phi_{x1}+\zeta_o)\exp(-\frac{\theta_n}{\tan(\zeta_o)})\Big] \\ & -\frac{1}{2}I_{x2}^2\ \tan(\zeta_o)\Big[\exp(-\frac{2\theta_{n+1}}{\tan(\zeta_o)}) - \exp(-\frac{2\theta_n}{\tan(\zeta_o)})\Big] \end{aligned} \tag{14.116}$$

Denoting the r.m.s. value of the current in a load phase by I_o and in a supply line by I_1

$$I_1 = I_o = \left[\frac{1}{\pi}\sum_{n=1,6} Q_n\right]^{1/2} \tag{14.117}$$

Since a thyristor conducts the load current during only one-half of a cycle, its r.m.s. value, I_{thrms}, is the load phase current divided by $\sqrt{2}$

$$I_{thrms} = \frac{I_o}{\sqrt{2}} \tag{14.118}$$

The r.m.s. values of the voltages are obtained in a similar way by considering each of the six subsegments in a half-cycle and summing the results

$$v_x = V_x \sin(\theta + \phi_x) \tag{14.119}$$

where the values of V_x and ϕ_x are obtained by reference to Table 14.3 or 14.4 as appropriate.

$$Q_n = \int_{\theta_n}^{\theta_{n+1}} {v_x}^2 \, \mathrm{d}\theta$$

i.e.

$$Q_n = \frac{1}{2}V_x^2\left\{(\theta_{n+1} - \theta_n) - \frac{1}{2}[\sin(2(\theta_{n+1} + \phi_x)) - \sin(2(\theta_n + \phi_x))]\right\} \tag{14.120}$$

and the r.m.s. value, V_y, is

$$V_y = \left[\frac{1}{\pi}\sum_{n=1,6} Q_n\right]^{1/2} \tag{14.121}$$

The line to line voltages are generally easier to measure than the line to neutral voltages. The r.m.s. value of the line to line voltage at the load terminals is derivable directly from the line to supply neutral and load neutral to supply neutral values because, as will be seen shortly, there are no Fourier components common to the line to line and the load neutral voltages. Under these circumstances

$$V_{oll} = \sqrt{3({V_o}^2 - {V_n}^2)} \tag{14.122}$$

where V_o is the r.m.s. value of the load line to supply neutral voltage, V_n is the r.m.s. value of the load neutral to supply neutral voltage, and V_{oll} is the r.m.s. value of the load line to line voltage.

14.13 The mean power, kVA, and power factor

Since the waveform of the power delivered to the load repeats in every 60° segment, averaging need only be done over two consecutive subsegments to obtain the mean power. For the 3-2 mode this involves integrating eqn 14.112 with respect to θ over the range α to $\gamma-120°$, integrating eqn 14.114 with respect to θ over the range $\gamma-120°$ to $\alpha+60°$, and dividing the sum of the two integrals by $\pi/3$. Denoting the mean power drawn from the supply by P_1 and the mean power delivered to the load by P_o

$$\begin{aligned} P_1 = P_o = \frac{3}{\pi}\Bigg\{ & P_u(\gamma - \frac{2\pi}{3} - \alpha) \\ & -\sqrt{2}V_s I_{bd}\ \sin(\zeta_o)\Big[\sin(\gamma - \frac{2\pi}{3} + \zeta_o)\exp(-\frac{\gamma - \frac{2\pi}{3}}{\tan(\zeta_o)}) \\ & - \sin(\alpha + \zeta_o)\exp(-\frac{\alpha}{\tan(\zeta_o)})\Big] \\ & -\sqrt{2}V_s I_{be}\sin(\zeta_o)\Big[\sin(\gamma - \frac{4\pi}{3} + \zeta_o)\exp(-\frac{\gamma - \frac{2\pi}{3}}{\tan(\zeta_o)}) \\ & - \sin(\alpha - \frac{2\pi}{3} + \zeta_o)\exp(-\frac{\alpha}{\tan(\zeta_o)})\Big] \\ & -\sqrt{2}V_s I_{bf}\sin(\zeta_o)\Big[\sin(\gamma + \zeta_o)\exp(-\frac{\gamma - \frac{2\pi}{3}}{\tan(\zeta_o)}) \\ & - \sin(\alpha + \frac{2\pi}{3} + \zeta_o)\exp(-\frac{\alpha}{\tan(\zeta_o)})\Big] \\ & + \frac{P_u}{2}(\alpha - \gamma + \pi) \\ & - \frac{3\sqrt{2}}{8}V_s\sqrt{2}I_s[\sin(2\alpha + \phi_c + \pi) - \sin(2\gamma + \phi_c - \pi)] \\ & -\sqrt{6}V_s I_{cd}\sin(\zeta_o)[\sin(\alpha + \frac{\pi}{2} + \zeta_o)\exp(-\frac{\alpha + \frac{\pi}{3}}{\tan(\zeta_o)}) \\ & - \sin(\gamma - \frac{\pi}{2} + \zeta_o)\exp(-\frac{\gamma - \frac{2\pi}{3}}{\tan(\zeta_o)})]\Bigg\} \end{aligned} \quad (14.123)$$

For the 2-0 mode the situation is considerably simpler, involving only the integral of eqn 14.114 over the range α to $\gamma - 60°$

$$P_1 = P_o = \frac{3}{\pi}\Bigg\{\frac{P_u}{2}(\gamma - \frac{\pi}{3} - \alpha)$$

$$+\frac{3\sqrt{2}}{8}V_s\sqrt{2}I_s\Big[\sin(2\gamma - \frac{\pi}{3} + \phi_c) - \sin(2\alpha + \phi_c + \frac{\pi}{3})\Big] \tag{14.124}$$

$$-\sqrt{6}V_s I_{cd}\sin(\zeta_o)\Big[\sin(\gamma - \frac{\pi}{6} + \zeta_o)\exp(-\frac{\gamma - \frac{\pi}{3}}{\tan(\zeta_o)})$$

$$-\sin(\alpha + \frac{\pi}{6} + \zeta_o)\exp(-\frac{\alpha}{\tan(\zeta_o)})\Big]\Bigg\}$$

The input kVA as conventionally measured, is $\sqrt{3}$ times the product of the r.m.s. line to line voltage of the supply and the r.m.s. line current. The r.m.s. value of the line to line voltage is $\sqrt{3}V_s$ and the r.m.s. value of the line current is given by eqn 14.117.

The input power factor as conventionally measured is the ratio of the mean input power and the input kVA.

Example 14.10

Determine the r.m.s. values of the currents and voltages, the mean power delivered to the load, the input kVA as conventionally measured, and the input power factor as conventionally measured for the example under the conditions of Examples 14.2 and 14.5.

Following the procedures given above, the values for the mode 3-2 case of Example 14.2 with a delay angle of 70° are

Supply line to line voltage = 440.00 V r.m.s.
Load line to supply neutral voltage = 242.57 V r.m.s.
Load neutral to supply neutral voltage = 14.18 V r.m.s.
Load line to line voltage = 419.43 V r.m.s.
Supply line current = 74.48 A r.m.s.
Load phase current = 74.48 A r.m.s.
Thyristor current = 52.66 A r.m.s.
Average power drawn from the supply = 36.06 kW
Average load power = 36.06 kW
Input kVA as conventionally measured = 56.76
Input power factor as conventionally measured = 0.6353

The corresponding quantities for the mode 2-0 case of Example 14.5 with a delay angle of 100° are

Supply line to line voltage = 440.00 V r.m.s.
Load line to supply neutral voltage = 237.64 V r.m.s.
Load neutral to supply neutral voltage = 14.43 V r.m.s.
Load line to line voltage = 410.85 V r.m.s.
Supply line current = 7.66 A r.m.s.
Load phase current = 7.66 A r.m.s.
Thyristor current = 5.42 A r.m.s.
Average power drawn from the supply = 2.16 kW
Average load power = 2.16 kW
Input kVA as conventionally measured = 5.84
Input power factor as conventionally measured = 0.3706

****

14.14 Fourier analysis

Because the determination of the extinction angle requires the solution of a transcendental equation, the problem must initially be solved in the time domain. However, once this has been done, Fourier analysis of the time domain waveforms is helpful because the spectral contents of the load and supply currents are important in many applications.

Fourier analysis can proceed in one of three ways. In the first method a complete time domain analysis is made and the Fourier components are found directly from the resulting analytic expressions for the waveforms. In the second method the time domain solution is carried as far as the determination of the waveform of the current in a load phase. This is then analyzed into its Fourier components and the components of all other quantities are derived from these. In the third method the time domain solution is carried only as far as the determination of the extinction angle. This defines the waveforms of the voltages at the load terminals. The Fourier analysis is performed on the potential of a load terminal and the spectra of all other components are derived by the application of a.c. circuit theory. This third approach will be used here.

14.14.1 *Spectrum of potential of terminal D*

The potential of load terminal D has half wave symmetry so that there are no even harmonics and its Fourier components can be determined by the use of the half range integrals of Section C.5.

The potential is composed of a series of subsegments of sinusoidal waveform, twelve subsegments per cycle, as defined in Table 14.3 for the 3-2 mode and Table 14.4 for the 2-0 mode. The potential during the nth sub-segment, which extends from θ_n to θ_{n+1}, can be expressed in general terms as

$$v_{dn} = V_{dn} \sin(\theta + \phi_{dvn}) \tag{14.125}$$

Thus, from eqns C.15 through C.17, the amplitudes, V_{dcm} and V_{dsm}, of the cosine and sine Fourier components of mth order, where m is odd, are

$$V_{dcm} = \frac{1}{\pi} \sum_{n=1,6} \int_{\theta_n}^{\theta_{n+1}} \sin(\theta + \phi_{dvn}) \cos(m\theta) \mathrm{d}\theta \tag{14.126}$$

and

$$V_{dsm} = \frac{1}{\pi} \sum_{n=1,6} \int_{\theta_n}^{\theta_{n+1}} \sin(\theta + \phi_{dvn}) \sin(m\theta) \mathrm{d}\theta \tag{14.127}$$

Expressions for these integrals are contained within the corresponding integrals for the general power electronics expression, eqns A1.33 through A.48. In these expressions X_{11}, X_{13}, and X_{14} are set to zero, X_{12} to V_{dn} and ϕ_{12} to ϕ_{dvn} to obtain, from eqn A.33 for $m \neq 1$

$$V_{dcm} = \sum_{n=1,6} \frac{V_{dn}}{2\pi} \left[-\frac{\cos((m+1)\theta + \phi_{dvn})}{m+1} + \frac{\cos((m-1)\theta - \phi_{dvn})}{m-1} \right]_{\theta_n}^{\theta_{n+1}} \tag{14.128}$$

and from eqn A.41 for $m \neq 1$

$$V_{dsm} = \sum_{n=1,6} \frac{V_{dn}}{2\pi} \left[-\frac{\sin((m+1)\theta + \phi_{dvn})}{m+1} + \frac{\sin((m-1)\theta - \phi_{dvn})}{m-1} \right]_{\theta_n}^{\theta_{n+1}} \tag{14.129}$$

These expressions are not valid when $m = 1$ because the $m - 1$ term in the denominator of the second part of the expressions is then zero. The correct expressions for this case are given in eqns A.40 and A.48 so that for $m = 1$

$$V_{dc1} = \sum_{n=1,6} \frac{V_{dn}}{2\pi} \left[-\frac{\cos(2\theta + \phi_{dvn})}{2} + \sin(\phi_{dvn})\, \theta \right]_{\theta_n}^{\theta_{n+1}} \tag{14.130}$$

and

$$V_{ds1} = \sum_{n=1,6} \frac{V_{dn}}{2\pi} \left[-\frac{\sin(2\theta + \phi_{dvn})}{2} + \cos(\phi_{dvn})\, \theta \right]_{\theta_n}^{\theta_{n+1}} \tag{14.131}$$

The amplitude and phase angle of the Fourier component is obtained from its cosine and sine parts by the use of eqns C.5 and either C.6 or C.7.

Since sine functions are being used, eqn C.7 is the appropriate one for the phase angle so that

$$v_d = \sum_{m=1,\infty,2} V_{dm}\sin(m\theta + \phi_{dvm}) \tag{14.132}$$

where

$$V_{dm} = \sqrt{V_{dcm}^2 + V_{dsm}^2)} \tag{14.133}$$

and

$$\phi_{dvm} = \arctan\left(\frac{V_{dcm}}{V_{dsm}}\right) \tag{14.134}$$

14.14.2 *Spectra of potentials of terminals E and F*

At the beginning of this chapter, eqns 14.7 through 14.9, it was seen that, because of the three phase symmetry, the potential of load terminal E is obtained by substituting $\theta - 2\pi/3$ for θ in the expression for v_d and the potential of load terminal F is obtained by substituting $\theta + 2\pi/3$ for θ in the expression for v_d. Thus, from eqn 14.133

$$v_e = \sum_{m=1,\infty,2} V_{dm}\sin[m(\theta - \frac{2\pi}{3}) + \phi_{dvm}] \tag{14.135}$$

and

$$v_f = \sum_{m=1,\infty,2} V_{dm}\sin[m(\theta + \frac{2\pi}{3}) + \phi_{dvm}] \tag{14.136}$$

The introduction of $\theta \pm 2\pi/3$ makes it convenient to subdivide the Fourier series into three components in which successive values of m change by 6 rather than 2, a procedure analogous to that introduced in Section 8.4. The values of m are then

$$m = 6k + 1 \text{ i.e. } 1,\ 7,\ 13,\ \text{etc.} \tag{14.137}$$

$$m = 6k + 3 \text{ i.e. } 3,\ 9,\ 15,\ \text{etc.} \tag{14.138}$$

$$m = 6k + 5 \text{ i.e. } 5,\ 11,\ 17,\ \text{etc.} \tag{14.139}$$

where k is any positive integer including zero.

Using the nomenclature of Section 8.4, the $6k + 1$ series is the positive sequence series, the $6k+3$ series is the zero sequence series, and the $6k+5$ series is the negative sequence series.

For the positive sequence series the angle $m2\pi/3$ can be written $2k \times 2\pi + 2\pi/3$, for the zero sequence series it can be written $(2k+1)2\pi$, and for the negative sequence series it can be written $(2k+1) \times 2\pi - 2\pi/3$. Since changing an angle by an integral multiple of 2π does not affect the circular functions of the angle, eqns 14.133, 14.136, and 14.137 may be rewritten as

$$\begin{aligned} v_d = & \sum_{m=1,\infty,6} V_{dm} \sin(m\theta + \phi_{dvm}) + \\ & \sum_{m=3,\infty,6} V_{dm} \sin(m\theta + \phi_{dvm}) + \\ & \sum_{m=5,\infty,6} V_{dm} \sin(m\theta + \phi_{dvm}) \end{aligned} \tag{14.140}$$

$$\begin{aligned} v_e = & \sum_{m=1,\infty,6} V_{dm} \sin(m\theta - \frac{2\pi}{3} + \phi_{dvm}) + \\ & \sum_{m=3,\infty,6} V_{dm} \sin(m\theta + \phi_{dvm}) + \\ & \sum_{m=5,\infty,6} V_{dm} \sin(m\theta + \frac{2\pi}{3} + \phi_{dvm}) \end{aligned} \tag{14.141}$$

$$\begin{aligned} v_f = & \sum_{m=1,\infty,6} V_{dm} \sin(m\theta + \frac{2\pi}{3} + \phi_{dvm}) + \\ & \sum_{m=3,\infty,6} V_{dm} \sin(m\theta + \phi_{dvm}) + \\ & \sum_{m=1,\infty,6} V_{dm} \sin(m\theta - \frac{2\pi}{3} + \phi_{dvm}) \end{aligned} \tag{14.142}$$

The triplens, i.e. Fourier components whose order is a multiple of three, which make up the zero sequence series, have the same phase angle in all three phases. Thus, for these components the potentials of the load terminals oscillate in synchronism. Because of the isolated neutral there can be no currents associated with these components and the zero sequence potential appears as the potential of the load neutral.

The positive and negative sequence systems are balanced three phase systems with the same amplitude in each phase and a phase difference of 120° between phases. As far as the load model is concerned there is no difference between the two systems. However, rotating electric machines present a different impedance to the two systems which must as a consequence be treated individually.

14.14.3 *The potential of the load neutral*

It has just been seen that the potential of the load neutral is represented by the zero sequence series, the triplens, and is therefore at three times the supply frequency, a fact already noted in connection with Figs 14.6 and 14.7

$$v_n = \sum_{m=3,\infty,6} V_{dm}\sin(m\theta + \phi_{dvm}) \tag{14.143}$$

14.14.4 *The voltage across a load phase*

The voltage across a load phase is the difference between the potential of the phase terminal and that of the load neutral. It is therefore the sum of the positive and negative sequence series

$$v_{dn} = \sum_{m=1,\infty,6} V_{dm}\sin(m\theta+\phi_{dvm}) + \sum_{m=5,\infty,6} V_{dm}\sin(m\theta+\phi_{dvm}) \tag{14.144}$$

14.14.5 *The voltage between load terminals.*

The voltage between a pair of load terminals is the difference between the potentials of those terminals and therefore does not contain any zero sequence components. As an example, v_{de} is given by

$$v_{de} = v_d - v_e = \sum_{m=1,\infty,6} V_{dm}[\sin(m\theta + \phi_{dvm}) - \sin(m\theta - \frac{2\pi}{3} + \phi_{dvm})] + \sum_{m=5,\infty,6} V_{dm}[\sin(m\theta + \phi_{dvm}) - \sin(m\theta + \frac{2\pi}{3} + \phi_{dvm})]$$

which reduces to

$$v_{de} = \sum_{m=1,\infty,6} \sqrt{3}V_{dm}\sin(m\theta + \phi_{dvm} + \frac{\pi}{6}) + \sum_{m=5,\infty,6} \sqrt{3}V_{dm}\sin(m\theta + \phi_{dvm} - \frac{\pi}{6}) \tag{14.145}$$

Example 14.11

Determine the amplitudes and phase angles of the Fourier components, up to and including the 35th order, of the potentials of load terminal D and the load neutral, and the voltages across load phase D and between load terminals D and E for the cases of Examples 14.2 and 14.5.

Table 14.5 *Expressions for the potential of load terminal D when $\alpha = 70°$ and the system is in mode 3-2 with $\gamma = 209.56°$*

Subseg.	Start	Finish	v_d
1	70.00°	89.56°	$359.26\sin(\theta)$
2	89.56°	130.00°	$359.26\sin(\theta)$
3	130.00°	149.56°	$359.26\sin(\theta)$
4	149.56°	190.00°	$359.26\sin(\theta)$
5	190.00°	209.56°	$359.26\sin(\theta)$
6	209.56°	250.00°	$312.90\sin(\theta - 0.0824)$

Expressions for the potential of load terminal D during the half cycle $\alpha < \theta < \alpha + \pi$ have been found in Example 14.8 following the prescriptions given in Tables 14.3 and 14.4. For the case of Example 14.2 when the delay angle is 70°, the extinction angle is 209.56°, and the system is in the 3-2 mode, the expressions are given in Table 14.5.

Substitution of these expressions into eqns 14.128 through 14.131 gives the amplitudes of the cosine and sine components. As an example, for the fifth order Fourier component $V_{dc5} = 4.0321$ V and $V_{ds5} = 12.4777$ V. Converting these to an amplitude and phase angle yields the fifth order component as

$$v_{d5} = 13.1130\sin(5\theta + 0.3126)$$

The amplitudes and phase angles of all the Fourier components up to order 35 are listed for the potential of load terminal D and for the load neutral in Table 14.6. Since there are no even harmonics these are omitted from the table. Phase angles are given in degrees normalized to the range ±180°.

The Fourier components of the voltage across load phase D are the components of the potential of terminal D less the triplens. The components of the voltage between load terminals D and E are obtained by following the procedure of eqn 14.145. Consider for example the fundamental and the fifth order components. The fundamental component of v_{dn} is 342.05 $\sin(\theta - 0.0426)$ and, as this is a positive sequence component,

$$v_{de1} = 342.05\sin(\theta - 0.0426) - 342.05\sin(\theta - 0.0426 - \frac{2\pi}{3})$$

i.e.

$$v_{de1} = 592.44\ \sin(\theta + 0.4810)$$

Table 14.6 *Fourier components of the potentials of load terminal D and of the load neutral when $\alpha = 70°$*

Order	Freq.	V_d		V_n	
		Amplitude	Phase	Amplitude	Phase
	Hz	A	degrees	A	degrees
1	60	342.05	−2.44		
3	180	19.22	119.04	19.22	119.04
5	300	13.11	17.91		
7	420	6.15	−85.19		
9	540	0.54	97.00	0.54	97.00
11	660	3.77	−93.76		
13	780	4.83	162.47		
15	900	3.60	59.67	3.60	59.67
17	1020	1.16	−50.57		
19	1140	1.32	53.60		
21	1260	2.70	−53.90	2.70	−53.90
23	1380	2.67	−156.67		
25	1500	1.48	98.80		
27	1620	0.27	−133.94	0.27	−133.94
29	1740	1.56	90.13		
31	1860	2.03	−13.38		
33	1980	1.54	−116.44	1.54	−116.44
35	2100	0.43	128.97		

The fifth order component of v_{dn} is $13.11\sin(5\theta + 0.3126)$ and, as this is a negative sequence component,

$$v_{de5} = 13.11\sin(5\theta + 0.3136) - 13.11\sin(5\theta + 0.3136 + \frac{2\pi}{3})$$

i.e.

$$v_{de5} = 22.71\ \sin(5\theta - 0.2100)$$

The amplitudes and phase angles of the components of v_{dn} and v_{de} are given in Table 14.7.

The spectrum of the potential of terminal D and its waveform are shown in Fig. 14.8. Fourier components beyond the fundamental are small but, even including components up to the 35th, the series does not accurately model the discontinuities in the voltage wave. Discontinuities always present problems to Fourier series since the Fourier components are continuous functions. However, as will soon be seen when deriving the current spectra, this difficulty is of very little consequence.

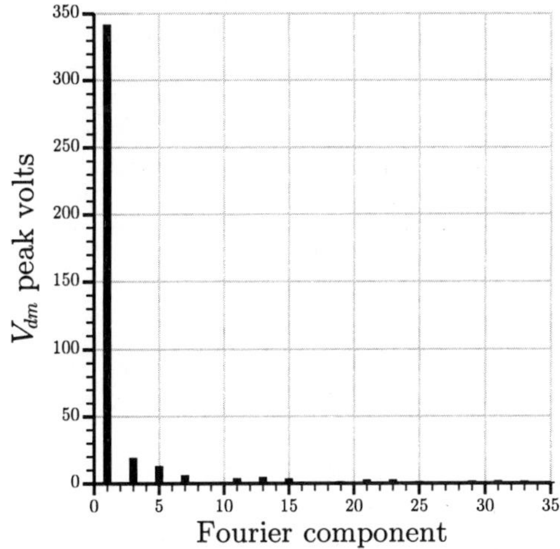

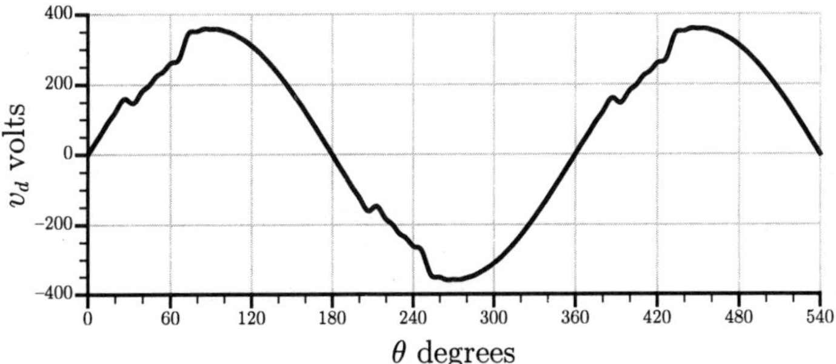

FIG. 14.8. The frequency spectrum and waveform of the potential of load terminal D.

An indication of the smallness of the harmonic content is provided by considering the r.m.s. values. In Example 14.10 the time domain solution gave the r.m.s. value of v_d as 242.572 V. The truncated Fourier series gives 242.543 V, a difference of only 0.01%, well within the uncertainty of all the calculations. From Table 14.6, the r.m.s. value of the fundamental is $342.05/\sqrt{2} = 241.87$ V. Thus, the r.m.s. value of all the harmonics is only $\sqrt{242.57^2 - 241.87^2} = 18.50$ V and the r.m.s. value of the remainder of the Fourier series which has been neglected is $\sqrt{242.57^2 - 242.54^2} = 3.74$ V.

For the case of Example 14.5 when the delay angle is 100°, the extinction angle is 200.23° and the system is in the 2-0 mode. Expressions for the potential of terminal D were found in Example 14.8 and are listed in Table 14.8. Substitution of these expressions in eqns 14.129 through 14.132 and conversion of the resulting cosine and sine component amplitudes into

Table 14.7 *Fourier components of the voltages across load phase D and between load terminals D and E when $\alpha = 70°$*

Order	$\mathbf{V}_{dn}$		$\mathbf{V}_{de}$	
	Amplitude V	Phase degrees	Amplitude V	Phase degrees
1	342.05	−2.44	592.44	27.56
5	13.11	17.91	22.71	−12.09
7	6.15	−85.19	10.65	−55.19
11	3.77	−93.76	6.53	−123.76
13	4.83	162.47	8.37	−167.53
17	1.16	−50.57	2.01	−80.57
19	1.32	53.60	2.29	83.60
23	2.67	−156.67	4.62	173.33
25	1.48	98.80	2.55	128.80
29	1.56	90.13	2.70	60.13
31	2.03	−13.38	3.52	16.62
35	0.43	128.97	0.74	98.97

amplitudes and phases gives the Fourier components for v_d and v_n listed in Table 14.9. The accuracy of this truncated series is just as high as that of the previous one, the r.m.s. value of v_d from the series being 233.2567 V as compared to the value of 233.2700 V derived from the time domain results.

****

14.14.6 *Fourier components of the currents*

Having obtained the Fourier components of the voltage across load phase D, the components of the current in this phase are easily obtained by the application of a.c. circuit theory.

The resistance of a phase is R_o and its equivalent reactance at the supply frequency is X_o. The equivalent reactance to the Fourier component of mth order is mX_o and the equivalent impedance, $\mathbf{Z}_{om}$, is

$$\mathbf{Z}_{om} = R_o + \mathrm{j}mX_o \tag{14.146}$$

i.e.

$$\mathbf{Z}_{om} = Z_{om}\underline{/\zeta_{om}} \tag{14.147}$$

where

$$Z_{om} = \sqrt{R_o{}^2 + (m\,X_o)^2} \tag{14.148}$$

Table 14.8 *Expressions for the potential of load terminal D when $\alpha = 100°$ and the system is in mode 2-0 with $\gamma = 200.23°$*

Subseg.	Start	Finish	v_d
1	100.00°	140.23°	$359.26\sin(\theta)$
2	140.23°	160.00°	$328.10\sin(\theta - 0.0524)$
3	160.00°	200.23°	$359.26\sin(\theta)$
4	200.23°	220.00°	$328.10\sin(\theta - 0.0524)$
5	220.00°	260.23°	$312.90\sin(\theta - 0.0824)$
6	260.23°	280.00°	$328.10\sin(\theta - 0.0524)$

and

$$\zeta_{om} = \arctan(\frac{mX_o}{R_o}) \tag{14.149}$$

When $m \neq 1$ the mth order current component is obtained by dividing the mth order component of v_{dn} by the mth order impedance. Thus, for $m \neq 1$

$$\mathbf{I}_{dm} = \frac{\mathbf{V}_{dnm}}{\mathbf{Z}_{om}} \tag{14.150}$$

where

$$\mathbf{V}_{dnm} = V_{dnm}\underline{/\phi_{dvm}} \tag{14.151}$$

and is found as described in eqns 14.130 and 14.131 with the proviso that m cannot be a multiple of three. Thus,

$$\mathbf{I}_{dm} = I_{dm}\underline{/\ \phi_{dcm}} \tag{14.152}$$

where

$$\mathbf{I}_{dm} = \frac{\mathbf{V}_{dnm}}{\mathbf{Z}_{om}} \tag{14.153}$$

and

$$\phi_{dcm} = \phi_{dvm} - \zeta_{om} \tag{14.154}$$

The fundamental, $m = 1$, is a special case since the driving voltage is then the difference between the fundamental component of v_{dn} and the load e.m.f., e_d, which has been called $v_{\delta r}$ and has been derived in eqn 14.25. Thus, for $m = 1$

$$\mathbf{I}_{dm} = \frac{\mathbf{V}_{\delta r}}{\mathbf{Z}_{o1}} \tag{14.155}$$

The currents in the a.c. lines are the same as the currents in the load phases and have the same Fourier series.

Table 14.9 *Fourier components of the potentials of load terminal D and of the load neutral when $\alpha = 100°$*

Order	$\mathbf{V}_d$		$\mathbf{V}_n$	
	Amplitude V	Phase degrees	Amplitude V	Phase degrees
1	329.21	−2.99		
3	19.56	89.47	19.56	89.47
5	3.56	−35.75		
7	3.52	26.73		
9	0.18	67.82	0.18	67.82
11	1.28	−27.23		
13	0.16	161.39		
15	3.73	−92.12	3.73	−92.12
17	1.40	146.80		
19	1.09	−150.64		
21	2.70	87.78	2.70	87.78
23	0.35	−43.49		
25	0.82	24.45		
27	0.10	−81.02	0.10	−81.02
29	0.54	−29.52		
31	0.05	95.81		
33	1.64	−94.18	1.64	−94.18
35	0.66	144.87		

The currents in the thyristors cannot be obtained directly but must be derived by multiplying the Fourier series for the phase current by the appropriate switching function as described in Section 8.19. This is an indirect method which can only produce an approximation to the series and it is more appropriate to analyze directly the time domain solution should these Fourier components be required.

14.14.7 *Mean power of the Fourier components*

The mean output power, P_{om}, associated with the mth order component is three times the triple product of the r.m.s. value of the phase voltage component, the r.m.s. value of the phase current component, and the cosine of their phase difference. Thus

$$P_{om} = 3\frac{V_{dnm}}{\sqrt{2}}\frac{I_{dm}}{\sqrt{2}}\cos(\phi_{dvm} - \phi_{dcm})$$

i.e.

$$P_{om} = \frac{3}{2} V_{dnm} I_{dm} \cos(\phi_{dvm} - \phi_{dcm}) \tag{14.156}$$

The mean input power, P_{11}, associated with the fundamental is

$$P_{11} = \frac{3}{2} \sqrt{2} V_s I_{d1} \cos(-\phi_{dc1}) \tag{14.157}$$

The mean input power associated with the higher order components is zero since a supply of negligible internal impedance with voltages of sinusoidal waveform is assumed.

Energy conservation requires that the mean input power equal the sum of the components of mean load power. Some of the fundamental component of input power is converted to components of higher frequency which are injected back into the supply.

This procedure only provides mean values. If the instantaneous values are required it is necessary to sum the products of the series for the load phase voltages with those for the load phase currents as described in Appendix D. As for the thyristor currents, it is both more efficient and more accurate to obtain this result directly from the time domain solution.

Example 14.12

Determine the Fourier series representing the current in load phase D up to and including the component of 35th order for the cases of Examples 14.2 and 14.5. Determine also the mean power associated with each component.

For the case of Example 14.2, the Fourier components of v_{dn} are obtained from Table 14.7. For the fundamental

$$v_{dn1} = 342.05 \ \sin(\theta - 0.0426)$$

and

$$e_d = 328.10 \ \sin(\theta - 0.0524)$$

so that

$$v_{\delta r} = v_{dn1} - e_d = 14.33 \ \sin(\theta - 0.1835)$$

The equivalent phase impedance at the fundamental frequency is

$$\mathbf{Z}_{o1} = 0.0700 + \mathrm{j}0.1206 = 0.1395 \underline{/ \ 1.0450}$$

where the phase angle is in radians.

Table 14.10 *Fourier components of the current in load phase D and the mean load powers associated with these components for the case when the delay angle is 70° and the system is in the 3-2 mode*

Order	$\mathbf{I}_{dm}$ Amplitude A	$\mathbf{I}_{dm}$ Phase degrees	Power kW
1	102.74	−49.36	36.003
5	21.59	−65.47	0.049
7	7.26	−170.45	0.006
11	2.84	179.26	0.001
13	3.08	75.03	0.001
17	0.56	−138.61	0.000
19	0.58	−34.65	0.000
23	0.96	−245.22	0.000
25	0.49	10.13	0.000
29	0.45	1.28	0.000
31	0.54	−102.30	0.000
35	0.10	39.92	0.000

The fundamental component of current is

$$\mathbf{I}_{d1} = \frac{14.33\underline{/0.1855}}{0.1395\underline{/1.0450}} = 102.74\underline{/-0.8615}$$

i.e.

$$i_{d1} = 102.74\sin(\theta - 0.8615)$$

The mean output power associated with this component is

$$P_{o1} = \frac{3}{2} \times 342.05 \times 102.74 \times \cos(-0.0426 + 0.8615) \div 1000.0 = 36.00 \text{ kW}$$

The fifth order voltage component is

$$v_{dn5} = 13.11\sin(5\theta + 0.3126)$$

and the impedance of this component is

$$\mathbf{Z}_{o5} = 0.0700 + \text{j}5 \times 0.1206 = 0.6072\underline{/1.4553}$$

The current component is therefore

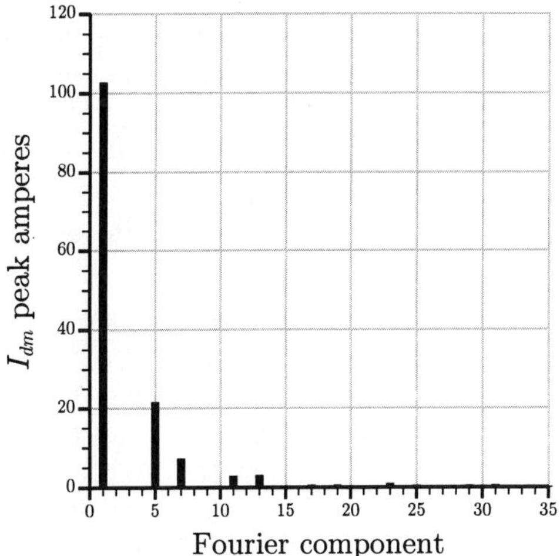

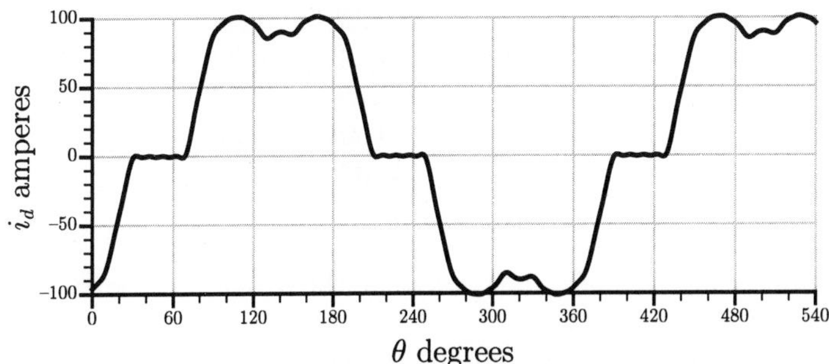

FIG. 14.9. The frequency spectrum and waveform of the current in load phase D.

$$\mathbf{I}_{d5} = \frac{13.11\underline{/0.3126}}{0.6072\underline{/1.4553}} = 21.59\underline{/-1.1427}$$

i.e.

$$i_{d5} = 21.59\ \sin(5\theta - 1.1427)$$

The mean output power associated with this component is

$$P_{o5} = 1.5 \times 13.11 \times 21.59 \times \cos(0.3126 + 1.1427) \div 1000.0 = 0.049\ \text{kW}$$

The complete set of values listed in Table 14.10 has been obtained following this method.

In Example 14.10 it was found that the mean power drawn from the supply was 36.0596 kW. The sum of the mean powers associated with the

Table 14.11 *Fourier components of the current in load phase D and the mean load power associated with this component for the case when the delay angle is 100° and the system is in the 2-0 mode*

Order	$\mathbf{I}_{dm}$		Power
	Amplitude A	Phase degrees	kW
1	7.99	−59.85	2.157
5	5.86	−119.13	0.004
7	4.16	−58.54	0.002
11	0.96	−114.21	0.000
13	0.10	73.95	0.000
17	0.68	−58.76	0.000
19	0.47	121.11	0.000
23	0.13	−132.05	0.000
25	0.27	−64.22	0.000
29	0.15	−118.37	0.000
31	0.01	6.88	0.000
35	0.16	55.82	0.000

Fourier components is 36.0597 kW, equal to the mean input power within the accuracy of the calculations.

The spectrum of the current and its waveform are shown in Fig. 14.9. In contrast to the voltage waveform of Fig. 14.8 the current waveform is indistinguishable from that obtained from the time domain solution and shown in Fig. 14.6, graph 5. This is an indication of the faster convergence of the current series due to the lack of current discontinuities.

A further indication of the accuracy of the truncated Fourier series is provided by considering the r.m.s. values. In Example 14.10 the time domain solution gave the r.m.s. value of i_d as 74.4786 A. The truncated Fourier series gives 74.4775 A, a difference of only 0.001%, well within the uncertainty of all the calculations. From Table 14.10 the r.m.s. value of the fundamental is $102.74/\sqrt{2} = 72.65$ A. Thus, the r.m.s. value of all the harmonics is only $\sqrt{74.48^2 - 72.65^2} = 16.42$ A and the r.m.s. value of the remaining terms of the Fourier series which has been neglected is $\sqrt{74.4786^2 - 74.4775^2} = 0.40$ A.

The components for the case of Example 14.5 when the system is in the 2-0 mode have been obtained by similar methods and are listed in Table 14.11. They show the same rapid convergence and high accuracy.

****

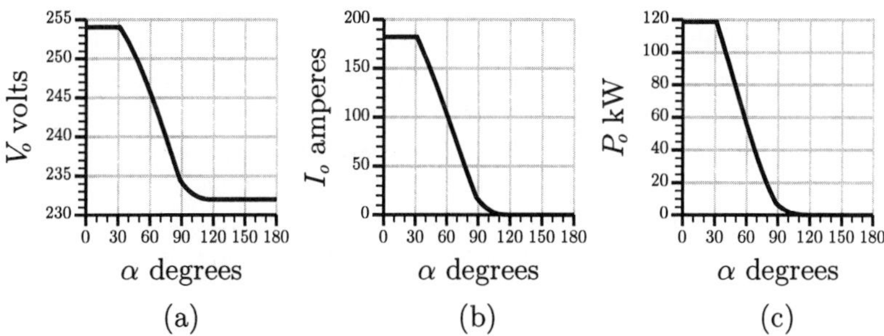

FIG. 14.10. Static control characteristics for the example. The r.m.s. values of the fundamental components of load phase voltage and current, and the mean power associated with the fundamental are shown as functions of delay angle.

14.15 The static control characteristics

A system such as this will often be used within a feedback control loop and the relationship between the controlling and the controlled quantities is of importance to the control system designer. The controlling variable is the delay angle. The controlled quantity depends on the type of system. In the case of a heating load it would be the mean load power. In the case of an electric motor it would be the fundamental component of either voltage, current, or power depending on the circumstances.

The controller itself does not have a unique control characteristic, the characteristic depending on interaction between the load and the controller. Figure 14.10 shows the control characteristics for the load which has been used for the example, the controlled quantities being the fundamental components of output voltage, current, and power. While being unique to this particular combination of a.c. regulator and load, their form is typical of the characteristics encountered in practice.

The characteristics have been drawn on the assumption that the delay angle controller will function correctly over a delay angle range of zero to 180°, a range suitable for the vast majority of loads. When the delay angle is less than the minimum effective value, 31.36° in this case, the thyristor gate pulses are wide enough to permit operation at the minimum delay. All three characteristics are therefore flat at the maximum value, for delay angles less than the minimum effective value. The gate pulses are also narrow enough to permit correct operation when the delay angle exceeds the cutoff value which in this case is 121.49°. Thus, the characteristics are flat at the minimum value above the cutoff delay, the minimum value being that of the load e.m.f. in the case of voltage and zero in the cases of current and power. The characteristics are more or less linear when the delay is between the minimum value and the critical value which is 88.41° in this case. The characteristics are highly nonlinear when the delay exceeds the critical value. The delay angle range from the minimum to the critical value covers most of the control range. In the example the power at the critical point is 6.17% of the maximum power, 7.32 kW as against 118.7 kW. While the delay angle range and the precise shape of the characteristics vary with the load, these observations apply to most situations and it would be rare for an actual system to enter the 2-0 mode.

APPENDIX A

THE BASIC POWER ELECTRONICS EXPRESSION

A.1 Introduction

The quantitative analysis of power electronic equipment frequently gives rise to expressions which are combinations of constant, sinusoidal, and exponential terms. A number of standard operations are performed on these expressions. If the expression represents a current and the voltage drop associated with its flow through an inductance is required, the derivative with respect to time of the expression is needed. The determination of the average value of the expression requires its integral with respect to time. If there is one such expression representing a voltage and another representing a current, their product is the instantaneous power. The integral of the product with respect to time is required for the average power. If the r.m.s. value of the expression is required it is necessary to obtain the integral of its square. Determination of the Fourier series representing the expression requires the integral of its product with $\cos(m\theta)$ and with $\sin(m\theta)$.

The above operations are facilitated if there is a general expression upon which the various algebraic manipulations have been performed and if there are subroutines which will return numerical values. The algebraic expressions for one of the commonest situations, a rectifier in an active circuit with resistance and inductance but no capacitance, are listed below. The slightly different case of a capacitor smoothed circuit is considered in Appendix B and that of a cycloconverter in Appendix E.

A.2 The general expression

When the circuit is free from capacitance, the most complex expression is that for a thyristor current during rectifier commutation, see eqn 5.67. It has four terms, a d.c. term, a sinusoidal term, and two exponentially decaying terms. All other expressions for rectifier circuits without capacitance can be derived from this expression by setting one or more of its coefficients to zero. In the text it has been called the basic power electronics expression. Here it is denoted by the symbol $f_1(\theta)$ and is expressed in the generalized form

$$f_1(\theta) = X_{11} + X_{12}\sin(\theta + \phi_{12}) + X_{13}\exp(-\frac{\theta}{T_{13}}) + X_{14}\exp(-\frac{\theta}{T_{14}}) \quad \text{(A.1)}$$

where $\theta = \omega_s t$, ω_s being the radian frequency of the supply and t being the time in seconds measured from some arbitrary origin.

Situations may be encountered, eqn 5.67 is one such, where it is convenient to use the cosine form of the second term. This is converted to the standard form of eqn A.1 by adding $\pi/2$ to the phase angle. Thus

$$\cos(\theta + \phi) = \sin[\theta + (\phi + \frac{\pi}{2})]$$

A.3 The derivative of the general expression

The derivative of the general expression with respect to θ is $f_1'(\theta)$ and is given by

$$f_1'(\theta) = X_{12}\ \cos(\theta + \phi_{12}) - \frac{X_{13}}{T_{13}}\exp(-\frac{\theta}{T_{13}}) - \frac{X_{14}}{T_{14}}\exp(-\frac{\theta}{T_{14}}) \quad \text{(A.2)}$$

The derivative with respect to time is obtained by multiplying this expression by the radian frequency of the supply, ω_s.

A.4 The integral of the general expression

The integral of the general expression with respect to θ is

$$\int f_1(\theta)\,\mathrm{d}\theta = X_{11}\theta - X_{12}\cos(\theta + \phi_{12}) - \\ T_{13}X_{13}\exp(-\frac{\theta}{T_{13}}) - T_{14}X_{14}\exp(-\frac{\theta}{T_{14}}) \quad \text{(A.3)}$$

The constant of integration has been omitted since the expression will always be used as a definite integral.

The integral with respect to time is obtained by dividing the expression by the radian frequency of the supply, ω_s.

A.5 The product of general expressions

The product of two general expressions is required in two situations, when determining power and when determining r.m.s. values. Given here is the product of two different general expressions, the first being that of eqn A.1 and the second $f_2(\theta)$ given by

$$f_2(\theta) = X_{21} + X_{22}\sin(\theta + \phi_{22}) + X_{23}\exp(-\frac{\theta}{T_{13}}) + X_{24}\exp(-\frac{\theta}{T_{14}}) \quad \text{(A.4)}$$

where the time constants have been set equal to those in the first expression since this will always be the case in practice.

For power, one general expression will represent a voltage and the other a current. For r.m.s. values, the two expressions will represent the same quantity and will therefore have identical coefficients.

The direct product has 16 terms but one of these is the product of sines and can be separated into the sum of a constant and a double frequency term. Some of the remaining terms can be combined. The result is the general product with thirteen terms, each term being listed separately below, the complete general expression being the sum of expressions A.6 through A.18, i.e.

$$f_1(\theta)\, f_2(\theta) = \sum_{n=1,13} g_n(\theta) \tag{A.5}$$

where

$$g_1(\theta) = X_{11}X_{21} + \frac{X_{12}X_{22}}{2}\cos(\phi_{12} - \phi_{22}) \tag{A.6}$$

$$g_2(\theta) = \frac{X_{12}X_{22}}{2}\sin(2\theta + \phi_{12} + \phi_{22} - \frac{\pi}{2}) \tag{A.7}$$

$$g_3(\theta) = X_{13}X_{23}\exp(-\frac{2\theta}{T_{13}}) \tag{A.8}$$

$$g_4(\theta) = X_{14}X_{24}\exp(-\frac{2\theta}{T_{14}}) \tag{A.9}$$

$$g_5(\theta) = [X_{11}X_{22}\sin(\phi_{22}) + X_{12}X_{21}\sin(\phi_{12})]\cos(\theta) \tag{A.10}$$

$$g_6(\theta) = [X_{11}X_{22}\cos(\phi_{22}) + X_{12}X_{21}\cos(\phi_{12})]\sin(\theta) \tag{A.11}$$

$$g_7(\theta) = (X_{11}X_{23} + X_{13}X_{21})\exp(-\frac{\theta}{T_{13}}) \tag{A.12}$$

$$g_8(\theta) = (X_{11}X_{24} + X_{14}X_{21})\exp(-\frac{\theta}{T_{c4}}) \tag{A.13}$$

$$g_9(\theta) = [X_{12}X_{23}\sin(\phi_{12}) + X_{13}X_{22}\sin(\phi_{22})]\cos(\theta)\exp(-\frac{\theta}{T_{13}}) \tag{A.14}$$

$$g_{10}(\theta) = [X_{12}X_{23}\cos(\phi_{12}) + X_{13}X_{22}\cos(\phi_{22})]\sin(\theta)\exp(-\frac{\theta}{T_{13}}) \tag{A.15}$$

$$g_{11}(\theta) = [X_{12}X_{24}\sin(\phi_{12}) + X_{14}X_{22}\sin(\phi_{22})]\cos(\theta)\exp(-\frac{\theta}{T_{14}}) \tag{A.16}$$

$$g_{12}(\theta) = [X_{12}X_{24}\cos(\phi_{12}) + X_{14}X_{22}\cos(\phi_{22})]\sin(\theta)\exp(-\frac{\theta}{T_{14}}) \tag{A.17}$$

$$g_{13}(\theta) = (X_{13}X_{24} + X_{14}X_{23})\exp(-(\frac{1}{T_{13}} + \frac{1}{T_{14}})\theta) \tag{A.18}$$

A.6 The integral of the product

The integral of the product of two basic expressions is required when deriving average power and r.m.s. values. The integral with respect to θ, omitting the constant of integration, is the sum of the thirteen expressions, eqns A.20 through A.32, i.e.

$$\int f_1(\theta)\, f_2(\theta)\, \mathrm{d}\theta = \sum_{n=1,13} h_n(\theta) \tag{A.19}$$

where

$$h_1(\theta) = [X_{11}X_{21} + \frac{X_{12}X_{22}}{2}\cos(\phi_{22} - \phi_{12})]\,\theta \tag{A.20}$$

$$h_2(\theta) = -\frac{X_{12}X_{22}}{4}\sin(2\theta + \phi_{12} + \phi_{22}) \tag{A.21}$$

$$h_3(\theta) = -\frac{T_{13}X_{13}X_{23}}{2}\exp(-\frac{2\theta}{T_{13}}) \tag{A.22}$$

$$h_4(\theta) = -\frac{T_{14}X_{14}X_{24}}{2}\exp(-\frac{2\theta}{T_{14}}) \tag{A.23}$$

$$h_5(\theta) = [X_{11}X_{22}\sin(\phi_{22}) + X_{12}X_{21}\sin(\phi_{21})]\sin(\theta) \tag{A.24}$$

$$h_6(\theta) = -[X_{11}X_{22}\cos(\phi_{22}) + X_{12}X_{21}\cos(\phi_{12})]\cos(\theta) \tag{A.25}$$

$$h_7(\theta) = -T_{13}(X_{11}X_{23} + X_{13}X_{21})\exp(-\frac{\theta}{T_{13}}) \tag{A.26}$$

$$h_8(\theta) = -T_{14}(X_{11}X_{24} + X_{14}X_{21})\exp(-\frac{\theta}{T_{14}}) \tag{A.27}$$

$$h_9(\theta) = \left\{ \frac{[T_{13}^2\sin(\phi_{12}) - T_{13}\cos(\phi_{12})]X_{12}X_{23}}{1 + T_{13}^2} + \frac{[T_{13}^2\sin(\phi_{22}) - T_{13}\cos(\phi_{22})]X_{13}X_{22}}{1 + T_{13}^2} \right\} \times \sin(\theta)\exp(-\frac{\theta}{T_{13}}) \tag{A.28}$$

$$h_{10}(\theta) = \left\{ -\frac{[T_{13}^2 \cos(\phi_{12}) + T_{13}\sin(\phi_{12})]X_{12}X_{23}}{1+T_{13}^2} + \frac{[T_{13}^2 \cos(\phi_{22}) + T_{13}\sin(\phi_{22})]X_{13}X_{22}}{1+T_{13}^2} \right\} \times \cos(\theta)\exp(-\frac{\theta}{T_{13}}) \qquad (A.29)$$

$$h_{11}(\theta) = \left\{ \frac{[T_{14}^2 \sin(\phi_{12}) - T_{14}\cos(\phi_{12})]X_{12}X_{24}}{1+T_{14}^2} + \frac{[T_{14}^2 \sin(\phi_{22}) - T_{14}\cos(\phi_{22})]X_{14}X_{22}}{1+T_{14}^2} \right\} \times \sin(\theta)\exp(-\frac{\theta}{T_{14}}) \qquad (A.30)$$

$$h_{12}(\theta) = \left\{ -\frac{[T_{14}^2 \cos(\phi_{12}) + T_{14}\sin(\phi_{12})]X_{12}X_{24}}{1+T_{14}^2} + \frac{[T_{14}^2 \cos(\phi_{22}) + T_{14}\sin(\phi_{22})]X_{14}X_{22}}{1+T_{14}^2} \right\} \times \cos(\theta)\exp(-\frac{\theta}{T_{14}}) \qquad (A.31)$$

$$h_{13}(\theta) = -\frac{T_{13}T_{14}(X_{13}X_{24} + X_{14}X_{23})}{T_{13}+T_{14}} \exp\left[-(\frac{1}{T_{13}} + \frac{1}{T_{14}})\theta\right] \qquad (A.32)$$

The integral with respect to time is obtained by dividing the above expressions by the radian frequency.

A.7 The Fourier cosine coefficients

Obtaining the amplitude of the mth Fourier cosine component requires the integral of the product of the general expression and $\cos(m\theta)$. This is the sum of the following five expressions, eqns A.34 through A.38, i.e.

$$\int f_1(\theta)\cos(m\theta)\,\mathrm{d}\theta = \sum_{n=1,5} j_n(\theta) \qquad (A.33)$$

where

$$j_1(\theta) = \frac{X_{11}}{m} \sin(m\theta) \tag{A.34}$$

$$j_2(\theta) = -\frac{X_{12}}{2(m+1)} \cos[(m+1)\theta + \phi_{12}] \tag{A.35}$$

$$j_3(\theta) = \frac{X_{12}}{2(m-1)} \cos[(m-1)\theta - \phi_{12}] \tag{A.36}$$

$$j_4(\theta) = \frac{X_{13}(m \sin(m\theta) - \cos(m\theta)/T_{13})}{m^2 + 1/T_{13}^2)]} \exp(-\frac{\theta}{T_{13}}) \tag{A.37}$$

$$j_5(\theta) = \frac{X_{14}(m \sin(m\theta) - \cos(m\theta)/T_{14})}{m^2 + 1/T_{14}^2)]} \exp(-\frac{\theta}{T_{14}}) \tag{A.38}$$

The expression for $j_1(\theta)$, eqn A.34, is indeterminate when $m = 0$. Its value for this particular case is obtained by applying l'Hôpital's rule as

$$j_1(\theta)_{m=0} = X_{11}\theta \tag{A.39}$$

The expression for $j_3(\theta)$, eqn A.36, is indeterminate when $m = 1$. l'Hôpital's rule gives the expression for this case as

$$j_3(\theta)_{m=1} = \frac{X_{12}}{2} \sin(\phi_{12})\,\theta \tag{A.40}$$

A.8 The Fourier sine coefficients

Obtaining the amplitude of the mth Fourier sine component requires the integral of the product of the general expression and $\sin(m\theta)$. This is the sum of the following five expressions, eqns A.42 through A.46, i.e.

$$\int f_1(\theta) \sin(m\theta)\, \mathrm{d}\theta = \sum_{n=1,5} k_n(\theta) \tag{A.41}$$

where

$$k_1(\theta) = -\frac{X_{11}}{m} \cos(m\theta) \tag{A.42}$$

$$k_2(\theta) = -\frac{X_{12}}{2(m+1)} \sin[(m+1)\theta + \phi_{12}] \tag{A.43}$$

$$k_3(\theta) = +\frac{X_{12}}{2(m-1)} \sin[(m-1)\theta - \phi_{12}] \tag{A.44}$$

$$k_4(\theta) = -\frac{X_{13}[m \cos(m\theta) + \sin(m\theta)/T_{13}]}{m^2 + 1/T_{13}^2} \exp(-\frac{\theta}{T_{13}}) \tag{A.45}$$

$$k_5(\theta) = -\frac{X_{14}[m\cos(m\theta) - \sin(m\theta)/T_{14}]}{m^2 + 1/T_{14}^2}\exp(-\frac{\theta}{T_{14}}) \tag{A.46}$$

When $m = 0$ the eqn A.42 contributes nothing to the definite integral and it is therefore, for this particular case, set to zero

$$k_1(\theta)_{m=0} = 0.0 \tag{A.47}$$

The third term is indeterminate when $m = 1$. l'Hôpital's rule gives the expression for this case as

$$k_3(\theta)_{m=1} = \frac{X_{12}}{2}\cos(\phi_{12})\theta \tag{A.48}$$

APPENDIX B

THE BASIC CAPACITOR SMOOTHING EXPRESSION

B.1 Introduction

In Chapter 7, where capacitor smoothing was considered, the solution of the rectifier differential equations produced an expression which contained an exponentially decaying oscillating term, a manifestation of energy flowing back and forth between the circuit inductance and capacitance. By analogy with the basic rectifier expression of earlier chapters, this was called the basic capacitor smoothing expression. The same kind of functions of this expression are required as were derived in Appendix A, i.e. its derivative, integral, square, etc. Generalized forms of the basic capacitor smoothing expression and its functions are listed here.

B.2 The basic capacitor smoothing expression

The basic expression as used in Chapter 7 has the general form of eqn B.1

$$\begin{aligned} f_{p1}(\theta) = {} & X_{11} + X_{12}\cos(\theta + X_{13}) + X_{14}\exp(X_{15}\theta) + \\ & X_{16}\cos(X_{17}\theta + X_{18})\exp(X_{19}\theta) \end{aligned} \tag{B.1}$$

where three changes from the form of the basic rectifier expression have been made. They are

1. Oscillating terms are expressed as cosine functions.
2. Phase angles are denoted by coefficients, X_{13} and X_{18}, rather than by the subscripted symbol ϕ.
3. The negative of the reciprocals of the time constants are employed in the exponential terms rather than the time constants.

B.2.1 *Cartesian form*

The polar form of $f_{p1}(\theta)$ with its amplitudes and phase angles is appropriate for visualizing and computing the waveform of the expression. However, for algebraic manipulation the Cartesian form, in which phase angles are eliminated by expanding the cosine terms into the sums of cosine and sine terms, is usually preferable. Making this change yields eqn B.2

$$f_{c1}(\theta) = X_{11} + X_{12c}\cos(\theta) + X_{12s}\sin(\theta) + X_{14}\exp(X_{15}\theta) + \\ X_{16c}\cos(X_{17}\theta)\exp(X_{19}\theta) + X_{16s}\cos(X_{17}\theta)\exp(X_{19}\theta) \tag{B.2}$$

where

$$X_{12c} = X_{12}\cos(X_{13}) \tag{B.3}$$

$$X_{12s} = -X_{12}\sin(X_{13}) \tag{B.4}$$

$$X_{16c} = X_{16}\cos(X_{18}) \tag{B.5}$$

$$X_{16s} = -X_{16}\sin(X_{18}) \tag{B.6}$$

B.3 The derivative with respect to θ

The derivative with respect to θ of the basic expression is,

$$f'_{p1}(\theta) = -X_{12}\sin(\theta + X_{13}) + X_{14}X_{15}\exp(X_{15}\theta) + \\ X_{16}\sqrt{X_{17}^2 + X_{19}^2}\cos\left[X_{27}\theta + X_{18} + \arctan(\frac{X_{17}}{X_{19}})\right]\exp(X_{19}\theta) \tag{B.7}$$

B.4 The integral with respect to θ

The integral with respect to θ of the basic expression is

$$\int f_{p1}(\theta)\,d\theta = X_{11}\theta + X_{12}\sin(\theta + X_{13}) + \frac{X_{14}}{X_{15}}\exp(X_{15}\theta) + \\ \frac{X_{16}X_{19}}{X_{17}^2 + X_{19}^2}\cos(X_{17}\theta + X_{18})\exp(X_{19}\theta) + \\ \frac{X_{16}X_{17}}{X_{17}^2 + X_{19}^2}\sin(X_{17}\theta + X_{18})\exp(X_{19}\theta) \tag{B.8}$$

Equation B.8 omits the constant of integration since the expression is always employed as a definite integral between given values of θ.

B.5 The product of two basic expressions

The product of two basic expressions, one for a current and the other for a voltage, is required when deriving an expression for power. The resulting expression is long and can be expressed in numerous ways, the form given below being among the simplest that can be obtained. The two basic expressions which are multiplied together are $f_{p1}(\theta)$, eqn B.1 and $f_{p2}(\theta)$, eqn B.9. Since the two expressions apply to the same circuit they have the same complex natural frequencies, i.e. $X_{25} = X_{15}$, $X_{27} = X_{17}$, and $X_{29} = X_{19}$ and these facts have been included in eqn B.9

$$f_{p2}(\theta) = X_{21} + X_{22}\cos(\theta + X_{23}) + X_{24}\exp(X_{15}\theta) + X_{26}\cos(X_{17}\theta + X_{28})\exp(X_{19}\theta) \tag{B.9}$$

The product is

$$f_{p1}(\theta)\, f_{p2}(\theta) = \sum_{n=1,18} g_n(\theta) \tag{B.10}$$

where the 18 expressions which are summed are given in eqns B.11 through B.28

$$g_1(\theta) = X_{11}X_{21} + \frac{1}{2}X_{12}X_{22}\cos(X_{13} - X_{23}) \tag{B.11}$$

$$g_2(\theta) = \frac{1}{2}X_{12}X_{22}\cos(2\theta + X_{13} + X_{23}) \tag{B.12}$$

$$g_3(\theta) = X_{14}X_{24}\exp(2X_{15}\theta) \tag{B.13}$$

$$g_4(\theta) = \frac{1}{2}X_{16}X_{26}\cos(X_{18} - X_{28})\exp(2X_{19}\theta) \tag{B.14}$$

$$g_5(\theta) = \frac{1}{2}X_{16}X_{26}\cos(2X_{17}\theta + X_{18} + X_{28})\exp(2X_{19}\theta) \tag{B.15}$$

$$g_6(\theta) = (X_{11}X_{24} + X_{14}X_{21})\exp(X_{15}\theta) \tag{B.16}$$

$$g_7(\theta) = [X_{11}X_{22}\cos(X_{23}) + X_{12}X_{21}\cos(X_{13})]\cos(\theta) \tag{B.17}$$

$$g_8(\theta) = -[X_{11}X_{22}\sin(X_{23}) + X_{12}X_{21}\sin(X_{13})]\sin(\theta) \tag{B.18}$$

$$g_9(\theta) = [X_{11}X_{26}\cos(X_{28}) + X_{16}X_{21}\cos(X_{18})] \times \cos(X_{17}\theta)\exp(X_{19}\theta) \tag{B.19}$$

$$g_{10}(\theta) = -[X_{11}X_{26}\sin(X_{28}) + X_{16}X_{21}\sin(X_{18})] \times \sin(X_{17}\theta)\exp(X_{19}\theta) \tag{B.20}$$

$$g_{11}(\theta) = [X_{12}X_{24}\cos(X_{13}) + X_{14}X_{22}\cos(X_{23})] \times \cos(\theta)\exp(X_{15}\theta) \tag{B.21}$$

$$g_{12}(\theta) = -[X_{12}X_{24}\sin(X_{13}) + X_{14}X_{22}\sin(X_{23})] \times \sin(\theta)\exp(X_{15}\theta) \tag{B.22}$$

$$g_{13}(\theta) = \frac{1}{2}[X_{12}X_{26}\cos(X_{13} + X_{28}) + X_{16}X_{22}\cos(X_{18} + X_{23})] \times \cos[(1 + X_{17})\theta]\exp(X_{19}\theta) \tag{B.23}$$

$$g_{14}(\theta) = -\frac{1}{2}[X_{12}X_{26}\sin(X_{13} + X_{28}) + X_{16}X_{22}\sin(X_{18} + X_{23})] \times \sin[(1 + X_{17})\theta]\exp(X_{19}\theta) \tag{B.24}$$

$$g_{15}(\theta) = \frac{1}{2}[X_{12}X_{26}\cos(X_{13} - X_{28}) + X_{16}X_{22}\cos(X_{18} - X_{23})] \times \cos[(1 - X_{17})\theta]\exp(X_{19}\theta) \tag{B.25}$$

$$g_{16}(\theta) = -\frac{1}{2}[X_{12}X_{26}\sin(X_{13} - X_{28}) - X_{16}X_{22}\sin(X_{18} - X_{23})] \times \sin[(1 - X_{17})\theta]\exp(X_{19}\theta) \tag{B.26}$$

$$g_{17}(\theta) = [X_{14}X_{26}\cos(X_{28}) + X_{16}X_{24}\cos(X_{18})] \times \cos(X_{17}\theta)\exp[(X_{15} + X_{19})\theta] \tag{B.27}$$

$$g_{18}(\theta) = -[X_{14}X_{26}\sin(X_{28}) + X_{16}X_{24}\sin(X_{18})] \times \sin(X_{17}\theta)\exp[(X_{15} + X_{19})\theta] \tag{B.28}$$

Once they are in numerical form, those of functions $g_1(\theta)$ through $g_{18}(\theta)$ which form cosine–sine pairs, can be combined into single expressions, either cosine or sine, with the appropriate amplitude and phase. As an example consider expressions $g_{13}(\theta)$, eqn B.23, and $g_{14}(\theta)$, eqn B.24. If A and B are the coefficients then

$$A = \frac{1}{2}[X_{12}X_{26}\cos(X_{13}+X_{28}) + X_{16}X_{22}\cos(X_{18}+X_{23})] \quad \text{(B.29)}$$

$$B = -\frac{1}{2}[X_{12}X_{26}\sin(X_{13}+X_{28}) + X_{16}X_{22}\sin(X_{18}+X_{23})] \quad \text{(B.30)}$$

and the combination of the two functions in cosine form is

$$\sqrt{A^2+B^2}\cos\left[(1+X_{17})\theta - \arctan(\frac{B}{A})\right]\exp(X_{19}\theta) \quad \text{(B.31)}$$

Among the 18 expressions, $g_1(\theta)$ through $g_{18}(\theta)$, there are six such pairs, $g_7(\theta)$ and $g_8(\theta)$, $g_9(\theta)$ and $g_{10}(\theta)$, $g_{11}(\theta)$ and $g_{12}(\theta)$, $g_{13}(\theta)$ and $g_{14}(\theta)$, $g_{15}(\theta)$ and $g_{16}(\theta)$, and $g_{17}(\theta)$ and $g_{18}(\theta)$. Combining each of these pairs reduces the number of expressions from 18 to 12 and simplifies numerical computations.

B.6 The integral of the product

The integral with respect to θ of the product of two basic expressions is required for the computation of average power and r.m.s. values. For the power computation, one expression is a voltage and the other is a current. For the r.m.s. value, the two expressions are identical

$$\int f_{p1}(\theta)\, f_{p2}(\theta)\, d\theta = \sum_{n=1,19} h_n(\theta) \quad \text{(B.32)}$$

where the nineteen $h_n(\theta)$ expressions are listed below in eqns B.33 through B.51. Again the constant of integration is omitted since the integral will always used as a definite integral between specified values of θ

$$h_1(\theta) = [X_{11}X_{21} + \frac{X_{12}X_{22}}{2}\cos(X_{13}-X_{23})]\,\theta \quad \text{(B.33)}$$

$$h_2(\theta) = \frac{X_{12}X_{22}}{4}\sin(2\theta + X_{13} + X_{23}) \quad \text{(B.34)}$$

$$h_3(\theta) = \frac{X_{14}X_{24}}{2X_{15}}\exp(2X_{15}\theta) \quad \text{(B.35)}$$

$$h_4(\theta) = \frac{X_{16}X_{26}}{4X_{19}}\cos(X_{18}-X_{28})\exp(2X_{19}\theta) \quad \text{(B.36)}$$

$$h_5(\theta) = \left\{ \frac{X_{16}X_{26}[X_{17}\sin(X_{18}+X_{28}) + X_{19}\cos(X_{18}+X_{28})]}{4(X_{17}^2 + X_{19}^2)} \right\} \times \cos(2X_{17}\theta + X_{18} + X_{28})\exp(2X_{19}\theta) \quad \text{(B.37)}$$

$$h_6(\theta) = \left\{ \frac{X_{16}X_{26}[X_{17}\cos(X_{18}+X_{28}) - X_{19}\sin(X_{18}+X_{28})]}{4(X_{17}^2 + X_{19}^2)} \right\} \times \sin(2X_{17}\theta + X_{18} + X_{28})\exp(2X_{19}\theta) \quad \text{(B.38)}$$

$$h_7(\theta) = [X_{11}X_{22}\sin(X_{23}) + X_{12}X_{21}\sin(X_{13})]\cos(\theta) \quad \text{(B.39)}$$

$$h_8(\theta) = [X_{11}X_{22}\cos(X_{23}) + X_{12}X_{23}\cos(X_{13})]\sin(\theta) \quad \text{(B.40)}$$

$$h_9(\theta) = \frac{X_{11}X_{24} + X_{14}X_{21}}{X_{15}}\exp(X_{15}\theta) \quad \text{(B.41)}$$

$$h_{10}(\theta) = \left\{ \frac{X_{11}X_{17}X_{26}\sin(X_{28}) + X_{11}X_{19}X_{26}\cos(X_{28})}{X_{17}^2 + X_{19}^2} + \frac{X_{16}X_{17}X_{21}\sin(X_{18}) + X_{16}X_{19}X_{21}\cos(X_{18})}{X_{17}^2 + X_{19}^2} \right\} \times \cos(X_{17}\theta)\exp(X_{19}\theta) \quad \text{(B.42)}$$

$$h_{11}(\theta) = \left\{ \frac{X_{11}X_{17}X_{26}\cos(X_{28}) - X_{11}X_{19}X_{26}\sin(X_{28})}{X_{17}^2 + X_{19}^2} + \frac{X_{16}X_{17}X_{21}\cos(X_{18}) - X_{16}X_{19}X_{21}\sin(X_{18})}{X_{17}^2 + X_{19}^2} \right\} \times \sin(X_{17}\theta)\exp(X_{19}\theta) \quad \text{(B.43)}$$

$$h_{12}(\theta) = \left\{ X_{12}X_{15}X_{24}\cos(X_{13}) + X_{14}X_{15}X_{22}\cos(X_{23}) + X_{12}X_{24}\sin(X_{13}) + X_{14}X_{22}\sin(X_{23}) \right\} \cos(\theta)\exp(X_{15}\theta) \quad \text{(B.44)}$$

$$h_{13}(\theta) = -\left\{ X_{12}X_{15}X_{24}\sin(X_{13}) - X_{14}X_{15}X_{22}\sin(X_{23}) - X_{12}X_{24}\cos(X_{13}) + X_{14}X_{22}\cos(X_{23}) \right\} \cos(\theta)\exp(X_{15}\theta) \quad \text{(B.45)}$$

$$h_{14}(\theta) = \left\{ \frac{X_{12}(1+X_{17})X_{26}\sin(X_{13}+X_{28})}{2[(1+X_{17})^2+X_{19}^2]} + \frac{X_{12}X_{19}X_{26}\cos(X_{13}+X_{28})}{2[(1+X_{17})^2+X_{19}^2]} + \frac{X_{16}(1+X_{17})X_{22}\sin(X_{18}+X_{23})}{2[(1+X_{17})^2+X_{19}^2]} + \frac{X_{16}X_{19}X_{22}\cos(X_{18}+X_{23})}{2[(1+X_{17})^2+X_{19}^2]} \right\} \times \cos[(1+X_{17})\theta]\exp(X_{19}\theta) \tag{B.46}$$

$$h_{15}(\theta) = \left\{ \frac{X_{12}(1+X_{17})X_{26}\cos(X_{13}+X_{28})}{2[(1+X_{17})^2+X_{19}^2]} - \frac{X_{12}X_{19}X_{26}\sin(X_{13}+X_{28})}{2[(1+X_{17})^2+X_{19}^2]} + \frac{X_{16}(1+X_{17})X_{22}\cos(X_{18}+X_{23})}{2[(1+X_{17})^2+X_{19}^2]} - \frac{X_{16}X_{19}X_{22}\sin(X_{18}+X_{23})}{2[(1+X_{17})^2+X_{19}^2]} \right\} \times \sin[(1+X_{17})\theta]\exp(X_{19}\theta) \tag{B.47}$$

$$h_{16}(\theta) = \left\{ \frac{X_{12}(1-X_{17})X_{26}\sin(X_{13}-X_{28})}{2[(1-X_{17})^2+X_{19}^2]} - \frac{X_{16}(1-X_{17})X_{22}\sin(X_{18}-X_{23})}{2[(1-X_{17})^2+X_{19}^2]} + \frac{X_{12}X_{19}X_{26}\cos(X_{13}-X_{28})}{2[(1-X_{17})^2+X_{19}^2]} - \frac{X_{16}X_{19}X_{22}\cos(X_{18}-X_{23})}{2[(1-X_{17})^2+X_{19}^2]} \right\} \times \cos[(1-X_{17})\theta]\exp(X_{19}\theta) \tag{B.48}$$

$$
\begin{aligned}
h_{17}(\theta) = \Bigg\{ & \frac{X_{12}(1 - X_{17})X_{26}\cos(X_{13} - X_{28})}{2[(1 - X_{17})^2 + X_{19}^2]} - \\
& \frac{X_{12}X_{19}X_{26}\sin(X_{13} - X_{28})}{2[(1 - X_{17})^2 + X_{19}^2]} - \\
& \frac{X_{16}(1 - X_{17})X_{22}\cos(X_{18} - X_{23})}{2[(1 - X_{17})^2 + X_{19}^2]} - \\
& \frac{X_{16}X_{19}X_{22}\sin(X_{18} - X_{23}}{2[(1 - X_{17})^2 + X_{19}^2]} \Bigg\} \times \\
& \sin[(1 - X_{17})\theta]\exp(X_{19}\theta)
\end{aligned}
\tag{B.49}
$$

$$
\begin{aligned}
h_{18}(\theta) = \Bigg\{ & \frac{X_{14}X_{17}X_{26}\sin(X_{28})}{(X_{15} + X_{19})^2 + X_{17}^2} + \\
& \frac{X_{14}(X_{15} + X_{19})X_{26}\cos(X_{28})}{(X_{15} + X_{19})^2 + X_{17}^2} + \\
& \frac{X_{16}X_{17}X_{24}\sin(X_{18})}{(X_{15} + X_{19})^2 + X_{17}^2} + \\
& \frac{X_{16}(X_{15} + X_{19})X_{24}\cos(X_{18})}{(X_{15} + X_{19})^2 + X_{17}^2} \Bigg\} \times \\
& \cos(X_{17}\theta)\exp[(X_{15} + X_{19})\theta]
\end{aligned}
\tag{B.50}
$$

$$
\begin{aligned}
h_{19}(\theta) = \Bigg\{ & \frac{X_{14}X_{17}X_{26}\cos(X_{28})}{(X_{15} + X_{19})^2 + X_{17}^2} - \\
& \frac{X_{14}(X_{15} + X_{19})X_{26}\sin(X_{28})}{X_{15} + X_{19})^2 + X_{17}^2} + \\
& \frac{X_{16}X_{17}X_{24}\cos(X_{18})}{(X_{15} + X_{19})^2 + X_{17}^2} - \\
& \frac{X_{16}(X_{15} + X_{19})X_{24}\sin(X_{18})}{(X_{15} + X_{19})^2 + X_{17}^2} \Bigg\} \times \\
& \sin(X_{17}\theta)\exp[(X_{15} + X_{19})\theta]
\end{aligned}
\tag{B.51}
$$

B.7 The Fourier cosine coefficients

The amplitude of the mth Fourier cosine coefficient requires the integral of the product of the general expression and $\cos(m\theta)$. This is sum of the eleven expressions, eqns B.53 through B.63, i.e.

$$\int f_p(\theta)\cos(m\theta)\,d\theta = \sum_{n=1,11} j_n(\theta) \tag{B.52}$$

where

$$j_1(\theta) = X_{11}\frac{\sin(m\theta)}{m} \tag{B.53}$$

$$j_2(\theta) = X_{14}X_{15}\frac{\cos(m\theta)\exp(X_{15}\theta)}{m^2+X_{15}^2} \tag{B.54}$$

$$j_3(\theta) = mX_{14}\frac{\sin(m\theta)\exp(X_{15}\theta)}{m^2+X_{15}^2} \tag{B.55}$$

$$j_4(\theta) = X_{12}\frac{\sin[(m+1)\theta+X_{13}]}{2(m+1)} \tag{B.56}$$

$$j_5(\theta) = X_{12}\frac{\sin[(m-1)\theta-X_{13}]}{2(m-1)} \tag{B.57}$$

$$j_6(\theta) = X_{16}X_{19}\frac{\cos[(m+X_{17})\theta+X_{18}]\exp(X_{19}\theta)}{2[(m+X_{17})^2+X_{19}^2]} \tag{B.58}$$

$$j_7(\theta) = X_{16}X_{17}\frac{\sin[(m+X_{17})\theta+X_{18}]\exp(X_{19}\theta)}{2[(m+X_{17})^2+X_{19}^2]} \tag{B.59}$$

$$j_8(\theta) = X_{16}X_{19}\frac{\cos[(m-X_{17})\theta-X_{18}]\exp(X_{19}\theta)}{2[(m-X_{17})^2+X_{19}^2]} \tag{B.60}$$

$$j_9(\theta) = -X_{16}X_{17}\frac{\sin[(m-X_{17})\theta-X_{18}]\exp(X_{19}\theta)}{2[(m-X_{17})^2+X_{19}^2]} \tag{B.61}$$

$$j_{10}(\theta) = X_{16}\frac{m\sin[(m+X_{17})\theta+X_{18}]\exp(X_{19}\theta)}{2[(m+X_{17})^2+X_{19}^2]} \tag{B.62}$$

$$j_{11}(\theta) = X_{16}\frac{m\sin[(m-X_{17})\theta-X_{18}]\exp(X_{19}\theta)}{2[(m-X_{17})^2+X_{19}^2]} \tag{B.63}$$

Two special cases must be considered, $j_1(\theta)$, eqn B.53, when $m = 0$ and $j_3(\theta)$, eqn B.55 when $m = 1$.

When $m = 0$, $j_1(\theta)$ becomes

$$j_1(\theta)_{m=0} = X_{11}\theta \tag{B.64}$$

and when $m = 1$, $j_3(\theta)$ becomes

$$j_1(\theta)_{m=1} = \frac{X_{12}}{2}\cos(X_{13})\,\theta \tag{B.65}$$

B.8 The Fourier sine coefficients

The amplitude of the mth Fourier sine component requires the integral of the product of the general expression and $\sin(m\theta)$. This is the sum of the eleven expression listed below, $k_m(\theta)$, eqn B.67, through $k_n(\theta)$, eqn B.77, i.e.

$$\int f_p(\theta)\sin(m\theta)\,d\theta = \sum_{n=1,11} k_n(\theta) \tag{B.66}$$

where

$$k_1(\theta) = -X_{11}\frac{\cos(m\theta)}{m} \tag{B.67}$$

$$k_2(\theta) = X_{14}X_{15}\frac{\sin(m\theta)\exp(X_{15}\theta)}{(m^2+X_{15}^2)} \tag{B.68}$$

$$k_3(\theta) = -X_{14}\frac{m\cos(m\theta)\exp(X_{15}\theta)}{(m^2+X_{15}^2)} \tag{B.69}$$

$$k_4(\theta) = -X_{12}\frac{\cos[(m+1)\theta+X_{13}]}{2(m+1)} \tag{B.70}$$

$$k_5(\theta) = -X_{12}\frac{\cos[(m-1)\theta-X_{13}]}{2(m-1)} \tag{B.71}$$

$$k_6(\theta) = X_{16}X_{19}\frac{\sin[(m+X_{17})\theta+X_{18}]\exp(X_{19}\theta)}{2[(m+X_{17})^2+X_{19}^2]} \tag{B.72}$$

$$k_7(\theta) = -X_{16}X_{17}\frac{\cos[(m+X_{17})\theta+X_{18}]\exp(X_{19}\theta)}{2[(m+X_{17})^2+X_{19}^2]} \tag{B.73}$$

$$k_8(\theta) = -X_{16}\frac{m\cos[(m+X_{17})\theta+X_{18}]\exp(X_{19}\theta)}{2[(m+X_{17})^2+X_{19}^2]} \tag{B.74}$$

$$k_9(\theta) = -X_{16}\frac{m\cos[(m - X_{17})\theta - X_{18}]\exp(X_{19}\theta)}{2[(m - X_{17})^2 + X_{19}^2]} \tag{B.75}$$

$$k_{10}(\theta) = X_{16}X_{19}\frac{\sin[(m - X_{17})\theta - X_{18}]\exp(X_{19}\theta)}{2[(m - X_{17})^2 + X_{19}^2]} \tag{B.76}$$

$$k_{11}(\theta) = X_{16}X_{17}\frac{\cos[(m - X_{17})\theta - X_{18}]\exp(X_{19}\theta)}{2[(m - X_{17})^2 + X_{19}^2]} \tag{B.77}$$

Two special cases must be considered, $k_1(\theta)$, eqn B.67, when $m = 0$ and $k_3(\theta)$, eqn B.69, when $m = 1$.

When $m = 0$, $k_1(\theta)$ contributes nothing to a definite integral and is therefore set to zero, i.e.

$$k_1(\theta)_{m=0} = 0.0 \tag{B.78}$$

When $m = 1$, $k_3(\theta)$ becomes

$$k_3(\theta)_{m=1} = -\frac{1}{2}X_{12}\sin(X_{13})\,\theta \tag{B.79}$$

APPENDIX C

FOURIER ANALYSIS

C.1 Introduction

Fourier analysis is a normal part of the engineering curriculum and this appendix does not attempt to repeat detailed treatment which has already been encountered elsewhere. Rather it collects together those concepts and formulas which are of particular value in power electronics.

Any repetitive waveform, such as that shown in Fig. C.1, can be expressed as the sum of its Fourier components. Thus if a wave is defined by a function $f(\theta)$ where $f(\theta + 2\pi) = f(\theta)$ then $f(\theta)$ can be expressed as

$$\begin{aligned} f(\theta) &= A_0 + A_1\cos(\theta) + A_2\cos(2\theta) + A_3\cos(3\theta) + \cdots \\ &\quad + B_1\sin(\theta) + B_2\sin(2\theta) + B_3\sin(3\theta) + \cdots \end{aligned} \tag{C.1}$$

or more briefly

$$f(\theta) = A_0 + \sum_{m=1,\infty} A_m\ \cos(m\theta) + B_m\ \sin(m\theta) \tag{C.2}$$

The form of eqn C.2 expresses each Fourier component as the sum of a cosine and a sine term. In some cases it is more convenient to have a form in which each component is represented by a single term having an appropriate amplitude and phase. This alternative form is given in cosine form in eqn C.3 and in sine form in eqn C.4, the amplitude and phase angles being defined by eqns C.5 through C.7

$$f(\theta) = A_0 + \sum_{m=1,\infty} C_m\ \cos(m\theta + \gamma_{cm}) \tag{C.3}$$

or alternatively

$$f(\theta) = A_0 + \sum_{m=1,\infty} C_m\ \sin(m\theta + \gamma_{sm}) \tag{C.4}$$

where

$$C_m = \sqrt{A_m^2 + B_m^2} \tag{C.5}$$

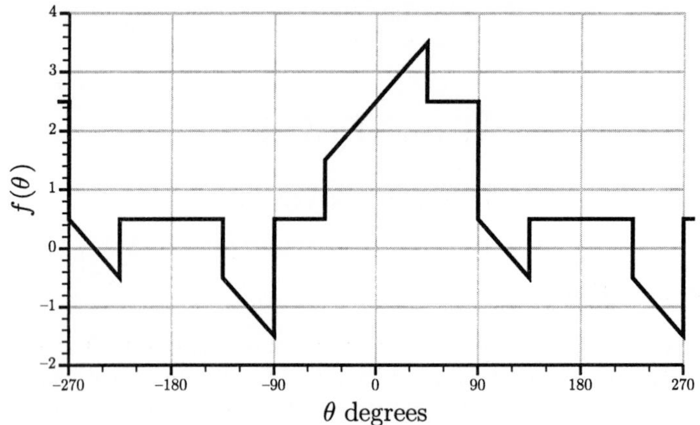

FIG. C.1. A wave with a period of 360°.

and, for the cosine form, eqn C.3,

$$\gamma_{cm} = \arctan(-\frac{B_m}{A_m}) \tag{C.6}$$

and, for the sine form, eqn C.4,

$$\gamma_{sm} = \arctan(\frac{A_m}{B_m}) \tag{C.7}$$

C.2 The amplitudes

The amplitudes of the cosine and sine components are given by the following integrals evaluated over any complete cycle of θ, i.e. for the range $\beta < \theta < \beta + 2\pi$, where β can have any value. For $m > 0$

$$A_m = \frac{1}{\pi} \int_{\beta}^{\beta+2\pi} f(\theta) \cos(m\theta)\, \mathrm{d}\theta \tag{C.8}$$

and

$$B_m = \frac{1}{\pi} \int_{\beta}^{\beta+2\pi} f(\theta) \sin(m\theta)\, \mathrm{d}\theta \tag{C.9}$$

For the special case of $m = 0$

$$A_m = \frac{1}{2\pi} \int_{\beta}^{\beta+2\pi} f(\theta) \mathrm{d}\theta \tag{C.10}$$

i.e. half the value given by direct use of eqn C.8 with $m = 0$.

While these integrals can be evaluated over any 2π interval it is usually most convenient to use either the interval 0 to 2π or $-\pi$ to $+\pi$.

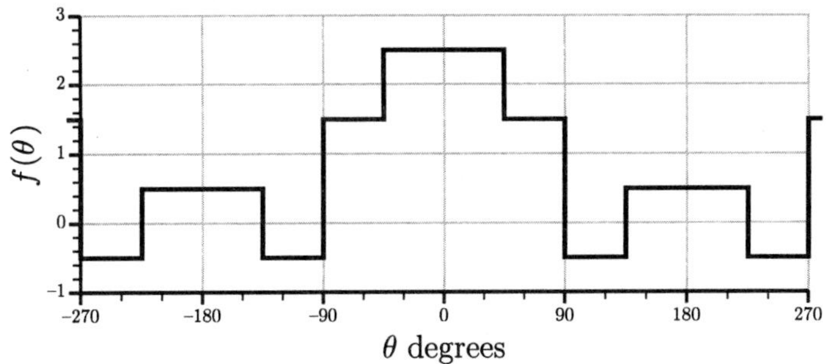

FIG. C.2. An even function, $f(-\theta) = f(\theta)$.

C.3 Some simplifications

Many particular cases occur in practice which are easily recognized visually and which permit the general expressions of eqns C.8 and C.9 to be simplified.

C.3.1 *The special case of an even function*

If $f(-\theta) = f(\theta)$, the shape of the function along the negative θ axis is the mirror image in the vertical axis of its shape along the positive θ axis. Such a function is said to be even, an example being shown in Fig. C.2. The usage of the word *even* in this context must be carefully differentiated from its usage in the context of Fourier components of even order.

Since sine functions are odd, the amplitudes of the sine components are zero and the Fourier series consists of only cosine components. Further, since cosine functions are even, $f(-\theta)\cos(-m\theta) = f(\theta)\cos(m\theta)$ and the full range integral from $-\pi$ *to* $+\pi$ may be replaced by twice the half range integral from 0 to π. Thus, for an even function

$$A_m = \frac{2}{\pi} \int_0^{\pi} f(\theta) \cos(m\theta)\, \mathrm{d}\theta \tag{C.11}$$

and

$$B_m = 0 \tag{C.12}$$

C.3.2 *The special case of an odd function*

In contrast to the above, if $f(-\theta) = -f(\theta)$, as is the case for the function shown in Fig. C.3, the function is said to be odd, the amplitudes of the

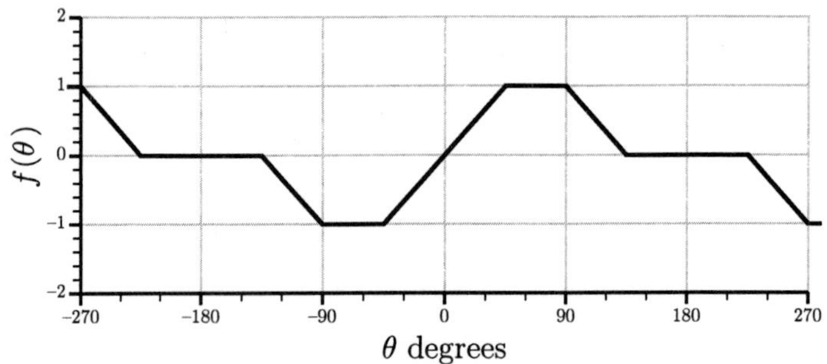

FIG. C.3. An odd function, $f(-\theta) = -f(\theta)$.

cosine components are zero, and the amplitudes of the sine components can be obtained by half range integrals. Thus, for an odd function

$$A_m = 0 \tag{C.13}$$

and

$$B_m = \frac{2}{\pi} \int_0^{\pi} f(\theta) \sin(m\theta)\, \mathrm{d}\theta \tag{C.14}$$

C.4 Offset even and odd functions

Sometimes a function which appears to be of general form can be converted to either the odd or even form by shifting the axes.

The upper graph of Fig. C.4 shows a function which can be converted to an even function by shifting the origin of the θ axis to $-30°$ i.e. by converting the function from $f(\theta)$ to $f(\psi)$ where $\psi = \theta + 30.0°$. The original θ axis is shown along the lower edge of the graph and the new ψ axis is shown along the top edge.

The lower graph of Fig. C.4 shows a function which can be converted to an odd function by moving the origin of its θ axis and removing its mean value, A_0. This is done by introducing a new angle ψ and a new function $h(\theta)$ which are the linear functions of the original angle θ and the original function $g(\theta)$

$$\psi = \theta - 45°$$

and

$$h(\theta) = g(\theta) - 0.5$$

The original scales are shown in Fig. C.4 along the bottom and left side of the graph and the modified scales along the upper and right side of the graph.

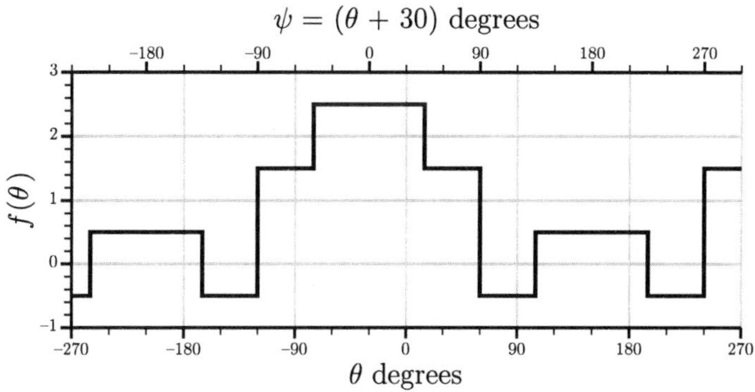

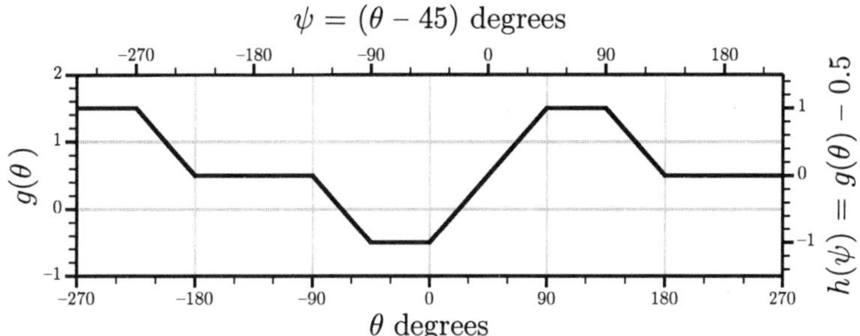

FIG. C.4. Displaced even and odd functions. The function shown in the upper graph can be converted to an even function by shifting the origin of the x axis. The function shown in the lower graph can be converted to an odd function by shifting the origins of both its x and y axes.

C.5 Absence of even harmonics

Electrical engineering waveforms, particularly those associated with rotating machines, are frequently symmetrical in the sense that the shape of a negative half wave is identical, except for the change in sign, to the shape of a positive half wave. A waveform of this type is shown in Fig. C.5. For such a wave

$$f(\theta + \pi) = -f(\theta) \tag{C.15}$$

For the Fourier components of even order, i.e. m even,

$$\cos[m(\theta + \pi)] = \cos(m\theta)\cos(m\pi) = \cos(m\theta)$$

and

$$\sin[m(\theta + \pi)] = \sin(m\theta)\cos(m\pi) = \sin(m\theta)$$

which are in conflict with the requirements of eqn C.15. Thus the amplitudes of all the even components are zero.

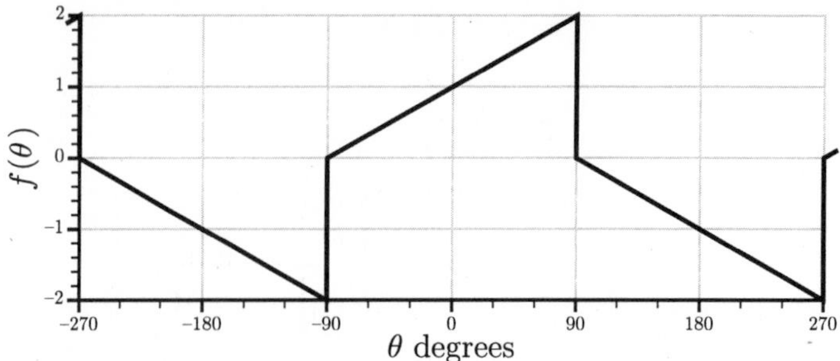

FIG. C.5. A function without even harmonics for which $f(\theta+\pi) = -f(\theta)$.

Again half range integrals can be employed so that, with the proviso that m be odd,

$$A_m = \frac{2}{\pi}\int_0^{\pi} f(\theta)\cos(m\theta)\,\mathrm{d}\theta \tag{C.16}$$

and

$$B_m = \frac{2}{\pi}\int_0^{\pi} f(\theta)\sin(m\theta)\,\mathrm{d}\theta \tag{C.17}$$

Further if the function is symmetrical and is also even, as is the case for the function shown in the upper graph of Fig. C.6, quarter range integrals can be employed. Thus for such functions

$$A_m = \frac{4}{\pi}\int_0^{\pi/2} f(\theta)\cos(m\theta)\,\mathrm{d}\theta \tag{C.18}$$

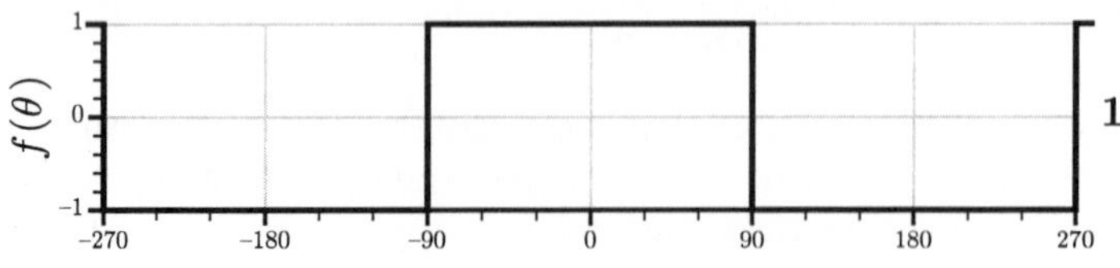

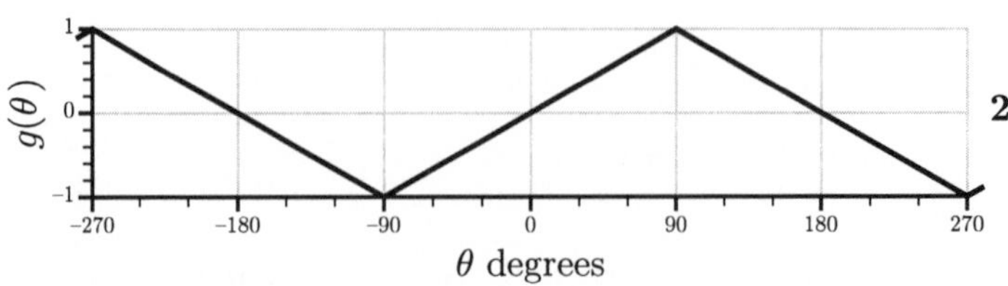

FIG. C.6. Even and odd functions without Fourier components of even order.

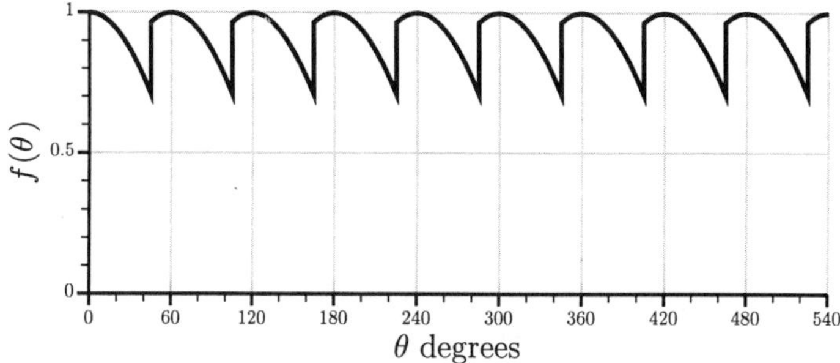

FIG. C.7. A wave which repeats six times per cycle.

Alternatively, if the function is symmetrical and is also odd, as is the case for the function shown in the lower graph of Fig. C.6

$$B_m = \frac{4}{\pi} \int_0^{\pi/2} f(\theta) \sin(m\theta)\, \mathrm{d}\theta \tag{C.19}$$

C.6 Waves which repeat more frequently

In power electronics, waves which repeat at a multiple of the basic frequency are often encountered. For example, the output of a semi-controlled bridge rectifier repeats at three times the supply frequency. Figure C.7 shows such a wave which repeats at six times the fundamental frequency.

If there are r repetitions per fundamental cycle then the Fourier spectrum contains only components whose orders are integral multiples of r, i.e.

$$m = kr \tag{C.20}$$

where k is any positive integer including zero.

For such waves, the integrals need only be evaluated over the repetition period. Equations C.8 and C.9 then become

$$A_m = \frac{r}{\pi} \int_0^{2\pi/r} f(\theta) \cos(m\theta)\, \mathrm{d}\theta \tag{C.21}$$

and

$$B_m = \frac{r}{\pi} \int_0^{2\pi/r} f(\theta) \sin(m\theta)\, \mathrm{d}\theta \tag{C.22}$$

C.7 R.m.s. values

The r.m.s. value of a wave is equal to the square root of the sum of the squares of the r.m.s. values of its Fourier components. Thus, if F_{rms} is the r.m.s. value of $f(\theta)$, then

$$F_{rms} = \sqrt{A_o^2 + \sum_{m=1,\infty} \frac{A_m^2 + B_m^2}{2}} \tag{C.23}$$

where the amplitudes A_m and B_m have been converted to r.m.s. values by dividing the sum of their squares by 2.

Alternatively, using the form of either eqn C.4 or eqn C.5

$$F_{rms} = \sqrt{A_o^2 + \sum_{m=1,\infty} \frac{C_m^2}{2}} \tag{C.24}$$

where again the amplitude C_m has been converted to an r.m.s. value by dividing its square by 2.

C.8 Useful integrals

A number of integrals which are frequently encountered when analyzing power electronics waveforms are given in eqns A.33 through A.48, in eqns B.52 through B.79, and in eqns E.37 through E.54.

APPENDIX D

THE PRODUCT OF FOURIER SERIES

D.1 Introduction

The Fourier analysis of power electronic waveforms regularly requires that the product of two series be obtained. An obvious example is the derivation of the instantaneous power as the product of a voltage series and a current series. Less obvious is the derivation of thyristor, diode, or input current from the output current. In general the Fourier series representing the output current of a rectifier, inverter, or chopper is easily obtained. Obtaining from this the series representing the current in an individual switching element or the input current to the device is more complex, requiring the product of the original series and the one for an appropriate switching function. The text, especially Section 8.19, has a number of examples of this procedure. A more detailed discussion of the process than is appropriate in the text is given here.

Obtaining the product of two Fourier series is a formidable task in which each term of the resulting series is the sum of an infinite number of product terms. It would not be practicable for manual computation and even the computer generated solution has a limit to the number of terms which can be generated and combined, thus limiting the accuracy of the result. If this limitation is unacceptable for a specific task there always remains the time domain solution which itself can be analyzed into Fourier components.

An indication of the accuracy of the product of series approach is provided by comparing the product of two rectangular waves with the exact result. This is a good test of the method since it is simple, all the waves involved being rectangular, and it is demanding because their Fourier series converge slowly.

D.2 The general product term

Consider the two Fourier series

$$f(\theta) = \sum_{m} F_{cm} \cos(m\theta) + F_{sm} \sin(m\theta) \tag{D.1}$$

and

$$g(\theta) = \sum_n G_{cn} \cos(n\theta) + G_{sn} \sin(n\theta) \tag{D.2}$$

The product, $h(\theta)$, of these two series is

$$\begin{aligned} h(\theta) = \sum_m \sum_n & F_{cm}G_{cn} \cos(m\theta)\cos(n\theta) + F_{sm}G_{sn} \sin(m\theta)\sin(n\theta) + \\ & F_{sm}G_{cn} \cos(m\theta)\sin(n\theta) + F_{sm}G_{cn} \sin(m\theta)\cos(n\theta) \end{aligned} \tag{D.3}$$

The products of sinusoids can be converted to sums and the resulting eight terms can be combined in pairs to yield

$$\begin{aligned} h(\theta) = \sum_m \sum_n & \frac{F_{cm}G_{cn} - F_{sm}G_{sn}}{2} \cos[(m+n)\theta] + \\ & \frac{F_{cm}G_{sn} + F_{sm}G_{cn}}{2} \sin[(m+n)\theta] + \\ & \frac{F_{cm}G_{cn} + F_{sm}G_{sn}}{2} \cos[(m-n)\theta] - \\ & \frac{F_{cm}G_{sn} - F_{sm}G_{cn}}{2} \sin[(m-n)\theta] \end{aligned} \tag{D.4}$$

Inevitably the $m - n$ term will give rise to negative frequencies. This can be avoided by using the magnitude of $m - n$ as the frequency and multiplying the sine term by its sign. Equation D.4 then becomes

$$\begin{aligned} h(\theta) = \sum_m \sum_n & \frac{F_{cm}G_{cn} - F_{sm}G_{sn}}{2} \cos[(m+n)\theta] + \\ & \frac{F_{cm}G_{sn} + F_{sm}G_{cn}}{2} \sin[(m+n)\theta] + \\ & \frac{F_{cm}G_{cn} + F_{sm}G_{sn}}{2} \cos(|m-n|\theta) - \\ & \frac{m-n}{|m-n|} \frac{F_{cm}G_{sn} - F_{sm}G_{cn}}{2} \sin(|m-n|\theta) \end{aligned} \tag{D.5}$$

The rth order component of $h(\theta)$ is the sum of many components of the product, specifically those components for which either

$$m + n = r$$

or

$$m - n = r$$

or

$$n - m = r$$

Rather than attempt to discover and sum these individual product components, it is simpler to first derive all the product terms and store them in temporary arrays. These arrays are then searched for terms of the same frequency and kind, cosine or sine, and these are summed to give the terms of the output array. This method is straightforward and effective but has the disadvantage that the size of the temporary arrays increases rapidly with the number of input terms. If one input series has M terms and the other has N terms, their product produces $2M \times N$ terms. Thus if there are twenty terms in each input series, temporary storage for 800 cosine and 800 sine terms is required. Of course many of these terms have the same frequency and, after sorting and summing, the final output series has far fewer terms.

This rapid expansion of the problem with the number of input terms is the weak link in this method. Many terms must be summed to obtain accurate estimates of the terms of the product series and yet if many terms are used the computational task grows enormously. A subroutine which accepts input series of up to 20 terms and limits the output series to 40 terms is satisfactory for most applications.

The application of the method and some indication of its accuracy is provided by Example D.1 which considers the product of the two rectangular waves shown in graphs 1 and 2 of Fig. D.1. Their product is also a rectangular wave and is shown in graph 3. The waveforms in Fig. D.1 are not exact but are derived from the truncated Fourier series, each wave being approximated by the first 20 nonzero Fourier components.

Example D.1

As an example of the process, consider the product of two rectangular waves $f(\theta)$ and $g(\theta)$. The waves are similar, having unit amplitude, unit frequency, and rectangular waveform, their only difference being that $g(\theta)$ lags $f(\theta)$ by 90°. An approximation to $f(\theta)$ is shown in graph 1 of Fig. D.1 and an approximation to $g(\theta)$ is shown in graph 2, the approximations being obtained from the first 20 nonzero terms of the Fourier series representing the functions.

The exact definition of the functions is that $f(\theta)$ has the value 1.0 for $0.0 < \theta < 180°$ and the value -1.0 for $180° < \theta < 360°$ and $g(\theta)$ has the value -1.0 for $0.0 < \theta < 90°$, the value 1.0 for $90° < \theta < 270°$, and the value -1.0 for $270° < \theta < 360°$.

The product $h(\theta) = f(\theta) \times g(\theta)$ also has a rectangular waveform of unity amplitude but its frequency is double that of the originating waves. An approximation to it is shown in graph 3 of Fig. D.1, the approximation being provided by the first 20 nonzero terms of its Fourier series, the Fourier components having been derived by taking the product of the series for $f(\theta)$ and $g(\theta)$. Since this is an approximation, the phrase *nonzero terms* needs qualification. In fact all Fourier components with amplitudes less than 0.000 05 have been ignored.

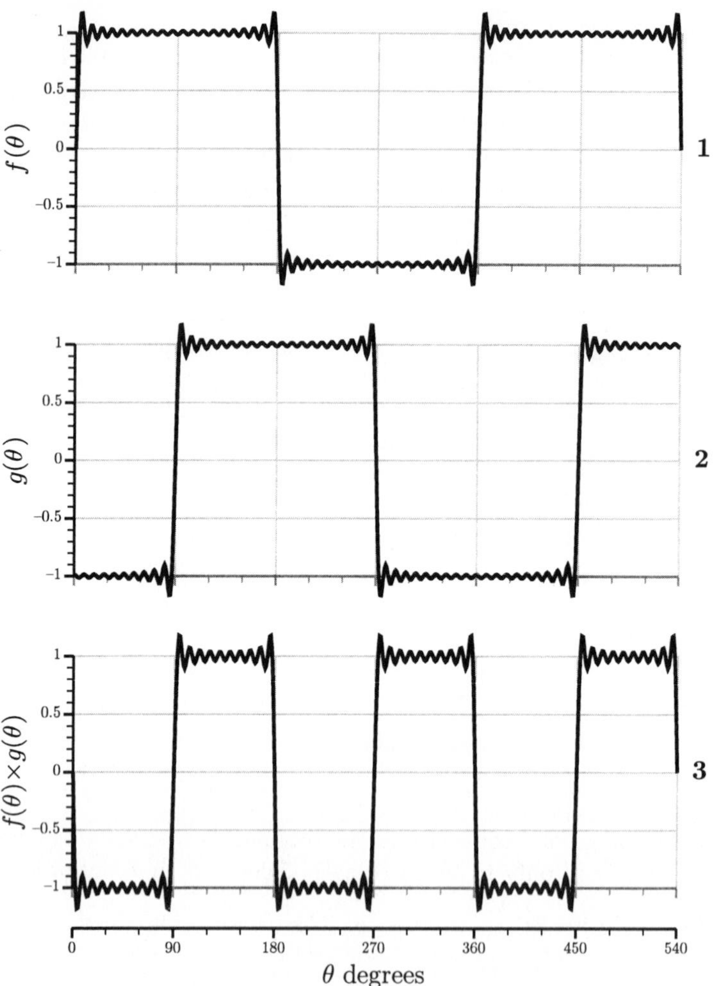

FIG. D.1. The product of rectangular waves. The first wave is shown in graph 1, the second in graph 2, and their product in graph 3. All waves are represented by their truncated Fourier series.

The approximation to $f(\theta)$ is

$$f(\theta) \approx \sum_{m=1,39,2} \frac{4}{m\pi} \sin(m\theta) \tag{D.6}$$

i.e.

$$f(\theta) \approx 1.2732 \sin(\theta) + 0.4244 \sin(3\theta) + 0.2546 \sin(5\theta) \cdots + 0.0326 \sin(39\theta)$$

The approximation to $g(\theta)$ is

$$g(\theta) \approx \sum_{m=1,39,2} -\sin(\frac{m\pi}{2})\frac{4}{m\pi}\cos(m\theta) \tag{D.7}$$

i.e.

$$g(\theta) \approx -1.2732\cos(\theta) + 0.4244\cos(3\theta) - 0.2546\cos(5\theta) \cdots + 0.0326\cos(39\theta)$$

The product of these two series initially gives rise to 400 terms which, after conversion of the products into sums results in 800 individual cosine and sine terms. Consider for example the product of the fifth order term of $f(\theta)$ with the third order term of $g(\theta)$. This is initially

$$0.2546\sin(5\theta) \times 0.4244\cos(3\theta)$$

which, after conversion to a sum, becomes

$$0.0540\sin(8\theta) + 0.0540\sin(2\theta)$$

Many other individual products give rise to components having frequencies of 2 or 8, e.g. the product of the first order term of $f(\theta)$ and the third order term of $g(\theta)$ gives rise to components having frequencies of 2 and 4, the product of the 29th order term of $f(\theta)$ and the 21st order term of $g(\theta)$ gives rise to components having frequencies of 8 and 50, etc. Consideration of the range of frequencies in the originating series shows that their product gives rise only to terms of even order and that the order ranges from a minimum of zero to a maximum of $2 \times 39 = 78$. Further, in this particular case the product gives rise only to sine terms, the cosine terms being absent because one of the originating series has only sine terms and the other only cosine terms.

The final step in the process is the combination of all terms of the same order. When this is done in this particular case it is found that terms whose order is an odd multiple of 2 are negligible, i.e. their amplitudes are less than 0.000 05.

Thus, in this case the product $f(\theta) \times g(\theta)$ gives rise to a series $h(\theta)$ of the type

$$h(\theta) \approx \sum_{m=2,78,4} H_m \sin(m\theta)$$

The order, m, and amplitude, H_m, thus derived are listed in columns 1 and 2 of Table D.1 and have been used to plot the waveform shown in graph 3 of Fig. D.1.

Table D.1 *Comparison of the product technique with the exact analytic solution for the product of the square waves of Fig. D.1*

Order	H_{approx}	H_{true}	H_{approx}/H_{true}
2	–1.2738	–1.2732	1.0004
6	–0.4250	–0.4244	1.0014
10	–0.2553	–0.2546	1.0026
14	–0.1827	–0.1819	1.0043
18	–0.1424	–0.1415	1.0065
22	–0.1169	–0.1157	1.0097
26	–0.0994	–0.0979	1.0147
30	–0.0869	–0.0849	1.0236
34	–0.0782	–0.0749	1.0438
38	–0.0756	–0.0670	1.1287
42	0.0088	–0.0606	–0.1446
46	0.0033	–0.0554	–0.0597
50	0.0020	–0.0509	–0.0395
54	0.0014	–0.0472	–0.0306
58	0.0011	–0.0439	–0.0256
62	0.0009	–0.0411	–0.0224
66	0.0008	–0.0386	–0.0202
70	0.0007	–0.0364	–0.0186
74	0.0006	–0.0344	–0.0173
78	0.0005	–0.0326	–0.0163

Fortunately the accuracy of this result is easily checked since the true product and its Fourier series are readily obtained. The true product has a rectangular waveform and alternates between +1 and −1 at twice the frequency of the originating waves. Its mean value is zero and its Fourier series is

$$h(\theta) = \sum_{m=2,\infty,4} -\frac{4}{m\pi}\sin(m\theta)$$

The amplitudes $-4/(m\pi)$ are listed in column 3 of Table D.1 and the ratios of the approximate amplitudes listed in column 2 to these true amplitudes are listed in column 4.

Inspection of Fig. D.1 indicates that the approximate Fourier series represents the true product quite well. In fact the accuracy appears to be nearly as good as that of the series representing the originating functions.

A more quantitative idea of the validity of the approximate series can be obtained from Table D.1. The comparison of the approximate amplitudes with the true amplitudes given in column 4 reveals that the accuracy of components whose order is in the range 2–38 is good, the ratio of the approximate to the true value gradually increasing from 1.0004 for the second order component to 1.1287 for the 38th order component. When the order exceeds 38 the accuracy falls precipitously, so much so that it can be said that the approximate amplitudes are of no value at these higher orders.

The drop in accuracy beyond the 38th order component is not peculiar to this particular example. The originating series were terminated at the 39th order component and the conclusion specific to this example can be generalized by saying that the accuracy of components of the product whose order exceeds the highest order of the originating series will be poor. The origin of this generalization is not difficult to find.

In general, the largest components in the originating series will be the ones of lowest order. Denoting the minimum order by m_{min} and the maximum order by n_{max}, the most significant contribution to the $n_{max} \pm m_{min}$ order components will be provided by the product of the m_{min} component of one originating series and the n_{max} order component of the other originating series. This most significant contribution is missing from product terms whose order is higher than n_{max}. This generalization, while of great value, is not absolutely reliable and it is wise to have the subroutine determine a few terms beyond this point.

APPENDIX E

THE BASIC CYCLOCONVERTER EXPRESSION

E.1 Introduction

This appendix provides various useful functions of the basic cycloconverter expression, eqn 11.25, which for the present purpose will be written in the generalized form

$$f_1(\theta) = X_{11}\cos(\theta + X_{12}) - X_{13}\cos(X_{14}\theta + X_{15}) + X_{16}\ \exp[X_{17}(\theta - X_{18})] \tag{E.1}$$

The determination of the voltage absorbed by an inductor requires a knowledge of the derivative of the basic expression. The determination of average values requires the integral of the basic expression. The determination of instantaneous power requires the product of two basic expressions. The determination of r.m.s. values and of mean power requires the integral with respect to θ of this product. Fourier analysis requires the integral with respect to θ of the product of the basic expression and $\cos(m\theta)$ and of the basic expression and $\sin(m\theta)$ where m is any positive integer including zero. These functions of the basic expression are given below. The integrals are given without the constant of integration as they will always be used to evaluate definite integrals.

E.2 The derivative of the basic expression

The derivative of the basic cycloconverter expression with respect to θ is $f_1'(\theta)$ where

$$f_1'(\theta) = -X_{11}\sin(\theta + X_{12}) + X_{13}X_{14}\sin(X_{14}\theta + X_{15}) + X_{16}X_{17}\exp[X_{17}(\theta - X_{18})] \tag{E.2}$$

E.3 The integral of the basic expression

The integral of the basic cycloconverter expression with respect to θ is given by

$$\int f_1(\theta)\,\mathrm{d}\theta = X_{11}\sin(\theta + X_{12}) - \frac{X_{13}}{X_{14}}\sin(X_{14}\theta + X_{15}) + \frac{X_{16}}{X_{17}}\exp[X_{17}(\theta - X_{18})] \tag{E.3}$$

E.4 The product of basic expressions

In deriving the product of two basic cycloconverter expressions the second expression is written as

$$f_2(\theta) = X_{21}\cos(\theta + X_{22}) - X_{23}\cos(X_{24}\alpha\theta + X_{25}) + X_{26}\exp[X_{27}(\theta - X_{28})] \tag{E.4}$$

The product is

$$f_1(\theta)\,f_2(\theta) = \sum_{n=1,9} g_n(\theta) \tag{E.5}$$

where

$$g_1(\theta) = X_{11}X_{21}\cos(\theta + X_{12})\cos(\theta + X_{22}) \tag{E.6}$$
$$g_2(\theta) = -X_{11}X_{23}\cos(\theta + X_{12})\cos(X_{24}\theta + X_{25}) \tag{E.7}$$
$$g_3(\theta) = X_{11}X_{26}\cos(\theta + X_{12})\exp[X_{27}(\theta - X_{28})] \tag{E.8}$$
$$g_4(\theta) = -X_{13}X_{21}\cos(X_{14}\theta + X_{15})\cos(\theta + X_{22}) \tag{E.9}$$
$$g_5(\theta) = X_{13}X_{23}\cos(X_{14}\theta + X_{15})\cos(X_{24}\theta + X_{25}) \tag{E.10}$$
$$g_6(\theta) = -X_{13}X_{26}\cos(X_{14}\alpha\theta + X_{15})\exp[X_{27}(\theta - X_{28})] \tag{E.11}$$
$$g_7(\theta) = X_{16}X_{21}\cos(\theta + X_{22})\exp[X_{17}(\theta - X_{18})] \tag{E.12}$$
$$g_8(\theta) = -X_{16}X_{23}\cos(X_{24}\alpha\theta + X_{25})\exp[X_{17}(\theta - X_{18})] \tag{E.13}$$
$$g_9(\theta) = X_{16}X_{26}\exp[X_{17}(\theta - X_{18}) + X_{27}(\theta - X_{28})] \tag{E.14}$$

E.5 The integral of the product

The integral with respect to θ of the product of two basic expressions is

$$\int f_1(\theta)\,f_2(\theta)\,\mathrm{d}\theta = \sum_{n=1,17} h_n(\theta) \tag{E.15}$$

where

$$h_1(\theta) = \frac{X_{11}X_{21}}{2}\cos(X_{12} - X_{22})\theta \tag{E.16}$$

$$h_2(\theta) = \frac{X_{11}X_{21}}{4}\sin(2\theta + X_{12} + X_{22}) \tag{E.17}$$

$$h_3(\theta) = -\frac{X_{11}X_{23}}{2(1+X_{24})}\sin[(1+X_{24})\theta + X_{12} + X_{25}] \tag{E.18}$$

$$h_4(\theta) = -\frac{X_{11}X_{23}}{2(1-X_{24})}\sin[(1-X_{24})\theta + X_{12} - X_{25}] \tag{E.19}$$

$$h_5(\theta) = \frac{X_{11}X_{26}X_{27}}{1+X_{27}^2}\cos(\theta + X_{12})\exp[X_{27}(\theta - X_{28})] \tag{E.20}$$

$$h_6(\theta) = \frac{X_{11}X_{26}}{1+X_{27}^2}\sin(\theta + X_{12})\exp[X_{27}(\theta - X_{28})] \tag{E.21}$$

$$h_7(\theta) = -\frac{X_{13}X_{21}}{2(1+X_{14})}\sin[(1+X_{14})\theta + X_{15} + X_{22}] \tag{E.22}$$

$$h_8(\theta) = -\frac{X_{13}X_{21}}{2(1-X_{14})}\sin[(1-X_{14})\theta - X_{15} + X_{22}] \tag{E.23}$$

$$h_9(\theta) = \frac{X_{13}X_{23}}{2(X_{14}+X_{24})}\sin[(X_{14}+X_{24})\theta + X_{15} + X_{25}] \tag{E.24}$$

$$h_{10}(\theta) = \frac{X_{13}X_{23}}{2(X_{14}-X_{24})}\sin[(X_{14}-X_{24})\theta + X_{15} - X_{25}] \tag{E.25}$$

$$h_{11}(\theta) = -\frac{X_{13}X_{26}X_{27}}{(X_{14}^2+X_{27}^2)}\cos(X_{14}\theta + X_{15})\exp[X_{27}(\theta - X_{28})] \tag{E.26}$$

$$h_{12}(\theta) = -\frac{X_{13}X_{14}X_{26}}{(X_{14}^2+X_{27}^2)}\sin(X_{14}\theta + X_{15})\exp[X_{27}(\theta - X_{28})] \tag{E.27}$$

$$h_{13}(\theta) = \frac{X_{16}X_{21}}{(1+X_{17}^2)}\sin(\theta + X_{22})\exp[X_{17}(\theta - X_{18})] \tag{E.28}$$

$$h_{14}(\theta) = \frac{X_{16}X_{17}X_{21}}{(1+X_{17}^2)}\cos(\theta + X_{22})\exp[X_{17}(\theta - X_{18})] \tag{E.29}$$

$$h_{15}(\theta) = -\frac{X_{16}X_{23}X_{24}}{(X_{17}^2+X_{24}^2)}\sin(X_{24}\theta + X_{25})\exp[X_{17}(\theta - X_{18})] \tag{E.30}$$

$$h_{16}(\theta) = -\frac{X_{16}X_{17}X_{23}}{(X_{17}^2+X_{24}^2)}\cos(X_{24}\theta + X_{25})\exp[X_{17}(\theta - X_{18})] \tag{E.31}$$

$$h_{17}(\theta) = \frac{X_{16}X_{26}}{(X_{17}+X_{27})}\exp[(X_{17}+X_{27})\theta - X_{17}X_{18} - X_{27}X_{28}] \tag{E.32}$$

Special situations arise in the case of component $h_4(\theta)$ when $X_{24} = 1$, in the case of component $h_8(\theta)$ when $X_{14} = 1$, in the case of component $h_{10}(\theta)$ when $X_{24} = X_{14}$, and in the case of component $h_{17}(\theta)$ when both X_{17} and X_{27} are zero. The appropriate expressions in the first three of these cases are obtained by application of l'Hôpital's method. When $X_{24} = 1$, component $h_4(\theta)$ becomes

$$h_4(\theta)_{(X_{24}=1)} = -\frac{X_{11}X_{23}}{2}\cos(X_{12} - X_{25})\theta \tag{E.33}$$

When $X_{14} = 1$, component $h_8(\theta)$ becomes

$$h_8(\theta)_{(X_{14}=1)} = -\frac{X_{13}X_{21}}{2}\cos(X_{15} - X_{22})\theta \tag{E.34}$$

When $X_{24} = X_{14}$, component $h_{10}(\theta)$ becomes

$$h_{10}(\theta)_{(X_{24}=X_{14})} = \frac{X_{13}X_{23}}{2}\cos(X_{15} - X_{25})\theta \tag{E.35}$$

In the last of these four special cases, component $h_{17}(\theta)$ when the exponents X_{17} and X_{27} are both zero, first substitute X_x for the sum $X_{17}+X_{27}$. l'Hôpital's method is now applied with X_x as the variable. This yields the value of the function in this special case as

$$h_{17}(\theta)_{(X_{27}=X_{17}=0)} = X_{16}X_{26}\,\theta \tag{E.36}$$

E.6 The Fourier cosine components

The determination of the amplitude of the mth Fourier cosine component requires the determination of the integral with respect to θ of the product of the basic expression and $\cos(m\theta)$. This is the sum of the six functions, $j_1(\theta)$ through $j_6(\theta)$ listed in eqns E.38 through E.43, i.e.

$$\int f_1(\theta)\cos(m\theta)\,\mathrm{d}\theta = \sum_{n=1,6} j_n(\theta) \tag{E.37}$$

where

$$j_1(\theta) = \frac{X_{11}}{2(1+m)}\sin[(1+m)\alpha\theta + X_{12}] \tag{E.38}$$

$$j_2(\theta) = \frac{X_{11}}{(1-m)} \sin[(1-m)\alpha\theta + X_{12}] \tag{E.39}$$

$$j_3(\theta) = -\frac{X_{13}}{2(X_{14}+m)} \sin[(X_{14}+m)\theta + X_{15}] \tag{E.40}$$

$$j_4(\theta) = -\frac{X_{13}}{2(X_{14}-m)} \sin[(X_{14}-m)\theta X_{15}] \tag{E.41}$$

$$j_5(\theta) = \frac{X_{16}}{(X_{17}{}^2+m^2)} X_{17} \cos(m\alpha\theta) \exp[X_{17}(\theta - X_{18})] \tag{E.42}$$

$$j_6(\theta) = \frac{mX_{16}}{(X_{17}{}^2+m^2)} \sin(m\theta) \exp[X_{17}(\theta - X_{18})] \tag{E.43}$$

Special situations arise in the case of component $j_2(\theta)$ when $m = 1$ and in the case of component $j_4(\theta)$ when $m = X_{14}$. The appropriate expressions in these cases are obtained by application of l'Hôpital's method. When $m = 1$, component $j_2(\theta)$ becomes

$$j_2(\theta)_{(m=1)} = \frac{X_{11}\cos(X_{12})}{2}\theta \tag{E.44}$$

When $m = X_{14}$ component $j_4(\theta)$ becomes

$$j_4(\theta)_{(m=X_{14})} = -\frac{X_{13}\cos(X_{15})}{2}\theta \tag{E.45}$$

E.7 The Fourier sine components

The determination of the amplitude of the mth Fourier sine component requires the determination of the integral with respect to θ of the product of the basic expression and $\sin(m\theta)$. This is the sum of the six functions, $k_1(\theta)$ through $k_6(\theta)$ listed in eqns E.47 through E.52, i.e.

$$\int f_1(\theta)\sin(m\theta)\,\mathrm{d}\theta = \sum_{n=1,6} k_n(\theta) \tag{E.46}$$

where

$$k_1(\theta) = -\frac{X_{11}}{2(1+m)} \cos[(1+m)\theta + X_{12}] \tag{E.47}$$

$$k_2(\theta) = \frac{X_{11}}{2(1-m)} \cos[(1-m)\alpha\theta + X_{12}] \tag{E.48}$$

$$k_3(\theta) = \frac{X_{13}}{2(X_{14}+m)} \cos[(X_{14}+m)\theta + X_{15}] \tag{E.49}$$

$$k_4(\theta) = -\frac{X_{13}}{2(X_{14}-m)} \cos[(X_{14}-m)\theta + X_{15}] \tag{E.50}$$

$$k_5(\theta) = \frac{X_{16}X_{17}}{(X_{17}{}^2+m^2)} \sin(m\alpha\theta) \exp[X_{17}(\theta - X_{18})] \tag{E.51}$$

$$k_6(\theta) = -\frac{mX_{16}}{(X_{17}{}^2+m^2)} \cos(m\theta) \exp[X_{17}(\theta - X_{18})] \tag{E.52}$$

Special situations arise in the case of component $k_2(\theta)$ when $m = 1$ and in the case of component $k_4(\theta)$ when $m = X_{14}$. The appropriate expressions in these cases are obtained by application of l'Hôpital's method.

When $m = 1$, component $k_2(\theta)$ becomes

$$k_2(\theta)_{(m=1)} = -\frac{X_{11}}{2} \sin(X_{12})\theta \tag{E.53}$$

When $m = X_{14}$, component $k_4(\theta)$ becomes

$$k_4(\theta)_{(m=X_{14})} = \frac{X_{13}}{2} \sin(X_{15})\theta \tag{E.54}$$

INDEX